Mathematics Study Resources

Volume 15

Series Editors

Kolja Knauer, Departament de Matemàtiques Informàtic, Universitat de Barcelona, Barcelona, Barcelona, Spain

Elijah Liflyand, Department of Mathematics, Bar-Ilan University, Ramat-Gan, Israel

This series comprises direct translations of successful foreign language titles, especially from the German language.

Powered by advances in automated translation, these books draw on global teaching excellence to provide students and lecturers with diverse materials for teaching and study.

Uwe Storch · Hartmut Wiebe

Fundamental Concepts of Mathematics

Set-Theoretic, Algebraic, and Topological Foundations as well as Real and Complex Numbers

Uwe Storch
Fakultät für Mathematik
Ruhr University Bochum
Bochum, Nordrhein-Westfalen, Germany

Hartmut Wiebe
Fakultät für Mathematik
Ruhr-Universität Bochum
Bochum, Nordrhein-Westfalen, Germany

ISSN 2731-3824   ISSN 2731-3832 (electronic)
Mathematics Study Resources
ISBN 978-3-662-69696-5   ISBN 978-3-662-69697-2 (eBook)
https://doi.org/10.1007/978-3-662-69697-2

This book is a translation of the original German edition "Grundkonzepte der Mathematik" by Uwe Storch, published by Springer-Verlag GmbH, DE in 2017. The translation was done with the help of an artificial intelligence machine translation tool. A subsequent human revision was done primarily in terms of content, so that the book will read stylistically differently from a conventional translation. Springer Nature works continuously to further the development of tools for the production of books and on the related technologies to support the authors.

Translation from the German language edition: "Grundkonzepte der Mathematik" by Uwe Storch and Hartmut Wiebe, © Springer-Verlag GmbH Deutschland 2017. Published by Springer Berlin Heidelberg. All Rights Reserved.

© The Editor(s) (if applicable) and The Author(s), under exclusive license to Springer-Verlag GmbH, DE, part of Springer Nature 2025

This work is subject to copyright. All rights are solely and exclusively licensed by the Publisher, whether the whole or part of the material is concerned, specifically the rights of translation, reprinting, reuse of illustrations, recitation, broadcasting, reproduction on microfilms or in any other physical way, and transmission or information storage and retrieval, electronic adaptation, computer software, or by similar or dissimilar methodology now known or hereafter developed.
The use of general descriptive names, registered names, trademarks, service marks, etc. in this publication does not imply, even in the absence of a specific statement, that such names are exempt from the relevant protective laws and regulations and therefore free for general use.
The publisher, the authors and the editors are safe to assume that the advice and information in this book are believed to be true and accurate at the date of publication. Neither the publisher nor the authors or the editors give a warranty, expressed or implied, with respect to the material contained herein or for any errors or omissions that may have been made. The publisher remains neutral with regard to jurisdictional claims in published maps and institutional affiliations.

This Springer imprint is published by the registered company Springer-Verlag GmbH, DE, part of Springer Nature.
The registered company address is: Heidelberger Platz 3, 14197 Berlin, Germany

If disposing of this product, please recycle the paper.

Preface

With the present volume, we begin a new edition of our Mathematics Textbook [18–21], which has also been published in four volumes by Springer Spectrum. We are particularly pleased that we were able to win Prof. Dr. C. Becker from RheinMain University of Applied Sciences in Wiesbaden, Prof. Dr. M. Kersken from Flensburg University of Applied Sciences, and Prof. Dr. F. Loose from the University of Tübingen as co-authors. We thank them very much for their willingness and commitment. Based on the existing editions, the material will be restructured and expanded. The volumes will deal more specifically with individual topics and cover them more comprehensively than was previously possible, so that they will contain the material of a regular mathematics study (without specializations). The individual volumes are less extensive than before. The following topics are planned:

1. Basic concepts of mathematics
2. Analysis of a variable
3. Linear Algebra I
4. Linear Algebra II
5. Differential calculus of several variables
6. Measure and integration theory
7. Analysis on manifolds
8. Functional analysis
9. Stochastics
10. Ordinary differential equations
11. Theory of functions
12. Differential Geometry/Differential Topology
13. Algebra

We ourselves are still designing the first two volumes, the remaining volumes will be independently edited by the three new authors mentioned above. M. Kersken is taking over Volume 3, F. Loose Volume 5, and C. Becker Volume 6.

The project was initiated by Dr. A. Rüdinger from Springer Spektrum publishing, to whom we express our heartfelt thanks for this and for the excellent support of our books so far.

Bochum Uwe Storch
March 2017 Hartmut Wiebe

Introduction

Mathematics displays its beautiful aspects
only to its patient followers.
M. Mirzakhani (1977–2017)

The present Vol. 1 of this textbook series deals with basic concepts from set theory, algebra, and topology in four chapters and provides an introduction to real and complex numbers. It serves as the basis for the subsequent volumes. This particularly applies to the terminology. Its content is based on chapters I, II, and IV of Volume 1 and chapter I of Volume 3 of our Mathematics Textbook [18, 20]. The chapter on the algebraic foundations summarizes the algebraic topics scattered across the four volumes [18–21] and has grown beyond the initially intended scope. However, the material has also been expanded in many places. Our goal was to create a conceptual foundation for mathematics, but also to present deeper results in detail. Although the presentation is systematic, a certain familiarity with natural, rational, and real numbers is expected from the outset in individual examples or exercises. The book is aimed at all those who want to engage more intensively with mathematics.

In the following, we describe the content of the individual chapters in more detail.

Chapter 1 is dedicated to the basic ways of speaking about sets, mappings, and relations. The order-theoretical concepts are discussed in detail. This includes an outline of the theory of cardinal and ordinal numbers and a proof of Zorn's Lemma with its immediate consequences such as the well-ordering theorem, comparability theorem for cardinal and ordinal numbers, and the product theorem for infinite cardinal numbers. Starting from the Peano axioms, we provide an introduction to natural numbers, for which complete induction is central. The basic methods and results of elementary combinatorics about counting finite sets are presented. The starting point of an introduction to elementary number theory is the Euclidean algorithm and the main theorem obtained from it about the unique prime factor decomposition of natural numbers.

As basic concepts of algebra, monoids and groups, rings and fields, as well as modules and algebras are discussed in Chap. 2, Homomorphisms play a decisive role as structure-preserving mappings and lead to standard constructions such as quotient formations, sums and products, and the discussion of free objects. We always emphasize the universal properties of the constructed objects and thus prepare for a more category-theoretical view in later volumes. The operation of monoids and groups provides a unified perspective in the treatment of various algebraic structures. Moreover, this revives the original meaning of groups as transformation groups. We develop group theory up to the Sylow theorems and concretize it on finite permutation groups. The simplicity of the alternating groups of sets with at least five elements is proven. Further applications include the quadratic reciprocity law according to Jacobi and Pólya's enumeration formula.

Rings, modules, and algebras are structures with multiple canonically connected links, which are considered under the already mentioned general aspects. With the general Chinese Remainder Theorem, the structure of the minimal rings, i.e., the residue class rings of $\mathbb{Z}$ and their unit groups, the prime residue class groups, are clarified. Modules and vector spaces are treated, including the concept of rank and dimension. The section on algebras discusses, among other things, very detailed (also non-commutative) polynomial algebras up to the Hilbert Basis Theorem and the prime factor decomposition. Principal ideal domains and in particular Euclidean domains with their specific module theory find their place. Application examples are finite fields and the two as well as the four squares theorem.

The real numbers form an ordered field. This fact, and thus the study of inequalities, are the starting point of Chap. 3. Central is the convergence of sequences, which in turn leads to the concept of completeness and to the characterization of $\mathbb{R}$ as a complete ordered field. Completeness is illuminated from various sides, from which also natural constructions of $\mathbb{R}$ result from the rational numbers. The transition to complex numbers is then a small step. However, trigonometric functions are already used here in anticipation of Vol. 2 for the polar coordinate representation of complex numbers. In the treatment of series, the concept of summability offers considerable methodological advantages. It is therefore consistently used. The continuity of real and complex functions on subsets of $\mathbb{R}$ or $\mathbb{C}$ is extensively treated, including the peculiarities in compact definition areas. As an application, one obtains the classical proof of the Fundamental Theorem of Algebra. The chapter concludes with the introduction of real exponential and logarithmic functions.

Topological structures play a crucial role in all areas of mathematics today. They are the subject of Chap. 4. Starting from the concept of distance, metric spaces are first introduced, which specifically include normed vector spaces. Metric spaces motivate the concept of the topological space with the associated homomorphisms, namely the continuous mappings. Relevant constructions such as image and pre-image topologies with their universal properties, especially quotients and products, are discussed in detail. The fundamental concepts of connectivity and compactness, which already played an important role in Chap. 3 for the spaces $\mathbb{R}$ and $\mathbb{C}$, are central objects of consideration. We discuss Tychonoff's theorem and the completeness of metric spaces as well as the various concepts of convergence in mapping spaces such as pointwise, uniform, locally uniform, and compact convergence up to the Arzelà-Ascoli theorem with the Hausdorff distance of closed sets in metric spaces as an application. Finally, summability in Hausdorff abelian topological groups as a generalization of summability in $\mathbb{R}$ and $\mathbb{C}$ is introduced.

The individual sections are supplemented by numerous exercises, the results of which are occasionally used in the text. Hints are given for the somewhat more difficult exercises. In addition, solutions to some exercises can be found in our workbooks [22] and [23]. The examples serve not only to illustrate the theory, but also to further it. We hope that they will structure the presentation more strongly and increase the overview. The end of examples and remarks is each marked by a $\diamond$ and the end of proofs by $\square$.

The book does not reflect the content of a single lecture. It does not have to be read page by page, rather the reader can select individual topics and use it as a reference book if needed. The detailed index is intended to help with this.

The second volume of this series deals with the analysis of functions of a real and complex variable, that is, with the differentiation and integration of such functions.

We sincerely thank Dr. A. Rüdinger and Ms. I. Ruhmann from Springer Spektrum publishing for their advice on the content direction of this volume, and Ms. A. Herrmann for the technical support.

Uwe Storch passed away in September 2017 due to a tragic bathing accident.
H.W.

Bochum   Uwe Storch
July 2017 Hartmut Wiebe

Contents

1 Foundations of Set Theory 1
   1.1   Sets 1
   1.2   Mappings and Families 7
   1.3   Relations 20
   1.4   Order Relations 28
   1.5   Natural Numbers and Complete Induction 40
   1.6   Finite Sets and Combinatorics 49
   1.7   Prime Factorization of Natural Numbers 64
   1.8   Infinite Sets and Cardinal Numbers 84

2 Algebraic Foundations 101
   2.1   Monoids and Groups 101
   2.2   Homomorphisms 126
   2.3   Induced Homomorphisms and Quotient Formation .. 154
   2.4   Operation of Monoids and Groups 180
   2.5   Permutation Groups 202
   2.6   Rings 229
   2.7   Ideals and Quotient Rings 246
   2.8   Modules and Vector Spaces 262
   2.9   Algebras 283
   2.10 Principal Ideal Domains and Factorial Integral Domains 314

3 Real and Complex Numbers 345
   3.1   Ordered Fields 345
   3.2   Convergent Sequences 355
   3.3   Real Number Fields 360
   3.4   Consequences of Completeness 380
   3.5   The complex numbers 392
   3.6   Series 410
   3.7   Summability 426
   3.8   Continuous Functions 439
   3.9   Continuous Functions on Compact Sets 466
   3.10 Real Exponential, Logarithm, and Power Functions 472

4	Topological Foundations	481
	4.1 Metric Spaces	481
	4.2 Topological Spaces and Continuous Mappings	492
	4.3 Connected Spaces	523
	4.4 Compact Spaces	537
	4.5 Complete Metric Spaces and Uniform Convergence	559

Literature 585

Foundations of Set Theory 1

1.1 Sets

Mathematics is formulated and taught today in the language of set theory. Objects that can be grouped into sets are considered. These can then, in turn, be elements of sets. The following classic and still current description is given by G. Cantor (1845–1918) in the work: Contributions to the Founding of the Transfinite Set Theory (First Article), Math. Ann. **46**, 481–512 (1895):

> By a **set** M we understand any collection of certain well-distinguished objects m of our intuition or our thinking (which are called the **elements** of M) into a whole.

H. Hermes (1912–2003) commented somewhat critically in his lectures on Cantor's definition of a set as follows: While people in a lecture hall are objects of our intuition and natural numbers are objects of our thinking, the classification of, for example, the Greek gods and goddesses is indeed questionable. (The ability to consider them as elements of a set is probably not doubted.)

Let A be a set and a an object. Then

$$a \in A \quad \text{resp.} \quad a \notin A,$$

means that a is an element of A or that a is not an element of A. A set is constituted by the elements it contains. Two sets are therefore exactly equal if they contain the same elements. This is the so-called **extensional viewpoint**.

By forming a set, a new object is always created. In particular, there are no sets $A_0, A_1, \ldots, A_n$ with $A_0 \in A_1 \in \cdots \in A_{n-1} \in A_n$ and $A_n = A_0$. Specifically, there are no sets A, B with $A \in A$ or with $A \in B \in A$. The set of all sets does not exist; it would have to contain itself. Thus, the basis for the so-called **Russell's Paradox** is also eliminated: B. Russell (1872–1970) considered the set B of all sets A that do not contain themselves and found that by definition B is an element of B exactly when B is not an element of B. This would be paradoxical. However, the set B does not exist, because it would be identical to the set of all sets for any set A due to $A \notin A$.

Remark 1.1.1 Set theory also exists as a mathematical theory. This attempts to axiomatize the methods of reasoning used in set theory. The only fundamental concept is the element-of relationship $a \in A$. A significant part of the axioms relates to the formation and existence of special sets, whereby a formal language is particularly used. One of the most commonly used axiom systems today is the **ZFC** system by E. Zermelo (1871–1953) and A. Fraenkel (1891–1965). The "C" stands for the **Axiom of Choice** (= Axiom of Choice), which we will address later, see for example Sect. 1.4. However, the reader should not be confused: Set theory is a mathematical theory like any other. To quote H. Hermes once more: Like all mathematicians, set theorists use the same dirty cloth (namely set theory) to polish their silver (namely set theory). For an introduction to axiomatic set theory, refer to books 7 and 5. ◊

Sets are described by explicitly listing their elements—this is called **enumerative notation**—or by characterizing their elements through a property that precisely applies to the objects belonging to the considered set. It is, for example

$$\{1, -1\} = \{1, -1, 1\} = \{x \mid x \text{ is a real number and it is } x^2 = 1\}$$
$$= \{x \in \mathbb{R} \mid x^2 = 1\}.$$

One and the same set can be described in many ways. As already mentioned, what is decisive for the equality of sets is that they contain the same elements. We use the following standard notations:

$\mathbb{N} = \{0, 1, 2, 3, \ldots\}$ Set of natural numbers (after DIN 5473),
$\mathbb{N}^* = \{1, 2, 3, \ldots\}$ Set of positive natural numbers,
$\mathbb{Z} = \{0, 1, -1, 2, -2, \ldots\}$ Set of entire numbers,
$\mathbb{Q} = \{\frac{a}{b} = a/b \mid a, b \in \mathbb{Z}, b \neq 0\}$ Set of rational numbers,
$\mathbb{R}$ Set of real numbers,
$\mathbb{R}^\times = \{x \in \mathbb{R} \mid x \neq 0\}$ Set of real numbers $\neq 0$,
$\mathbb{R}_+ = \{x \in \mathbb{R} \mid x \geq 0\}$ Set of nonnegative real numbers,
$\mathbb{R}_- = \{x \in \mathbb{R} \mid x \leq 0\}$ Set of nonpositive real numbers,
$\mathbb{R}_+^\times = \{x \in \mathbb{R} \mid x > 0\}$ Set of positive real numbers,
$\mathbb{C} = \{a + bi \mid a, b \in \mathbb{R}\}$ Set of complex numbers,
$\mathbb{C}^\times = \{z \in \mathbb{C} \mid z \neq 0\}$ Set of complex numbers $\neq 0$.

Elementary arithmetic in these number domains is used from the beginning. However, we will discuss real and complex numbers in detail in Chap. 3. Initially, complex numbers do not play a major role.

1.1 Sets

A and B are sets. A is called a **subset** of B, if every element of A is also an element of B. It is then written

$$A \subseteq B \quad \text{or} \quad B \supseteq A$$

and we speak of the **inclusion** of set A in set B. If $A \neq B$, the inclusion is called **proper**. In this case, we also write

$$A \subset B \quad \text{or} \quad B \supset A.$$

$A \subset B$ thus means that every element of set A also belongs to set B, but there is at least one element in B that does not belong to A.[1] It is apparent that from the inclusions $A \subseteq B$ and $B \subseteq C$ follows the inclusion $A \subseteq C$. Exactly then $A = B$ holds, when $A \subseteq B$ and $B \subseteq A$ apply. Therefore, one usually proves the equality $A = B$ of two sets A and B by verifying these two inclusions, that is, by showing for all x: *From $x \in A$ follows $x \in B$ and from $x \in B$ follows $x \in A$.* The set that contains no element at all is called the **empty set** and is denoted by

$$\emptyset$$

It is a subset of every set.[2]—We briefly describe the most important set operations. Let A and B be sets. Then the sets are called

$A \cup B := \{x \mid x \in A \text{ or } x \in B\}$	the **conjunction**,
$A \cap B := \{x \mid x \in A \text{ and } x \in B\}$	the **intersection** und
$A - B := \{x \mid x \in A \text{ and } x \notin B\}$	the **difference**

of the sets A and B. Here, as in the following, the colon before the equals sign means that the left side of the equation is defined by the right side (and conversely, the right side by the left, if the colon is after the equals sign). If B is a subset of A, then the difference $A - B$ is called the **complement** of B in A. It is also denoted by $\complement B = \complement_A B$. Note that in the definition of the union, the word "or" is used in the non-exclusive sense, as always in mathematics, i.e., not in the sense of "either-or." $A \cap B$ is thus a subset of $A \cup B$. The set

$$A \triangle B := \{x \mid \text{either } x \in A \text{ or } x \in B\} = (A - B) \cup (B - A)$$
$$= (A \cup B) - (A \cap B)$$

is called the **symmetric difference** of A and B. Two sets with an empty intersection are called **disjoint**. Thus, $A \triangle B = A \cup B$ holds if and only if the sets A and B are

---

[1] The inclusion symbols $\subset$ and $\supset$ are not used consistently in the literature. They occasionally denote the not necessarily proper inclusion.

[2] A beginner can easily overlook that a set can be empty. If one wants to pick an element a_0 from a given set A, one should ensure that A is not empty.

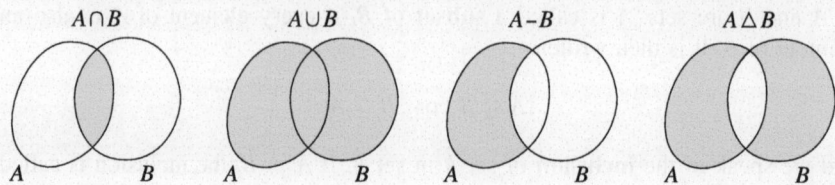

Fig. 1.1 Venn diagrams

disjoint. These and similar concept formations are often illustrated by so-called **Euler** or **Venn diagrams** (after L. Euler (1707–1783) and J. Venn (1834–1923), respectively), see Fig. 1.1.

We note some calculation rules for the described operations:

Proposition 1.1.2 *A, B and C are sets. Then the following holds:*

(1) $A \cup \emptyset = A$, $A \cap \emptyset = \emptyset$.
(2) $A \cup B = B \cup A$, $A \cap B = B \cap A$. **(commutativity)**
(3) $A \cup (B \cup C) = (A \cup B) \cup C$,
    $A \cap (B \cap C) = (A \cap B) \cap C$. **(associativity)**
(4) $A \cup (B \cap C) = (A \cup B) \cap (A \cup C)$,
    $A \cap (B \cup C) = (A \cap B) \cup (A \cap C)$. **(distributivity)**
(5) $A - (B \cup C) = (A - B) \cap (A - C)$,
    $A - (B \cap C) = (A - B) \cup (A - C)$. **(rules of De Morgan** (1806 – 1871))
(6) $A - (A - B) = A \cap B$.

Proof We exemplarily prove the first of the equations from (4) and first show the inclusion $A \cup (B \cap C) \subseteq (A \cup B) \cap (A \cup C)$. Let $x \in A \cup (B \cap C)$. Then $x \in A$ or $x \in B \cap C$, thus $x \in A$ or simultaneously $x \in B$ and $x \in C$. Then the statements "$x \in A$ or $x \in B$" and "$x \in A$ or $x \in C$" are both true. Therefore, $x \in A \cup B$ and $x \in A \cup C$ and thus $x \in (A \cup B) \cap (A \cup C)$.

To prove the inclusion $(A \cup B) \cap (A \cup C) \subseteq A \cup (B \cap C)$, let $x \in (A \cup B) \cap (A \cup C)$. Then $x \in A \cup B$ and $x \in A \cup C$, so it holds simultaneously "$x \in A$ or $x \in B$" and "$x \in A$ or $x \in C$". It follows $x \in A$ or $x \in B \cap C$, i.e., $x \in A \cup (B \cap C)$. □

Regarding the associative laws 1.1.2 (3), it should be noted that the set $A \cap B \cap C := (A \cap B) \cap C = A \cap (B \cap C)$ or the set $A \cup B \cup C := (A \cup B) \cup C = A \cup (B \cup C)$ is simply the set of those elements that lie in all or at least one of the sets A, B, C. In this way, intersection and union can be defined for any number of sets, see Sect. 1.2. For subsets of a set, the rules (5) from Proposition 1.1.2 can be formulated as follows: *The complement of the union is the intersection of the complements, and the complement of the intersection is the union of the complements.* Finally, *the complement of the complement is the original subset.* We will occasionally use more

1.1 Sets

Fig. 1.2 Cartesian Product

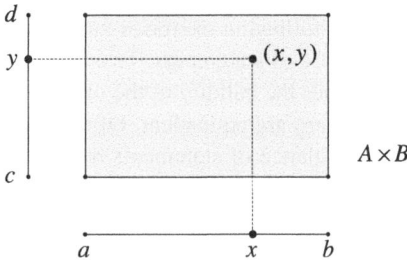

of such elementary set relationships without always justifying them in detail. Some examples can also be found in the exercises for this section.

Let A be an arbitrary set again. The set of subsets of A is called the **power set** of A and is denoted by

$$\mathfrak{P}(A)$$

The power set $\mathfrak{P}(\{\Box, \Diamond, \triangle\})$ of the set $\{\Box, \Diamond, \triangle\}$, for example, is the set $\{\emptyset, \{\Box\}, \{\Diamond\}, \{\triangle\}, \{\Box, \Diamond\}, \{\Box, \triangle\}, \{\Diamond, \triangle\}, \{\Box, \Diamond, \triangle\}\}$.[3] Intersection, union, and difference of sets from $\mathfrak{P}(A)$ again belong to $\mathfrak{P}(A)$. Note that the power set $\mathfrak{P}(\emptyset) = \{\emptyset\}$ of the empty set $\emptyset$ is *not* empty.[4]

For sets A and B, the set of **pairs** (x, y), $x \in A$, $y \in B$, is called the **Cartesian product** or the **cross product** of A and B and is denoted by

$$A \times B$$

Two pairs (x, y) and (v, w) are equal if and only if their corresponding **components** match, i.e., if $x = v$ and $y = w$. Thus, it is necessary to distinguish between the pair (x, y) and the set $\{x, y\}$. For $x \neq y$, $(x, y) \neq (y, x)$ is, of course, $\{x, y\} = \{y, x\}$.[5]

Example 1.1.3 If A and B are the real intervals $[a, b] = \{x \in \mathbb{R} \mid a \leq x \leq b\}$ and $[c, d] = [y \in \mathbb{R} \mid c \leq y \leq d]$ with $a < b$ and $c < d$, then $A \times B$ is the "rectangle"

$$[a, b] \times [c, d] = \{(x, y) \in \mathbb{R} \times \mathbb{R} \mid a \leq x \leq b, c \leq y \leq d\} \subseteq \mathbb{R} \times \mathbb{R},$$

see Fig. 1.2. The Cartesian product of two sets is generally illustrated as such a rectangle. Cross products with any number of factors are discussed in Sect. 1.2. ◊

---

[3] Even if the elements of the set A are concrete objects of our perception, its power set as a whole is only an object of thought.
[4] This is, in a way, a creation out of nothing.
[5] According to K. Kuratowski (1896–1980), the concept of the pair can be directly traced back to basic concepts of set theory by setting, for example, $(x, y) := \{x, \{x, y\}\}$.

In the following exercises and also later, we will frequently use the concept of equivalence of statements. Two statements are called **equivalent**, if the validity of one implies the validity of the other. Several statements are called equivalent if any two of them are equivalent. Often (though not always), it is useful to demonstrate the equivalence of statements $\alpha_1, \ldots, \alpha_n$ through a so-called **circular argument**, where (possibly after renumbering) the implications $\alpha_1 \Rightarrow \alpha_2, \alpha_2 \Rightarrow \alpha_3, \ldots, \alpha_{n-1} \Rightarrow \alpha_n$ and finally $\alpha_n \Rightarrow \alpha_1$ are shown. If the statements α, β are equivalent, we write $\alpha \Leftrightarrow \beta$.

Exercises

Exercise 1.1.1 For a set A, the following statements are equivalent: (i) $A = \emptyset$. (ii) For every set B, $A - B = A \cap B$. (iii) There exists a set B with $A - B = A \cap B$. (iv) For every set B, $B - A = B \cup A$. (v) There exists a set B with $B - A = B \cup A$.

Exercise 1.1.2 For sets A, B, the following statements are equivalent: (i) $A \subseteq B$. (ii) $A \cap B = A$. (iii) $A \cup B = B$. (iv) $A - B = \emptyset$. (v) $B - (B - A) = A$. (vi) For every set C, $A \cup (B \cap C) = (A \cup C) \cap B$. (vii) There exists a set C with $A \cup (B \cap C) = (A \cup C) \cap B$.

Exercise 1.1.3 For sets A, B, it holds that $(A \cap B) \cap (A - B) = \emptyset$ and $(A \cap B) \cup (A - B) = A$.

Exercise 1.1.4 For sets A, B, C, the following holds:

a) $A \cap (B - C) = (A \cap B) - (A \cap C)$.
b) $(A \cup B) - C = (A - C) \cup (B - C)$.
c) $(A - B) - C = A - (B \cup C)$.
d) $A - (B - C) = (A - B) \cup (A \cap C)$.

Exercise 1.1.5 The sets A, B are equal if and only if $A \cup B = A \cap B$ is true.

Exercise 1.1.6 For sets A, B, C, show:

a) $A \triangle A = \emptyset$, $A \triangle \emptyset = A$.
b) Exactly then is $A = B$, when $A \triangle B = \emptyset$ is.
c) Exactly then is $A \cap B = \emptyset$, when $A \triangle B = A \cup B$ is.
d) $(A \triangle B) \cap (A \cap B) = \emptyset$, $(A \triangle B) \cup (A \cap B) = A \cup B$.
e) $(A \triangle B) \cap C = (A \cap C) \triangle (B \cap C)$.
f) From $A \triangle B = A \triangle C$ follows $B = C$.
g) $(A \triangle B) \triangle C = A \triangle (B \triangle C)$.

Exercise 1.1.7 The *non-empty* sets A, B are exactly then equal when $A \times B = B \times A$ is.

1.2 Mappings and Families

Exercise 1.1.8 For sets A, B, C, show:

a) $A \times (B \cup C) = (A \times B) \cup (A \times C)$.
b) $A \times (B \cap C) = (A \times B) \cap (A \times C)$.
c) $A \times (B - C) = (A \times B) - (A \times C)$.

Exercise 1.1.9 Let A, B, C, D be sets. From $A \subseteq C$ and $B \subseteq D$ follows $A \times B \subseteq C \times D$. In the case of $A \neq \emptyset$ and $B \neq \emptyset$ the reverse also holds.

Exercise 1.1.10 For sets A, B, C, D, the following holds:

a) $(A \times B) \cap (C \times D) = (A \cap C) \times (B \cap D)$.
b) $(A \times B) \cup (C \times D) \subseteq (A \cup C) \times (B \cup D)$. The equality holds here if and only if both the condition "$A \subseteq C$ or $D \subseteq B$" and the condition "$C \subseteq A$ or $B \subseteq D$" is fulfilled. $\square$

1.2 Mappings and Families

The concept of mapping is one of the most fundamental in all of mathematics. The reader should become familiar with it as early as possible.

A and B be sets. A **mapping from A to B** is understood as a rule f, which assigns to each element $x \in A$ exactly one element from B, which we generally denote by $f(x)$. A mapping f from A to B is briefly written in the form

$$f : A \longrightarrow B \quad \text{or} \quad A \xrightarrow{f} B$$

and on the element level through

$$x \longmapsto f(x), \quad x \in A.$$

A is called the **domain** and B the **range** or also the **codomain** of f. Both sets are constitutive for the mapping f. For $x \in A f(x) \in B$ is called the **image** of x under f or also the **value** of f for the **argument** x or at the **position** x. The **graph** of f is the subset

$$\Gamma := \Gamma(f) := \Gamma_f := \{(x, f(x)) \mid x \in A\} \subseteq A \times B$$

of $A \times B$. Together with the domain and range, the graph $\Gamma(f)$ fully characterizes the mapping f. In principle, mappings are to be identified with their graphs. The definition of a mapping as an assignment (rule) is generally too vague. A mapping f with domain A and range B is thus a subset Γ of $A \times B$ with the following property: For every $x \in A$ there is exactly one $y \in B$ (namely $y = f(x)$) with $(x, y) \in \Gamma$. We note that mappings f and g are equal if and only if their domains and ranges respectively coincide and if for all elements x from the common domain

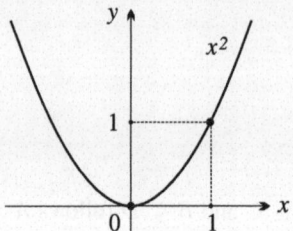 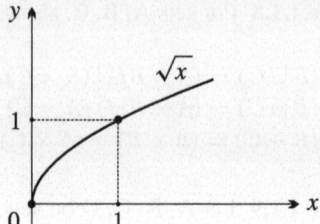

Fig. 1.3 Quadratic function and square root function

$f(x) = g(x)$ holds. The manner in which the values are described is irrelevant. This is again the extensional standpoint , see Sect. 1.1. The set of all mappings from A to B is denoted by [6]

$$\mathrm{Abb}(A, B) \quad \text{or} \quad B^A.$$

Instead of mappings, one often speaks of **functions**. This terminology is particularly common when the values lie in numerical domains. Sum, product, and similar operations of functions f, g on the set A with values in the same numerical domain, such as the real numbers $\mathbb{R}$, are then defined by the corresponding operations for the values. For example,

$$(f+g)(x) := f(x)+g(x) \quad \text{and} \quad (fg)(x) := f(x)g(x)$$

for all x from the common domain A of f and g. Similarly, multiplication by a number a is explained:

$$(af)(x) := af(x), \quad x \in A.$$

The **constant function** a assigns the fixed value a to each argument x.

Example 1.2.1 The quadratic function $\mathbb{R} \to \mathbb{R}$ is given by $x \mapsto x^2$. The square root initially does not provide a function of $\mathbb{R}$ in $\mathbb{R}$, since negative numbers do not have a real square root. Even on $\mathbb{R}_+$ the square root still does not provide a function $\mathbb{R}_+ \to \mathbb{R}$, since a positive real number has two square roots in $\mathbb{R}$. However, through $x \mapsto \sqrt{x}$ a function $\mathbb{R}_+ \to \mathbb{R}$ is defined, if—as is generally customary—*under $\sqrt{x}$ for $x \in \mathbb{R}_+$ the non-negative square root is understood.* The corresponding graphs are thus the subsets of $\mathbb{R} \times \mathbb{R}$ or $\mathbb{R}_+ \times \mathbb{R}$ shown in Fig. 1.3. ◊

Example 1.2.2 We mention some other frequently used functions, see Fig. 1.4.

---

[6] The power notation B^A is motivated by the general products of sets introduced further below.

1.2 Mappings and Families

(1) The mapping $\mathbb{R} \to \mathbb{R}$ with $x \mapsto x$ is called the **identity** of $\mathbb{R}$. It is defined in an analogous manner for any set A and is then denoted by $\mathrm{id} = \mathrm{id}_A$. Its graph is the **diagonal** $\Delta_A := \{(x, x) \mid x \in A\} \subseteq A \times A$. If $f : A \to A$ is any mapping of the set A into itself, then a point $x \in A$ with $f(x) = x$ is called a **fixed point** of f. The mapping f has a fixed point if and only if its graph $\Gamma(f)$ intersects the diagonal $\Delta_A = \Gamma(\mathrm{id}_A)$. The identity of A is the mapping of A into itself for which every point of A is a fixed point. The set of fixed points of $f : A \to A$ we denote by

$$\mathrm{Fix}\, f = \mathrm{Fix}(f, A) = \{x \in A \mid f(x) = x\}.$$

Many important theorems in mathematics are theorems about the existence of fixed points, so-called **fixed point theorems**.

(2) The function $\mathbb{R} \to \mathbb{R}$ with $x \mapsto |x| := \begin{cases} x, & \text{if } x \geq 0, \\ -x, & \text{if } x < 0, \end{cases}$ is called **(absolute) value** or **absolute value function** (on $\mathbb{R}$).

(3) The function $\mathbb{R} \to \mathbb{R}$ with $x \mapsto \mathrm{Sign}\, x := \begin{cases} 1, & \text{if } x > 0, \\ 0, & \text{if } x = 0, \\ -1, & \text{if } x < 0, \end{cases}$ is called **sign (function)** or **signum** (on $\mathbb{R}$).

(4) The function $\mathbb{R} \to \mathbb{R}$ with $x \mapsto [x]$, where $[x]$ is the largest integer that is $\leq x$, is called the **Gauss bracket** (after C. F. Gauss (1777–1855)). For $x \in \mathbb{R}$ is thus $[x]$ determined by the following two conditions: (1) $[x] \in \mathbb{Z}$ and (2) $[x] \leq x < [x] + 1$. For example, $[\pi] = 3$ and $[-\pi] = -4$. The Gauss bracket of x is often also denoted by $\lfloor x \rfloor$ and is called the **floor** of x. The smallest integer that is $\geq x$ is denoted by $\lceil x \rceil$. It is called the **ceiling** of x. For all $x \in \mathbb{R}$ is $\lceil x \rceil = -\lfloor -x \rfloor$, and for $x \notin \mathbb{Z}$ is $\lceil x \rceil = \lfloor x \rfloor + 1$, see Fig. 1.4. ◊

Let $f : A \to B$ be a mapping. For $A' \subseteq A$ or $B' \subseteq B$ are called

$$f(A') := \{f(x) \mid x \in A'\} \subseteq B \quad \text{and} \quad f^{-1}(B') := \{x \in A \mid f(x) \in B'\} \subseteq A$$

the **image** of A' or the **preimage** of B' under f. The image $f(A)$ is called the **image** (not to be confused with the image range) of f per se and is denoted by $\mathrm{Im}\, f$. It is a subset of the image range of f (and generally different from it). For

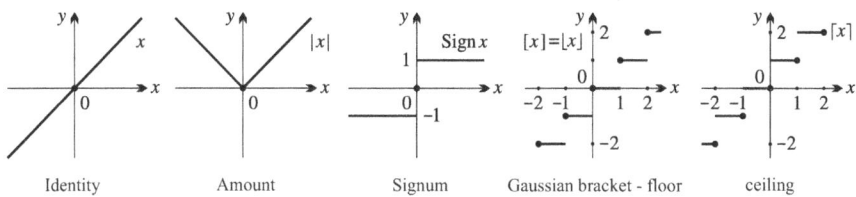

Fig. 1.4 Examples of figures $\mathbb{R} \to \mathbb{R}$

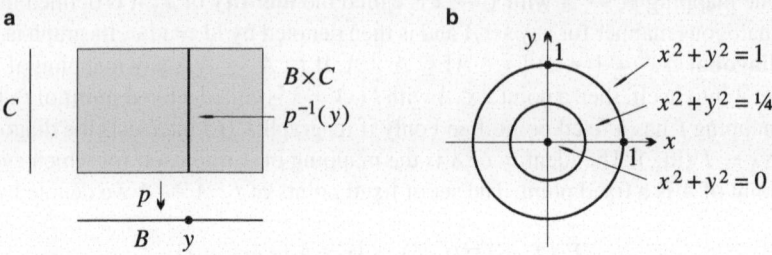

Fig. 1.5 Fibers of a projection, fibers of $x^2 + y^2$

$y \in B$ is called $f^{-1}(y) := f^{-1}(\{y\}) = \{x \in A \mid f(x) = y\}$ the **fiber of f over** y. For $x \in A$ is $f^{-1}(f(x))$ the **fiber of f through** x. If $f : A \to A$ is a mapping from A into itself, then $A' \subseteq A$ is called f-**invariant**, if $f(A') \subseteq A'$ holds. Then f induces a mapping by restriction $f' = f|A' : A' \to A'$, $f'(x) := f(x)$ for $x \in A'$, from A' into itself. For example, $\operatorname{Im} f = f(A)$ is always f-invariant.

Example 1.2.3 If B and C are sets, then the fibers of the **projection** $p : (y, z) \mapsto y$ from $B \times C$ onto B are the sets $\{y\} \times C \subseteq B \times C$, $y \in B$, see Fig. 1.5a. The fiber $f^{-1}(a)$ of the function $f : (x, y) \mapsto x^2 + y^2$ from $\mathbb{R} \times \mathbb{R}$ in $\mathbb{R}$ is empty for $a < 0$, contains only the point $0 = (0, 0)$ for $a = 0$, and is the circle around 0 with radius $\sqrt{a}$ for $a > 0$, see Fig. 1.5b. ◇

Let $f : A \to B$ be a mapping and $A' \subseteq A$ or $B' \subseteq B$ subsets with $f(A') \subseteq B'$. Then f naturally defines a mapping $A' \to B'$ by restricting the domain to A' and the range to B', and mapping the arguments $x \in A'$ as in f. This **restriction** is denoted by $f|A'$, noting that this does not express the (usually less significant) restriction of the image range from B to B'.

The terms introduced below are fundamental:

Definition 1.2.4 Let $f : A \to B$ be a mapping.

(1) f is called **injective**, if for every $y \in B$ there is at most one $x \in A$ with $f(x) = y$.
(2) f is called **surjective**, if for every $y \in B$ there is (at least) one $x \in A$ with $f(x) = y$.
(3) f is called **bijective**, if for every $y \in B$ there is exactly one $x \in A$ with $f(x) = y$.

Exactly then is $f : A \to B$ injective if every fiber of f contains at most one element. This means: If $x, x' \in A$ are elements with $f(x) = f(x')$, then it follows $x = x'$. *Exactly then is f surjective if every fiber of f contains at least one element*, i.e., if $\operatorname{Im} f = B$ is. An injective mapping $f : A \to B$ we occasionally also write in the form $f : A \hookrightarrow B$ and a bijective mapping in the form $f : A \xrightarrow{\sim} B$. *Only then is f bijective if each fiber of f contains exactly one element*, i.e., *iff f is injective and surjective*. If $f : A \to B$ is a surjective mapping, it is also referred to as a mapping

1.2 Mappings and Families

from A **onto** B. A bijective mapping of a set A onto itself is called a **permutation** of A. The set of all permutations of A is denoted by

$$\mathfrak{S}(A), \quad \text{in particular} \quad \mathfrak{S}(\{1, 2, \ldots, n\}) \quad \text{with} \quad \mathfrak{S}_n, n \in \mathbb{N}.$$

The set $\mathfrak{S}_0 = \mathfrak{S}(\emptyset) = \{\emptyset\}$ contains the empty mapping and is therefore non-empty.

Example 1.2.5 The elements f in $\mathfrak{S}_n$ are conveniently given with a **value table**

$$\begin{pmatrix} 1 & 2 & \cdots & n \\ f(1) & f(2) & \cdots & f(n) \end{pmatrix}$$

where one also omits the first row if it is clear that the arguments are listed in their natural order. Thus, $\binom{123}{312}$ or briefly $(3, 1, 2)$ is the permutation $1 \mapsto 3, 2 \mapsto 1, 3 \mapsto 2$ from $\mathfrak{S}_3$. ◇

Example 1.2.6 The simplest non-trivial permutations of a set A (with more than one element) are the transpositions of A. A **transposition** τ swaps two different elements $a, b \in A$ and leaves the remaining elements of A fixed. Thus, $\tau(a) = b, \tau(b) = a$ and $\tau(x) = x$ for $x \in A - \{a, b\}$. This transposition is denoted by $\langle a, b \rangle$. For convenience, one also sets $\langle a, a \rangle := \mathrm{id}_A$) for $a \in A$. ◇

Example 1.2.7 The mapping $\mathbb{R} \to \mathbb{R}$ with $x \mapsto x^2$ is neither injective nor surjective. The mapping $\mathbb{R}_+ \to \mathbb{R}$ with $x \mapsto x^2$ is injective but not surjective. The mapping $\mathbb{R} \to \mathbb{R}_+$ with $x \mapsto x^2$ is surjective but not injective. Finally, the mapping $\mathbb{R}_+ \to \mathbb{R}_+$ with $x \mapsto x^2$ and also the mapping $\mathbb{R}_+ \to \mathbb{R}_+$ with $x \mapsto \sqrt{x}$ are bijective. See Fig. 1.3. ◇

Definition 1.2.8 Let $f : A \to B$ be a bijective mapping. The mapping $B \to A$, which assigns to each $y \in B$ the (uniquely determined) element $x \in A$ with $f(x) = y$ is called the f **inverse mapping** or the **inverse function** of f. It is denoted by f^{-1}.

Bijective mappings are also called **invertible** mappings. For any set A, the identity $\mathrm{id}_A : A \to A, x \mapsto x$, a bijective mapping that is inverse to itself. If $f : A \to B$ is any bijective mapping, then $f^{-1}(f(x)) = x$ holds for all $x \in A$ and $f(f^{-1}(y)) = y$ for all $y \in B$. With f, f^{-1} is also bijective and it holds

$$(f^{-1})^{-1} = f.$$

If $\Gamma(f)(\subseteq A \times B)$ is the graph of the bijective mapping $f : A \to B$, then the graph $\Gamma(f^{-1})$ ($\subseteq B \times A$) of the inverse mapping to $f^{-1} : B \to A$ is the image of $\Gamma(f)$ under the bijective mapping $(x, y) \mapsto (y, x)$ from $A \times B$ to $B \times A$, which swaps the components and whose inverse mapping $B \times A \to A \times B$ also swaps the components (but is a different mapping when $A \neq B$). If f is bijective, then for a

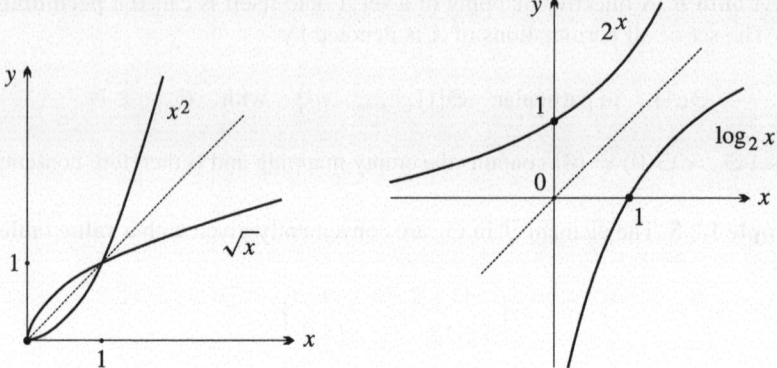

Fig. 1.6 x^2 and $\sqrt{x}$ or 2^x and $\log_2 x$ as inverse functions of each other

subset $B' \subseteq B$ the image of B' under the inverse mapping $f^{-1}: B \to A$ and the preimage of B' under $f: A \to B$ coincide, so that the common designation $f^{-1}(B')$ does not lead to misunderstandings.

Example 1.2.9 The inverse mapping of the bijective square function $\mathbb{R}_+ \to \mathbb{R}_+$ with $x \mapsto x^2$ is the square root function $\mathbb{R}_+ \to \mathbb{R}_+$ with $x \mapsto \sqrt{x}$, and, conversely, the inverse function of the square root function is the square function.—Let $a \in \mathbb{R}_+^\times$, $a \neq 1$. The inverse mapping of the bijective exponential function $\mathbb{R} \to \mathbb{R}_+^\times$ with $x \mapsto a^x$ is the logarithm function $\mathbb{R}_+^\times \to \mathbb{R}$ with $x \mapsto \log_a x$. See Fig. 1.6. ◊

Definition 1.2.10 A, B, and C are sets, $f: A \to B$ and $g: B \to C$ are mappings. Then the mapping $A \to C$ with $x \mapsto g(f(x))$ is called the **composition** or **concatenation** of f and g. It is denoted by $g \circ f$ or simply by gf.

Compositions are clearly illustrated with **commutative diagrams**. The commutativity of the diagram in Fig. 1.7 means, for example, that $f = h \circ g$ is. In this case, the fiber condition $g^{-1}(g(a)) \subseteq f^{-1}(f(a))$ holds for every $a \in A$, i.e., (the restriction of) f is constant on the fibers of g respectively. Namely, if $a' \in g^{-1}(g(a))$, then $f(a') = h(g(a')) = h(g(a)) = f(a)$. Moreover, for an element $c = g(a) \in \text{Im}\, g$ it is necessarily $h(c) = h(g(a)) = f(a)$, i.e., the mapping is uniquely determined on the image of g by f.

Conversely, if $f: A \to B$ and $g: A \to C$ are mappings with $g^{-1}(g(a)) \subseteq f^{-1}(f(a))$ for all $a \in A$ and if g is surjective, then there is exactly one mapping $h: C \dashrightarrow B$ with $f = h \circ g$, with which the diagram in Fig. 1.7 is commutative. It

Fig. 1.7 Induced mapping

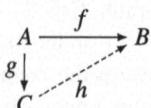

1.2 Mappings and Families

is called the f-**induced** mapping by means of g. As already mentioned, $h(c)$ is to be defined for $c \in C$ by $h(c) := f(a)$, where a is any element of the fiber $g^{-1}(c)$. The fact that this value is independent of the choice of a is precisely the assumed fiber condition that f is constant on the fibers of g. The image of the f-induced mapping h is evidently identical to the image of f. h is injective if and only if for all $a \in A$ the equality of the fibers $g^{-1}(g(a)) = f^{-1}(f(a))$ holds.

According to the definition, two mappings f and g can initially only be composed in sequence if the range of f and the domain of g are the same. However, the composition can already be performed if $\mathrm{Im} f$ is only a subset of the domain of g. Generally, this does not lead to misunderstandings. Trivially, $\mathrm{id}_B \circ f = f = f \circ \mathrm{id}_A$ is valid for every mapping $f: A \to B$. If f is bijective, then, as already mentioned, $f^{-1} \circ f = \mathrm{id}_A$ and $f \circ f^{-1} = \mathrm{id}_B$. Through these equations, f^{-1} is uniquely determined. More generally, the following statement holds, the proof of which we leave to the reader.

Proposition 1.2.11 *Let $f: A \to B$ and $g: B \to A$ be mappings with $gf = \mathrm{id}_A$ and $fg = \mathrm{id}_B$. Then f and g are bijective, and it holds that $f^{-1} = g$, $g^{-1} = f$.*

Fundamental is also the **associativity** of the composition of mappings:

Theorem 1.2.12 *Let $f: A \to B$, $g: B \to C$, and $h: C \to D$ be mappings. Then $(hg)f = h(gf)$.*

Proof For all $a \in A$ it holds: $((hg)f)(a) = (hg)(f(a)) = h(g(f(a))) = h((gf)(a)) = (h(gf))(a)$. Thus, as claimed, $(hg)f = h(gf)$. □

According to Theorem 1.2.12, one can omit the parentheses when composing mappings. This also applies to the sequential composition of more than three mappings:

Theorem 1.2.13 (General Associative Law) *For mappings $f_i: A_i \to A_{i-1}$, $i = 1, \ldots, n$, ($n \in \mathbb{N}^*$) the composition $f_1 \circ \cdots \circ f_n: A_n \to A_0$ is independent of the bracketing.*

Proof We conclude by induction on n and refer to Theorem 1.5.1 or Theorem 1.5.5. For $n = 1, 2$ there is nothing to show, for $n = 3$ it is Theorem 1.2.12. So let $n \geq 3$. Then, by the induction hypothesis, for all $k = 1, \ldots, n-1$ the products $f_1 \circ \cdots \circ f_k$, $f_{k+1} \circ \cdots \circ f_n$ are independent of the bracketing. We have $(f_1 \circ \cdots \circ f_k) \circ (f_{k+1} \circ \cdots \circ f_n) = f_1 \circ (f_2 \circ \cdots \circ f_n)$ to show for all $k = 1, \ldots, n-1$. This is also done by induction (now on k). Let $k > 1$. Then, however, $f_1 \circ \cdots \circ f_k = (f_1 \circ \cdots \circ f_{k-1}) \circ f_k$ and $f_k \circ (f_{k+1} \circ \cdots \circ f_n) = f_k \circ \cdots \circ f_n$, thus

$$(f_1 \circ \cdots \circ f_k) \circ (f_{k+1} \circ \cdots \circ f_n) = ((f_1 \circ \cdots \circ f_{k-1}) \circ f_k) \circ (f_{k+1} \circ \cdots \circ f_n)$$
$$= (f_1 \circ \cdots \circ f_{k-1}) \circ (f_k \circ (f_{k+1} \circ \cdots \circ f_n))$$

$$= (f_1 \circ \cdots \circ f_{k-1}) \circ (f_k \circ f_{k+1} \circ \cdots \circ f_n) = f_1 \circ (f_2 \circ \cdots \circ f_n),$$

where in the last equation the induction hypothesis for $k-1$ was used. □

According to Theorem 1.2.13, in particular, the n-fold composition, $n \in \mathbb{N}^*$, of the same mapping $f : A \to A$ is well-defined. We denote this **iterate** $f \circ \cdots \circ f$ (n-times) of f briefly with f^n and also set $f^0 := \mathrm{id}_A$.

Remark 1.2.14 For functions with values in $\mathbb{R}$ or $\mathbb{C}$ (or more generally in multiplicative semigroups), f^n usually means the function $x \mapsto (f(x))^n$, which is generally different from the just introduced n-fold composition $x \mapsto f(\cdots(f(x))\cdots)$, if this is even defined (i.e., if the domain and range of f coincide). For example, the square of the identity $\mathbb{R} \to \mathbb{R}$ is the function $x \mapsto x^2 = x \cdot x$, but the composition of the identity with itself is again the identity $x \mapsto x$. Similarly, with a bijective function $f : A \to A$, $A \subseteq \mathbb{R}^\times$ (or $A \subseteq \mathbb{C}^\times$), one must distinguish between the inverse function f^{-1} and the reciprocal $1/f : x \mapsto 1/f(x)$, which is also often denoted by f^{-1}. Usually, the context makes it clear what is meant. ◊

Example 1.2.15 Even if the compositions $f \circ g$ and $g \circ f$ are both defined, they generally do not agree. For the functions $f : x \mapsto x+1$ and $g : x \mapsto x^2$ of $\mathbb{R}$ in itself, for example, is $f \circ g : x \mapsto x^2 + 1$ and $g \circ f : x \mapsto (x+1)^2 = x^2 + 2x + 1$. Thus, *the composition of mappings does not generally satisfy a commutation law.* ◊

If $f : A \to B$ and $g : B \to C$ are injective (or surjective or bijective), then this also holds for $gf : A \to C$. In the bijective case, we explicitly note:

Proposition 1.2.16 *If $f : A \to B$ and $g : B \to C$ are bijective mappings, then $gf : A \to C$ is bijective with $(gf)^{-1} = f^{-1}g^{-1}$.*

Proof With the associativity of composition, it holds that $(f^{-1}g^{-1})(gf) = f^{-1}(g^{-1}g)f = f^{-1}\mathrm{id}_B f = f^{-1}f = \mathrm{id}_A$ and $(gf)(f^{-1}g^{-1}) = g(ff^{-1})g^{-1} = g\mathrm{id}_B g^{-1} = gg^{-1} = \mathrm{id}_C$. Now, Proposition 1.2.11 is applied. □

Example 1.2.17 The set $\mathfrak{S}(A)$ of permutations of a set A is, according to Prop. 1.2.16, closed under composition $\circ$ and also under the formation of inverses. Furthermore, the identity id_A belongs to the permutations of A. Since the composition is also associative, $\mathfrak{S}(A) = (\mathfrak{S}(A), \circ)$ is a group in the sense of Definition 2.1.5, namely the **permutation group** of A, see also Example 2.1.13. We have already encountered simple permutations in the transpositions $\langle a, b \rangle$, $a, b \in A$, $a \neq b$, in Example 1.2.6. These each coincide with their inverse mapping. Generally, a mapping $f : A \to A$ with $f^2 = \mathrm{id}_A$ is a permutation of A with $f = f^{-1}$. Such a mapping f is called an **involution** or a **reflection** of A and its set of fixed points $\mathrm{Fix}(f, A)$ the **mirror** of f (which can be empty). ◊

1.2 Mappings and Families

Mappings allow elements of a set to be identified by indices, to be indexed. We briefly address the general terminology associated with this. Let $f : I \to A$ be a mapping. Then we say the set $\operatorname{Im} f$ is indexed by f through I, and call I the corresponding **index set**. For $i \in I$ we often write $f(i)$ as f_i and instead of referring to the mapping $f : I \to A$ we speak of the **family** $(f_i)_{i \in I}$ or f_i, $i \in I$. Such families are also called I-**tuples** (of elements from A or with components from A). In the I-tuple $(f_i)_{i \in I}$ the element f_i is called the i-**th component**. We emphasize that for different indices i and j the components f_i and f_j may coincide. The **indexing** f therefore does not need to be injective. Specifically, if I is the set $\mathbb{N}$ of natural numbers, then the families with the index set $I = \mathbb{N}$ are the (infinite) **sequences** $(f_n)_{n \in \mathbb{N}}$ or $(f_0, f_1, f_2, \ldots)$. In $I = \{1, 2, \ldots, n\}$ these are the n-**tuples** $(f_1, f_2, \ldots, f_n)$. The 2-tuple (f_1, f_2), i.e., the mapping $1 \mapsto f_1, 2 \mapsto f_2$, can be identified with the pair (f_1, f_2); 3-tuples are called **triples**, etc. An infinite sequence $(f_n)_{n \in \mathbb{N}}$ is called **stationary** if it is constant from a certain point, that is, if there is a $n_0 \in \mathbb{N}$ with $f_n = f_{n_0}$ for all $n \geq n_0$. The value f_{n_0} is then called the **limit value** or **limit** of the sequence.

If $(A_i)_{i \in I}$ is a family of sets A_i with $I \neq \emptyset$, then

$$\bigcap_{i \in I} A_i := \{x \mid x \in A_i \text{ for all } i \in I\} \qquad \text{the \textbf{intersection} and}$$

$$\bigcup_{i \in I} A_i := \{x \mid x \in A_i \text{ for (at least) one } i \in I\} \qquad \text{the \textbf{conjunction}}$$

of the sets A_i, $i \in I$. If $A_1, \ldots, A_n$ is a finite sequence of sets, we also write

$$A_1 \cap \cdots \cap A_n := \bigcap_{i=1}^{n} A_i \quad \text{resp.} \quad A_1 \cup \cdots \cup A_n := \bigcup_{i=1}^{n} A_i$$

for their intersection or their union. The union of the empty family of sets ($I = \emptyset$) we set equal to $\emptyset$. However, we define the intersection for the empty family of sets only if a fixed set A is given, of which all sets to be considered are subsets. *In this case, A is the intersection over the empty family of sets.*

Furthermore, let $(A_i)_{i \in I}$ be a family of sets and $A := \bigcup_{i \in I} A_i$ their union. Then the set of I-tuples $(a_i)_{i \in I}$ of elements from A with $a_i \in A_i$ for all $i \in I$ is called the **Cartesian product** or the **cross product** of the sets A_i. It is denoted by

$$\prod_{i \in I} A_i.$$

In the case of $I = \{1, \ldots, n\}$ the notation $A_1 \times \cdots \times A_n := \prod_{i=1}^{n} A_i$ is also common. For two sets A, B, we thus obtain the product $A \times B$ already introduced earlier (when pairs are identified with 2-tuples). If $A_i \neq \emptyset$ for each $i \in I$, then the cross product is also $\prod_{i \in I} A_i \neq \emptyset$. This is the so-called axiom of choice in set theory. We will discuss it in more detail in Sect. 1.4. In the case $A_i = A$ for all $i \in I$ the product $\prod_{i \in I} A_i$ is simply the set $A^I = \operatorname{Abb}(I, A)$ of all mappings from I to A. In $I = \{1, \ldots, n\}$ is the set of n-tuples of elements from A. It is denoted by A^n.

Thus, it is $A^n = \{(a_1, \ldots, a_n) \mid a_i \in A, i = 1, \ldots, n\}$. $A^0 = \{\emptyset\}$ contains as its only element the empty mapping $\emptyset \to A$. The Cartesian product $\prod_{i \in I} A_i$ includes the **canonical projections**, namely, for a $j \in I$ the mapping $\prod_i A_i \to A_j$, which assigns to an I-tuple $(a_i) \in \prod_i A_i$ the j-th component a_j, the j- **th projection**. In the case $A_i = A$ for all $i \in I$ the j-th projection assigns to a mapping $f \in A^I$ the value $f(j) \in A$ of f at the point $j \in I$.

For a subset $J \subseteq I$ denotes e_J the function $I \to \{0, 1\}$ with

$$e_J(i) := \begin{cases} 1, & \text{if } i \in J, \\ 0, & \text{if } i \notin J. \end{cases}$$

e_J is called the **indicator function** of J (with respect to I).[7] If $J = \{j\}, j \in I$, a single-element subset, then we write more briefly e_j for $e_{\{j\}}$. Thus, e_j is the I-tuple that has the value 1 at position j and the value 0 at all other positions:

$$e_j = (\delta_{ij})_{i \in I} \quad \text{with} \quad \delta_{ij} := \begin{cases} 1, & \text{if } i = j, \\ 0, & \text{if } i \neq j. \end{cases}$$

δ_{ij} is called the **Kronecker symbol**. For $I = \{1, \ldots, n\}$ and $j \in I$ e_j is the n-tuple

$$e_j = (0, \ldots, 0, 1, 0, \ldots, 0),$$

where the one is at the j-th position. Often, n-tuples are also noted as columns instead of rows (especially when calculating with matrices).

Exercises

Exercise 1.2.1 Sketch the graphs of the following functions $\mathbb{R} \to \mathbb{R}$:

a) $x \mapsto \{x\} := x - [x]$ (**sawtoothcurve**).

b) $x \mapsto \begin{cases} x - |x|, & \text{if } [x] \leq x < [x] + \dfrac{1}{2}, \\ [x] + 1 - x, & \text{if } [x] + \dfrac{1}{2} \leq x < [x] + 1 \end{cases}$ (**Distance to the nearest whole number**). The restriction of twice this function to the unit interval $[0, 1]$ is called the **tent function**. It maps the unit interval onto itself. Also sketch the graphs of some of the iterates of this function.

c) $x \mapsto \begin{cases} 0, & \text{if } x \leq 0, \\ 1, & \text{if } x > 0 \end{cases}$ (**Heaviside-function**).

---

[7] Some authors also call indicator functions **characteristic functions**. Besides e_J, the notations $\chi_J, 1\!\!1_J$ etc. are common. The elements $0, 1$ may lie in any ring, which must be specified if necessary. In general and without specification, this is (the ring) $\mathbb{Z}$.

1.2 Mappings and Families

d) $x \mapsto \lceil x \rceil$ (cf. Example 1.2.2 (4)).
e) $x \mapsto x|x| = (\text{Sign } x)x^2$.
f) $x \mapsto x + |x - 1|$.
g) $x \mapsto |x^2 - 4|$.

Exercise 1.2.2 Let f, g, and h be the functions defined by $f(x) := 1/(1+x^2)$, $g(x) := |x|$ and $h(x) := x + 1$ from $\mathbb{R}$ into itself. Form the compositions fg, fh, gh, gf, hg, hf and check which of these functions coincide.

Exercise 1.2.3 For which $a, b, c \in \mathbb{R}$ is the function $f: \mathbb{R} \to \mathbb{R}$ with $f(x) := ax^2 + bx + c$ bijective?

Exercise 1.2.4 Check which of the following mappings from $\mathbb{R} \times \mathbb{R}$ into itself (where the value for $(x, y) \in \mathbb{R} \times \mathbb{R}$ is given) is injective, surjective, or bijective. In the bijective case, provide the inverse mapping.

$(y, 3); \quad (x + y^2, y + 2); \quad (xy, x + 1); \quad (xy, x + y); \quad (2x^2 - y, x + y);$

$(x - y, x^2 - y^2); \quad (xy, x^2 - y^2); \quad \left(x/\sqrt{1 + x^2 + y^2}, y/\sqrt{1 + x^2 + y^2}\right).$

Solve the corresponding exercise for the mapping from $\mathbb{R} \times \mathbb{R} - \{(0,0)\}$ into itself with $(x, y) \mapsto \left(x/(x^2 + y^2), y/(x^2 + y^2)\right)$.

Exercise 1.2.5 Let $a, b, c, d \in \mathbb{R}$ and $f: \mathbb{R} \to \mathbb{R}$, $g: \mathbb{R} \to \mathbb{R}$ be the functions defined by $f(x) := ax + b$ and $g(x) := cx + d$. Under what conditions are f and g interchangeable, i.e., when does $f \circ g = g \circ f$ hold?

Exercise 1.2.6 The fibers of real-valued functions are also called **level sets**. Sketch the level sets of the following functions $\mathbb{R} \times \mathbb{R} \to \mathbb{R}$ for the values $-2, -1, 0, 1, 2$:

a) $(x, y) \mapsto x + y$, $(x, y) \mapsto xy$.
b) $(x, y) \mapsto |x - 1| + |y + 2|$.
c) $(x, y) \mapsto |y - x|$.
d) $(x, y) \mapsto x^2 - 4x + y^2$.
e) $(x, y) \mapsto \sqrt[3]{x + 1} - \sqrt{|y|}$.
f) $(x, y) \mapsto xy - (x + y)$.

Exercise 1.2.7 Let $f: A \to B$ and $g: B \to C$ be mappings and $gf: A \to C$ their composition.

a) If gf is injective, then f is injective.
b) If gf is surjective, then g is surjective.
c) If gf is bijective, then f is injective and g is surjective. (Provide an example where gf is bijective, but neither f nor g is.)

d) If gf is bijective and f (or g) is bijective, then g (or f) is also bijective.

Exercise 1.2.8 Prove Prop. 1.2.11.

Exercise 1.2.9

a) Let $f: A \to B, g: B \to C$ and $h: C \to D$ be mappings. If gf and hg are bijective, then f, g, and h are bijective.
b) Let $f: A \to B, g: B \to A$ and $h: A \to B$ be mappings. From $gf = \mathrm{id}_A$ and $hg = \mathrm{id}_B$ it follows that $f = h$, i.e., the bijectivity of g and $g^{-1} = f = h$.

Exercise 1.2.10 Let $f: A \to B$ a mapping.

a) The following statements are equivalent: (i) f is injective. (ii) For all sets C and all mappings $g_1: C \to A$ and $g_2: C \to A$ it follows from $fg_1 = fg_2$ already $g_1 = g_2$. Moreover, if $A \neq \emptyset$ (and thus also $B \neq \emptyset$), these two conditions are equivalent to: (iii) There exists a mapping $g: B \to A$ with $gf = \mathrm{id}_A$. (Such a mapping g is called a **retraction** to f.)
b) The following statements are equivalent: (i) f is surjective. (ii) For all sets D and all mappings $h_1: B \to D$ and $h_2: B \to D$ it follows from $h_1 f = h_2 f$ already $h_1 = h_2$. (iii) There exists a mapping $h: B \to A$ with $fh = \mathrm{id}_B$. (Such a mapping h is called a **section** to f. For each $b \in B$ $h(b)$ belongs to the fiber $f^{-1}(b)$.—To construct a section h, one uses the axiom of choice.)

Exercise 1.2.11 Let $f: A \to A$ be a mapping. Then $\mathrm{Fix} f \subseteq \mathrm{Im} f$ holds, and the equality holds if and only if $f^2 = f$ is. (It is then said that f is a **projection**.)

Exercise 1.2.12 Let $p: A \to B$ and $q: C \to D$ be mappings. For mappings $f: A \to C$ and $g: B \to D$, the following conditions are equivalent: (i) It is $q \circ f = g \circ p$, i.e., the diagram in Fig. 1.8 is commutative. (ii) It is $f(p^{-1}(b)) \subseteq q^{-1}(g(b))$ for all $b \in B$.

Exercise 1.2.13 Let $f: A \to B$ be a mapping. For the mappings induced by f $f_*: \mathfrak{P}(A) \to \mathfrak{P}(B)$, $A' \mapsto f(A')$, and $f^*: \mathfrak{P}(B) \to \mathfrak{P}(A)$, $B' \mapsto f^{-1}(B')$, the following holds:

a) Equivalent are: (i) f is injective. (ii) f_* is injective. (iii) f^* is surjective.
b) Equivalent are: (i) f is surjective. (ii) f_* is surjective. (iii) f^* is injective.
c) If f is bijective, then f_* and f^* are inverse bijective mappings to each other.

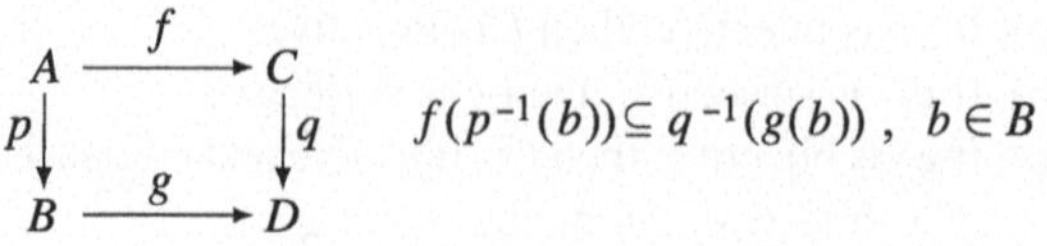

Fig. 1.8 Fiber condition for commuting mappings

$$\begin{array}{ccc} A & \xrightarrow{f} & C \\ p \downarrow & & \downarrow q \\ B & \xrightarrow{g} & D \end{array} \qquad f(p^{-1}(b)) \subseteq q^{-1}(g(b)), \; b \in B$$

1.2 Mappings and Families

Exercise 1.2.14 Let $f: A \to B$ be a mapping.

a) For all $A' \subseteq A$ is $f^{-1}(f(A')) \supseteq A'$.
b) For all $B' \subseteq B$ is $f(f^{-1}(B')) \subseteq B'$.
c) For all $A', A'' \subseteq A$ is $f(A' \cup A'') = f(A') \cup f(A'')$, $f(A' \cap A'') \subseteq f(A') \cap f(A'')$ and $f(A' - A'') \supseteq f(A') - f(A'')$.
d) For all $B', B'' \subseteq B$ is $f^{-1}(B' \cup B'') = f^{-1}(B') \cup f^{-1}(B'')$, $f^{-1}(B' \cap B'') = f^{-1}(B') \cap f^{-1}(B'')$ and $f^{-1}(B' - B'') = f^{-1}(B') - f^{-1}(B'')$. In particular, $f^{-1}(\complement_B B'') = \complement_A(f^{-1}(B''))$ is.
e) For all $A' \subseteq A$, $B' \subseteq B$ is $f(A' \cap f^{-1}(B')) = f(A') \cap B'$.
f) Equivalent are: (i) f is surjective. (ii) For all $B' \subseteq B$ is $f(f^{-1}(B')) = B'$.
g) Equivalent are: (i) f is injective. (ii) For all $A' \subseteq A$ is $f^{-1}(f(A')) = A'$. (iii) For all $A', A'' \subseteq A$ is $f(A' \cap A'') = f(A') \cap f(A'')$. (iv) For all $A', A'' \subseteq A$ with $A' \cap A'' = \emptyset$ is $f(A') \cap f(A'') = \emptyset$. (v) For all $A', A'' \subseteq A$ is $f(A' - A'') = f(A') - f(A'')$.

Exercise 1.2.15 Let $f: A \to B$ and $g: B \to C$ be mappings. For all $A' \subseteq A$ or $C' \subseteq C$ holds $(gf)(A') = g(f(A'))$ and $(gf)^{-1}(C') = f^{-1}(g^{-1}(C'))$. Thus, it is $(gf)_* = g_* f_*$ and $(gf)^* = f^* g^*$. (See Ex. 1.2.13.—It is said that the transition from f to f_* is **covariant** and that from f to f^* **contravariant**.)

Exercise 1.2.16 Let A, B be sets, $B \neq \emptyset$. There is a surjective mapping from A onto B if and only if there is an injective mapping from B into A. (See Ex. 1.2.10b).)

Exercise 1.2.17 Let $A_i, i \in I$, and $B_j, j \in J$, be families of sets with $I \neq \emptyset \neq J$. Then the following holds:

a) $(\bigcap_{i \in I} A_i) \cup (\bigcap_{j \in J} B_j) = \bigcap_{(i,j) \in I \times J} (A_i \cup B_j)$, $(\bigcap_{i \in I} A_i) \cup (\bigcup_{j \in J} B_j) = \bigcap_{i \in I} (\bigcup_{j \in J} A_i \cup B_j)$.

b) $(\bigcup_{i \in I} A_i) \cap (\bigcup_{j \in J} B_j) = \bigcup_{(i,j) \in I \times J} (A_i \cap B_j)$, $(\bigcup_{i \in I} A_i) \cap (\bigcap_{j \in J} B_j) = \bigcup_{i \in I} (\bigcap_{j \in J} A_i \cap B_j)$.

c) $(\bigcup_{i \in I} A_i) - (\bigcup_{j \in J} B_j) = \bigcup_{i \in I} (\bigcap_{j \in J} (A_i - B_j))$, $(\bigcup_{i \in I} A_i) - (\bigcap_{j \in J} B_j) = \bigcup_{(i,j) \in I \times J} (A_i - B_j)$.

If A is another set, then $A - (\bigcup_{j \in J} B_j) = \bigcap_{j \in J} (A - B_j)$ and $A - (\bigcap_{j \in J} B_j) = \bigcup_{j \in J} (A - B_j)$. In particular, the complement of a union is equal to the intersection of the complements, and the complement of an intersection is equal to the union of the complements.

Exercise 1.2.18 Let $f: A \to B$ be a mapping and $A_i, i \in I$, respectively $B_j, j \in J$, be families of subsets $A_i \subseteq A$ respectively $B_j \subseteq B$. Then the following holds:

a) $f(\bigcup_{i \in I} A_i) = \bigcup_{i \in I} f(A_i)$ and $f(\bigcap_{i \in I} A_i) \subseteq \bigcap_{i \in I} f(A_i)$.

b) $f^{-1}(\bigcup_{i \in I} B_i) = \bigcup_{i \in I} f^{-1}(B_i)$ and $f^{-1}(\bigcap_{i \in I} B_i) = \bigcap_{i \in I} f^{-1}(B_i)$.

Exercise 1.2.19 Let A, I, and J be sets. The mapping

$$f \mapsto \left(j \mapsto (i \mapsto f_{ij} = f(i,j))\right)$$

is a bijective mapping from $M_{I,J}(A) := A^{I \times J}$ to $(A^I)^J$. (The elements $f : I \times J \to A$ of $M_{I,J}(A)$ are the so-called $I \times J$-**matrices** with coefficients in A. The **partial mappings** $f_{\bullet j}: I \to A$, $i \mapsto f_{ij} = f(i,j), j \in J$, respectively $f_{i\bullet}: J \to A, j \mapsto f_{ij} = f(i,j), i \in I$, are called the *j*-**th column** respectively the *i*-**th row** of the matrix $f = (f_{ij}) \in M_{I,J}(A)$. The mapping $f \mapsto (i \mapsto f_{i\bullet})$ is a bijective mapping from $M_{I,J}(A) = A^{I \times J}$ to $(A^J)^I$.—If $I = J$, then we write $M_I(A)$ for $M_{I,I}(A)$.)

Exercise 1.2.20 Let I be a set. The mapping $J \mapsto e_J$, which assigns the indicator function e_J of J to a subset $J \subseteq I$, is a bijective mapping of the power set $\mathfrak{P}(I)$ of I onto the set $\{0,1\}^I$ of all mappings $I \to \{0,1\}(\subseteq \mathbb{N})$. The inverse mapping is $e \mapsto e^{-1}(1)$.

Exercise 1.2.21 Let I be a set and J, K subsets of I with the complements $J' = I - J$ and $K' = I - K$ in I.

a) Prove the following equations about indicator functions (which are to be understood as functions with values in $\mathbb{Z}$): $e_\emptyset = 0$, $e_I = 1$, $e_{J \cap K} = e_J e_K$, $e_{J \cup K} = e_J + e_K - e_J e_K$, $e_{J-K} = e_J(1 - e_K)$. In particular, $e_{J'} = 1 - e_J$ and $e_{J \triangle K} = e_J + e_K - 2e_J e_K$.
b) In general, let $K_1 := J \cap K$, $K_2 := J \cap K'$, $K_3 := J' \cap K$, $K_4 := J' \cap K'$. Then, for each of the 16 subsets $S \subseteq \{1,2,3,4\}$, provide the indicator function of $\bigcup_{i \in S} K_i$ (using the indicator functions of J and K).

1.3 Relations

Who is married to whom can be most easily and clearly described by specifying the set of married couples. In general, one defines:

Definition 1.3.1 Let A and B be sets. A **relation between** A **and** B is a subset R of $A \times B$. If $A = B$, we speak of a **relation on** A.

If R is a relation between A and B and $(x, y) \in R$, then one also writes more suggestively xRy. Instead of a letter like R, specific symbols that already suggest certain properties of this relation are often used. For example, the equality sign $=$ describes the equality relation (on a set A), that is, the **diagonal** $\Delta_A = \{(a,a) \mid a \in A\} \subseteq A \times A$. For $x, y \in A$ namely $x = y$ is equivalent to $(x, y) \in \Delta_A$. This diagonal is the graph of the identity id_A. Every mapping $f: A \to B$ defines a relation between A and B, namely its graph $\Gamma_f \subseteq A \times B$. In general: If it is to be emphasized that a relation R between A and B is considered as a subset of $A \times B$—and it is nothing else—one also speaks of the **graph of the relation** R.

1.3 Relations

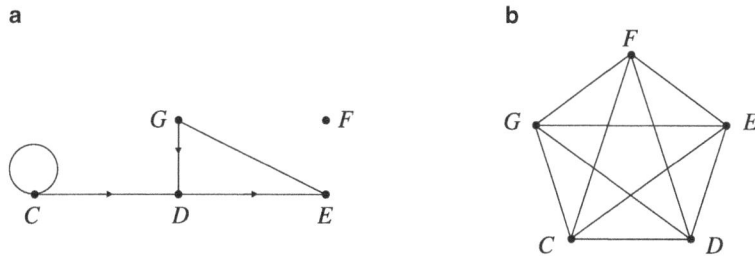

Fig. 1.9 Examples of graphs

Remark 1.3.2 (Graphs—Quivers) A relation R on a set A is occasionally also called a **directed graph** on A. Such a directed graph is illustrated in such a way that the elements of A are represented by **vertices** in the plane (or in space) and two vertices P, Q are connected with an **arrow** from P to Q if the pair belonging to (P, Q) from $A \times A$ lies in R. If both (P, Q) and (Q, P) are pairs in R, then P and Q are simply connected with an **edge** instead of a double arrow when $P \neq Q$. If (P, P) belongs to R, then a **loop** is attached to the vertex P. For example, the diagram in Fig. 1.9a represents the relation $\{(C, C), (C, D), (D, E), (E, G), (G, E), (G, D)\}$ on the set $\{C, D, E, F, G\}$.

If the relation R is **symmetric**, i.e., if with (P, Q) also (Q, P) belongs to R, then the graph contains no arrows, and it is called an **undirected graph**. An undirected graph without loops is simply called a **graph**. Every directed graph defines an undirected graph by ignoring the arrow directions and a simple graph by omitting the loops. A simple graph on a set A is determined by a subset K of the set $\mathfrak{E}_2(A)$ of the two-element subsets $\{P, Q\}$, $P, Q \in A$, $P \neq Q$, of A. Two vertices P, Q, $P \neq Q$, of the graph are connected by an edge if and only if $\{P, Q\}$ belongs to K. K is now called the set of edges of the graph. The graph in Fig. 1.9b is the **complete graph** for the vertex set $A = \{C, D, E, F, G\}$, which is defined by *all* two-element subsets of A.

Often, one has to consider generalized graphs or so-called **quivers** (English: quiver), where two vertices $P, Q \in A$ can be connected by multiple edges. This situation is most easily described by a mapping $\Gamma : K \to A \times A$, which is equivalent to the two component mappings $\alpha : K \to A$ and $\omega : K \to A$ with $\Gamma(k) = (\alpha(k), \omega(k))$. $\alpha(k)$ is then the **starting** and $\omega(k)$ the **endpoint** of the **arrow** $k \in K$. Directed graphs are thus special quivers. An **(oriented) path** of length $n \in \mathbb{N}^*$ in the quiver Γ is a sequence $\gamma = (k_1, \ldots, k_n)$ of arrows $k_i \in K$ with $\omega(k_i) = \alpha(k_{i+1})$ for $i = 1, \ldots, n-1$. The **starting point** $\alpha(\gamma)$ of such a path is $\alpha(k_1)$, its **endpoint** is $\omega(\gamma) = \omega(k_n)$. Furthermore, each vertex $P \in A$ of the quiver defines a path of length 0 with starting and endpoint P. Two paths γ_1 and γ_2 of lengths n_1 and n_2 with $\omega(\gamma_1) = \alpha(\gamma_2)$ can be naturally combined into a path $\gamma_1\gamma_2$ of length $n_1 + n_2$. A path in a graph, in general, is simply a sequence $(P_0, \ldots, P_n)$ of vertices, where $\{P_i, P_{i+1}\}$, $i = 0, \ldots, n-1$, is an edge. If $P_0 = P_n$ and $n > 0$, it is called a **loop**. A graph without loops is simply called a **forest**. A graph is called **connected** if it is not empty and any two vertices can be connected by a path. A connected forest is called a **tree**. The undirected paths of a directed graph are by definition the paths of the corresponding graph in general.

If one does not distinguish between the individual arrows in a quiver, which each connect points P, Q, and is only interested in their number $|\Gamma^{-1}(P, Q)|$, this leads to the concept of a **weighted graph**. This is generally a relation $R \subseteq A \times A$ together with a mapping $v: R \to L$, which is usually written as a square matrix $(\ell_{PQ})_{P,Q \in A} \in M_A(L)$, where ℓ_{PQ} for $(P, Q) \notin R$ is a fixed element not belonging to the image of v from L (usually this is 0, mostly L is a set of numbers). A relation $R \subseteq A \times A$ is thus described by its so-called **adjacency matrix**$(e_{PQ} = e_R(P, Q))_{P,Q \in A}$ with $e_{PQ} = 1$ if and only if $(P, Q) \in R$ is, and $e_{PQ} = 0$ otherwise. If R is a graph per se (without loops), then the adjacency matrix is symmetric (i.e., it is $e_{PQ} = e_{QP}$) and its diagonal elements e_{PP}, $P \in A$, are all equal to 0. The number of edges containing the vertex P of such a graph is called the **degree**v_P of P. The elements $\neq P$ of these edges are the v_P **neighbors** of P.[8] $\diamond$

We note some frequently occurring properties of relations $R \subseteq A \times A$ on a set A. R is called

(1) **reflexive**, if for all $a \in A$ holds: $(a, a) \in R$;
(2) **antireflexive**, if for all $a \in A$ holds: $(a, a) \notin R$;
(3) **symmetric**, if for all $a, b \in A$ from $(a, b) \in R$ always $(b, a) \in R$ follows;
(4) **antisymmetric**, if for all $a, b \in A$ from $(a, b) \in R$ and $(b, a) \in R$ always $a = b$ follows;
(5) **asymmetric**, if for all $a, b \in A$ from $(a, b) \in R$ always $(b, a) \notin R$ follows;
(6) **transitive**, if for all $a, b, c \in A$ from $(a, b) \in R$ and $(b, c) \in R$ always $(a, c) \in R$ follows.

The identification of things that one does not want to distinguish in a given situation to simplify considerations is captured by the concept of equivalence relation, which is fundamental to all of mathematics:

Definition 1.3.3 A relation $\sim$ on the set A is called an **equivalence relation** on A, if it is reflexive, symmetric, and transitive, i.e., if for all $a, b, c \in A$ holds:

(1) $a \sim a$; (2) from $a \sim b$ follows $b \sim a$; (3) from $a \sim b$ and $b \sim c$ follows $a \sim c$.

Two elements $a, b \in A$ are called **equivalent** (with respect to the equivalence relation $\sim$), if $a \sim b$ and thus also $b \sim a$ holds. For an element $a \in A$ the set of elements equivalent to a from A is called the **equivalence class** of the element a.

The equivalence class of a is generally denoted by $[a] = [a]_\sim$ or $\bar{a}$ (or in a similar manner). Thus, it is $[a] = \{x \in A \mid x \sim a\}$. The transition from the elements a to

---

[8] We note that the terminology for graphs in the literature is not uniform (and often must be inferred from the context).

1.3 Relations

their equivalence classes $[a]$ is an abstraction or identification process (specified by the respective equivalence relation $\sim$), which is described by the following theorem.

Theorem 1.3.4 *Let $\sim$ be an equivalence relation on the set A.*

(1) *For each $a \in A$ is $a \in [a]$ and in particular $[a] \neq \emptyset$. It is $A = \bigcup_{a \in A}[a]$.*
(2) *For all $a, b \in A$ the following three conditions are equivalent:*

$$\text{(i) } [a] = [b]. \quad \text{(ii) } [a] \cap [b] \neq \emptyset. \quad \text{(iii) } a \sim b.$$

Proof (1) results from $a \sim a$. To prove (2), we conduct a circular argument.

(i) $\Rightarrow$ (ii): For $[a] = [b]$ is $a \in [a] = [b]$, thus $a \in [a] \cap [b]$.
(ii) $\Rightarrow$ (iii): Let $c \in [a] \cap [b]$, i.e., $c \sim a$ and $c \sim b$. From $a \sim c$ and $c \sim b$ follows, due to transitivity, $a \sim b$.
(iii) $\Rightarrow$ (i): Due to symmetry, it suffices to show $[a] \subseteq [b]$. Let $x \in [a]$, thus $x \sim a$. Together with $a \sim b$ follows $x \sim b$, thus $x \in [b]$. $\square$

Let $R \subseteq A \times A$ be an equivalence relation $\sim$ on the set A. The set of equivalence classes with respect to $\sim$ is a subset of the power set $\mathfrak{P}(A)$ and is called the **quotient set** or the **quotient space** of A with respect to $\sim$. It is denoted by

$$A/R \quad \text{or} \quad A/\sim \quad \text{or} \quad \overline{A} \quad \text{or} \quad [A] \quad \text{or} \quad [A]_R \quad \text{or} \quad [A]_\sim \quad \text{and so on}$$

and the sets $C \times C$, $R = \bigcup_{C \in A/R}(C \times C)$, are pairwise disjoint, thus forming a partition of R. The surjective mapping $\pi = \pi_R \colon a \mapsto [a]$ from A onto $C \in A/R$, which assigns to an element $a \in A$ equivalence class $[a] \in \overline{A}$ with respect to $\sim$, is called the **canonical projection** from A onto $\overline{A}$. Its fibers are precisely the equivalence classes with respect to $\sim$. It is said that $\overline{A}$ arises from A by **identifying** the elements equivalent with respect to $\sim$. An element of an equivalence class is called a **representative** of this equivalence class. If one selects exactly one representative from each equivalence class (axiom of choice!), these together form a **complete system of representatives** or a so-called **fundamental domain** for the set of all equivalence classes. A subset $B \subseteq A$ is called **saturated with respect to the equivalence relation** R on A or R-**saturated**, if B contains with each element also every element equivalent to it with respect to R, that is, if B is a union of equivalence classes with respect to R. B is exactly then R-saturated, if $B = \pi_R^{-1}(\pi_R(B))$ is. The canonical projection $\pi_R \colon A \to A/R$ has the following universal property: If $f \colon A \to B$ is an arbitrary mapping, then there is a mapping $\overline{f} \colon A/R \to B$ with $f = \overline{f} \circ \pi_R$, if f is constant on the fibers of π_R, that is, on the equivalence classes with respect to R, respectively. $\overline{f} \colon A \to B$ is then **the mapping induced by** f.

Fig. 1.10 R is finer than S

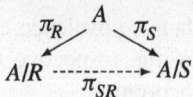

If R and S are equivalence relations on the set A, then R is called **finer** than S (or S **coarser** than R), if $R \subseteq S$ is, i.e., if from aRb always aSb follows or, equivalently, if for every $a \in A$ always $[a]_R \subseteq [a]_S$ holds, i.e., if the S-equivalence classes of A are R-saturated. The equality relation (with the diagonal Δ_A as graph) is the finest equivalence relation on A. Its equivalence classes are the single-element subsets of A. The coarsest equivalence relation on A is the **universal relation** (with $A \times A$ as graph). In $A \neq \emptyset$ A is its only equivalence class. If R is finer than S, then the fibers $[a]_R$ of the canonical projection $\pi_R \colon A \to A/R$ are contained in the fibers $[a]_S$ of the canonical projection $\pi_S \colon A \to A/S$ contained. Consequently, π_S induces a uniquely determined surjective mapping $\pi_{SR} \colon A/R \to A/S$ with $\pi_S = \pi_{SR} \circ \pi_R$, see Fig. 1.10. For $a \in A$, the fiber $\pi_{SR}^{-1}([a]_S)$ is the set of R-equivalence classes contained in $[a]_S$.

Each mapping $f \colon A \to B$ defines the equivalence relation

$$R_f \quad \text{by} \quad aR_fb \iff f(a) = f(b).$$

The equivalence classes of R_f are the non-empty fibers of f, more precisely: The equivalence class of $a \in A$ with respect to R_f is the fiber $f^{-1}(f(a))$ of f through a. If R is an equivalence relation on A that is finer than R_f, then f induces a mapping $\overline{f} \colon A/R \to B$ with $f = \overline{f} \circ \pi_A$, i.e., with $\overline{f}([a]_R) = f(a)$ for all $a \in A$, which is injective if and only if $R = R_f$ is. It is surjective if and only if f is surjective.

The equivalence classes with respect to an equivalence relation on a set A form a partition of the set A. In this context, a subset of the power set $\mathfrak{P}(A)$ of A is called a **partition** of A, if its elements are pairwise disjoint and their union is the entire A. In the case of equivalence classes with respect to an equivalence relation, the elements $\neq \emptyset$, it is also referred to as a **partition** of A. More generally, a family A_i, $i \in I$, of subsets of A is called a **partition** of A, if $A_i \cap A_j = \emptyset$ is for $i \neq j$ and $\bigcup_{i \in I} A_i = A$. In this case, we also write

$$A = \biguplus_{i \in I} A_i .$$

We do not require—as some authors do—that all $A_i \neq \emptyset$ are. However, if all sets A_i of a partition are non-empty, it is called a **proper partition** of A. Partitions are thus special proper partitions. *The partitions A_i, $i \in I$, of A correspond bijectively to the mappings $f \colon A \to I$:* The partition A_i, $i \in I$, defines on the one hand the mapping $f \colon A \to I$ with $f(a) := i$, if $a \in A_i$, and on the other hand the mapping $F \colon A \to I$ the partition $A_i := F^{-1}(i)$, $i \in I$, of A. The surjective mappings correspond to the proper partitions. Holds $\bigcup_{i \in I} A_i = A$, without necessarily $A_i \cap A_j = \emptyset$ is for all $i \neq j$, then the family A_i, $i \in I$, is called a **covering** of A.

1.3 Relations

Example 1.3.5 (Connected components of a graph) Let Γ be a graph with vertex set A and edge set $K \subseteq \mathfrak{E}_2(A) \subseteq \mathfrak{P}(A)$. Two points $P, Q \in A$ are called **connectable**, if there is a path $(P = P_0, P_1, \ldots, P_n = Q)$ with starting point P and endpoint Q in Γ. The connectivity is evidently an equivalence relation on A. The return path $(Q = P_n, \ldots, P_1, P_0 = P)$ connects Q with P. The graph Γ is thus connected if and only if there is exactly one equivalence class. Therefore, the equivalence classes with respect to connectivity are generally called the **connected components** of Γ. The connected components of a directed graph or a quiver are by definition the connected components of the associated graph in general. ◇

Example 1.3.6 (Congruence relations) Let $n \in \mathbb{N}$ be a natural number. Two integers a and b are called **congruent modulo** n, if their difference $b - a$ is divisible by n, that is, if there exists an $k \in \mathbb{Z}$ with $b = a + kn$ exists. One then writes

$$a \equiv b \bmod n \quad \text{or} \quad a \equiv b(n).$$

This relation is an equivalence relation: (1) It is $a \equiv a \bmod n$ because of $a - a = 0 = 0 \cdot n$; (2) from $a \equiv b \bmod n$, i.e., $b - a = kn$ with $k \in \mathbb{Z}$, follows $a - b = (-k)n$, and finally (3) from $a \equiv b$ and $b \equiv c \bmod n$, i.e., $b - a = kn$ and $c - b = \ell n$ with $k, \ell \in \mathbb{Z}$, follows $c - a = (c - b) + (b - a) = kn + \ell n = (k + \ell)n$.—For $n = 0$, this relation is the equality relation, for $n = 1$, the universal relation. Now let $n > 0$. Then two integers are congruent modulo n if and only if they leave the same remainder (between 0 and $n - 1$) when divided by n, see Theorem 1.7.5. *In this case, the numbers $0, \ldots, n-1$ form a canonical complete system of representatives.* Generally, any n consecutive integers form a complete system of representatives. Thus, there are exactly n equivalence classes, the so-called **residue classes modulo** n. The set of residue classes modulo n is denoted by

$$\mathbb{Z}/\mathbb{Z}n$$

For $a \in \mathbb{Z}$,

$$[a]_n = a + \mathbb{Z}n = a + [0]_n = \{a + sn \mid s \in \mathbb{Z}\}$$

is the residue class of a in $\mathbb{Z}/\mathbb{Z}n$. In the case $n = 2$, the residue class $\overline{0} = [0]_2$ is the set of even numbers and the residue class $\overline{1} = [1]_2$ is the set of odd numbers. The congruence relation modulo n is finer than the congruence relation modulo m if and only if $[0]_n = \mathbb{Z}n \subseteq [0]_m = \mathbb{Z}m$, i.e., if n is a multiple of m or—which is the same—if m is a divisor of n. In this case, the canonical projection $\pi_n \colon \mathbb{Z} \to \mathbb{Z}/\mathbb{Z}n$ induces the surjective mapping $\pi_{m,n} \colon \mathbb{Z}/\mathbb{Z}n \to \mathbb{Z}/\mathbb{Z}m$ with $\pi_{m,n}([a]_n) = [a]_m$ for all $a \in \mathbb{Z}$.

The addition $+$ and the multiplication $\cdot$ on $\mathbb{Z}$ induce corresponding operations on $\mathbb{Z}/\mathbb{Z}n$. For $a, b \in \mathbb{Z}$, one sets

$$(a + \mathbb{Z}n) + (b + \mathbb{Z}n) := (a + b) + \mathbb{Z}n, \quad (a + \mathbb{Z}n) \cdot (b + \mathbb{Z}n) := (a \cdot b) + \mathbb{Z}n.$$

To define this sum and this product, we have used the specific representatives a, b of the residue classes. It must be shown that the results do not depend on the choice of these representatives, i.e., that from $a \equiv a'$ mod n, thus $a = a' + kn$, and $b \equiv b'$ mod n, thus $b = b' + \ell n$, $k, \ell \in \mathbb{Z}$, follows $a + b \equiv a' + b'$ mod n as well as $ab \equiv a'b'$ mod n. However, it is $a + b = a' + b' + (k + \ell)n$ and $ab = a'b' + (kb' + a'\ell + k\ell n)n$. The necessity to verify such **well-definedness**, i.e., the **independence from the choice of representative**, is typical for calculations with equivalence classes and will repeatedly encounter the reader. The two operations + and $\cdot$ on $\mathbb{Z}/\mathbb{Z}n$ are thus well-defined and chosen such that for the canonical projection $\pi_n \colon \mathbb{Z} \to \mathbb{Z}/\mathbb{Z}n$ holds:

$$\pi_n(a + b) = [a + b]_n = [a]_n + [b]_n = \pi_n(a) + \pi_n(b),$$
$$\pi_n(ab) = [ab]_n = [a]_n[b]_n = \pi_n(a)\pi_n(b).$$

Thus, the relevant calculation rules from $\mathbb{Z}$ transfer to $\mathbb{Z}/\mathbb{Z}n$, for example, the associative laws of addition and multiplication

$$([a]\dot{+}[b])\dot{+}[c] = [a\dot{+}b]\dot{+}[c] = [(a\dot{+}b)\dot{+}c] = [a\dot{+}(b\dot{+}c)] = [a]\dot{+}[b\dot{+}c] = [a]\dot{+}([b]\dot{+}[c]),$$

as well as the commutative laws, etc. The tables for addition and multiplication in $\mathbb{Z}/\mathbb{Z}5$ or $\mathbb{Z}/\mathbb{Z}6$ are given in Fig. 1.11. In doing so, we have again denoted the residue class $[a]$ for simplicity with a.[9] In the case $n = 2$, we obtain the known calculation rules for **parities**:

$$\text{pair} + \text{pair} = \text{pair} = \text{impair} + \text{impair}, \quad \text{impair} + \text{pair} = \text{impair};$$

$$\text{pair} \cdot \text{pair} = \text{pair} = \text{impair} \cdot \text{pair}, \quad \text{impair} \cdot \text{impair} = \text{impair}.$$

The sets of residue classes $\mathbb{Z}/\mathbb{Z}n$ with these additions and multiplications are like $\mathbb{Z}$ commutative rings and fundamental objects of mathematics. We will discuss them in a more general context in Chap. 2 in more detail. The congruence relations $\equiv$ mod n were first systematically used by C. F. Gauss in the "Disquisitiones Arithmeticae" (1801).

In general, for a real number $T \neq 0$, one writes

$$a \equiv b \bmod T \quad \text{or} \quad a \equiv b(T),$$

when a and b are real numbers whose difference $b - a$ is an *integer* multiple of T. This is an equivalence relation on $\mathbb{R}$. Proof! For $a \in \mathbb{R}$ the equivalence class

---

[9] $\mathbb{Z}/\mathbb{Z}5$ is, by the way, a field, in $\mathbb{Z}/\mathbb{Z}6$ however, only the residue classes 1 and 5 possess an inverse with respect to multiplication.

1.3 Relations

+	0 1 2 3 4
0	0 1 2 3 4
1	1 2 3 4 0
2	2 3 4 0 1
3	3 4 0 1 2
4	4 0 1 2 3

·	0 1 2 3 4
0	0 0 0 0 0
1	0 1 2 3 4
2	0 2 4 1 3
3	0 3 1 4 2
4	0 4 3 2 1

+	0 1 2 3 4 5
0	0 1 2 3 4 5
1	1 2 3 4 5 0
2	2 3 4 5 0 1
3	3 4 5 0 1 2
4	4 5 0 1 2 3
5	5 0 1 2 3 4

·	0 1 2 3 4 5
0	0 0 0 0 0 0
1	0 1 2 3 4 5
2	0 2 4 0 2 4
3	0 3 0 3 0 3
4	0 4 2 0 4 2
5	0 5 4 3 2 1

Fig. 1.11 Addition and multiplication in $\mathbb{Z}/\mathbb{Z}5 = \{0, 1, 2, 3, 4\}$ or in $\mathbb{Z}/\mathbb{Z}6 = \{0, 1, 2, 3, 4, 5\}$

$\bar{a} = a + \mathbb{Z}T$ of a contains exactly the elements $a + kT, k \in \mathbb{Z}$. T and $-T$ define the same relation. The numbers of the half-open interval

$$[0, |T|[:= \{x \in \mathbb{R} \mid 0 \leq x < |T|\}$$

form a complete system of representatives for the set

$$\mathbb{R}/\mathbb{Z}T$$

of equivalence classes. The uniquely determined representative of $\bar{a} = a + \mathbb{Z}T$ in $[0, |T|[$ is $a - [a/|T|] \cdot |T|$, where $[-]$ denotes the Gauss bracket. One defines the functions $x \mapsto x$ DIV T and $x \mapsto x$ MOD T on $\mathbb{R}$ with values in $\mathbb{Z}$ or in the interval $[0, |T|[$ through the equation

$$x = (x \text{ DIV } T) \cdot T + (x \text{ MOD } T) \quad \text{with} \quad x \text{ DIV } T \in \mathbb{Z}, \ 0 \leq x \text{ MOD } T < |T|,$$

thus x DIV $T = \text{Sign } T \cdot [x/|T|]$.[10] In $T = n \in \mathbb{N}^*$ is $\mathbb{Z}/\mathbb{Z}n$ the set of those equivalence classes in $\mathbb{R}/\mathbb{Z}n$ that have an integer representative. If $S \neq 0$ is another real number $\neq 0$, then the congruence relation modulo T is finer than that modulo S if and only if T is an *integer* multiple of S, i.e., $T/S \in \mathbb{Z}$ holds. On the set $\mathbb{R}/\mathbb{Z}T$, as on $\mathbb{Z}/\mathbb{Z}n$, the addition

$$(a + \mathbb{Z}T) + (b + \mathbb{Z}T) := (a + b) + \mathbb{Z}T, \quad a, b \in \mathbb{R},$$

is well-defined, but *not* a corresponding multiplication. Why not? ◇

---

[10] Note that the "integer quotient" x DIV T can be different from the integer part $[x/T]$. The functions DIV and MOD are primarily defined in computer languages and occasionally in a different manner there.

Exercises

Exercise 1.3.1 Every partition of a set A defines an equivalence relation, whose equivalence classes are exactly the sets of the given partition of A.

Exercise 1.3.2 Provide examples of relations that satisfy two of the three properties of an equivalence relation, but not the third. How many relations are there on a set with n elements and how many of them are reflexive, antireflexive, symmetric, antisymmetric, or asymmetric, see 1.6.2 and 1.6.3? (The number of equivalence relations is called the n-**th Bell number** β_n and is determined in Ex. 1.6.14.)

1.4 Order Relations

The great importance of general (not necessarily total) orders was first recognized by F. Hausdorff (1868–1942) and described in his work "Grundzüge der Mengenlehre" from 1914. Today, they and the related concepts are fundamental to all areas of mathematics.

Definition 1.4.1 A is a set and $\leq$ a relation on A. Then $\leq$ is called an **order (relation)** on A and $A = (A, \leq)$ an **ordered set**, if $\leq$ is reflexive, antisymmetric, and transitive, i.e., if for all $a, b, c \in A$ holds:

(1) $a \leq a$; (2) from $a \leq b, b \leq a$ follows $a = b$

(3) from $a \leq b, b \leq c$ follows $a \leq c$.

Moreover, if always one (and then for $a \neq b$ exactly one) of the relations $a \leq b$ or $b \leq a$ holds, then $\leq$ is called a **complete** or **total order (relation)** on A.[11]

An order on A induces an order on every subset of A. A totally ordered set is also called a **chain**. Subsets of chains are again chains. With an order $\leq$, instead of writing "$a \leq b$ and $a \neq b$", one writes more concisely "$a < b$". The relation $<$ is antireflexive, asymmetric, and transitive. The relations $\leq$ and $<$ determine each other. Instead of $a \leq b$ (or $a < b$), one also writes $b \geq a$ (or $b > a$). This again provides an order. It is called the **opposite order** to $\leq$ and is also denoted by $\leq^{op}$ Two elements $a, b \in A$ are called **comparable** with respect to the order $\leq$, if $a \leq b$ or $a \geq b$ holds. For $a \in A$ we denote by $A_{\leq a}$ and $A_{<a}$ the **closed** and **open initial segment**

$$A_{\leq a} := \{x \in A \mid x \leq a\} \quad \text{resp.} \quad A_{<a} := \{x \in A \mid x < a\} = A_{\leq a} - \{a\}.$$

---

[11] In the literature, one also finds the term "partially ordered set" for an ordered set (in English "poset"). Then an "ordered set" is often understood to be a totally ordered set.

1.4 Order Relations

Analogously, the **final segments** $A_{\geq a}$ or $A_{>a}$ defined. These are the initial sections of the opposite order $\leq^{\mathrm{op}}=\geq$. This also allows the convenient description of the **closed**, **open**, and **half-open intervals** in A. For $a, b \in A$ is

$$[a,b] := A_{\geq a} \cap A_{\leq b}, \quad]a,b[:= A_{>a} \cap A_{<b},$$

$$]a,b] := A_{>a} \cap A_{\leq b} \quad \text{and} \quad [a,b[:= A_{\geq a} \cap A_{<b}.$$

The elements of $A_{>a}$ are also called the **successors** and the elements of $A_{<a}$ the **predecessors** of a. If b is a successor of a, i.e., $a < b$, and there is no element $c \in A$ with $a < c < b$, then $]a, b[= \emptyset$, then b is called a **direct** or **immediate successor** of a and a a **direct** or **immediate predecessor** of b. The elements a,b or b,a are then (direct) **neighbors** or **adjacent**. By definition, for (arbitrary) the elements of $a, b \in A$ (or of $[a, b] \cup [b, a]$) **between** (or **strictly between**) a and b. The between relation is thus symmetric. There is an element between a and b if and only if a and b are comparable. a and b are adjacent if and only if they are comparable but different, and moreover, no element lies strictly between a and b. All these definitions are thus adapted to natural language usage.

An element $a_0 \in A$ is called a **greatest element** or a **maximal element** of A, if $a \leq a_0$ holds for all $a \in A$, if thus $A_{\leq a_0} = A$ is, and a **maximal element** of A, if $a \leq a_0$ holds for all $a \in A$ comparable with a_0, if thus $A_{>a_0} = \emptyset$ is. Correspondingly, **smallest elements** or **minimal elements** of A are defined. These are the maxima or the maximal elements with respect to the respective opposite order. *Maxima and minima are, if they exist, each uniquely determined.* Namely, if M and M' are maxima in A with respect to $\leq$, then it holds $M \leq M'$, since M' is a maximum, and $M' \leq M$, since M is a maximum, thus overall $M = M'$ due to the antireflexivity of $\leq$. The maximum or minimum of an ordered set A is denoted, if it exists, by

$$\operatorname{Max} A \quad \text{resp.} \quad \operatorname{Min} A.$$

More generally, $\operatorname{Max}(a_i, i \in I)$ or $\operatorname{Min}(a_i, i \in I)$ denotes for a family $a_i, i \in I$, of elements of an ordered set A the largest or smallest element of the (ordered) subset $\{a_i \mid i \in I\} \subseteq A$ (if it exists). For a finite non-empty family in a *chain*, these elements always exist, as can be easily proven by induction on the number of elements in the family, see Ex. 1.5.10. An ordered set A is called **well-ordered** if *every* non-empty subset of A has a smallest element. Then A is particularly totally ordered, because for $a, b \in A$ the element$\{a, b\} \subseteq A$ has a smallest element. If the ordered set A has a smallest element $\operatorname{Min} A$, then every minimal element in $A - \{\operatorname{Min} A\}$ is called an **atom** of A. If A has a largest element $\operatorname{Max} A$, each maximal element in $A - \{\operatorname{Max} A\}$ is called an **antiatom** of A. The antiatoms of $(A, \leq)$ are the atoms of $(A, \leq^{\mathrm{op}})$.

Example 1.4.2 Equality is (with respect to inclusion) the smallest order on a set A. It is the only order on A that is simultaneously an equivalence relation, and is characterized by the fact that any two different elements of A are *not* comparable. This is referred to as an **antichain** or also—somewhat inconsistently—as a **totally unordered set**. Each element of an antichain is minimal and maximal, but neither a smallest nor a largest element, if the antichain contains more than one element. ◊

Example 1.4.3 The natural order $\leq$ is a total order on $\mathbb{R}$ and thus on every subset of $\mathbb{R}$. There is neither a largest nor a smallest element in $\mathbb{R}$. The subset $\mathbb{N} \subseteq \mathbb{R}$ has a smallest element, namely 0, but no largest element. 1 is an atom in $\mathbb{N}$, and indeed the only one. ◊

Example 1.4.4 Let A be a set. On the power set $\mathfrak{P}(A)$ of A, the inclusion $\subseteq$ defines an order that is not a total order if A contains at least two elements. Namely, if $a, b \in A$ are different, then $\{a\}, \{b\} \in \mathfrak{P}(A)$ are not comparable. A is the largest element in $\mathfrak{P}(A)$ and $\emptyset$ the smallest. The single-element subsets $\{a\}$, $a \in A$, are the atoms of $\mathfrak{P}(A)$. What are the anti-atoms of $\mathfrak{P}(A)$? The inclusion orders on power sets and their subsets are the prototypes for orders. ◊

Example 1.4.5 (Hasse Diagrams) In illustrating an order using a directed graph, whose vertices are points of the drawing plane, the following simplifications are generally used: Care is taken to ensure that the arrows always run from bottom to top. Then the arrowheads can be omitted. Additionally, all loops and all connecting lines that can be deduced based on the transitivity of the order relation are removed. Such a diagram for an order relation is called a **Hasse diagram** (after H. Hasse (1898–1979)). A typical example of a Hasse diagram is the left figure in Fig. 1.12. The other two Hasse diagrams are the diagram for the natural order on the set $\{0, 1, \ldots, n\}$ or the one for inclusion on the set $\mathfrak{P}(\{1, 2, 3\})$. ◊

Example 1.4.6 (Sum Order) Let I be an *ordered* index set and $A_i, i \in I$, be a family of ordered sets indexed by I. On the disjoint union $\biguplus_{i \in I} A_i = \bigcup_{i \in I} \{i\} \times A_i$, an order is defined by

$$(i, a_i) \leq (j, a_j) \text{ if and only if } i < j \text{ or if } i = j \text{ and } a_i \leq a_j \text{ is,}$$

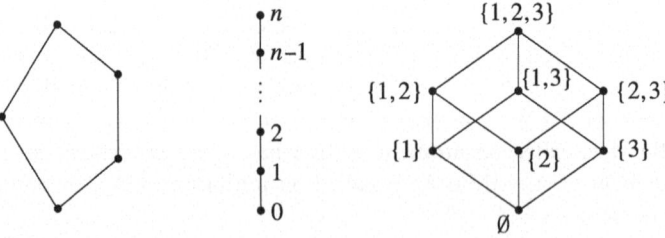

Fig. 1.12 Hasse Diagrams

1.4 Order Relations

This so-called **sum order** generally depends not only on the orders on the sets A_i, $i \in I$, but also on the order on the index set I. Usually, it is clear from the context which order is chosen on I. The sum order is complete if and only if all A_i and the subset $I' := \{i \in I \mid A_i \neq \emptyset\} \subseteq I$ are completely ordered. In particular, $\biguplus_{i=1}^{n} A_i = A_1 \uplus \cdots \uplus A_n$ is completely ordered if all $A_1, \ldots, A_n$ are completely ordered and $\{1, \ldots, n\}$ carries the natural order. If I is an antichain, then elements of A_i and A_j are not comparable for $i \neq j$. In this case, one speaks of the **sum** of the ordered sets A_i, $i \in I$, in the absolute sense. ◊

Example 1.4.7 (Product Order) Let A_i, $i \in I$, be a family of ordered sets. On the Cartesian product $\prod_{i \in I} A_i$, an order is obviously defined by the rule

$$(a_i)_{i \in I} \leq (b_i)_{i \in I} \text{ if and only if } a_i \leq b_i \text{ for all } i \in I,$$

which is called the **product order** on $\prod_{i \in I} A_i$. In particular, for an ordered set A and an arbitrary set I, the set of all mappings from I to A carries the product order. For two mappings $f, g : I \to A$, $f \leq g$ holds if and only if for all $i \in I$ the values $f(i)$ and $g(i)$ satisfy the condition $f(i) \leq g(i)$. If all $A_i \neq \emptyset$ and there are two different indices $i, j \in I$ such that A_i and A_j contain more than one element, then the product order is not complete. ◊

Example 1.4.8 (Lexicographic Order) Let $A_1, \ldots, A_n$ be ordered sets. Then on the product $A_1 \times \cdots \times A_n$ an order is defined by the rule

$$(a_1, \ldots, a_n) < (b_1, \ldots, b_n) \text{ if and only if}$$

$(a_1, \ldots, a_n) \neq (b_1, \ldots, b_n)$ and for the smallest index i with $a_i \neq b_i$ is $a_i < b_i$.

It is called the **lexicographic order** on $A_1 \times \cdots \times A_n$. With respect to the lexicographic order, $(1, 2) < (2, 1)$. (In the dictionary, the word "ab" comes before "ba".) With respect to the product order, $(1, 2)$ and $(2, 1)$ are not comparable. If all $A_1, \ldots, A_n$ are completely ordered, then so is $A_1 \times \cdots \times A_n$ with respect to the lexicographic order. See Fig. 1.13. The lexicographic order is also defined on a product $\prod_{i \in I} A_i$ of ordered sets when the index set I is infinite and *well-ordered*. Occasionally, we use further variants of the lexicographic order.

The monotonic mappings of ordered sets are those mappings of ordered sets that are compatible with the respective orders. We define:

Definition 1.4.9 Let $f : A \to B$ be a mapping of ordered sets $(A, \leq_A)$ and $(B, \leq_B)$.

(1) f is **monotonically increasing** (respectively **monotonically decreasing**), if for all $x, y \in A$ with $x \leq_A y$ it holds: $f(x) \leq_B f(y)$ (respectively $f(x) \geq_B f(y)$).

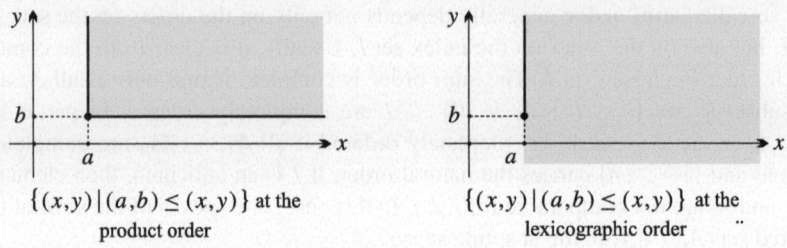

Fig. 1.13 Product order and lexicographic order

(2) f is **strictly monotonically increasing** (respectively **strictly monotonically decreasing**), if for all $x, y \in A$ with $x <_A y$ it holds: $f(x) <_B f(y)$ (respectively $f(x) >_B f(y)$).

(3) f is **monotonic** (respectively **strictly monotonic**), if f is monotonically increasing or monotonically decreasing (respectively if f is strictly monotonically increasing or strictly monotonically decreasing).

In particular, (strictly) monotonically increasing and (strictly) monotonically decreasing sequences in an ordered set A are defined. A mapping $f: A \to B$ is monotonically decreasing (respectively strictly monotonically decreasing) if and only if f is considered as a mapping from $(A, \leq_A)$ in $(B, \leq_B^{op})$ or also from $(A, \leq_A^{op})$ in $(B, \leq_B)$ monotonically increasing (or strictly monotonically increasing). Injective monotone mappings are strictly monotone. Conversely, if $f: A \to B$ is strictly monotone *and A completely ordered*, then f is injective. If f is even bijective, then B is also completely ordered and $f^{-1}: B \to A$ is strictly monotone (of the same monotonicity type as f). In general, $f: A \to B$ is called an **(order) isomorphism** of ordered sets, if f is bijective and both f and f^{-1} are (strictly) monotonically increasing. A bijective (strictly) monotonically increasing mapping is generally not yet an order isomorphism. However, this is the case if A is completely ordered. If there is an isomorphism $f: A \xrightarrow{\sim} B$, then A and B are called of the **same order type** or **(order) isomorphic**.$(\mathbb{Z}, \leq)$ and $(\mathbb{Z}, \leq^{op})$ for example are of the same order type, $(\mathbb{N}, \leq)$ and $(\mathbb{N}, \leq^{op})$ however, are not. If $(A, \leq)$ is an ordered set, then the mapping $f: (A, \leq) \to (\mathfrak{P}(A), \subseteq)$, $a \mapsto A_{\leq a}$, strictly monotonically increasing, and induces an order isomorphism from A to $\text{Im} f \subseteq \mathfrak{P}(A)$. For each set A, the mapping $(\mathfrak{P}(A), \subseteq) \xrightarrow{\sim} (\{0, 1\}^A, \leq)$, $B \mapsto e_B$, is an order isomorphism (where $\leq$ is the product order on $\{0, 1\}^A$).—For ordered sets A and B, the sums $A \uplus B$ and $B \uplus A$ are order isomorphic. However, if the index set $\{1, 2\}$ carries the natural order, then $A \uplus B$ and $B \uplus A$ are generally not isomorphic. For example, $\{x\} \uplus \mathbb{N}$ is order isomorphic to $\mathbb{N}$, but $\mathbb{N} \uplus \{x\}$ is order isomorphic to $\overline{\mathbb{N}} = \mathbb{N} \uplus \{\infty\}$. Analogously, the product orders on $A \times B$ and $B \times A$ are always isomorphic. However, the lexicographic orders on $A \times B$ and $B \times A$ are generally not isomorphic. For example, $\mathbb{N} \times \{1, 2\}$ is order isomorphic to $\mathbb{N}$, $\{1, 2\} \times \mathbb{N}$ however, to the also

1.4 Order Relations

well-ordered set $\mathbb{N} \uplus \mathbb{N}$. (The index set $\{1, 2\}$ again carries the natural order with $1 < 2$.)

For the following considerations, we have to introduce some additional order-theoretical terms, which are of general significance. Let $A = (A, \leq)$ be an ordered set and $B \subseteq A$ a subset of A. Then an element $S \in A$ is called an **upper bound of B in A**, if $b \leq S$ holds for all $b \in B$. The set of upper bounds of B in A is denoted by $\mathrm{UB}(B) = \mathrm{UB}_A(B)$. If the smallest upper bound Min $\mathrm{UB}(B)$ of B in A exists, it is called the **least upper bound** or the **supremum of B in A** and is denoted by

$$\mathrm{Sup}\, B = \mathrm{Sup}_A B \; (= \mathrm{Min}\, \mathrm{UB}_A(B)).$$

Accordingly, an element $s \in A$ a **lower bound of B in A**, if $s \leq b$ holds for all $b \in B$. The set of lower bounds of B in A we denote by $\mathrm{LB}(B) = \mathrm{LB}_A(B)$. If the greatest lower bound Max $\mathrm{LB}(B)$ of B in A exists, it is called the **lower limit** or the **infimum of B in A**. We then denote it by

$$\mathrm{Inf}\, B = \mathrm{Inf}_A B \; (= \mathrm{Max}\, \mathrm{LB}_A(B)).$$

Note $\mathrm{Sup}_A A = \mathrm{Max}\, A$, $\mathrm{Sup}_A \emptyset = \mathrm{Min}\, A$, $\mathrm{Inf}_A A = \mathrm{Min}\, A$ and $\mathrm{Inf}_A \emptyset = \mathrm{Max}\, A$ (where one side of the equations exists if and only if the other side exists). A subset $B \subseteq A$ is called **bounded above** (or **bounded below**), if it has an upper (or lower) bound in A, that is, if $\mathrm{UB}_A(B) \neq \emptyset$ (or $\mathrm{LB}_A(B) \neq \emptyset$) holds. It is called **bounded** if it is both bounded above and bounded below.

Example 1.4.10 (Lattices) For every subset $\mathcal{B} \subseteq \mathfrak{P}(A) = (\mathfrak{P}(A), \subseteq)$ the supremum and the infimum exist. It is $\mathrm{Sup}\, \mathcal{B} = \bigcup_{B \in \mathcal{B}} B$ and $\mathrm{Inf}\, \mathcal{B} = \bigcap_{B \in \mathcal{B}} B$. An ordered set in which—as here in $(\mathfrak{P}(A), \subseteq)$—*every* subset has a supremum and an infimum is called a **complete lattice**. For example, $\overline{\mathbb{N}} = (\mathbb{N} \uplus \{\infty\}, \leq)$ is also a complete lattice, but not $\mathbb{N} = (\mathbb{N}, \leq)$. A **lattice** in general is by definition an ordered set in which any two elements x, y have a supremum and an infimum, which are usually denoted by $x \vee y$ or $x \sqcup y$ respectively with $x \wedge y$ or $x \sqcap y$ are denoted. A totally ordered set is always a lattice. Conversely, a finite *non-empty* lattice A is due to $\mathrm{Sup}(a_1, \ldots, a_n, a_{n+1}) = \mathrm{Sup}(\mathrm{Sup}(a_1, \ldots, a_n), a_{n+1})$ and $\mathrm{Inf}(a_1, \ldots, a_n, a_{n+1}) = \mathrm{Inf}(\mathrm{Inf}(a_1, \ldots, a_n), a_{n+1})$, $a_1, \ldots, a_{n+1} \in A$, $n \in \mathbb{N}^*$, always a complete lattice. An ordered set in which any two elements have an upper bound (or a lower bound) is called **directed upwards** (or **directed downwards**). In such a set A, every finite subset has an upper (or a lower) bound, *if $A \neq \emptyset$ is*. Lattices are directed upwards and downwards. ◊

A fundamental theorem about the existence of maximal elements is Zorn's Lemma 1.4.15. First, we define:

Definition 1.4.11 An ordered set A is called **inductively ordered** (or **strictly inductively ordered**) if every chain in A has an upper bound (or an upper limit) in A.

Strictly inductively ordered sets are particularly inductively ordered. Inductively ordered sets are not empty, as in them the empty chain has an upper bound. Strictly inductively ordered sets always have a minimum as the upper limit of the empty chain. Every finite ordered set with a minimal element is strictly inductively ordered. The fundamental lemma for strictly inductively ordered sets is the following fixed point theorem, whose proof can be skipped without loss of understanding on the first reading:

Lemma 1.4.12 *Let A be a strictly inductively ordered set. Then every mapping $f : A \to A$ with $x \leq f(x)$ for all $x \in A$ has a fixed point.* In other words: *There is no mapping $g : A \to A$ with $x < g(x)$ for all $x \in A$.*

Proof The statement is trivial if A is a chain. Then, namely, $S := \operatorname{Sup} A = \operatorname{Max} A$ is a fixed point of A because $S \leq f(S) \leq S$. The general case is reduced to this special case.

In the further proof, we call a subset $B \subseteq A$ *admissible* if B is invariant under f (i.e., if $f(B) \subseteq B$) and if for every chain $K \subseteq B$ the upper bound $\operatorname{Sup} K = \operatorname{Sup}_A K$ belongs to B. Then B is also strictly inductively ordered. A itself is admissible, and arbitrary intersections of admissible subsets of A are again admissible. In particular, there is a smallest admissible subset of A. By replacing A with this, we can henceforth assume *that A has no proper admissible subsets*. We show that A is then a chain, thus concluding the proof according to the preliminary remark.

We call a point $x \in A$ *separating* (with respect to f) if for all $y \in A$ with $y < x$ it holds $f(y) \leq x$. First, we show:

(*) *Let $x \in A$ be a separating point. Then $y \leq x$ or $f(x) \leq y$ for all $y \in A$. In particular, x (due to the general assumption $x \leq f(x)$) is comparable with every $y \in A$.*

It suffices to prove that the set $B := \{y \in A \mid y \leq x \text{ or } f(x) \leq y\}$ is admissible. B clearly contains x and $f(x)$. Now let $K \subseteq B$ be a chain in B and $S := \operatorname{Sup}_A K$. If $y \leq x$ for all $y \in K$, then x is an upper bound of K and $S \leq x$, thus $S \in B$. Otherwise, there exists a $y \in K$ with $f(x) \leq y \leq S$ and thus also $S \in B$. Finally, we show that B is invariant under f. For this, let $y \in B$. If $y < x$, then $f(y) \leq x$, since x is a separating point with respect to f, and consequently $f(y) \in B$. If $y = x$, then $f(y) = f(x) \in B$ as already noted. However, if $f(x) \leq y$, then $f(x) \leq y \leq f(y)$ and therefore again $f(y) \in B$.

We now show *that every point $x \in A$ is separating with respect to f*. Then A according to what has already been shown—as claimed—a chain. Again, it suffices to show that the set T of points separating with respect to f is permissible. Let $K \subseteq T$ be a chain and $S := \operatorname{Sup}_A K$. Then S is also a separating point. Namely, if $y \in A$ and $y < S$, then y is not an upper bound of K and there exists a $x \in K$ with $x \not\leq y$. Then, due to $x \leq f(x)$ also $f(x) \not\leq y$. Since x is separating, $y \leq x$ according to (*). Since $y = x$ is not possible, $y < x$ and consequently $f(y) \leq x \leq S$. Thus, S is indeed separating. It remains to show $f(T) \subseteq T$: Let $t \in T$. We have to prove that $f(y) \leq f(t)$ is for $y \in A$

1.4 Order Relations

with $y < f(t)$. Since t is separating, however, $y \le t$ or $f(t) \le y$ again according to(∗). The second condition is not possible due to $y < f(t)$. Therefore, $y \le t$. If even $y < t$, then $f(y) \le t \le f(t)$ according to the definition of a separating point. However, if $y = t$, then $f(y) = f(t)$. □

The following somewhat weaker version 1.4.13 of Zorn's Lemma 1.4.15 (after M. Zorn (1906–1993)) is sufficient in many cases. In the proof, the already mentioned so-called axiom of choice is decisively used, the validity of which we do not question.[12]

Axiom of Choice Let A_i, $i \in I$, be an arbitrary family of *non-empty* sets. Then there exists a mapping $f : I \to \bigcup_{i \in I} A_i$ with $f(i) \in A_i$ for all $i \in I$, i.e., the Cartesian product $\prod_{i \in I} A_i$ is non-empty.

Theorem 1.4.13 *Every strictly inductively ordered set A has a maximal element.*

Proof Assume A contains no maximal element. Then for every $x \in A$ the set $A_{>x} := \{y \in A \mid x < y\}$ is non-empty. According to the axiom of choice, there exists a mapping $g: A \to A$ with $g(x) \in A_{>x}$, i.e., with $x < g(x)$ for all $x \in A$, in contradiction to Lemma 1.4.12.

The following **Maximal Chain Theorem** 1.4.14 was already stated in 1914 in the book "Grundzüge der Mengenlehre" by F. Hausdorff, cited at the beginning of this section.

Theorem 1.4.14 (Hausdorff's Theorem) *Every ordered set A possesses a maximal chain (with respect to inclusion).*

Proof The set $\mathcal{K} \subseteq \mathfrak{P}(A) = (\mathfrak{P}(A), \subseteq)$ of chains in A is strictly inductively ordered. Namely, if $\mathcal{C} \subseteq \mathcal{K}$ is a chain in $\mathcal{K}$, then $K := \bigcup_{C \in \mathcal{C}} C \subseteq A$ is also a chain in A and thus an upper bound of $\mathcal{C}$ in $\mathcal{K}$. To demonstrate that K is a chain, let $x, x' \in K$. Then there exist $C, C' \in \mathcal{C}$ with $x \in C$ and $x' \in C'$. Since $\mathcal{C}$ is a chain, $C \subseteq C'$ or $C' \subseteq C$ holds. In any case, x and y both lie in one of the chains C or C' and are thus comparable. Proposition 1.4.13 now provides the assertion. □

The general Zorn's Lemma from 1935 is an immediate consequence of Hausdorff's Maximal Chain Theorem.

---

[12] The validity of the axiom of choice presupposes—especially for uncountable index sets I — great confidence in the capabilities of the human mind and is not accepted by all mathematicians (or only in a limited way). Therefore, one tries to avoid the application of the axiom of choice where possible, particularly the consequences expressed in the following four theorems.

Theorem 1.4.15 (Zorn's Lemma) *Every inductively ordered set A possesses a maximal element.*

Proof Let K be a maximal chain in A according to Hausdorff's maximal chain theorem. By assumption, K has an upper bound S. This is a maximal element of A. If there were an element $a \in A$ with $S < a$, then $K \uplus \{a\}$ would be a chain in A that is truly larger than K, contradicting the maximality of K. □

As a first application of Zorn's lemma or theorem 1.4.13, we demonstrate the **well-ordering theorem of Zermelo**, which E. Zermelo (1871–1953) already proved in 1904 and which G. Cantor still considered self-evident. Zermelo's second proof from 1908 already contains many arguments used in the proof of the crucial fixed-point theorem 1.4.12. We repeat the following definition:

Definition 1.4.16 An ordered set A is called **well-ordered**, if every non-empty subset of A contains a smallest element.

The set $\mathbb{N}$ of natural numbers with the standard order is the non-trivial exemplary case of a well-ordered set, see Sect. 1.5. Thus, every completely ordered set whose elements can be enumerated in a strictly ascending sequence $a_0 < a_1 < a_2 < a_3 < \cdots$ is well-ordered. The set $\mathbb{Z}$ of integers is not well-ordered. Every non-empty well-ordered set has a smallest element. If A_i, $i \in I$, is a family of well-ordered sets with a well-ordered index set I, then the disjoint union $\uplus_{i \in I} A_i$ with respect to the sum order is also well-ordered, see Example 1.4.6. If $A_1, \ldots, A_n$ are well-ordered, then the product $A_1 \times \cdots \times A_n$ is well-ordered with respect to the lexicographic order, see Example 1.4.8. (As with the sum orders, one should distinguish between the two products $A_1 \times A_2 = \uplus_{i \in A_1} \{i\} \times A_2$ and $A_2 \times A_1 = \uplus_{j \in A_2} \{j\} \times A_1$.) Products $\prod_{i \in I} A_i$ with an infinite well-ordered index set I are completely ordered with respect to the lexicographical order, but not well-ordered if $A_i \neq \emptyset$ is for every i and infinitely many of the A_i contain more than one element. Every element a of a well-ordered set A, which is not the greatest element of A, has an immediate successor, namely the smallest element of the open final segment $A_{>a} \neq \emptyset$. If $A' \subset A$ is a proper subset of A that contains with every element x all elements $\leq x$, then A' is an open initial segment $A_{<a}$ with a (uniquely determined) $a \in A$, namely the smallest element of $A - A' \neq \emptyset$. It is then $A - A' = A_{\geq a}$.

Theorem 1.4.17 (Well-Ordering Theorem) *Every set A has a well-ordering.*

Proof We consider the set $\mathcal{W}$ of pairs $(B, \leq)$, where B is a subset of A and $\leq$ is a well-ordering on B. For two such pairs $(B_1, \leq_1)$ and $(B_2, \leq_2)$ we set $(B_1, \leq_1) \leq (B_2, \leq_2)$, if $B_1 \subseteq B_2$ holds, $\leq_2$ induces the order $\leq_1$ on B_1 and $x < y$ for all $x \in B_1, y \in B_2 - B_1$ is. Then $(\mathcal{W}, \leq)$ is (even strictly) inductively ordered. Namely, if $\mathcal{K} \subseteq \mathcal{W}$ is a chain, then $M := \bigcup_{B \in \mathcal{K}} B$ can be ordered in a natural way. Namely, if $x, y \in M$, then there is a $(B_1, \leq_1) \in \mathcal{W}$ with $x, y \in B_1$ and we set $x \leq y$ exactly when $x \leq_1 y$ holds.

This determination is evidently independent of the choice of $(B_1, \leq_1)$ and provides a well-ordering on M. To see the latter, let $L \subseteq M$ be a non-empty subset of M. Then there is a $K \in \mathcal{K}$ with $K \cap L \neq \emptyset$, and $K \cap L$ contains a smallest element m, since K is well-ordered. m is a smallest element of L. Namely, if $x \in L$ arbitrary and $x \in B \in \mathcal{K}$, then $B \subseteq K$ or $K \subseteq B$, since $\mathcal{K}$ is a chain. Thus, $x \in K$ or $x \in B - K$ holds. In the first case, $x \in L \cap K$, and by choice of m, $m \leq x$ holds. In the second case, $\mathcal{K} \subset B$ and $m < x$ by definition of the order on $\mathcal{W}$. $(M, \leq)$ is thus an upper bound of $\mathcal{K}$.

According to Zorn's Lemma 1.4.15 (or already according to Theorem 1.4.13), there exists a maximal element $(N, \leq) \in \mathcal{W}$. It suffices to show that $N = A$. However, if there were $a_n \in A - N$, then $(N' := N \uplus \{a\}, \leq')$ would be with the well-ordering $\leq'$, which extends $\leq$ from N to N' and for which $x <' a$ is for all $x \in N$, a greater element than $(N, \leq)$ in $\mathcal{W}$. Contradiction! □

The Well-Ordering Theorem 1.4.17 in turn implies the Axiom of Choice. Namely, if A_i, $i \in I$, is a family of non-empty sets and $\leq$ a well-ordering on $A := \bigcup_{i \in I} A_i$, then $i \mapsto \operatorname{Min} A_i$, $i \in I$, is a choice function $I \to A$. *The Axiom of Choice, Zorn's Lemma, and the Well-Ordering Theorem are thus equivalent principles of set theory.* Finally, it should be noted that to this day, no uncountable set (cf. Def. 1.8.1) with an explicit well-ordering is known.

The natural extension of the concept of well-ordering to arbitrary ordered sets is the concept of Artinian order.

Definition 1.4.18 An ordered set A is called **Artinian** (or **Noetherian**) ordered if every non-empty subset of A has a minimal (or a maximal) element.[13]

An ordered set $(A, \leq)$ is thus Noetherian ordered if and only if it is Artinian with respect to the opposite order $\geq$. The following criterion is very useful:

Lemma 1.4.19 *Let $A = (A, \leq)$ be an ordered set. Then the following are equivalent:*

(i) *A is Artinian (or Noetherian) ordered.*
(ii) *There is no infinite strictly monotonically decreasing (or no infinite strictly monotonically increasing) sequence in A.*
(iii) *Every infinite monotonically decreasing sequence $a_0 \geq a_1 \geq a_2 \geq \cdots$ (or every infinite monotonically increasing sequence $a_0 \leq a_1 \leq a_2 \leq \cdots$) in A is stationary.*

Proof It suffices to treat the Artinian case. (ii) and (iii) are trivially equivalent. For the proof of (i) $\Rightarrow$ (ii) let $a_0 > a_1 > a_2 > \cdots$ be a strictly monotonically decreasing sequence in A. Then the set $\{a_i \mid i \in \mathbb{N}\}$ has no minimal element. Contradiction! For the proof of (ii) $\Rightarrow$ (i) let conversely $B \subseteq A$ be a non-empty subset without a minimal element. Then we construct recursively a strictly monotonically decreasing

---

[13] According to E. Artin (1898–1962) and E. Noether (1882–1935).

infinite sequence $a_0 > a_1 > a_2 > \cdots$ of elements from B, which again results in a contradiction. $a_0 \in B$ be arbitrary. If $a_0, \ldots, a_n \in B$ with $a_0 > \cdots > a_n$ are already constructed, then a_n is not a minimal element of B, and there are elements in B that are smaller than a_n, and we choose one of these as a_{n+1}. Here, too, we use the axiom of choice (albeit in a weaker form). □

For Artinian or Noetherian ordered sets, the proof principle of Artinian or Noetherian induction can be formulated:

Theorem 1.4.20 (Artinian and Noetherian Induction) *Let A be an Artinian (respectively, Noetherian) ordered set. For each $a \in A$ let a statement $S(a)$ be given. Then, for each $a \in A$ the statement $S(a)$ holds under the assumption that $S(b)$ holds for all $b \in A$ with $b < a$ (respectively, with $b > a$), then $S(a)$ holds for all $a \in A$.*

Proof If $B := \{a \in A \mid S(a) \text{ is not true}\} \neq \emptyset$, then B would contain a minimal (respectively, a maximal) element a_0. For all $b \in A$ with $b < a_0$ (respectively, with $b > a_0$) then $S(b)$ holds and thus by assumption also $S(a_0)$. Contradiction! □

If the set A in Theorem 1.4.20 is infinite and well-ordered, it is also referred to as **transfinite induction**. The ordinary induction in the case $A = \mathbb{N}$ is included, see Sect. 1.5, in particular Theorem 1.5.5.

Exercises

Exercise 1.4.1 The present exercise uses the elementary theory of divisibility in the domain of positive natural numbers, see Sect. 1.7. On the set $\mathbb{N}^*$ of positive natural numbers, let $|$ be the divisibility relation, i.e., $x \mid y$ holds if and only if x is a divisor of y. Show that $|$ is an order relation on $\mathbb{N}^*$ with 1 as the smallest element. On $\mathbb{N}^* - \{1\}$, exactly the prime numbers are the minimal elements with respect to $|$, i.e., the atoms in $(\mathbb{N}^*, |)$. Draw the Hasse diagrams for the set of divisors of 12 and of 30, respectively. (The second is order-isomorphic to $\mathfrak{P}(\{2, 3, 5\})$.) The chains in $\mathbb{N}^*$ with respect to $|$ correspond bijectively to the finite or infinite sequences $(q_0, q_1, q_2, \ldots)$ with $q_n \in \mathbb{N}^*$ and $q_n \geq 2$ for $n \geq 1$. The corresponding chain $\{q_0, q_0 q_1, q_0 q_1 q_2, \ldots\}$ is maximal if and only if the sequence is infinite and $q_0 = 1$ is, and the remaining q_n are prime numbers. □

Exercise 1.4.2 Let $\preceq$ be a reflexive and transitive relation on the set A; thus, it holds that $a \preceq a$ and from $a \preceq b \preceq c$ follows $a \preceq c$ for all $a, b, c \in A$. Such relations are called **quasi-orders**. Every order is a quasi-order. For quasi-orders, one often uses analogous terms as for orders, e.g., upper bound, lower bound, directed upwards or downwards, etc. They are usually self-explanatory.

a) By "$a \sim b$ if and only if $a \preceq b$ and $b \preceq a$", an equivalence relation $\sim$ is defined on A. On the set $\overline{A}$ of equivalence classes of A with respect to $\sim$, an order relation

1.4 Order Relations

is well-defined by "$[a] \leq [b]$ if and only if $a \preceq b$". (It is particularly to be shown that the $\leq$ relationship for two equivalence classes does not depend on the representatives used for the definition.)

b) If $f : A' \to A$ is an arbitrary mapping, then $a' \preceq_f b'$ is defined as a quasi-order on A' if and only if $f(a') \preceq f(b')$. (Part a) shows that in this way every quasi-order on a set A can be obtained from an order by choosing the canonical projection $A \to \overline{A}$ for f in the situation of a) and applying the order defined in a) on $\overline{A}$.)

c) Let R be an arbitrary relation on A. Then we define $a \preceq_R b$ if and only if there is a finite sequence $a_0, \ldots, a_n$ in A with $a_0 = a$, $a_n = b$ and $a_i R a_{i+1}$ for $i = 0, \ldots, n-1$. (Such a sequence is called a **directed path of length n** in A with respect to R from a to b, see also Remark 1.3.2.) Show that $\preceq_R$ is a quasi-order on A. When is it even an order?

Exercise 1.4.3 Let A be an inductively ordered set. If $a \in A$, then there is a maximal element $M \in A$ with $a \leq M$. (Consider the section $A_{\geq a} \subseteq A$, which is also inductively ordered.)

Exercise 1.4.4 Let $A = (A, \leq)$ be an ordered set.

a) If a, b are incomparable elements in A, then $x \leq' y$ is defined exactly when $x \leq y$ or when $x \leq a$ and $b \leq y$ holds, an order $\leq'$ on A, for which $a \leq' b$ is. It follows: An order on A is maximal with respect to inclusion (orders considered as subsets of $A \times A$) if and only if it is complete.
b) The set of orders on A is strictly inductively ordered with respect to inclusion.
c) Every order on A is contained in a complete order. More precisely: A given order on A is the intersection of all complete orders containing it. (Every ordered set with n elements, $n \in \mathbb{N}$, is thus order-isomorphic to a suborder of the natural order of $\mathbb{N}^*_{\leq n} = \{1, \ldots, n\}$.)

Exercise 1.4.5 Provide an example of a (necessarily infinite) inductively ordered set with a smallest element that is not strictly inductively ordered.

Exercise 1.4.6 Let A be an ordered set. We call a subset C of A **cofinal** in A, if for every $x \in A$ there exists a $y \in C$ with $x \leq y$. C is called **weakly cofinal** in A, if there is no $x \in A$ with $y < x$ for all $y \in C$.

a) If $C \subseteq A$ is cofinal, then C is weakly cofinal. If A is completely ordered, then the converse also holds.
b) A has a well-ordered, weakly cofinal subset. (Conclude similarly as in the proof of 1.4.17.) In particular, it holds: A completely ordered set always has a well-ordered cofinal subset. (For additional remarks, see Exercise 1.8.12.)

1.5 Natural Numbers and Complete Induction

In the following, we want to assume the set $\mathbb{N}$ of natural numbers with their natural order and the usual addition and multiplication as known. In particular, we use the so-called **successor function** $S: n \mapsto n' := n+1$ with the following characteristic properties:

(Peano 1) S is injective with $\mathbb{N}^* = \mathbb{N} - \{0\}$ as image.
(Peano 2) $M = \mathbb{N}$ is the only S-invariant subset $M \subseteq \mathbb{N}$ with $0 \in M$.

The essential condition (Peano 2) means: If M is a subset of the set $\mathbb{N}$ of natural numbers with the two properties

$$(1)\ 0 \in M; \quad (2)\ \text{for all } n \in M \text{ is also } S(n) = n+1 \in M,$$

then $M = \mathbb{N}$. Thus, one reaches every natural number, starting from 0, by repeatedly adding 1. (Peano 1) and (Peano 2) are called the **Peano axioms** (after G. Peano (1858–1932)). The preimage $S^{-1}(n) = n-1$ of $n \in \mathbb{N}^* = \mathbb{N} - \{0\}$ is called the **predecessor** of n. How one can conversely derive the fundamental properties and arithmetic operations of $\mathbb{N}$ from the Peano axioms, we will explain in more detail at the end of this section in Remark 1.5.7. The axiom (Peano 2) is the basis for the following so-called **principle of induction**.

Theorem 1.5.1 (Complete Induction) *To each natural number $n \in \mathbb{N}$ is assigned a statement $A(n)$. The following conditions are fulfilled:*

(1) **Base case:** $A(0)$ *holds.*
(2) **Inductive step:** *For each $n \in \mathbb{N}$ the validity of $A(n)$ implies the validity of $A(n+1)$.*

Then $A(n)$ holds for all $n \in \mathbb{N}$.

Proof Let $M := \{n \in \mathbb{N} \mid A(n) \text{ gilt}\} \subseteq \mathbb{N}$. By assumption (1) is $0 \in M$, and by assumption (2) contains M with each n also $S(n) = n+1$. Thus, $M = \mathbb{N}$ is according to the Peano axiom (Peano 2), and that is the assertion. □

In the inductive step from n to $n+1$ according to 1.5.1 (2) one calls the validity of $A(n)$ the **induction hypothesis** and the validity of $A(n+1)$ the **induction claim**. Of course, in the induction step, one can also conclude from the predecessor $n-1$ to n, $n \in \mathbb{N}^*$. Often, the following variant is used: Let $n_0 \in \mathbb{N}$, and to each natural number $n \geq n_0$ a statement $A(n)$ be assigned. If then $A(n_0)$ holds and for every $n \geq n_0$ the validity of $A(n)$ always also implies that of $A(n+1)$, then $A(n)$ holds for all $n \geq n_0$. To see this, consider the set

$$M := \{n \in \mathbb{N} \mid n < n_0\} \cup \{n \in \mathbb{N} \mid n \geq n_0 \text{ and } A(n) \text{ is true}\}.$$

1.5 Natural Numbers and Complete Induction

Before we discuss some examples of complete induction, we briefly explain the use of summation and product symbols in $\mathbb{R}$ or $\mathbb{C}$.[14] Let a_i, $i \in I$, be a finite family of numbers. Then we denote by

$$\sum_{i \in I} a_i \quad \text{or} \quad \prod_{i \in I} a_i$$

the sum or the product of all these numbers a_i. The fact that these sums and products are well-defined is based on the commutativity of addition and multiplication in $\mathbb{R}$ and $\mathbb{C}$. In the case of a finite sequence $a_m, a_{m+1}, \ldots, a_n$, one also writes

$$\sum_{i=m}^{n} a_i = a_m + a_{m+1} + \cdots + a_n \quad \text{or} \quad \prod_{i=m}^{n} a_i = a_m a_{m+1} \cdots a_n$$

for their sum or product. For the empty index set, the sum is defined to be 0 and the product to be 1. If $I = \{i\}$ contains only one element i, then the sum and product are equal to the number a_i. The sum and product do not change with the so-called **reindexing**, i.e., if $\sigma : J \to I$ is a bijective mapping, then

$$\sum_{j \in J} a_{\sigma(j)} = \sum_{i \in I} a_i, \quad \text{or} \quad \prod_{j \in J} a_{\sigma(j)} = \prod_{i \in I} a_i.$$

For example, one obtains $\sum_{j=m-k}^{n-k} a_{j+k} = \sum_{i=m}^{n} a_i$ by shifting the index set by $k \in \mathbb{Z}$. On occasion, we use other self-evident calculation rules for sum and product. If, for instance, I is the disjoint union of I' and I'', then $I = I' \uplus I''$, so $\sum_{i \in I} a_i = \sum_{i \in I'} a_i + \sum_{i \in I''} a_i$, for $1 \leq m \leq n$ specifically $\sum_{i=1}^{n} a_i = \sum_{i=1}^{m} a_i + \sum_{i=m+1}^{n} a_i$.

Example 1.5.2 (Some Arithmetic Series) For all $n \in \mathbb{N}$ is

$$\sum_{k=1}^{n} k = 1 + 2 + \cdots + n = \frac{n(n+1)}{2}.$$

We prove this by induction over n. The base case for $n = 0$ holds, since $\sum_{k=1}^{0} k$ as an empty sum is equal to 0 and also $0(0+1)/2 = 0$ holds. In the induction step from n to $n+1$, we may assume $\sum_{k=1}^{n} k = n(n+1)/2$ and then have to show $\sum_{k=1}^{n+1} k = (n+1)(n+2)/2$. However, it is indeed

$$\sum_{k=1}^{n+1} k = \left(\sum_{k=1}^{n} k \right) + (n+1) = \frac{n(n+1)}{2} + (n+1) = \frac{(n+1)(n+2)}{2}.$$

$\square$

---

[14] For generalizations, we refer to Sect. 2.1.

Similarly, one proves the following formulas (which one should also know):

$$\sum_{k=1}^{n} k^2 = \frac{n(n+1)(2n+1)}{6}, \quad \sum_{k=1}^{n} k^3 = \left(\frac{n(n+1)}{2}\right)^2 = \left(\sum_{k=1}^{n} k\right)^2.$$ ◊

Example 1.5.3 (Finite Geometric Series) For any real (or complex) number q different from 1 and any $n \in \mathbb{N}$ holds

$$\sum_{k=0}^{n} q^k = 1 + q + \cdots + q^n = \frac{q^{n+1} - 1}{q - 1}.$$

For $n = 0$, both sides are equal to 1. The step from n to $n+1$ follows from

$$\sum_{k=0}^{n+1} q^k = \left(\sum_{k=0}^{n} q^k\right) + q^{n+1} = \frac{q^{n+1} - 1}{q - 1} + q^{n+1} = \frac{q^{n+1} - 1 + q^{n+1}(q - 1)}{q - 1}$$
$$= \frac{q^{n+2} - 1}{q - 1}.$$

For $q = 1$, of course, $\sum_{k=0}^{n} q^k = \sum_{k=0}^{n} 1 = n + 1$. ◊

From the principle of induction 1.5.1 follows the well-ordering property:

Theorem 1.5.4 (Well-ordering principle for $\mathbb{N}$) *The set $\mathbb{N}$ of natural numbers is well-ordered, i.e., every non-empty subset M of $\mathbb{N}$ contains a smallest element, thus an $m_0 \in M$ with $m_0 \leq m$ for all $m \in M$.*

Proof For $n \in \mathbb{N}$ let $A(n)$ be the following statement: If M contains a natural number m with $m \leq n$, then M has a smallest element. We demonstrate the validity of the statement $A(n)$ by induction over n, thereby proving the assertion 1.5.4.

$A(0)$ holds: If M contains a natural number $m \leq 0$, then necessarily $m = 0$ and 0 is the smallest element of M.

In the step from n to $n+1$, we assume the validity of $A(n)$. If M even contains an element $m \leq n$, then it also contains a smallest element by the induction hypothesis. Otherwise, M contains the number $n+1$, since M contains an element $m \leq n+1$ by assumption. In this case, $n+1$ is the smallest element of M. □

The statement 1.5.4 together with Theorem 1.4.20 allows the following induction scheme:

Theorem 1.5.5 (Generalized Principle of Induction) *For each $n \in \mathbb{N}$ let a statement $A(n)$ be given. Then, if for each $n \in \mathbb{N}$ the statement $A(n)$ holds under the assumption that $A(m)$ holds for all $m < n$, then $A(n)$ holds for all $n \in \mathbb{N}$.*

1.5 Natural Numbers and Complete Induction

With the principle of induction, one proves the possibility of the **recursive definition** of sequences. Let A be a set and $(h_n)_{n \in \mathbb{N}}$ be a sequence of mappings $h_n : A^n \to A$, $(x_0, \ldots, x_{n-1}) \mapsto h_n(x_0, \ldots, x_{n-1})$. h_0 is thus given by an element $a_0 \in A$(**base case**). *Then there is exactly one sequence* $(a_n)_{n \in \mathbb{N}}$ *with* $a_n = h_n(a_0, \ldots, a_{n-1})$, $n \in \mathbb{N}^*$, thus

$$a_0, \quad a_1 = h_1(a_0), \quad a_2 = h_2(a_0, a_1), \quad \ldots, \quad a_n = h_n(a_0, a_1, \ldots, a_{n-1}), \quad \ldots$$

One shows for the *proof* by complete induction over $n \in \mathbb{N}$: There are uniquely determined functions $H_n : \mathbb{N}_{\leq n} \to A$, $n \in \mathbb{N}$, with the properties $H_0(0) = a_0$, $H_n(n) = h_n(H_0(0), \ldots, H_{n-1}(n-1))$, $n \in \mathbb{N}^*$, and $H_m = H_n|_{\mathbb{N}_{\leq m}}$ for all $m, n \in \mathbb{N}$, $m \leq n$. Then one sets $a_n := H_n(n)$, $n \in \mathbb{N}^*$. □

The value a_n often depends only on the preceding sequence element a_{n-1}, and often h_n is always the same mapping $h : A \to A$. The recursion scheme then simplifies to

$$a_0, \quad a_1 = h(a_0), \quad a_2 = h(a_1), \quad \ldots, \quad a_n = h(a_{n-1}), \quad \ldots$$

Apparently, in this case, $a_n = h^n(a_0)$ is for all $n \in \mathbb{N}$, where $h^n = h \circ \cdots \circ h$ (*n*-times) is the *n*-th iterate of h. Occasionally, we will use other recursion schemes without comment, as long as they are self-explanatory or can be easily reduced to the above scheme. Already $\sum_{i=1}^n a_i$ and $\prod_{i=1}^n a_i$ are—strictly speaking—to be defined recursively:

$$\sum_{i=1}^0 a_i = 0, \quad \sum_{i=1}^n a_i = \left(\sum_{i=1}^{n-1} a_i\right) + a_n; \quad \prod_{i=1}^0 a_i = 1, \quad \prod_{i=1}^n a_i = \left(\prod_{i=1}^{n-1} a_i\right) a_n.$$

Example 1.5.6 (Fibonacci Sequence) The sequence defined recursively by $F_0 = 0$, $F_1 = 1$, $F_n = F_{n-1} + F_{n-2}$, $n \geq 2$, $(F_n)_{n \in \mathbb{N}}$ is called the **Fibonacci Sequence** and F_n the *n*-th **Fibonacci number**. The first 12 terms $F_0, \ldots, F_{11}$ of the Fibonacci sequence are thus

$$0, 1, 1, 2, 3, 5, 8, 13, 21, 34, 55, 89.$$

For the *n*-th Fibonacci number, the explicit representation is valid

$$F_n = \frac{1}{\sqrt{5}}\left(\left(\frac{1+\sqrt{5}}{2}\right)^n - \left(\frac{1-\sqrt{5}}{2}\right)^n\right), \quad n \in \mathbb{N} \quad \textbf{(Binetsche Formel)}.$$

We prove this by induction over n. For $n = 0$ and $n = 1$, the formula is obviously correct. The induction step for $n \geq 2$ follows from

$$F_n = F_{n-1} + F_{n-2}$$

$$= \frac{1}{\sqrt{5}}\left(\left(\frac{1+\sqrt{5}}{2}\right)^{n-1} - \left(\frac{1-\sqrt{5}}{2}\right)^{n-1}\right) + \frac{1}{\sqrt{5}}\left(\left(\frac{1+\sqrt{5}}{2}\right)^{n-2} - \left(\frac{1-\sqrt{5}}{2}\right)^{n-2}\right)$$

$$= \frac{1}{\sqrt{5}}\left(\left(\frac{1+\sqrt{5}}{2}\right)^{n-2}\left(\frac{1+\sqrt{5}}{2}+1\right) - \left(\frac{1-\sqrt{5}}{2}\right)^{n-2}\left(\frac{1-\sqrt{5}}{2}+1\right)\right)$$

$$= \frac{1}{\sqrt{5}}\left(\left(\frac{1+\sqrt{5}}{2}\right)^{n-2}\left(\frac{1+\sqrt{5}}{2}\right)^{2} - \left(\frac{1-\sqrt{5}}{2}\right)^{n-2}\left(\frac{1-\sqrt{5}}{2}\right)^{2}\right)$$

$$= \frac{1}{\sqrt{5}}\left(\left(\frac{1+\sqrt{5}}{2}\right)^{n} - \left(\frac{1-\sqrt{5}}{2}\right)^{n}\right).$$

One sets

$$\Phi := \frac{1}{2}\left(1+\sqrt{5}\right) = 1{,}618033988749894848204\ldots$$

Then $\Phi^2 = \Phi + 1$ is, thus $\frac{1}{2}(\sqrt{5}-1) = \Phi - 1 = \Phi^{-1}$ and

$$\mathsf{F}_n = (\Phi^n - (-1)^n \Phi^{-n})/\sqrt{5}.$$

F_n, $n \in \mathbb{N}$, *is thus the integer that* $\Phi^n/\sqrt{5}$ *is closest.* Φ is called the **number of the golden ratio**. If $a, b \in \mathbb{R}_+^\times$ and the point a divides the line segment $[0, a+b] \subseteq \mathbb{R}$ in the ratio of the golden ratio, i.e., $(a+b)/a = a/b$, then $a/b = 1 + b/a = \Phi$.[15] ◊

In the present section, we have verified given statements using complete induction. More interesting and important, of course, is to develop methods by which such results can be obtained. (For some of the preceding formulas, this is done in the subsequent volume on analysis of a variable.)

Remark 1.5.7 (Natural Numbers according to Peano) According to a communication from H. Weber (1842–1913), L. Kronecker (1823–1891) is said to have remarked about the natural numbers in 1886:

The whole numbers[16] were made by the dear God, everything else is the work of man.

We do not want to dissuade the reader—especially the beginner—from this naive attitude towards the natural numbers. Nevertheless, in this remark, which the reader can initially skip, it will be explained in more detail how the elementary properties of the natural numbers can be derived solely from the Peano axioms.

---

[15] The choice of the letter Φ is meant to recall $\Phi\varepsilon\iota\delta\iota\alpha\varsigma$ (5th century BC). The number Φ of the Golden Ratio is often also denoted by τ.—For $\alpha := \pi/5$, the equations $0 = \sin 3\alpha - \sin 2\alpha = (4\cos^2\alpha - 1 - 2\cos\alpha)\sin\alpha$ follow from $4\cos^2\alpha - 2\cos\alpha - 1 = 0$ and $2\cos\alpha = 2\cos(\pi/5) = \Phi$. Thus, one can construct $\cos(\pi/5)$ and consequently the regular decagon as well as the regular pentagon with the Golden Ratio. See the representation of ζ_5 in Ex. 3.5.28.

[16] Kronecker means the natural numbers.

1.5 Natural Numbers and Complete Induction

According to Peano, a system of natural numbers, as already noted at the beginning, is a triple $(\mathbb{N}, 0, S)$ consisting of a (necessarily infinite) set $\mathbb{N}$, an element $0 \in \mathbb{N}$ and a so-called **successor function** $S: \mathbb{N} \to \mathbb{N}$ with the following properties:

(Peano 1) S is injective and $0 \notin \operatorname{Im} S$.
(Peano 2)   If $M \subseteq \mathbb{N}$ is a subset of $\mathbb{N}$ with $0 \in M$ and $S(M) \subseteq M$, then $M = \mathbb{N}$.

Since $M := \{0\} \cup S(\mathbb{N})$ is a 0-containing S-invariant subset of $\mathbb{N}$, according to the **Induction Axiom** (Peano 2) $M = \mathbb{N}$, i.e., $S(\mathbb{N}) = \mathbb{N}^* := \mathbb{N} - \{0\}$ and consequently *S is a bijective mapping* $\mathbb{N} \to \mathbb{N}^*$. One usually sets $1 := S(0)$, $2 := S(1) = S(S(0))$, $3 := S(2) = S(S(S(0)))$ etc. *A model for such a triple $\mathbb{N} = (\mathbb{N}, 0, S)$ is already obtained from any injective mapping $f: X \to X$ that is not surjective.* For this, one chooses an arbitrary element 0 in X, which does not belong to the image $f(X)$ of f, and considers the set of subsets I of X that are **inductive** with respect to f and 0. These are by definition the f-invariant subsets $I \subseteq X$ with $0 \in I$. Since obviously any arbitrary intersection of inductive subsets of X and in particular X itself are inductive, there is a smallest inductive subset $\mathbb{N} \subseteq X$, namely the intersection of all inductive subsets of X. Then $(\mathbb{N}, 0, S)$ with $S := f|\mathbb{N}$ is a system of natural numbers in the sense of Peano. The axiom (Peano 1) holds by construction. Furthermore, if $M \subseteq \mathbb{N}$ is invariant under S with $0 \in M$, then $M \subseteq X$ is invariant under f with $0 \in M$, consequently inductive and thus equal to $\mathbb{N}$, since $\mathbb{N}$ is the smallest inductive subset of X. Therefore, the induction axiom (Peano 2) also holds.

The **model of J. v. Neumann** (1903–1957) for the natural numbers is obtained according to this principle in the following way: For sets, one considers the assignment $f: A \mapsto A \cup \{A\} = A \uplus \{A\}$, see Sect. 1.1. Then $\emptyset \notin \operatorname{Im} f$. Furthermore, f is injective, i.e.: If A, B are sets with $A \uplus \{A\} = B \uplus \{B\}$, then $A = B$. From $A \uplus \{A\} = B \uplus \{B\}$ it initially follows $A \in B$ or $A = B$ and analogously $B \in A$ or $B = A$. Since not both $A \in B$ and $B \in A$ can hold simultaneously, see Sect. 1.1, $A = B$ must be the case. If there is now a set of sets that is invariant under f (which is required in axiomatic set theory as the **Infinity Axiom**), we obtain a model of the natural numbers with $0 = \emptyset$ and successor function $S(n) = n \uplus \{n\}$, $n \in \mathbb{N}$. The first natural numbers in this model are thus $0 = \emptyset$, $1 = 0 \uplus \{0\} = \{\emptyset\}$, $2 = 1 \uplus \{1\} = \{\emptyset, \{\emptyset\}\}$, $3 = 2 \uplus \{2\} = \{\emptyset, \{\emptyset\}, \{\emptyset, \{\emptyset\}\}\}$ etc. The number n is thus a set with exactly n elements here, which corresponds to the usual representation of a natural number n (already proposed by B. Russell).—The **Zermelo model** for the natural numbers uses the injective mapping $A \mapsto \{A\}$, for which $\emptyset$ also does not belong to the image. Thus, the sets $\emptyset, \{\emptyset\}, \{\{\emptyset\}\}, \{\{\{\emptyset\}\}\}, \ldots$ (which, except for $\emptyset$, are all 1-element) also a model for the natural numbers.

Now let $\mathbb{N} = (\mathbb{N}, 0, S)$ be again a general system of natural numbers. The Peano axiom (Peano 2) immediately implies the validity of the principle of induction 1.5.1 and the possibility of recursive definition. In particular, the iterates f^n, $n \in \mathbb{N}$, of a mapping $f: A \to A$ are recursively defined by

$$f^0 = \operatorname{id}_A, \quad f^{S(n)} = f \circ f^n, \ n \in \mathbb{N}.$$

Then it holds

(1) $f^m \circ f^n = f^n \circ f^m$ and (2) $(f^m)^n = (f^n)^m$ for all $m, n \in \mathbb{N}$,

which we want to prove by induction over n. (After addition and multiplication are defined in $\mathbb{N}$, naturally $f^m \circ f^n = f^n \circ f^m = f^{m+n}$ and $(f^m)^n = (f^n)^m = f^{mn}$.)

Proof (1) For $n = 0$, it is $f^m \circ f^0 = f^m \circ \mathrm{id} = f^m = \mathrm{id} \circ f^m = f^0 \circ f^m$. For $n = 1$, we prove the equation $f^m \circ f = f \circ f^m = f^{S(m)}$ by induction over m, where the conclusion from m to $S(m)$ results from $f^{S(m)} \circ f = f \circ f^m \circ f = f \circ f^{S(m)}$. The general conclusion from n to $S(n)$ is thus obtained because of $f^m \circ f^{S(n)} = f^m \circ f \circ f^n = f \circ f^m \circ f^n = f \circ f^n \circ f^m = f^{S(n)} \circ f^m$, where the penultimate equation uses the induction hypothesis.

In the proof of (2), we use that for every mapping $g: A \to A$ with $f \circ g = g \circ f$ also $f^m \circ g = g \circ f^m$ and $(fg)^m = f^m \circ g^m$ hold for all $m \in \mathbb{N}$, which again results from a simple induction over m. For $n = 0$, one now obtains $(f^m)^0 = \mathrm{id} = (\mathrm{id})^m = (f^0)^m$. In the conclusion from n to $S(n)$, $f \circ f^n = f^n \circ f$ holds because of (1) and thus $(f^m)^{S(n)} = f^m \circ (f^m)^n = f^m \circ (f^n)^m = (f \circ f^n)^m = (f^{S(n)})^m$.

We now explain how addition and multiplication on $\mathbb{N}$ can be introduced solely with the help of the successor function S. The addition of $1 = S(0)$ is S itself, i.e., it is $m + 1 := S(m)$. The **predecessor** $S^{-1}(m)$ of a natural number $\neq 0$ is then $m - 1$. The **addition** of any number $n \in \mathbb{N}$ is obtained by n-fold addition of 1, i.e., we set

$$m + n := S^n(m), \quad m, n \in \mathbb{N}.$$

It is $m = S^m(0)$ (induction over m) and $m + n = S^n(m) = S^n(S^m(0)) = S^m(S^n(0)) = n + m$ (cf. (1)), i.e., *addition is commutative*. Furthermore, the *cancellation rule of addition* holds: From $m + k = n + k$, i.e., $S^k(m) = S^k(n)$ follows $m = n$, since S^k is injective like S (induction over k). Moreover, $m + n$ is only 0 if m and n are both 0. Why? Furthermore, it results in $m + (n + 1) = S^{n+1}(m) = S(S^n(m)) = (n + m) + 1$. For any mapping $f: A \to A$ one obtains by induction over n the already mentioned equation $f^m \circ f^n = f^{m+n}$. In the conclusion from n to $S(n) = n + 1$ it is namely

$$f^m f^{n+1} = f^m \circ f \circ f^n = f \circ f^m \circ f^n = f \circ f^{m+n} = f^{(m+n)+1} = f^{m+(n+1)}.$$

Now directly follows the general *associative law of addition*: For $k, m, n \in \mathbb{N}$ it is

$$(k + m) + n = S^n(k + m) = S^n(S^m(k)) = S^{m+n}(k) = k + (m + n).$$

The **Product** of two natural numbers m, n is the sum $m + \cdot s + m$ with n summands, i.e., we set

$$m \cdot n = mn := (S^m)^n(0).$$

1.5 Natural Numbers and Complete Induction

Specifically, $0 \cdot n = n \cdot 0 = 0$ and $1 \cdot n = n \cdot 1 = n$. Only then is the product mn equal to 0, if one of the two factors m or n is equal to 0. Why? Because of $(S^m)^n = (S^n)^m$, see (2), multiplication is *commutative*. For an arbitrary mapping $f : A \to A$ one obtains by induction over n the already mentioned equation $(f^m)^n = f^{mn}$. In the induction step from n to $n + 1$, namely $(f^m)^{n+1} = f^m \circ (f^m)^n = f^m \circ f^{mn} = f^{m+mn} = f^{m(n+1)}$ because of $m(n + 1) = (S^m)^{n+1}(0) = (S^m \circ (S^m)^n)(0) = S^m(mn) = mn + m$. The obtained equations imply the *associative law of multiplication* and also the *distributive law*. For $k, m, n \in \mathbb{N}$ namely $(km)n = (S^{km})^n(0) = ((S^k)^m)^n(0) = (S^k)^{mn}(0) = k(mn)$ and $k(m + n) = (S^k)^{m+n}(0) = ((S^k)^m \circ (S^k)^n)(0) = (S^{km} \circ S^{kn})(0) = km + kn$.

With the help of addition, the natural order on $\mathbb{N}$ can be obtained as follows. For $m, n \in \mathbb{N}$ is

$$m \le n \text{ if and only if there is a } k \in \mathbb{N} \text{ with } m + k = n.$$

It is thus $\{n \in \mathbb{N} \mid m \le n\} = \operatorname{Im} S^m$ (because $k = S^k(0)$ for all $k \in \mathbb{N}$). This indeed provides an order. The antisymmetry of $\le$ follows from the fact that a sum $k + \ell$ in $\mathbb{N}$ can only be 0 if both summands are equal to 0. *The order $\le$ is even complete*, i.e., for arbitrary $m, n \in \mathbb{N}$ holds $n \in \operatorname{Im} S^m$ or $m \in \operatorname{Im} S^n$. This is shown by induction over n using the following identity: $\operatorname{Im} S^n = \operatorname{Im} S^{n+1} \uplus \{n\}$. It is thus $\mathbb{N} - \operatorname{Im} S^n = \{m \in \mathbb{N} \mid m < n\} = \{0, \ldots, n-1\}$ the n-element set of natural numbers $<n$. Now the **well-ordering property** of $(\mathbb{N}, \le)$ can also be proven as executed in theorem 1.5.4.

There is essentially only one system of natural numbers. Namely, if $\mathbb{N} = (\mathbb{N}, 0, S)$ and $\mathbb{N}' = (\mathbb{N}', 0', S')$ are triples that both satisfy the Peano axioms, then there are uniquely determined mappings $f : \mathbb{N} \to \mathbb{N}'$ and $f' : \mathbb{N}' \to \mathbb{N}$ with $f(0) = 0'$, $f \circ S = S' \circ f$ or $f'(0') = 0$, $f' \circ S' = S \circ f'$. They are recursively defined by $f(0) = 0'$, $f(S(n)) = S'(f(n))$, $n \in \mathbb{N}$, and analogously for f'. Then necessarily applies $f' \circ f = \operatorname{id}_\mathbb{N}$ and $f \circ f' = \operatorname{id}_{\mathbb{N}'}$, since $\operatorname{id}_\mathbb{N}$ and $\operatorname{id}_{\mathbb{N}'}$ fulfill the same recursions as $f' \circ f$ or $f \circ f'$. f and f' are therefore inverse mappings to each other, which map the structures given by $0, S$, and $0', S'$ onto $\mathbb{N}$ or $\mathbb{N}'$. ◇

Exercises

Exercise 1.5.1 Prove the formulas given at the end of Example 1.5.2.

Exercise 1.5.2 For all $n \in \mathbb{N}$ holds:

a) $\sum_{k=1}^{n}(-1)^{k-1}k = \frac{1}{4}\bigl(1 + (-1)^{n-1}(2n+1)\bigr)$.
b) $\sum_{k=1}^{n}(-1)^{k-1}k^2 = (-1)^{n+1} \cdot \frac{n(n+1)}{2}$.
c) $\sum_{k=1}^{n}(2k-1) = n^2$.
d) $\sum_{k=1}^{n}k(k+1) = \frac{1}{3}n(n+1)(n+2)$.
e) $\sum_{k=1}^{n}(2k-1)^2 = \frac{n}{3}(4n^2 - 1)$.

Exercise 1.5.3 For all $n \in \mathbb{N}$ holds:

a) $\sum_{k=1}^{n} \frac{1}{k(k+1)} = 1 - \frac{1}{n+1}$.
b) $\sum_{k=1}^{n} \frac{1}{4k^2-1} = \frac{1}{2}\left(1 - \frac{1}{2n+1}\right)$.
c) $\sum_{k=1}^{n} \frac{1}{k(k+1)(k+2)} = \frac{1}{4} - \frac{1}{2(n+1)(n+2)}$.
d) $\sum_{k=1}^{n} \frac{k-1}{k(k+1)(k+2)} = \frac{1}{4} - \frac{2n+1}{2(n+1)(n+2)}$.

Exercise 1.5.4 For all $n \geq 1$, the following holds:

a) $\prod_{k=2}^{n}\left(1 - \frac{1}{k^2}\right) = \frac{1}{2}\left(1 + \frac{1}{n}\right)$.

b) $\prod_{k=2}^{n}\left(1 - \frac{2}{k(k+1)}\right) = \frac{1}{3}\left(1 + \frac{2}{n}\right)$.

c) $\prod_{k=2}^{n} \frac{k^3-1}{k^3+1} = \prod_{k=2}^{n}\left(1 - \frac{2}{k^3+1}\right) = \frac{2}{3}\left(1 + \frac{1}{n(n+1)}\right)$.

Exercise 1.5.5 For all $n \in \mathbb{N}$ and all $q \in \mathbb{R}$, $q \neq 1$, the following holds:

a) $\prod_{k=0}^{n}\left(1 + q^{2^k}\right) = \frac{q^{2^{n+1}} - 1}{q - 1}$.

b) $\sum_{k=1}^{n} k q^k = \frac{nq^{n+2} - (n+1)q^{n+1} + q}{(q-1)^2}$.

Exercise 1.5.6 For all $n \in \mathbb{N}$ applies:

a) 5 divides $2^{n+1} + 3 \cdot 7^n$.
b) 3 divides $n^3 + 2n$.
c) 6 divides $n^3 - n$.
d) 7 divides $5^{2n+1} + 2^{2n+1}$.
e) 30 divides $n^5 - n$.
f) 3 divides $2^{2n} - 1$.
g) 15 divides $3n^5 + 5n^3 + 7n$.
h) 133 divides $11^{n+2} + 12^{2n+1}$.
i) 5 divides $3^{n+1} + 2^{3n+1}$.

Exercise 1.5.7 For the recursively defined sequences (a_n) in a) to e), prove the given explicit representation in each case.

a) $a_0 = 2$, $a_n = 2 - 1/a_{n-1}$, $n \geq 1$. Then $a_n = (n+2)/(n+1)$ is for all $n \in \mathbb{N}$.
b) $a_0 = 0$, $a_1 = 1$, $a_n = \frac{1}{2}(a_{n-1} + a_{n-2})$, $n \geq 2$. Then

$$a_n = \frac{2}{3}\left(1 - \frac{(-1)^n}{2^n}\right), \quad n \in \mathbb{N}.$$

c) $a_0 = 1$, $a_n = 1 + (1/a_{n-1})$, $n \geq 1$. Then $a_n = F_{n+2}/F_{n+1}$ is for all $n \in \mathbb{N}$, where F_k for $k \in \mathbb{N}$ is the k-th Fibonacci number.
d) $a_0 = 0$, $a_1 = 1$, $a_n = a_{n-1} + 2a_{n-2}$, $n \geq 2$. Then $a_n = \frac{1}{3}(2^n - (-1)^n)$, $n \in \mathbb{N}$.
e) $a_0 = 0$, $a_1 = 1$, $a_n = 2a_{n-1} + a_{n-2}$, $n \geq 2$. It is

$$a_n = \frac{\left(1+\sqrt{2}\right)^n - \left(1-\sqrt{2}\right)^n}{2\sqrt{2}}, \quad n \in \mathbb{N}.$$

Exercise 1.5.8 The sequence (a_n) is recursively defined by $a_0 = 1$, $a_n = \sum_{k=0}^{n-1} a_k$, $n \geq 1$.

a) It is $a_n = 2^{n-1}$ for all $n \geq 1$.
b) The number of finite sequences (of variable length) of positive natural numbers with sum n is equal to a_n. (For example, one has $a_3 = 4$ and $3 = 2 + 1 = 1 + 2 = 1 + 1 + 1$.—For the case where the sequence length is fixed, see Ex. 1.6.19.)

Exercise 1.5.9 Prove by induction the following equations for the Fibonacci numbers F_n, $n \in \mathbb{N}$.

a) It is $F_{n+m} = F_{n-1}F_m + F_n F_{m+1}$ for all $m \geq 0$ and all $n \geq 1$. Specifically, for all $n \geq 1$, the equation $F_{2n} = F_n(F_{n-1} + F_{n+1}) = F_{n+1}^2 - F_{n-1}^2$ holds.
b) For all $n \geq 1$, it is $F_n^2 = F_{n-1}F_{n+1} + (-1)^{n+1}$.
c) For $\Phi = \frac{1}{2}(1+\sqrt{5})$ holds $\Phi^n = F_{n-1} + F_n\Phi$, $n \in \mathbb{N}^*$. (Through these identities, the Fibonacci numbers F_n can be defined for all $n \in \mathbb{Z}$. Then the equations $F_n = F_{n-1} + F_{n-2}$ and $F_n = (-1)^{n+1}F_{-n}$ as well as the formulas in a) and b) hold for all $m, n \in \mathbb{Z}$.—Since the powers Φ^n can be well calculated with the equations

$$(a + b\Phi)^2 = (a^2 + b^2) + (2a+b)b\Phi, \quad a, b \in \mathbb{Z},$$

and the method of fast exponentiation (cf. Remark (2) in Example 2.2.23), one also obtains a method for calculating F_n for a given $n \in \mathbb{N}$, without having to determine all Fibonacci numbers F_k, $0 \leq k < n$.)

Exercise 1.5.10 Let $a_1, \ldots, a_n$, $n \in \mathbb{N}^*$, be elements of a totally ordered set. Prove by induction on n the existence of Min$(a_1, \ldots, a_n)$ and Max$(a_1, \ldots, a_n)$. □

1.6 Finite Sets and Combinatorics

In this section, we want to derive some counting formulas for finite sets. These belong to elementary **combinatorics** and are used repeatedly. More comprehensive information can be found in books 1 and 17.

Let $n \in \mathbb{N}$. The prototype for a finite set with n elements is the set

$$\mathbb{N}^*_{\leq n} = \{1, \ldots, n\} = \{x \in \mathbb{N}^* \mid x \leq n\}$$

of the first n positive natural numbers. It should be remembered that $\mathbb{N}^* = \mathbb{N} - \{0\}$ is. In particular, $\mathbb{N}^*_{\leq 0} = \emptyset$. We say a set A is a **finite set with n elements**, if there is a bijective mapping from $\mathbb{N}^*_{\leq n}$ onto A, meaning the elements of A can be numbered with the numbers $1, \ldots, n$, where different elements receive different numbers. It then holds that $A = \{a_1, \ldots, a_n\}$ with $a_i \neq a_j$ for $i \neq j$. The number n is uniquely determined and is called the **number of elements** or the **cardinal number** of A. It is denoted by[17]

$$|A| \quad \text{or} \quad \text{Card } A.$$

This uniqueness of the number of elements of a finite set is by no means self-evident and results from the following lemma:

Lemma 1.6.1 *If $m, n \in \mathbb{N}$ and is $f : \mathbb{N}^*_{\leq n} \to \mathbb{N}^*_{\leq m}$ a bijective mapping, then $n = m$.*

Proof We use induction on n. Since $\mathbb{N}^*_{\leq m}$ is not empty for $m > 0$, the base case $n = 0$ is clear. In the induction step from n to $n + 1$, the induction hypothesis follows in the case $f(n+1) = m$ due to $\mathbb{N}^*_{\leq n+1} = \mathbb{N}^*_{\leq n} \uplus \{n+1\}$ and $\mathbb{N}^*_{\leq m} = \mathbb{N}^*_{\leq m-1} \uplus \{m\}$ directly from the induction assumption. Otherwise, let $k := f^{-1}(m) \leq n$ and $\sigma = \langle k, n+1 \rangle$ be the transposition of $\mathbb{N}^*_{\leq n+1}$, which swaps k and $n + 1$ and has the remaining elements of $\mathbb{N}^*_{\leq n+1}$ as fixed points. Then $f \circ \sigma : \mathbb{N}^*_{\leq n+1} \to \mathbb{N}^*_{\leq m}$ is bijective with $(f \circ \sigma)(n+1) = f(k) = m$, and one is in the already mentioned trivial case. □

We denote the set of finite subsets of a set A by $\mathfrak{E}(A)$ and the set of n-element subsets of A by $\mathfrak{E}_n(A)$, $n \in \mathbb{N}$. Thus,

$$\mathfrak{E}(A) = \biguplus_{n \in \mathbb{N}} \mathfrak{E}_n(A),$$

and $\mathfrak{E}(A) = \mathfrak{P}(A)$ holds if and only if A is a finite set.

If $A \to B$ is a bijective mapping, then A is finite if and only if B is finite. In this case, $|A| = |B|$. If the set A is partitioned into the finite subsets $A_1, \ldots, A_m$ (which are pairwise disjoint), then $A = \biguplus_{i=1}^{m} A_i$ is also finite and it holds

$$|A| = \left| \biguplus_{i=1}^{m} A_i \right| = \sum_{i=1}^{m} |A_i| = |A_1| + \cdots + |A_m|.$$

---

[17] In English literature, the designation #A for $|A|$ is also common.

1.6 Finite Sets and Combinatorics

This is proven by induction over n. *Thus, the addition of natural numbers is realized by the union of disjoint finite sets.* If $A_1, \ldots, A_m$ are arbitrary finite sets, then the cross product $\prod_{i=1}^{m} A_i = A_1 \times \cdots \times A_m$ is also finite with

$$\left| \prod_{i=1}^{m} A_i \right| = \prod_{i=1}^{m} |A_i| = |A_1| \cdots |A_m|.$$

In the *proof* of this equation (by induction over m), it suffices to treat the case $m = 2$. However, it is $A_1 \times A_2$ the union of the pairwise disjoint sets $\{a\} \times A_2$, $a \in A_1$, and consequently

$$|A_1 \times A_2| = \sum_{a \in A_1} |\{a\} \times A_2| = \sum_{a \in A_1} |A_2| = |A_1| \cdot |A_2|.$$

Thus, the multiplication of natural numbers is realized by the Cartesian product of finite sets. For each finite set A and each $m \in \mathbb{N}$ it follows in particular $|A^m| = |A|^m$. It is A^m the set of m-tuples of elements from A, i.e., the set of mappings from $\mathbb{N}^*_{\leq m}$ into A. Replacing the set $\mathbb{N}^*_{\leq m}$ with any finite set I with m elements, one obtains:

Theorem 1.6.2 *If I and A are finite sets with m and n elements respectively, then the set A^I of mappings from I into A is finite with n^m elements. Thus, it is $|A^I| = |A|^{|I|}$.*

Note $A^\emptyset = \{\emptyset\}$ for any set A. Thus, there are exactly 2^n 0-1 sequences of length $n \in \mathbb{N}$. If one assigns to a sequence $(k_1, \ldots, k_s)$, $s \in \mathbb{N}$, positive natural numbers with sum $k_1 + \cdots + k_s = n$ the 0-1 sequence in which alternately k_1 zeros, k_2 ones, k_3 zeros, ... appear, then for $n \geq 1$ one obtains a bijective mapping of the set of these sequences onto the 2^{n-1}-element set of 0-1 sequences of length n that start with a zero. This solves Ex. 1.5.8b) combinatorially.[18]

Let I be a set. The mapping that assigns to each subset $J \subseteq I$ the indicator function e_J is a bijective mapping of the power set $\mathfrak{P}(I)$ onto the set $\{0,1\}^I$. From Proposition 1.6.2 it follows:

Corollary 1.6.3 *Let I be a finite set with m elements. Then $\mathfrak{P}(I)$ is a finite set with 2^m elements. Thus, it holds that $|\mathfrak{P}(I)| = 2^{|I|}$.*

The following statement gives the number of *injective* mappings between finite sets.

---

[18] We generally recommend looking for combinatorial justifications for equations whose terms can be interpreted combinatorially. For example, one can derive the simple equations $2^0 + 2^1 + 2^2 + \cdots + 2^n = 2^{n+1} - 1$, $n \in \mathbb{N}$, by appropriately counting the 0-1 sequences of length $n+1$ that differ from the 0-tuple.

Theorem 1.6.4 *Let I and A be finite sets with m and n elements, respectively, then there are exactly*

$$[n]_m := \prod_{k=0}^{m-1}(n-k) = n(n-1)\cdots(n-m+1)$$

injective mappings from I to A.

Proof We prove the statement by induction on m. For $m = 0$, I is empty, and there is exactly one mapping from I to A (namely the empty mapping); $[n]_0$ is also 1 as an empty product. Furthermore, the statement is also obvious for $m = 1$: There are n injective mappings from $I = \{i\}$ to A, and it is $[n]_1 = n$.

In the step from m to $m+1$ elements, let I be a set with $m+1$ elements. First, let $m \leq n$ and let i_0 be a fixed element from I. By the induction hypothesis, there are $[n]_m$ injective mappings from $I' := I - \{i_0\}$ to A. If $f : I' \to A$ is such a mapping, then there are $n - m$ injective mappings from I to A, whose restriction to I' coincides with f: As the image of i_0, due to injectivity, only one of the $n - m$ elements in $A - f(I')$ is possible. Therefore, in total, there are $[n]_m(n - m) = [n]_{m+1}$ injective mappings from I to A. However, if $m > n$, then by the induction hypothesis, there are no injective mappings from I' to A and thus certainly none from I to A. Moreover, in this case, with $[n]_m = 0$ also $[n]_{m+1} = [n]_m(n-m) = 0$. □

The symbol $[n]_m$ is called the **generalized factorial** or **descending factorial**. It is defined for any number (or even any ring element) α and each $m \in \mathbb{N}$:

$$[\alpha]_m := \prod_{k=0}^{m-1}(\alpha - k) = \alpha(\alpha - 1)\cdots(\alpha - m + 1).$$

Often, the so-called **ascending factorial** or the **Pochhammer symbol** $(\alpha)_m := [\alpha + m - 1]_m = \alpha(\alpha + 1)\cdots(\alpha + m - 1)$ is used.

Remark 1.6.5 (Combinatorial Principle) For $I = \{1, \ldots, m\}$ it follows from Theorem 1.6.4: If A is a set with n elements, then there are exactly $[n]_m$ tuples $(a_1, \ldots, a_m)$ with pairwise distinct elements $a_i \in A$, $i = 1, \ldots, m$. This is a special case of the following general **combinatorial principle**: *Let there be a set $C \subseteq A^m$ of m-tuples. Furthermore, there are natural numbers $n_1, \ldots, n_m$ with the following properties: If $k < m$ and $(a_1, \ldots, a_k)$ is the beginning of a sequence from C, then there are exactly n_{k+1} elements $a_{k+1} \in A$, for which $(a_1, \ldots, a_k, a_{k+1})$ is also the beginning of a sequence from C. Then C has exactly $n_1 \cdots n_m$ elements.* This is also easily proven by induction over m. In the induction step from $m - 1$ to m, one uses the induction hypothesis for the set

$$C' := \{(a_1, \ldots, a_{m-1}) \in A^{m-1} \mid \text{there is an } a_m \in A \text{ with } (a_1, \ldots, a_{m-1}, a_m) \in C\}.$$

1.6 Finite Sets and Combinatorics

Example 1.6.6 Let B and C be finite sets with $|B| > |C|$. Then, of course, as already mentioned, there is no injective mapping from B to C. For every mapping $f : B \to C$ there are thus elements $b, b' \in B$ with $b \neq b'$ but $f(b) = f(b')$. This statement is also known as the **(Dirichlet's) pigeonhole principle** (after L. Dirichlet (1805–1859)): If you distribute $m := |B|$ objects into $n := |C|$ pigeonholes and $m > n$, then there is more than one object in at least one pigeonhole. *More generally, if $m > nr$, $r \in \mathbb{N}$, then there are more than r objects in at least one pigeonhole.*

Now let B and C be finite sets with the same number of elements. For a mapping $f : B \to C$ the following are equivalent: (i) f is injective. (ii) f is surjective. (iii) f is bijective.

If f is injective, then $f(B)$ has the same number of elements as B and thus as C, therefore it exhausts C completely. Conversely, if f is not injective, then $f(B)$ has fewer elements than B and thus than C. Then $f(B) \neq C$ and f is not surjective.

From Proposition 1.6.4 it follows directly:

Corollary 1.6.7 *B and C are finite sets with the same number of elements $|B| = |C| = n$. Then there are exactly*

$$n! := [n]_n = \prod_{k=1}^{n} k = 1 \cdots n$$

bijective mappings from B to C. In particular, there are exactly $n!$ permutations of the n-element set B.

If $n \in \mathbb{N}$, then one reads n-**factorial** for the number $n!$. One can determine $n!$ recursively through $0! = 1$ and $n! = (n-1)! \, n$. For natural numbers m and n with $m \leq n$ it is apparent that

$$[n]_m = \frac{n!}{(n-m)!}.$$

Remark 1.6.8 For larger values of n, one uses for $n!$ the important estimation

$$\sqrt{2\pi n}\left(\frac{n}{e}\right)^n e^{1/(12n+1)} < n! < \sqrt{2\pi n}\left(\frac{n}{e}\right)^n e^{1/12n}.$$

In particular, one has for $n \to \infty$ the good approximation

$$n! \sim \sqrt{2\pi n}\left(\frac{n}{e}\right)^n.$$

Here, the Euler's number e = 2,7182818284590... is the base of the natural logarithms.[19] We will prove this so-called **Stirling's formula** in Vol. 2. ◊

The following counting formula is one of the most important results of elementary combinatorics.

Theorem 1.6.9 *Let A be a finite set with n elements. For every natural number $m \leq n$, the number of m-element subsets of A is then*

$$\binom{n}{m} := \frac{[n]_m}{[m]_m} = \frac{n!}{m!(n-m)!} = \frac{n(n-1)\cdots(n-m+1)}{1 \cdot 2 \cdots m}.$$

Proof Let $I := \{1, \ldots, m\}$. By definition, every m-element subset of A is the image of an injective mapping $I \to A$. Two such mappings f and g have the same image if and only if there is a permutation σ of I with $f = g \circ \sigma$. Consequently, according to Corollary 1.6.7, each $[m]_m = m!$ of these mappings has the same image. Since, according to Theorem 1.6.4, there are a total of $[n]_m$ injective mappings from $I \to A$, the assertion follows. □

The numbers $\binom{n}{m}$—pronounced: n choose m (in English: n choose m)—in Theorem 1.6.9 are called **binomial coefficients**. They are defined for arbitrary (real or complex) numbers α and any $m \in \mathbb{N}$:

$$\binom{\alpha}{m} := \frac{[\alpha]_m}{m!} = \frac{\alpha(\alpha-1)\cdots(\alpha-m+1)}{1 \cdot 2 \cdots m}.$$

Note that with this definition, the formula in Theorem 1.6.9 also holds for $m > n$. Furthermore, one sets $\binom{\alpha}{m} = 0$ for negative integers m. This immediately results in the following calculation rules:

Proposition 1.6.10 *For arbitrary α and all $m \in \mathbb{Z}$ the following holds:*

(1) $\binom{\alpha}{0} = 1$. (2) $\binom{\alpha}{m+1} = \binom{\alpha}{m} \cdot \frac{\alpha-m}{m+1} = \binom{\alpha-1}{m} \cdot \frac{\alpha}{m+1}$, $m \neq -1$.

(3) $\binom{-\alpha}{m} = (-1)^m \binom{m+\alpha-1}{m}$. (4) $\binom{\alpha+1}{m} = \binom{\alpha}{m} + \binom{\alpha}{m-1}$.

Example 1.6.11 (1) The values $\binom{\alpha}{m}$, $m \in \mathbb{N}$, are generally calculated most quickly recursively with a fixed α using Proposition 1.6.10 (2). For $\alpha \in \mathbb{N}$ and $\alpha \geq m \geq$

---

[19] We use the asymptotic equality $\sim$ in the following sense: If (a_n) and (b_n) are sequences of real (or complex) numbers, then $a_n \sim b_n$ means that the sequence (a_n/b_n) is defined for sufficiently large n (i.e., $b_n \neq 0$ is for sufficiently large n) and converges to 1, see Ex. 3.2.9.

1.6 Finite Sets and Combinatorics

0, this recursion formula can also be understood combinatorially: If $\mathfrak{E}_m$ and $\mathfrak{E}_{m+1}$ are the sets of m- and $(m+1)$-element subsets of $I := \{1, \ldots, \alpha\}$ and $\mathfrak{X}_m$ is the set of pairs (C, c) with $C \in \mathfrak{E}_m$, $c \in I - C$, then $\mathfrak{X}_m \to \mathfrak{E}_{m+1}$, $(C, c) \mapsto C \cup \{c\}$, is a surjective mapping whose fibers all contain exactly $m + 1$ elements. Due to $|\mathfrak{X}_m| = \binom{\alpha}{m} \cdot (\alpha - m)$, the recursion $\binom{\alpha}{m} \cdot \frac{\alpha - m}{m+1} = \binom{\alpha}{m+1}$ results with the Schäfer rule from Ex. 1.6.6.

If n, m, and $n - m$ are large natural numbers, then Stirling's formula from Remark 1.6.8 provides the approximation

$$\binom{n}{m} = \frac{n!}{m!(n-m)!} \approx \frac{1}{\sqrt{2\pi n x(1-x)}\left(x^x(1-x)^{1-x}\right)^n}, \quad x := m/n.$$

The function $x^x(1-x)^{(1-x)}$ in the interval $[0, 1]$ is very well approximated by the lower semicircular arc $1 - \sqrt{x(1-x)}$ of the circle with center $(1/2, 1)$ and radius $1/2$. The logarithm of the reciprocal of this function, i.e., $-x \ln x - (1-x) \ln(1-x)$, $x \in]0, 1[$, is called the **Boltzmann-Gibbs entropy (function)**.

(2) (**Pascal's Triangle**) The formula from Proposition 1.6.10 (4) provides a method for calculating the binomial coefficients $\binom{n}{m}$ for $m, n \in \mathbb{N}$. For this, these binomial coefficients are noted in the form of the so-called **Pascal's Triangle**, see Fig. 1.14.

In it, the binomial coefficients $\binom{n+1}{m}$ of the $(n+1)$-th row are obtained by starting this row with 1 and then adding two adjacent binomial coefficients of the n-th row (which have already been calculated).

(3) For $0 \leq m \leq n$, it follows directly from the definition $\binom{n}{m} = \binom{n}{n-m}$, so that in this case, one can replace the number m with $n - m$ for the calculation of $\binom{n}{m}$ (which is useful when $m > n/2$). With Theorem 1.6.9, this can also be seen as follows: If the set A contains n elements, then the mapping $B \mapsto A - B$ is a bijective mapping of the set of m-element subsets B of A onto the set of $(n-m)$-element subsets of A.

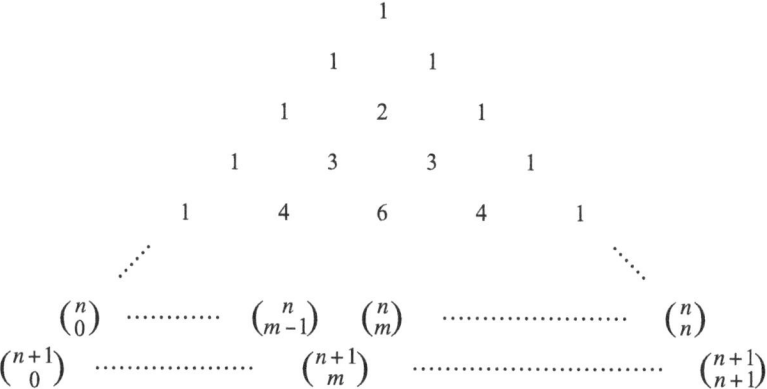

Fig. 1.14 Pascal's Triangle

Incidentally, the formula in Proposition 1.6.10 (4) for $\alpha = n \in \mathbb{N}$ can also be proven with Theorem 1.6.9. For this, let A be a $(n+1)$-element set and $a \in A$ a fixed element. The $\binom{n+1}{m}$ m-element subsets of A are then on the one hand the $\binom{n}{m}$ m-element subsets of $A - \{a\}$ and on the other hand the $\binom{n}{m-1}$ sets $B \uplus \{a\}$, where B runs through the $(m-1)$-element subsets of $A - \{a\}$. ◊

Example 1.6.12 Let A be a finite set with n elements. According to Theorem 1.6.9, A has exactly $\binom{n}{m}$ subsets with m elements for every natural number m with $0 \leq m \leq n$, thus $\sum_{m=0}^{n} \binom{n}{m}$ subsets in total. With Corollary 1.6.3, it follows that

$$\sum_{m=0}^{n} \binom{n}{m} = 2^n,$$

which can also be easily confirmed by induction using Proposition 1.6.10 (4). ◊

Example 1.6.13 *Let m and n be natural numbers. The number of m-tuples $(x_1, \ldots, x_m)$ of natural numbers with $\sum_{i=1}^{m} x_i \leq n$ is equal to $\binom{n+m}{m}$.*

Proof The mapping $(x_1, x_2, \ldots, x_m) \longmapsto \{x_1 + 1, x_1 + x_2 + 2, \ldots, x_1 + \cdots + x_m + m\}$ is a bijective mapping of this set of m-tuples onto the set of m-element subsets of the set $\{1, 2, \ldots, n+m\}$. □

The number of m-tuples $(x_1, \ldots, x_m) \in \mathbb{N}^m$ with $\sum_{i=1}^{m} x_i = n$ is equal to $\binom{n+m-1}{m-1}$, if one exceptionally sets $\binom{-1}{-1} := 1$ here. For the *proof*, it should only be noted that the mapping $(x_1, \ldots, x_{m-1}, x_m) \longmapsto (x_1, \ldots, x_{m-1})$ for $m \geq 1$ is a bijective mapping of the set of m-tuples $(x_1, \ldots, x_m)$ with $\sum_{i=1}^{m} x_i = n$ onto the set of $(m-1)$-tuples $(x_1, \ldots, x_{m-1})$ with $\sum_{i=1}^{m-1} x_i \leq n$.—The number of m-tuples $(x_1, \ldots, x_m) \in \mathbb{N}^m$ with $\sum_{i=1}^{m} x_i = n$ can also be interpreted as the number of ways to distribute n objects into m compartments, if two such distributions are identified when they result from a permutation of the objects.

Example 1.6.14 (Polynomial Coefficients) Let $A_1, \ldots, A_r$ be pairwise disjoint finite sets with $m_1, \ldots, m_r$ elements and $A := \biguplus_{i=1}^{r} A_i$. Then $\sigma \mapsto (\sigma|A_1, \ldots, \sigma|A_r)$ a bijective mapping of the set $\mathfrak{S}'$ of those permutations σ of A with $\sigma(A_i) = A_i$ for $i = 1, \ldots, r$ onto the set $\mathfrak{S}(A_1) \times \cdots \times \mathfrak{S}(A_r)$. Therefore, $|\mathfrak{S}'| = m_1! \cdots m_r!$. Generally, for an r-tuple $m = (m_1, \ldots, m_r)$ of natural numbers, one sets

$$m! := m_1! \cdots m_r!.$$

Now let A be a finite set with n elements, and $m = (m_1, \ldots, m_r)$ an r-tuple of natural numbers with

$$n = |m| = m_1 + \cdots + m_r.$$

1.6 Finite Sets and Combinatorics

The number of mappings $f : A \to \{1, \ldots, r\}$ with $|f^{-1}(i)| = m_i$ for $i = 1, \ldots, r$ is then

$$\binom{n}{m} := \frac{n!}{m!} = \frac{n!}{m_1! \cdots m_r!}.$$

Proof The desired number is equal to the number of partitions $(A_1, \ldots, A_r)$ of A with (pairwise disjoint) subsets $A_i \subseteq A$, for which $|A_i| = m_i$ holds, $i = 1, \ldots, r$. According to the combinatorial principle formulated in Remark 1.6.5, this number is equal to

$$\binom{n}{m_1}\binom{n-m_1}{m_2}\cdots\binom{n-m_1-\cdots-m_{r-1}}{m_r}$$

$$= \frac{n!}{m_1!(n-m_1)!} \cdot \frac{(n-m_1)!}{m_2!(n-m_1-m_2)!} \cdots \frac{(n-m_1-\cdots-m_{r-1})!}{m_r!(n-m_1-\cdots-m_r)!}$$

$$= \frac{n!}{m_1! m_2! \cdots m_r!} = \binom{n}{m}. \qquad \square$$

For example, there are $\binom{32}{10,10,10,2} = 2.753.294.408.504.640$ possible card distributions in the game of Skat.—The numbers $\binom{n}{m}$ are called **polynomial coefficients** or also **multinomial coefficients**. For $r = 2$, the binomial coefficients result. Generally, for a number α and an r-tuple $m = (m_1, \ldots, m_r) \in \mathbb{N}^r$ with $|m| = m_1 + \cdots + m_r$

$$\binom{\alpha}{m} = \binom{\alpha}{m_1, \ldots, m_r} = \frac{[\alpha]_{|m|}}{m!} = \frac{\alpha(\alpha - 1) \cdots (\alpha - |m| + 1)}{m_1! \cdots m_r!}.$$

◊

An important application of the binomial coefficients is the binomial theorem. We formulate it here only for real or complex numbers. However, it holds (with the same proof) for *commutative* elements a, b of any ring, see Theorem 2.6.3.

Theorem 1.6.15 (Binomial Theorem) *For real (or complex) numbers a and b and any natural number n holds*

$$(a+b)^n = \sum_{m=0}^{n} \binom{n}{m} a^m b^{n-m} = \sum_{m=0}^{n} \binom{n}{m} a^{n-m} b^m$$

$$= a^n + na^{n-1}b + \binom{n}{2}a^{n-2}b^2 + \cdots + nab^{n-1} + b^n.$$

Proof (by induction on n)) For $n \leq 2$, the statement is well-known as the "first binomial formula" (and trivial). The transition from n to $n+1$ follows with Proposition

1.6.10 (4) (and the distributive law) as follows:

$$(a+b)^{n+1} = (a+b)^n(a+b) = \sum_{m=0}^{n} \binom{n}{m} a^m b^{n-m}(a+b)$$

$$= \sum_{m=0}^{n} \binom{n}{m} a^{m+1} b^{n-m} + \sum_{m=0}^{n} \binom{n}{m} a^m b^{n+1-m}$$

$$= \sum_{m=1}^{n+1} \binom{n}{m-1} a^m b^{n+1-m} + \sum_{m=0}^{n} \binom{n}{m} a^m b^{n+1-m}$$

$$= \sum_{m=0}^{n+1} \left(\binom{n}{m-1} + \binom{n}{m}\right) a^m b^{n+1-m} = \sum_{m=0}^{n+1} \binom{n+1}{m} a^m b^{n+1-m}. \quad \square$$

The binomial theorem can also be proven less formally as follows: For any real or complex numbers (or pairwise commutative ring elements) $a_1, \ldots, a_n, b_1, \ldots, b_n$ is evidently

$$\prod_{i=1}^{n} (a_i + b_i) = \sum_{H} a^H b^{H'},$$

where H runs through all subsets of $\{1, \ldots, n\}$ and H' is respectively the complement of H in $\{1, \ldots, n\}$. Furthermore, we have $a^H := \prod_{i \in H} a_i$ and analogously $b^{H'} := \prod_{i \in H'} b_i$ set. If now $a_1 = \cdots = a_n = a$ and $b_1 = \cdots = b_n = b$, then $a^H = a^{|H|}$ and $b^{H'} = b^{n-|H|}$. According to theorem 1.6.9, there are exactly $\binom{n}{m}$ subsets H of $\{1, \ldots, n\}$ with $|H| = m$. From this follows theorem 1.6.15.—With a completely analogous argument, using example 1.6.14, one obtains the following generalization of the binomial theorem.

Theorem 1.6.16 (Polynomial Theorem) *For real (or complex) numbers $a_1, \ldots, a_r$ and any natural number n holds*

$$(a_1 + \cdots + a_r)^n = \sum_{m \in \mathbb{N}^r, |m|=n} \binom{n}{m} a^m = \sum_{\substack{(m_1, \ldots, m_r) \in \mathbb{N}^r \\ m_1 + \cdots + m_r = n}} \frac{n!}{m_1! \cdots m_r!} a_1^{m_1} \cdots a_r^{m_r}.$$

More generally, for arbitrary real or complex numbers $a_{\rho i}$, $1 \leq \rho \leq r$, $1 \leq i \leq n$,

$$\prod_{i=1}^{n} (a_{1i} + \cdots + a_{ri}) = \sum_{(\rho_1, \ldots, \rho_n) \in \{1, \ldots, r\}^n} a_{\rho_1 1} \cdots a_{\rho_n n} = \sum_{(H_1, \ldots, H_r)} a_1^{H_1} \cdots a_r^{H_r},$$

where finally, the sum is taken over all decompositions $(H_1, \ldots, H_r)$ of $\{1, \ldots, n\}$ and $a_\rho^{H_\rho} := \prod_{i \in H_\rho} a_{\rho i}$ was set.

1.6 Finite Sets and Combinatorics

Exercises

Exercise 1.6.1

a) $2^n \leq n!$ for all $n \in \mathbb{N}$, $n \neq 1, 2, 3$.
b) $n^2 \leq 2^n$ for all $n \in \mathbb{N}$, $n \neq 3$.
c) $3^n \leq (n+1)!$ for all $n \in \mathbb{N}$, $n \neq 1, 2, 3$.
d) $(n+1)^2 \leq 3^n$ for all $n \in \mathbb{N}$, $n \neq 1$.
e) $n^3 \leq 2^n$ for all $n \in \mathbb{N}$, $n \geq 10$.
f) $n! \leq n^{n-1}$ for all $n \in \mathbb{N}^*$.

Exercise 1.6.2

a) $\binom{-1}{n} = (-1)^n$, $n \in \mathbb{N}$.
b) $\binom{-1/2}{n} = (-1)^n \frac{1 \cdot 3 \cdots (2n-1)}{2 \cdot 4 \cdots 2n} = \left(\frac{-1}{4}\right)^n \binom{2n}{n}$, $n \in \mathbb{N}$.
c) $\binom{1/2}{n} = \frac{1}{2n}\binom{-1/2}{n-1} = \frac{(-1)^{n-1}}{2n} \frac{1 \cdot 3 \cdots (2n-3)}{2 \cdot 4 \cdots (2n-2)} = \frac{-1}{2n-1}\left(\frac{-1}{4}\right)^n \binom{2n}{n}$, $n \in \mathbb{N}^*$.

Exercise 1.6.3 For all $\alpha \in \mathbb{R}$ (or $\mathbb{C}$) and $n \in \mathbb{N}$ the following holds:

a) $\alpha \binom{\alpha-1}{n} = (n+1)\binom{\alpha}{n+1}$.
b) $n\binom{\alpha}{n} + (n+1)\binom{\alpha}{n+1} = \alpha \binom{\alpha}{n}$.
c) $\frac{\alpha+1+n}{\alpha+1}\binom{\alpha+1}{n} = \frac{\alpha+n}{\alpha}\binom{\alpha}{n} + \frac{\alpha+n-1}{\alpha}\binom{\alpha}{n-1}$, $\alpha \neq 0, -1$.

Exercise 1.6.4 Prove by induction for all $n \in \mathbb{N}$:

a) $\sum_{k=0}^{n} k \cdot (k!) = (n+1)! - 1$.
b) $\sum_{k=0}^{n} 2^{n-k}\binom{n+k}{k} = 4^n$. (See also Ex. 1.6.21.)
c) $\sum_{k=m}^{n} \binom{k}{m} = \binom{n+1}{m+1}$, $m \in \mathbb{N}$, $m \leq n$.
d) $\sum_{k=0}^{n} \binom{\alpha+k}{k} = \binom{\alpha+n+1}{n}$, $\sum_{k=0}^{n}(-1)^k \binom{\alpha}{k} = n(-1)^n \binom{\alpha-1}{n}$, $\alpha \in \mathbb{R}$ (or $\mathbb{C}$).

Exercise 1.6.5

a) Let A be a *non-empty* finite set. The number of subsets of A with an even number of elements is equal to the number of subsets of A with an odd number of elements. (For finite $A \neq \emptyset$ it is therefore $\sum_{H \subseteq A}(-1)^{|H|} = 0$.—Let $a \in A$ be fixed. To each $B \subseteq A$ assign the subset $B \triangle \{a\}$.)
b) It is $\sum_{m=0}^{n}(-1)^m \binom{n}{m} = 0$ for $n \in \mathbb{N}^*$. (Use a) or $(1-1)^n = 0$ for $n \in \mathbb{N}^*$.)
c) It is $\sum_{k=0}^{n} \binom{2n+1}{2k} = 4^n = \sum_{k=0}^{n} \binom{2n+1}{2k+1}$ for $n \in \mathbb{N}$.
d) It is $\sum_{k=0}^{n} \binom{2n}{2k} = \frac{1}{2} 4^n = \sum_{k=0}^{n-1} \binom{2n}{2k+1}$ for $n \in \mathbb{N}^*$.

Exercise 1.6.6 Let $f: A \to B$ be a mapping of finite sets.

a) It is $|A| = \sum_{y \in B} |f^{-1}(y)|$. If specifically all fibers of f have the same number of elements m, then $|A| = m|B|$ holds (**Shepherd's Rule**).[20]
b) If f is surjective, then $|B| = \sum_{x \in A} |f^{-1}(f(x))|$.

Exercise 1.6.7 Let A be a set with n elements and B a subset of A with k elements. The number of m-element subsets of A that include B is $\binom{n-k}{m-k}$.

Exercise 1.6.8 For $m, n \in \mathbb{N}$ with $m \leq n$ show $\sum_{k=0}^{m} \binom{n}{k}\binom{n-k}{m-k} = 2^m \binom{n}{m}$. (Calculate the sum of the numbers in Ex. 1.6.7, where B runs through all k-element subsets of A, in two different ways, or use the formula $\binom{n}{k}\binom{n-k}{m-k} = \binom{n}{m}\binom{m}{k}$.)

Remark Let S, T be finite sets, $R \subseteq S \times T$ and $p: R \to S$ respectively $q: R \to T$ the natural projections. Then it holds $|R| = \sum_{s \in S} |p^{-1}(s)| = \sum_{t \in T} |q^{-1}(t)|$. This **principle of double counting** used in the first solution hint (and also before) provides many interesting counting formulas and serves as a model for related reasoning in other areas of mathematics.

Exercise 1.6.9

a) For $m, n, k \in \mathbb{N}$ prove

$$\sum_{j=0}^{k} \binom{m}{j}\binom{n}{k-j} = \binom{m+n}{k}.$$

(Count the k-element subsets of an $(m+n)$-element set $\{x_1, \ldots, x_m, y_1, \ldots, y_n\}$ in two different ways.—Incidentally, this so-called **Vandermonde's identity** also holds for arbitrary real or complex numbers m, n, which can be derived from

---

[20] n sheep together have $4n$ legs. It can occasionally be more convenient to count the legs first instead of the sheep, see for example the proof of Theorem 1.6.9.

1.6 Finite Sets and Combinatorics

the above special case using the identity theorem for polynomials or directly obtained using the binomial series, see Vol. 2. Often, the validity of (combinatorially proven) identities for binomial coefficients of natural numbers can be extended to arbitrary real or complex numbers using the identity theorem for polynomials 2.9.32.

b) For $n \in \mathbb{N}$ is $\sum_{j=0}^{n} \binom{n}{j}^2 = \binom{2n}{n}$.

Exercise 1.6.10 Let V be an association with n members.

a) The number of ways to elect a board from m association members and from them a 1st, 2nd, ..., k-th chairperson is $\binom{n}{m} \cdot [m]_k = n!/k!(n-m)!$.
b) The number of ways to elect a 1st, 2nd, ..., k-th chairperson and to complement the set of these chairpersons to a board with a number of members $\leq n$ is $[n]_k \cdot 2^{n-k}$.

Exercise 1.6.11 Using Ex. 1.6.10 show $\sum_{m=k}^{n} [m]_k \binom{n}{m} = 2^{n-k}[n]_k$ for $k, n \in \mathbb{N}$, $k \leq n$.

Exercise 1.6.12 Let A be a set. The n-tuples $(a_1, \ldots, a_n) \in W_n(A) := A^n$ of elements from A are also called **words** of length n over the **alphabet** A, $n \in \mathbb{N}$. The empty word is the only word of length 0. $W(A) := \biguplus_{n \in \mathbb{N}} W_n(A)$ is the set of all words over the alphabet A. Now let A be a finite alphabet with $m := |A| \geq 2$ letters and $n \in \mathbb{N}$. Then there are exactly m^n words of length n over A and $(m^{n+1} - m)/(m - 1)$ non-empty words with a length $\leq n$. Thus, there are $2(2^n - 1)$ non-empty words of length $\leq n$ over the two-element Morse alphabet $-$, $\cdot$.

Exercise 1.6.13 On a set with $n \in \mathbb{N}$ elements, there are exactly 2^{n^2} relations.

Exercise 1.6.14 Let A be a finite set with n elements. The number of equivalence relations on A is called the n-th **Bell number** β_n, $n \in \mathbb{N}$.

a) The numbers β_n satisfy the recursion scheme $\beta_0 = 1$, $\beta_{n+1} = \sum_{k=0}^{n} \binom{n}{k} \beta_k$.
b) Let $\beta_{m,n} := \sum_{i=0}^{m} \binom{m}{i} \beta_{n-i}$, $0 \leq m \leq n$. Then $\beta_{0,n} = \beta_n$, $n \in \mathbb{N}$, and the $\beta_{m,n}$ satisfy the recursion

$$\beta_{0,0} = 1, \quad \beta_{0,n+1} = \beta_{n,n}, \quad \beta_{m+1,n+1} = \beta_{m,n} + \beta_{m,n+1}, \quad m, n \in \mathbb{N}, \ m \leq n.$$

c) With b) confirm the table from Fig. 1.15.
   (**Note** The Bell number β_n can also be interpreted as the number of ways to distribute n objects into n boxes, if two such distributions are identified when they result from a permutation of the boxes. For example, there are only 2 ($=\beta_2$) and not 4 ($=2^2$) essentially different ways to assign the beds of a double room to a woman and a man. For n objects and m boxes, the corresponding number is $\beta(n, m) := \sum_{k=0}^{m} S(n, k)$. Here, $S(n, k)$ is the so-called **Stirling number of the second kind**. By definition, it is the number of equivalence relations with

Fig. 1.15 The first 11 Bell numbers

n	0	1	2	3	4	5	6	7	8	9	10
β_n	1	1	2	5	15	52	203	877	4140	21.147	115.975

exactly k equivalence classes on a set with n elements. Thus, it is $\beta_n = \beta(n,n)$. The $S(n,k)$ satisfy the initial conditions $S(n,0) = \delta_{n,0}$, $S(0,k) = \delta_{0,k}$, $n, k \in \mathbb{N}$, and the recursion equation

$$S(n+1,k) = kS(n,k) + S(n,k-1), \quad n \in \mathbb{N}, k \in \mathbb{N}^*.$$

The $\beta(n,m)$ satisfy the initial conditions $\beta(0,m) = 1$, $m \in \mathbb{N}$, $\beta(n,0) = 0$, $n \in \mathbb{N}^*$ as well as the recursion equation

$$\beta(n+1, m+1) = \sum_{k=0}^{n} \binom{n}{k} \beta(k,m), \quad n, m \in \mathbb{N}.$$

For the recursive calculation of the $\beta(n,m)$, use the numbers

$$\beta_k(n,m) := \sum_{i=0}^{k} \binom{k}{i} \beta(n-i, m),$$

for the $\beta_{k+1}(n+1, m) = \beta_k(n, m) + \beta_k(n+1, m)$, $\beta_0(n,m) = \beta(n,m)$, $\beta_n(n,m) = \beta(n+1, m+1)$ holds.)

Exercise 1.6.15 Let A be a set with n elements.

a) The number of pairs (B,C) of disjoint subsets B, C of A is 3^n.
b) The number of m-tuples of pairwise disjoint subsets of A is $(m+1)^n$. (If $A_1, \ldots, A_m \subseteq A$ are pairwise disjoint, then $(A_1, \ldots, A_m, A - \biguplus_{\mu=1}^{m} A_\mu)$ is a partition of A.—The case $m=1$ is Corollary 1.6.3.)

Exercise 1.6.16 Let $n, r \in \mathbb{N}$. Then it holds that $\sum_{m \in \mathbb{N}^r, |m|=n} \binom{n}{m} = r^n$. (Use $r^n = (1 + \cdots + 1)^n$ or Ex. 1.6.15b).—For $r=2$ see Example 1.6.12.)

Exercise 1.6.17 For real (or complex) numbers $\alpha \neq 0, -1, -2, \ldots$ and $m, n \in \mathbb{N}$ the following holds:

a) $\sum_{k=0}^{n} (-1)^k \frac{1}{\alpha+k} \binom{n}{k} = \frac{n!}{[\alpha+n]_{n+1}}$. In particular ($\alpha = m+1$)

$$\frac{(m+n+1)!}{m!n!} = (m+1)\binom{m+n+1}{n}$$

is a divisor of kgV$(m+1, \ldots, m+n+1)(=$ kgV$(1, \ldots, m+n+1)$, if $m \leq n+1)$, see Sect. 1.7.

1.6 Finite Sets and Combinatorics

b) $\sum_{k=0}^{n}(-1)^k \frac{[\beta+k-1]_k}{[\alpha+k-1]_k}\binom{n}{k} = \frac{[\alpha-\beta+n-1]_n}{[\alpha+n-1]_n}$, $\beta \in \mathbb{R}$ (or $\mathbb{C}$).

Exercise 1.6.18 The product of k consecutive integers is divisible by $k!$.

Exercise 1.6.19 Let $m \in \mathbb{N}^*$. The number of sequences $(a_1, \ldots, a_m)$ of length m of positive natural numbers a_i with sum $n \in \mathbb{N}^*$ is $\binom{n-1}{m-1}$. (See Example 1.6.13.)

Exercise 1.6.20 Let $n, k \in \mathbb{N}$ with $k \leq n$. Then the number of k-element subsets A of the n-element set $\{1, \ldots, n\}$, which for no $x \in A$ also contains the successor $x + 1$, is equal to $\binom{n-k+1}{k}$. (Use a similar trick as in Example 1.6.13.—The probability that in the lottery "6 out of 49" two adjacent numbers are drawn is thus $\left(\binom{49}{6} - \binom{44}{6}\right) / \binom{49}{6} \approx 0,4951 \ldots$) It holds $\sum_{k=0}^{n}\binom{n-k+1}{k} = F_{n+2}$, where F_k, $k \in \mathbb{N}$, are the Fibonacci numbers, see Example 1.5.6.

Remark This summation formula also results from the following **Zeckendorf's theorem**, which the reader can easily prove by induction: *Every natural number $m < F_{n+2}$, $n \in \mathbb{N}$, can be uniquely (apart from the order of the summands) written as a sum of pairwise distinct n Fibonacci numbers $F_2 = 1, F_3 = 2, \ldots, F_{n+1}$, where no two summands are adjacent Fibonacci numbers, and every such sum is $< F_{n+2}$.* One obtains this Zeckendorf representation of m through the following greedy algorithm: Determine for $m \geq 1$ the *largest* Fibonacci number $F_k \leq m, k \geq 2$, and then continue with $m - F_k$ instead of m continues in the same way. For example, $100 = 89 + 11 = 89 + 8 + 3 = F_{11} + F_6 + F_4$.—The Zeckendorf theorem is the basis of a strategy for the so-called **Fibonacci Nim game**. In this game, a position is given by a pair $(m, k) \in \mathbb{N} \times \mathbb{N}$ and m is realized by a pile of m game pieces. The player who is to move in this position removes ℓ game pieces with $1 \leq \ell \leq k$ (if this is possible, i.e., $m \geq \ell$ is). The opponent may then remove at most 2ℓ stones from the remaining game pieces on the next move. Thus, he finds the position $(m - \ell, 2\ell)$ before him. The one who can no longer move loses (which is exactly the case for the positions with $m = 0$ or with $k = 0$). *The winning positions are exactly the favorable positions*, where a position (m, k) is called "favorable" if $m \neq 0$ and $k \geq Z(m)$ for the smallest Fibonacci number $Z(m)$, which appears in the Zeckendorf representation of m. This follows directly from the following two statements, which the reader can easily confirm: (1) If (m, k) is favorable, then $(m - Z(m), 2Z(m))$ is not favorable. (2) If (m, k) is not favorable and $m \neq 0$, then every possible subsequent position is favorable.—A standard starting position $(m, m - 1), m \in \mathbb{N}^*$, is therefore only a losing position for the player who moves if m is a Fibonacci number $F_k, k \geq 2$.

Exercise 1.6.21 Let $n \in \mathbb{N}$. M denote the set of subsets with at least $(n+1)$ elements of a $(2n + 1)$-element set $A := \{x_1, \ldots, x_{2n+1}\}$, and for $k = 0, \ldots, n$ let $M_k \subseteq M$ be the set of those subsets of A that contain exactly n of the elements $x_1, \ldots, x_{n+k}$ as well as the element x_{n+k+1}. (Additionally, in the subsets of A, which are elements of M_k, there can also be some of the elements $x_{n+k+2}, \ldots, x_{2n+1}$.) Solve exercise 1.6.4b) again by counting $M = \biguplus_{k=0}^{n} M_k$ in two different ways.

Exercise 1.6.22 Let A, B be completely ordered finite sets with m and n elements, respectively.

a) The number of monotonically increasing mappings $A \to B$ is $\binom{m+n-1}{n-1}$.
b) The number of strictly monotonically increasing mappings $A \to B$ is $\binom{n}{m}$.
c) The number of surjective monotonically increasing mappings $A \to B$ is $\binom{m-1}{n-1}$.

(See example 1.6.13 and exercise 1.6.19.—Also in the present exercise, let $\binom{-1}{-1} := 1$.) □

Exercise 1.6.23 (Sieve Formula) Let $A_1, \ldots, A_n$ be subsets of the finite set A. For $I \subseteq \mathbb{N}^*_{\leq n} = \{1, \ldots, n\}$ let $A_I := \bigcap_{i \in I} A_i$ (with $A_\emptyset = A$) and $B := \bigcup_{i \in I} A_i$. One shows by induction over n:

$$\sum_{I \subseteq \mathbb{N}^*_{\leq n}} (-1)^{|I|} |A_I| = |A - B| \, (= |A| - |B|) \quad \text{bzw.} \quad |B| = \sum_{I \subseteq \mathbb{N}^*_{\leq n}, \, I \neq \emptyset} (-1)^{|I|-1} |A_I|.$$

(One can also conclude combinatorially: Is B_k for $k \in \mathbb{N}$ the set of $x \in A$, which lie in exactly k of the sets $A_1, \ldots, A_n$, it holds that $A = \biguplus_{k=0}^{n} B_k$ and $x \in B_k$ belongs exactly to the 2^k of the sets A_I, for which $I \subseteq \{i \in \mathbb{N}^*_{\leq n} \mid x \in A_i\}$ is. Due to $\sum_{\ell=0}^{k}(-1)^\ell \binom{k}{\ell} = \delta_{0,k}$ for $k \in \mathbb{N}$, see Ex. 1.6.5b), it follows that $x \in A$ is indeed counted on the left side of the first formula exactly once, if x is not in B, i.e., if $x \in B_0$ is.—Often, even for $n = 2$, the formulas $|A| \geq |A_1 \cup A_2| = |A_1| + |A_2| - |A_1 \cap A_2|$ and in particular $|A_1 \cap A_2| \geq |A_1| + |A_2| - |A|$ are of great use.)

Exercise 1.6.24 Let X be a set and $\mathfrak{P}(X)$ the power set of X with inclusion as order.

a) The set $\mathfrak{E}(X) \subseteq \mathfrak{P}(X)$ of the finite subsets of X is artinianly ordered.
b) The following conditions are equivalent: (i) X is finite. (ii) $\mathfrak{P}(X)$ is finite. (iii) $\mathfrak{P}(X)$ is artinianly ordered. (iv) $\mathfrak{P}(X)$ is noetherianly ordered. (v) $\mathfrak{E}(X)$ is noetherianly ordered.

1.7 Prime Factorization of Natural Numbers

In this section, we address the multiplicative structure of the set $\mathbb{N}^*$ of positive natural numbers. Let $a, b \in \mathbb{N}^*$. a is called a **divisor** of b or b a **multiple** of a, if there is a $c \in \mathbb{N}^*$ with $b = ac$. We then say b is **divisible** by a or a **divides** b, and we write $a \mid b$. The set of divisors of a in $\mathbb{N}^*$ is denoted by $T(a) = T_{\mathbb{N}^*}(a)$ and their (finite) number by $\tau(a) = \tau_{\mathbb{N}^*}(a)$. The multiples of a in $\mathbb{N}^*$ form the set $\mathbb{N}^* a = \{na \mid n \in \mathbb{N}^*\}$. In an analogous manner, the divisibility concepts in $\mathbb{Z}$ are defined. It is $T_\mathbb{Z}(0) = \mathbb{Z}$.

1.7 Prime Factorization of Natural Numbers

Definition 1.7.1 A natural number p is called a **prime number** or **prime**, if $p \geq 2$ and 1 and p are the only divisors of p in $\mathbb{N}^*$, that is, if $\tau_{\mathbb{N}^*}(p) = 2$ is true. The set of prime numbers in $\mathbb{N}^*$ we denote by

$$\mathbb{P}.$$

If n is a natural number ≥ 2, then the smallest natural number $p \geq 2$, which divides n is evidently a prime number; because every divisor of p also divides n. From this immediately follows the following statement, which was already proven by Euclid (ca. 325–265 BC) in this way:

Theorem 1.7.2 *There are infinitely many prime numbers.*

Proof Assume there are only finitely many prime numbers, say $p_1, \ldots, p_r$. Then $p_1 \cdots p_r + 1$ is in $\mathbb{N}^*$ and is ≥ 2, therefore, as we have just noted, it has a prime divisor p. This prime number p is different from $p_1, \ldots, p_r$, since p would otherwise also divide 1. □

Example 1.7.3 The sequence of prime numbers begins with

$$2, 3, 5, 7, 11, 13, 17, 19, 23, 29, \ldots$$

The largest prime number explicitly known at the time (2017) is the number $2^{74.207.281} - 1$. It has $[\log_{10} 2^{74.207.281}] + 1 = [74.207.281 \cdot \log_{10} 2] + 1 = 22.338.618$ digits in the decimal system and is of the form $M(p) = 2^p - 1$, where p itself is prime. Such numbers $M(p)$ are called **Mersenne numbers** (after M. Mersenne (1588–1648)), see also Ex. 1.7.9. All very large known prime numbers are Mersenne or related numbers. This is because there are relatively convenient primality tests for them. For a Mersenne number $M(p)$, for example, the so-called **Lucas test** (after É. Lucas (1842–1891)) is used: One recursively forms the sequence r_n, $n \geq 1$, where $r_1 = -4$ and r_{n+1} for $n \in \mathbb{N}^*$ is the smallest non-negative remainder when dividing $r_n^2 - 2$ by $M(p)$, see 1.7.5. *Exactly then is the Mersenne number $M(p)$, $p \geq 3$ prime, a prime number, if $r_{p-1} = 0$ is*, see example 2.10.35. The smallest Mersenne number that is not a prime number is $M(11) = 2047 = 23 \cdot 89$. This is no coincidence: If p is a prime number $\equiv 3 \mod 4$ and $2p + 1$ is also prime, then $2p + 1$ is a divisor of $M(p)$, see Ex. 2.7.10.

The **prime number function** $\pi : \mathbb{R}_+ \to \mathbb{N}$ gives for $x \in \mathbb{R}_+$ the number $\pi(x)$ of prime numbers $p \in \mathbb{P}$ with $p \leq x$. According to Theorem 1.7.2, π is unbounded. If p_n is the n-th prime number in the above list, $n \in \mathbb{N}^*$, then $\pi(p_n) = n$. ◊

Proposition 1.7.4 *Every positive natural number is a product of prime numbers.*

Proof The statement holds for the number 1, which is represented as an empty product of prime numbers. Now let n be a natural number ≥ 2, and assume the claim holds for all smaller natural numbers. As already noted, n then has a prime divisor $p \in \mathbb{P}$, and it is $n = pm$ with a $m \in \mathbb{N}^*$, $m < n$, which according to the induction hypothesis can be written in the form $m = p_1 \cdots p_r$ with prime numbers p_i, $i = 1, \ldots, r$, can be written. Thus, $n = pp_1 \cdots p_r$ is also a product of prime numbers. □

We want to show that the prime factorization of a natural number according to Proposition 1.7.4 is unique up to the order of the factors. For this, we recall the division with remainder and the derived **Euclidean (division) algorithm**.

Theorem 1.7.5 (Division with Remainder) *Let a and b be integers with $b > 0$. Then there exist uniquely determined integers q and r such that*

$$a = qb + r \quad \text{and} \quad 0 \leq r < b.$$

Proof The existence of q and r is proven for $a \geq 0$ by induction over a, by applying the induction hypothesis to $a - b$ when $a \geq b$: From $a - b = \tilde{q}b + r$, $\tilde{q} \in \mathbb{Z}$, $0 \leq r < b$, one obtains $a = qb + r$ with $q := \tilde{q} + 1$. If $a < 0$, then one has $-a = q'b + r'$ with $0 \leq r' < b$ and

$$a = \begin{cases} (-q' - 1)b + (b - r'), & \text{if } r' \neq 0, \\ (-q')b, & \text{if } r' = 0. \end{cases}$$

To prove uniqueness, let also $a = q_1 b + r_1$ with $0 \leq r_1 < b$. Then $0 = a - a = (q - q_1)b + (r - r_1)$, and b divides $|r_1 - r| < b$. It follows $|r_1 - r| = 0$, thus $r = r_1$ and then also $q = q_1$. □

The number r in Proposition 1.7.5 is the (smallest non-negative) **remainder** of a when divided by b. Due to $a/b = q + (r/b)$, q is the **integer part** of a/b, i.e., $q = [a/b]$, where $[-]$ is the Gauss bracket. With the notations from Example 1.3.6, $q = a$ DIV b and $r = a$ MOD b.

Example 1.7.6 (*g*-al development) Let g be a natural number ≥ 2. For every natural number $n \geq 1$, there are then uniquely determined natural numbers r and $a_0, \ldots, a_r$ with $a_r \neq 0$ and $0 \leq a_i < g$ as well as

$$n = a_0 + a_1 g + \cdots + a_r g^r = \sum_{i=0}^{r} a_i g^i.$$

One obtains the **digits** a_i of this so-called *g*-**adic expansion** of n recursively through continuous division with remainder according to the following scheme:

$$n = q_0 = q_1 g + a_0, \quad 0 \leq a_0 < g,$$
$$q_1 = q_2 g + a_1, \quad 0 \leq a_1 < g,$$

1.7 Prime Factorization of Natural Numbers

$$\vdots$$
$$q_{r-1} = q_r g + a_{r-1}, \quad 0 \leq a_{r-1} < g,$$
$$q_r = a_r, \quad 0 < a_r < g.$$

The uniqueness of the division with remainder provides the uniqueness of these digits. One also writes briefly $n = (a_r \ldots a_0)_g$. For $g = 2$, $g = 3$, $g = 10$, or $g = 16$, one speaks of the **binary-** or **ternary-** or **decimal-** or **hexadecimal-** or **sedecimal expansion** of n. In the last system, the digits $10, \ldots, 15$ are usually denoted by the letters A, ..., F. From the g-adic expansion $n = a_0 + a_1 g + \cdots + a_r g^r$, one calculates the number n in reverse most quickly recursively through

$$n_0 = a_r,$$
$$n_1 = n_0 g + a_{r-1} \ (= a_r g + a_{r-1}),$$
$$\vdots$$
$$n_{r-1} = n_{r-2} g + a_1 \ (= a_r g^{r-1} + a_{r-1} g^{r-2} + \cdots + a_2 g + a_1),$$
$$n_r = n_{r-1} g + a_0 = n.$$

This is a special case of the so-called **Horner's scheme**, see example 2.9.19.

Occasionally, the following generalization of the g-adic expansion is used: Let h_j, $j \in \mathbb{N}^*$, be a sequence of natural numbers >1 and let $g_i := h_1 \cdots h_i, i \in \mathbb{N}$, (also $g_0 = 1$). Then each $n \in \mathbb{N}$ has a unique representation $n = \sum_{i \in \mathbb{N}} a_i g_i$ with $0 \leq a_i < h_{i+1}$ for all i and $a_i = 0$ for almost all i.[21] The a_i are recursively determined by $q_0 = n$, $q_i = q_{i+1} h_{i+1} + a_i$ for $0 \leq a_i < h_{i+1}$, and conversely, $n = n_r$ can be expressed for $a_i = 0$ for $i > r$ through the recursion $n_0 = a_r, n_{i+1} = n_i h_{r-i} + a_{r-i-1}$ gain. In the above g-al development, $h_j = g$ is for all $j \in \mathbb{N}^*$. The Old Babylonian **sexagesimal system** is actually the mixed system with $h_1 = 10, h_2 = 6; h_3 = 10, h_4 = 6; \ldots$, from which one obtains the pure sexagesimal system with $g = h_{2k-1} h_{2k} = 60$, $k \in \mathbb{N}^*$, by combining every two digits. Our decimal system is also occasionally treated as a mixed system (e.g., in tally lists) $(5, 2; 5, 2; \ldots)$. The Mayans had a pure base-20 system. ◊

For $a, b \in \mathbb{N}^* d \in \mathbb{N}^*$ is called the **greatest common divisor (gcd)** of a and b, if d is a common divisor of a and b and if every other common divisor of a and b is a divisor of d. It is uniquely determined by a and b and is denoted by

$$\gcd(a, b)$$

---

[21] A property holds for **almost all** members of a family $(x_i)_{i \in I}$, if it holds for all x_i with at most finitely many exceptions, i.e., if there is a *finite* subset $J \subseteq I$ such that all $x_i, i \in I - J$, possess the property in question. This extremely useful and suggestive manner of speaking was probably first used in the textbook "Grundzüge der Differential- und Integralrechnung," p. 13, by G. Kowalewski.

Thus, it is $T(a) \cap T(b) = T(\gcd(a, b))$.[22] If $\gcd(a, b) = 1$, then 1 is the only common divisor of a and b in $\mathbb{N}^*$, then a and b are called **coprime**. a and b are coprime if and only if they have no common prime divisor. If a, b have the greatest common divisor d, then a / d and b / d are evidently coprime. Furthermore, if $p \in \mathbb{P}$ is a prime number and $b \in \mathbb{N}^*$ is a number not divisible by p, then p and b are coprime.

The existence and a fast calculation method of $\gcd(a, b)$ is provided by the **Euclidean Algorithm**. One sets $r_0 := a$ and $r_1 := b$ and successively performs the following divisions with remainder:

$$r_0 = q_1 r_1 + r_2, \qquad 0 < r_2 < r_1,$$
$$r_1 = q_2 r_2 + r_3, \qquad 0 < r_3 < r_2,$$
$$\vdots$$
$$r_{k-1} = q_k r_k + r_{k+1}, \qquad 0 < r_{k+1} < r_k,$$
$$r_k = q_{k+1} r_{k+1}.$$

The procedure ends if r_{k+1} is a divisor of r_k. Since the remainders become successively smaller, this case will certainly occur after a finite number of steps.

Theorem 1.7.7 *It is* $\gcd(a, b) = r_{k+1}$. *In particular, two positive natural numbers always have a greatest common divisor.*

Proof First, $r_{k+1} = \gcd(r_k, r_{k+1})$, since r_{k+1} is a divisor of r_k. It is therefore sufficient to show that for every i with $1 \le i \le k$ it holds: If $\gcd(r_i, r_{i+1})$, exists, then so does $\gcd(r_{i-1}, r_i)$ and both are equal. However, from the equation $r_{i-1} = q_i r_i + r_{i+1}$ it follows that the numbers r_i and r_{i+1} have the same common divisors as the numbers r_{i-1} and r_i. $\square$

Parallel to the Euclidean algorithm, the remainders r_i can be represented in the form

$$r_i = s_i a + t_i b$$

with $s_i, t_i \in \mathbb{Z}$, $i = 0, \ldots, k + 1$. In particular,

$$r_{k+1} = \gcd(a, b) = s_{k+1} a + t_{k+1} b$$

with the integers s_{k+1} and t_{k+1}.. For this, one sets recursively

$$s_0 = 1, \quad t_0 = 0; \quad s_1 = 0, \quad t_1 = 1;$$

---

[22] Note that the gcd of a and b is *not* defined by the natural order of $\mathbb{N}^*$, but by divisibility $|$ (which also defines an order on $\mathbb{N}^*$). If $\gcd(a, b)$ exists, then it is naturally also the greatest common divisor of a and b with respect to the natural order of $\mathbb{N}^*$.

1.7 Prime Factorization of Natural Numbers

Fig. 1.16 Example of the Euclidean Algorithm

i	0	1	2	3	4
q_i		2	1	164	
s_i	1	0	1	-1	165
t_i	0	1	-2	3	-494

$$s_{i+1} = s_{i-1} - q_i s_i, \quad t_{i+1} = t_{i-1} - q_i t_i, \quad i = 1, \ldots, k.$$

Then indeed $r_0 = s_0 a + t_0 b$, $r_1 = s_1 a + t_1 b$ and

$$r_{i+1} = r_{i-1} - q_i r_i = s_{i-1} a + t_{i-1} b - q_i s_i a - q_i t_i b = s_{i+1} a + t_{i+1} b, \quad i = 1, \ldots, k.$$

We summarize once again:

Theorem 1.7.8 (Bezout's Lemma) *If a and b are positive natural numbers, then there exist integers s and t such that $\gcd(a, b) = sa + tb$. Specifically: If a and b are coprime positive natural numbers, then there exist integers s and t such that $1 = sa + tb$.*

Example 1.7.9 Let $a := 36.667$ and $b := 12.247$. Then one obtains

$$36.667 = 2 \cdot 12.247 + 12.173, \quad 12.247 = 1 \cdot 12.173 + 74,$$

$$12.173 = 164 \cdot 74 + 37, \quad 74 = 2 \cdot 37.$$

Thus, it is $37 = \gcd(36.667, 12.247) = 165 \cdot 36.667 - 494 \cdot 12.247$. The s_i and t_i, $i = 0, \ldots, 4$, are derived from the table in Fig. 1.16. ◊

A direct consequence of Bezout's lemma is:

Theorem 1.7.10 (Euclid's Lemma) *If a prime number $p \in \mathbb{P}$ divides a product $b_1 \cdots b_r$ of positive natural numbers, then p divides at least one of the factors $b_1, \ldots, b_r$.*

Proof Without loss of generality, let $r = 2$ (induction over r). By assumption, $b_1 b_2 = pc$ with a $c \in \mathbb{N}^*$. Assume p does not divide b_1. Then p and b_1 are coprime, and according to Theorem 1.7.8, there exist numbers $s, t \in \mathbb{Z}$ with $1 = sp + tb_1$. It follows $b_2 = spb_2 + tb_1 b_2 = p(sb_2 + tc)$, i.e., p divides b_2. □

Trivially, among the natural numbers ≥ 2, only prime numbers can possess the divisibility property specified in Euclid's Lemma 1.7.10. As the proof of the following uniqueness statement shows, it is the essential property of prime numbers. With regard to the defining property in Definition 1.7.1, one should initially speak of **indivisible** numbers and only after the proof of Euclid's Lemma 1.7.10 of prime numbers, see Ex. 1.7.36.

Theorem 1.7.11 (Fundamental Theorem of Elementary Number Theory) *Every positive natural number can be uniquely represented as a product of prime numbers, except for the order of the factors.*

Proof Due to Proposition 1.7.4, we only need to show uniqueness. Let

$$n = p_1 \cdots p_r = q_1 \cdots q_s$$

be representations of $n \in \mathbb{N}^*$ as a product of prime numbers $p_1, \ldots, p_r$ and $q_1, \ldots, q_s$, respectively. By induction over n, we show that $r = s$ and after renumbering $p_i = q_i$, $i = 1, \ldots, r$, holds. Let $r \geq 1$ and thus also $s \geq 1$. Then p_1 divides the product $q_1 \cdots q_s$ and thus, according to Euclid's Lemma 1.7.10, one of the factors. We can assume after renumbering that p_1 is a divisor of q_1. Since q_1 is prime, it is necessarily $p_1 = q_1$, and after canceling $p_1 = q_1$ follows $p_2 \cdots p_r = q_2 \cdots q_s =: m$. The induction hypothesis, applied to $m < n$, then provides the assertion of the theorem. □

Let $n \in \mathbb{N}^*$. In the prime factorization of n, identical prime factors are combined into powers. This yields the **canonical prime factorization**

$$n = \prod_{p \in \mathbb{P}} p^{v_p(n)}.$$

In this product, $\mathbb{P} = \{2, 3, 5, \ldots\} \subseteq \mathbb{N}^*$ is the set of all prime numbers, and the so-called p-**exponents**

$$v_p(n) \in \mathbb{N}$$

are only different from 0 for finitely many $p \in \mathbb{P}$, so that in the above product only finitely many factors $\neq 1$ need to be considered. In the specific case, only these essential factors are noted, for example, $1001 = 7 \cdot 11 \cdot 13$ and $10.200 = 2^3 \cdot 3 \cdot 5^2 \cdot 17$. One also sets $v_p(0) = \infty$ for all $p \in \mathbb{P}$. Apparently, the p-exponent v_p is **additive**, i.e., it holds

$$v_p(ab) = v_p(a) + v_p(b) \quad \text{for all } a, b \in \mathbb{N}$$

(where we set $\alpha + \infty := \infty + \alpha := \infty$ for all $\alpha \in \overline{\mathbb{N}} = \mathbb{N} \cup \{\infty\}$). If

$$a = \prod_{p \in \mathbb{P}} p^{v_p(a)} \quad \text{und} \quad b = \prod_{p \in \mathbb{P}} p^{v_p(b)}$$

are the canonical prime factorizations of two numbers $a, b \in \mathbb{N}^*$, then a is a divisor of b if and only if $v_p(a) \leq v_p(b)$ for all $p \in \mathbb{P}$. From this, the representation for the greatest common divisor results

$$\gcd(a, b) = \prod_{p \in \mathbb{P}} p^{\operatorname{Min}(v_p(a), v_p(b))}.$$

1.7 Prime Factorization of Natural Numbers

Analogously,

$$\mathrm{lcm}(a, b) = \prod_{p \in \mathbb{P}} p^{\mathrm{Max}\,(v_p(a), v_p(b))}$$

is the **least common multiple (lcm)** of a and b. By definition, this is the common multiple of a and b that divides every other common multiple of a and b, i.e., it is $\mathbb{N}^* a \cap \mathbb{N}^* b = \mathbb{N}^* \mathrm{lcm}(a, b)$.[23]

For an *integer* $a \in \mathbb{Z}^* := \mathbb{Z} - \{0\}$, one obtains a canonical prime factorization by providing the prime factorization of $|a| = -a$ with the sign $\varepsilon(a) = \mathrm{Sign}\,a = -1$ when $a < 0$. We set $\gcd(a, b) = \gcd\left(|a|, |b|\right)$ as well as $\mathrm{lcm}(a, b) = \mathrm{lcm}\left(|a|, |b|\right)$ for $a, b \in \mathbb{Z}^*$. $\gcd(a, b)$ is again the integer uniquely determined up to the sign, which divides a and b and is divided by every common divisor of a and b. Accordingly, $\mathrm{lcm}(a, b)$ is the integer uniquely determined up to the sign, which is a common multiple of a and b and divides every common multiple of a and b.

For a non-zero *rational* number $x = a / b$, $a, b \in \mathbb{Z}^*$, one obtains the **canonical representation**

$$x = \varepsilon \prod_{p \in \mathbb{P}} p^{v_p(x)}$$

by combining the prime factorizations of a and b with p-**exponents** $v_p(x) = v_p(a) - v_p(b) \in \mathbb{Z}$, of which only finitely many are $\neq 0$, and the sign $\varepsilon = \varepsilon(x) = \mathrm{Sign}\,x \in \{1, -1\}$ of x. x is an integer if and only if all multiplicities $v_p(x)$ are non-negative. If a and b are *coprime* integers, i.e., $\gcd(a, b) = 1$, then a / b is called a **reduced representation** of x. From any representation $x = a'/b'$, $a', b' \in \mathbb{Z}$, $b' \neq 0$, such a representation $x = a / b$, $a := a'/\gcd(a', b')$, $b := b'/\gcd(a', b')$, can always be obtained by reducing. *It is uniquely determined by x except for the signs of a and b*. For normalization, one usually chooses $b > 0$. This makes the representation unique. The reduced representation of 0 is then $0 / 1$.

Example 1.7.12 (Enumeration of $\mathbb{Q}_+^\times$—Calkin-Wilf Tree) According to the last remark, is $(a, b) \mapsto a/b$ a bijective mapping of the set T of coprime pairs (a, b) of positive natural numbers onto the set $\mathbb{Q}_{>0} = \mathbb{Q}_+^\times$ of positive rational numbers. With the help of the Euclidean algorithm, more precisely, the classical Euclidean algorithm by **successive subtraction** (= antepheiresis[24]), the set T can be naturally endowed with the structure of a (directed) graph (cf. Remark 1.3.2).

The **classical Euclidean algorithm**, which is described in Book VII, § 2 of Euclid's Elements, proceeds from a pair $(a, b) \in (\mathbb{N}^*)^2$ to the pair $(a, b - a)$ if $a < b$, and to $(a - b, b)$ if $a > b$, and finally ends at $(\gcd(a, b), \gcd(a, b))$, that is, at $(1, 1)$,

---

[23] For the lcm of positive natural numbers, a similar remark applies as in the previous footnote for the gcd.
[24] This term already dates back to Aristotle (384–322 BC). Successive subtraction was also used by the Greeks for comparing line segments, see Example 3.3.11 on continued fractions.

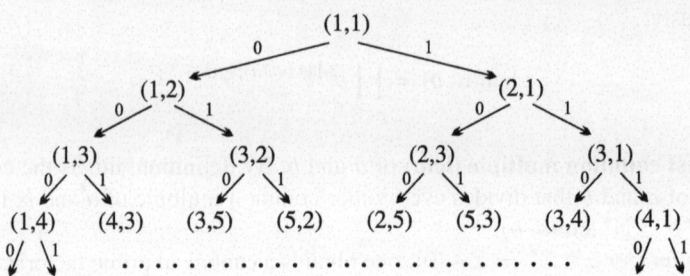

Fig. 1.17 Calkin-Wilf tree

when the initial pair (a, b) was coprime.[25] In the reverse direction, one thus obtains the pair $(a, b) \in \mathbf{T}$, starting from $(1, 1)$, through a *unique* sequence

$$(1, 1) = (a_0, b_0) \to (a_1, b_1) \to \cdots \to (a_i, b_i) \to$$
$$(a_{i+1}, b_{i+1}) \to \cdots \to (a_r, b_r) = (a, b),$$

where (a_{i+1}, b_{i+1}) is one of the pairs $(a_i, a_i + b_i)$ or $(a_i + b_i, b_i), 0 \le i < r$. We therefore connect two pairs $(c, d), (e, f) \in \mathbf{T}$ with an arrow $\xrightarrow{0}$ of type 0, if $(e, f) = (c, c+d)$ is, and with an arrow $\xrightarrow{1}$ of type 1, if $(e, f) = (c+d, d)$ is. We denote these arrows with $0_{(c,d)}$ or $1_{(c,d)}$ or also briefly with 0 or 1, if the origin of the arrow is clear from the context. Then each pair $(a, b) \in \mathbf{T}$ is reachable by a uniquely determined finite path $(\varepsilon_1, \ldots, \varepsilon_r) \in \{0, 1\}^r, r \in \mathbb{N}$, from $(1, 1)$. If we arrange the 0–1 sequences of a fixed length $r \in \mathbb{N}$ lexicographically and furthermore set $(\varepsilon_1, \ldots, \varepsilon_r) < (\eta_1, \ldots, \eta_s)$, if $r < s$, we obtain a canonical bijective enumeration. $(c_1, d_1), (c_2, d_2), (c_3, d_3), \ldots$ of the elements of T and thus a bijective enumeration $c_1/d_1, c_2/d_2, c_3/d_3, \ldots$ of $\mathbb{Q}_+^\times$ (which, however, has little to do with the natural order of $\mathbb{Q}_+^\times$). Overall, we obtain the so-called **Calkin-Wilf tree**. It begins as indicated in Fig. 1.17.

The enumeration of $\mathbb{Q}_+^\times$ thus starts with

1/1, 1/2, 2/1, 1/3, 3/2, 2/3, 3/1, 1/4, 4/3, 3/5, 5/2, 2/5, 5/3, 3/4, 4/1, ...

The pair (c_n, d_n), which is reached from $(1, 1)$ on the path $(\varepsilon_1, \ldots, \varepsilon_r) \in \{0, 1\}^r$ of length $r \in \mathbb{N}$ is at the position $n = 2^r + \sum_{i=1}^{r} \varepsilon_i 2^{r-i} = (1\varepsilon_1 \ldots \varepsilon_r)_2 \in \mathbb{N}^*$. The binary expansion of $n \in \mathbb{N}^*$ thus determines the path to (c_n, d_n). For $n = (90.317)_{10} = (10110000011001101)_2$, for example, it results in (c_n, d_n) from

$(1, 1) \xrightarrow{0} (1, 2) \xrightarrow{1} (3, 2) \xrightarrow{1} (5, 2) \xrightarrow{0} \cdots \xrightarrow{0} (5, 27) \xrightarrow{1} (32, 27) \xrightarrow{1} (59, 27)$
$\xrightarrow{0} (59, 86) \xrightarrow{0} (59, 145) \xrightarrow{1} (204, 145) \xrightarrow{1} (349, 145) \xrightarrow{0} (349, 494) \xrightarrow{1} (843, 494)$

---

[25] The ordinary Euclidean algorithm combines several similar steps of successive subtraction, which are performed consecutively, into one step.

1.7 Prime Factorization of Natural Numbers

to $(c_{90.317}, d_{90.317}) = (843, 494)$ or the fraction $c_{90.317}/d_{90.317}$ to 843 / 494 (>1, since n is odd). Which pair is reached with an alternating path $(0, 1, 0, 1, \ldots)$ of length $r \in \mathbb{N}$, and at which position is it? Which number is the largest in the pairs of the r-th row, $r \in \mathbb{N}$? The Calkin-Wilf tree is a so-called (infinite) **binary decision tree with root** $(1, 1)$. The sequence $(c_n, d_n), n \in \mathbb{N}^*$, can be easily determined recursively. It is $(c_1, d_1) = (1, 1)$ and evidently $(c_n, d_n) = (c_m, c_m + d_m)$, if $n = 2m$ is even, or $(c_n, d_n) = (c_m + d_m, d_m)$, if $n = 2m + 1 > 1$ is odd.

A sequence $(\varepsilon_1, \ldots, \varepsilon_r) \in \{0, 1\}^r$ can also be represented by a tuple $[k_1, \ldots, k_s; \varepsilon]$ *of positive* integers $k_1, \ldots, k_s$, provided with a marking $\varepsilon \in \{0, 1\}$ for $s > 0$, characterize, where $k_1, \ldots, k_s \in \mathbb{N}^*$ indicate the numbers of zeros and ones that alternate in $(\varepsilon_1, \ldots, \varepsilon_r)$ (thus it is $r = k_1 + \cdots + k_s$) and ε is the last component, see the remark on Proposition 1.6.2. $[k_1, \ldots, k_s; 1-\varepsilon]$ is then the sequence in which the zeros and ones are swapped. For example, the sequence $(0, 1, 1, 0, 0, 0, 0, 0, 1, 1, 0, 0, 1, 1, 0, 1)$ from above is noted in this way with $[1, 2, 5, 2, 2, 2, 1, 1; 1]$. If $(a, b) \in \mathbf{T}$ with $a > b$ and $q_1, \ldots, q_{k+1}$ are the quotients in the usual Euclidean algorithm for the division of a by b with remainder, then $q_{k+1} > 1$ and $[q_{k+1} - 1, q_k, \ldots, q_1; 1]$ is the path (of length $q_1 + \cdots + q_k + q_{k+1} - 1$) from $(1, 1)$ to (a, b) and $[q_{k+1} - 1, q_k, \ldots, q_1; 0]$ the path from $(1, 1)$ to (b, a). For the numbers $a := 36.667/37 = 991$ and $b := 12.247/37 = 331$ from example 1.7.9, for instance, $k = 3$ and $q_1 = 2$, $q_2 = 1$, $q_3 = 164$, $q_4 = 2$, and one obtains the path $[1, 164, 1, 2; 1] = (0, 1, \ldots, 1, 0, 1, 1)$ of length 168 from $(1, 1)$ to $(991, 331)$. This pair or the fraction 991 / 331 is thus at position $2^{168} + 2^{166} + \cdots + 2^3 + 2^1 + 2^0 = (2^{169} - 1) - 2^{167} - 2^2 = 3 \cdot 2^{167} - 5$ in the enumeration of T or an enumeration of $\mathbb{Q}_+^\times$. At which position is the pair $(331, 991)$?

In the sequence $(c_1, d_1), (c_2, d_2), (c_3, d_3), \ldots$ of the elements of T, it is apparent that $d_i = c_{i+1}$ holds for all $i \in \mathbb{N}^*$. It is thus sufficient to specify the first components of the pairs. The sequence thus obtained

$$(c_n)_{n \in \mathbb{N}^*} = (1, 1, 2, 1, 3, 2, 3, 1, 4, 3, 5, 2, 5, 3, 4, \ldots)$$

is called the **Stern-Brocot sequence**. The pair $(c_n, d_n) \in \mathbf{T}$ is then $(c_n, c_{n+1}), n \in \mathbb{N}^*$, and $c_1/c_2, c_2/c_3, c_3/c_4, \ldots$ is the corresponding bijective enumeration of the positive rational numbers. The recursion from above for the pairs (c_n, d_n) yields for the c_n, $n \in \mathbb{N}^*$, the following recursion: *It is $c_1 = 1$ and $c_n = c_m$, if $n = 2m$ is even, respectively $c_n = c_m + c_{m+1}$, if $n = 2m + 1 > 1$ is odd.* From $n = 2$, (c_n) is a zigzag sequence with $1 = c_2 <; c_3 > c_4 < c_5 > \cdots$. The sequence elements c_n can also be interpreted combinatorially. Let b_n for $n \in \mathbb{N}$ the number of representations of n as a sum of powers of 2, where each power of 2 may be used at most twice and the order of the summands is disregarded, thus $b_0 = 1$ (empty sum), $b_1 = 1$, $b_2 = 2$ (because $2 = 2 = 1 + 1$), $b_3 = 1$ (because $3 = 2 + 1$), $b_4 = 3$ (because $4 = 4 = 2 + 2 = 2 + 1 + 1$), …Then it is $b_n = c_{n+1}$, $n \in \mathbb{N}$, because the sequence $c'_n := b_{n-1}, n \in \mathbb{N}^*$, obviously satisfies the same recursion as the sequence $c_n, n \in \mathbb{N}^*$. (One distinguishes how often the $1 = 2^0$ is used as a summand.) Furthermore, the term following the term (c_n, d_n) can be easily specified directly. It is

$$(c_{n+1}, d_{n+1}) = (d_n, (1 + 2[c_n/d_n])d_n - c_n), \ n \in \mathbb{N}^*,$$

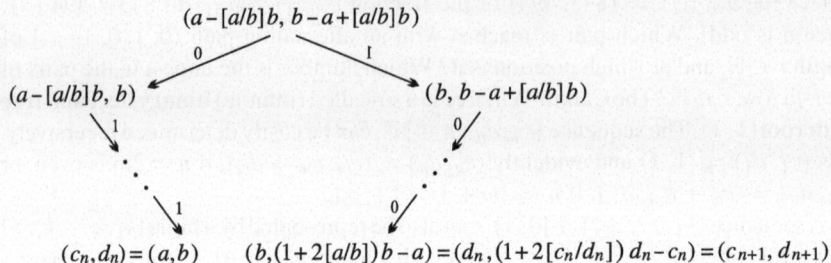

Fig. 1.18 Section from the Calkin-Wilf tree

as with $n \notin \{2^{r+1} - 1 \mid r \in \mathbb{N}\}$ the section of the decision tree shown in Fig. 1.18 shows, but also with $n = 2^{r+1} - 1$, i.e., with $(c_n, d_n) = (r + 1, 1)$, trivially applies.

We leave it to the reader as a combinatorial exercise to discover further details in the Calkin-Wilf tree. Furthermore, we refer to the works of N. Calkin, H. S. Wilf, Recounting the Rationals, Am. Math. Monthly **167**, 360–363 (2000) and J.-P. Delahaye, The Misunderstood Sister of the Fibonacci Sequence, Spektrum der Wiss., 64–69 (May 2015) as well as the literature cited therein. ◇

Example 1.7.13 (Irrationality of Roots) Let a and n be natural numbers ≥ 2. There is a rational number x with $x^n = a$ if and only if a is already the n-th power of a natural number, i.e., if x itself is an integer. In other words: *If a is not the n-th power of a natural number, then $\sqrt[n]{a} = a^{1/n}$ is irrational.*[26] For the *proof* we can assume $x > 0$. If then $x = \prod_{p \in \mathbb{P}} p^{v_p(x)}$ is the canonical representation of x, then $a = x^n = \prod_{p \in \mathbb{P}} p^{nv_p(x)} \in \mathbb{N}^*$ is the canonical representation of a. From this follows $nv_p(x) \geq 0$ and thus $v_p(x) \geq 0$ for all $p \in \mathbb{P}$, i.e., $x \in \mathbb{N}^*$, as claimed. ◇

Exercises

Exercise 1.7.1 For any two integers $a, b \in \mathbb{Z}$ with $b \neq 0$, there are integers q and r such that $a = qb + r$ and $|r| \leq \frac{1}{2}|b|$. If b is odd, then q and r are uniquely determined; if b is even, then there are two possibilities for $|r| = \frac{1}{2}|b|$. (r is called the **absolutely smallest remainder** modulo b.)

Exercise 1.7.2 (Bachet's Weight Problem) Let $r \in \mathbb{N}$.

a) With a balance scale and weights of $1, 2, 2^2, \ldots, 2^r$ available exactly once each, any weight $n \in \mathbb{N}$ with $n \leq 2^{r+1} - 1$ can be weighed. The weights may only

---

[26] For $n = 2$ see also footnote 34 in Sect. 1.8 further below.

be placed on one of the two pans. There is no other set of weights with $\leq r+1$ weights that makes this possible.

b) If weights can be placed on *both* pans, then with a balance scale and the set of weights $1, 3, 3^2, \ldots, 3^r$, any weight $n \in \mathbb{N}$ with $n \leq \frac{1}{2}(3^{r+1}-1)$ can be weighed. This is not possible with any other set of weights with $\leq r+1$ weights.

Exercise 1.7.3 Determine the binary and sexagesimal representation of 10^6. What decimal number is $(ABCDEF)_{16}$? How can one generally derive the sexagesimal representation from the binary representation and vice versa?

Exercise 1.7.4

a) For $n \in \mathbb{N}^*$, the prime divisors of $n! + 1$ and for $n > 1$ also the prime divisors of $n! - 1$ are all $> n$. (This again shows that there are infinitely many prime numbers. It is unknown whether infinitely many of the numbers $n! + 1$ or $n! - 1$ are themselves prime.)

b) If $m \in \mathbb{N}^*$, $m > 1$, $m \neq 4$, is given, then the largest prime number $\leq m$ is the smallest number $n \in \mathbb{N}^*$ such that $n!$ is divisible by all $k \in \mathbb{N}^*$ with $k \leq m$. Or, to put it slightly differently: For $n \in \mathbb{N}^*$, $n \neq 3$, $n!$ is divisible by all $k \in \mathbb{N}^*$ with $k < p$, where p is the smallest prime number $> n$. (For the proof, conveniently use the (proven by Chebyshev (1821–1894)) **Bertrand's Postulate:** *If $n \in \mathbb{N}^*$, then there is a prime number q with $n < q \leq 2n$.*—The so-called Platonic number $7! = 5040$ is divisible by all numbers ≤ 10. However, the least common multiple $\text{kgV}(1, 2, \ldots, 10) = 2^3 \cdot 3^2 \cdot 5 \cdot 7 = 2520$ is smaller.)

Exercise 1.7.5 For $n \in \mathbb{N}^*$ none of the n numbers $(n+1)! + 2, \ldots, (n+1)! + n + 1$ is prime. Thus, there are arbitrarily large prime gaps.

Exercise 1.7.6 Show for $a = 3, 4, 6$, that there are infinitely many prime numbers in the sequence $an + (a-1)$, $n \in \mathbb{N}$, by arguing similarly to Theorem 1.7.2 with $ap_1 \cdots p_r + (a-1)$. (In general, it holds: If a, b are coprime positive natural numbers, then there are infinitely many prime numbers of the form $an + b$ with $n \in \mathbb{N}$ (**Dirichlet's Theorem**), see Vol. 2, Theorem 2.7.22.)

Exercise 1.7.7 Let $n, r \in \mathbb{N}^*$, $n \geq 2$. If n has no prime divisor $\leq \sqrt[r+1]{n}$, then n is the product of at most r (not necessarily distinct) prime numbers. In particular, n is prime if n has no prime divisor $\leq \sqrt{n}$. (This is the basis of the so-called **Sieve of Eratosthenes** (after Eratosthenes (276–194 BC)): To obtain the prime numbers $< N^2$, $N \in \mathbb{N}^*$, one successively strikes out from the list of natural numbers from 2 to $N^2 - 1$ the true multiples of the next number not yet struck out, as long as this is $< N$. The remaining numbers are then the prime numbers $< N^2$.)

Exercise 1.7.8 Due to $m^4 + 4^m = (m^2 - 2^{(m+1)/2}m + 2^m)(m^2 + 2^{(m+1)/2}m + 2^m)$ for odd $m \in \mathbb{N}^*$ (cf. the identity of Sophie Germain in Ex. 3.5.5) none of the numbers

$n^4 + 4^n$, $n \in \mathbb{N}$, $n \geq 2$, is prime. Determine the canonical prime factorization of these numbers for $2 \leq n \leq 10$.

Exercise 1.7.9 Let $a, n \in \mathbb{N}$ with a,n ≥ 2. If $a^n - 1$ is prime, then $a = 2$ and n is prime, thus $a^n - 1$ is a Mersenne prime, cf. Example 1.7.3.

Exercise 1.7.10 Let $a, n \in \mathbb{N}^*$, $a \geq 2$. If $a^n + 1$ is prime, then a is even and n is a power of 2. (The numbers $F_m := 2^{2^m} + 1$, $m \in \mathbb{N}$, are called **Fermat numbers** (after P. Fermat (1601(?)–1665)). $F_0 = 3$, $F_1 = 5$, $F_2 = 17$, $F_3 = 257$, $F_4 = 65.537$ are prime. Whether there are more Fermat prime numbers is unknown. A primality test for the Fermat numbers F_m can be found in Theorem 2.5.29. Due to $F_{m+1} = 2 + F_0 \cdots F_m$ (Proof!) two different Fermat numbers are coprime.—It is not known whether there are infinitely many prime numbers of the form $a^2 + 1$. For the significance of Fermat numbers in circle division, see Vol. 13 on Algebra.)

Exercise 1.7.11 For $x, m, n \in \mathbb{N}^*$ with $x \geq 2$ and $d := \gcd(m, n)$ is

$$\gcd(x^m - 1, x^n - 1) = x^d - 1.$$

(One can easily reduce to the case $d = 1$. Then are

$$\frac{x^m - 1}{x - 1} = x^{m-1} + \cdots + x + 1 \quad \text{and} \quad \frac{x^n - 1}{x - 1} = x^{n-1} + \cdots + x + 1$$

coprime.) In particular, the Mersenne numbers $M(p) = 2^{p-1} - 1$, $p \in \mathbb{P}$, are pairwise coprime.—From $m = qn + r$, $q, r \in \mathbb{N}$, follows $x^m - 1 = x^{nq}x^r - 1 = Q \cdot (x^n - 1) + (x^r - 1)$ with $Q = Q(x) := (x^{n(q-1)} + \cdots + x^n + 1)x^r$. Thus, the Euclidean algorithm for $x^m - 1$ and $x^n - 1$ runs parallel to the Euclidean algorithm for m and n. Thus, there are polynomials $R, S \in \mathbb{Z}[X]$ with $R \cdot (X^m - 1) + S \cdot (X^n - 1) = X^d - 1$. In other words: $X^m - 1$ and $X^n - 1$ generate in the polynomial ring $\mathbb{Z}[X]$ the principal ideal $\mathbb{Z}[X](X^d - 1)$, see Sect. 2.7. Determine R and S for $m = 11$ and $n = 7$.

Exercise 1.7.12

a) Determine the prime factorization of 81,057,226,635,000.
b) If $n = p_1^{\alpha_1} \cdots p_r^{\alpha_r}$ is the prime factorization of the positive natural number n with pairwise distinct prime numbers $p_1, \ldots, p_r$, then $\tau(n) = (\alpha_1 + 1) \cdots (\alpha_r + 1)$ is the number of divisors of n in $\mathbb{N}^*$. How many divisors does the number given in a) have?

1.7 Prime Factorization of Natural Numbers

Exercise 1.7.13

a) Let $a \in \mathbb{N}^*$. For how many $x \in \mathbb{N}^*$ is $x(x+a)$ a square number? Determine these x for $a \in \{15, 30, 60, 120\}$.
b) Let $n \in \mathbb{N}^*$. The number of pairs $(u, v) \in \mathbb{N}^2$ with $u^2 - v^2 = n$ is $\lceil \tau(n)/2 \rceil$, if n is odd, $\lceil \tau(n/4)/2 \rceil$, if $4 \mid n$, and equal to 0 otherwise. Provide all four representations of $u^2 - v^2 = 1000$ with $u, v \in \mathbb{N}$. The number of (pairwise incongruent) **Pythagorean triangles** (i.e., the right-angled triangles with positive integer side lengths), one leg of which is equal to the given number $a \in \mathbb{N}^*$ is $\lfloor \tau(a^2)/2 \rfloor$, if a is odd, and $\lfloor \tau(a^2/4)/2 \rfloor$, if a is even.

Remarks (1) A triple $(a, b, c) \in (\mathbb{N}^*)^3$ is called a **Pythagorean number triple** if $a^2 + b^2 = c^2$ holds, i.e., if a, b, c are the side lengths of a Pythagorean triangle. If these side lengths are coprime (which is exactly the case when they are pairwise coprime), the triple is called **primitive**. In a Pythagorean number triple (a, b, c) it is always $a \neq b$ (why?). The two Pythagorean triples (a, b, c) and (b, a, c) determine congruent (but not actually congruent) Pythagorean triangles. In a primitive Pythagorean triple (a, b, c) exactly one of the leg lengths a, b is even (why?). If this is b, then there is a uniquely determined coprime pair $(u, v) \in (\mathbb{N}^*)^2$, where one component is even, with $u > v$ and $a = u^2 - v^2$, $b = 2uv$ and $c = u^2 + v^2$. (**Indian formulas**—Is $b = 2b'$ with $b' \in \mathbb{N}^*$, then $b'^2 = \frac{1}{2}(c+a) \cdot \frac{1}{2}(c-a)$ with odd coprime $a, c \in \mathbb{N}^*$. Now use Ex. 1.7.23. See also Ex. 3.5.12b).)

(2) More difficult than the present exercise is to find the number of Pythagorean triples (a, b, c) for a given integer hypotenuse length $c > 0$. The sought number of triples with $a < b$ is $\lfloor \tau'(c^2)/2 \rfloor$, where $\tau'(c^2)$ is the number of those natural divisors of c^2 whose prime divisors are all $\equiv 1 \bmod 4$. For $c > 1$, the number of primitive triples among them is equal to $2^{\omega(c)-1}$, if $\omega(c)$ is the number of different prime divisors of c and these are all $\equiv 1 \bmod 4$, and 0 otherwise, see the two-square theorem 2.10.42. The smallest Pythagorean triple is the so-called **Egyptian triple** $(3, 4, 5)$ for $c = 5$. For $c = 39 = 3 \cdot 13$ there is (up to swapping the first two components) exactly one such triple, which is not primitive (namely the so-called **Indian triple** $(15, 36, 39)$ with the associated primitive triple $(5, 12, 13)$), which we want to call the primitive Indian triple), for $c = 65 = 5 \cdot 13$ however 4 (which ones?), including 2 primitive ones, and for $c = 57 = 3 \cdot 19$ none.

Exercise 1.7.14 (Egyptian Rope-Stretchers Problem) Let $s \in \mathbb{N}^*$. The primitive Pythagorean triples (a, b, c) (cf. the remarks on the preceding exercise) with even b and perimeter $s = a + b + c$ correspond bijectively to the representations $s = e \cdot 2f$ with coprime e and $2f$, $e < 2f < 2e$, where $e, f \in \mathbb{N}^*$ are. The corresponding triple is

$$(a, b, c) = (e(2f - e), 2f(e - f), (e - f)^2 + f^2).$$

(One can use the Indian formulas from Ex. 1.7.13 or conclude directly.—How does one obtain all Pythagorean triples with given s with $s = a + b + c$?)

Exercise 1.7.15 Determine all pairs $(a, b) \in (\mathbb{N}^*)^2$ with $(a^2 + b^2)/ab \in \mathbb{N}^*$.

Exercise 1.7.16

a) Let $n \in \mathbb{N}^*$ and $p \in \mathbb{P}$. Then

$$v_p(n!) = \left[\frac{n}{p}\right] + \left[\frac{n}{p^2}\right] + \left[\frac{n}{p^3}\right] + \cdots .$$

(Due to $[x/m] = [[x]/m]$ for all $x \in \mathbb{R}$ and all $m \in \mathbb{N}^*$ one calculates the summands conveniently recursively. One easily proves that for each $g \in \mathbb{N}^*, g \geq 2$, the equation $\sum_{i \geq 1}[n/g^i] = (n - \sum_{i \geq 0} a_i)/(g-1)$ holds, where the a_i are the digits in the g-adic expansion of $n = \sum_{i \geq 0} a_i g^i$, see example 1.7.6. In particular, n is congruent modulo $g - 1$ to the (g-adic) **digit sum** $\sum_{i \geq 0} a_i$ of n. For $g = 10$, this is known as the **nines test**.—More generally, one shows: If $n_i, i \in I$, is a finite family of positive natural numbers, then the p-exponent of the product $\prod_{i \in I} n_i$ is equal to $\sum_{k \in \mathbb{N}^*} v_k$, where v_k for $k \in \mathbb{N}^*$ is the number of $i \in I$ for which n_i is divisible by p^k.)

b) Let $n, k \in \mathbb{N}^*, k \leq n$. Every prime power that divides $\binom{n}{k}$ is $\leq n$.

c) For every prime power $p^\alpha > 1$ and every $k \in \mathbb{N}^*, 1 \leq k \leq p^\alpha$, is

$$v_p\left(\binom{p^\alpha}{k}\right) = \alpha - v_p(k) = v_p(p^\alpha) - v_p(k).$$

Exercise 1.7.17

a) Determine the canonical prime factorization of 50! as well as of the product $\prod_{k=1}^{50}(2k-1) = 100!/2^{50} \cdot 50! = 1 \cdot 3 \cdot 5 \cdots 99$ of the first 50 odd natural numbers.

b) Determine the canonical prime factorization of lcm$(1, 2, 3, \ldots, 50)$.

c) Let $n \in \mathbb{N}^*$. For $p \in \mathbb{P}$ determine $v_p(B(n))$, where $B(n) := \text{kgV}(1, 2, 3, \ldots, n)$ is the lcm of the first n positive natural numbers.

Exercise 1.7.18 Let $n, k \in \mathbb{N}^*$ be coprime. Show that $\binom{n}{k}$ is divisible by n and $\binom{n-1}{k-1}$ is divisible by k. (Consider the formula $k\binom{n}{k} = n\binom{n-1}{k-1}$.)

Exercise 1.7.19 Let $p \in \mathbb{P}$. For $r, k \in \mathbb{N}$ with $r < k < p$, $\binom{p+r}{k}$ is divisible by p. In particular, $\binom{p}{k}$ is divisible by p for $0 < k < p$.

Exercise 1.7.20 Let $p \in \mathbb{P}$. By induction over n, prove the **Little Fermat's Theorem**: For $n \in \mathbb{N}$, p is a divisor of $n^p - n$, i.e., $n^p \equiv n \bmod p$ and $n^{p-1} \equiv 1 \bmod p$, if $p \nmid n$. (Use Ex. 1.7.19.)

1.7 Prime Factorization of Natural Numbers

Exercise 1.7.21 For every natural number n, $n^8 - n^2$ is divisible by $4 \cdot 7 \cdot 9 = 252$. Also discuss the divisibility statements from Ex. 1.5.6 again.

Exercise 1.7.22 Let $r \in \mathbb{N}^*$, $m = (m_1, \ldots, m_r) \in \mathbb{N}^r$ and $n := |m| = \sum_{i=1}^{r} m_i$. All prime numbers p, for which Max $(m_1, \ldots, m_r) < p \leq n$ holds, divide $\binom{n}{m} = n!/m_1! \cdots m_r!$. □

Exercise 1.7.23 Let $n \in \mathbb{N}^*$. The product of two coprime natural numbers a and b is exactly then the n-th power of a natural number, if this holds for a and b individually.

Exercise 1.7.24 Let $a, b \in \mathbb{N}^*$. Then $\gcd(a,b) \cdot \text{lcm}(a,b) = ab$. (This provides a convenient method for calculating the LCM via the Euclidean algorithm.)

Exercise 1.7.25 Let $v \in \mathbb{N}^*$ be a common multiple of the numbers $a_1, \ldots, a_n \in \mathbb{N}^*$, $n \geq 1$.

a) Equivalent are: (i) It is $\text{lcm}(a_1, \ldots, a_n) = v$. (ii) It is $\gcd(v/a_1, \ldots, v/a_n) = 1$. (iii) There exists $s_1, \ldots, s_n \in \mathbb{Z}$ with $1/v = s_1/a_1 + \cdots + s_n/a_n$. (lcm and gcd of integers $b_1, \ldots, b_n$ are explained analogously as in the case $n = 2$. If $\gcd(b_1, \ldots, b_n) = 1$ holds, then $b_1, \ldots, b_n$ are called **coprime**. This concept is to be clearly distinguished from that of pairwise coprimality.)
b) Let $a := a_1 \cdots a_n$ and $a_i' := \text{lcm}(a_j, j \neq i)$ be for $i = 1, \ldots, n$. Equivalent are: (i) The numbers $a_1, \ldots, a_n$ are *pairwise* coprime. (ii) It is $\text{lcm}(a_1, \ldots, a_n) = a$. (iii) For $i = 1, \ldots, n$, the numbers a_i and a_i' are each coprime. (iv) The numbers $a_1', \ldots, a_n'$ are coprime. (v) The numbers $a/a_1, \ldots, a/a_n$ are coprime. (vi) There exists $s_1, \ldots, s_n \in \mathbb{Z}$ with $1/a = s_1/a_1 + \cdots + s_n/a_n$.

Exercise 1.7.26 Let $a_1, \ldots, a_n \in \mathbb{N}^*$. There exist numbers $u_1, \ldots, u_n \in \mathbb{Z}$ with

$$\gcd(a_1, \ldots, a_n) = u_1 a_1 + \cdots + u_n a_n.$$

In particular, $a_1, \ldots, a_n$ are coprime if and only if there exist integers $u_1, \ldots, u_n$ with $1 = u_1 a_1 + \cdots + u_n a_n$. The coefficients $u_1, \ldots, u_n$ are obtained algorithmically by successive application of the method described before Proposition 1.7.8, utilizing the relation $\gcd(a_1, \ldots, a_{n-1}, a_n) = \gcd(\gcd(a_1, \ldots, a_{n-1}), a_n)$. However, this algorithm often yields coefficients $u_1, \ldots, u_n$ that are disproportionately large in magnitude. A better approach is as follows: First, number the numbers $a_1, \ldots, a_n$ such that a_1 is minimal among the a_i, and then proceed to the tuple $(a_1, r_2, \ldots, r_n)$, where r_j is the remainder of a_j when a_j is divided by a_1, strike out the zeros among the r_j and calculate with the new tuple as at the beginning. In doing so, one must check how the coefficients of the constructed tuples can be represented as linear combinations of the $a_1, \ldots, a_n$, starting with $a_i = \sum_{k=1}^{n} \delta_{ik} a_k$. (For this, also compare the proof of the elementary divisor theorem in Vol. 3: Linear Algebra 1.) Determine integers u_1, u_2, u_3 with $1 = u_1 \cdot 88 + u_2 \cdot 152 + u_3 \cdot 209$.

Exercise 1.7.27 Let $a_1, \ldots, a_n \in \mathbb{N}^*$ be coprime. Then there is a number $f \in \mathbb{N}$ such that every natural number $b \geq f$ has a representation $b = u_1 a_1 + \cdots + u_n a_n$ with *natural* numbers $u_1, \ldots, u_n$. The smallest such number f is called the **conductor** of the additive submonoid generated by $a_1, \ldots, a_n$

$$\mathbb{N}a_1 + \cdots + \mathbb{N}a_n = \{u_1 a_1 + \cdots + u_n a_n \mid u_1, \ldots, u_n \in \mathbb{N}\}$$

of $\mathbb{N}$. For $n = 2$, the conductor of $M := \mathbb{N}a_1 + \mathbb{N}a_2$ is equal to $f := (a_1 - 1)(a_2 - 1)$. In this case, there are exactly $f/2$ elements in $\mathbb{N} - M$. (**Sylvester's theorem** (after J. Sylvester (1814–1897))—For $0 \leq c \leq f - 1$ exactly one of the two numbers c and $f - 1 - c$ is an element of M.—One can also conclude by induction over $a_1 + a_2$. From a_1, a_2 proceed to $a_1 - a_2, a_2$ if $a_1 > a_2$.)

Exercise 1.7.28 Let $a, b \in \mathbb{N}^*$ and $d := \gcd(a, b) = sa + tb$ with $s, t \in \mathbb{Z}$. Then $d = s'a + t'b$ also holds for $s', t' \in \mathbb{Z}$, if there is a $k \in \mathbb{Z}$ with $s' = s - kb/d$, $t' = t + ka/d$.

Exercise 1.7.29

a) Let $p_1 = 2, p_2 = 3, p_3 = 5, \ldots$ be the (infinite) sequence of prime numbers. Furthermore, let A be an alphabet with a (finite or infinite) enumeration $A = \{a_1, a_2, a_3, \ldots\}$, $a_i \neq a_j$ for $i \neq j$, of its letters. Then, through $(a_{i_1}, \ldots, a_{i_n}) \mapsto p_1^{i_1} \cdots p_n^{i_n}$, an injective mapping of the set $W(A) = \biguplus_{n \in \mathbb{N}} A^n$ of words over A into the set $\mathbb{N}^*$ of positive natural numbers is given.

(**Remark** Such an encoding of words over A is called a **Gödelization** (after K. Gödel (1906–1978)). The natural number assigned to a word in this process is called the **Gödel number** of this word. Which numbers $m \in \mathbb{N}^*$ appear as the Gödel number of a word $W \in W(A)$ and how is the corresponding word $W \in W(A)$ determined, if applicable?)

b) Let A be the finite alphabet $\{a_1, a_2, \ldots, a_g\}$ with g letters, $g \geq 2$, and $a_0 \notin A$ an additional letter. A word $W = (a_{i_1}, \ldots, a_{i_n})$ over A we identify after padding with a_0 with the infinite sequence $(a_{i_1}, \ldots a_{i_n}, a_0, a_0, \ldots)$. Then the mapping $(a_{i_\nu})_{\nu \in \mathbb{N}^*} \mapsto \sum_{\nu=1}^{\infty} i_\nu g^{\nu-1}$ is a *bijective* mapping of the set $W(A)$ of words over A onto the set $\mathbb{N}$ of natural numbers and in particular a Gödelization of $W(A)$. (It is a variant of the g-adic development.)

Exercise 1.7.30 Let $a, b \in \mathbb{Q}_+^\times$. Exactly then is $\sqrt{a} + \sqrt{b} \in \mathbb{Q}_+^\times$, if and only if both a and b are squares of a rational number.[27]

---

[27] The irrational number $\sqrt{2} + \sqrt{3} = 3,14626\ldots$ was given by Plato (427–347 BC) as an approximation of the circle number $\pi = 3,14159\ldots$. At least, K. Popper suspects this in: The Open Society and Its Enemies 1, Munich 1980, note 9 [4] to chapter 6.

1.7 Prime Factorization of Natural Numbers

Exercise 1.7.31

a) Let $x := a/b \in \mathbb{Q}$ be a *reduced* fraction, $a, b \in \mathbb{Z}$, $b > 0$. For integers $a_0, \ldots, a_n$ and $a_n \neq 0$, $n \geq 1$, let $a_n x^n + \cdots + a_1 x + a_0 = 0$ hold, i.e., x is a root of the polynomial $a_n X^n + \cdots + a_1 X + a_0$. Then a is a divisor of a_0 and b is a divisor of a_n. In particular, $x \in \mathbb{Z}$, if the highest coefficient is $a_n = 1$ (**Gauss's Lemma**).
b) Determine all rational roots of the polynomials

$$X^3 + \frac{3}{4}X^2 + \frac{3}{2}X + 3 \quad \text{and} \quad 3X^7 + 4X^6 - X^5 + X^4 + 4X^3 + 5X^2 - 4.$$

Exercise 1.7.32

a) Let $x, y \in \mathbb{Q}_+^\times$ and $y = c/d$ be a reduced representation of y with $c, d \in \mathbb{N}^*$. Then x^y is rational if and only if x is the d-th power of a rational number.
b) Let $x \in \mathbb{Q}_+^\times$ and a be a natural number ≥ 2, which is not of the form b^d with $b, d \in \mathbb{N}^*$, $d \geq 2$. Then $\log_a x$ is integer or irrational. (According to a theorem by Gelfond-Schneider, $\log_a x$ is even transcendental if $\log_a x$ is irrational.)

Exercise 1.7.33 The pairs $(x, y) \in (\mathbb{Q}_+^\times)^2$ of positive rational numbers with $x < y$ and $x^y = y^x$ are $\left((1 + \frac{1}{n})^n, (1 + \frac{1}{n})^{n+1}\right)$, $n \in \mathbb{N}^*$. In particular, $(2, 4)$ is the only pair $(x, y) \in (\mathbb{N}^*)^2$ with $x < y$ and $x^y = y^x$. (The given pairs form the well-known interval nesting for Euler's number $e = 2{,}718281828\ldots$, see example 3.3.8. For every *real* positive number x with $1 < x < e$, there is exactly one real number $y > x$ with $x^y = y^x$. Then $y > e$. To prove these statements, note that $x^y = y^x$ is equivalent to $(\ln x)/x = (\ln y)/y$, and discuss the function $f(x) = (\ln x)/x$ on $\mathbb{R}_+^\times$, see Fig. 1.19. For the exponential and logarithm function, see Sect. 3.10.)

Exercise 1.7.34 Let $m, n \in \mathbb{N}^*$ be coprime. The sequence $a_0, a_1, \ldots$ is recursively defined by $a_0 = n$, $a_{i+1} = a_0 \cdots a_i + m$, $i \in \mathbb{N}$. For $i \geq 1$, it is $a_{i+1} = (a_i - m)a_i + m = a_i^2 - ma_i + m$.

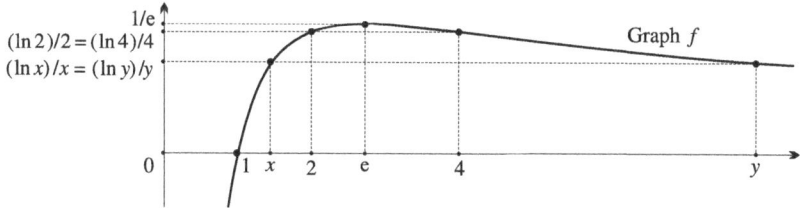

Fig. 1.19 The function $f(x) = (\ln x)/x$

a) It is ggT$(a_i, a_j) = 1$ for all $i, j \in \mathbb{N}$ with $i \neq j$. The prime divisors of the a_i, $i \in \mathbb{N}$, yield infinitely many different prime numbers. (The a_i are often well-suited for testing prime factorization methods.)

b) For all $i \in \mathbb{N}$ is

$$\frac{1}{a_0} + \frac{m}{a_1} + \cdots + \frac{m^i}{a_i} = \frac{m+1}{n} - \frac{m^{i+1}}{a_{i+1} - m}.$$

One deduces

$$\sum_{i=0}^{\infty} \frac{m^i}{a_i} = \frac{m+1}{n}.$$

c) For $m = 2$ and $n = 1$ is $a_{i+1} = F_i = 2^{2^i} + 1$ the i-th Fermat number, $i \in \mathbb{N}$. From b) follows $\sum_{i=0}^{\infty} 2^i/F_i = 1$. Determine the first terms of the sequence (a_i) for $m = n = 1$, i.e., with $a_0 = 1$, $a_1 = 2$ and $a_{i+1} = a_i^2 - a_i + 1 = (a_i - 1)^2 + a_i$, $i \in \mathbb{N}^*$, and provide their prime factorization.

Exercise 1.7.35 For each $s \geq 2$, $(m_s, n_s) := \left(2(2^{s-1} - 1), 2^{s+1}(2^{s-1} - 1)\right)$ is a pair (m, n) of positive natural numbers such that $m < n$ and m and n as well as $m + 1$ and $n + 1$ have the same prime divisors respectively. (There are other such pairs (m, n), e.g., (75, 1215), see Makowski: Ens. Math. **14**, 193 (1968).)

Exercise 1.7.36 The uniqueness of the decomposition of a positive natural number as a product of irreducible numbers according to Theorem 1.7.11 is far less obvious than the existence of such a decomposition, see Proposition 1.7.4 and the remark before Theorem 1.7.11. Let $q \in \mathbb{N}^*$ be any prime number (e.g., $q = 2$ or $q = 12{,}345{,}678{,}901{,}234{,}567{,}891$ and $N := \mathbb{N}^* - \{q\}$). N is multiplicatively closed, and every element in N is a product of irreducible elements of N. However, such a decomposition is generally no longer unique. Show more precisely: The irreducible elements in N are, besides the usual prime numbers $p \neq q$ and their products pq with q, the two elements $q_2 := q^2$ and $q_3 := q^3$. The element $n := q^6 \in N$ has the two essentially different decompositions $n = q_2 \cdot q_2 \cdot q_2 = q_3 \cdot q_3$ as a product of irreducible elements of N. The irreducible element q_3 divides in N the product $q_2 \cdot q_2 \cdot q_2$, but none of the factors. Similarly, q_2 divides in N the product $q_3 \cdot q_3$, but not q_3 (cf. however, Proposition 1.7.10). Similarly, $m := pq^3 = (pq)q^2$ in N has two essentially different decompositions (p prime number $\neq q$).—One should also discuss the Hilbertian multiplicative monoids $H_{\mathbb{N}^*} := \{n \in \mathbb{N}^* \mid n \equiv 1 \bmod 4\} \subseteq \mathbb{N}^*$ or $H_{\mathbb{Z}^*} := \{a \in \mathbb{Z}^* \mid a \equiv 1 \bmod 4\} \subseteq \mathbb{Z}^* = \mathbb{Z} - \{0\}$ and determine all irreducible and all prime elements therein. (The monoids $H_{\mathbb{N}^*}$ and $H_{\mathbb{Z}^*}$ appear as monoids of discriminants of quadratic $\mathbb{Z}$-algebras, cf. Example 2.10.37. In $H_{\mathbb{Z}^*}$, the theorem of unique prime factorization holds as in $\mathbb{N}^*$, but not in $H_{\mathbb{N}^*}$.—A prime element is an element that divides a product only if it divides at least one factor.—See also the general discussion of divisibility concepts in Sect. 2.1.)

1.7 Prime Factorization of Natural Numbers

Exercise 1.7.37 Let N be a subset of $\mathbb{N}$ with $N^* := N \cap \mathbb{N}^* \neq \emptyset$, which contains exactly two elements $r_1, r_2, r_1 \geq r_2$, also $r_1 + r_2$ and $r_1 - r_2$. Then there is (exactly) one element $k \in N^*$ with $N = \mathbb{N}k = \{nk \mid n \in \mathbb{N}\}$. (Consider the smallest element in N^*.)

Exercise 1.7.38 An (arbitrary) sequence $(x_i)_{i \in \mathbb{N}}$ is called **periodic**, if there are numbers $t \in \mathbb{N}$ and $r \in \mathbb{N}^*$ with $x_{i+r} = x_i$ for all $i \geq t$. (t, r) is then called a **period pair** for (x_i). Show: If $(x_i)_{i \in \mathbb{N}}$ is periodic, then there is a uniquely determined period pair.$(m, k) \in \mathbb{N} \times \mathbb{N}^*$ for (x_i) such that for each period pair (t, r) of (x_i) holds: It is $m \leq t$ and k divides r.[28] (To prove the existence of k, use Ex. 1.7.37.—m is called the (smallest) **pre-period length**, k the (smallest) **period length**, and the smallest period pair (m, k) the **periodicity type** of $(x_i)_{i \in \mathbb{N}}$. $(x_0, \ldots, x_{m-1})$ is called the **pre-period** and $(x_m, \ldots, x_{m+k-1})$ the **period** of $(x_i)_{i \in \mathbb{N}}$. A non-periodic sequence is defined to have the type $(\infty, 0)$. A sequence of the periodicity type $(m, 1)$ is stationary with the single-element period x_m as a limit value. A periodic sequence of the periodicity type $(0, k)$ is called **purely periodic**.) If $(x_i)_{i \in \mathbb{N}}$ and $(y_i)_{i \in \mathbb{N}}$ are sequences of the periodicity type (m, k) or (n, ℓ), then the pair sequence $(x_i, y_i)_{i \in \mathbb{N}}$ has the periodicity type $(\text{Max}(m, n), \text{lcm}(k, \ell))$.

Exercise 1.7.39 The division with remainder of natural numbers gives rise to the so-called **Euclidean Nim game**: In this game, a game position is given by a pair $(a, b) \in \mathbb{N}^* \times \mathbb{N}^*$, represented by two heaps of a and b game pieces, respectively. If $a \neq b$, the player whose turn it is may and must take away a positive multiple of the number of pieces in the smaller heap from the heap with the larger number of pieces, without removing the larger heap entirely. A game with the initial position (a, b) ends with the position $(\gcd(a, b), \gcd(a, b))$.[29] The player who was able to make the last move wins. Show that: $(a, b) \in \mathbb{N}^* \times \mathbb{N}^*$ is a winning position for the first player if and only if $a\,/b$ or—equivalently—if $b\,/\,a$ is *not* in the open interval $]\Phi^{-1} = \Phi - 1, \Phi[$ ($\Phi = \frac{1}{2}(1+\sqrt{5})$).—The assertion follows immediately from the following two statements, which in turn are a consequence of the fact that $]\Phi^{-1}, \Phi[$ is the (only) open interval of length 1 in $\mathbb{R}_+^{\times}$ that is invariant under inversion[30] : (1) If $1 < a/b < \Phi$, then $(a-b)/b = a/b - 1 < \Phi - 1 = \Phi^{-1}$. (2) If $a/b > \Phi$, then there is (exactly) one $q \in \mathbb{N}^*$ with $(a-qb)/b = a/b - q \in]\Phi^{-1}, \Phi[$.—A Fibonacci pair $(\mathsf{F}_{n+1}, \mathsf{F}_n), n \geq 1$, represents a winning position if and only if n is even. The

---

[28] (m, k) is thus the smallest element in the set of all period pairs of (x_i), where $\mathbb{N} \times \mathbb{N}^* = (\mathbb{N}, \leq) \times (\mathbb{N}^*, |)$ carries the product order. The order on $\mathbb{N}^*$ is divisibility, see Ex. 1.4.1.

[29] The Euclidean Nim games playfully determine the gcd of two positive natural numbers.

[30] Note that $\Phi = \frac{1}{2}(1 + \sqrt{5})$ is irrational. $a\,/b$ therefore cannot be equal to one of the boundary points Φ^{-1}, Φ of the interval. If $\Phi = a/b$ or $\Phi^{-1} = b/a = \Phi - 1$, $a, b \in \mathbb{N}^*$, were rational, then a game with the initial position (a, b) would never end due to $(a-b)/b = a/b - 1 = b/a$, which is absurd. Because of this simple argument, it is assumed that Φ (and not $\sqrt{2} = $ length of the diagonal in the unit square) was the first number recognized by the Greeks (specifically by the Pythagorean Hippasus of Metapontum (around 450 BC)) as irrational, see K. von Fritz: The discovery of incommensurability by Hippasus of Metapontum, Ann. of Math. **48**, 242–264 (1945).

pair (36.667, 12.247) from Example 1.7.9 describes a winning position. How does the game proceed if the player with the first move plays optimally from this starting position? (It ends after a total of 5 moves.)

1.8 Infinite Sets and Cardinal Numbers

The simplest infinite sets are those whose elements can be numbered using natural numbers: $a_0, a_1, a_2, \ldots$

Definition 1.8.1 Let A be a set.

(1) A is called (at most) **countable**, if A is empty or there exists a surjective mapping from $\mathbb{N}$ onto A.
(2) A is called **countably infinite**, if there exists a bijective mapping from $\mathbb{N}$ onto A.
(3) A is called **uncountable**, if A is not countable.

A is countable if and only if there exists an injective mapping from A into $\mathbb{N}$, see Exercise 1.2.16. Images and subsets of countable sets are again countable. Every countable set is either finite or countably infinite. Every infinite set has a countably infinite subset, see Remark 1.5.7.

Lemma 1.8.2 $\mathbb{N} \times \mathbb{N}$ *is countable.*

Proof The mapping $g : \mathbb{N} \times \mathbb{N} \to \mathbb{N}$ with $(m, n) \mapsto 2^m(2n + 1) - 1$ is bijective. m is namely the 2-exponent of $g(m, n) + 1$ and $2n + 1$ the odd residual factor. □

Remark 1.8.3 One can easily recognize the countability of $\mathbb{N} \times \mathbb{N}$ also with the help of the **first** or **Cauchy's diagonal method**, in which the pairs $(m, n) \in \mathbb{N} \times \mathbb{N}$ are counted one after the other in the diagonals according to the scheme in Fig. 1.20. ◊

Theorem 1.8.4 *The union of countably many countable sets is countable.*

Fig. 1.20 Cauchy's diagonal method

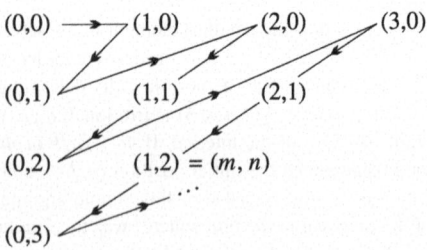

1.8 Infinite Sets and Cardinal Numbers

Proof Let A_i, $i \in I$, be countable sets with the countable index set I and A their union. We can readily assume that I and all A_i are non-empty. Then there are surjective mappings $f_i \colon \mathbb{N} \to A_i$, $i \in I$, and $h \colon \mathbb{N} \to I$, and the mapping $\mathbb{N} \times \mathbb{N} \to A$ with $(m, n) \mapsto f_{h(m)}(n)$ is also surjective. Since $\mathbb{N} \times \mathbb{N}$ is countable according to Lemma 1.8.2, this also applies to A. □

Corollary 1.8.5 *If the sets $A_1, \ldots, A_n$ are countable, then $A_1 \times \cdots \times A_n$ is also countable.*

Proof The statement is reduced by induction to the case $n = 2$. $A_1 \times A_2 = \bigcup_{a \in A_1} (\{a\} \times A_2)$ is countable as a countable union of countable sets. □

Corollary 1.8.6 *The set $\mathbb{Q}$ of rational numbers is countable.*

Proof $\mathbb{Z}$ is countable as a union of the two countable sets $\mathbb{N}$ and $-\mathbb{N} := \{-n \mid n \in \mathbb{N}\}$. Therefore, $\mathbb{Z} \times (\mathbb{Z} - \{0\})$ is also countable. The mapping $(a, b) \mapsto a/b$ from $\mathbb{Z} \times (\mathbb{Z} - \{0\})$ to $\mathbb{Q}$ is surjective. Consequently, $\mathbb{Q}$ is also countable. □

Remark 1.8.7 A more number-theoretically motivated *bijective* enumeration $\mathbb{N}^* \to \mathbb{Q}_+^\times$ of the positive rational numbers is described in Example 1.7.12. ◇

Example 1.8.8 *Let A be a countable set. Then the set $\mathfrak{E}(A)$ of the finite subsets of A is also countable.* For $n \in \mathbb{N}^*$ the set of non-empty subsets of A with at most n elements is the image of the mapping from A^n into $\mathfrak{P}(A)$ with $(a_1, \ldots, a_n) \mapsto \{a_1, \ldots, a_n\}$ and therefore countable (cf. Corollary 1.8.5). Now the assertion follows directly from Theorem 1.8.4.—The dual development $E \mapsto \sum_{n \in E} 2^n$ provides an explicit bijective mapping $\mathfrak{E}(\mathbb{N}) \xrightarrow{\sim} \mathbb{N}$, cf. Example 1.7.6. ◇

However, there are sets that are not countable. This is one of the great discoveries of G. Cantor. For example, the full power set of a countably infinite set is no longer countable. This is a special case of the following famous Cantor's theorem:

Theorem 1.8.9 *A be a set. Then there is no surjective mapping from A onto the power set $\mathfrak{P}(A)$ of A.—In particular, the power set of an infinite set is uncountable.*

Proof Let $f \colon A \to \mathfrak{P}(A)$ be an arbitrary mapping. We consider the subset $B := \{a \in A \mid a \notin f(a)\}$ of A and claim that B does not belong to the image of f. Suppose there is $B = f(b)$ with a $b \in A$. If $b \in B$, then according to the definition of B, the contradiction $b \notin f(b) = B$ immediately follows. However, if $b \notin B = f(b)$, then $b \in B$ is according to the definition of B, which is also a contradiction. □

Theorem 1.8.9 shows that, for example, the infinite sets $\mathbb{N}$, $\mathfrak{P}(\mathbb{N})$, $\mathfrak{P}(\mathfrak{P}(\mathbb{N}))$, ... continuously become significantly larger.

Remark 1.8.10 The proof principle used in the proof of Theorem 1.8.9 is the so-called **second** or **Cantor's diagonal argument**. If one identifies the elements $J \in \mathfrak{P}(A)$ with their indicator functions $e_J \in \{0,1\}^A$, then the function $e := e_B$ in the proof of Theorem 1.8.9 is defined by

$$e(a) := \begin{cases} 1, & \text{if } f(a)(a) = 0, \\ 0, & \text{if } f(a)(a) = 1, \end{cases}$$

i.e., with the help of the "diagonal values" $f(a)(a)$, $a \in A$. By construction, $e(a) \neq f(a)(a)$ is for all $a \in A$, thus certainly $e = e_B \notin \operatorname{Im} f$. Cantor's diagonal argument turns Russell's paradox, see Sect. 1.1, into a positive: The set of all sets that do not contain themselves cannot exist. ◇

Example 1.8.11 *Let A be an uncountable set and B a countable subset of A. Then there is a bijective mapping from A onto $A - B$. Proof.* With A, $A - B$ is also uncountable. Let C be a countably infinite subset of $A - B$. Then the sets $B \cup C$ and C are both countably infinite. Therefore, there is a bijective mapping $g : B \cup C \longrightarrow C$. Now we define the bijective mapping $f : A \longrightarrow A - B$ by $f(x) := x$ for $x \notin B \cup C$ and $f(x) := g(x)$ for $x \in B \cup C$. ◇

The set $\mathbb{R}$ of real numbers is also uncountable. There is, as we will show shortly, a bijective mapping from $\mathfrak{P}(\mathbb{N})$ to $\mathbb{R}$. In general, two sets A and B are called **equipotent** or of **equal cardinality**, if there is a bijective mapping from A to B. One then writes $|A| = |B|$ or also $\operatorname{Card} A = \operatorname{Card} B$. $\mathfrak{P}(\mathbb{N})$ and $\mathbb{R}$ are therefore equipotent. The cardinality of $\mathbb{R}$ is called the **cardinality of the continuum** and is denoted following Cantor with

$$\aleph.$$

($\aleph$ (= Aleph) is the first letter of the Hebrew alphabet.) The cardinality of $\mathbb{N}$ is denoted with

$$\aleph_0.$$

On any set of sets, equipotency is an equivalence relation, since the identity, the inverse of a bijective mapping, and the composition of bijective mappings are bijective.

Theorem 1.8.12 *The sets $\mathfrak{P}(\mathbb{N})$ and $\mathbb{R}$ are equinumerous, i.e., there exists a bijective mapping from $\mathfrak{P}(\mathbb{N})$ to $\mathbb{R}$. In particular, $\mathbb{R}$ is not countable.*

Proof Let $\mathcal{U}$ be the set of *infinite* subsets of $\mathbb{N}$ different from $\mathbb{N}$. According to examples 1.8.8 and 1.8.11, $\mathcal{U}$ and $\mathfrak{P}(\mathbb{N})$ are equinumerous. The mapping

$$f : \mathcal{U} \longrightarrow\,]0,1[:= \{x \in \mathbb{R} \mid 0 < x < 1\} \quad \text{with} \quad A \longmapsto \sum_{n \in A} \frac{1}{2^{n+1}} = \sum_{n=0}^{\infty} \frac{e_A(n)}{2^{n+1}}$$

1.8 Infinite Sets and Cardinal Numbers

is bijective, since each $x \in \,]0, 1[$ has a unique *infinite* binary fraction expansion according to example 3.3.10.[31] Furthermore, $\mathbb{R}$ is equinumerous to the interval $]0, 1[$. The mapping $x \mapsto x/(1 + |x|)$ maps $\mathbb{R}$ bijectively onto the open interval $\,]-1, 1[$, which in turn is mapped bijectively onto $]0, 1[$ by the mapping $x \mapsto \frac{1}{2}(x + 1)$. Overall, we obtain that $\mathfrak{P}(\mathbb{N})$ and $\mathbb{R}$ are equinumerous. □

Remark 1.8.13 The statement, which is somewhat weaker compared to theorem 1.8.12, that $\mathbb{R}$ is uncountable, can also be obtained by applying a variant of Cantor's diagonal argument in the following way to decimal fractions: With $\mathbb{R}$, the interval $[0, 1]$ would also be countable. Assume

$$0, a_{11}a_{12}a_{13}a_{14} \ldots$$
$$0, a_{21}a_{22}a_{23}a_{24} \ldots$$
$$0, a_{31}a_{32}a_{33}a_{34} \ldots$$
$$0, a_{41}a_{42}a_{43}a_{44} \ldots$$
$$\vdots$$

is an enumeration of these numbers, represented as finite or infinite decimal fractions with the digits $a_{ij} \in \{0, 1, \ldots, 9\}$. Then, for example, the number $0, a_1 a_2 a_3 a_4 \ldots$ with $a_i := 4$, if $a_{ii} \neq 4$, and $a_i := 6$, if $a_{ii} = 4$, does not appear in this enumeration. Contradiction!—The first proof that Cantor gave for the uncountability of $\mathbb{R}$ can be found in example 3.3.6. ◊

Example 1.8.14 *If I is non-empty and countable, then $\mathbb{R}^I$ has the cardinality of the continuum. Proof.* According to Theorem 1.8.12 and Exercise 1.2.20, it suffices to show that $(\{0, 1\}^{\mathbb{N}})^I$ and $\{0, 1\}^{\mathbb{N}}$ are equinumerous. According to Exercise 1.2.19, however, the sets $(\{0, 1\}^{\mathbb{N}})^I$ and $\{0, 1\}^{\mathbb{N} \times I}$ are equinumerous, and according to Corollary 1.8.5, the sets $\mathbb{N} \times I$ and $\mathbb{N}$. Overall, therefore, $\{0, 1\}^{\mathbb{N} \times I}$ and $\{0, 1\}^{\mathbb{N}}$ are equinumerous.—*The spaces $\mathbb{R}, \mathbb{R}^2, \mathbb{R}^3, \ldots$ and likewise the sequence space $\mathbb{R}^{\mathbb{N}}$ therefore all have the same cardinality $\aleph$.* ◊

Example 1.8.15 Let A be a subset of the x-axis in the (x, y)-coordinate plane $\mathbb{R} \times \mathbb{R}$. To each point $a \in A$ a circle[32] should be attached, which lies in the upper half-plane and touches the x-axis at a, with the condition that any two of these circles should not have a common point. Under what conditions on A is this exercise solvable?

Let K_1 and K_2 be two circles with radii R_1 and R_2, which lie in the upper half-plane and touch the x-axis, see Fig. 1.21. The distance of their points of contact on the x-axis is d. According to the Pythagorean theorem, the circles touch each other if

---

[31] We recall that e_A is the indicator function of $A \subseteq \mathbb{N}$. The mapping f is even an order isomorphism if $\mathcal{U} \subseteq \mathfrak{P}(\mathbb{N}) = \{0, 1\}^{\mathbb{N}}$ carries the lexicographical order. Note $\sum_{n \in \mathbb{N}} 1/2^{n+1} = 1$.
[32] By a circle, we mean here (and in the exercises of this section) a circular disk with a positive radius including its periphery.

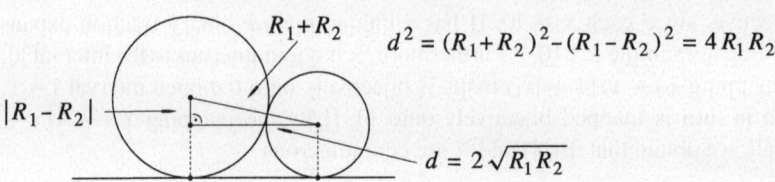

Fig. 1.21 Condition for circles to touch

$d = 2\sqrt{R_1 R_2}$ is true. Thus, they have no common point exactly when $d > 2\sqrt{R_1 R_2}$ is true.[33]

Now let $A = \mathbb{Q}$. If we choose for each rational number p/q (where $q > 0$ is chosen minimally) a radius $< 1/2q^2$, then the corresponding circles do not intersect pairwise. Namely, if p/q and r/s are two different (reduced) rational numbers with $p, q, r, s \in \mathbb{Z}, q, s > 0$, then it is (If one chooses for p/q each the radius $1/2q^2$, then the circles can touch, and they do so precisely when $|p/q - r/s| = 1/qs$, i.e., $ps - qr = \pm 1$. A pair $p/q, r/s$ of fractions with this property is called a **Farey pair**.[34]

In general, the initially posed problem is always solvable if A is countable.

If namely $a_0, a_1, a_2, \ldots$ is an enumeration of the elements of A, then one chooses recursively the radius R_i to the point a_i so small that the circle with the contact point a_i does not intersect with any of the already constructed circles with the contact points $a_0, a_1, \ldots, a_{i-1}$.

Conversely, if there is a solution for the given set A, then A is necessarily countable. In particular, the posed problem for $A = \mathbb{R}$ is not solvable. According to Theorem 1.8.4, it suffices to show that $A \cap [n, n+1]$ is countable for every $n \in \mathbb{Z}$. Let A_q for $q \in \mathbb{N}^*$ be the set of points in $A \cap [n, n+1]$, for which the radius of

---

[33] The reader should "see" the inequality $R_1 + R_2 \geq 2\sqrt{R_1 R_2}$ for the arithmetic and geometric mean of the radii R_1, R_2 (where the equality sign holds only for $R_1 = R_2$).

[34] The estimate $|b - c| \geq 1/qs$ (or better $|b - c| \geq 1/\mathrm{lcm}(q, s)$) for *different* rational numbers b,c with denominators $q, s \in \mathbb{N}^*$ is very frequently used. We provide here the following simple example for a so-called **diophantine approximation**, which is often the basis for proofs of irrationality and even transcendence (see also the example 3.3.11 on continued fractions): From

$$0 < \left|1 - \sqrt{2}\right|^{2^m} = \left|p_m - q_m\sqrt{2}\right| < 1/2^{2^m}, \quad m \in \mathbb{N},$$

and $p_{m+1} - q_{m+1}\sqrt{2} = (p_m - q_m\sqrt{2})^2 = p_m^2 + 2q_m^2 - 2p_m q_m \sqrt{2}$, thus $p_0 = q_0 = 1, p_{m+1} = p_m^2 + 2q_m^2 \in \mathbb{N}^*$ and $q_{m+1} = 2p_m q_m \in \mathbb{N}^*$, follows not only the irrationality of $\sqrt{2}$, but also the excellent approximations $\sqrt{2} \approx p_m/q_m$ with

$$0 < p_m/q_m - \sqrt{2} < 1/q_m 2 2^{2^m}, \quad m \in \mathbb{N}^*.$$

By the way: Because of $p_{m+1}/q_{m+1} = (p_m^2 + 2q_m^2)/2p_m q_m = \frac{1}{2}(p_m/q_m + 2q_m/p_m)$ is $x_m := p_m/q_m$, $m \in \mathbb{N}^*$, the sequence of the Babylonian method of extracting roots for $a := 2$ according to example 3.3.7 and exercise 3.5.45, which begins with the initial value $x_0 := p_0/q_0 = 1/1 = 1$.—Similarly, one can conclude for any root $\sqrt{n}$, $n \in \mathbb{N}^*$ is not a square number.

1.8 Infinite Sets and Cardinal Numbers

the corresponding circle is $\geq 1/q$. In A_q, due to the above intersection condition, there are only finitely many points (namely at most $(q+1)/2$). Consequently, $A \cap [n, n+1] = \bigcup_{q \geq 1} A_q$ is again countable according to Theorem 1.8.4. The idea of this method, to show the countability of A, underlies many proofs of countability.— *By the way, every set $\mathcal{K}$ of pairwise disjoint circular disks in the plane $\mathbb{R}^2$ is necessarily countable.* A circular disk $K \in \mathcal{K}$ always contains a point $(x_K, y_K) \in \mathbb{Q}^2 \cap K$, and $\mathcal{K} \to \mathbb{Q}^2, K \mapsto (x_K, y_K)$, is injective and $\mathbb{Q}^2$ countable. $\diamond$

Every set A represents a **cardinal number** or a **Cardinality**, which we denote with

$$|A| \quad \text{or} \quad \text{Card}\, A$$

where two sets A and B represent the same cardinal number if there is a bijective mapping $A \xrightarrow{\sim} B$. This condition indeed defines an equivalence relation on any system of sets. The cardinal numbers of finite sets are represented by the initial segments $\mathbb{N}^*_{\leq n}$, $n \in \mathbb{N}$, of $\mathbb{N}^*$ and can be identified with the natural numbers. We set

$$|A| \leq |B|,$$

if there is an *injective* mapping $A \to B$. If $|A| \leq |B|$, but $|A| \neq |B|$, then we write

$$|A| < |B|.$$

In particular, $|A| \leq |B|$ for $A \subseteq B$. For finite B, the equality sign holds here only if $A = B$. However, if B is not finite, there are always *proper* subsets $A \subset B$ with $|A| = |B|$, see Ex. 1.8.3. Typical examples are the sets $\mathbb{N}^* \subset \mathbb{N} \subset \mathbb{Z} \subset \mathbb{Q}$, which all have the same cardinality $\aleph_0 = |\mathbb{N}|$. A proper part can thus be exactly as large as the whole. This phenomenon has long caused irritation in the treatment of infinite sets and even led (as still with Gauss) to the exclusion of actually infinite sets. It is $\aleph_0 = |\mathbb{N}| < |\mathbb{R}| = \aleph$, see theorem 1.8.12.

The $\leq$-relation for cardinal numbers is trivially reflexive and transitive. That it is also antisymmetric and thus an order relation is the content of the following famous Bernstein equivalence theorem, which is also called the **Bernstein-Schröder theorem** (named after F. Bernstein (1878–1956) and E. Schröder (1841–1902)) and can be proven in an astonishingly elementary way.

Theorem 1.8.16 (Bernstein Equivalence Theorem) *A and B be sets. If $|A| \leq |B|$ and also $|B| \leq |A|$, then $|A| = |B|$, i.e., the sets A and B are equinumerous.*

Proof Let $f : A \to B$ and $g : B \to A$ be injective mappings. For $x \in A$ we now set $x_0 := x$; $x_1 := g^{-1}(x_0)$, if $x_0 \in \text{Im}\, g$; $x_2 := f^{-1}(x_1)$, if $x_1 \in \text{Im}\, f$, etc. We define $\ell(x) := n \in \mathbb{N}$, if the sequence thus obtained $x_0, x_1, \ldots, x_n$ can no longer be extended, and $\ell(x) := \infty$, if an infinite sequence can be constructed in this way. Let

$$A_0 := \{x \in A \mid \ell(x) \text{ even}\}, \quad A_1 := \{x \in A \mid \ell(x) \text{ uneven}\},$$

$$A_\infty := \{x \in A \mid \ell(x) = \infty\}.$$

Then $A = A_0 \uplus A_1 \uplus A_\infty$. Similarly, $B = B_0 \uplus B_1 \uplus B_\infty$ with correspondingly defined sets $B_0, B_1, B_\infty \subseteq B$. It is $f(A_1) \subseteq B_0, f(A_0) = B_1$ and $f(A_\infty) = B_\infty$, furthermore $g(B_0) = A_1$. Thus, the mapping $h\colon A \to B$ with

$$h(x) := \begin{cases} f(x), & \text{if } x \in A_0 \uplus A_\infty, \\ g^{-1}(x), & \text{if } x \in A_1, \end{cases}$$

is bijective, and A and B are equinumerous. $\square$

Example 1.8.17 The following statement is a typical application of Bernstein's equivalence theorem: *The set $C = C_\mathbb{R}(\mathbb{R})$ of continuous functions $\mathbb{R} \to \mathbb{R}$ has the cardinality of the continuum. Proof.* The mapping $\mathbb{R} \to C$, which assigns to each real number a assigns the constant function a, is injective. Thus, $|\mathbb{R}| = \aleph \leq |C|$. The mapping $C \to \mathbb{R}^\mathbb{Q}$, which assigns to each continuous function $f\colon \mathbb{R} \to \mathbb{R}$ its restriction $f|\mathbb{Q}$ to $\mathbb{Q}$, is also injective, since a continuous function is uniquely determined by its values on $\mathbb{Q}$ (cf. Exercise 3.8.21: if $x = \lim x_n$ with $x \in \mathbb{R}$ and $x_n \in \mathbb{Q}, n \in \mathbb{N}$, then $f(x) = \lim f(x_n)$). Due to $|\mathbb{R}^\mathbb{Q}| = \aleph$, cf. Example 1.8.14, it also holds that $|C| \leq \aleph$. With Bernstein's equivalence theorem, it follows overall $|C| = \aleph$, as claimed.

We note that the set $\mathbb{R}^\mathbb{R}$ of all mappings from $\mathbb{R}$ into itself, due to $\{0,1\}^\mathbb{R} \subseteq \mathbb{R}^\mathbb{R}$ according to 1.8.9, has a greater cardinality than $\mathbb{R}$. From $|\mathbb{R}| = |\{0,1\}^\mathbb{N}|$ and $|\mathbb{N} \times \mathbb{R}| = |\mathbb{R}|$ follows however $|\mathbb{R}^\mathbb{R}| = |\{0,1\}^{\mathbb{N} \times \mathbb{R}}| = |\{0,1\}^\mathbb{R}| = |\mathfrak{P}(\mathbb{R})|$. $\Diamond$

The order $\leq$ on cardinal numbers is complete, i.e., any two cardinal numbers are comparable. In other words:

Theorem 1.8.18 (Comparison Theorem for Cardinal Numbers) *Let A and B be sets. Then there exists an injective mapping $A \to B$ or an injective mapping $B \to A$, i.e., it holds $|A| \leq |B|$ or $|B| \leq |A|$.*

Proof The proof is simple, but it uses Zorn's Lemma 1.4.15. We consider the set $\mathfrak{M}$ of triples, (M, f, N), where $M \subseteq A$ and $N \subseteq B$ is as well as $f\colon M \xrightarrow{\sim} N$ a bijective mapping. For $(M, f, N), (R, g, S) \in \mathfrak{M}$ we set

$$(M, f, N) \leq (R, g, S) \iff M \subseteq R,\ N \subseteq S \text{ and } f = \Gamma_f \subseteq \Gamma_g = g \text{ (that is } f = g|M\text{)}.$$

1.8 Infinite Sets and Cardinal Numbers

Trivially, this is an order on $\mathcal{M}$. This order is even (strictly) inductive: Namely, if (M_i, f_i, N_i), $i \in I$, a chain in $\mathcal{M}$, then (M, f, N) with $M := \bigcup_{i \in I} M_i$, $N := \bigcup_{i \in I} N_i$ and $\Gamma_f := \bigcup_{i \in I} \Gamma_{f_i}$ is the upper bound for this chain. According to Zorn's Lemma 1.4.15, $\mathcal{M}$ has a maximal element (M_0, f_0, N_0). It suffices to show that $M_0 = A$ or $N_0 = B$ is. However, if there were elements $a \in A - M_0$, $b \in B - N_0$, then (M_0', f_0', N_0') with $M_0' := M_0 \uplus \{a\}$, $N_0' := N_0 \uplus \{b\}$ and $\Gamma_{f_0'} := \Gamma_{f_0} \uplus \{(a, b)\}$ a truly larger element in $\mathcal{M}$ than (M_0, f_0, N_0). Contradiction! □

Set operations can be transferred to cardinal numbers. If $\alpha = |A|$ and $\beta = |B|$ are cardinal numbers, then one sets

$$\alpha + \beta := |A \uplus B|, \quad \alpha \cdot \beta := |A \times B| \quad \text{and} \quad \beta^\alpha = \left|B^A\right| \ (= |\text{Abb}(A, B)|).$$

For finite cardinal numbers, these operations coincide with the usual arithmetic operations on $\mathbb{N}$. For example, $\aleph = 2^{\aleph_0}$, $\aleph^n = \aleph^{\aleph_0} = \aleph$ for all $n \in \mathbb{N}^*$ and $\aleph^\aleph = 2^\aleph$, see the examples 1.8.14 and 1.8.17. In general, the sum and product of cardinal numbers are commutative and associative. Furthermore, the power rule $(\alpha^\beta)^\gamma = \alpha^{\beta \cdot \gamma}$ holds (due to $|(A^B)^C| = |A^{B \times C}|$ for arbitrary sets A, B, C, see Ex. 1.2.19). For every cardinal number α, $\alpha < 2^\alpha$ holds according to Theorem 1.8.9. Sum and product can be defined for arbitrary families $\alpha_i = |A_i|$, $i \in I$, of cardinal numbers:

$$\sum_{i \in I} \alpha_i := \left|\biguplus_{i \in I} A_i\right| \quad \text{and} \quad \prod_{i \in I} \alpha_i := \left|\prod_{i \in I} A_i\right|.$$

All these arithmetic operations are well-defined, i.e., independent of the chosen representatives A, B, A_i, $i \in I$, of the involved cardinal numbers.

We have already proven the equations $\aleph_0 \cdot \aleph_0 = \aleph_0$ and $\aleph \cdot \aleph = \aleph$, see Lemma 1.8.2 and Example 1.8.14. In general, the following important and perhaps surprising statement holds:

Theorem 1.8.19 (Product theorem for infinite sets) *For any infinite cardinal number α, it holds that $\alpha \cdot \alpha = \alpha$.*

Proof We have to show: If A is an infinite set, then $|A| \cdot |A| = |A \times A| = |A|$. For this, we consider the set $\mathcal{M}$ of pairs (M, f), where $M \subseteq A$ is an infinite subset of A and $f : M \to M \times M$ is a bijective mapping. Since A possesses a countably infinite subset, such pairs exist. For $(M, f), (N, g) \in \mathcal{M}$ we set

$$(M, f) \leq (N, g) \quad \text{exactly if } M \subseteq N \text{ and } g|M = f.$$

This evidently defines an order on $\mathcal{M}$. If (M_i, f_i), $i \in I$, is a non-empty chain in $\mathcal{M}$ and $M := \bigcup_{i \in I} M_i$, then $M \times M = \bigcup_{i \in I} M_i \times M_i$ and the bijective mappings $f_i : M_i \to M_i \times M_i$ define a bijective mapping $f : M \to M \times M$ with $f|M_i = f_i$, $i \in I$.

Consequently, (M, f) is an upper bound of the considered chain. $\mathcal{M}$ is thus strictly inductively ordered. According to Zorn's Lemma 1.4.15, $\mathcal{M}$ now has a maximal element (M_0, f_0). In particular, $|M_0| \cdot |M_0| = |M_0|$. However, in general, $M_0 \neq A$. (Construct examples for this!) However, it suffices to show that $|A - M_0| \leq |M_0|$; for then $|M_0| \leq |A| = |M_0| + |A - M_0| \leq 2 \cdot |M_0| \leq |M_0| \cdot |M_0| = |M_0|$, thus $|A| = |M_0|$ (since $\leq$ is an order for cardinal numbers according to Bernstein's equivalence theorem 1.8.16) and $|A \times A| = |A|$.

However, if $|A - M_0| \not\leq |M_0|$ were the case, then according to the comparison theorem 1.8.18 $|M_0| \leq |A - M_0|$ would hold. That is, there would be a subset $N_0 \subseteq A - M_0$ with $|N_0| = |M_0|$. Then

$$(M_0 \uplus N_0) \times (M_0 \uplus N_0) = (M_0 \times M_0) \uplus \big((M_0 \times N_0) \uplus (N_0 \times M_0) \uplus (N_0 \times N_0)\big)$$

and $|(M_0 \times N_0) \uplus (N_0 \times M_0) \uplus (N_0 \times N_0)| = 3 \cdot |N_0 \times M_0| = |N_0|$ would be the case. Thus, there would be a bijective mapping $f_1 : N_0 \to (M_0 \times N_0) \uplus (N_0 \times M_0) \uplus (N_0 \times N_0)$, which together with the mapping $f_0 : M_0 \to M_0 \times M_0$ would define a bijective extension $g_0 : M_0 \uplus N_0 \to (M_0 \uplus N_0) \times (M_0 \uplus N_0)$ of f_0, which contradicts the maximality of (M_0, f_0). □

Cardinal numbers are even well-ordered by $\leq$. We formulate this as follows:

Theorem 1.8.20 (Well-ordering of Cardinal Numbers) *Let α_i, $i \in I$, be a nonempty family of cardinal numbers. Then there exists an $i_0 \in I$ with $\alpha_{i_0} \leq \alpha_i$ for all $i \in I$.*

Proof Let $\alpha_i = |A_i|$ and A be a set with $A_i \subseteq A$ and $|A_i| < |A|$ for all $i \in I$. We choose a well-ordering on A according to Theorem 1.4.17. According to the comparison theorem below 1.8.25 for ordinal numbers, there is for each $i \in I$ a (uniquely determined) element $a_i \in A$ such that A_i (with respect to the well-ordering induced by A) is order-isomorphic and in particular equinumerous to the initial segment $A_{<a_i} = \{a \in A \mid a < a_i\}$. Since A is well-ordered, there is a $i_0 \in I$ with $a_{i_0} \leq a_i$ for all $i \in I$. Then $A_{<a_{i_0}} \subseteq A_{<a_i}$ and consequently $\alpha_{i_0} = |A_{i_0}| = |A_{<a_{i_0}}| \leq |A_{<a_i}| = \alpha_i$ for all $i \in I$. □

Remark 1.8.21 (Continuum Hypothesis) $\aleph_0 = |\mathbb{N}|$ is the smallest infinite cardinal number, and it is $\aleph_0 < \aleph (= 2^{\aleph_0} = |\mathbb{R}|)$. The so-called **(Cantor's) Continuum Hypothesis** states that there is no cardinal number α with $\aleph_0 < \alpha < \aleph$, in other words, that every uncountable subset of $\mathbb{R}$ is equinumerous to $\mathbb{R}$. Denote $\aleph_1$ as the successor of $\aleph_0$, i.e., the smallest cardinal number $> \aleph_0$, as it exists according to Theorem 1.8.20, then $\aleph_0 < \aleph_1 \leq \aleph$ and the Continuum Hypothesis states $\aleph_1 = \aleph$. According to K. Gödel (1906–1978), there is a model of axiomatic set theory, see Remark 1.1.1, in which the Continuum Hypothesis holds, and according to P. J. Cohen (1934–2007), there is also a model in which it does not hold. It is still a problem to find a *natural* (easily acceptable) axiom system for set theory in which the validity or invalidity of the Continuum Hypothesis can be proven. The **general Continuum**

1.8 Infinite Sets and Cardinal Numbers

Hypothesis states that generally 2^α is the successor of any infinite cardinal number α. Regarding the validity or invalidity of this general Continuum Hypothesis, the same applies as for the specific Continuum Hypothesis. ◊

Remark 1.8.22 (Ordinal Numbers) In this remark, which can be skipped on the first reading, we provide a brief introduction to the theory of ordinal numbers. It uses well-ordered sets, for whose basic properties we refer to the end of Sect. 1.4. Due to its great importance, we formulate the proof principle of Artinian induction 1.4.20 once again specifically for well-ordered sets: ◊

Theorem 1.8.23 (Transfinite Induction) *Let A be a well-ordered set. To each $a \in A$ an assertion $S(a)$ is assigned. Then, if for every $a \in A$ the assertion $S(a)$ holds under the assumption that $S(b)$ holds for all $b < a$, then $S(a)$ holds for all $a \in A$.*

An **ordinal number** is by definition the order type of a well-ordered set A, which is denoted by

$$\operatorname{Ord} A .$$

Two well-ordered sets A and B thus define the same ordinal number if A and B are order isomorphic. In particular, A and B are then equinumerous and define the same cardinal number. The mapping $\sigma = \operatorname{Ord} A \mapsto \operatorname{Card} A =: \operatorname{Card} \sigma$ from the ordinal numbers to the cardinal numbers is therefore well-defined and moreover surjective, since according to Theorem 1.4.17 every set can be well-ordered. Because two *finite* well-ordered (i.e., totally ordered) sets with the same number of elements are order isomorphic, this mapping provides a bijective correspondence of the set of finite ordinal numbers to the set of finite cardinal numbers, which can therefore both be identified with the set $\mathbb{N}$ of natural numbers.[35] One sets

$$\omega := \operatorname{Ord} \mathbb{N}.$$

It is thus $\operatorname{Card} \omega = \aleph_0$. Fundamental is the following lemma:

Lemma 1.8.24 *If A and B are isomorphic well-ordered sets, then there is exactly one order isomorphism $f : A \to B$. In particular, id_A is the only order isomorphism $A \to A$.*

Proof The proof is based on a lemma that is interesting in its own right:

Lemma *If A is a well-ordered set and $h : A \to A$ is a strictly monotonically increasing mapping, then $x \leq h(x)$ for all $x \in A$.—In particular, A is not order isomorphic to any of its open initial segments $A_{<a}$, $a \in A$.*

---

[35] This is probably the reason why in everyday life cardinal numbers and ordinal numbers are not always clearly distinguished.

For the *proof of the lemma*, let $C := \{x \in A \mid h(x) < x\}$. If $C \neq \emptyset$, then C would contain a smallest element c. By definition of C, it holds that $h(c) < c$. Since h is strictly monotonically increasing, it is then $h(h(c)) < h(c)$. On the other hand, since $h(c) \notin C$ is (due to the minimality of c), it is $h(c) \leq h(h(c))$. Contradiction!—To prove the addition in the lemma, we assume that there is a $a \in A$ strictly monotonically increasing mapping $h': A \to A_{<a}$. The composition $h := \iota \circ h': A \to A$ with the canonical injection $\iota: A_{<a} \to A$ is then a strictly monotonically increasing mapping with $h(a) < a$, which cannot exist according to what was shown before.

To prove Lemma 1.8.24, let $f, g: A \to B$ be order isomorphisms, i.e., strictly monotonically increasing bijective mappings. Then $f^{-1}g, g^{-1}f: A \to A$ are also strictly monotonically increasing. According to the auxiliary theorem, $x \leq (f^{-1}g)(x)$ and $x \leq (g^{-1}f)(x)$ hold for all $x \in A$. It follows $f(x) \leq f((f^{-1}g)(x)) = g(x)$ and analogously $g(x) \leq f(x)$ Thus, $f(x) = g(x)$ holds for all $x \in A$.

The following statement is the basis for the comparison of ordinal numbers.

Theorem 1.8.25 (Comparison Theorem for Ordinal Numbers) *Let A and B be well-ordered sets. Then exactly one of the following three cases occurs:*

(1) *A and B are order isomorphic.*
(2) *There exists a $b \in B$ and an order isomorphism $A \xrightarrow{\sim} B_{<b}$. Here, b and the isomorphism are uniquely determined.*
(3) *There exists a $a \in A$ and an order isomorphism $B \xrightarrow{\sim} A_{<a}$. Here, a and the isomorphism are uniquely determined.*

Proof The uniqueness statements are clear according to Lemma 1.8.24 and the auxiliary theorem used therein. To prove the existence statements, we consider the set A' of those $a' \in A$, for which there exists a $b' \in B$ and an order isomorphism $A_{<a'} \to B_{<b'}$. The element $f(a') := b' \in B$ is then (like the isomorphism) uniquely determined by a'. The mapping $f: A' \to B$ defined in this way is strictly monotonically increasing, and A' contains with each element a' all elements $\leq a'$. Thus, it is either $A' = A$ or $A' = A_{<a}$ for a $a \in A$. The same applies to the $\operatorname{Im} B' := f(A')$. If $A' = A$, then (1) or (2) holds. Similarly, (1) or (3) holds if $B' = B$ is the case. However, if $A' = A_{<a}$ and $B' = B_{<b}$, then $f: A' \to B'$ is an order isomorphism and $a \in A'$. Contradiction! □

If A and B are well-ordered sets that represent the ordinal numbers $\sigma = \operatorname{Ord} A$ and $\tau = \operatorname{Ord} B$, then we set

$$\sigma \leq \tau,$$

if A is order isomorphic to B or an open initial segment $B_{<b}$ of B. According to the comparison theorem 1.8.25, this is exactly the case if there is a strictly monotonically increasing mapping $A \to B$. The comparison theorem further shows

1.8 Infinite Sets and Cardinal Numbers

that this provides a complete order for ordinal numbers. $\omega = \operatorname{Ord} \mathbb{N}$ is the smallest infinite ordinal number. If A is a well-ordered set, then $\operatorname{Ord}(A_{<a}) < \operatorname{Ord} A$ for every $a \in A$. If A is infinite, then there are indeed proper subsets $A' \subset A$ with $\operatorname{Ord} A' = \operatorname{Ord} A$. For finite ordinal numbers, the order defined here coincides with the natural order on $\mathbb{N}$. Analogous to the well-ordering theorem 1.8.20 for cardinal numbers, one proves the following well-ordering theorem for ordinal numbers, which we leave to the reader.

Theorem 1.8.26 (Well-ordering of Ordinal Numbers) *Let σ_i, $i \in I$, be a nonempty family of ordinal numbers. Then there exists a $i_0 \in I$ with $\sigma_{i_0} \leq \sigma_i$ for all $i \in I$.*

Certain operations for well-ordered sets can be transferred to ordinal numbers. If $\sigma = \operatorname{Ord} A$ and $\tau = \operatorname{Ord} B$ are ordinal numbers, then one defines

$$\sigma + \tau := \operatorname{Ord}(A \uplus B), \quad \sigma \cdot \tau := \operatorname{Ord}(B \times A).$$

The disjoint union $A \uplus B$ carries the sum order with $a < b$ for $a \in A$, $b \in B$ and the product $B \times A$ the lexicographic order. Note the changed order of factors in the definition of the product (which already goes back to Cantor). Addition and multiplication of ordinal numbers are evidently associative. For every ordinal number σ, $\sigma + 1 (\neq \sigma)$ is the direct successor of σ. If σ is infinite, then $1 + \sigma = \sigma$. *Thus, the addition of ordinal numbers is not commutative. It is $\sigma < \sigma + \tau$ for every ordinal number $\tau \neq 0$.* Also, multiplication is not commutative. Thus, $2 \cdot \omega = \omega$, but $\omega \cdot 2 = \omega + \omega \neq \omega$:

$$2 \cdot \omega \colon (0, 1) < (0, 2) < (1, 1) < (1, 2) < (2, 1) < (2, 2) < \cdots$$
$$\omega \cdot 2 \colon (1, 0) < (1, 1) < (1, 2) < \cdots < (2, 0) < 2, 1) < (2, 2) < \cdots.$$

The equation $\omega = 2 \cdot \omega = (1 + 1) \cdot \omega \neq 1 \cdot \omega + 1 \cdot \omega = \omega + \omega$ shows that the right distributive law does not hold. On the other hand, the left distributive law $\rho \cdot (\sigma + \tau) = \rho \cdot \sigma + \rho \cdot \tau$ holds for all ordinal numbers σ, τ. Proof!

Powers of ordinal numbers are initially not defined, because the power $B^A = \operatorname{Abb}(A, B)$ of well-ordered sets with the lexicographic order is no longer well-ordered if A is infinite and B contains at least two elements. However, one can define the powers through transfinite recursion. For this, one first needs the concept of the **limit** $\lim_{i \in I} \sigma_i$ of a family σ_i, $i \in I$, of ordinal numbers. We can represent the σ_i by initial segments $A_{<a_i}$ of one and the same well-ordered set A. Then

$$\lim_{i \in I} \sigma_i := \operatorname{Ord} \bigcup_{i \in I} A_{<a_i}.$$

For example, $\omega = \lim_{n \in \mathbb{N}} n$. More generally, if the well-ordered set $A \neq \emptyset$ has no greatest element, then $\sigma = \operatorname{Ord} A = \lim_{a \in A} \operatorname{Ord} A_{<a} = \lim_{\tau < \sigma} \tau$. Ordinal numbers $\sigma \neq 0$ of this kind are called **limit numbers**. They are necessarily infinite. Every non-limit number $\sigma \neq 0$ is a direct successor $\sigma = \sigma' + 1$.

We can now define the **powers** σ^τ of ordinal numbers recursively as follows:

$$\sigma^0 = 1; \quad \sigma^{\tau+1} = \sigma^\tau \cdot \sigma; \quad \sigma^\tau = \lim_{0 < \rho < \tau} \sigma^\rho, \text{ if } \tau \text{ limit number.}$$

If $\tau = n$ is finite, then $\sigma^\tau = \sigma^n$ is the n-fold product of σ with itself. One should try to develop a conception of $\sigma^\omega = \lim_{n \in \mathbb{N}} \sigma^n$. The ordinal number ω^ω is evidently countable. It must not be confused with the order type of $\mathbb{N}^\mathbb{N}$ with respect to the lexicographic order. The latter is uncountable and equal to the order type of the half-open real interval $[0, 1[$ or of $\mathbb{R}_+$ (and in particular not a well-order), see Ex. 1.8.14b).

We have used the principle of transfinite recursion for the definition of powers. It proceeds completely analogously to the recursive definition of sequences $\mathbb{N} \to X$ with values in a set X, see Sect. 1.5, and is now proven analogously as there through transfinite induction.

Theorem 1.8.27 (Transfinite Recursion) *Let A be a well-ordered set. Furthermore, let $h_a \colon X^{A_{<a}} \to X$, $a \in A$, be a family of mappings* (**recursion equations**). *In particular, if $A \neq \emptyset$ is, for the smallest element $0 \in A$ the mapping $h_0 \colon \{\emptyset\} \to X$ is given by an element $x_0 \in X$* (**recursion start**). *Then there is exactly one mapping $H \colon A \to X$ with $H(a) = h_a(H|A_{<a})$ for all $a \in A$.*

One could also define the product $\sigma \cdot \tau$ of ordinal numbers recursively, namely through the following scheme:

$$\sigma \cdot 0 = 0; \quad \sigma \cdot (\tau+1) = \sigma \cdot \tau + \sigma; \quad \sigma \cdot \tau = \lim_{\rho < \tau} \sigma \cdot \rho, \text{ if } \tau \text{ limit number.}$$

We have already noted several times that an infinite set possesses multiple non-isomorphic well-orderings. It even holds:

Theorem 1.8.28 *Let α be an infinite cardinal number and the cardinal number β the direct successor of α. Then the order type of the (well-ordered) set of well-orderings with cardinal number α is the smallest ordinal number with cardinal number β. Consequently, there are exactly β pairwise non-isomorphic well-orderings on a set with cardinal number α.*

Proof Let B be a set with cardinality β, equipped with the smallest well-ordering for the cardinality β, see Proposition 1.8.26. (This is a limit number.) Then $|B_{<b}| < \beta$ for all $b \in B$, and every well-ordering for the cardinality α is isomorphic to exactly one segment $B_{<b}$, $b \in B$, of B, see Proposition 1.8.25. Therefore, the ordered set of ordinals with cardinality α is isomorphic to the set $B_\alpha := \{b \in B \mid |B_{<b}| = \alpha\}$, equipped with the order induced by B. This is, again according to the comparison proposition 1.8.25, isomorphic to a segment of B or to B itself. If c is the smallest element of B_α, then $|B_{<c}| = \alpha$ and $B_\alpha = B_{\geq c}$. Because $\alpha < \beta$ and $B = B_{<c} \uplus B_{\geq c}$ is $\beta = |B| = |B_{<c}| + |B_{\geq c}| = \text{Max}(|B_{<c}|, |B_{\geq c}|) = |B_{\geq c}|$, see Exercise 1.8.15a), and B_α cannot be isomorphic to a segment of B, as this has a cardinality $< \beta$. □

1.8 Infinite Sets and Cardinal Numbers

The set of ordinal numbers for a fixed cardinal number α is called the **number class** of α, and its smallest element is called the **initial number** of α. The number class of $\aleph_0$ is called the **first number class**. According to theorem 1.8.28, its cardinal number is the successor $\aleph_1$ of $\aleph_0$. The initial number of $\aleph_0$ is ω. The continuum hypothesis $\aleph = \aleph_1$, see remark 1.8.21, can thus be formulated as follows: The set $\mathbb{N}$ has continuum many pairwise non-isomorphic well-orderings.

Exercises

Exercise 1.8.1 Show that the mapping $\mathbb{R} \to]-1, 1[$, $x \mapsto x/(1+|x|)$, which was used in the proof of theorem 1.8.12, is bijective.

Exercise 1.8.2 Justify that the set $\mathbb{R}$ of all real numbers and the set $\mathbb{R} - \mathbb{Q}$ of irrational numbers are equinumerous (Example 1.8.11). □

Exercise 1.8.3 For a set A, the following are equivalent: (i) A is infinite (i.e., not finite). (ii) There exists a proper subset of A that is equinumerous with A. (iii) There exists an injective mapping $A \to A$ that is not surjective. (iv) There exists a surjective mapping $A \to A$ that is not injective. (R. Dedekind (1831–1916) used these characterizations for the *definition* of infinite sets.)

Exercise 1.8.4 The set of (infinite) sequences with elements from a set with at least two elements is uncountable.

Exercise 1.8.5 Let $f : A \to B$ be a mapping. If B is countable and all fibers of f are countable, then A is also countable.

Exercise 1.8.6 The mapping $\mathbb{N} \times \mathbb{N} \to \mathbb{N}$ of the Cauchy diagonal method is explicitly given by $(m, n) \mapsto \frac{1}{2}(m+n+1)(m+n) + n$. One proves directly that this mapping is bijective.

Exercise 1.8.7 In generalization of Ex. 1.8.6, one proves for $k \geq 1$ the bijectivity of the mapping $f_k : \mathbb{N}^k \to \mathbb{N}$ with

$$(m_1, m_2, \ldots, m_k) \mapsto \binom{m_1}{1} + \binom{m_1 + m_2 + 1}{2} + \cdots + \binom{m_1 + \cdots + m_k + k - 1}{k}.$$

One obtains the preimage $f_k^{-1}(n)$ for $n \in \mathbb{N}$ recursively through the following greedy algorithm: One determines the largest $\ell \in \mathbb{N}$ with $\binom{\ell+k-1}{k} \leq n$ and then $f_{k-1}^{-1}\left(n - \binom{\ell+k-1}{k}\right)$.

Exercise 1.8.8 One shows constructively without using Bernstein's equivalence theorem that the sets $\mathbb{R}$ and $\mathbb{R} \times \mathbb{N}$ are equinumerous. □

Exercise 1.8.9 Let $B \subseteq \mathbb{R}^2$ be a countable set and R_n, $n \in \mathbb{N}$, a sequence of positive real numbers with $\lim R_n = 0$.[36] Then there is a sequence K_n of pairwise disjoint circles (as in Footnote 32 to Example 1.8.15) in $\mathbb{R}^2$ with the radii R_n, $n \in \mathbb{N}$, whose union encompasses B. (For example, if one chooses $B := \mathbb{Q}^2$, then there is no longer a circle that lies entirely in the complement of $\bigcup_{n \in \mathbb{N}} K_n$. However, it is not possible to cover the entire plane $\mathbb{R}^2$ with pairwise disjoint circles, see Exercise 3.4.20.)

Exercise 1.8.10 The set of polynomials with rational coefficients is countable, while that of polynomials with arbitrary real (or complex) coefficients has the cardinality of the continuum.

Exercise 1.8.11 A barber claims he shaves exactly the men in his village who do not shave themselves. Show that the barber is lying with this statement or is a woman. (See Proposition 1.8.9.)

Exercise 1.8.12

a) The set $\mathfrak{E}(\mathbb{N})$ of finite subsets of $\mathbb{N}$ possesses a well-ordered cofinal subset (with respect to inclusion), see Exercise 1.4.6.
b) If X is an uncountable infinite set, then there is no cofinal chain in the set $\mathfrak{E}(X)$ of finite subsets of X. (Every chain $\mathcal{K} \subseteq \mathfrak{E}(X)$ is well-ordered and countable; because $K \mapsto |K|$ is an injective strictly monotone mapping $\mathcal{K} \to \mathbb{N}$.—Note that $\mathfrak{E}(X)$ is a lattice and in particular directed upwards. Every infinite chain in $\mathfrak{E}(X)$ is weakly cofinal.)
c) If A is a countable lattice, then A possesses a well-ordered cofinal subset. (If $A = \{a_0, a_1, a_2, \ldots\}$, then $\{\mathrm{Sup}(a_0, \ldots, a_n) = a_0 \vee \cdots \vee a_n \mid n \in \mathbb{N}\}$ is such a subset.)
d) If the ordered set A possesses a countable totally ordered cofinal subset, then there exists in A a strictly monotonically increasing (finite or infinite) cofinal sequence $a_0 < a_1 < a_2 < \cdots$.
e) There are well-ordered sets without countable cofinal subsets. (Examples are the well-ordered sets that represent the initial number of a cardinal number α', where α' is the successor of an infinite cardinal number α, see the remarks following the proof of Theorem 1.8.28.) □

Exercise 1.8.13 Let $(A, \leq)$ be an ordered set. A subset $D \subseteq A$ is called **dense in A**, if for any two elements $a, b \in A$ with $a < b$ there exists a $d \in D$ with $a < d < b$. The order $\leq$ is called **dense** if A is dense in A. If D is dense in A, then D is also dense in every subset of A that includes D. If D is dense in A and D' is dense in D, then D' is dense in A. If A is dense and A has two distinct comparable elements, then A is infinite. $\mathbb{Q}$ and (e.g.) also the set $\mathbb{Z}_2 := \{a/2^n \mid a \in \mathbb{Z}, n \in \mathbb{N}\}$ is dense in $\mathbb{R}$.

---

[36] For each $\varepsilon > 0$ there is thus a $n_0 \in \mathbb{N}$ with $R_n \leq \varepsilon$ for all $n \geq n_0$, see Definition 3.2.1.

1.8 Infinite Sets and Cardinal Numbers

a) For every countable totally ordered set A, there is a strictly monotonically increasing mapping $f : A \to \mathbb{Q}$, i.e., every countable totally ordered set is order-isomorphic to a subset of $\mathbb{Q}$. (Use an enumeration $a_0, a_1, a_2, \ldots$ of the elements of A and define f recursively.)

b) Exactly then is A order-isomorphic to $\mathbb{Q}$, if A possesses the following properties: (1) A is non-empty and countable. (2) A is dense. (3) A is completely ordered. (4) A has neither a smallest nor a largest element.—In particular, every countable subset of $\mathbb{R}$, which includes $\mathbb{Q}$, is order-isomorphic to $\mathbb{Q}$.

c) If A is order-isomorphic to $\mathbb{Q}$ and $E \subseteq A$ is a finite subset of A, then $A - E$ is also order-isomorphic to $\mathbb{Q}$.

d) It follows that $\mathbb{R}$ is not countable. (If $\mathbb{R}$ were countable, then $\mathbb{R}$ would be order-isomorphic to $\mathbb{Q}$ and thus also to $\mathbb{R} - \{0\}$. Show that this is not possible, using the definition 3.3.1.)

Exercise 1.8.14

a) The mapping $2^{\mathbb{N}} \to [0, 1]$, $(\varepsilon_n) \mapsto \sum_{n=0}^{\infty} 2\varepsilon_n/3^{n+1}$ is a strictly monotonically increasing mapping of the sequence space $2^{\mathbb{N}}$, equipped with the lexicographical order, into the real unit interval $[0, 1]$ (with the natural order), whose image is the so-called Cantor discontinuum $\mathcal{C}$, see Ex. 3.4.19. $\mathcal{C}$ and $2^{\mathbb{N}}$ thus have the same order type.

b) The set $\mathbb{N}^{\mathbb{N}}$ of infinite sequences of natural numbers with the lexicographical order is of the same order type as the real half-open interval $[0, 1[$. (For example, $(c_n)_{n \in \mathbb{N}} \mapsto \sum_{n=0}^{\infty} 1/2^{c_0 + \cdots + c_n + n + 1}$ is, according to the dual development of real numbers, see Example 3.3.10, a strictly monotonically decreasing bijective mapping $\mathbb{N}^{\mathbb{N}} \to]0, 1]$.)

Exercise 1.8.15

a) If α and β are cardinal numbers $\neq 0$, of which at least one is infinite, then $\alpha + \beta = \alpha \cdot \beta = \mathrm{Max}\,(\alpha, \beta)$.

b) If A is an infinite set and $B \subseteq A$ is a subset with $|B| < |A|$, then $|A| = |A - B|$. (See example 1.8.11 for the case where B is countable.)

Exercise 1.8.16

a) For every infinite cardinal number α and $n \in \mathbb{N}^*$ it holds that $\alpha^n = \alpha$.

b) For every infinite set A it holds that $|A| = |\mathfrak{E}(A)|$. (In the case that A is countably infinite, see example 1.8.8.)

Exercise 1.8.17 Let $f : A \to B$ be a mapping with infinite B and $|f^{-1}(b)| \leq |B|$ for all $b \in B$. Then $|A| \leq |B|$. It follows: If $A = \bigcup_{i \in B} A_i$ with $|A_i| \leq |B|$, then $|A| \leq \sum_{i \in B} |A_i| \leq |B|$.

Exercise 1.8.18 If A is infinite and I is a non-empty set with $|I| \leq |A|$, then there is a partition A_i, $i \in I$, of A with $|A_i| = |A|$ for all $i \in I$.

Exercise 1.8.19 Let α, β be cardinal numbers with $2 \leq \alpha \leq \beta$ and β infinite. Then $\alpha^\beta = 2^\beta$.

Exercise 1.8.20 Let σ, τ, ρ be ordinal numbers. Prove by transfinite induction over ρ the power calculation rule $(\sigma^\tau)^\rho = \sigma^{\tau \cdot \rho}$.

Algebraic Foundations

2.1 Monoids and Groups

Becoming familiar with addition and multiplication in the number ranges $\mathbb{N}, \mathbb{Z}, \mathbb{Q}$ (and $\mathbb{R}$) is a desirable goal already in elementary education. Sets on which one or more arithmetic operations (such as addition and multiplication) are defined with certain calculation rules play a significant role in all areas of mathematics. It is an important exercise of algebra to study such sets with connections from a general perspective. First, we define:

Definition 2.1.1 An **operation** on a set A is a mapping $A \times A \to A$.

For arbitrary connections, we use neutral connection symbols such as $*$ or $\diamond$ etc. and suggestively write $a * b$ or $a \diamond b$ for the image of $(a, b) \in A \times A$ under the given connection $A \times A \to A$. However, we often leave out the connection symbol altogether and use the so-called **multiplicative notation** ab for $a * b$. Then one generally speaks of a **multiplication** and calls ab the **product** of a and b as well as a and b themselves its **factors.** Unless otherwise stated, these are our standard notations for a connection. When using the sum symbol $+$ as a connection symbol, the connection is called an **addition** and $a + b$ the **sum** with the **addends** a and b (**additive notation**). A set A with a connection $*$ is called a **magma** (or also a **connection structure**) $A = (A, *)$.

Let $(A, *)$ be a magma and $A' \subseteq A$ a subset of A. If $a * b \in A'$ for all $a, b \in A'$, then $*$ defines by restriction to $A' \times A'$ a composition on A', which is called the **induced composition** and is again denoted by $*$. In this case, $A' = (A', *)$ is called a **submagma** of $A = (A, *)$.

The reader should always consider a link $A \times A \to A$ on a set A also as a square $A \times A$ matrix with coefficients in A and the partial mappings

$$L: A \to A^A, \ a \mapsto (L_a: x \mapsto ax) \quad \text{or} \quad R: A \to A^A, \ b \mapsto (R_b: x \mapsto xb),$$

Fig. 2.1 Linkage table

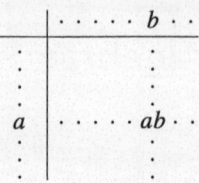

cf. exercise 1.2.19. The row mapping $L_a\colon A \to A$ is called the **left translation** with $a \in A$ in A and the column mapping $R_b\colon A \to A$ the **right translation** with $b \in A$ in A. The matrix $(ab)_{a,b \in A}$ is also called the associated **linkage table**. In its a-th row and b-th column is the product ab, cf. Fig. 2.1.

In general, only links that satisfy certain calculation rules are of interest. In the following definition, we introduce some of the most important of these rules.

Definition 2.1.2 Let $*$ be a combination on the set A.

(1) The combination $*$ is called **associative**, if for all $a, b, c \in A$ applies:

$$(a * b) * c = a * (b * c).$$

(2) Elements $a, b \in A$ **commute**, if $a * b = b * a$ is. The combination $*$ is called **commutative**, if any two elements of A commute, i.e. if for all $a, b \in A$ applies:

$$a * b = b * a$$

(3) An element $e \in A$ is called a **neutral element** (with respect to $*$), if for all $a \in A$ applies:

$$e * a = a * e = a.$$

Apparently applies:

(1) The following statements are equivalent: (i) The operation $*$ is associative. (ii) It holds $L_{a*b} = L_a \circ L_b$ for all $a, b \in A$. (iii) It holds $R_c \circ R_b = R_{b*c}$ for all $b, c \in A$. (iv) It holds $R_c \circ L_a = L_a \circ R_c$ for all $a, c \in A$.

(2) The operation $*$ is commutative if and only if $L_a = R_a$ holds for all $a \in A$. If the commutative operation is written additively, the translation $L_a = R_a$ is occasionally also denoted by T_a and referred to as the **shift** by a. It is then $T_a(x) = a + x = x + a$ for all $a, x \in A$.

(3) $e \in A$ is exactly a neutral element when $L_e = R_e = id_A$ is. If $L_e = id_A$ applies for a $e \in A$, then e is called **left-neutral.** Correspondingly, e is called **right-neutral** when $R_e = id_A$ is. e is exactly a neutral element when it is both left-neutral and right-neutral. *If A has a neutral element e, it is uniquely determined.* If $e' \in A$ is another neutral element, then $e' = ee' = e$. The first equation applies because e is left-neutral, and the second because e' is right-neutral. If it exists, we also denote the uniquely determined neutral element of A with

2.1 Monoids and Groups

$$e_A,$$

when the reference to A should be made clear.

Definition 2.1.3 Let $A = (A, *)$ be a magma, i.e., a set A with an operation $*$.

(1) A is called a **semigroup** if the operation $*$ is associative. If it is also commutative, we speak of a **commutative** or **abelian semigroup**.
(2) A is called a **monoid** if A is a semigroup with a neutral element. If the operation is also commutative, we speak of a **commutative** or **abelian monoid**.

A submagma of a (commutative) semigroup is again a (commutative) semigroup. This is then referred to as a **subsemigroup**. A submagma of a monoid is a subsemigroup, but not necessarily a monoid. However, it is a monoid if it contains the neutral element of the monoid. In this case, it is referred to as a **submonoid**. A submonoid M' of a monoid M therefore always has the same neutral element by definition as M. Note that a magma or a semigroup always contains the empty magma or the empty semigroup as a subobject. If the semigroup A is commutative, the translations are pairwise interchangeable: $L_a \circ L_b = L_{a*b} = L_{b*a} = L_b \circ L_a$ for all $a, b \in A$. If A is a monoid, the reverse also applies. However, this is generally not the case for arbitrary semigroups. (Example?)

In multiplicative notation, the neutral element of a monoid M is also called the **identity element** of M and is denoted by

$$1 = 1_M$$

The additive notation is usually only used for commutative semigroups. (An exception is the addition of ordinal numbers in Remark 1.8.22.) The neutral element of a commutative monoid is usually called the **zero element**

$$0 = 0_M$$

of M. So it follows that $0 + a = a = a + 0$ for all $a \in M$.

Definition 2.1.4 Let M be a (multiplicative) monoid with neutral element e and $a \in M$. An element $a' \in M$ is called an a **inverse element** if

$$aa' = a'a = e$$

is. a' is called **right-inverse** to a, if $aa' = e$ is, and **left-inverse**, if $a'a = e$ is. An element is thus an inverse element to a if it is both right- and left-inverse.

The inverse element to a as in a' is uniquely determined by Definition 2.1.4 *through $a \in M$. If $a'' \in M$ also fulfills the condition for an inverse element to $a \in M$, then $a' = a'e = a'(aa'') = (a'a)a'' = ea'' = a''$. The second equation holds because a'' is a right inverse to a, and the fourth because a' is a left inverse to a. It follows that: If $a \in M$ has a right inverse $a'' \in M$ and a left inverse $a' \in M$, then $a' = a''$ and a has an inverse.*

Let M continue to be a monoid. $a \in M$ has a right inverse (or a left inverse) if and only if the left translation L_a (or the right translation R_a) is surjective. a has an inverse *only if* a', *if the left translation L_a or the right translation R_a is bijective. In this case*, $L_a^{-1} = L_{a'}$ *and* $R_a^{-1} = R_{a'}$. If $a' \in M$ is inverse to a, then $id_M = L_e = L_{aa'} = L_a L_{a'}$ and likewise $id_M = L_e = L_{a'a} = L_{a'} L_a$. Conversely, if L_a is bijective, then a has a right inverse $a' = L_a^{-1}(e)$, and it is $id_M = L_e = L_{aa'} = L_a L_{a'}$. Since L_a is bijective, it is necessarily $L_{a'} = L_a^{-1}$, and consequently $a'a = L_{a'}(a) = L_a^{-1}(a) = e$. The same conclusion is drawn for the right translations.

In multiplicative notation, the inverse to $a \in M$ (if it exists) is denoted by

$$a^{-1}$$

and in additive notation, it is usually denoted by

$$-a.$$

$-a$. The **negative** of a is thus denoted. So it is $(-a) + a = 0 = a + (-a)$. For any $b \in M$, in additive notation, we write briefly

$$b - a := b + (-a)$$

and call this element the **difference** of b and a.[1]

If $a, b \in M$ commutes and a is invertible, then a^{-1} and b also commute. From $ab = ba$ it follows first $b = eb = a^{-1}ab = a^{-1}ba$ and then $ba^{-1} = (a^{-1}b)aa^{-1} = a^{-1}b$. If a' is inverse to a, then naturally a is inverse to a', so $(a^{-1})^{-1} = a$ in multiplicative and $-(-a) = a$ in additive notation. Moreover, the neutral element e is invertible with inverse e. If $a, b \in M$ is invertible, then so is the product ab with

$$(ab)^{-1} = b^{-1}a^{-1}.$$

This is simply the calculation rule $L_{ab}^{-1} = (L_a L_b)^{-1} = L_b^{-1} L_a^{-1} = L_{b^{-1}a^{-1}}$ for the corresponding invertible left translations L_a and L_b, cf. Proposition 1.2.16. Overall, the invertible elements of the monoid M form a submonoid of M, which we denote with

$$M^\times$$

The elements of $M^\times$ are called the **units** of M. The monoid M is called **sharp**, if $M^\times = \{e\}$ is, i.e., if the neutral element of M is the only unit in M. In the submonoid $M^\times \subseteq M$, every element is invertible. This situation deserves its own name:

Definition 2.1.5 A **group** is a monoid in which every element is invertible. A set $G = (G, *)$ with an operation $*$ is therefore exactly a group when the following applies:

---

[1] We remind you that additive notation is usually only used for commutative operations.

2.1 Monoids and Groups

(1) The operation is associative (i.e., it is $(a*b)*c = a*(b*c)$ for all $a, b, c \in G$).
(2) G has a neutral element e (with $e*a = a = a*e$ for all $a \in G$).
(3) Every element $a \in G$ has an inverse a^{-1} (with $a*a^{-1} = e = a^{-1}*a$).

The cardinal number $|G|$ of a group G is called the **order** of G and is also denoted by Ord G. The invertible elements of a monoid M form the **group** $M^\times$ **of invertible elements** or the **unit group** of M. The monoid M is a group if and only if $M = M^\times$ is. If the operation of a group is commutative, we speak of a **commutative** or **abelian group**. A submonoid of a group G, which is itself a group (with the induced operation), is called a **subgroup** of G. We prove the following simple criterion for this:

Proposition 2.1.6 (Subgroup Criterion) *Let G be a (multiplicatively written) group with neutral element e. A subset H of G is a subgroup of G if and only if the following holds:*

(1) $e \in H$.   (2) *With $a, b \in H$ it is also $ab \in H$ and $a^{-1} \in H$.*

Proof It is clear that H is a subgroup of G if H fulfills conditions (1) and (2). Conversely, let H be a subgroup of G. We have to show that the neutral element e_H of H coincides with e and that the inverse a' of an element $a \in H$ in the subgroup H necessarily is the inverse a^{-1} of a in the group G. But it is $e = e_H e_H^{-1} = e_H e_H e_H^{-1} = e_H e = e_H \in H$. Furthermore, from $aa' = e = a^{-1}a$ the equations $a' = ea' = (a^{-1}a)a' = a^{-1}(aa') = a^{-1}e_H = a^{-1}e = a^{-1}$ follow, so $a' = a^{-1}$. □

Condition (2) in 2.1.6 can obviously be formulated as follows if (1) is valid:

(2′) *With $a, b \in H$ it is also $ab^{-1} \in H$.*

Here, condition (1) can be replaced by the condition "(1′) $H \neq \emptyset$". From Proposition 2.1.6 it immediately follows:

Lemma 2.1.7 *The intersection of any family of subgroups of a group is again a subgroup.*

The subgroups of a given group G thus form a complete lattice (with respect to inclusion). The supremum of a family F_i, $i \in I$, of subgroups is the intersection of those subgroups of G that include all F_i (to which G itself belongs). The same naturally applies to the set of submagmas or subsemigroups or submonoids of a magma or a semigroup or a monoid.

We continue to discuss semigroups, monoids, and groups, preferring the multiplicative notation as before. The associative law states that in a product with three factors a, b, c from a semigroup, the brackets can be omitted: $abc := (ab)c = a(bc)$. This also applies to products with any finite number of factors: *If* $x_1, \ldots, x_n$, $n \geq 1$, *elements of a semigroup H, then the product is* $p_n = x_1 \cdots x_n$ *independent of how it is bracketed when calculating.* This **general associative law** follows from the special case $n = 3$ by induction over n. The proof proceeds analogously to the proof of Theorem 1.2.13, replacing the images $f_1, \ldots, f_n$ by the elements $x_1, \ldots, x_n$. However, the statement can also be traced back directly to Theorem 1.2.13: After possible adjunction of a neutral element e (cf. exercise 2.1.7) one can assume that H is a monoid. Then $L_x(e) = x$ for all $x \in M$ and the statement follows from the fact that $L_x \circ L_y = L_{xy}$ for all $x, y \in H$ and $L_{x_1} \circ \cdots \circ L_{x_n}$ is independent of the bracketing according to Sentence 1.2.13. For the above n-fold product p_n, the product symbol

$$\prod_{i=1}^{n} x_i = x_1 \cdots x_n.$$

is used. In additive notation, the sum symbol

$$\sum_{i=1}^{n} x_i = x_1 + \cdots + x_n.$$

is used. If the considered semigroup M is even a monoid, thus possessing a neutral element e, the so-called **empty product** is defined as the neutral element e for $n = 0$. So,

$$\prod_{i=1}^{0} x_i = e \quad \text{(or if additively written } \sum_{i=1}^{0} x_i = 0\text{)}.$$

holds. For a *completely ordered* finite index set I and a family x_i, $i \in I$, one sets

$$x^I = \prod_{i \in I} x_i := x_{i_1} \cdots x_{i_n},$$

when $i_1 < \cdots < i_n$, $n \in \mathbb{N}$, are the elements of I in the given order. In additive notation, it is common to write the sum in the form $x_I = \sum_{i \in I} x_i$. If the elements $x_1, \ldots, x_n$, $n \in \mathbb{N}$, are invertible in M, then obviously (induction over n)

$$(x_1 \cdots x_n)^{-1} = \left(\prod_{i=1}^{n} x_i\right)^{-1} = \prod_{i=1}^{n} x_{n+1-i}^{-1} = x_n^{-1} \cdots x_1^{-1}.$$

holds. If the semigroup or the monoid is abelian, then the product $x_1 \cdots x_n$ does not only not depend on the bracketing, but also not on the order of the factors $x_1, \ldots, x_n$. So, for every permutation $\sigma \in \mathfrak{S}_n$ it holds.

$$\prod_{i=1}^{n} x_i = \prod_{i=1}^{n} x_{\sigma i}.$$

This **general commutative law** is proven in the following steps: Initially, the statement is true by definition of commutativity if σ is a transposition that swaps

2.1 Monoids and Groups

$$
\begin{array}{llll}
x_{11}+ & x_{12}+\cdots+ & x_{1n} & \sum_{j=1}^{n}x_{1j} \\
+x_{21}+ & x_{22}+\cdots+ & x_{2n} & \sum_{j=1}^{n}x_{2j} \\
\cdots\cdots\cdots\cdots\cdots \\
+x_{m1}+ & x_{m2}+\cdots+ & x_{mn} & \sum_{j=1}^{n}x_{mj} \\
\sum_{i=1}^{m}x_{i1}+\sum_{i=1}^{m}x_{i2}+\cdots+\sum_{i=1}^{m}x_{in}=\sum_{j=1}^{n}(\sum_{i=1}^{m}x_{ij}) & = \sum_{i=1}^{m}(\sum_{j=1}^{n}x_{ij})
\end{array}
$$

Fig. 2.2 Swapping of summations

two adjacent indices (and leaves the others fixed). It also holds for the composition $\sigma\tau$ of two permutations in $\mathfrak{S}_n$, if it holds for σ and τ individually. Finally, it is used that every permutation in $\mathfrak{S}_n$ is a composition of transpositions of the mentioned type, see for example exercise 2.5.10. In the case of a commutative M, for a finite family x_i, $i \in I$, of elements from M, the product or the sum

$$ x^I = \prod_{i\in I} x_i \quad \text{or} \quad x_I = \sum_{i\in I} x_i $$

can also be defined for any index set I. It is simply the product or the sum of the family obtained by completely ordering I in any way. In general, the product $\prod_{i\in I} x_i$ is already defined when the elements x_i, $i \in I$, commute pairwise. Moreover, the product of pairwise commuting elements $x_i \in M$, $i \in I$, of a monoid M is already defined when I is potentially infinite, but x_i is equal to the neutral element $e \in M$ for almost all $i \in I$. In this case, the product is equal to the product over the finite subfamily x_i, $i \in \{i \in I \mid x_i \neq e\} \subseteq I$.

Example 2.1.8 Let I, J be finite sets and x_{ij}, $(i,j) \in I \times J$, a family of elements of a commutative monoid written additively. Then by bracketing

$$ \sum_{(i,j)\in I\times J} x_{ij} = \sum_{i\in I}\Big(\sum_{j\in J} x_{ij}\Big) = \sum_{j\in J}\Big(\sum_{i\in I} x_{ij}\Big); $$

especially with $I := \mathbb{N}^*_{\leq m} = \{1,\ldots,m\}$, $J := \mathbb{N}^*_{\leq n} = \{1,\ldots,n\}$

$$ \sum_{\substack{1\leq i\leq m \\ 1\leq j\leq n}} x_{ij} = \sum_{i=1}^{m}\Big(\sum_{j=1}^{n} x_{ij}\Big) = \sum_{j=1}^{n}\Big(\sum_{i=1}^{m} x_{ij}\Big). $$

The last equations are well overviewed with the scheme from Fig. 2.2. ◊

Products with the same factors are written as **powers**. If x is an element of a semigroup and $n \in \mathbb{N}^*$, then x^n is the n-fold product of x with itself. If $x^2 = x$, then $x^n = x$ for all $n \in \mathbb{N}^*$. Such an element x is called **idempotent** or a **projection** (since this property characterizes the projections in the monoid A^A of the mappings of a set A into itself). We denote the set of idempotent elements of a semigroup H with

$$\mathrm{Idp}(H).$$

In a monoid M with neutral element e one also defines $x^0 := e$ and $x^{-n} := (x^{-1})^n = (x^n)^{-1}$, if $x \in M^\times$ is invertible. In additive notation, we have the corresponding **multiples** nx. The following elementary calculation rules apply:

$$x^{m+n} = x^m x^n, \quad (x^m)^n = x^{mn},$$

for *commuting* elements x, y and

$$(xy)^m = x^m y^m,$$

for all $m, n \in \mathbb{Z}$ as far as the terms that appear are defined. In additive notation, these rules are:

$$(m+n)x = mx + nx, \quad n(mx) = (mn)x = (nm)x, \quad m(x+y) = mx + my.$$

In a group G, for every family a_i, $i \in I$, of elements of G, there is a smallest subgroup

$$\mathrm{H}(a_i, i \in I) \subseteq G,$$

that contains these elements (namely the intersection of all subgroups of G that contain all a_i, $i \in I$). This group is called the **subgroup generated by** a_i, $i \in I$, of G, and a_i, $i \in I$, is called a **generating system** of this subgroup. It consists precisely of all finite products of elements of the set $\{a_i \mid i \in I\} \cup \{a_i^{-1} \mid i \in I\}$. In the case of $I = \{1\}$, $a := a_1$ these are exactly the powers a^n, $n \in \mathbb{Z}$, of a. So it is

$$\mathrm{H}(a) = \{a^n \mid n \in \mathbb{Z}\}.$$

A group G is called **cyclic**, if it is generated by a single element, i.e. if $G = \mathrm{H}(a)$ with a (suitable) $a \in G$ is. Because of $a^n a^m = a^{n+m} = a^{m+n} = a^m a^n$ for all $m, n \in \mathbb{Z}$ every cyclic group $\mathrm{H}(a)$ is commutative. In general, in an *abelian* group G, the subgroup generated by the family $a = (a_i)_{i \in I}$ contains exactly the products

$$a^n := \prod_{i \in I} a_i^{n_i}, \quad n := (n_i) \in \mathbb{Z}^{(I)} := \{(z_i) \in \mathbb{Z}^I \mid z_i = 0 \text{ for almost all } i \in I\}.$$

It is therefore the set of all finite products of integer powers of the a_i, $i \in I$. In additive notation, the subgroup generated by the family $a = (a_i)_{i \in I}$ contains the sums

$$na := \sum_{i \in I} n_i a_i, \quad n := (n_i) \in \mathbb{Z}^{(I)}.$$

It is denoted by $\sum_{i \in I} \mathbb{Z} a_i$, for $I = \{1, \ldots, n\}$ also with $\mathbb{Z} a_1 + \cdots + \mathbb{Z} a_n$, for $I = \{1\}$, $a := a_1$ thus with $\mathbb{Z} a$.

Let again in general M be a (multiplicative, not necessarily commutative) monoid with neutral element e. If $a \in M^\times$ is invertible in M, then the left translations $L_a: x \mapsto ax$ and the right translation $R_a: y \mapsto ya$ are permutations of M with the inverse mappings $L_{a^{-1}}$ or $R_{a^{-1}}$. In other words:

Theorem 2.1.9 *Let a be an invertible element in the monoid M. Then for each $b \in M$ there are uniquely determined elements x and y in M with*

$$ax = b \quad \text{and} \quad ya = b, \quad \text{and particularily} \quad x = a^{-1}b \quad \text{and} \quad y = ba^{-1}.$$

If M is an abelian monoid, then naturally the solutions x and y in Theorem 2.1.9 coincide. In additive notation, this is the difference $x = y = b + (-a) = b - a$. In a group $M = M^\times$ all equations $ax = b$ and $ya = b$, $a, b \in M$, have unique solutions $x, y \in M$.

Invertible elements a can be shortened, i.e., from $ab = ac$ or $ba = ca$ it follows that $b = c$. This is the injectivity of the left or right translation with a. We generally define:

Definition 2.1.10 Let H be a semigroup.

(1) The element $a \in H$ is called **left regular** (or **right regular**) in H, if the left translation $L_a: H \to H$ (or the right translation $R_a: H \to H$) with a on H is injective.
(2) The element $a \in H$ is called **regular** in H, if a is both left- and right-regular in H. We denote the set of regular elements of H with

$$H^*.$$

(3) The semigroup H is called **regular**, if $H = H^*$ is.

A regular element $a \in H^*$ can be shortened: For $b, c \in H$ the equality $b = c$ follows both from $ab = ac$ and from $ba = ca$. In regular semigroups, one can generally shorten. Therefore, regular semigroups are also called **semigroups with shortening rule**. In groups, the shortening rule always applies. In general, $M^\times \subseteq M^*$ applies for every monoid M. Since the composition of injective mappings is again injective, *is H^* always a regular subgroup of the semigroup H and M^* a regular submonoid of the monoid M*. In a monoid M the neutral element e_M is the only element that is both regular and idempotent. If $a \in M$ is regular and idempotent, then from $a^2 = aa = a = e_M a$ after shortening of a the equation $a = e_M$ follows. (It was sufficient to require that a is left- or right-regular.)

In the following, we discuss the relevant **divisibility terms** and limit ourselves to *abelian* monoids. The model for this is the divisibility in $(\mathbb{N}, \cdot)$ or $(\mathbb{Z}, \cdot)$. So let M be a multiplicative *abelian* monoid with neutral element 1 and $a, b \in M$. Then a is called a **divisor** of b or b a **multiple** of a, if there is a $c \in M$ gives with $b = ac$, i.e. when b is in the image $aM = Ma$ of the translation $L_a = R_a$ or – equivalently – when $Ma \supseteq Mb$ is.[2] We then write

---

[2] In non-abelian monoids, one has to distinguish between left and right divisibility.

$a \mid b$ or more exactly $a \mid_M b$.

The set of all divisors of b in M we denote with

$$T(b) = T_M(b).$$

$T(1)$ is the group $M^\times$ of units of M. The set of multiples of a is simply Ma.[3] So it holds

$$a \mid b \iff a \in T(b) \iff T(a) \subseteq T(b) \iff b \in Ma \iff Mb \subseteq Ma.$$

The divisibility relation is reflexive and transitive, i.e. a quasi-order, see exercise 1.4.2. The associated equivalence relation

$$\| = {}_M\|_M$$

is given by

$$a \parallel b \quad \text{if and only if} \quad a \mid b \quad \text{and} \quad b \mid a,$$

i.e. when $Ma = Mb$ or $T(a) = T(b)$ is. On the quotient set $\overline{M} := M/\|$ is then by

$$[a] \leq [b] \quad \text{if and only if} \quad a \mid b$$

(so $T(a) \subseteq T(b)$ or $Ma \supseteq Mb$) an order well-defined, and the mapping $\overline{M} \to \mathfrak{P}(M)$, $[a] \mapsto Ma$, induces an order isomorphism from $(\overline{M}, \leq)$ to the image (ordered by anti-inclusion) $\supseteq \{Ma \mid a \in M\} \subseteq \mathfrak{P}(M)$. Moreover, $\overline{M}$ is a monoid by virtue of the well-defined operation $[a] \cdot [b] := [ab]$. Furthermore, $[a][c] \leq [b][c]$ for all $a, b, c \in \overline{M}$ with $[a] \leq [b]$, i.e., the translations in $\overline{M}$ are monotonically increasing, and $\overline{M}$ is an *ordered monoid with* $[1] = M^\times$ *as the smallest element*, cf. Exercise 2.1.16 for the concept of the ordered monoid. The divisibility relation in $\overline{M}$ is the order relation $\leq$, and its study is essentially equivalent to the study of divisibility in M. Note that the unit group $(\overline{M})^\times = \{[1]\}$ of $\overline{M}$ is trivial, i.e., $\overline{M}$ *is a pointed monoid*. As already mentioned, it is often more convenient to calculate in the submonoid $\{Ma \mid a \in M\} \subseteq \mathfrak{P}(M)$ isomorphic to $\overline{M}$ with the complex multiplication $(Ma) \cdot (Mb) = Mab$, in which divisibility is the anti-inclusion $\supseteq$.[4] The equivalence classes with respect to $\|$, i.e., the elements of $\overline{M}$, is called **divisor classes**. Each divisor class $[a]$ is a subset of the set $T(a)$ of all divisors of a. A divisor b of a is called a **proper divisor** of a if b does not belong to the divisor class of a – or equivalently – if $T(b) \subset T(a)$ or $Ma \subset Mb$ applies. The **trivial divisors** of $a \in M$ are by definition the units of M and the elements of the divisor class $[a]$ of a. The number of divisor classes in $T(a)$, i.e., the number of elements of the section $\overline{M}_{\leq [a]}$ in $\overline{M}$ we denote with

$$\tau(a) = \tau_M(a).$$

---

[3] Because of this clear description of the multiples, it is often more convenient to operate with the concept of multiples rather than the concept of divisors. Compared to the sets of multiples Ma, the sets of divisors $T_M(a)$ are usually much harder to overlook (already in $\mathbb{N}^*$).

[4] For the concept of isomorphism, see Sect. 2.2.

2.1 Monoids and Groups

The divisor class $[a] = [a]_{M \| M}$ of a always includes the elements associated with a. The element $b \in M$ is called **associated** to $a \in M$, if there is an $e \in M^\times$ with $b = ea$. Associativity is an equivalence relation on M with the equivalence classes $M^\times a \subseteq [a]$, $a \in M$. It is generally finer than mutual divisibility $\|$. Example? *But if $a \in M^*$ is a regular element of M, then $[a] = M^\times a$.* If $a \mid b$ and $b \mid a$, then $b = ac$ and $a = bd$, so $a = acd$ and thus, since a is regular, $e = cd$, i.e. $d, c \in M^\times$. a and b are therefore associated. The quotient set with respect to associativity is denoted by $M/M^\times$. On $M/M^\times$, the complex multiplication $(M^\times b) \cdot (M^\times c) = M^\times bc$ provides a monoid structure, cf. Example 2.1.17. The neutral element $M^\times$ is the only unit in $M/M^\times$, i.e. $M/M^\times$ is also a pointed monoid.

An element $q \in M$ is called **indecomposable** or **irreducible**, if q is not a unit in M and if from a decomposition $[q] = [c][d] = [cd]$ into $\overline{M}$ always $[q] = [c]$ or $[q] = [d]$ follows. If the non-unit q only has the trivial divisors (i.e., $\tau(q) = 2$), i.e., the divisor class of q is an atom in the ordered set $\overline{M} \subseteq \mathfrak{P}(M)$ of all divisor classes of M or – equivalently – is $T(q)$ an atom with respect to the inclusion in $\{T(a) \mid a \in M\} \subseteq \mathfrak{P}(M)$ or Mq an anti-atom (again with respect to the inclusion) in $\{Ma \mid a \in M\} \subseteq \mathfrak{P}(M)$, then q is obviously indecomposable. Conversely, a *regular* indecomposable element q only has the elements of the divisor classes $M^\times$ and $[q] = M^\times q$ as divisors. Proof!

Prime elements are to be distinguished from indecomposable elements. An element $p \in M$ is called **prime** in M or a **prime element,** if p is not a unit in M and p only divides a product ab of elements $a, b \in M$ if p divides one of the factors a or b.[5] Only then is an element of M irreducible or prime, if the same applies to all elements of its divisor class. *A prime element $p \in M$ is always indecomposable.* If $[p] = [c][d]$, $c, d \in M$, then $[p]$ divides the product $[c][d]$ and thus $[c]$ or $[d]$ and conversely, $[c]$ and $[d]$ also divide $[p]$. So $[c] = [p]$ or $[d] = [p]$. With

$$\mathbb{I}_M \quad \text{resp.} \quad \mathbb{P}_M$$

we always denote a full representative system for the divisor classes of the irreducible or prime elements of M. We always choose $\mathbb{P}_M \subseteq \mathbb{I}_M$. Often, $\mathbb{I}_M$ and $\mathbb{P}_M$ indicates canonical representatives, such as in $M = \mathbb{Z}^*$ the set $\mathbb{I}_{\mathbb{Z}^*} = \mathbb{P}_{\mathbb{Z}^*} = \mathbb{P} \subseteq \mathbb{N}^*$ of positive prime numbers. In $(\mathbb{N}, \cdot)$ or $(\mathbb{Z}, \cdot)$ 0 is also prime.

Suprema and infima with respect to the quasi-order $\mid_M$ of M or the associated order of $\overline{M} = M/M \|_M$ are called **least common multiples** (**LCM**) or **greatest common divisors** (**GCD**). If a_i, $i \in I$, is a family of elements of M, then each element $a \in M$, which represents a supremum (or an infimum) of the divisor classes $\overline{a}_i$, $i \in I$, is called a least common multiple $v = \text{lcm}(a_i, i \in I)$ (or a greatest common divisor $d = \gcd(a_i, i \in I)$) of the a_i, $i \in I$. *These are thus only defined up to equivalence with respect to mutual divisibility $\|$.* v is a common multiple of the a_i

---

[5] As with irreducible elements, we do not require (as many authors do) that prime elements are regular. In general, it should be noted that irreducible elements or prime elements are defined differently in the literature. For regular elements, however, the definitions are usually equivalent.

∧	T	F		∨	T	F		△	T	F		⇒	T	F		⇔	T	F		∣	T	F
T	T	F		T	T	T		T	F	T		T	T	F		T	T	F		T	F	T
F	F	F		F	T	F		F	T	F		F	T	T		F	F	T		F	T	T

Fig. 2.3 Logical combinations

and divides every other common multiple of the $a_i, i \in I$. Similarly, d is a common divisor of the $a_i, i \in I$, and a multiple of every other common divisor.

As already noted, associativity and mutual divisibility $_M\|_M$ coincide when the monoid M is regular. *For a regular monoid M, therefore, $M/_M\|_M = M/M^\times$.* Furthermore, for a divisor a of b with $b = ac$, the element $c \in M$ is uniquely determined by a and b due to the cancellation rule. It is called the **quotient** b/a of b and a. This is also a divisor of b and is called the **complementary divisor** of b. Since with M also $M/M^\times$ is regular, it follows: *If M is a regular abelian monoid, then $M/M^\times$ is a pointed ordered monoid with [1] as the smallest element,* specifically, *if M is additionally pointed, then M itself is an ordered monoid with respect to divisibility with 1 as the smallest element.* For example, for the additive-monoid $(\mathbb{N}, +)$, divisibility is the natural order on $\mathbb{N}$. If M is regular and pointed, then $\mathbb{I}_M$ and $\mathbb{P}_M$ are simply the sets of all irreducible and all prime elements of M. Elements $a, b \in M$ with $\gcd(a,b) = 1$ are called **coprime**. The elements $a, b \in M$ are therefore coprime if and only if the units in M are the only common divisors of a and b.

We will use the divisibility concepts discussed here, in particular for the regular multiplicative monoid A^* of the non-zero elements of an integral domain A, see Sect. 2.10. In this case, the addition in A provides additional means for studying the multiplicative monoid A^*. An example of this is already the monoid $\mathbb{Z}^* = (\mathbb{Z}^*, \cdot)$ with $\mathbb{Z}^*/\mathbb{Z}^\times = \mathbb{N}^*$. The proof of the main theorem of elementary number theory makes essential use of the addition in $\mathbb{Z}$, see Sect. 1.7.

We now discuss some examples related to the concepts introduced in this section so far.

Example 2.1.11 On a finite set A with $n \in \mathbb{N}$ elements, there are exactly $|A^{A \times A}| = |A|^{|A \times A|} = n^{n^2}$ combinations, cf. theorem 1.6.2. Thus, there are already $2^4 = 16$ combinations on a two-element set. Many conjunctions of a language connect two statements in such a way that the truth value "T = True" or "F = False" of the overall statement depends only on the truth values of the individual statements. For example, the conjunctions ∧ (= and = both-and), ∨ (= or), △ (= either-or), ⇒ (= if, then), ⇔ (= exactly when), and ∣ (= not both) have the combination tables (= **truth tables**) from Fig. 2.3. Except for "⇒", all these combinations are commutative: Their combination tables are symmetric to the main diagonal, which leads from top left to bottom right. Which of these combinations are associative and which have a neutral element? The combination "∣" is the negation of "∧" and is called **Sheffer's stroke**. It is abbreviated as "nand". The combination "neither-nor" (briefly "nor") is the negation of "or". One should specify its combination table. ◊

2.1 Monoids and Groups

Example 2.1.12 (Opposite Magma) Let $A = (A, *)$ be a magma with the operation $(a, b) \mapsto a * b$. Then $A^{\mathrm{op}} = (A, *^{\mathrm{op}})$ with the opposite operation $(a, b) \mapsto a *^{\mathrm{op}} b := b * a$ is also a magma. It is called the A **opposite** magma. It is $(A^{\mathrm{op}})^{\mathrm{op}} = A$. A is commutative if and only if $A = A^{\mathrm{op}}$ is. A is a semigroup, or a monoid, or a group, if and only if the same applies for A^{op}. We then speak of the opposite semigroup, the opposite monoid, or the opposite group. The neutral element and the respective inverses coincide in A and A^{op}. For a monoid M, $(M^{\mathrm{op}})^{\times} = (M^{\times})^{\mathrm{op}}$ is therefore. For a semigroup H, $(H^{\mathrm{op}})^{*} = (H^{*})^{\mathrm{op}}$ is. $\Diamond$

Example 2.1.13 Let X be a set. The set X^X of all mappings from X to itself, with the composition $\circ$ of mappings as a link, is a monoid with the identity id_X as a neutral element. Its invertible elements are the permutations of X. They form the so-called **permutation groups** $\mathfrak{S}(X) := \left(X^X\right)^{\times}$ of X. The inverse element to a permutation $\sigma \in \mathfrak{S}(X)$ is the mapping σ^{-1} inverse to σ. If X is a finite set with n elements, then X^X and $\mathfrak{S}(X)$ are also finite with n^n and $n!$ elements, see Theorem 1.6.2 or corollary 1.6.7. In this case, $\mathrm{Ord}\,\mathfrak{S}(X) = n!$ is true. In particular, the permutation group $\mathfrak{S}_n = \mathfrak{S}(\{1, \ldots, n\})$ has the order $n!$ for each $n \in \mathbb{N}$. If X is infinite, then $\mathrm{Ord}\,\mathfrak{S}(X) = 2^{|X|}$ is true, see Exercise 2.1.4b). We will deal more extensively with the finite permutation groups in Sect. 2.5. The left-regular elements in the mapping monoid X^X are exactly the injective mappings and the right-regular elements are exactly the surjective mappings, see Exercise 1.2.10. The regular elements in X^X thus coincide with the invertible elements: $\left(X^X\right)^{*} = \left(X^X\right)^{\times} = \mathfrak{S}(X)$. $\Diamond$

Example 2.1.14 Let X be a set. On the power set $\mathfrak{P}(X)$ of X, the union $(A, B) \mapsto A \cup B = \mathrm{Sup}(A, B)$ and the intersection $(A, B) \mapsto A \cap B = \mathrm{Inf}(A, B)$ are associative and commutative operations with the neutral elements $\emptyset$ and X.[6] These neutral elements are the only invertible elements in $(\mathfrak{P}(X), \cup)$ and $(\mathfrak{P}(X), \cap)$. For the set $\mathfrak{E}(X) \subseteq \mathfrak{P}(X)$ of finite subsets of X, $(\mathfrak{E}(X), \cup)$ is a submonoid of $(\mathfrak{P}(X), \cup)$ and $(\mathfrak{E}(X), \cap)$ is a subgroup of $(\mathfrak{P}(X), \cap)$ (but not a submonoid, if X is infinite). – Also, the symmetric difference $(A, B) \mapsto A \triangle B = (A \cup B) - (A \cap B)$ is an associative and commutative operation on $\mathfrak{P}(X)$ (for associativity see exercise 1.1.6f)). The empty set $\emptyset$ is a neutral element. Furthermore, $A \triangle A = \emptyset$ for every $A \in \mathfrak{P}(X)$. $(\mathfrak{P}(X), \triangle)$ *is therefore an abelian group in which every element is its own inverse.* Such groups are called **elementary 2-groups**, cf. Ex. 2.1.6. $(\mathfrak{E}(X), \triangle)$ is a subgroup of $(\mathfrak{P}(X), \triangle)$. The groups $(\mathfrak{E}(X), \triangle)$ are, by the way, all elementary 2-groups up to isomorphism, cf. Ex. 2.3.6. – In general, an element a of a monoid M is called **involutory** or an **involution**, if $a^2 = e_M$ is. This is exactly the case when a is invertible with $a^{-1} = a$. In the mapping monoid X^X the involutions are exactly the reflections $\sigma \in \mathfrak{S}(X)$ with $\sigma^2 = \mathrm{id}_X$, cf. Example 1.2.17. Therefore, in any monoid, the involutory elements are often called **reflections**.

The operations $\cup = \mathrm{Sup}$ and $\cap = \mathrm{Inf}$ in $\mathfrak{P}(X)$ are defined analogously for any lattices. We recall that a **lattice** is an ordered set $V = (V, \leq)$ in which every two

---

[6] Sup and Inf are understood with respect to inclusion.

elements have a supremum and an infimum, cf. Example 1.4.10. On such a lattice are

$$a \sqcup b := \operatorname{Sup}(a,b) \quad \text{and} \quad a \sqcap b := \operatorname{Inf}(a,b), \quad a,b \in V,$$

associative and commutative operations. All elements of V are idempotent with respect to both operations, i.e., it is $a \sqcup a = a \sqcap a = a$ for all $a \in A$. $a \leq b$ if and only if $a \sqcup b = b$ or if $a \sqcap b = a$ is. Furthermore, one has the so-called **fusion rules**

$$a \sqcup (a \sqcap b) = a = a \sqcap (a \sqcup b), \quad a,b \in V.$$

An element $0 \in V$ is a neutral element with respect to $\sqcup$ if and only if 0 is the smallest element in V, and $1 \in V$ is a neutral element with respect to $\sqcap$ if and only if 1 is the largest element in V. If the zero element 0 exists in V, it is the only invertible element with respect to $\sqcup$. The same applies to the one element 1. If V is replaced by the oppositional lattice $V^{\mathrm{op}} = (V, \leq^{\mathrm{op}}) = (V, \geq)$, the operations $\sqcup$ and $\sqcap$ swap their roles and a lattice is obtained again (**Duality principle for lattices**). – Conversely, if $V = (V, \sqcup, \sqcap)$ is a set with two associative and commutative operations $\sqcup$ or $\sqcap$, for which the above merging rules apply, then by "$a \leq b$ exactly when $a \sqcup b = b$" an order is defined on V, with respect to which V is a lattice with $\operatorname{Sup}(a,b) = a \sqcup b$ and $\operatorname{Inf}(a,b) = a \sqcap b$ for all $a,b \in V$. Proof! A subset V' of a lattice $V = (V, \sqcup, \sqcap)$ is called a **sublattice**, if V' is closed with respect to the two operations $\sqcup$ and $\sqcap$, if with $a,b \in V'$ always also $\operatorname{Sup}_V(a,b)$ and $\operatorname{Inf}_V(a,b)$ belong to V'. Note that a subset *can be a lattice with respect to the order induced by V without V' being a sublattice of V*. Example?   ◊

Example 2.1.15 The classical number ranges provide a wealth of important examples for monoids and groups. $\mathbb{N}$ is an abelian monoid with respect to addition and multiplication, with neutral elements 0 and 1 respectively. These are also the only invertible elements in $(\mathbb{N}, +)$ and in $(\mathbb{N}, \cdot)$. The designation $\mathbb{N}^*$ for the set $\mathbb{N} - \{0\}$ of regular elements in the multiplicative monoid $(\mathbb{N}, \cdot)$ we have already anticipated. The subgroup $\{0\}$ of the multiplicative monoid $\mathbb{N} = (\mathbb{N}, \cdot)$ is of course a monoid, but not a submonoid, since its neutral element 0 is not the neutral element of $(\mathbb{N}, \cdot)$. $(\mathbb{Z}, +)$ is an abelian group with neutral element 0, which includes $(\mathbb{N}, +)$ as a submonoid. $(\mathbb{Z}, \cdot)$ is an abelian monoid with neutral element 1 and $\mathbb{Z}^\times = \{1, -1\}$ as a group of invertible elements. $\mathbb{Z}^* = \mathbb{Z} - \{0\}$ is the submonoid of regular elements of $(\mathbb{Z}, \cdot)$. The divisibility in $\mathbb{Z}^*$ and $\mathbb{N}^* = \mathbb{Z}^*/\mathbb{Z}^\times$ is dominated by the main Theorem of elementary number Theory 1.7.11.

$(\mathbb{Z}, +)$ in turn is a subgroup of the additive abelian group $(\mathbb{Q}, +)$. $(\mathbb{Q}, \cdot)$ is an abelian monoid, but not a group; the only element in $(\mathbb{Q}, \cdot)$ that is not invertible is 0. (0 cannot be shortened!) So it is $\mathbb{Q}^\times = (\mathbb{Q} - \{0\}, \cdot) = \mathbb{Q}^*$. The positive rational numbers form a subgroup $\mathbb{Q}_+^\times$. The same applies to $\mathbb{R}$: $(\mathbb{R}, +)$ is an abelian group, $(\mathbb{R}, \cdot)$ an abelian monoid with $\mathbb{R}^\times = (\mathbb{R} - \{0\}, \cdot)$ as the group of invertible elements. Again, $\mathbb{R}^\times = \mathbb{R}^*$ and $\mathbb{R}_+^\times = \{x \in \mathbb{R} \mid x > 0\}$ is a subgroup of $\mathbb{R}^\times$.   ◊

2.1 Monoids and Groups

Example 2.1.16 (Products) The formation of products is an important method in all areas of mathematics to create new structures from given ones and to trace back given structures to already known ones. Let A_i, $i \in I$, be a family of (multiplicatively written) magmas. Then the product set $\prod_{i \in I} A_i$ with **component-wise combination**

$$(a_i)_{i \in I} \cdot (b_i)_{i \in I} := (a_i \cdot b_i)_{i \in I}, \quad (a_i)_{i \in I}, (b_i)_{i \in I} \in \prod_{i \in I} A_i$$

is a magma. It is called the **product** of the magmas A_i, $i \in I$. If all A_i are semigroups or monoids or groups, the same applies to the product. More generally, for a family of monoids M_i with neutral elements e_i, $i \in I$, the element $e := (e_i)_{i \in I}$ is the neutral element of the product, and it holds

$$\left(\prod_{i \in I} M_i\right)^{\times} = \prod_{i \in I} M_i^{\times} \quad \text{with} \quad (a_i)_{i \in I}^{-1} = \left(a_i^{-1}\right)_{i \in I} \quad \text{for} \quad (a_i)_{i \in I} \in \prod_{i \in I} M_i^{\times}.$$

Similarly, $\left(\prod_{i \in I} H_i\right)^* = \prod_{i \in I} H_i^*$ holds for a family of semigroups H_i, $i \in I$. For monoids M_i, the following submonoid of the product monoid is often used:

$$\prod'_{i \in I} M_i := \left\{ (a_i)_{i \in I} \in \prod_{i \in I} M_i \,\middle|\, a_i = e_i \text{ for almost all } i \in I \right\} \subseteq \prod_{i \in I} M_i.$$

It is called the **restricted product** of the M_i, $i \in I$. Here too, $\left(\prod'_{i \in I} M_i\right)^{\times} = \prod'_{i \in I} M_i^{\times}$ applies. In $\prod'_{i \in I} M_i$ (and thus also in $\prod_{i \in I} M_i$), the individual factors M_i, $i \in I$, are embedded as submonoids. We identify an element $a_i \in M_i$ with the I-tuple, which at the position i has the value a_i and at the other positions the respective neutral element. If all monoids M_i are equal to the same monoid M, we denote the restricted I-fold product with

$$M^{(I)}.$$

It is the submonoid of those I-tuples in M^I, for which almost all components are equal to the neutral element of M. Note that according to the above, the monoid M in $M^{(I)}$ (at $M \neq \{e_M\}$) is canonically embedded in $|I|$ in different ways. These embeddings must not (at $I \neq \emptyset$) be confused with the **diagonal embedding** $M \to M^I$, which assigns to each $a \in M$ the constant tuple (a_i) with $a_i = a$ for all $i \in I$. $M^{(I)}$ is a submonoid of the full I-fold product M^I and coincides with it when I is finite. $M^{(I)}$ is the monoid of mappings $f: I \to M$ with $f(i) = e_M$ for almost all $i \in I$. – If all M_i, $i \in I$, are additively written abelian monoids, then

$$\bigoplus_{i \in I} M_i = \{(a_i)_{i \in I} \mid a_i \in M_i \text{ for all } i \in I \text{ and } a_i = 0_i \text{ for almost all } i \in I\}$$

their restricted product. It is called the **direct sum** of the M_i, $i \in I$. ◊

Example 2.1.17 (Complex multiplication) Let M be a *non-empty* multiplicative magma. For subsets $A, B \subseteq M$ we denote by AB the so-called **complex product**

$$AB := \{ab \mid a \in A, b \in B\}$$

of A and B. AB is thus the image of $A \times B$ under the operation $M \times M \to M$ of M. We write aB or Ab for $\{a\}B$ or $A\{b\}$. With this **complex multiplication**, the power set $\mathfrak{P}(M)$ of M is also a magma. We consider M itself with respect to the canonical embedding $a \mapsto \{a\}$ appears as a submagma of $\mathfrak{P}(M)$. The empty set is an absorbing element in $\mathfrak{P}(M)$ with $\emptyset \cdot A = A \cdot \emptyset = \emptyset$ for all $A \in \mathfrak{P}(M)$.[7] $\mathfrak{P}(M) - \{\emptyset\}$ is a submagma of $\mathfrak{P}(M)$. In additive notation, the complex product

$$A + B = \{a + b \mid a \in A, b \in B\}$$

is called the **Minkowski sum** of A and B. For example, the Minkowski sum of two intervals $[a,b]$ and $[c,d]$, $a \le b$, $c \le d$, in $(\mathbb{R},+)$ the interval $[a+c, b+d]$. Apparently, $\mathfrak{P}(M)$ is associative or commutative exactly when the same applies for M. Furthermore, $\mathfrak{P}(M)$ has a neutral element exactly when M has a neutral element e_M. In this case, $e_{\mathfrak{P}(M)} = \{e_M\} = e_M$. $\mathfrak{P}(M)$ is a monoid exactly when M is a monoid. In this case, M is a submonoid of $\mathfrak{P}(M)$ and $\mathfrak{P}(M)^\times = M^\times$. $\mathfrak{P}'(M) := \mathfrak{P}(M) - \{\emptyset\}$ is a group only when M is the trivial monoid $\{e_M\}$.

If $A \subseteq M^\times$ is any set of invertible elements of the monoid M, then the set A^{-1} usually refers to the set $A^{-1} := \{a^{-1} \mid a \in A\}$, i.e., the image of A under the inverse formation $M^\times \to M^\times$. However, A^{-1} *is only the inverse of A in $P(M)$, if A is a singleton*. Also note the calculation rule $(AB)^{-1} = B^{-1}A^{-1}$ for $A, B \subseteq M^\times$.

Now let M be a group and F, H subgroups of M. Then $(FH)^{-1} = H^{-1}F^{-1} = HF$. It follows: *The complex product FH is also a subgroup of M, if $FH = HF$ is.* This last condition is certainly met when M is abelian.—In the group $\mathfrak{S}_3$ with the cyclic subgroups E, F, H, which are generated by the transpositions $\langle 1, 2 \rangle$, $\langle 1, 3 \rangle$ and $\langle 2, 3 \rangle$, $EFH = \mathfrak{S}_3$ holds, but the product of any two different of these subgroups is not a subgroup. ◊

The additive monoid $(\mathbb{N}, +)$ and the additive group $(\mathbb{Z}, +)$ are of universal significance. This is mainly due to the fact that for any monoid M and an element $a \in M$ or $a \in M^\times$ the exponential mappings $\mathbb{N} \to M$ or $\mathbb{Z} \to M$ with $n \mapsto a^n$ with the operations in $\mathbb{N}$ or $\mathbb{Z}$ and in M are compatible. This is exactly what the power calculation rule $a^{m+n} = a^m a^n$ states. The group $\mathbb{Z} = (\mathbb{Z}, +)$ is cyclic and is generated by both 1 and -1 (and no other element). Fundamental is the following theorem about the subgroups of $\mathbb{Z}$:

---

[7] In general, an element 0 of a magma N is called an **absorbing element**, if $0 \cdot x = x \cdot 0 = 0$ is for all $x \in N$. Do not confuse the absorbing element 0 with the zero element of an additive monoid. Zero is an absorbing element in the multiplicative monoids $(\mathbb{N}, \cdot)$ and $(\mathbb{Z}, \cdot)$ (and in the multiplicative monoid of every ring). An absorbing element of an additive monoid is often denoted by ∞ or with $-\infty$.

2.1 Monoids and Groups

Theorem 2.1.18 *For every subgroup H of $\mathbb{Z} = (\mathbb{Z}, +)$ there is exactly one $n \in \mathbb{N}$ with $H = \mathbb{Z}n$. In particular, all subgroups of $\mathbb{Z}$ are cyclic.*

Proof In the case of $H = \{0\}$, $n = 0$. In the case of $H \neq \{0\}$, H contains positive numbers; because at $a \in H$, $a \neq 0$, a or $-a (\in H)$ is positive. For the smallest number $n \in H \cap \mathbb{N}^*$, cf. Theorem 1.5.4, then $H = \mathbb{Z}n$ applies. Because of $n \in H$, $\mathbb{Z}n \subseteq H$ is also the case. Conversely, let $a \in H$ be arbitrary. Division with remainder provides a representation $a = qn + r$, $q, r \in \mathbb{Z}$, $0 \leq r < n$. Because of $r = a + (-q)n \in H \cap \mathbb{N}$, $r = 0$ is necessarily the case after choosing n and thus $a = qn \in \mathbb{Z}n$. The uniqueness of n is trivial. □

For the subgroup $H = \mathbb{Z}n \subseteq \mathbb{Z}$, $-n$ can also be used as a generating element (but no further). A generating element of H is therefore only determined up to the sign when $H \neq 0$. As an example to Theorem 2.1.18, the reader should prove the following proposition:

Proposition 2.1.19 *Let there be $a_1, \ldots, a_n \in \mathbb{N}^*$. Then the following applies*

$$\mathbb{Z}a_1 + \cdots + \mathbb{Z}a_n = \mathbb{Z}\gcd(a_1, \ldots, a_n) \quad \text{and} \quad \mathbb{Z}a_1 \cap \cdots \cap \mathbb{Z}a_n = \mathbb{Z}\operatorname{lcm}(a_1, \ldots, a_n).$$

For an *arbitrary* family $a_i \in \mathbb{Z}$, $i \in I$, of whole numbers, the greatest common divisor and the least common multiple are the generating elements of $\sum_{i \in I} \mathbb{Z}a_i$ or $\bigcap_{i \in I} \mathbb{Z}a_i$. Both are again determined only up to the sign.

A subgroup H of any group G defines two important decompositions of G. We use the complex multiplication from Example 2.1.17. In particular, we use for the subgroup H the so-called **left cosets** $aH = \{a\}H$ and the **right cosets** $Ha = H\{a\}$, $a \in G$. The left translation $L_a : x \mapsto ax$ and the right translation $R_a : y \mapsto ya$ induce bijective mappings $L_a|H : H \to aH$ or $R_a|H : H \to Ha$. In particular,

$$|aH| = |H| = |Ha| \quad \text{for all } a \in G.$$

Furthermore, $(aH)^{-1} = H^{-1}a^{-1} = Ha^{-1}$ and $(Ha)^{-1} = a^{-1}H^{-1} = a^{-1}H$ hold for all $a \in G$. The set of left cosets (or the set of right cosets) of G with respect to H is denoted by

$$G/H \quad (\text{resp. } G\backslash H).$$

According to the last remark, the inverse formation of G induces mutually inverse bijections $G/H \to G\backslash H$ and $G\backslash H \to G/H$. The common cardinal number of G/H and $G\backslash H$ is called the **index** of H in G and is denoted by

$$[G : H] := \operatorname*{Ind}_G H := |G/H| = |G\backslash H|$$

For abelian groups, the left and right cosets coincide and are simply called **cosets**. Fundamental is the following general lemma:

Lemma 2.1.20 *If H is a subgroup of the group G, then $G/H \subseteq \mathfrak{P}(G)$ and also $G\backslash H \subseteq \mathfrak{P}(G)$ are partitions of G.*

Proof It is sufficient to consider the left cosets. Due to $a = ae \in aH$ is $G = \bigcup_{a \in G} aH$. It remains to show $aH = bH$ if $aH \cap bH \neq \emptyset$ is for $a, b \in G$. Let $c = ah_1 = bh_2$ with $h_1, h_2 \in H$. For $h \in H$ is then $ah = bh_2 h_1^{-1} h \in bH$, thus $aH \subseteq bH$. Similarly, $bH \subseteq aH$ holds. □

The equivalence relations $_H\equiv$ and $\equiv_H$ on G to the partitions G/H (or $G\backslash H$) of G can apparently be described as follows: For $a, b \in G$ applies $a_H\equiv b$, i.e. $aH = bH$ (or $a \equiv_H b$, i.e. $Ha = Hb$) exactly when there is a $h \in H$ with $ah = b$ (or with $ha = b$). These relations are called the **congruence relations** on G with respect to the subgroup H. The fibers of the canonical projection $G \to G/H$ (or $G \to G\backslash H$) are exactly the elements of G/H (or $G\backslash H$). Since all of these have the same cardinal number $|H|$, we obtain:

Theorem 2.1.21 (Lagrange's Theorem) *If H is a subgroup of the group G, then*

$$|G| = |H| \cdot [G : H].$$

In a finite group G, the order $|H|$ of every subgroup H of G is a divisor of the order $|G|$ of G with the quotient $|G|/|H| = [G : H]$.

For example, if $|G|$ is a prime number, then G only has the trivial subgroups $\{e_G\}$ and G. *In particular, then $G = \mathrm{H}(a)$ for every $a \in G$, $a \neq e_G$, and G is cyclic.*

The decompositions G/H and $G\backslash H$ of a group into its cosets with respect to a subgroup $H \subseteq G$ usually do not coincide. As we will see in the next two sections, the case $G/H = G\backslash H$ is of particular importance. We therefore define it here:

Definition 2.1.22 A subgroup N of a group G is called a **normal subgroup** or a **normal divisor** of G, if $G/N = G\backslash N$ is, if thus $aN = Na$ for all $a \in N$ is. In this case, the elements of $G/N = G\backslash N$ are simply the **cosets** (or **residue classes**) of G modulo N.

If N is a normal divisor, then $AN = NA$ holds for every subset $A \subseteq G$. In particular, then $HN = NH \subseteq G$ is a subgroup of G for every subgroup $H \subseteq G$.

Example 2.1.23 If G is an abelian group, then — as already mentioned — every subgroup $H \subseteq G$ is normal in G. The cosets with respect to the subgroup $\mathbb{Z}m \subseteq \mathbb{Z}$ of $\mathbb{Z}$, $m \in \mathbb{N}^*$, are the m pairwise different residue classes $a + \mathbb{Z}m$, $a \in \mathbb{N}$, $0 \leq a < m$. The elements of $a + \mathbb{Z}m$ are those whole numbers that have the remainder a when divided by m. Compare this with the already mentioned Example 1.3.6. With this example in mind, in abelian groups—especially when they are written additively — a coset is also called a "residue class". — $\{e_G\}$ and G itself are trivially normal subgroups of any group G. The simplest example of a non-normal subgroup is provided by the permutation group $G = \mathfrak{S}_3 = \mathfrak{S}(\{1, 2, 3\})$ of a set of three elements and its subgroup H, which consists of the identity id and a transposition, for example $\binom{123}{213} = \langle 1, 2\rangle$. The set $\mathfrak{S}_3/H$ of left cosets then contains the elements

$$H, \quad \langle 1, 3\rangle H = \left\{\langle 1, 3\rangle, \binom{123}{231}\right\}, \quad \langle 2, 3\rangle H = \left\{\langle 2, 3\rangle, \binom{123}{312}\right\}$$

Fig. 2.4 Monoid for the projection p of a set X

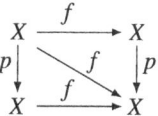

and the set $\mathfrak{S}_3 \backslash H$ of right cosets the elements

$$H, \quad H\langle 1,3\rangle = \left\{\langle 1,3\rangle, \begin{pmatrix}123\\312\end{pmatrix}\right\}, \quad H\langle 2,3\rangle = \left\{\langle 2,3\rangle, \begin{pmatrix}123\\231\end{pmatrix}\right\}.$$

The two sets $\mathfrak{S}_3/H$ and $\mathfrak{S}_3\backslash H$ do *not* coincide. The cyclic subgroup N of $\mathfrak{S}_3$ with index 2, which contains the identity, the permutation $\binom{123}{231}$ and its inverse $\binom{123}{312}$, is normal. Its cosets are N and the complement $S_3 - N$. In general, a subgroup of index 2 in any group G is normal. Its two cosets are necessarily the subgroup itself and its complement in G. ◊

Exercises

Exercise 2.1.1 Provide all associative operations on the set $\{W, F\}$ of the two different elements W,F, see Example 2.1.11.

Exercise 2.1.2 Let M be a semigroup and $a \in M$. Then the elements of M that commute with a form a subsemigroup $Z(a) = Z_M(a)$ of M. If M is a monoid, then $Z(a)$ is a submonoid of M with $Z(a)^\times = Z(a) \cap M^\times$. In particular, $Z(a)$ is a subgroup when M is a group. ($Z(a)$ is called the **centralizer** of a in M. For any subsets $A \subseteq M$, $Z(A) = Z_M(A) := \bigcap_{a \in A} Z(a)$ is called the **centralizer** of A in M, and again $Z(A)^\times = Z(A) \cap M^\times$ holds if M is a monoid. $Z(M)$ is called the **center** of M. M is commutative exactly when $Z(M) = M$ is. If M is a group, its center $Z(M)$ is a normal subgroup of M.)

Exercise 2.1.3 Let X be a set. If X has at least two elements, then the monoid X^X is not commutative. If X has at least three elements, then the permutation group $\mathfrak{S}(X)$ is not commutative.

Exercise 2.1.4 Let X be a set.

a) If $|X| > 1$, then X has a permutation σ without a fixed point. (If X is infinite, σ can be chosen as an involution due to $|X| = |X| + |X|$. For finite X, there is an involution on X without a fixed point exactly when X is even.)

b) If X is infinite, then $|\mathfrak{S}(X)| = 2^{|X|}$. ($|\mathfrak{S}(X)| \leq 2^{|X|}$ follows from

$$|\mathfrak{S}(X)| \leq |\mathfrak{P}(X \times X)| = 2^{|X \times X|} = 2^{|X|}.$$

For the reverse inequality, consider the mapping $\mathfrak{S}(X) \to \mathfrak{P}(X)$, $\sigma \mapsto \text{Fix}\,\sigma$, and use a).)

Exercise 2.1.5 Let H be a multiplicative semigroup and $\mathrm{Idp}(H) = \{x \in H \mid x^2 = x\}$ the set of idempotent elements (= set of projections) of H.

a) If $h \in \mathrm{Idp}(H)$, then $hHh = \{hxh \mid x \in H\}$ is a subsemigroup of H with neutral element h.
b) For $x, y \in \mathrm{Idp}(H)$, the following conditions are equivalent: (i) $xHx \subseteq yHy$. (ii) $x \in yHy$. (iii) $x = yxy$. (iv) $x = yx = xy$.
c) Through "$x \leq y$ exactly when $x, y \in H$ fulfills the equivalent conditions from b)" an order is defined on the set $\mathrm{Idp}(H)$. If $1 \in H$ is a neutral element of H, then 1 is the greatest element of $\mathrm{Idp}(H)$. If $0 \in H$ is an absorbing element of H (i.e., $0 \cdot x = x \cdot 0 = 0$ holds for all $x \in H$), then 0 is the smallest element of H. If H is commutative, then $\mathrm{Idp}(H)$ is an (ordered, cf. Exercise 2.1.16) subgroup of H and $xy = \mathrm{Inf}(x, y)$ holds for $x, y \in \mathrm{Idp}(H)$.
d) Let X be a set and $p \in X^X$ be a projection (i.e., let $p^2 = p \circ p = p$). Describe set-theoretically the elements of the monoid pX^Xp (according to a). (These are, for example, exactly the mappings $f: X \to X$ with $\mathrm{Im} f \subseteq \mathrm{Im} p$ and $f(p^{-1}(x)) \subseteq p^{-1}(f(x))$, for which the diagram from Fig. 2.4 is commutative.) If $p, q \in X^X$ are comparable projections with respect to the canonical order on $\mathrm{Idp}(X^X)$ with $\mathrm{Im} \, p = \mathrm{Im} \, q$, then $p = q$

Exercise 2.1.6 Let M be a monoid.

a) If every element of M is involutory, then M is an abelian group, a so-called **elementary (abelian) 2-group**.(For the structure of such groups, see Exercise 2.3.6a).)
b) If M is a regular monoid and $(xy)^2 = x^2 y^2$ holds for all $x, y \in M$, then M is abelian. Provide a non-abelian monoid M (with 3 elements) for which $(xy)^2 = x^2 y^2$ holds for all $x, y \in M$.

Exercise 2.1.7 Let A be a multiplicative magma and e an element that is not in A. Then $A' := A \uplus \{e\}$ with respect to the operation $*$ with

$$a * b := ab, \quad \text{if} \quad a, b \in A; \quad e * c := c * e := c \quad \text{for all} \quad c \in A',$$

is a magma with neutral element e, which includes A as a submagma. If A is a semigroup, then A' is a pointed monoid (even if A is already a monoid). (It is said that A' arises from A by **adjoining a neutral element**. Through this construction, problems about semigroups can occasionally be reduced to problems about monoids. Construct in an analogous way to a magma (or a semigroup) a magma (or a semigroup) with an absorbing element by **adjoining an absorbing element**.)

Exercise 2.1.8 Let M be a monoid, in which every equation of the form $ax = b$ with $a, b \in M$ is solvable, for which therefore *all* left translations L_a, $a \in M$, are surjective. Then M is a group. Provide a non-empty semigroup (with two elements) in which all left translations are bijective, but which is not a group.

2.1 Monoids and Groups

Exercise 2.1.9 Let H be a non-empty semigroup, in which *all* left- *and all* right translations are surjective. Then H is a group. In particular, every non-empty regular and finite semigroup is a group, and every non-empty finite subsemigroup of a group is a subgroup.

Exercise 2.1.10 Let M be a monoid and $x \in M$ be an element with $x^d = e$ for a $d \in \mathbb{N}^*$. Then $x \in M^\times$, and for all $m, n \in \mathbb{Z}$ applies $x^m = x^n$, when m and n are congruent modulo d.

Exercise 2.1.11 For $a, b \in \mathbb{R}$ let $f_{a,b}: \mathbb{R} \to \mathbb{R}$ be defined by $f_{a,b}(x) := ax + b$, $x \in \mathbb{R}$. Then $M := \{f_{a,b} \mid a, b \in \mathbb{R}\}$ is a submonoid of $\mathbb{R}^\mathbb{R}$ with

$$M^\times = \{f_{a,b} \mid a, b \in \mathbb{R}, a \neq 0\}$$

as the group of units. ($M^\times$ is the so-called **affine group** $A_1(\mathbb{R})$, see the end of Example 2.6.19.)

Exercise 2.1.12 Let G be a finite group with n elements and $(a_1, \ldots, a_n) \in G^n$. Then there are indices r,s with $0 \leq r < s \leq n$ and $a_{r+1} \cdots a_s = e_G$. (The $n + 1$ products $a_1 \cdots a_s$, $s = 0, \ldots, n$, cannot be pairwise different.)

Exercise 2.1.13 Examine whether the specified subsets H are each subgroups of the group G:

$G := (\mathbb{Q}, +);$ $H := \{a/b \mid a, b \in \mathbb{Z}, b \text{ impair}\}.$
$G := (\mathbb{Q}^\times, \cdot);$ $H := \{x^n \mid x \in \mathbb{Q}^\times\}$ ($n \in \mathbb{Z}$ fixed),
  $H := \{a/b \mid a, b \in \mathbb{Z}, a \text{ and } b \text{ not divisble by } 5\}.$
$G := (\mathbb{C}^\times, \cdot);$ $H := \{z = a + b\mathrm{i} \in \mathbb{C} \mid |z|^2 = a^2 + b^2 = 1\},$
  $H := \{z = a + b\mathrm{i} \in \mathbb{C} \mid \Re z = a > 0\}.$
$G := \mathfrak{S}(A)$ (A a fixed set);
  $H := \{\sigma \in \mathfrak{S}(A) \mid \sigma(b) = b \text{ for all } b \in B\}$ ($B \subseteq A$ fixed),
  $H := \{\sigma \in \mathfrak{S}(A) \mid \sigma(a) = a \text{ for almost all } a \in A\},$
  $H := \{\sigma^n \mid \sigma \in \mathfrak{S}(A)\}$ ($n \in \mathbb{N}$ fixed),
  $H := \{\sigma \in \mathfrak{S}(A) \mid \sigma^n = \mathrm{id}_A\}$ ($n \in \mathbb{N}$ fixed).

Exercise 2.1.14 Let G be a group.

a) Let $H_1, H_2 \subseteq G$ be subgroups. $H_1 \cup H_2$ is also a subgroup if and only if $H_1 \subseteq H_2$ or $H_2 \subseteq H_1$ is.
b) If $A \subseteq G$, then A or the complement $G - A$ generates the group G.
c) If G is finite and $A, B \subseteq G$ are subsets with $|A| + |B| > |G|$, then $AB = G$.

Exercise 2.1.15

a) Every subgroup H of $\mathbb{Q} = (\mathbb{Q}, +)$ that has a finite generating system is already cyclic. (H is a subgroup of $\mathbb{Z}(1/m)$, where m is a common denominator (i.e., a common multiple of the denominators) of finitely many generators of H. – If $a_1, \ldots, a_n \in \mathbb{Q}$, $n \in \mathbb{N}^*$, then the uniquely determined non-negative generating elements of the subgroups $\sum_{i=1}^{n} \mathbb{Z} a_i$ or $\bigcap_{i=1}^{n} \mathbb{Z} a_i$ of $\mathbb{Q}$ are called the **greatest common divisor** or the **least common multiple** of the rational numbers $a_1, \ldots, a_n$. They are denoted by $\gcd(a_1, \ldots a_n)$ or $\operatorname{lcm}(a_1, \ldots, a_n)$. These designations are compatible with the terminology already introduced for integers due to Theorem 2.1.19. Show $\gcd(a_1, \ldots, a_n) \cdot \operatorname{lcm}(a_1^{-1}, \ldots, a_n^{-1}) = 1$, if $a_1, \ldots, a_n \in \mathbb{Q}^\times$.)

b) The group $\mathbb{Q} = (\mathbb{Q}, +)$ does not have a finite generating system. Moreover, $\mathbb{Q}$ does not have a minimal generating system, i.e., every generating system of $\mathbb{Q}$ can be reduced.

Exercise 2.1.16 Let $M = (M, \leq)$ be a (multiplicatively written) magma with an order $\leq$. M is called an **ordered magma**, if the so-called **monotonicity laws** apply, i.e., if all left and right translations in M are monotonically increasing. Then $(M, \geq)$ is also an ordered magma. For example, for every magma N, the power set $\mathfrak{P}(N)$ with complex multiplication with respect to inclusion $\subseteq$ and thus also with respect to anti-inclusion $\supseteq$ is an ordered magma with N as the largest or $\emptyset$ smallest element. A submagma of an ordered magma is again an ordered magma. $(\mathbb{R}, +)$ with the natural order and thus also its sub-semigroups or subgroups, such as $\mathbb{N}, \mathbb{Z}, \mathbb{Q}$ etc., are ordered semigroups or groups. The same applies to the multiplicative monoid $(\mathbb{R}_+, \cdot)$. In particular, $(\mathbb{R}_+^\times, \cdot)$ is an ordered group. The group $(\mathbb{R}^\times, \cdot)$ is not (with respect to the natural order) an ordered group. If M is an ordered monoid, then $M_{\geq 1}$ and $M_{\leq 1}$ are sharp(!) submonoids of M due to the monotonicity laws.

a) Let M be a commutative monoid. Then the monoid $\overline{M} = M/\|\|$ of the divisor classes of M with respect to the order given by divisibility on M is an ordered pointed monoid with [1] as the smallest element, and the natural injection $\overline{M} \to \mathfrak{P}(M)$, $[a] \mapsto Ma$, is compatible with both the operations and the orders $\leq$ or $\supseteq$ (thus an isomorphism of ordered monoids from $\overline{M}$ to its image). (Note that for $a, b \in M$ the inclusion $T_M(a) T_M(b) \subseteq T_M(ab)$ in $\mathfrak{P}(M)$ is generally strict. Example? The multiples sets are usually more manageable than the divisor sets.) This construction can be generalized: Let $N \subseteq M$ be a submonoid. We set

$$a \mid_N b \quad \text{if and only if} \quad Nb \subseteq Na.$$

Then $\mid_N$ is again a quasi-order on M, and $M/_N\|\|_N$ is an ordered monoid, where the equivalence relation $_N\|\|_N$ is defined by "$a \,_N\|\|_N b$ if and only if $Na = Nb$", and the order by "$[a]_N \leq [b]_N$ if and only if $a \mid_N b$ are given. The submonoid $\overline{N} := N/_N\|\|_N \subseteq M/_N\|\|_N$ is the submonoid of the elements $\geq_N [1]_N$. If M is regular and sharp, then $M = M/_N\|\|_N$ itself is an ordered monoid with $M_{\geq 1} = N$.

2.1 Monoids and Groups

(Suprema and infima with respect to the order $\leq_N$ are denoted with lcm_N and $\gcd_N$ respectively.)

b) **(Ordered Groups)** Let G be an (not necessarily abelian) ordered group. Then $P := G_{\geq 1}$ is a submonoid of G with $P^{-1} = G_{\leq 1}$. Furthermore, (1) $aPa^{-1} = P$ holds for all $a \in G$ as well as (2) $P \cap P^{-1} = \{1\}$. G is a completely ordered group if and only if $P \cup P^{-1} = G$ is. Conversely, if $P \subseteq G$ is a submonoid of G with (1) and (2), then there is exactly one order $\leq_P$ on G such that $(G, \leq_P)$ is an ordered group with $G_{\geq 1} = P$ is. It is $a \leq_P b$ exactly when $a^{-1}b \in P$ is (or when $ba^{-1} = a(a^{-1}b)a^{-1} \in P$ is). (**Note** If G is an additively written abelian ordered group, then $P := G_{\geq 0}$ (somewhat abusively) is also called the **positivity domain** of G. It is a pointed submonoid of G with $P \cap (-P) = \{0\}$. For example, the product group $(\mathbb{R} \times \mathbb{R}, +)$ is an ordered group with respect to both the product order and the lexicographic order. The positivity domains are sketched in Fig. 1.13, where for (a, b) the point $(0, 0)$ is to be chosen. The lexicographic order is complete.)

Exercise 2.1.17 Let M be a commutative regular monoid with neutral element 1 and $a, b, c \in M$. Prove the following calculation rules for the gcd and the lcm. (Note that the gcd and lcm in M are each uniquely determined only up to associativity.)

a) If $\gcd(ac, bc)$ exists, then $\gcd(a, b)$ also exists and $\gcd(ac, bc) = \gcd(a, b)c$ applies. Provide an example where $\gcd(a, b)$ exists without $\gcd(ac, bc)$ existing.
b) $\text{lcm}(ac, bc)$ exists precisely when $\text{lcm}(a, b)$ exists, and $\text{lcm}(ac, bc) = \text{lcm}(a, b)c$ applies. (Note $M(ac) \cap M(bc) = (Ma \cap Mb)c$ and $M\,\text{kgV}(a,b) = Ma \cap Mb$.)
c) If $\text{lcm}(a, b)$ exists, then $\gcd(a, b)$ also exists and $\text{lcm}(a, b)\gcd(a, b) = ab$ applies (up to associativity). If the elements a,b are coprime, i.e. if $\gcd(a, b) = 1$ and *exists* $\text{lcm}(a, b)$, then $\text{lcm}(a, b) = ab$. Conversely, from $\text{lcm}(a, b) = ab$ it already follows that a and b are coprime. Provide an example of two coprime elements whose least common multiple does not exist. (**Note** Some authors define the coprimality of elements a,b of a commutative regular monoid by the condition $\text{lcm}(a, b) = ab$, which is more restrictive than our definition.)
d) If $\gcd(ac, bc)$ exists for all $c \in M$, then $\text{lcm}(a, b)$ also exists and thus $\text{lcm}(ac, bc)$ for all $c \in M$. (One reduces to the case $\gcd(a, b) = 1$. Then ab is a least common multiple of a,b.)
e) $\gcd(a, b)$ exists precisely when $a, b \in M$ for any $\text{lcm}(a, b)$, if $a, b \in M$ exists for any.

Exercise 2.1.18 Let M be a regular commutative monoid. For elements $a \in M - M^\times$ and $x \in M$ let

$$v_a(x) \in \overline{\mathbb{N}} = \mathbb{N} \uplus \{\infty\}$$

the so-called a-**exponent** of x, that is the supremum in $\overline{\mathbb{N}}$ of the exponents $m \in \mathbb{N}$ with $a^m \mid x$. If $\mathsf{v}_a(x)$ is finite, then x has the representation $x = a^{\mathsf{v}_a(x)} x'$, where x' is the so-called a-**free factor** of x with $\mathsf{v}_a(x') = 0$. If $x = a^m b$ with $m \in \mathbb{N}$ and $\mathsf{v}_a(b) = 0$, then $m = \mathsf{v}_a(x)$ and $b = x'$ is the a-free factor of x. Provide an example that $\mathsf{v}_a(x) = \infty$ can be. (In the additive sharp monoid $(\mathbb{Z} \times \mathbb{N}^*) \uplus (\mathbb{N} \times \{0\}) \subseteq (\mathbb{Z} \times \mathbb{N}, +)$ for example, the element $p := (1, 0)$ is the only prime element and also the only irreducible element. It divides every element $\neq (0, 0)$, and it is $\mathsf{v}_p(x) = \infty$ for every $x \in \mathbb{Z} \times \mathbb{N}^*$.)

a) For $a \in M - M^\times$, the a-exponent $\mathsf{v}_a \colon M \to \overline{\mathbb{N}}$ is **superadditive**, i.e., it is $\mathsf{v}_a(xy) \geq \mathsf{v}_a(x) + \mathsf{v}_a(y)$ for all $x, y \in M$. Exactly then is v_a additive (i.e., exactly then the equality sign always applies in the inequality), when a is prime.

b) Let $p \notin M^\times$. The following statements are equivalent: (i) p is prime. (ii) For every element $a \in M$ with $p \nmid a$, $\mathrm{lcm}(a, p) = ap$ applies. (iii) The p-exponent $\mathsf{v}_p \colon M \to (\overline{\mathbb{N}}, +)$, $x \mapsto \mathsf{v}_p(x)$, is additive. (Note: If $q \in M$ is irreducible and $q \nmid a$ applies, then $\gcd(a, q) = 1$. So if q is irreducible and $\mathrm{lcm}(a, q)$ exists for all $a \in M$, then q is prime according to Exercise 2.1.17c).

c) Let $a \in M - M^\times$ be a product of prime elements. Then $(\mathsf{v}_p(a))_{p \in \mathbb{P}_M} \in \mathbb{N}^{(\mathbb{P}_M)}$ and

$$a = \varepsilon \prod_{p \in \mathbb{P}_M} p^{\mathsf{v}_p(a)} \quad \text{with a unit } \varepsilon \in M^\times.$$

If $a = \eta \prod_{p \in \mathbb{P}_M} p^{\alpha_p}$ with $(\alpha_p) \in \mathbb{N}^{(\mathbb{P}_M)}$ and $\eta \in M^\times$ is another such representation, then $(\alpha_p) = (\mathsf{v}_p(a))$ and $\eta = \varepsilon$ (**uniqueness of prime factorization**). The divisors of a are exactly the elements $\eta \prod_{p \in \mathbb{P}_M} p^{\alpha_p}$ with $\eta \in M^\times$, $(\alpha_p) \in \mathbb{N}^{(\mathbb{P}_M)}$, $\alpha_p \leq \mathsf{v}_p(a)$ for all $p \in \mathbb{P}_M$ (induction over $\sum_p \mathsf{v}_p(a)$). In particular, $\tau_M(a) = \prod_{p \in \mathbb{P}_M} (\mathsf{v}_p(a) + 1)$, see Problem 1.7.12b). a has only finitely many divisors in M if and only if $M^\times$ is finite. It is $|T_M(a)| = |M^\times| \cdot \tau_M(a)$.

Exercise 2.1.19 Let G be a group, N a normal subgroup in G and F, H subgroups of G with $F \subseteq H$.

a) $HN = NH$ is a subgroup of G.
b) If F is a normal subgroup in H, then FN is a normal subgroup in HN. In particular, FN is a normal subgroup in G if F is normal in G.
c) $H \cap N$ is a normal subgroup in H.

Exercise 2.1.20 The normal subgroups of a group G form a complete sublattice of the lattice of all subgroups of G. If N_i, $i \in I$, is a family of normal subgroups of G, then $N_i N_j = N_j N_i$ for all $i, j \in I$ and $\bigcap_{i \in I} N_i$ is the infimum or $\mathrm{H}\left(\bigcup_{i \in I} N_i\right) = \bigcup_{J \in \mathfrak{E}(I)} N_J$ the supremum of the N_i, $i \in I$, where $N_J := \prod_{i \in J} N_i$ was set for each *finite* subset $J \subseteq I$.

Exercise 2.1.21 Let H and F be subgroups of the group G with $H \subseteq F$. If $a_i, i \in I$, or $b_j, j \in J$, are each complete representative systems for F/H or G/F, then $a_i b_j$,

2.1 Monoids and Groups

$(i,j) \in I \times J$, is a complete representative system for G/H. – In particular, we get the so-called **index theorem**

$$[G:H] = [G:F][F:H].$$

Exercise 2.1.22 Let H,F be subgroups of the group G.

a) It is $[F:(F \cap H)] \leq [G:H]$. Every complete representative system of $F/(F \cap H)$ is part of a complete representative system for G/H.

b) If FH is a subgroup of G, i.e. $FH = HF$, then

$$[FH:H] = [F:(F \cap H)] \quad \text{and} \quad [G:F][G:H] = [G:FH][G:(F \cap H)].$$

applies. What are these formulas in the case of the subgroups $F := \mathbb{Z}m$, $H := \mathbb{Z}n$ of $G := \mathbb{Z}$, $m, n \in \mathbb{N}^*$?

c) If the subgroups $H_1, \ldots, H_n$ of the group G all have a finite index, so does their intersection $H := H_1 \cap \cdots \cap H_n$ and

$$[G:H] \leq [G:H_1]\cdots[G:H_n].$$

applies. If the indices $[G:H_i]$, $i = 1, \ldots, n$, are pairwise coprime, then the equality sign applies.

Exercise 2.1.23 The **Möbius µ-function** $\mu: \mathbb{N}^* \to \mathbb{Z}$ (named after A. Möbius (1790–1868)) is defined by

$$\mu(n) := \begin{cases} (-1)^r, & \text{if } n \in \mathbb{N}^* \text{ is squarefree with exactly } r \text{ prime faktors,} \\ 0, & \text{otherwise.} \end{cases}$$

The values $\mu(1), \ldots, \mu(10)$ are in order $1, -1, -1, 0, -1, 1, -1, 0, 0, 1$. For $n \in \mathbb{N}^*$, let $\mathbb{P}_n$ denote the set of prime divisors of n and for a finite set $H \subseteq \mathbb{P}$ of prime numbers, let $p^H = \prod_{p \in H} p$.

a) Prove the following fundamental property of the µ-function: For $n \in \mathbb{N}^*$ is

$$\sum_{d|n} \mu(d) = \delta_{1,n} = \begin{cases} 1, & \text{if } n = 1, \\ 0, & \text{otherwise.} \end{cases}$$

(It is $\sum_{d|n} \mu(d) = \sum_{H \subseteq \mathbb{P}_n} (-1)^{|H|} = 1$, if $\mathbb{P}_n = \emptyset$, and $= 0$, if $\mathbb{P}_n \neq \emptyset$, cf. Exercise 1.6.5a).

b) Let H be an additively written abelian group. For a **number-theoretical function** $f: \mathbb{N}^* \to H$ is called $F = S(f): \mathbb{N}^* \to H$, $n \mapsto \sum_{d|n} f(d)$, the **summatory function** of f. According to a), for example, $n \mapsto \delta_{1,n}$ is the summatory function of the µ-function. The following **Möbius inversion formula** applies:

$$f(n) = \sum_{d|n} \mu(d) F(n/d) = \sum_{H \subseteq \mathbb{P}_n} (-1)^{|H|} F(n/p^H), \quad n \in \mathbb{N}^*.$$

(It is $\sum_{d|n} \mu(d) F(n/d) = \sum_{(d,e),\, de|n} \mu(d) f(e) = \sum_{e|n} \left(\sum_{d|(n/e)} \mu(d) \right) f(e) = f(n)$.) Formulate this inversion formula also for multiplicatively written abelian groups.

2.2 Homomorphisms

We have already encountered mappings between magmas that are compatible with the respective operations. Such structure-compatible mappings are called homomorphisms (and in more general situations also morphisms). They are of utmost importance in all of mathematics—not just in algebra—for comparing structures. Here we initially focus on sets with operations, which we — unless otherwise stated — write multiplicatively.

Definition 2.2.1 A mapping $\varphi: M \to N$ between magmas M and N is called a **homomorphism** or a **multiplicative** mapping (of magmas), if for all $a, b \in M$ the following holds:

$$\varphi(ab) = \varphi(a)\varphi(b).$$

Bijective homomorphisms are called **isomorphisms**. In the case of additive notation of the operations, instead of homomorphisms, we also speak of **additive** mappings.

A mapping $\varphi: M \to N$ between magmas is exactly a homomorphism when φ commutes with the left or right translations in M or N, i.e., if for all a or b in M the following holds: $\varphi \circ L_a = L_{\varphi(a)} \circ \varphi$ or $\varphi \circ R_b = R_{\varphi(b)} \circ \varphi$. A homomorphism $\varphi: M^{\mathrm{op}} \to N$ (or – equivalently – a homomorphism $\varphi: M \to N^{\mathrm{op}}$), for which $\varphi(ab) = \varphi(b)\varphi(a)$ holds for all $a, b \in M$, is called an **antihomomorphism.**

If $\psi: L \to M$ is next to $\varphi: M \to N$ another homomorphism, then the composition $\varphi\psi: L \to N$ is obviously also a homomorphism. Furthermore, $\varphi^{-1}: N \to M$ is an isomorphism, if φ is an isomorphism. If namely $c = \varphi(a)$ and $d = \varphi(b)$, $a, b \in M$, are arbitrary elements of N, then $\varphi^{-1}(cd) = \varphi^{-1}(\varphi(a)\varphi(b)) = \varphi^{-1}(\varphi(ab)) = ab = \varphi^{-1}(c)\varphi^{-1}(d)$.

The set of homomorphisms or isomorphisms from M to N we denote with

$$\mathrm{Hom}(M,N) \quad \text{resp.} \quad \mathrm{Iso}(M,N).$$

Homomorphisms from M to itself ($M = N$) are called **endomorphisms,** bijective endomorphisms, i.e., isomorphisms from M onto itself, are also called **automorphisms.** Since the identity id_M is trivially an automorphism of M, the endomorphisms or automorphisms of M form a submonoid or a subgroup

$$\mathrm{End}\, M = \mathrm{Hom}(M,M) \quad \text{resp.} \quad \mathrm{Aut}\, M = \mathrm{Iso}(M,M) = (\mathrm{End}\, M)^{\times}$$

of the monoid M^M or the permutation group $\mathfrak{S}(M) = (M^M)^{\times}$. Two magmas M and N are called **isomorphic,** if there is an isomorphism from M to N. In this case, we write

2.2 Homomorphisms

$$M \cong N.$$

The isomorphism of magmas is reflexive, symmetric, and transitive, and thus an equivalence relation on any set of magmas. An equivalence class with respect to isomorphism is called an **isomorphism class.** For example, the single-element magmas form an isomorphism class. Isomorphic magmas have the same structure. They only differ in the naming of their elements. If M is a magma and $\varphi: M \to X$ is a bijective mapping to any *set* X, then X has exactly one operation $*$, with respect to which φ is an isomorphism, namely $x * y := \varphi(\varphi^{-1}(x)\varphi^{-1}(y))$, $x, y \in X$. It is said that *the structure of M has been transferred to X by means of φ.*

Even the following simple proposition shows the importance of homomorphisms.

Proposition 2.2.2 *For a* surjective *homomorphism* $\varphi: M \to M'$ *of magmas, the following applies:*

(1) *If M is associative, i.e. if M is a semigroup, then so is M'.*
(2) *If M is commutative, then so is M'.*
(3) *If M has a neutral element, then so does M'. More precisely: If e_M is a neutral element of M, then $\varphi(e_M)$ is a neutral element of M'.*
(4) *If M is a monoid, then so is M', and for $a \in M^\times$ is $\varphi(a) \in (M')^\times$ with $(\varphi(a))^{-1} = \varphi(a^{-1})$. In particular, M' is a group, if this applies to M.*

Proof For any elements $a', b', c' \in M'$ there are $a, b, c \in M$ with $\varphi(a) = a'$, $\varphi(b) = b'$, $\varphi(c) = c'$. If M is now associative, then $(a'b')c' = (\varphi(a)\varphi(b))\varphi(c) = \varphi(ab)\varphi(c) = \varphi((ab)c) = \varphi(a(bc)) = \varphi(a)\varphi(bc) = \varphi(a)(\varphi(b)\varphi(c)) = a'(b'c')$ applies.

The proof of (2) results from commutative M from $a'b' = \varphi(a)\varphi(b) = \varphi(ab) = \varphi(ba) = \varphi(b)\varphi(a) = b'a'$. If e_M is a neutral element of M, then $\varphi(e_M)a' = \varphi(e_M)\varphi(a) = \varphi(e_M a) = \varphi(a) = a'$ and analogously $a'\varphi(e_M) = a'$, $\varphi(e_M)$ thus a neutral element of M'. (4) finally follows from (1) and (3) as well as from $e_{M'} = \varphi(e_M) = \varphi(aa^{-1}) = \varphi(a)\varphi(a^{-1})$ and $e_{M'} = \varphi(e_M) = \varphi(a^{-1}a) = \varphi(a^{-1})\varphi(a)$ for $a \in M^\times$. ∎

If M and N are monoids, then a homomorphism $\varphi: M \to N$ in the sense of Definition 2.2.1 does not necessarily map the neutral element e_M of M to the neutral element e_N of N. However, φ is only called a **monoid homomorphism** if additionally $\varphi(e_M) = e_N$ holds. According to Proposition 2.2.2, this is automatically the case if φ is surjective, and also when M has only a single idempotent element, for example when N is a regular monoid, and *especially when N is a group*. In this case, $\varphi(e_M) = \varphi(e_M^2) = (\varphi(e_M))^2$ is also idempotent, so it equals e_N by assumption. A semigroup homomorphism from groups is therefore always a **group homomorphism**. The constant mapping $M \to N$, $a \mapsto e_N$, is a monoid homomorphism, the so-called **trivial homomorphism** from M to N. For monoids M,N, Hom(M, N) is therefore never empty. If M is a group, then the inverse formation

$a \mapsto a^{-1}$ in M is an involutional anti-isomorphism of M due to $(ab)^{-1} = b^{-1}a^{-1}$ and $(a^{-1})^{-1} = a$. In the case of commutative M, it is an automorphism of M. A group and its opposite group are therefore always isomorphic. (This is generally not the case for monoids. Example?)

Let $\varphi \colon M \to N$ be a homomorphism of semigroups. Then φ is also compatible with multiple products (induction over n):

$$\varphi(a_1 \cdots a_n) = \varphi\left(\prod_{i=1}^{n} a_i\right) = \prod_{i=1}^{n} \varphi(a_i) = \varphi(a_1) \cdots \varphi(a_n), \quad a_1, \ldots, a_n \in M,\ n \in \mathbb{N}^*.$$

If φ is a monoid homomorphism, this also applies for $n = 0$. For powers, specifically:

Proposition 2.2.3 *Let $\varphi \colon M \to N$ be a monoid homomorphism and $a \in M$. Then the following applies:*

(1) *It is $\varphi(a^n) = \varphi(a)^n$ for all $n \in \mathbb{N}$.*
(2) *If $a \in M^\times$, then $\varphi(a) \in N^\times$ and it applies $\varphi(a^n) = \varphi(a)^n$ for all $n \in \mathbb{Z}$.*

Proof (1) follows directly from the preliminary remark. (2) is obtained from (1) and the following: If $a \in M^\times$, then $\varphi(a) \in N^\times$ and $\varphi(a^{-1}) = \varphi(a)^{-1}$ holds. The latter, however, follows from $e_N = \varphi(e_M) = \varphi(aa^{-1}) = \varphi(a)\varphi(a^{-1})$ and $e_N = \varphi(e_M) = \varphi(a^{-1}a) = \varphi(a^{-1})\varphi(a)$. □

A monoid homomorphism $\varphi \colon M \to N$ induces a group homomorphism $M^\times \to N^\times$, which is usually denoted by $\varphi^\times$. For magmas M,N, the set $\mathrm{Hom}(M, N)$ is a subset of the product magma $N^M = \mathrm{Abb}(M, N)$ (with $(fg)(a) = f(a)g(a)$ for $f, g \in N^M, a \in M$). For associative operations, even the following holds:

Proposition 2.2.4 *Let M, N be semigroups (or monoids or groups) and N be commutative. Then $\mathrm{Hom}(M, N)$ is a subsemigroup (or a submonoid or a subgroup) of N^M.*

Proof Let $f, g \in \mathrm{Hom}(M, N)$. For $a, b \in M$ applies

$$(fg)(ab) = f(ab)g(ab) = f(a)f(b)g(a)g(b) = f(a)g(a)f(b)g(b)$$
$$= (fg)(a)(fg)(b).$$

Thus, $\mathrm{Hom}(M, N)$ is a subgroup of N^M. If M,N are monoids, then the neutral element of N^M is the trivial homomorphism $a \mapsto e_N$, thus $\mathrm{Hom}(M, N)$ is a submonoid of N^M. If M,N are groups, then $\mathrm{Hom}(M, N)$ contains with f also the inverse f^{-1} (which should not be confused with a possibly existing homomorphism inverse to $f\, N \to M$). This inverse is namely the composition of f with the inverse formation $b \mapsto b^{-1}$ of N, which (since N is commutative) is an (involutory) automorphism of N. According to Proposition 2.1.6, $\mathrm{Hom}(M, N)$ is then a subgroup of N^M. □

2.2 Homomorphisms

Let L, M, N be additively written abelian groups. According to Proposition 2.2.4, $\mathrm{Hom}(L,M)$ and $\mathrm{Hom}(M,N)$ are also abelian groups, and for the composition

$$\mathrm{Hom}(M,N) \times \mathrm{Hom}(L,M) \to \mathrm{Hom}(L,N), \quad (g,f) \mapsto g \circ f,$$

the following **distributive laws** obviously apply:

$$(g_1 + g_2) \circ f = g_1 \circ f + g_2 \circ f, \quad g \circ (f_1 + f_2) = g \circ f_1 + g \circ f_2,$$

$f_1, f_2, f \in \mathrm{Hom}(L,M)$, $g, g_1, g_2 \in \mathrm{Hom}(M,N)$. It is said that the composition of homomorphisms of abelian groups is **biadditive**. $\mathrm{End}\, N = \mathrm{Hom}(N,N)$ thus has the composition $\circ$ as a natural operation in addition to the addition $+$. With respect to the addition, $\mathrm{End}\, N$ is a group and with respect to the composition a monoid with id_N as the neutral element and $\mathrm{Aut}\, N = \mathrm{End}(N,N)^\times$ as the group of invertible elements. Both operations are also connected by the distributive laws. $\mathrm{End}\, N = (\mathrm{End}\, N, +, \circ)$ *is thus for an abelian group N a ring*, see Definition 2.6.1.

Let $\varphi \colon M \to N$ be a homomorphism of magmas and M' or N' submagmas of M or N. Apparently, the image $\varphi(M')$ and the preimage $\varphi^{-1}(N')$ are submagmas of N or M. Analogous statements apply to semigroups, monoids, and groups. (In the case of groups, Proposition 2.2.3 (2) is used.) In particular, for a monoid homomorphism $\varphi \colon M \to N$, the preimage

$$\ker \varphi := \varphi^{-1}(e_N) = \{x \in M \mid \varphi(x) = e_N\}$$

of the trivial submonoid $\{e_N\} \subseteq N$ is a submonoid of M, the so-called **kernel of** φ. If M is a group, then $\ker \varphi$ is a subgroup of M. For groups, the following applies more precisely:

Theorem 2.2.5 *Let $\varphi \colon G \to H$ be a homomorphism of groups and $a \in G$. Then*

$$\varphi^{-1}(\varphi(a)) = a \ker \varphi = (\ker \varphi)a.$$

In particular, $\ker \varphi$ is a normal subgroup of G, and the non-empty fibers of φ are the cosets of $\ker \varphi$ in G. It holds that

$$|G| = |\mathrm{Im}\, \varphi| \cdot |\ker \varphi|.$$

All elements of the fiber $\varphi^{-1}(\varphi(a))$ are obtained through $a \in G$, by multiplying a with all elements of the kernel of φ from the right or the left.

Proof We only show $\varphi^{-1}(\varphi(a)) = a \ker \varphi = \{ay \mid y \in \ker \varphi\}$. One proves $\varphi^{-1}(\varphi(a)) = (\ker \varphi)a$ analogously. So let $x \in \varphi^{-1}(\varphi(a))$ be, i.e. $\varphi(x) = \varphi(a)$. Then $\varphi(a^{-1}x) = \varphi(a)^{-1}\varphi(x) = e_H$ is, thus $y := a^{-1}x \in \ker \varphi$, $x = ay \in a \ker \varphi$. This proves the inclusion $\varphi^{-1}(\varphi(a)) \subseteq a \ker \varphi$. Conversely, if $x = ay$ with $y \in \ker \varphi$, then $\varphi(x) = \varphi(a)\varphi(y) = \varphi(a)e_H = \varphi(a)$ is, which also proves the inclusion $a \ker \varphi \subseteq \varphi^{-1}(\varphi(a))$. □

The following corollary directly results:

Corollary 2.2.6 (Injectivity Criterion) *A homomorphism $\varphi \colon G \to H$ of groups is injective if and only if $\ker \varphi = \{e_G\}$ is trivial.*

The injectivity criterion does not generally apply to monoid homomorphisms. Of course, an injective monoid homomorphism has a trivial kernel, but the converse is not usually true. Example?

Every normal subgroup $N \subseteq G$ *of a group* G *appears as the kernel of a group homomorphism* $G \to H$. The set $G/N = G\backslash N$ of the cosets with respect to N is then a subgroup of $\mathfrak{P}(G)$ due to $(aN)(bN) = aNbN = abNN = abN$ and the canonical projection $\pi_N \colon G \to G/N$, $a \mapsto aN$, is a homomorphism of semigroups. *Therefore*, see Proposition 2.2.2, G/N *is a group with neutral element* $\pi_N(e_G) = N$ *and* $(aN)^{-1} = \pi_N(a)^{-1} = \pi_N(a^{-1}) = a^{-1}N$, $a \in G$, *and the kernel of* π_N *is* N.[8] In the situation of Theorem 2.2.5 the homomorphism $\varphi \colon G \to H$ induces an injective mapping $\overline{\varphi} \colon G/\ker \varphi \to H$ with $\overline{\varphi}(a \ker \varphi) = \varphi(a)$. The groups $G/\ker \varphi$ and $\operatorname{Im} \varphi$ are therefore canonically isomorphic. In the next section, we will discuss these so-called **Quotient groups** G/N, $N \subseteq G$ normal subgroup, and discuss the just mentioned isomorphism in more detail and in a more general context.

For a subgroup F of G and a group homomorphism $\varphi \colon G \to H$ with $N := \ker \varphi$ is exactly then $\varphi^{-1}(\varphi(F)) = F$, when $N \subseteq F$ applies. In addition, we mention:

Proposition 2.2.7 *Let* $\varphi \colon G \to H$ *be a surjective homomorphism of groups. Then the mappings* $G' \mapsto \varphi(G')$ *and* $H' \mapsto \varphi^{-1}(H')$ *are inverse bijective mappings of the set of subgroups* $G' \subseteq G$ *with* $\ker \varphi \subseteq G'$ *and the set of subgroups* $H' \subseteq H$. *The normal subgroups of* G, *which include* $\ker \varphi$, *correspond to the normal subgroups of* H.

Note that in the situation of Proposition 2.2.7, the φ image of *every* normal subgroup of G is a normal subgroup of H; because $\varphi(aG') = \varphi(a)\varphi(G')$ and $\varphi(G'a) = \varphi(G')\varphi(a)$ for every subgroup $G' \subseteq G$ and every $a \in G$.

Example 2.2.8 (Cayley Mapping) Let M be a magma. Then the mapping

$$L \colon M \to M^M, \ a \mapsto (L_a \colon x \mapsto ax),$$

is a homomorphism of the magma M *into the mapping monoid* M^M, *if* $L_{ab} = L_a \circ L_b$ *is for all* $a, b \in M$, *i.e. if the operation on* M *is associative,* M *is therefore a semigroup. If* M *is a monoid, then* L *is because of* $L(e_M) = \operatorname{id}_M$ *even a monoid homomorphism and* L *is also injective because of* $L_a(e_M) = ae_M = a$. (For a semigroup, L is generally not injective.) Thus, L induces an isomorphism from M to the submonoid $L(M) \subseteq M^M$. In the case of a group G, the image of L lies in the group of invertible elements of the monoid G^G, i.e. in the permutation group $\mathfrak{S}(G)$ of G, and L induces a group isomorphism from G to the subgroup $L(G)$ of $\mathfrak{S}(G)$.

---

[8] Note: G/N is not a submonoid of the monoid $e_{G/N} = N \neq \{e_G\} = e_{\mathfrak{P}(G)}$ at $\mathfrak{P}(G)$.

2.2 Homomorphisms

Theorem 2.2.9 (Cayley's Representation Theorem) *Every group G is isomorphic to a subgroup of the permutation group $\mathfrak{S}(G)$ of G, and every finite group with $\operatorname{Ord} G = n$, $n \in \mathbb{N}^*$, is isomorphic to a subgroup of the group $\mathfrak{S}_n$.*

The addition results from the following: If $g: X \to Y$ is a bijective mapping of arbitrary sets, then the so-called **conjugation** $\kappa_g: f \mapsto gfg^{-1}$ **with** g is an isomorphism of the monoid X^X onto the monoid Y^Y, which therefore induces a group isomorphism $\kappa_g: \mathfrak{S}(X) \to \mathfrak{S}(Y)$.

A homomorphism $\varphi: M \to N$ is also called a **representation** of M in N. The representation is called **faithful**, if φ is injective. In this case, φ induces an isomorphism from M to $\varphi(M) \subseteq N$. The Cayley representation theorem thus states that $L: G \to \mathfrak{S}(G)$ is a faithful representation of the group G in the permutation group $\mathfrak{S}(G)$. It is also called the **regular representation** of the group G. Generally, a representation of a group G in the permutation group $\mathfrak{S}(X)$ of a set X is called an **operation** of G on X. Such operations are an essential tool for studying groups. We will deal with them in more detail in Sect. 2.4.

Because of $R_a \circ R_b = R_{ba}$ for elements a,b of a semigroup M, $R: a \mapsto R_a$ is an antihomomorphism from M to M^M, thus a representation of the oppositional semigroup M^{op} in the mapping monoid M^M (which is generally not faithful). It is again faithful if M and thus M^{op} is a monoid or a group. This is referred to as an **anti-representation**.

However, an arbitrary semigroup M can also be faithfully represented in a mapping monoid. First of all, M can be faithfully represented in a monoid $M' := M \uplus \{e\}$ by adjoinment of a neutral element e, cf. Ex. 2.1.7, and M' can then be faithfully represented in $M'^{M'}$. ◊

Example 2.2.10 (Order of a group element – Isomorphism classes of cyclic groups) Let G be a group, $a \in G$ and $\mathrm{H}(a)$ the cyclic subgroup of G generated by a. Then the mapping

$$\varphi: n \mapsto a^n, \quad n \in \mathbb{Z},$$

is a surjective homomorphism of the group $\mathbb{Z} = (\mathbb{Z}, +)$ onto $\mathrm{H}(a)$. The kernel of φ is the subgroup

$$\ker \varphi = \{n \in \mathbb{Z} \mid a^n = e\} \subseteq \mathbb{Z}.$$

According to Theorem 2.1.18 this is the group $\mathbb{Z}m$ of integer multiples of a uniquely determined number $m \in \mathbb{N}$. This m is called the **order** of a and is denoted by

$$\operatorname{Ord} a.$$

So it is

$$\ker \varphi = \mathbb{Z} \operatorname{Ord} a.$$

and thus $a^n = e$ for $n \in \mathbb{Z}$ exactly when n is a multiple of $\operatorname{Ord} a$ is.

Fig. 2.5 Cyclic group Z_m of order $m \in \mathbb{N}^*$ (here Z_{16})

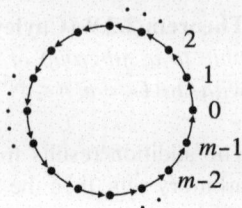

If $m = \text{Ord } a = 0$ is the case, then φ is injective and consequently bijective according to 2.2.6, i.e., all powers a^n, $n \in \mathbb{Z}$, of a are pairwise different and φ is an *isomorphism from* $\mathbb{Z}$ *to* $H(a)$.[9]

Let $m = \text{Ord } a > 0$. According to Theorem 2.2.5, $\varphi^{-1}(a^n) = n + \ker \varphi = n + \mathbb{Z}m$ is the remainder class of n modulo m. Therefore,

$$a^n = a^{n'} \quad \text{if and only if} \quad n \equiv n' \mod m = \text{Ord } a,$$

applies, i.e., if the difference $n - n'$ of the exponents n, n' is divisible by $m = \text{Ord } a$. Thus,

$$e_G = a^0,\ a = a^1, \ldots, a^{m-1}$$

are the pairwise different powers of a, and *the cyclic group* $H(a)$ *has the finite (positive) order* m. In any case, the induced mapping

$$\overline{\varphi}\colon \mathbb{Z}/\mathbb{Z}m \to H(a), \quad r + \mathbb{Z}m \mapsto a^r,$$

is well-defined and bijective and even a group isomorphism of the quotient group $\mathbb{Z}/\mathbb{Z}m$ onto $H(a)$ because of $\overline{\varphi}([r]_m + [s]_m) = \overline{\varphi}([r+s]_m) = a^{r+s} = a^r a^s = \overline{\varphi}([r]_m)\overline{\varphi}([s]_m)$. Note that we have already introduced the group $(\mathbb{Z}/\mathbb{Z}m, +)$ in Example 1.3.6. We have proven:

Theorem 2.2.11 *Every cyclic group G is isomorphic to exactly one of the cyclic residue class groups $\mathbb{Z}/\mathbb{Z}m$, $m \in \mathbb{N}$. If a is a generating element of G with $\text{Ord } a = m$, then*

$$\mathbb{Z}/\mathbb{Z}m \xrightarrow{\sim} G, \quad r + \mathbb{Z}m \mapsto a^r,$$

is an isomorphism of groups. In particular, G is infinite if and only if $m = 0$, and $\text{Ord } G = \text{Ord } a = m$, if $m \in \mathbb{N}^*$ is.

The residue class groups $\mathbb{Z}/\mathbb{Z}m$, $m \in \mathbb{N}$, thus represent all cyclic groups up to isomorphism. We use

$$\mathbf{Z}_m := (\mathbb{Z}/\mathbb{Z}m, +),\ m \in \mathbb{N}^*, \quad \text{resp.} \quad \mathbf{Z}_0 := (\mathbb{Z}, +)\ (= \mathbb{Z}/\mathbb{Z}0)$$

---

[9] Often in this case, it is also said that the order of a is ∞, since $H(a)$ then has infinitely many elements.

2.2 Homomorphisms

as standard models for a finite cyclic group of order $m \in \mathbb{N}^*$ or for an infinite cyclic group. If we identify the elements of Z_m with their standard representatives $0, \ldots, m - 1$ for $m > 0$, the Cayley representation $\mathbb{Z}_m \to \mathfrak{S}(\mathbb{Z}_m)$, cf. Example 2.2.8, maps the canonical generating element $1 \in \mathbb{Z}_m$ to the permutation

$$\begin{pmatrix} 0 & 1 & 2 & \cdots & m-2 & m-1 \\ 1 & 2 & 3 & \cdots & m-1 & 0 \end{pmatrix}$$

which is also briefly denoted by $\langle 0, 1, 2, \ldots, m-2, m-1 \rangle$, cf. Sect. 2.5. It cyclically swaps the m elements $0, 1, 2, \ldots, m-2, m-1$, cf. Fig. 2.5. The subgroup generated by it in $\mathfrak{S}(\mathbb{Z}_m) \cong \mathfrak{S}_m$ is thus also a natural model for the cyclic group of order m.

To prove sentences about cyclic groups, these standard models are often used. However, note that *an isomorphism $\mathbb{Z}_m \xrightarrow{\sim} G$ is only uniquely determined in the case $m = 1$ and $m = 2$*, since for $m \neq 1, 2$ with a always also $a^{-1} \neq a$ is a generating element of the group $G = H(a)$ and $r + \mathbb{Z}m \mapsto a^r$ or $r + \mathbb{Z}m \mapsto a^{-r}$ are two different isomorphisms $\mathbb{Z}_m \xrightarrow{\sim} G$. More precisely:

Theorem 2.2.12 *Let G be a cyclic group with generating element a and $\operatorname{Ord} a = m$. Exactly then is a^r, $r \in \mathbb{Z}$, also a generating element of G, when $\gcd(r, m) = 1$ is. In particular, G has for $m \in \mathbb{N}^*$ exactly the pairwise different generating elements a^n with m coprime n, $0 \leq n < m$.*

Proof We switch to additive notation and can assume $G = \mathbb{Z}/\mathbb{Z}m$ and $a = 1 + \mathbb{Z}m$. Let $\pi_m \colon \mathbb{Z} \to \mathbb{Z}/\mathbb{Z}m$ with $\pi_m(r) = r + \mathbb{Z}m$ be the canonical projection. The subgroup generated by $n \cdot (1 + \mathbb{Z}m) = n + \mathbb{Z}m \; U \subseteq \mathbb{Z}/\mathbb{Z}m$ contains exactly the elements $rn + \mathbb{Z}m$, $r \in \mathbb{Z}$, and coincides with $\mathbb{Z}/\mathbb{Z}m$ exactly when its preimage $\pi_m^{-1}(U) = \mathbb{Z}n + \mathbb{Z}m$ is equal to $\mathbb{Z}$. According to Proposition 2.1.19, this is equivalent to $\gcd(n, m) = 1$. – The addition results from the fact that $0 + \mathbb{Z}m = 0 \cdot (1 + \mathbb{Z}m), \ldots, (m-1) + \mathbb{Z}m = (m-1) \cdot (1 + \mathbb{Z}m)$ at $m > 0$ are the m different elements of G. □

Let $m \in \mathbb{N}^*$. The generating elements of the group $\mathbb{Z}/\mathbb{Z}m$ are then also represented by the elements of the set $U_m := \{n \in \mathbb{N}^*_{\leq m} = \{1, \ldots, m\} \mid \gcd(n, m) = 1\} \subseteq \mathbb{N}^*_{\leq m}$. Their number is denoted with

$$\varphi(m)$$

$\varphi \colon \mathbb{N}^* \to \mathbb{N}$ is called the **Eulerian (φ-)Function**. *A finite cyclic group of order m therefore has exactly $\varphi(m)$ generating elements.* If $m = p^\alpha$, $\alpha > 0$, is a power of the prime number $p \in \mathbb{P}$, then the multiples $1 \cdot p, \ldots, p^{\alpha-1} p = p^\alpha$ of p are exactly the elements of $\mathbb{N}^*_{\leq p^\alpha} - U_{p^\alpha}$. Consequently,

$$\varphi(p^\alpha) = |U_{p^\alpha}| = p^\alpha - p^{\alpha-1} = p^{\alpha-1}(p-1).$$

For any $m \in \mathbb{N}^*$ with the canonical prime factor decomposition $m = p_1^{\alpha_1} \cdots p_k^{\alpha_k}$, $p_1 < \cdots < p_k$, $\alpha_1, \ldots, \alpha_k > 0$, is $\mathbb{N}_{\leq m}^* - U_m = \bigcup_{i=1}^k V_i$, $V_i := \{\ell \in \mathbb{N}_{\leq m}^* \mid p_i \mid \ell\}$. With the sieve formula from exercise 1.6.23 it follows due to $V_I := \bigcap_{i \in I} V_i = \{\ell \in \mathbb{N}_{\leq m}^* \mid p^I \mid \ell\}$, $p^I = \prod_{i \in I} p_i$, $I \subseteq \mathbb{N}_{\leq k}^*$, specifically $V_\emptyset = \mathbb{N}_{\leq m}^*$, the so-called **Euler's formula**

$$\varphi(m) = |U_m| = \left| \mathbb{N}_{\leq m}^* - \bigcup_{i=0}^k V_i \right| = \sum_{I \subseteq \mathbb{N}_{\leq k}^*} (-1)^{|I|} \frac{m}{p^I} = m \prod_{i=1}^k \left(1 - \frac{1}{p_i}\right)$$

$$= m \prod_{p \in \mathbb{P}, p \mid m} \left(1 - \frac{1}{p}\right).$$

In particular, $\varphi(m_1 m_2) = \varphi(m_1)\varphi(m_2)$ is for coprime $m_1, m_2 \in \mathbb{N}^*$, from which also $\varphi(p_1^{\alpha_1} \cdots p_k^{\alpha_k}) = p_1^{\alpha_1 - 1}(p_1 - 1) \cdots p_k^{\alpha_k - 1}(p_k - 1)$ follows. See also Exercise 2.2.11. Furthermore, for $m, \alpha \in \mathbb{N}^*$ the formula $\varphi(m^\alpha) = m^{\alpha-1}\varphi(m)$, e.g. $\varphi(10^\alpha) = 4 \cdot 10^{\alpha-1} = 2^{\alpha+1} \cdot 5^{\alpha-1}$.

Let $m \in \mathbb{N}^*$. The preimage of a subgroup $U \subseteq \mathbb{Z}/\mathbb{Z}m$ with respect to the canonical projection $\pi : \mathbb{Z} \to \mathbb{Z}/\mathbb{Z}m$ is a subgroup of $\mathbb{Z}$, which includes $\ker \pi = \mathbb{Z}m$, thus $\pi^{-1}(U) = \mathbb{Z}d \supseteq \mathbb{Z}m$, $d \mid m$, and $U = \pi(\pi^{-1}(U)) \subseteq \mathbb{Z}/\mathbb{Z}m$ is the cyclic subgroup of order m/d generated by $d + \mathbb{Z}m$. If U is the subgroup of $\mathbb{Z}/\mathbb{Z}m$ generated by $r + \mathbb{Z}m \in \mathbb{Z}/\mathbb{Z}m$, $r \in \mathbb{Z}$, then $\pi^{-1}(U) = \mathbb{Z}r + \mathbb{Z}m = \mathbb{Z}\,\mathrm{ggT}(r,m)$, see Theorem 2.1.19. We have proven in analogy to Theorem 2.1.18:

Theorem 2.2.13 *Let G be a finite cyclic group of order $m \in \mathbb{N}^*$ with generating element $a \in G$. Then for every divisor n of m there is exactly one subgroup of order n (and index m/n). This is also cyclic and is generated by $a^{m/n}$. Furthermore, the following holds*

$$\mathrm{Ord}(a^r) = m/\mathrm{ggT}(r,m) \quad \text{and} \quad \mathrm{H}(a^r) = \mathrm{H}(a^{\mathrm{ggT}(r,m)}), \quad r \in \mathbb{Z}.$$

Sentence 2.2.12 is a special case of the general formula in Sentence 2.2.13. Since each of the elements of a cyclic group of order m generates one of the $\tau(m)$ cyclic subgroups of $\mathbb{Z}/\mathbb{Z}m$ ($\tau(m)$ = number of divisors of m in $\mathbb{N}^*$) and each of these cyclic subgroups of order d according to sentence 2.2.12 exactly has $\varphi(d)$ generating elements,

$$\sum_{d \mid m} \varphi(d) = m, \quad m \in \mathbb{N}^*.$$

$\mathrm{id}_{\mathbb{N}^*} : m \mapsto m$, $m \in \mathbb{N}^*$, is therefore the summation function of the φ-function. With the Möbius inversion formula, cf. exercise 2.1.23, this results once again in the Euler formula

$$\varphi(m) = \sum_{H \subseteq \mathbb{P}_m} (-1)^{|H|} \frac{m}{p^H} = m \prod_{p \in \mathbb{P}_m} \left(1 - \frac{1}{p}\right), \quad m \in \mathbb{N}^*.$$

2.2 Homomorphisms

In any finite group G, an element $a \in G$ generates a cyclic subgroup $H(a)$ of order $\operatorname{Ord} a$, which according to Lagrange's Theorem 2.1.21 divides the order $|G| = \operatorname{Ord} G$ of G. This results in the following important sentence:

Theorem 2.2.14 (Fermat's Little Theorem) *Let G be a finite group. For every $a \in G$ the following holds*
$$a^{\operatorname{Ord} G} = e_G.$$

We mention the following elegant proof of Fermat's Little Theorem for finite *abelian* groups G, given by Euler: If $a \in G$, then the translation $L_a : x \mapsto ax$ is bijective and consequently $\prod_{x \in G} x = \prod_{x \in G}(ax) = a^{|G|} \prod_{x \in G} x$, thus $e_G = a^{|G|}$ after simplification.

With the above formula $\sum_{d|m} \varphi(d) = m$ the following frequently used criterion for the cyclicity of finite groups is easily derived, see for example the proof of Theorem 2.6.22.

Theorem 2.2.15 *Let G be a finite group of order m. For every divisor d of m there are at most d elements $x \in G$ with $x^d = e_G$. Then G is cyclic.*

Proof For $d \mid m$ let $\alpha(d)$ be the number of $x \in G$ with $\operatorname{Ord} x = d$. We show $\alpha(d) \leq \varphi(d)$ for all $d|m$. Because of $\sum_{d|m} \alpha(d) = m = \sum_{d|m} \varphi(d)$ then $\alpha(d) = \varphi(d)$ applies for all $d|m$ and in particular $\alpha(m) = \varphi(m) \geq 1$, i.e. G is cyclic. But if $\alpha(d) = 0$, then certainly $\alpha(d) \leq \varphi(d)$. Now let $\alpha(d) > 0$ and $a \in G$ be an element of order d. By assumption, the d elements of $H(a)$ are the only elements $x \in G$ with $x^d = e_G$. In particular, there is no element of order d in the complement $G - H(a)$, and it is $\alpha(d) = \varphi(d)$. □

Let G be an arbitrary group. Elements $a \in G$ with $\operatorname{Ord} a > 0$ are called **torsion elements** of G. We denote the set of torsion elements of G with TG. It is
$$TG = \bigcup_{n \in \mathbb{N}^*} T_n G,$$
where
$$T_n G := \{x \in G \mid x^n = e_G\}$$
is the so-called *n*-**torsion** of G. It is often simply denoted by $_n G$. It is the set of those elements of G whose order is a divisor of n, or according to Theorem 2.2.14 also the union of all finite subgroups of G whose order divides n. For completeness, we also set $T_0 G := \{x \in G \mid x^0 = e_G\} = G$. The group G is called a **torsion group** if $G = TG$ is, and **torsion-free** if $TG = \{e_G\}$ is. The image of the **power mapping** $x \mapsto x^n$ of G, $n \in \mathbb{N}$, we denote with
$$^n G := \{x^n \mid x \in G\}.$$
In additive notation, this is the subgroup nG of the *n*-times of the elements of G. If all power mappings $x \mapsto x^n$, $n \in \mathbb{N}^*$, of G are surjective, i.e. is $G = {}^n G$ for all

Fig. 2.6 Unit circle

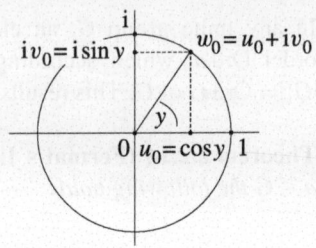

$n \in \mathbb{N}^*$, then the group G is called **divisible**. This is exactly the case when $G = {}^pG$ for all $p \in \mathbb{P}$ is. The group $\mathbb{Q} = (\mathbb{Q}, +)$ is torsion-free and divisible. Every homomorphic image of a divisible group is again divisible.

If G is abelian and $n \in \mathbb{N}^*$, then the power mapping $G \to G$, $x \mapsto x^n$, is even a group endomorphism with $\mathrm{T}_n G = {}_n G$ as kernel and ${}^n G$ as image. In particular, then $\mathrm{T}_n G$ and also $\mathrm{T}G$ as well as ${}^n G$ are subgroups of G. According to theorem 2.2.5, $|G| = |{}_n G| \cdot |{}^n G|$ holds for every *abelian* group G and every $n \in \mathbb{N}^*$. If $n = p$ is a prime number, then the p-torsion ${}_p G$ of a (any) group G is also called the p-**base** of G. If G is abelian, then the p-base ${}_p G$ is an elementary abelian p-group, cf. Ex. 2.3.6. *For a finite abelian group G the groups are ${}_n G$ and $G/{}^n G$ (and of course – as for any abelian groups – the groups ${}^n G$ and $G/{}_n G$) are isomorphic*. This is trivial if G is cyclic, as then both groups are also cyclic of the same order, and generally follows from the main theorem for finite abelian groups, cf. Ex. 2.2.25. ◊

Example 2.2.16 (Exponential mappings and the multiplicative groups $\mathbb{R}_+^\times, \mathbb{C}^\times$) In this example, we discuss some important groups from analysis that we will significantly expand later. Anticipating Sect. 3.5 (and the next volume), we use some less elementary properties for $\mathbb{C}$.

(1) Let $a \in \mathbb{R}$, $a > 0$, $a \neq 1$. *The real exponential mapping*

$$x \mapsto a^x, \quad x \in \mathbb{R},$$

is an isomorphism of the additive group $(\mathbb{R}, +)$ of $\mathbb{R}$ onto the multiplicative group $(\mathbb{R}_+^\times, \cdot)$ of positive real numbers. That it is a group homomorphism is the so-called **addition theorem $a^{x+y} = a^x a^y$ of the exponential function**. *The inverse isomorphism $\mathbb{R}_+^\times \xrightarrow{\sim} \mathbb{R}$ is the* **logarithm**

$$y \mapsto \log_a y, \quad y \in \mathbb{R}_+^\times,$$

to the base a. We refer to Sect. 3.10. For the Euler number $\mathrm{e} = 2{,}71828\ldots$ is $\mathbb{R} \xrightarrow{\sim} \mathbb{R}_+^\times$, $x \mapsto \exp x = \mathrm{e}^x$, the real exponential function par excellence with the natural logarithm $\mathbb{R}_+^\times \xrightarrow{\sim} \mathbb{R}$, $y \mapsto \log_\mathrm{e} y = \ln y$, as inversion. – The group $\mathbb{R}^\times$ is isomorphic to the product group $\{\pm 1\} \times \mathbb{R}_+^\times$. The mapping $\{\pm 1\} \times \mathbb{R}_+^\times \xrightarrow{\sim} \mathbb{R}^\times$, $(\varepsilon, x) \mapsto \varepsilon x$, is a group isomorphism with $y \mapsto (\mathrm{Sign}\, y, |y|)$ as inverse isomorphism.

2.2 Homomorphisms

(2) The complex e-function $\mathbb{C} \to \mathbb{C}^\times$, $z \mapsto \exp z = e^z$, can be defined as follows:

$$e^z = e^{x+iy} = e^x e^{iy} = e^x(\cos y + i \sin y), \quad z = x+iy, \quad x = \Re z, \quad y = \Im z \in \mathbb{R},$$

where cos and sin are the real trigonometric functions already known from school, whose basic properties we want to use here. They will be discussed in detail only in the second volume. From the addition theorem for the real e-function $x \mapsto e^x$ and the addition theorems for the trigonometric functions

$$\cos(\alpha + \beta) = \cos \alpha \cos \beta - \sin \alpha \sin \beta,$$

$$\sin(\alpha + \beta) = \sin \alpha \cos \beta + \cos \alpha \sin \beta, \quad \alpha, \beta \in \mathbb{R},$$

the **addition theorem for the complex e-function**, i.e. $\exp: z \mapsto e^z$ *is a group homomorphism from* $(\mathbb{C}, +)$ *into* $(\mathbb{C}^\times, \cdot)$. In other words: It is

$$e^{z+w} = e^z e^w, \quad z, w \in \mathbb{C}.$$

For $z = x + iy$ and $w = u + iv$, $x, y, u, v \in \mathbb{R}$, the following applies

$$e^{z+w} = e^{(x+u)+i(y+v)} = e^{x+u}\big(\cos(y+v) + i\sin(y+v)\big)$$
$$= e^x e^u \big(\cos y \cos v - \sin y \sin v\big) + i e^x e^u \big(\sin y \cos v + \cos y \sin v\big),$$
$$e^z e^w = e^x(\cos y + i \sin y)e^u(\cos v + i \sin v)$$
$$= e^x e^u \big(\cos y \cos v - \sin y \sin v\big) + i e^x e^u \big(\cos y \sin v + \sin y \cos v\big).$$

$\exp: \mathbb{C} \to \mathbb{C}^\times$ *is surjective.* Is $w = u + iv \in \mathbb{C}^\times$, then $w = |w| w_0$ with

$$|w| = \sqrt{u^2 + v^2} \in \mathbb{R}_+,$$

where $w_0 = u_0 + i v_0$, $|w_0|^2 = u_0^2 + v_0^2 = 1$ is. Then there is an $x \in \mathbb{R}$ with $e^x = |w|$, cf. (1), and due to the properties of the trigonometric functions — the point $(u_0, v_0) \in \mathbb{R}^2$ lies on the unit circle! — a $y \in \mathbb{R}$ with $w_0 = \cos y + i \sin y$, cf. Fig. 2.6. Then $w = e^{x+iy}$.

The kernel of exp is the subgroup

$$U := \{z = x + iy \in \mathbb{C} \mid e^z = e^x(\cos y + i \sin y) = 1\}.$$

For $z = x + iy \in U$ is then necessarily $e^x = 1$, i.e. $x = 0$, and $\cos y = 1$, $\sin y = 0$, i.e. $y \in \mathbb{Z}2\pi$. Therefore,

$$U = \ker \exp = \mathbb{Z}2\pi i.$$

For $w \in \mathbb{C}^\times$ every number $z \in \mathbb{C}$ with $\exp z = w$ is a **(natural) logarithm**

$$z = \log w$$

of w. A complex number $\neq 0$ therefore has infinitely many natural logarithms, and any two of these logarithms differ by an element from ker exp, i.e. by an integer multiple of $2\pi i$. The representation

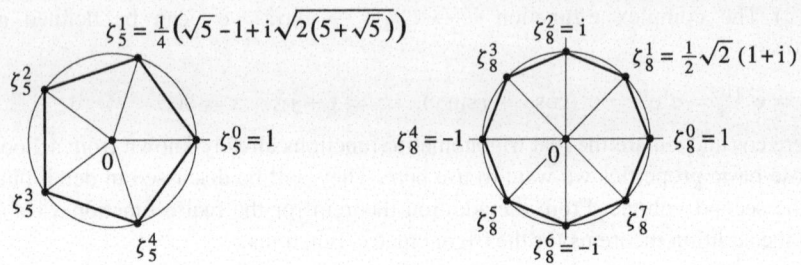

Fig. 2.7 Groups $_5\mathbb{E}$ and $_8\mathbb{E}$ of the complex 5-th and 8-th roots of unity

$$w = e^{\log w} = e^{\Re \log w}\bigl(\cos(\Im \log w) + i\sin(\Im \log w)\bigr) = |w|\bigl(\cos(\Im \log w) + i\sin(\Im \log w)\bigr)$$

is also called the **polar coordinate representation** of w and the angle

$$\mathrm{Arg}\, w := \Im \log w$$

an **argument** of w. This representation is not unique, as $\log w$ is not unique. However, there is exactly one logarithm with $-\pi < \Im \log w \le \pi$. This is called the **principal value** of the natural logarithm or the **natural logarithm** par excellence and is denoted by

$$\ln w.$$

The corresponding argument $\mathrm{Arg}\, w = \Im \ln w \in\,]-\pi, \pi]$ is the **standard argument** of w. The (bijective) function $\ln\colon \mathbb{C}^\times \to \mathbb{C}$ extends the real logarithm function $\ln\colon \mathbb{R}_+^\times \to \mathbb{R}$, but is no longer a group homomorphism. Because of $e^{i\pi} = \cos\pi + i\sin\pi = -1$ [10] for example, $\ln(-1) = i\pi$, but $\ln\bigl((-1)^2\bigr) = \ln 1 = 0 \ne 2i\pi$. The equation holds true

$$\ln(w_1 w_2) = \ln w_1 + \ln w_2,$$

if $\mathrm{Arg}\, w_1 + \mathrm{Arg}\, w_2 \in\,]-\pi, \pi]$ is fulfilled for the standard arguments of w_1 and w_2. If $a \in \mathbb{C}^\times$ and $a \ne 1$, then $\ln a \ne 0$ and one defines

$$a^z := e^{z \ln a}, \quad z \in \mathbb{C}.$$

This **exponential function to the base** a also defines a surjective group homomorphism $(\mathbb{C}, +) \to (\mathbb{C}^\times, \cdot)$. It is the composition of the isomorphism $(\mathbb{C}, +) \xrightarrow{\sim} (\mathbb{C}, +)$, $z \mapsto z \ln a$, with the exponential homomorphism $\exp\colon (\mathbb{C}, +) \to (\mathbb{C}^\times, \cdot)$. Its kernel is therefore the subgroup $\{z \in \mathbb{C} \mid e^{z \ln a} = 1\} = \mathbb{Z} 2\pi i / \ln a \subseteq \mathbb{C}$.

---

[10] $e^{i\pi} + 1 = 0$ is considered the most beautiful formula in mathematics.

2.2 Homomorphisms

Now let

$$U := \left\{ z = x + iy \in \mathbb{C} \,\middle|\, |z| = \sqrt{x^2 + y^2} = 1 \right\} \subseteq \mathbb{C}^\times$$

be the so-called **circle group** of complex numbers of magnitude 1. They form the unit circle in $\mathbb{C} = \mathbb{R}^2$. The pre-image $\exp^{-1}(U)$ is the imaginary axis $\mathbb{R}i \subseteq \mathbb{C}$. Consequently, exp induces a surjective homomorphism $\mathbb{R}i \to U$, $i\varphi \mapsto e^{i\varphi} = \cos\varphi + i\sin\varphi,$[11] with kernel $\mathbb{Z}2\pi i$, and $\mathbb{R} \to U$, $t \mapsto e^{2\pi i t} = \cos 2\pi t + i \sin 2\pi t$ is a surjective group homomorphism with kernel $\mathbb{Z}$. Its fibers are the residue classes $t + \mathbb{Z}$, $t \in \mathbb{R}$, for which the half-open interval $[a, a+1[$ or also $]a, a+1]$, for each $a \in \mathbb{R}$ forms a full representative system. *The induced mapping* $\mathbb{R}/\mathbb{Z} \xrightarrow{\sim} U$, $t + \mathbb{Z} \mapsto e^{2\pi i t}$, *is a group isomorphism*, where $\mathbb{R}/\mathbb{Z}$ is the quotient group with the Minkowski addition $(t + \mathbb{Z}) + (s + \mathbb{Z}) = (t + s) + \mathbb{Z}, t, s \in \mathbb{R}$, see already Example 1.3.6. The group

$$\mathbb{T} := \mathbb{R}/\mathbb{Z} \;(\cong U)$$

is called the (1-dimensional) **Torus Group**.[12]

The group isomorphism $\mathbb{R}_+^\times \times U \to \mathbb{C}^\times$, $(r, u) \mapsto ru$, provides the group isomorphism

$$\mathbb{R} \times U \xrightarrow{\sim} \mathbb{C}^\times, \quad (t, u) \mapsto e^t u,$$

with the real exponential isomorphism $\mathbb{R} \to \mathbb{R}_+^\times$, $t \mapsto e^t$. The set $\mathbb{R} \times U \subseteq \mathbb{R} \times \mathbb{C} = \mathbb{R}^3$ is a cylinder. Therefore, the group $\mathbb{R} \times U \cong \mathbb{C}^\times$ is often called the **cylinder group**. Because of $U \cong \mathbb{T} = \mathbb{R}/\mathbb{Z}$, it is also isomorphic to $\mathbb{R} \times \mathbb{T}$.

The torsion subgroup $T\mathbb{C}^\times$ of $\mathbb{C}^\times$ is the group

$$\mathbb{E} := \{\xi \in \mathbb{C} \mid \xi^n = 1 \text{ for } n \in \mathbb{N}\}$$

of the so-called complex **roots of unity**. It corresponds to the torsion subgroup by virtue of the isomorphism $\mathbb{R} \times (\mathbb{R}/\mathbb{Z}) \xrightarrow{\sim} \mathbb{C}^\times$, $(x, t + \mathbb{Z}) \mapsto e^x e^{2\pi i t}$, with the torsion subgroup $T(\mathbb{R} \times (\mathbb{R}/\mathbb{Z})) = T\mathbb{R} \times T(\mathbb{R}/\mathbb{Z}) = \{0\} \times \mathbb{Q}/\mathbb{Z} \cong \mathbb{Q}/\mathbb{Z}$. The figure $\mathbb{Q}/\mathbb{Z} \xrightarrow{\sim} \mathbb{E}$, $t + \mathbb{Z} \mapsto e^{2\pi i t}$, is therefore a group isomorphism. For each $m \in \mathbb{N}^*$ the m-torsion subgroup $T_m(\mathbb{Q}/\mathbb{Z})$ is the only subgroup of order m of $\mathbb{Q}/\mathbb{Z}$. It is cyclic and is generated by $(1/m) + \mathbb{Z}$. Consequently, the m-torsion $T_m\mathbb{C}^\times = {}_m\mathbb{E} \subseteq \mathbb{E}$ of $\mathbb{C}^\times$, that is the group of so-called m-**th complex unit roots**, is the *cyclic group*

$$_m\mathbb{E} = \{\zeta_m^k \mid 0 \leq k < m\}$$

of order m, which is generated by

$$\zeta_m := e^{2\pi i/m} = \cos(2\pi/m) + i\sin(2\pi/m)$$

---

[11] $e^{i\varphi} = \cos\varphi + i\sin\varphi$ is also called the **Euler's equation**.
[12] The product group $\mathbb{T}^n = (\mathbb{R}/\mathbb{Z})^n$ is the so-called n-**dimensional torus group**, $n \in \mathbb{N}$.

is generated. An element of order m in $\mathbb{C}^\times$ is called a **primitive m-th unit root**. Each of them generates $_m\mathbb{E}$. According to Theorem 2.2.12 these are exactly the numbers $\zeta_m^k = e^{2\pi ik/m}$, $0 \le k < m$, $\gcd(k,m) = 1$. The cases $m = 5$ and $m = 8$ are shown in Fig. 2.7. If $t = a/b \in \mathbb{Q}$, $a \in \mathbb{Z}$, $b \in \mathbb{N}^*$, $\mathrm{ggT}(a,b) = 1$, the canonical reduced representation of any rational number with numerator a and denominator b (see the remarks before Example 1.7.12), then $e^{2\pi ia/b} = \left(e^{2\pi i/b}\right)^a$ is a primitive b-th root of unity. ◊

Example 2.2.17 (Homomorphisms into direct products) Let N_i, $i \in I$, be an arbitrary family of magmas and $N := \prod_{i\in I} N_i$ their product, see Example 2.1.16. Then the canonical projections $p_i : N \to N_i$, $(y_j)_{j\in I} \mapsto y_i$, are homomorphisms, $i \in I$. If M is any magma, then a mapping $\varphi : M \to N$ is uniquely determined by the compositions $\varphi_i = p_i\varphi$, $i \in I$. It is $\varphi(x) = (\varphi_i(x))_{i\in I}$ for all $x \in M$. One writes $\varphi = (\varphi_i)_{i\in I}$. Apparently, φ is a homomorphism exactly when all φ_i are homomorphisms, i.e., *the canonical mapping*

$$\mathrm{Hom}\left(M, \prod_{i\in I} N_i\right) \xrightarrow{\sim} \prod_{i\in I} \mathrm{Hom}(M, N_i), \quad \varphi \mapsto (p_i\varphi)_{i\in I},$$

is bijective. This fact is also called the **universal property of the product** of the N_i, $i \in I$. If the family $\varphi_i : N \to N_i$, $i \in I$, defines an isomorphism $N \xrightarrow{\sim} \prod_{i\in I} N_i$, then one says, N together with the homomorphisms φ_i, $i \in I$, represents the product of the N_i, $i \in I$. A completely analogous statement applies to semigroups, monoids, and groups and the associated sets of homomorphisms. In the case of groups, it is clear that

$$\ker \varphi = \bigcap_{i\in I} \ker \varphi_i.$$

The image of φ is usually not so easy to specify. As an example, consider the following situation: Let G be a group and $\varphi_i : G \to H_i$ be group homomorphisms with $N_i := \ker \varphi_i$, $i = 1, \ldots, n$, and

$$\varphi : G \longrightarrow H_1 \times \cdots \times H_n, \quad x \mapsto (\varphi_1(x), \ldots, \varphi_n(x)),$$

the associated homomorphism from G into the product $H_1 \times \cdots \times H_n$ with

$$\ker \varphi = N_1 \cap \cdots \cap N_n.$$

Then N_i for each $i = 1, \ldots, n$ is a normal subgroup in G and N_iF for each subgroup $F \subseteq G$ because of $N_iF = FN_i$ is also a subgroup of G. φ can only be surjective if all $\varphi_1, \ldots, \varphi_n$ are surjective. It now follows:

Theorem 2.2.18 (Chinese Remainder Theorem) *The group homomorphisms $\varphi_i : G \to H_i$ are surjective, and let $N_i := \ker \varphi_i$, $i = 1, \ldots, n$. The homomorphism $\varphi = (\varphi_i) : G \to H_1 \times \cdots \times H_n$ is also surjective if and only if for all $i = 1, \ldots, n$ the following holds:*

2.2 Homomorphisms

$$N_i \cdot N_i' = G \quad \text{with} \quad N_i' := \bigcap_{j \neq i} N_j.$$

In particular, G represents with the homomorphisms $\varphi_1, \ldots, \varphi_n$ exactly the product of the H_i, if $N_i \cdot N_i' = G$, $i = 1, \ldots, n$, and $N := N_1 \cap \cdots \cap N_n = \{e_G\}$ holds.

Proof Let φ be surjective and $i \in \{1, \ldots, n\}$. We denote the neutral elements of $H_1, \ldots, H_n$ with $e_1, \ldots, e_n$. For a given $y \in G$ there is an $x \in G$ with

$$\varphi(x) = (e_1, \ldots, e_{i-1}, \varphi_i(y), e_{i+1}, \ldots, e_n).$$

Therefore, $\varphi_i(x) = \varphi_i(y)$ and $\varphi_j(x) = e_j$ for $j \neq i$, thus $y \in N_i x$, see Theorem 2.2.5, and $x \in \bigcap_{j \neq i} N_j$.

Conversely, let the given condition for the N_i be fulfilled. It suffices to show that for $i = 1, \ldots, n$ and all $y \in H_i$ the elements $y' := (e_1, \ldots, e_{i-1}, y, e_{i+1}, \ldots, e_n)$ belong to Im φ. Due to the surjectivity of φ_i there exists for a given $y \in H_i$ a $x \in G$ with $\varphi_i(x) = y$. According to the assumption, there exists a $z \in N_i$ and a $w \in N_i' = \bigcap_{j \neq i} N_j$ with $x = zw$. Then

$$\varphi(w) = \varphi(z^{-1}x) = \big(\varphi_1(w), \ldots, \varphi_{i-1}(w), \varphi_i(z^{-1}x), \varphi_{i+1}(w), \ldots, \varphi_n(w)\big) = y'.$$

holds. □

By the way, from $N_i \cdot N_i' = G$, $i = 1, \ldots, n$, even $N_1' \cdots N_n' = G$, see exercise 2.2.15. Standard examples for the Chinese Remainder Theorem 2.2.18 are the canonical projections $\pi_i : \mathbb{Z} \to \mathbb{Z}/\mathbb{Z}m_i$ (of additive groups) with $m_i \in \mathbb{N}^*$, $i = 1, \ldots, n$. Then it is $N_i = \mathbb{Z}m_i$, $N_i' = \bigcap_{j \neq i} \mathbb{Z}m_j = \mathbb{Z}m_i'$, $m_i' = \text{kgV}(m_j, j \neq i)$ and $N_i + N_i' = \mathbb{Z}\text{ggT}(m_i, m_i')$, $i = 1, \ldots, n$, and $N = N_1 \cap \cdots \cap N_n = \mathbb{Z}m$ with $m := \text{kgV}(m_1, \ldots, m_n)$. The core of the induced homomorphism

$$\pi : \mathbb{Z} \longrightarrow \mathbb{Z}/\mathbb{Z}m_1 \times \cdots \times \mathbb{Z}/\mathbb{Z}m_n$$

is $\mathbb{Z}m \subseteq \mathbb{Z}$. According to Theorem 2.2.18, π is surjective if m_i and m_i' are coprime for all $i = 1, \ldots, n$. This is obviously the case when $m_1, \ldots, m_n$ are pairwise coprime, see exercise 1.7.25b). In this case, $m = m_1 \cdots m_n$. We obtain:

Corollary 2.2.19 *If $m_1, \ldots, m_n \in \mathbb{N}^*$ are pairwise coprime and is $m = m_1 \cdots m_n$, then the canonical homomorphism*

$$\overline{\pi} : \mathbb{Z}/\mathbb{Z}m \xrightarrow{\sim} \mathbb{Z}/\mathbb{Z}m_1 \times \cdots \times \mathbb{Z}/\mathbb{Z}m_n, \quad r + \mathbb{Z}m \longmapsto (r + \mathbb{Z}m_1, \ldots, r + \mathbb{Z}m_n),$$

is an isomorphism (of additive cyclic groups of order m).

Corollary 2.2.19 therefore states: If $m_1, \ldots, m_n \in \mathbb{N}^*$ are pairwise coprime and $r_1, \ldots, r_n$ are arbitrary integers, then there exist integers r with $r \equiv r_i \mod m_i$ for all $i = 1, \ldots, n$. Furthermore, any two such r are congruent modulo $m = m_1 \cdots m_n$.

We note that in the situation of Corollary 2.2.19 the injectivity of $\overline{\pi}$ is trivial due to $\ker \pi = \mathbb{Z}m$. The surjectivity then also follows from the fact that both groups contain the same number of elements, namely $m = m_1 \cdots m_n$. Since the $m_i' := m/m_i$, $i = 1, \ldots, n$, are coprime, there are $c_i \in \mathbb{Z}$ with

$1 = m'_1 c_1 + \cdots + m'_n c_n$, cf. Proposition 2.1.19. *The inverse mapping to* $\bar{\pi}$ *is then explicitly given by*

$$(r_1 + \mathbb{Z}m_1, \ldots, r_n + \mathbb{Z}m_n) \longmapsto (r_1 m'_1 c_1 + \cdots + r_n m'_n c_n) + \mathbb{Z}m.$$

Special cases of Corollary 2.2.19 were already known in ancient China 2000 years ago. From Sun Tse (ca. 150 BC) comes the following exercise; Determine all natural numbers r with $r \equiv 2 \bmod 3$, $r \equiv 3 \bmod 5$ and $r \equiv 2 \bmod 7$. The reader should solve this exercise.

For $n = 2$, Theorem 2.2.18 states: The group G with the surjective homomorphisms $\varphi_i \colon G \to H_i$, $i = 1,2$, represents the product $H_1 \times H_2$, if $N_1 N_2$ ($= N_2 N_1$) $= G$ and $N_1 \cap N_2 = \{e_G\}$ are for the normal subgroups $N_i := \ker \varphi_i$, $i = 1,2$. In this case, φ_1 induces an isomorphism $\varphi_1 | N_2 \colon N_2 \xrightarrow{\sim} H_1$ and φ_2 an isomorphism $\varphi_2 | N_1 \colon N_1 \xrightarrow{\sim} H_2$. Thus, we obtain the isomorphism $N_2 \times N_1 \xrightarrow{\sim} G$, $(x_2, x_1) \mapsto x_2 x_1$. In particular, for $x_1 \in N_1$ and $x_2 \in N_2$ elementwise $x_1 x_2 = x_2 x_1$. One says, G is the **product of the normal subgroups** N_1 and N_2. However, note that in a group G there can be subgroups F_1 and F_2 with $F_1 F_2 = F_2 F_1 = G$ and $F_1 \cap F_2 = \{e_G\}$ without G being isomorphic to the product $F_1 \times F_2$. An example is provided by the group $G = \mathfrak{S}_3$. ◊

Example 2.2.20 (Homomorphisms on restricted products) In this example, we limit ourselves to homomorphisms of monoids (and groups). Let M_i, $i \in I$, be a family of monoids. Homomorphisms on the product $\prod_{i \in I} M_i$ are difficult to describe when the index set I is infinite. Such a homomorphism is usually not already determined by its values on the factors M_i, which are canonically embedded in $\prod_{i \in I} M_i$, cf. Example 2.1.16. However, a monoid homomorphism $\varphi \colon \prod'_{i \in I} M_i \to N$ on the restricted product $M := \prod'_{i \in I} M_i$ is determined by its values on the submonoids $M_i \subseteq M$, $i \in I$, cf. loc. cit. For clarity, we denote the canonical embedding $M_i \to M$ with ι_i. Since the elements of M_i commute with those of M_j for $i \neq j$, it is clear that: *For any monoid N, the canonical mapping*

$$\mathrm{Hom}\left(\prod_{i \in I}{}' M_i, N\right) \longrightarrow \prod_{i \in I} \mathrm{Hom}(M_i, N), \quad \varphi \longmapsto (\varphi \circ \iota_i)_{i \in I},$$

is injective. Its image contains exactly the I-tuples $(\varphi_i) \in \prod_{i \in I} \mathrm{Hom}(M_i, N)$, for which $\varphi_i(x_i)\varphi_j(x_j) = \varphi_j(x_j)\varphi_i(x_i)$ for all $x_i \in M_i$, $x_j \in M_j$, $i,j \in I$, applies. For such a tuple (φ_i) and the associated homomorphism $\varphi \colon \prod'_{i \in I} M_i \to N$ *then applies*

$$\varphi\big((x_i)_{i \in I}\big) = \prod_{i \in I} \varphi_i(x_i).$$

In particular, the specified mapping is bijective when N is commutative. In this case, the specified mapping is even an isomorphism of monoids, cf. Proposition 2.2.4. This statement is called the **universal property of restricted products**. It applies completely analogously to the restricted products of groups. (One also constructs a restricted product for semigroups H_i, $i \in I$, with the corresponding universal property and uses the monoids $H'_i := H_i \uplus \{e_i\}$ resulting from the

2.2 Homomorphisms

adjunction of neutral elements e_i. Then $\prod'_{i\in I} H_i := \prod'_{i\in I} H'_i - \{e\}$ with $e := (e_i)_{i\in I}$ and the canonical embeddings $\iota_i \colon H_i \to \prod'_{i\in I} H_i$, $i \in I$, the desired universal property.) In the case of additively written abelian monoids or groups M_i, $i \in I$, and N the universal property of the **direct sums** simply reads: *The canonical mapping*

$$\mathrm{Hom}\left(\bigoplus_{i\in I} M_i, N\right) \longrightarrow \prod_{i\in I} \mathrm{Hom}(M_i, N), \quad \varphi \longmapsto (\varphi \circ \iota_i)_{i\in I},$$

is an isomorphism of monoids or groups. For the tuple $(\varphi_i) \in \prod_{i\in I} \mathrm{Hom}(M_i, N)$ is the original image of the homomorphism

$$\bigoplus_{i\in I} M_i \longrightarrow N, \quad (x_i)_{i\in I} \longmapsto \sum_{i\in I} \varphi_i(x_i).$$

An important example is the following: Let M_i, $i \in I$, be a family of submonoids (or subgroups) of the additively written monoid (or the additively written group) M. The canonical inclusions $M_i \hookrightarrow M$, $i \in I$, then define the homomorphism

$$\bigoplus_{i\in I} M_i \longrightarrow M, \quad (x_i)_{i\in I} \longmapsto \sum_{i\in I} x_i,$$

whose image is the submonoid (or the subgroup) $\sum_{i\in I} M_i \subseteq M$ of M that is generated by the M_i. If this homomorphism is injective, then the sum $\sum_{i\in I} M_i$ is said to be direct and one speaks of the **(internal) direct sum**

$$\sum_{i\in I}^{\oplus} M_i$$

of the M_i, $i \in I$. For groups, this situation can be characterized as follows:

Lemma 2.2.21 (Direct Sums of Subgroups) *The sum $\sum_{i\in I} H_i \subseteq G$ of the subgroups H_i, $i \in I$, of the additive abelian group G is direct exactly when the following condition applies: For each $i \in I$ is*

$$H_i \cap \sum_{j\neq i} H_j = \{0\}.$$

If I is completely ordered, then this condition is also equivalent to the following: It applies $H_i \cap \sum_{j<i} H_j = \{0\}$ for all $i \in I$.

Proof Let the sum of the H_i be direct and $x = x_i = \sum_{j\neq i} x_j \in H_i \cap \sum_{j\neq i} H_j$. Then $0 = x_i + \sum_{j\neq i}(-x_j)$, so $x_k = 0$ for all $k \in I$ and in particular $x = x_i = 0$. – If, conversely, the average condition is fulfilled and $0 = \sum_{k\in I} x_k = 0$ for $(x_k) \in \bigoplus_{k\in I} H_k$, then $x_i = \sum_{j\neq i}(-x_j) \in H_i \cap \sum_{j\neq i} H_j = \{0\}$, so $x_i = 0$ for each $i \in I$. – We leave the proof of the addition to the reader. □

As an application, we prove the so-called primary decomposition of an abelian torsion group. Let G be any group and $p \in \mathbb{P}$ a prime number. The p-**primary component** of G is understood to be the part *set*

$$G(p) := \bigcup_{n \in \mathbb{N}} T_{p^n} G \subseteq TG$$

of those elements of G, whose order is a power of p. G is called a p-**group**, if $G = G(p)$ is, if therefore the order of every element of G is a p-power. If G is finite and $|G|$ is a power of p, then G is a p-group. The converse also holds, since for every prime divisor q of the order of a finite group G, there is an element of order q in G. This is simple if G is abelian, see Ex. 2.2.20, and is generally the Theorem 2.4.8 of Cauchy. The p-primary components of an *abelian* group are obviously subgroups. It even holds:

Theorem 2.2.22 (Primary decomposition of abelian torsion groups) *Every additively written abelian torsion group G is the (internal) direct sum of its primary components, i.e. it holds*

$$G = TG = \sum_{p \in \mathbb{P}}^{\oplus} G(p).$$

Proof It is $G = \sum_{p \in \mathbb{P}} G(p)$. For this, it is sufficient to show that every finite cyclic group $\mathbb{Z}/\mathbb{Z}m$ is the sum of its primary components, where $m = p_1^{\alpha_1} \cdots p_n^{\alpha_n}$ is the canonical prime factor decomposition of $m \in \mathbb{N}^*$. According to the Chinese Remainder Theorem 2.2.19, however, $\mathbb{Z}/\mathbb{Z}m \cong \mathbb{Z}/\mathbb{Z}p_1^{\alpha_1} \oplus \cdots \oplus \mathbb{Z}/\mathbb{Z}p_n^{\alpha_n}$. The sum $\sum_{p \in \mathbb{P}} G(p)$ is direct: We use the criterion from Lemma 2.2.21. Let $x \in G(p) \cap \sum_{q \neq p} G(q)$. Then the order of x is on the one hand a power of p and on the other hand a product of q powers, $q \in \mathbb{P} - \{p\}$. However, this is only possible for $\text{Ord } x = 1$, i.e. for $x = 0$. □

When studying Abelian torsion groups G, one should always keep their primary decompositions in mind. This is particularly true for finite Abelian groups. Following the remark after the Chinese Remainder Theorem 2.2.19, the primary components $x_i \in G(p_i)$ of an element $x \in G$ of order $m = p_1^{\alpha_1} \cdots p_n^{\alpha_n}$ can be directly specified: If $m'_i = m/p_i^{\alpha_i}$ and $1 = m'_1 c_1 + \cdots + m'_n c_n$ with $c_i \in \mathbb{Z}$, then $x = x_1 + \cdots + x_n$ with $x_i := m'_i c_i x \in G(p_i)$, $i = 1, \ldots, n$. ◊

Example 2.2.23 (Exponent of a group) Let G be a group with neutral element e. For $n \in \mathbb{Z}$, let $\chi_n : G \to G$ be the power mapping $x \mapsto x^n$. Because of $\chi_m(\chi_n(x)) = (x^n)^m = x^{mn} = \chi_{mn}(x)$, the mapping $\chi : \mathbb{Z} \to G^G$, $n \mapsto \chi_n$, is a monoid homomorphism of the multiplicative monoid $(\mathbb{Z}, \cdot)$ into the mapping monoid G^G (with composition as the operation). Exactly then is $\chi_m = \chi_n$, when $x^{m-n} = e$ is for all $x \in G$. The set of $k \in \mathbb{Z}$ with $x^k = e$ for all $x \in G$ is evidently a subgroup of $\mathbb{Z} = (\mathbb{Z}, +)$. The uniquely determined generating element ≥ 0 of this subgroup is called the **exponent** of G and is denoted with

$$\text{Exp } G$$

$\text{Exp } G$ is the least common multiple of the orders $\text{Ord } a$, $a \in G$. So it follows $\chi_m = \chi_n$ exactly when $m - n \in \mathbb{Z}\,\text{Exp } G$ is, i.e. when $m \equiv n$ mod $\text{Exp } G$ is. The

2.2 Homomorphisms

fibers of the mapping χ are the residue classes $k + \mathbb{Z}\operatorname{Exp} G$, $k \in \mathbb{Z}$, and χ induces an injective homomorphism

$$\overline{\chi} : (\mathbb{Z}/\mathbb{Z}\operatorname{Exp} G, \cdot) \to (G^G, \circ)$$

multiplicative monoids. The multiplication on $\mathbb{Z}/\mathbb{Z}\operatorname{Exp} G$ is the canonical one (through multiplication of the representatives) and was already introduced in Example 1.3.6. In particular, $\chi_n = \operatorname{id}_G = \chi_1$ is exactly when $n \equiv 1 \bmod \operatorname{Exp} G$ is, and exactly then are χ_m and χ_n inverse to each other, when $mn \equiv 1 \bmod \operatorname{Exp} G$ is.

Let G now be an abelian group, which we now write additively. Then χ_n is an endomorphism of G for each $n \in \mathbb{Z}$ due to $\chi_n(x+y) = n(x+y) = nx + ny = \chi_n(x) + \chi_n(y)$ and χ is a mapping $\chi : \mathbb{Z} \to \operatorname{End} G$. Due to $\chi_{m+n}(x) = (m+n)x = mx + nx = \chi_m(x) + \chi_n(x)$, χ is also a homomorphism of the additive group $(\mathbb{Z}, +)$ into the additive group $\operatorname{End} G$. Overall, $\chi : \mathbb{Z} \to \operatorname{End} G$ *is therefore a ring homomorphism with* $\ker \chi = \mathbb{Z}\operatorname{Exp} G$, cf. Proposition 2.2.4 and the remark on it. χ induces an injective homomorphism $\overline{\chi} : \mathbb{Z}/\mathbb{Z}\operatorname{Exp} G \to \operatorname{End} G$. This is not only a homomorphism of the additive groups, but also a homomorphism of the multiplicative monoids $(\mathbb{Z}/\mathbb{Z}\operatorname{Exp} G, \cdot)$ and $(\operatorname{End} G, \circ)$, thus also a ring homomorphism.

Let $G = \mathrm{H}(a) = \mathbf{Z}_m$ even be cyclic with $m = \operatorname{Exp} G$. Then $\overline{\chi}$ *is also surjective and thus a canonical ring isomorphism*

$$\mathbb{Z}/\mathbb{Z}m \xrightarrow{\sim} \operatorname{End} \mathbf{Z}_m, \quad [n]_m \mapsto (\chi_n : x \mapsto nx).$$

If $\varphi : \mathbf{Z}_m \to \mathbf{Z}_m$ is an endomorphism with $\varphi(a) = na$, then $\varphi = \chi_n$. In particular, $\overline{\chi}$ induces the group isomorphism

$$(\mathbb{Z}/\mathbb{Z}m, \cdot)^\times \xrightarrow{\sim} (\operatorname{Aut} \mathbf{Z}_m, \circ).$$

If G is finite, i.e. $m > 0$, then the group $(\mathbb{Z}/\mathbb{Z}m)^\times$ consists of the residue classes $[n]_m = n + \mathbb{Z}m$ with $\gcd(n,m) = 1$, cf. Theorem 2.2.12. Its order is $\varphi(m)$, and according to the Little Fermat's Theorem 2.2.14 is

$$n^{\varphi(m)} \equiv 1 \bmod m, \text{ if } \gcd(n,m) = 1 \quad \textbf{(Euler's formula)}.$$

We will study the unit groups $(\mathbb{Z}/\mathbb{Z}m, \cdot)^\times$ in more detail in Example 2.7.10. It is called the **prime residue class group** modulo m. Here we already introduce the following notation: If $a \in \mathbb{Z}$, $\operatorname{ggT}(a,m) = 1$, then let

$$\operatorname{Ord}_m a$$

be the order of $[a]_m = a + \mathbb{Z}m$ in $(\mathbb{Z}/\mathbb{Z}m, \cdot)^\times$. $\operatorname{Ord}_m a$ *divides* $\operatorname{Ord}(\mathbb{Z}/\mathbb{Z}m)^\times = \varphi(m)$.

Let now G be a multiplicatively written, not necessarily abelian torsion group. Then χ_n is bijective exactly when n and $\operatorname{Ord} a$ are coprime for every $a \in G$. Because χ_n maps for every $a \in G$ the cyclic group $\mathrm{H}(a)$ onto itself. If $\operatorname{Exp} G > 0$, then this condition is equivalent to n and $\operatorname{Exp} G$ being coprime (since $\operatorname{Exp} G$ is the lcm of the orders $\operatorname{Ord} a$, $a \in G$). In this case, as already noted, χ_k is the inverse mapping of χ_n, where $nk \equiv 1 \bmod \operatorname{Exp} G$ is. For a finite group G, the condition $\gcd(n, \operatorname{Exp} G) = 1$ is equivalent with $\gcd(n, \operatorname{Ord} G) = 1$, because $\operatorname{Exp} G$ and $\operatorname{Ord} G$ have for a finite group G have the same prime divisors. First of all, $\operatorname{Exp} G$ is due to $x^{\mathrm{Ord}\, G} = e$ for all $x \in G$ a divisor of Ord G. On the other hand, every prime divisor p of Ord G is a divisor of Exp G, since G then has an element of order p. For abelian G, this is — as already noted — simple, see exercise 2.2.20, in the general case this is the Theorem 2.4.8 by Cauchy. — Give an example of a torsion group G with Exp $G = 0$, for which an exponent $n > 1$ exists such that $\chi_n : G \to G$ is bijective. (It should be noted that then n and Exp G due to $\gcd(n, 0) = n$ are *not* coprime.)

Remarks (1) Power mappings of groups are often used in cryptography for the construction of **public-key cryptosystems** in the sense of W. Diffie, M. Hellman, and R. Merkle. In these cryptosystems, a message is encrypted with a public key **(public key)** made accessible by the recipient. To decrypt, the recipient uses a key **(private key)** known only to him, which should be difficult to find for outsiders with disproportionately large effort. In the present case, the public key is the finite group G and the exponent n coprime to Ord G (or Exp G). A message is encoded as element $x \in G$, and the easily calculable power x^n is transmitted, see the following remark (2). To recover x, one needs an element k with $nk \equiv 1 \bmod \text{Exp}\, G$ (which is guaranteed by $nk \equiv 1 \bmod \text{Ord}\, G$). Then $(x^n)^k = x$ is the original message. Finding the inverse k to n modulo Ord G or modulo Exp G should be very difficult, for example, because the order of G is unknown. For an explicit example, see the RSA codes at the end of Example 2.7.10.

Bijective mappings $f : A \to B$ (like here the power mappings $x \mapsto x^n$), whose values are easy to calculate, while determining the values of the inverse mappings $f^{-1} : B \to A$ is very difficult, but then becomes easy when additional information (like here the element k) is available, are called **trapdoor functions** or **one-way functions**[13].

(2) In this context, **rapid exponentiation** is important. Let M be any monoid, $x \in M$ and $n \in \mathbb{N}^*$. If then $n = (a_r \ldots a_1 a_0)_2 = a_0 + a_1 2 + \cdots + a_r 2^r$ with $a_i \in \{0, 1\}$, $i = 0, \ldots, r$, the dual development of n, then $x^n = x^{a_0} (x^2)^{a_1} \cdots (x^{2^r})^{a_r}$. To determine x^n one therefore calculates recursively, with $y_0 = x$, $x_0 = y_0$, if $a_0 = 1$, or $x_0 = e_M$, if $a_0 = 0$, starting with the elements

$$y_{i+1} = y_i^2, \quad x_{i+1} = \begin{cases} x_i y_{i+1}, & \text{if } a_{i+1} = 1, \\ x_i, & \text{if } a_{i+1} = 0, \end{cases} \quad 0 \le i < r.$$

Then $x^n = x_r$. So at most $2r$ multiplications in M are needed (instead of $n - 1$ as with naive exponentiation $x^0 = e_M, x, x^2 = x \cdot x, \ldots, x^{n-1} = x^{n-2} x, x^n = x^{n-1} x$).

(3) We also mention the so-called **discrete logarithm problems** (**DLP**). Let a, x be elements of a group G. The search is for an exponent $t \in \mathbb{N}$ with $x = a^t$, if it exists. t is then called a (discrete) **logarithm of x to the base** a and writes

$$t = \mathrm{Log}_a x.$$

---

[13] In English, trapdoor functions or one-way functions.

2.2 Homomorphisms

In Ord $a = m \in \mathbb{N}^*$, $\mathrm{Log}_a x$ is only determined modulo m, and then for $\mathrm{Log}_a x$ the representative t with $0 \leq t < m$ is chosen. If a factor decomposition $m = m_1 \cdots m_s$ of m with as small factors as possible $m_1, \ldots, m_s$ is known, then according to Pohlig and Hellman $\mathrm{Log}_a x$ is obtained by solving discrete logarithm problems for group elements of the orders $m_1, \ldots, m_s$. We note the sought t with $0 \leq t < m$ in the form $t = t_0 + t_1 m_1 + \cdots + t_{s-1} m_1 \cdots m_{s-1}$ with uniquely determined t_i, $0 \leq t_i < m_{i+1}$, cf. the remark in Example 1.7.6, and now use the additive notation. Then $x = ta$, and it follows

$$x_i := x - (t_0 + \cdots + t_{i-1} m_1 \cdots m_{i-1})a = m_1 \cdots m_i (t_i + \cdots + t_{s-1} m_{i+1} \cdots m_{s-1})a,$$

as well as recursion for determining the "digits" $t_0, \ldots, t_{s-1}$:

$$x_0 = x, \quad m_2 \cdots m_s x_0 = t_0 (m/m_1)a;$$

$$x_i = x_{i-1} - t_{i-1} m_1 \cdots m_{i-1} a, \quad m_{i+2} \cdots m_s x_i = t_i (m/m_{i+1})a, \quad 0 < i < s.$$

Since the elements $(m/m_i)a$ have the order m_i, the assertion is proven. If a recursion step cannot be executed, the given logarithm problem has no solution. — To solve discrete logarithm problems with relatively small $m = $ Ord a (about $m \leq 10^{20}$?), one often proceeds as follows: For a $k \geq 1$, the powers $1 = a^0, \ldots, a^{k-1}$ are stored and the elements $x, xa^{-k}, x(a^{-k})^2, \ldots$ are successively calculated until an $j \in \mathbb{N}$ is found for which an $i < k$ with $x(a^{-k})^j = a^i$ exists. Then $x = a^{i+jk}$. If the discrete logarithm problem is solvable, there is such a j with $j < \lceil m/k \rceil$. As a rule, k is chosen — as long as the memory allows — close to $\sqrt{m}$ (**Babystep-Giantstep method**).

If the group G is finite and initially only the prime factor decomposition of the order $n = p_1^{\alpha_1} \cdots p_r^{\alpha_r}$ of G is known, then $m = $ Ord $a = p_1^{\alpha_1 - \gamma_1} \cdots p_r^{\alpha_r - \gamma_r}$, where γ_ρ is the maximum of the $\gamma \in \mathbb{N}$ with $\gamma \leq \alpha_\rho$ and $a^{n/p_\rho^\gamma} = e_G$ is, $\rho = 1, \ldots, r$. Proof! ◊

Exercise

Exercise 2.2.1 Let M and N be magmas, semigroups, monoids or groups and $\varphi: M \to N$ a mapping with the graph $\Gamma_\varphi = \{(x, \varphi(x)) \mid x \in M\} \subseteq M \times N$.

a) φ is a homomorphism if and only if Γ_φ is a submagma, a subsemigroup, a submonoid or a subgroup of the product $M \times N$. In this case, $x \mapsto (x, \varphi(x))$ is an isomorphism from M to Γ_φ.
b) Let M and N be groups. Γ_φ is a normal subgroup of $M \times N$ if and only if φ is a homomorphism whose image lies in the center $Z(N)$ of N, cf. Exercise 2.1.2. In this case, $(x, y) \mapsto \varphi(x) y^{-1}$ is a surjective group homomorphism $M \times N \to N$, whose kernel is equal to Γ_φ.

Exercise 2.2.2 For a group G, the following are equivalent: (i) G is abelian. (ii) The squaring $x \mapsto x^2$ is an endomorphism of G. (iii) The inverse formation $x \mapsto x^{-1}$ is an automorphism of G. (iv) The "hyperbola"

$$H_G := \{(x,y) \in G \times G \mid xy = e_G\} = \{(x,x^{-1}) \mid x \in G\}$$

of G is a subgroup of $G \times G$. (In this case, H_G is isomorphic to G.) (v) The multiplication mapping $\mu: G \times G \to G$, $\mu(x,y) = xy$, is a group homomorphism. (In this case, μ is surjective with $\ker \mu = H_G$ (cf. (iv)).) (vi) The diagonal $\Delta_G = \{(x,x) \mid x \in G\}$ is a normal subgroup of G. (vii) The quotient formation $\nu: G \times G \to G$, $\nu(x,y) = xy^{-1}$, is a group homomorphism. (In this case, ν is surjective with $\ker \nu = \Delta_G$ (cf. (vi)).)

Exercise 2.2.3 Let M be a monoid and $a \in M^\times$. The **conjugation** of M with a is by definition the mapping

$$\kappa_a: M \to M, \quad x \mapsto axa^{-1}.$$

a) For each $a \in M^\times$, κ_a is an automorphism of M with $(\kappa_a)^{-1} = \kappa_{a^{-1}}$.
b) The mapping $\kappa: M^\times \to \mathrm{Aut}\, M$, $a \mapsto \kappa_a$, is a group homomorphism. Its kernel is the subgroup $Z(M)^\times = Z(M) \cap M^\times = \{a \in M^\times \mid ax = xa \text{ for all } x \in M\}$ (cf. exercise 2.1.2), and its image is a normal subgroup of $\mathrm{Aut}\, M$. (It is $\varphi \kappa_a \varphi^{-1} = \kappa_{\varphi(a)}$ for $a \in M^\times$ and $\varphi \in \mathrm{Aut}\, M$. – Note that the inclusion $Z(M)^\times \subseteq Z(M^\times)$ can be strict. Example?)

Note The automorphisms κ_a, $a \in M^\times$, are the so-called **inner automorphisms** of M, and the quotient group $\mathrm{Out}\, M := \mathrm{Aut}\, M / \mathrm{Inn}\, M$, $\mathrm{Inn}\, M := \mathrm{Im}\, \kappa \cong M^\times / Z(M)^\times$, is called the **group of outer automorphisms** of M.

Exercise 2.2.4 For a subgroup N of a group G, the following are equivalent: (i) N is a normal subgroup in G (i.e., $aN = Na$ for all $a \in G$). (i') For all $a \in G$, $aN \subseteq Na$ is true. (i'') For all $a \in G$, $Na \subseteq aN$ is true. (ii) For all $a \in G$, $aNa^{-1} = N$ is true. (ii') For all $a \in G$, $aNa^{-1} \subseteq N$ is true (i.e., all inner automorphisms of G map N onto itself).

Exercise 2.2.5 Let M and N each be magmas, semigroups, monoids or groups and $\varphi, \psi: M \to N$ homomorphisms.

a) The set of $x \in M$ with $\varphi(x) = \psi(x)$ is a corresponding subobject of M.
b) Infer from a): Two homomorphisms $\varphi, \psi: M \to N$ already agree if they agree on a generating system of M.
c) If $\psi: M \xrightarrow{\sim} N$ is an isomorphism, then $\mathrm{Iso}(M,N) = \psi \circ (\mathrm{Aut}\, M) = (\mathrm{Aut}\, N) \circ \psi$ and the conjugation $\chi \mapsto \psi \circ \chi \circ \psi^{-1}$ with ψ is a group isomorphism $\mathrm{Aut}\, M \xrightarrow{\sim} \mathrm{Aut}\, N$.

Exercise 2.2.6 Let $\varphi: G \to H$ be a homomorphism of groups and $a, b \in G$.

a) For $n \in \mathbb{N}$ is $\varphi(T_n G) \subseteq T_n H$, i.e., if $\mathrm{Ord}\, a$ is a divisor of n, then $\mathrm{Ord}\, \varphi(a)$ is also a divisor of n. Furthermore, $\varphi(TG) \subseteq TH$.

2.2 Homomorphisms

b) If φ is injective, then Ord $\varphi(a) = $ Ord a.
c) It is Ord$(ab) = $ Ord(ba). (Note $ba = b(ab)b^{-1} = \kappa_b(ab)$.)

Exercise 2.2.7 The only endomorphisms of the additive group $\mathbb{Q} = (\mathbb{Q}, +)$ are the stretches $L_a \colon x \mapsto ax$, $a \in \mathbb{Q}$. (Consider that an endomorphism $\varphi \colon \mathbb{Q} \to \mathbb{Q}$ is determined by its value $\varphi(1)$. It is $\varphi = L_{\varphi(1)}$.)

Exercise 2.2.8 Let G be a group and H a cyclic group with generating element a of order $n \in \mathbb{N}$ (e.g. $H = \mathbb{Z}/\mathbb{Z}n$ and $a = 1 + \mathbb{Z}n$). Then the mapping Hom$(H, G) \to T_n G$, $\varphi \mapsto \varphi(a)$, is bijective. If G is abelian, then it is an isomorphism of groups.

Exercise 2.2.9 Let G and H be finite cyclic groups of orders m and n respectively.

a) The group Hom(G, H) is also cyclic and Hom$(G, H) \cong \mathbb{Z}/\mathbb{Z}\gcd(m, n)$. List the elements of Hom(G, H) explicitly. (For the case $G = H$ see also Example 2.2.23. – Which group is Hom$(G, \mathbb{Z})$?)
b) Let m be a divisor of n and $\varphi \colon F \to H$ a homomorphism of a subgroup $F \subseteq G$ into H. Then φ can be extended to a homomorphism $G \to H$. (One can assume $m = n$ and then $G = H$. Then φ is an endomorphism of F, thus a power mapping $x \mapsto x^k$ with a $k \in \mathbb{N}$. – Show that without the condition $m \mid n$ the statement generally does not hold.)
c) Under the conditions of b), let φ be injective. Then φ can be extended to an injective homomorphism $G \to H$. (It can be assumed that m and n are powers of a prime number p.)

Exercise 2.2.10 A group of order 4 is cyclic, thus isomorphic to Z_4, or isomorphic to $\mathbb{Z}_2 \times \mathbb{Z}_2$. (In the second case, we speak of a **Klein four-group**. — With respect to how many of the $4^{16} = 2^{32}$ operations on a set A with 4 elements is A a group? There are $24/2 + 24/6 = 16$. Note $|\text{Aut } \mathbb{Z}_4| = 2$, $|\text{Aut}(\mathbb{Z}_2 \times \mathbb{Z}_2)| = 6$.)

Exercise 2.2.11 Let $G_1, \ldots, G_n$ be groups and $G := G_1 \times \cdots \times G_n$ their product. Furthermore, let $a = (a_1, \ldots, a_n)$ be an element in G.

a) It is Ord $a = \text{lcm}(\text{Ord } a_1, \ldots, \text{Ord } a_n)$.
b) G is cyclic exactly when the factors G_i are all cyclic and the orders $m_1, \ldots, m_n \in \mathbb{N}$ of generating elements of $G_1, \ldots, G_n$ are pairwise coprime. In particular, G is exactly a finite cyclic group when the G_i, $i = 1, \ldots, n$, are finite cyclic groups with pairwise coprime orders. In this case, Ord $G = $ Ord $G_1 \cdots$ Ord G_n and a is a generating element of G exactly when the component a_i is a generating element of G_i, $i = 1, \ldots, n$. (**Note** The last statement provides once again the Chinese Remainder Theorem 2.2.19 and very clearly the multiplicativity of the Euler's φ function: $\varphi(m_1 \cdots m_n) = \varphi(m_1) \cdots \varphi(m_n)$ for pairwise coprime $m_1, \ldots, m_n \in \mathbb{N}^*$.)

Exercise 2.2.12 Let F and H be subgroups of the group G and $\varphi \colon F \times H \to G$ the multiplication mapping $(x, y) \mapsto xy$, whose image is the complex product $FH \subseteq G$. Im $\varphi = FH$ is a subgroup of G if and only if $FH = HF$.

a) $\varphi^{-1}(xy) = \{(xz, z^{-1}y) \mid z \in F \cap H\}$ is given. In particular, all non-empty fibers of φ have the same cardinal number $|F \cap H|$, and it follows
$$|F| \cdot |H| = |F \cap H| \cdot |FH|.$$
b) φ is injective if and only if $F \cap H = \{e_G\}$ is given.
c) φ is bijective if and only if $F \cap H = \{e_G\}$ and $FH = G$ are given.
d) φ is a group homomorphism if and only if F and H are elementwise interchangeable. In this case, the normal subgroup $\ker \varphi$ is the abelian subgroup $\{(z, z^{-1}) \mid z \in F \cap H\} \subseteq F \times H$ (which is isomorphic to $F \cap H$).

Exercise 2.2.13 Let F and H be subgroups of the group G, for which the multiplication mapping $\varphi \colon F \times H \to G$, $(x, y) \mapsto xy$, is bijective, for which therefore $F \cap H = \{e_G\}$ and $FH = G$ applies, see the preceding exercise. In this case, every element $w \in G$ has a representation $w = p_{F,H}(w) q_{F,H}(w)$ with uniquely determined elements $p_{F,H}(w) \in F$ and $q_{F,H}(w) \in H$. The mappings $p_{F,H} \colon G \to G$ and $q_{F,H} \colon G \to G$ are projections of G (i.e., $p_{F,H}^2 = p_{F,H}$ and $q_{F,H}^2 = q_{F,H}$) with $\mathrm{Im}\, p_{F,H} = \mathrm{Fix}(p_{F,H}, G) = F$ or $\mathrm{Im}\, q_{F,H} = \mathrm{Fix}(q_{F,H}, G) = H$. $p_{F,H}$ is called the **projection of** G **onto** F **along** H and $q_{F,H}$ the **projection of** G **along** F **onto** H. Because of $w = (w^{-1})^{-1} = (q_{F,H}(w^{-1}))^{-1}(p_{F,H}(w^{-1}))^{-1}$ applies $p_{H,F}(w) = (q_{F,H}(w^{-1}))^{-1}$ and $q_{H,F}(w) = (p_{F,H}(w^{-1}))^{-1}$. In the abelian case, of course, $p_{H,F} = q_{F,H}$ and $q_{H,F} = p_{F,H}$.

a) The following conditions are equivalent: (i) The projection $q_{F,H} \colon G \to G$ is a group homomorphism. (ii) It is $q_{F,H} = p_{H,F}$. (iii) F is a normal subgroup of G. (**Note** If these conditions are met, then $\ker q_{F,H} = F = \ker p_{H,F}$ and it is said that H is a **weak complement** to F in G and $G = FH$ is the **semidirect product** of F and H. Note for $u, x \in F$ and $v, y \in H$ the equation $(uv)(xy) = u(vxv^{-1})vy = (u\kappa_v(x))(vy)$. In particular, $p_{F,H}(vx) = \kappa_v(x)$ and $q_{F,H}(vx) = v = p_{H,F}(vx)$. The multiplication in G is thus determined by the multiplications in F and H as well as the conjugation homomorphism $\kappa|H \colon H \to \mathrm{Aut}\, F$. For semidirect products, we also refer to Example 2.4.12.)

b) The following conditions are equivalent: (i) φ is a group isomorphism. (ii) The projections $p_{F,H}$ and $q_{F,H}$ are group homomorphisms. (iii) It is $p_{F,H} = q_{H,F}$ and $q_{F,H} = p_{H,F}$. (iv) F and H are normal subgroups of G. (v) The subgroups F and H commute elementwise. (**Note** If these conditions are met, it is said that G is the **direct product** of the subgroups F and H and calls H a **(strong) complement** to F in G (as well as F a (strong) complement to H). A semidirect product $G = FH$ (cf. a)) is exactly a direct product when the conjugation homomorphism $\kappa|H \colon H \to \mathrm{Aut}\, F$ is trivial. If G is abelian, the given conditions are always met.)

2.2 Homomorphisms

c) Provide an example where in the considered situation neither F nor H is a normal subgroup in G. (Such examples can be found in $G := \mathfrak{S}_4$.)

(For the above exercise, also see the Examples 2.2.17 and 2.2.20.)

Exercise 2.2.14 Let F be a normal subgroup of the group G. F is called a **(strong) direct factor** or a **weak direct factor** of G, if F has a (strong) complement or a weak complement in G, cf. the above exercise. Every strong direct factor is also a weak direct factor. For abelian groups, both terms coincide.

a) Only then is F a (strong) factor of G, if there is a projection $p: G \to G$ with $\operatorname{Im} p = F$ that is a group homomorphism. In this case, $\ker p$ is a (strong) complement of F in G.
b) F is a weak factor of G if and only if there is a projection $p: G \to G$ with $\ker p = F$ that is a group homomorphism. In this case, $\operatorname{Im} p$ is a weak complement of F in G.
c) Let G' be a subgroup of G with $F \subseteq G'$. If H is a strong (or weak) complement to F in G, then $H \cap G'$ is a strong (or weak) complement to F in G'.
d) If H is a weak complement to F in G, then the restriction of the canonical projection $\pi: G \to G/F$ to H is a group isomorphism $\pi|H: H \xrightarrow{\sim} G/F$. In particular, all weak complements to F in G are isomorphic.
e) If G/F is abelian (i.e., is $[G, G] \subseteq F$, see Example 2.3.10), then F is exactly a (strong) direct factor of G when $F \cap Z(G)$ is a (strong = weak) direct factor of $Z(G)$. In this case, every complement of $F \cap Z(G)$ in $Z(G)$ is a (strong) complement of F in G.

Exercise 2.2.15 For normal divisors $N_1, \ldots, N_n$ of a group G with $N'_i := \bigcap_{j \neq i} N_j$, $i = 1, \ldots, n$, the following conditions are equivalent: (i) $N_i N'_i = G$ for $i = 1, \ldots, n$. (ii) $N'_1 \cdots N'_n = G$. (iii) The homomorphism $\pi := (\pi_{N_1}, \ldots, \pi_{N_n}): G \to G/N_1 \times \cdots \times G/N_n$ is surjective. (See Theorem 2.2.18.) — If the indices $[G : N_i]$, $i = 1, \ldots, n$, are finite and pairwise coprime, then these conditions are fulfilled.

Exercise 2.2.16 Let $\varphi_i: G \to H_i$ be surjective homomorphisms of the group G into the finite groups H_i with pairwise coprime orders, $i = 1, \ldots, n$. Then the homomorphism $\varphi = (\varphi_1, \ldots, \varphi_n): G \to H_1 \times \cdots \times H_n$ is also surjective.

Exercise 2.2.17 Let F, H be subgroups of the abelian group G with $FH = G$ and let $\varphi: F \to L$ or $\psi: H \to L$ be homomorphisms into the abelian group L. There exists a homomorphism $\chi: G \to L$ with $\chi|F = \varphi$ and $\chi|H = \psi$, if $\varphi|(F \cap H) = \psi|(F \cap H)$ is. Show with an example that a similar statement does not generally hold for arbitrary groups G.

Exercise 2.2.18 Let G be an abelian group with $^{n^k}G = \{e_G\}$ for a $n \in \mathbb{N}^*$ and a $k \in \mathbb{N}$. If $F \subseteq G$ is a subgroup of G with $G = F \cdot {}^nG$, then $G = F$. (It is $G = F \cdot {}^{n^m}G$ for all $m \in \mathbb{N}$.)

Exercise 2.2.19 Let $p \in \mathbb{P}$ and G be a p-group.

a) If $\varphi: G \to H$ is a group homomorphism and if $\varphi|_{p}G$ is injective, then φ is also injective.
b) If $G = {}^pG$, then G is divisible.

Exercise 2.2.20 Let G be a finite abelian group. For each divisor d of Ord G there is a subgroup of G of order d. In particular, for each prime divisor p of Ord G there is an element of order p in G. (Because of Theorem 2.2.22 one can assume that Ord G is a prime power p^α. Then one concludes by induction over α. If $a \in G$ is an element of order p, then apply the induction hypothesis to the group $G/\mathrm{H}(a)$ whose order is $p^{\alpha-1}$ and then use Proposition 2.2.7. – Even in non-abelian finite groups G, there is an element of order p for every prime divisor p of Ord G, see Cauchy's Theorem 2.4.8. For abelian finite groups, this can also be easily proven as follows: Let $G = \mathrm{H}(a_1, \ldots, a_n)$. Then $\mathrm{H}(a_1) \times \cdots \times \mathrm{H}(a_n) \to G$, $(x_1, \ldots, x_n) \mapsto x_1 \cdots x_n$, is a surjective group homomorphism and consequently Ord G is a divisor of Ord $\bigl(\mathrm{H}(a_1) \times \cdots \times \mathrm{H}(a_n)\bigr) = \mathrm{Ord}\, a_1 \cdots \mathrm{Ord}\, a_n$, and every prime divisor of Ord G divides at least one of the orders Ord $a_1, \ldots,$ Ord a_n. In non-abelian finite groups, there is not always a subgroup of order d for every divisor d of the group order. For example, the permutation group $\mathfrak{S}_5$ has no subgroup of order 15. (This would have to be cyclic.)

Exercise 2.2.21 Let G be a finite group. G is cyclic if and only if G has at most one subgroup of order d for every divisor d of $|G|$. (Show that the d-torsion $\mathrm{T}_d G$ of G has at most d elements and then use Theorem 2.2.15.)

Exercise 2.2.22 Let G be a finite abelian group.

a) The following conditions are equivalent: (i) G is cyclic. (ii) Every primary component of G is cyclic. (iii) For every prime divisor p of $|G|$, the p-base $_pG$ is cyclic. (For the proof of (iii) $\Rightarrow$ (ii), let the order of G be a prime power p^α, $\alpha > 0$. Then $|{}^pG| = |G|/|_pG|$ is true. In $|_pG| = p$, every element of the complement $G - {}^pG$ is a generating element of G, see exercise 2.2.18. – The implication (iii) $\Rightarrow$ (i) strengthens the cyclicality criterion 2.2.15 for finite abelian groups. The quaternion group Q_4 of order 8 is not cyclic, although $|_2Q_4| = 2$ is true, see exercise 2.4.7. A finite p-group G with $p > 2$ and $|_pG| = p$ is already cyclic.)
b) If the order of G is square-free, then G is cyclic.

Exercise 2.2.23 Let G be a finite abelian group with neutral element e.

Fig. 2.8 $\overline{\varphi}$ is induced by φ by means of ψ.

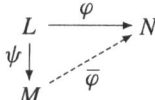

a) There is an element $a \in G$ with $\operatorname{Ord} a = \operatorname{Exp} G$, see Example 2.2.23. (According to Theorem 2.2.22, one can assume that $\operatorname{Ord} G$ is a power of a prime number. — The statement does not generally apply to finite groups, for example not for $G = \mathfrak{S}_3$.)
b) G is cyclic if and only if $\operatorname{Ord} G = \operatorname{Exp} G$ is. (This once again follows the cyclicity criterion 2.2.15 for finite *abelian*groups.)

Exercise 2.2.24 Let G be a finite abelian group and $a_0 \in G$ be an element with $\operatorname{Ord} a_0 = \operatorname{Exp} G$. Then $\mathrm{H}(a_0)$ is a direct factor of G. (Let $G = \mathrm{H}(a_0, a_1, \ldots, a_n)$. Using exercise 2.2.9b) and exercise 2.2.17, the identity of $\mathrm{H}(a_0)$ is successively set to homomorphisms $\mathrm{H}(a_0, a_1) = \mathrm{H}(a_0) \mathrm{H}(a_1) \to \mathrm{H}(a_0)$, $\ldots$, $G = \mathrm{H}(a_0) \mathrm{H}(a_1) \cdots \mathrm{H}(a_n) \to \mathrm{H}(a_0)$ and so on. – With the Zorn's Lemma 1.4.15, the statement is also proven for infinite abelian groups with $\operatorname{Exp} G > 0$.)

Exercise 2.2.25 Prove the so-called **Fundamental Theorem of Finite Abelian Groups** : Every finite abelian group G is a direct product of cyclic groups $H_1, \ldots, H_r$, where furthermore the relations $1 < \operatorname{Ord} H_r \mid \operatorname{Ord} H_{r-1} \mid \cdots \mid \operatorname{Ord} H_1$ can be assumed. (According to Exercise 2.2.23a) and Exercise 2.2.24, G is isomorphic to $H_1 \times G'$, where H_1 is a cyclic group of order $\operatorname{Exp} G$.)

Remark The orders of the H_i are uniquely determined by G through the given divisibility properties and are called the **elementary divisors** of the group G. For proof, one can assume that G is a p-group. Then the elementary divisors are, for example, determined by the orders of the groups $_p G, _{p^2} G, \ldots$.

Exercise 2.2.26 Let $p \in \mathbb{P}$ be a prime number. $\mathbb{Z}_p$ is the subgroup $\{a/p^n \mid a \in \mathbb{Z}, n \in \mathbb{N}\} \subseteq \mathbb{Q} = (\mathbb{Q}, +)$ and $\mathbb{Z}_{(p)}$ is the subgroup $\{a/b \mid a, b \in \mathbb{Z}, p \nmid b\}$. With π_p the canonical surjective homomorphism $\mathbb{Q}/\mathbb{Z} \to \mathbb{Q}/\mathbb{Z}_{(p)}$, $x + \mathbb{Z} \mapsto x + \mathbb{Z}_{(p)}$, is denoted.

a) $\mathbb{Q}/\mathbb{Z}$ is a torsion group, and $\mathbb{Z}_p/\mathbb{Z}$ is the p-primary component of $\mathbb{Q}/\mathbb{Z}$. So it is $\mathbb{Q}/\mathbb{Z} = \sum_{p \in \mathbb{P}}^{\oplus} \mathbb{Z}_p/\mathbb{Z}$. For each $m \in \mathbb{N}^*$ there is exactly one subgroup of order m in $\mathbb{Q}/\mathbb{Z}$. It is generated by the residue class $m^{-1} + \mathbb{Z}$ and is equal to the m-torsion $\mathrm{T}_m(\mathbb{Q}/\mathbb{Z})$ of $\mathbb{Q}/\mathbb{Z}$.
b) $\mathbb{Q}/\mathbb{Z}_{(p)}$ is a p-group. The homomorphism induced by the π_p, $p \in \mathbb{P}$, $\pi : \mathbb{Q}/\mathbb{Z} \to \prod_{p \in \mathbb{P}} \mathbb{Q}/\mathbb{Z}_{(p)}$ is injective, its image is the direct sum $\bigoplus_{p \in \mathbb{P}} \mathbb{Q}/\mathbb{Z}_{(p)} \subseteq \prod_{p \in \mathbb{P}} \mathbb{Q}/\mathbb{Z}_{(p)}$. So $\pi : \mathbb{Q}/\mathbb{Z} \xrightarrow{\sim} \bigoplus_{p \in \mathbb{P}} \mathbb{Q}/\mathbb{Z}_{(p)}$ is an isomorphism.
c) π induces an isomorphism $\mathbb{Z}_p/\mathbb{Z} \xrightarrow{\sim} \mathbb{Q}/\mathbb{Z}_{(p)}$. (The group $\mathbb{Z}_p/\mathbb{Z}$ and any group isomorphic to it is called the **Prüfer p-group** and is denoted by $\mathrm{I}(p)$.)

Exercise 2.2.27 The isomorphism type of the group $\mathbb{Q}/\mathbb{Z}$ (and any group isomorphic to it) is also denoted by $\mathbf{Z}_\infty$. This designation is motivated by the following characterization: A torsion group G is isomorphic to $\mathbb{Q}/\mathbb{Z}$, if it contains exactly one subgroup of order n for each $n \in \mathbb{N}^*$ (which is necessarily cyclic, cf. exercise 2.2.21).

Exercise 2.2.28 Let G be a group. The mapping $\varphi \mapsto \ker\varphi$ is a bijective mapping of the set of surjective homomorphisms $\varphi\colon G \to \mathbf{Z}_2$ onto the set $\mathcal{N}_2 = \mathcal{N}_2(G)$ of subgroups of index 2 in G (which are automatically normal in G). The intersection $\bigcap_{N \in \mathcal{N}_2} N$ is equal to the normal subgroup generated by the squares in G $\mathrm{H}(^2G)$. If $\mathcal{N}_2$ is finite, then $|\mathcal{N}_2| = 2^n - 1$ and $G/\mathrm{H}(^2G) \cong \mathbf{Z}_2^n$ with a $n \in \mathbb{N}$. (Use exercise 2.3.6.) For each $n \in \mathbb{N}^*$ the set ^nG of the n-th powers obviously generates a normal subgroup $\mathrm{H}(^nG)$ of G, and the exponent of the quotient group $G/\mathrm{H}(^nG)$ divides n. Furthermore, $\mathrm{H}(^nG)$ is in the intersection $\bigcap_{N \in \mathcal{N}_n} N$ of the normal subgroups N of index n in G. ($\mathcal{N}_n$ is the set of these normal subgroups.) If $n = p$ is a prime number and G is abelian, then the equality $\mathrm{H}(^pG) = {}^pG = \bigcap_{N \in \mathcal{N}_p} N$ applies again. For a prime number $p > 2$ and arbitrary groups, however, this does not generally apply.

Exercise 2.2.29 Let G be a finite abelian group, $n \in \mathbb{N}^*$ and $d := \gcd(n, |G|)$. For each prime divisor p of n, let the p-primary component $G(p)$ of G is cyclic, i.e., the p-base $_pG$ of G is cyclic. Exactly then is $x \in G$ an n-th power in G, when $x^{|G|/d} = e_G$ is (**Euler's criterion for n-th powers**). (The statement can be reduced to the case that G is cyclic. – Euler himself formulated the criterion only for the case $n = 2$. Therefore, it reads: *If G has exactly one element f of order 2, then an element $x \in G$ is exactly a square in G, when $x^{|G|/2} = e_G$ is. Otherwise, $x^{|G|/2} = f$.* – To generally find an $x \in G$ that fulfills the above condition for an n-th power, a $y \in G$ with $y^n = x$, one can proceed as follows: If n is coprime to $|G|$, then $y = x^m$ is the only solution, where $mn \equiv 1 \bmod |G|$ is. Otherwise, based on the primary decomposition of G, one can assume that G is a cyclic group of order p^α, $\alpha > 0$, see Theorem 2.2.22 and the remark following it. Through trial and error, one then finds a generating element a of G. (The probability that a randomly chosen element from G is a generating element of G is $\varphi(p^\alpha)/p^\alpha = 1 - p^{-1}$, so at least $1/2$.) Let $n = kp^\beta$ be with $p \nmid k$. By assumption, $x^{p^{\alpha-\mathrm{Min}(\alpha,\beta)}} = e_G$. So we can assume $\beta \leq \alpha$ and $x^{p^{\alpha-\beta}} = e_G$. First, we determine a $z \in G$ with $z^k = x$. It is then sufficient to find a y with $y^{p^\beta} = z$. To do this, we calculate the logarithm $\ell := \mathrm{Log}_a z$ of z to the base a, for example using the method described in Remark (3) to Example 2.2.23. By assumption, ℓ is a multiple of p^α, and $y := a^{\ell/p^\alpha}$ is a p^αth root of z.)

2.3 Induced Homomorphisms and Quotient Formation

In this section, we prove homomorphism and isomorphism theorems and discuss typical constructions (including quotient formations), which are of exemplary importance for the entire algebra and beyond.

2.3 Induced Homomorphisms and Quotient Formation

Let L, M, N be magmas and $\psi: L \to M$, $\varphi: L \to N$ be (initially) arbitrary mappings, where ψ is surjective. Then is the equivalence relation induced by ψ R_ψ (with the fibers $\psi^{-1}(\psi(a))$ as equivalence classes $[a]_{R_\psi}$, $a \in L$) finer than the equivalence relation induced by φ R_φ (with the equivalence classes $\varphi^{-1}(\varphi(a)) = [a]_{R_\varphi}$), is therefore $[a]_{R_\psi} \subseteq [a]_{R_\varphi}$ for all $a \in L$, so φ induces a mapping $\overline{\varphi}: M \dashrightarrow N$, which is uniquely determined by the condition $\varphi = \overline{\varphi} \circ \psi$, i.e., the commutativity of the diagram from Fig. 2.8, is uniquely determined. It is $\overline{\varphi}(c) = \varphi(a)$, where a is any element from the (non-empty) fiber $\psi^{-1}(c), c \in M$, is. Exactly then is $\overline{\varphi}$ surjective, when φ is surjective, and exactly then injective, when $R_\varphi = R_\psi$ is. Now if ψ and φ are homomorphisms, then also $\overline{\varphi}$ a homomorphism. If $c, d \in M$ and $a \in \psi^{-1}(c)$, $b \in \psi^{-1}(d)$, then $ab \in \psi^{-1}(cd)$, since ψ is a homomorphism, and $\overline{\varphi}(cd) = \varphi(ab) = \varphi(a)\varphi(b) = \overline{\varphi}(c)\overline{\varphi}(d)$. We have proven:

Theorem 2.3.1 (Theorem of the induced homomorphism) *Let $\psi: L \to M$ and $\varphi: L \to N$ be homomorphisms of magmas. ψ is surjective, and it holds $\psi^{-1}(\psi(a)) \subseteq \varphi^{-1}(\varphi(a))$ for all $a \in L$, i.e. it holds $\varphi(a) = \varphi(b)$ for all $a, b \in L$ with $\psi(a) = \psi(b)$. Then there is exactly one homomorphism $\overline{\varphi}: M \to N$ with $\varphi = \overline{\varphi} \circ \psi$. – Exactly then is $\overline{\varphi}$ surjective, when φ is surjective. – Exactly then is $\overline{\varphi}$ injective, when $\psi^{-1}(\psi(a)) = \varphi^{-1}(\varphi(a))$ is for all $a \in L$. – Exactly then is $\overline{\varphi}$ an isomorphism, when φ is surjective and $\psi^{-1}(\psi(a)) = \varphi^{-1}(\varphi(a))$ for all $a \in L$ applies.*

If in the situation of Theorem 2.3.1 ψ and φ are monoid homomorphisms, then it is also clear that $\overline{\varphi}$ is a monoid homomorphism. In the case of groups, $\psi^{-1}(\psi(a)) = a \ker \psi = (\ker \psi)a$ and $\varphi^{-1}(\varphi(a)) = a \ker \varphi = (\ker \varphi)a$, $a \in L$, the fibers, see Theorem 2.2.5, and we obtain the following important result for groups:

Theorem 2.3.2 (Theorem of the Induced Homomorphism for Groups) *Let $\psi: G \to F$ and $\varphi: G \to H$ be homomorphisms of groups. Let ψ be surjective, and let $\ker \psi \subseteq \ker \varphi$ hold. Then there exists exactly one homomorphism $\overline{\varphi}: F \to H$ with $\varphi = \overline{\varphi} \circ \psi$.*

Exactly then is $\overline{\varphi}$ surjective, if φ is surjective. – Exactly then is $\overline{\varphi}$ injective, if $\ker \psi = \ker \varphi$ is. – Exactly then is $\overline{\varphi}$ an isomorphism, if φ is surjective and $\ker \psi = \ker \varphi$ holds **(Isomorphism theorem for groups)**.

Let $\psi: L \to M$ be a surjective homomorphism of magmas again. Then the inclusion $\psi^{-1}(\psi(a)) \cdot \psi^{-1}(\psi(b)) \subseteq \psi^{-1}(\psi(ab))$ applies for all $a, b \in L$, i.e., the equivalence relation R_ψ induced by ψ is compatible in the following sense:

Definition 2.3.3 Let R be an equivalence relation on the magma L. Then R is called **compatible** (with the given operation on L), if $[a]_R[b]_R \subseteq [ab]_R$ for all $a, b \in L$ is, i.e., if for all $a, b, c, d \in L$ applies: From aRc and bRd follows $(ab)R(cd)$.

The compatibility of the equivalence relation R can be checked in two steps: R is compatible exactly when for all $a,b,c \in L$ applies: From aRc follows $(ab)R(cb)$ and $(ba)R(bc)$. The universal relation $R = L \times L$ is always compatible on L. If $R \subseteq L \times L$ is any relation on L, then the smallest compatible equivalence relation on L that includes R (which is the intersection of all compatible equivalence relations on L that include R) is called the **compatible equivalence relation generated by R**. We denote it by $\langle R \rangle$.

Theorem 2.3.4 (Quotients) *Let R be a compatible equivalence relation on the magma L. Then by*

$$[a]_R \cdot [b]_R := [ab]_R, \quad a,b \in L$$

an operation on L/R is defined, with respect to which the canonical projection

$$\pi_R \colon L \to L/R, \quad a \mapsto [a]_R,$$

*is a homomorphism. L/R is called the **quotient of L with respect to** R. If L is a semigroup or a monoid or a group, the same applies to the quotient L/R.*

Proof It only needs to be shown that the operation on L/R is well-defined, i.e., that from $[a]_R = [c]_R$ and $[b]_R = [d]_R$ always follows $[ab]_R = [cd]_R$. But this is exactly the assumed compatibility of R. The additions follow from Proposition 2.2.2. □

Let R be a compatible equivalence relation on the *group* L with neutral element e. Then $\pi_R \colon L \to L/R$ is a surjective group homomorphism with $N := \ker \pi_R = [e]_R$ and $[a]_R = aN = Na$. In other words:

Theorem 2.3.5 (Quotient Groups) *The compatible equivalence relations on a group G are precisely the congruence relations with respect to the normal subgroups of G. If N is a normal subgroup of G, then the operation on the quotient group is $G/N = G\backslash N$ the complex multiplication of the cosets with respect to N: $(aN) \cdot (bN) = (ab)N$, $a,b \in G$, and the canonical projection $\pi_N \colon G \to G/N$ is the mapping $a \mapsto aN = Na$, $a \in G$.*

In the situation of Theorem 2.3.5, the group G/N is also called the **quotient group** or the **factor group** of G according to the normal subgroup N of G. We have already introduced these groups following Theorem 2.2.5. Note that unlike groups for arbitrary compatible equivalence relations R on a magma L, the inclusion $[a]_R[b]_R \subseteq [ab]_R$ can be strict, i.e., *the operation on L/R is not necessarily the complex multiplication of* $\mathfrak{P}(L)(\supseteq L/R)$. For example, as already noted, the multiplication of $\mathbb{Z}$ and the congruence relation mod m on $\mathbb{Z}$, $m \in \mathbb{N}$, are compatible (due to $(r + \mathbb{Z}m) \cdot (s + \mathbb{Z}m) \subseteq rs + \mathbb{Z}m$ for all $r,s \in \mathbb{Z}$). With respect to the multiplication $(r + \mathbb{Z}m)(s + \mathbb{Z}m) = rs + \mathbb{Z}m$, $\mathbb{Z}/\mathbb{Z}m$ is therefore a monoid. For $m \geq 2$, however, for the complex product, for example, $(0 + \mathbb{Z}m) \cdot (0 + \mathbb{Z}m) = \mathbb{Z}m^2 \subset \mathbb{Z}m = 0 + \mathbb{Z}m$ applies. — From Theorem 2.3.1 it follows immediately:

2.3 Induced Homomorphisms and Quotient Formation

Theorem 2.3.6 (Universal Property of the Quotient) *Let L and N be magmas or semigroups or monoids or groups, R a compatible equivalence relation on L and $\varphi \colon L \to N$ a homomorphism. If R is finer than the equivalence relation R_φ, then there exists exactly one homomorphism $\overline{\varphi} \colon L/R \to N$ with $\varphi = \overline{\varphi} \circ \pi_R$. Exactly then is $\overline{\varphi}$ surjective or injective or an isomorphism, when φ is surjective or $R = R_\varphi$ or when φ is surjective and $R = R_\varphi$ is. Specifically, the isomorphism applies*

$$L/R_\varphi \cong \mathrm{Im}\, \varphi.$$

For groups, we formulate explicitly, see Theorem 2.3.5:

Theorem 2.3.7 *Let G and H be groups, N a normal subgroup and $\varphi \colon G \to H$ a homomorphism. If $N \subseteq \ker \varphi$ and $\pi_N \colon G \to G/N$ is the canonical homomorphism, then there exists exactly one homomorphism*

$$\overline{\varphi} \colon G/N \to H$$

with $\varphi = \overline{\varphi} \circ \pi_N$. It is $\overline{\varphi}(aN) = \varphi(a)$ for all $a \in G$. Exactly then is $\overline{\varphi}$ surjective or injective or an isomorphism, when φ is surjective or $N = \ker \varphi$ applies or when both φ is surjective and $N = \ker \varphi$ applies.

In particular, we obtain:

Corollary 2.3.8 (Isomorphism Theorem for Groups) *Let $\varphi \colon G \to H$ be a homomorphism of groups. Then it induces φ an isomorphism*

$$\overline{\varphi} \colon G/\ker \varphi \xrightarrow{\sim} \mathrm{Im}\, \varphi.$$

Example 2.3.9 (Cyclic Groups) We consider once again a cyclic group G with generating element a. The surjective exponential homomorphism $\varphi_a \colon \mathbb{Z} \to G$, $n \mapsto a^n$, with $\ker \varphi_a = \mathbb{Z}\, \mathrm{Ord}\, a$ then induces the isomorphism

$$\mathbb{Z}/\mathbb{Z}\, \mathrm{Ord}\, a \xrightarrow{\sim} G,$$

which we have already discussed in detail in Example 2.2.10. ◊

Example 2.3.10 (Commutator Group) Let G be a group, H a commutative group and $\varphi \colon G \to H$ a homomorphism from G to H. For any elements $a, b \in G$, the so-called **commutator**

$$[a, b] := aba^{-1}b^{-1}$$

is obviously an element of $\ker \varphi$. Therefore, $\ker \varphi$ contains the subgroup of G generated by the commutators $[a, b]$, $a, b \in G$. This is denoted with

$$[G, G] \quad \text{or} \quad D(G)$$

and is called the **commutator group** or the **derived group** of G. It is a normal subgroup of G. *Proof.* For every group homomorphism $\psi \colon G \to G'$ is $\psi([a, b]) = [\psi(a), \psi(b)]$. In particular, for $a, b, c \in G$

$$c[a, b]c^{-1} = [cac^{-1}, cbc^{-1}],$$

results when choosing for ψ the conjugation $\kappa_c \colon x \mapsto cxc^{-1}$ with the element $c \in G$. For all $c \in G$ follows $c[G,G]c^{-1} \subseteq [G,G]$ and thus $c[G,G] = [G,G]c$.

G is exactly abelian when $[G,G]$ consists only of the neutral element. The residue class group $G/[G,G]$ is abelian, and according to Theorem 2.3.7 the given homomorphism $\varphi \colon G \to H$ induces a group homomorphism $\overline{\varphi} \colon G/[G,G] \to H$. The **abelianization**

$$G_{\mathrm{ab}} := G/[G,G]$$

of G with the canonical projection $\pi_{\mathrm{ab}} \colon G \to G_{\mathrm{ab}}$ thus has the following universal property: *For every abelian group H the mapping is* $\mathrm{Hom}(G_{\mathrm{ab}}, H) \to \mathrm{Hom}(G, H)$, $\overline{\varphi} \mapsto \overline{\varphi} \circ \pi_{\mathrm{ab}}$, *an isomorphism of abelian groups*. In particular, every homomorphism from G into an abelian group is trivial if and only if the group G_{ab} is trivial, i.e. when $G = [G,G]$ is. Such groups G are called **perfect**.

For an arbitrary magma M, the Abelianization M_{ab} is the quotient magma of M with respect to the compatible equivalence relation generated by the pairs (ab, ba), $a, b \in M$. ◊

Example 2.3.11 (Simple Groups) A group G is called **simple** if $G \neq \{e_G\}$ is and G and $\{e_G\}$ are the only normal divisors of G, i.e. if $G \neq \{e_G\}$ is and every homomorphism from G into a group H is trivial or injective, see Theorem 2.3.5. *An abelian group is simple if and only if it is cyclic of prime order,* see Exercise 2.3.5b). Further examples of simple groups are the alternating groups $\mathfrak{A}_n$, $n \geq 5$, see Theorem 2.5.19. Non-abelian simple groups are perfect, see Example 2.3.10. The classification of all finite simple groups is an important and difficult problem of group theory, but today it is considered essentially solved.[14] ◊

Example 2.3.12 (Generating Systems of Finite Abelian Groups) Let H be a finite abelian group. The group H is trivial if and only if $^p H = H$ or $_p H = \{e\}$ is for every prime number $p \in \mathbb{P}$. This follows, for example, from the formula $|H| = |_p H| \cdot |^p H|$, i.e. $|H/^p H| = |_p H|$, and from $_p H \neq \{e_H\}$ for every prime divisor p of $|H|$. We want to generalize this consideration somewhat.

Lemma 2.3.13 *The elements $x_1, \ldots, x_r \in H$ generate the finite abelian group H exactly when for every prime number p the residue classes of $x_1, \ldots, x_r$ generate the elementary abelian p-group $H/^p H$.*

Proof The elements $x_1, \ldots, x_r$ may satisfy the given condition. Then let $F = \mathrm{H}(x_1, \ldots, x_r)$ be the subgroup generated by $x_1, \ldots, x_r$. By assumption, $H = F \cdot {}^p H$, so $^p(H/F) = (F \cdot {}^p H)/F = H/F$ for every prime number p. According to the preliminary remark, the group H/F is trivial and thus $H = F$. □

---

[14] See for example Gorenstein, D.; Lyons, R. Solomon, R: The classification of the finite simple groups, Mathematical Surveys and Monographs, vol. 40, Number 1–6. Providence, R.I. 1994–2004.

2.3 Induced Homomorphisms and Quotient Formation

Let now $p_1,\ldots,p_n$ be the different prime divisors of $|H|$. Then H is the direct product $H = H(p_1) \times \cdots \times H(p_n)$ of its primary components $H(p_i)$, $i = 1,\ldots,n$, cf. Theorem 2.2.22. According to Lemma 2.3.13 and Exercise 2.3.6, the minimum number of elements of a generating system of $H(p_i)$ is equal to α_i, if $|H(p_i)/^{p_i}H(p_i)| = |_{p_i}H(p_i)| = |_{p_i}H| = p_i^{\alpha_i}$ is. It follows with the Chinese Remainder Theorem, cf. Exercise 2.2.11b):

Theorem 2.3.14 *The minimum number for the number of elements of a generating system of a finite abelian group H is* $\text{Max}(\alpha_1,\ldots,\alpha_n)$, *if for the different prime divisors* $p_1,\ldots,p_n$ *of $|H|$ the following holds: The order of the p_i-core $_{p_i}H$ of H, i.e. the number of solutions of the equation $x^{p_i} = e_H$, is $p_i^{\alpha_i}$, $i = 1,\ldots,n$.*

From the last sentence, the following cyclicity criterion for finite abelian groups H arises again: H is exactly cyclic when the equation $x^p = e_H$ for every prime divisor p of $|H|$ only p solutions in H has. See also exercise 2.2.22a). Theorem 2.3.14 also naturally arises from the main theorem for finite abelian groups. If H has the m elementary divisors $e_1,\ldots,e_m (> 1)$, then m is the minimum number of elements of a generating system of H, see exercise 2.2.25. ◊

Example 2.3.15 (Further Isomorphism Theorems) Let H and N be subgroups of the group G, N is a normal divisor. The composition of the canonical injection $H \to HN$ and the canonical projection $HN \to (HN)/N$ is surjective with $H \cap N$ as the kernel. The general isomorphism Theorem 2.3.8 thus implies the following so-called **Noether's Isomorphism Theorem**: *The mapping*

$$H/(H \cap N) \xrightarrow{\sim} (HN)/N, \quad a(H \cap N) \mapsto aN, \ a \in H,$$

is a group isomorphism.

Now let H also be a normal divisor in G with $N \subseteq H$. Then the identity of G induces a surjective homomorphism according to Theorem 2.3.7 $G/N \to G/H$, whose kernel is obviously H/N. The isomorphism Theorem 2.3.8 now provides: *The mapping*

$$(G/N)/(H/N) \xrightarrow{\sim} G/H, \quad \bar{a}(H/N) \mapsto aH, \ \bar{a} = aN \in G/N, \ a \in G,$$

is an isomorphism of groups. — The reader should understand the above statements only as examples of many similar applications of induced homomorphisms. ◊

Example 2.3.16 (Fraction Monoids – Grothendieck Groups) Compatible equivalence relations on monoids (and not just on groups) have already been encountered in Sect. 2.1. Thus, the divisibility relation $a \mid_N b$ if and only if $aN \supseteq bN$" on an abelian monoid M with submonoid $N \subseteq M$ is a quasi-order on M, whose associated equivalence relation "$a_N \|_N b$ if and only if $aN = bN$" is a compatible equivalence relation on M. From $a_N \|_N c$ and $b_N \|_N d$ it follows that $aN = cN$, $bN = dN$ and thus $(ab)N = (aN)(bN) = (cN)(dN) = (cd)N$, i.e. $ab_N \|_N cd$.

The associated quotient monoid $\overline{M} = M/_N\|_N$ we have already introduced in Sect. 2.1, see also exercise 2.1.16. $a \mid_N b$ holds if and only if $\overline{a} \mid_{\overline{N}} \overline{b}$ holds, where $\overline{N} = N/_N\|_N$ is the canonical image of N in $\overline{M}$. Since the divisibility $\mid_{\overline{N}}$ in $\overline{M}$ is even an order, divisibility considerations become often simpler and clearer by transition to $\overline{M}$. If $N = M^\times$, then the divisibility $\mid_{M^\times}$ is already an equivalence relation and the corresponding equivalence classes are the associativity classes $aM^\times$, $a \in M$. The quotient monoid $M/_{M^\times}\|_{M^\times} = M/M^\times$ is sharp, i.e., its unit group is trivial.

Let now $S \subseteq M$ be a submonoid of the (not necessarily commutative) monoid M and furthermore $\varphi: M \to L$ a monoid homomorphism such that the image $\varphi(S) \subseteq L^*$ only contains regular elements of L. If there are elements $a, b \in M$ for $s, t \in S$ with $\varphi(sat) = \varphi(sbt)$, then it follows $\varphi(s)\varphi(a)\varphi(t) = \varphi(s)\varphi(b)\varphi(t)$, thus $\varphi(a) = \varphi(b)$, since $\varphi(s)$ and $\varphi(t)$ are regular in L. From $(sat)R_\varphi(sbt)$ it always follows $aR_\varphi b$, where R_φ the compatible equivalence relation defined by φ on M. If $\approx_S$ is the finest compatible equivalence relation R on M with the additional property "from $(sat)R(sbt)$ for $a, b \in M$, $s, t \in S$ always follows aRb" — it exists as the intersection of all R with this property -, then $\pi_S(S) \subseteq (M/\approx_S)^*$, where $\pi_S: M \to M/\approx_S$ is the canonical projection. If $s, t \in S$ and $\pi_S(s)\pi_S(a)\pi_S(t) = \pi_S(s)\pi_S(b)\pi_S(t)$ applies, then $sat \approx_S sbt$ and consequently $a \approx_S b$, i.e. $\pi_S(a) = \pi_S(b)$. Together with the universal property 2.3.6 of the quotient monoids, we obtain the following universal property of $M/\approx_S$:

Proposition 2.3.17 *Let $S \subseteq M$ be a submonoid of M and $\pi_S: M \to M/\approx_S$ the canonical projection. Then $\pi_S(S) \subseteq (M/\approx_S)^*$ and for every monoid homomorphism $\varphi: M \to L$ with $\varphi(S) \subseteq L^*$ there exists exactly one monoid homomorphism $\overline{\varphi}: M/\approx_S \to L$ with $\varphi = \overline{\varphi} \circ \pi_S$.*

In the situation of Proposition 2.3.17, $M/\approx_S$ is called the **regularization of M with respect to** $S \subseteq M$. When $S = M$, we speak of the regularization of M per se. It is a regular monoid that is canonically associated with M. For any subset $T \subseteq M$, the regularization of M with respect to T is understood as the regularization of M with respect to the submonoid generated by T, which we denote by $\langle T \rangle$.

The relation $\approx_S$ is generally difficult to oversee. It becomes much clearer, *if S is in the center $Z(M)$ of M. Then the following holds*

$$a \approx_S b \iff \text{there is } s \in S \text{ with } sa = sb, \quad a, b \in M.$$

Proof The relation R_0 defined by the right side is trivially an equivalence relation. Moreover, it is compatible; because from aR_0c and bR_0d it follows that $sa = sc$ and $tb = td$ with $s, t \in S$ and consequently $(st)(ab) = (st)(cd))$ with $st \in S$, i.e. $(ab)R_0(cd)$. Furthermore, from $(sat)R_0(sbt)$ for $a, b \in M$, $s, t \in S$, i.e. $usta = ustb$ with a $u \in S$, always aR_0b due to $ust \in S$. Finally, R_0 is finer than any compatible equivalence relation R with the property "from $(sat)R(sbt)$ for $a, b \in M$, $s, t \in S$ always follows aRb". From aR_0b it follows that $sa = sb$ with $s \in S$ and therefore $(sa)R(sb)$ and thus aRb (one chooses $t = e_M$). □

2.3 Induced Homomorphisms and Quotient Formation

We are now looking for a monoid for $S \subseteq M$ that fulfills the analogous universal property from Proposition 2.3.17 for units instead of regular elements, *while we maintain the convenience assumption* $S \subseteq Z(M)$. Since units are regular elements, we initially assume that S is even a submonoid of M^*, thus $S \subseteq Z(M) \cap M^*$ applies. If then $\varphi \colon M \to L$ is a monoid homomorphism with $\varphi(S) \subseteq L^\times$, then $\varphi' \colon M \times S \to L$, $(a,s) \mapsto \varphi(a)\varphi(s)^{-1}$, is also a monoid homomorphism, see Example 2.2.20, since the elements $\varphi(s)^{-1}$, $s \in S$, like the $\varphi(s)$ with all elements $\varphi(a)$, $a \in M$, are interchangeable. The associated (compatible) equivalence relation $R_{\varphi'}$ is given by

$$(a,s) R_{\varphi'} (b,t) \iff \varphi(a)\varphi(s)^{-1} = \varphi(b)\varphi(t)^{-1}$$
$$\iff \varphi(at) = \varphi(a)\varphi(t) = \varphi(b)\varphi(s) = \varphi(bs),$$

$a,b \in M$. This suggests considering the relation

$$(a,s) \equiv_S (b,t) \iff at = bs, \quad a,b \in M, \ s,t \in S,$$

on the product monoid $M \times S$. This is indeed an equivalence relation. To prove transitivity, let $(a,s) \equiv_S (b,t) \equiv_S (c,u)$. Then $at = bs$, $bu = ct$ and consequently $aut = bus = cst$, so $au = cs$, since $t \in S$ is regular, and thus $(a,s) \equiv_S (c,u)$. The relation $\equiv_S$ is also compatible. From $(a,s) \equiv_S (c,u)$ and $(b,t) \equiv_S (d,v)$ it follows that $au = cs$ and $bv = dt$ and consequently $(ab)(uv) = (cd)(st)$, i.e. $(ab, st) \equiv_S (cd, uv)$. According to the universal property 2.3.6 of the quotient $(M \times S)/\equiv_S$ with the canonical projection $\pi'_S \colon M \times S \to (M \times S)/\equiv_S$, $(a,s) \mapsto [a,s]$, there is exactly one homomorphism $\varphi \colon M \to L$ with $\varphi(S) \subseteq L^\times$ or its extension $\varphi' \colon M \times S \to L$ to the homomorphism $\overline{\varphi'} \colon (M \times S)/\equiv_S \to L$ with $\varphi' = \overline{\varphi'} \circ \pi'_S$. Moreover, π'_S maps the elements of $S = S \times \{e_M\} \subseteq M \times S$ to central units of $(M \times S)/\equiv_S$. It is indeed $[s, e_M] \cdot [e_M, s] = [s,s] = [e_M, e_M]$ and $[s, e_M][a,t] = [sa, t] = [as, t] = [a, t][s, e_M]$ for all $(a, t) \in M \times S$. More generally, $[a, s]$ is the quotient $[a, e_M]/[s, e_M]$. It is also simply referred to as

$$a/s \quad \text{or} \quad \frac{a}{s}.$$

If S is any arbitrary submonoid of the center $Z(M) \subseteq M$, we first form the regularization $\overline{M}$ of M with respect to S with canonical projection $\overline{\pi} \colon M \to \overline{M}$. Then $\overline{S} := \overline{\pi}(S) \subseteq \overline{M}$ is a submonoid of $Z(\overline{M}) \cap \overline{M}^*$, and the monoid

$$S^{-1}M := M_S := (\overline{M} \times \overline{S})/\equiv_{\overline{S}} \quad \text{with}$$

$$\frac{a}{s} = a/s := [\overline{a}, \overline{s}] = [\overline{a}, \overline{e}_M] \cdot [\overline{s}, \overline{e}_M]^{-1}, \ a \in M, \ s \in S,$$

is defined. It is called the **monoid of fractions** or the **fraction monoid** of M **with denominators in** S.[15] $S^{-1}M$ can also be directly defined as a quotient monoid of the product monoid $M \times S$ with respect to the compatible equivalence relation "$(a,s) \equiv_S (b,t) \iff$ there is a $u \in S$ with $atu = bsu$". In other words: In $S^{-1}M$ applies

$$a/s = b/t \iff \text{there is a } u \in S \text{ with } atu = bsu.$$

The multiplication in $S^{-1}M$ is given by

$$(a/s) \cdot (b/t) = (ab)/(st).$$

With the canonical homomorphism $\iota_S : M \to S^{-1}M$, $a \mapsto a/e_M$, the monoid of fractions $S^{-1}M$ has the following universal property due to the universal properties of $\overline{M}$ and $\overline{M} \times \overline{S}/\equiv_{\overline{S}}$:

Theorem 2.3.18 *Let $S \subseteq M$ be a submonoid of the monoid M with $S \subseteq Z(M)$. Also, let $\iota_S : M \to S^{-1}M$ be the canonical homomorphism $a \mapsto a/e_M$ from M into the monoid $S^{-1}M$ of fractions of M with denominators in S. Then $\iota_S(S) \subseteq Z(S^{-1}M)^\times$ holds, and for every monoid homomorphism $\varphi : M \to L$ with $\varphi(S) \subseteq L^\times$ there exists exactly one homomorphism $S^{-1}\varphi : S^{-1}M \to L$ with $\varphi = (S^{-1}\varphi) \circ \iota_S$. It is*

$$(S^{-1}\varphi)(a/s) = \varphi(a)\varphi(s)^{-1} = \varphi(s)^{-1}\varphi(a), \quad a \in M, s \in S.$$

Apparently, ι_S is injective exactly when $S \subseteq Z(M) \cap M^*$ applies. In this case, one usually identifies the elements $a \in M$ with the elements $a/e_M \in S^{-1}M$. Then a/s is simply the quotient $as^{-1} = s^{-1}a$. If $T \subseteq M$ is any subset of $Z(M)$, then by definition $T^{-1}M := M_T := \langle T \rangle^{-1}M = M_{\langle T \rangle}$, where $\langle T \rangle$ is the submonoid of M generated by T.

If M is commutative, then the monoid $Q(M) := (M^*)^{-1}M$ of fractions of M with all regular elements of M as denominators is called the **total fraction monoid** of M. If $S \subseteq M^*$ is any submonoid of regular elements of M, then $S^{-1}M$ can be identified with the submonoid $\{a/s \in Q(M) \mid a \in M, s \in S\} \subseteq Q(M)$ of $Q(M)$.

These constructions are often applied to additively written abelian monoids M. They are then referred to as **difference monoids**. If M is such an additive monoid with neutral element 0 and $S \subseteq M$ is a submonoid, then the monoid of differences of M with respect to S is the quotient monoid of the direct sum $M \oplus S$ according to the compatible equivalence relation

$$(a,s) \equiv_S (b,t) \iff \text{there is a } u \in S \text{ with } a+t+u = b+s+u.$$

The element $[a,s]$ in the difference monoid is the difference $[a,0] - [s,0]$. If S contains only regular elements, then simply "$(a,s) \equiv_S (b,t) \iff a+t = b+s$" applies. In this case, the canonical homomorphism $\iota_S : a \mapsto [a,0]$ is injective and

---

[15] Often one also speaks of a quotient monoid. However, this terminology can easily be confused with that of quotient monoids with respect to compatible equivalence relations.

2.3 Induced Homomorphisms and Quotient Formation

M is identified with a submonoid of the difference monoid. Then $[a, s]$ is the difference $a - s$. In the case $S = M$, the difference monoid is an abelian group and is called the **difference group** or also the **Grothendieck group** of M (after A. Grothendieck (1928–2014) and is denoted by

$$G(M).$$

With the canonical homomorphism $\iota_M : M \to G(M)$, $a \mapsto [a, 0]$, it has the following universal property:

Theorem 2.3.19 (Universal Property of the Grothendieck Group) *Let M be a commutative monoid and $\iota_M : M \to G(M)$ the canonical homomorphism from M into its Grothendieck group $G(M)$. If $\varphi : M \to L$ is a homomorphism from M into a group L, then there exists exactly one group homomorphism $\widetilde{\varphi} : G(M) \to L$ with $\varphi = \widetilde{\varphi} \circ \iota_M$. Exactly then is ι_M injective, $\widetilde{\varphi}$ thus an extension of φ, when M is a regular monoid.*

The monoids $S^{-1}M$ or the Grothendieck groups $G(M)$ also exist when S is not in the center of the monoid M or when M is not commutative. They are then, however, more difficult and confusing to describe, see the end of Example 2.3.34.

The prime example of a Grothendieck group is the additive group $\mathbb{Z} = (\mathbb{Z}, +)$ of the integers as the Grothendieck group of the additive monoid $\mathbb{N} = (\mathbb{N}, +)$ of the natural numbers. In general, it seems to hold: If M and $S \subseteq M$ are submonoids of the monoid L with $S \subseteq Z(M) \cap L^\times$ and $\iota_M : M \to L$ the canonical embedding, then the canonical homomorphism $S^{-1}\iota_M : S^{-1}M \to L$, $a/s \mapsto as^{-1} = s^{-1}a$ is injective. If $S = M$, M is therefore commutative, and $M \subseteq L^\times$, then its image is the subgroup of M in $L^\times$ generated by $\{ab^{-1} = b^{-1}a \mid a, b \in M\}$, which is thus isomorphic to the Grothendieck group of M.

The total fraction monoid $(\mathbb{Z}^*)^{-1}\mathbb{Z}$ of the multiplicative monoid $\mathbb{Z} = (\mathbb{Z}, \cdot)$ is, according to the last remark, the multiplicative monoid $\mathbb{Q} = (\mathbb{Q}, \cdot)$. The tautological mapping $(\mathbb{Z}^*)^{-1}\mathbb{Z} \to \mathbb{Q}$, $a/s \mapsto a/s$, is an isomorphism. The Grothendieck group $G(\mathbb{Z}^*) = (\mathbb{Z}^*)^{-1}\mathbb{Z}^*$ of $\mathbb{Z}^*$ is therefore $\mathbb{Q}^\times$. The addition on $\mathbb{Q}$ can be directly obtained with the help of $\mathbb{Z} \times \mathbb{Z}^*$. The rules for fractions motivate the following addition on $\mathbb{Z} \times \mathbb{Z}^*$:

$$(a, s) + (b, t) := (at + bs, st), \quad a, b \in \mathbb{Z},\ s, t \in \mathbb{Z}^*.$$

It is commutative and associative with neutral element $(0, 1)$, as one can directly verify. Furthermore, the equivalence relation $\equiv_{\mathbb{Z}^*}$ on $\mathbb{Z} \times \mathbb{Z}^*$ is compatible with this addition. From $(a, s) \equiv_S (c, u)$ and $(b, t) \equiv_S (d, v)$, i.e., $au = cs$ and $bv = dt$, it follows that $(at + bs)uv = atuv + bsuv = cstv + dtsu = (cv + du)st$, thus $(a, t) + (b, s) \equiv_S (c, u) + (d, v)$. The addition on $\mathbb{Z} \times \mathbb{Z}^*$ thus defines a commutative monoid structure on $(\mathbb{Z}^*)^{-1}\mathbb{Z} = \mathbb{Q}$ with $(a/s) + (b/t) = (at + bs)/st$. It is even a group, the negative of a/s is $(-a)/s$. The additive monoid $(\mathbb{Z} \times \mathbb{Z}^*, +)$ is regular, but not a group.

Determine the Grothendieck group of $(\mathbb{Z} \times \mathbb{Z}^*, +)$ and the kernel of the canonical surjective homomorphism $G(\mathbb{Z} \times \mathbb{Z}^*, +) \to ((\mathbb{Z}^*)^{-1}\mathbb{Z}, +) = (\mathbb{Q}, +)$.

Furthermore, one can easily verify the validity of the distributive laws for the operations $+$ and $\cdot$ on $(\mathbb{Z}^*)^{-1}\mathbb{Z}$. (On $\mathbb{Z} \times \mathbb{Z}^*$ the distributive laws for $+$ and $\cdot$ do not generally apply!) We will discuss these last constructions more generally for rings (instead of $\mathbb{Z}$) in Sect. 2.6.

Example 2.3.20 (Free commutative monoids – Factorial monoids) Let I be an index set. The elements of the direct sum $\mathbb{N}^{(I)}$ are I-tuples of natural numbers $n := (n_i)_{i \in I}$, whose components almost all vanish. The addition is done component-wise. Thus, $n = \sum_{i \in I} n_i e_i$ with $e_i := (\delta_{ji})_{j \in I}$, where the Kronecker delta δ_{ji} vanishes for $j \neq i$ and is equal to 1 for $j = i$. So, $\mathbb{N}^{(I)} = \sum_{i \in I} \mathbb{N} e_i = \sum_{i \in I}^{\oplus} \mathbb{N} e_i = \bigoplus_{i \in I} \mathbb{N} e_i$. Often, the base element e_i is simply denoted by i. So, I is considered as a subset of $\mathbb{N}^{(I)}$. The elements of $\mathbb{N}^{(I)}$ are then simply the formal sums $\sum_{i \in I} n_i i$, where almost all $n_i \in \mathbb{N}$ vanish. The commutative monoid $\mathbb{N}^{(I)}$ with the canonical injection $\iota_I : I \to \mathbb{N}^{(I)}, i \mapsto i = e_i$, has the following universal property, see Example 2.2.20: ◊

Theorem 2.3.21 (Universal Property of Free Commutative Monoids) *Let I be an index set and $x = (x_i)_{i \in I} \in M^I$ an arbitrary family of elements of the commutative monoid M. Then there exists exactly one monoid homomorphism $\varphi \colon \mathbb{N}^{(I)} \to M$ with $\varphi(i) = \varphi(e_i) = x_i, i \in I$. In other words: The mapping*

$$\mathrm{Hom}(\mathbb{N}^{(I)}, M) \xrightarrow{\sim} M^I = \mathrm{Abb}(I, M), \quad \varphi \mapsto \varphi \circ \iota_I = \varphi | I,$$

is an isomorphism of commutative monoids. It is

$$\varphi(n) = x^n = \prod_{i \in I} x_i^{n_i}, \quad n = (n_i)_{i \in I} = \sum_{i \in I} n_i e_i \in \mathbb{N}^{(I)}.$$

The image of φ is the submonoid of M generated by the family $x = (x_i)_{i \in I}$.

In additive notation, $\varphi(n) = nx = \sum_{i \in I} n_i x_i$, $n \in \mathbb{N}^{(I)}$, and $\mathrm{Im}\,\varphi = \sum_{i \in I} \mathbb{N} x_i$. If $x_i, i \in I$, is a generating system of M, then φ is surjective and according to the isomorphism Theorem 2.3.6 consequently $\mathbb{N}^{(I)}/R_\varphi \xrightarrow{\sim} M$, $[n]_{R_\varphi} \mapsto x^n$, an isomorphism. *With the quotient monoids of $\mathbb{N}^{(I)}$ and specifically with the quotient monoids of $\mathbb{N}^m$, $m \in \mathbb{N}$, we thus know all commutative monoids or all finitely generated commutative monoids up to isomorphism.* Often, only a generating system T is given for $R = R_\varphi$, i.e., a subset $T \subseteq R \subseteq \mathbb{N}^{(I)} \times \mathbb{N}^{(I)}$ such that $R = \langle T \rangle$ is the smallest compatible equivalence relation on $\mathbb{N}^{(I)}$ that includes T.

Monoids that are isomorphic to a monoid of the type $\mathbb{N}^{(I)} = (\mathbb{N}^{(I)}, +)$, I any index set, are called **free commutative monoids**. If $\varphi \colon \mathbb{N}^{(I)} \xrightarrow{\sim} M$ is an isomorphism onto a multiplicative monoid M, the images of the base elements are usually denoted by capital letters, such as $e_i \in \mathbb{N}^{(I)}$, usually with capital letters, such as X_i, $i \in I$. Every element of M then has a unique representation of the form

$$X^n = \prod_{i \in I} X_i^{n_i}, \quad n = (n_i)_{i \in I} \in \mathbb{N}^{(I)},$$

2.3 Induced Homomorphisms and Quotient Formation

and M has, with respect to the embedding $\iota_I : I \to M$, $i \mapsto X_i$, the analogous universal property as $\mathbb{N}^{(I)}$. The elements $X^n \in M$, $n \in \mathbb{N}^{(I)}$, are also called **monomials** in the **indeterminates** X_i, $i \in I$. The mapping

$$\deg : M \to \mathbb{N} = (\mathbb{N}, +), \quad X^n \mapsto \deg X^n := |n| = \sum_{i \in I} n_i,$$

is a monoid homomorphism and is called **degree (function)**. If $I = \{i\}$ is a singleton and $X := X_i$, it simply deals with the powers $X^n = X_i^n$ with $\deg X^n = n$, $n \in \mathbb{N}$. For any function $\gamma : I \to N$, $i \mapsto \gamma_i$, with values in an additive monoid N, the homomorphism $\deg_\gamma : M \to N$, $X^n \mapsto \deg_\gamma X^n := n\gamma = \sum_{i \in I} n_i \gamma_i$ is called the γ-**degree** with respect to the **weights** $\gamma_i = \deg_\gamma X_i \in N$ of the indeterminates X_i, $i \in I$.

In the free commutative monoid M, the set of indeterminates X_i, $i \in I$, which we identify as before with the set I itself, is uniquely determined. It is the set $\mathbb{P}_M$ of prime elements in M, which is also equal to the set of irreducible elements of M, cf. Exercise 2.1.18. *Every irreducible element in M is therefore prime.* In particular, the number $|I|$ is uniquely determined. It is called the **rank** of the free commutative monoid M. Furthermore, $\operatorname{Aut} M \xrightarrow{\sim} \mathfrak{S}(I)$, $\varphi \mapsto \varphi|I$, is an isomorphism of groups, since every automorphism of M necessarily permutes the prime elements of M. With the help of Exercise 2.1.18c), the free commutative monoids can be characterized as follows:

Proposition 2.3.22 *A commutative monoid M is free if and only if it is regular and sharp and is generated by the set $\mathbb{P}_M$ of its prime elements.*

Since units play a subordinate role in divisibility questions, we want to generalize the concept of the free commutative monoid somewhat.

Definition 2.3.23 A commutative monoid M is called **factorial**, if M is regular and $M/M^\times$ is free.

So if M is factorial and $\mathbb{P}_M$ is a representative system for the associativity classes of the prime elements of M, then every element $a \in M$ has the **unique prime factor decomposition**

$$a = \varepsilon \prod_{p \in \mathbb{P}_M} p^{\nu_p(a)} \quad \text{with } \varepsilon \in M^\times \text{ and } (\nu_p(a))_{p \in \mathbb{P}_M} \in \mathbb{N}^{(\mathbb{P}_M)}.$$

In other words: *The mapping* $M^\times \times \mathbb{N}^{(\mathbb{P}_M)} \xrightarrow{\sim} M$, $(\varepsilon, (\alpha_p)) \mapsto \varepsilon \prod_p p^{\alpha_p}$, *is a monoid isomorphism.* The free commutative (isomorphic to $\mathbb{P}_M$) submonoid of M generated by $\mathbb{N}^{(\mathbb{P}_M)}$ is a representative system for the associativity classes of the elements of M. For $a, b \in M$ are

$$\gcd(a, b) = \prod_{p \in \mathbb{P}_M} p^{\operatorname{Min}(\nu_p(a), \nu_p(b))} \quad \text{and} \quad \operatorname{lcm}(a, b) = \prod_{p \in \mathbb{P}_M} p^{\operatorname{Max}(\nu_p(a), \nu_p(b))}$$

the (determined by the choice of $\mathbb{P}_M$) natural representatives for the greatest common divisors or the least common multiples of a and b. We note the following criteria for factorial monoids:

Lemma 2.3.24 *Let M be a regular commutative monoid and M' the subgroup $M' := M - M^\times \subseteq M$ of non-units of M. The following conditions are equivalent:*

(i) *M is factorial.*
(ii) *Every element of M' is a product of prime elements.*
(iii) *Every element of M' is a product of irreducible elements, and every irreducible element of M is prime.*
(iv) *Every element of M' is a product of irreducible elements, and for any two elements $a, b \in M$ the greatest common divisor exists $\gcd(a, b)$.*
(v) *Every element of M' is a product of irreducible elements, and for any two elements $a, b \in M$ the least common multiple exists $\mathrm{lcm}(a, b)$.*

For the equivalence of (iv) or (v) with (iii), see exercise 2.1.18c). The equivalence of (ii) and (iii) is tautological, but is often used because it is often easy to check the decomposability into irreducible elements. For example, this is the case when the number $\tau(a)$ of divisor classes for each $a \in M'$ is finite (which is a necessary condition for factoriality) or when there is no sequence $a_0, a_1, a_2, \ldots$ of elements in M such that a_{i+1} is always a proper divisor of a_i or — equivalently — when the set $M/M^\times$ of associativity classes is Artinian or $\{Ma \mid a \in M\}$ Noetherian ordered, see exercise 2.3.16. Then remains the usually more difficult exercise of proving the prime element property of irreducible elements.

If M is a factorial monoid and $S \subseteq M$ is a submonoid, then the fraction monoid $S^{-1}M$ is also factorial, see exercise 2.3.10. A certain reversal is provided by the following lemma by M. Nagata (1927–2008):

Lemma 2.3.25 (Nagata's Lemma) *Let M be a regular commutative monoid and $S \subseteq M$ a submonoid with the following properties: (1) Every non-unit of M is a product of irreducible elements of M. (2) S is generated by prime elements of M. If then $S^{-1}M$ is factorial, then so is M.*

Proof We can assume that S is generated by a set $\mathbb{P}_S$ of pairwise non-associated prime elements of M, so that each element of S is a (unique) product $\prod_{p \in \mathbb{P}_S} p^{\alpha_p}$, $(\alpha_p) \in \mathbb{N}^{(\mathbb{P}_S)}$, is. It suffices to show that every irreducible element $q \in M$, which is not associated with any element from $\mathbb{P}_S$, is prime in M. Initially, such a q is not a unit in $S^{-1}M$. If $qb/s = 1$, $b \in M$, $s \in S$, then $qb = s$ would be a product of prime elements from $\mathbb{P}_S$ and thus the irreducible element q itself would be associated with an element of $\mathbb{P}_S$. Contradiction! q is therefore a product of prime elements in $S^{-1}M$, i.e., $q = q_1 \cdots q_n/s$ holds with $n \in \mathbb{N}^*$, $s \in S$ and elements $q_1, \ldots q_n \in M$, which are prime in $S^{-1}M$. It follows $sq = q_1 \cdots q_n$ and consequently $q = q'_1 \cdots q'_m$ with elements $q'_1, \ldots, q'_m \in M$, which are prime in $S^{-1}M$ (because of $v_p(s) \leq v_p(sq) = v_p(q_1) + \cdots + v_p(q_n)$ for every prime element $p \in \mathbb{P}_S$). Since q

2.3 Induced Homomorphisms and Quotient Formation

is irreducible in M, $m = 1$ and q is prime in $S^{-1}M$. Then q is also prime in M. If q is a divisor of ab in M, then q is also a divisor of ab in $S^{-1}M$ and consequently $qc = as$ or $qd = bt$ with $c, d \in M$ and $s, t \in S$. It follows $s|c$ or $t|d$ and thus $q|a$ or $q|b$ in M. □

The prerequisite in Lemma 2.3.25 by Nagata, that in M every element is a product of irreducible elements, cannot be completely eliminated. An example is provided by the additive monoid $M = (\mathbb{Z} \times \mathbb{N}^*) \uplus (\mathbb{N} \times \{0\}) \subseteq (\mathbb{Z} \times \mathbb{N}, +)$ from Exercise 2.1.18 with the only prime element $p = (1, 0)$. If then $S = \mathbb{N}p = \mathbb{N} \times \{0\}$ is the submonoid generated by p, then the difference monoid of M with respect to S is the factorial monoid $\mathbb{Z} \times \mathbb{N}$. However, M itself is not factorial.

Example 2.3.26 (Free abelian groups) Let I be an index set. The Grothendieck group of the free abelian monoid $(\mathbb{N}^{(I)}, +)$ is the abelian group $G(\mathbb{N}^{(I)}) = \mathbb{Z}^{(I)} = (\mathbb{Z}^{(I)}, +)$ with the canonical injection $\mathbb{N}^{(I)} \hookrightarrow \mathbb{Z}^{(I)}$. Every element $n = (n_i)_i \in \mathbb{Z}^{(I)}$ has the representation $n = \sum_{i \in I} n_i e_i$, where the coefficients $n_i \in \mathbb{Z}$ are uniquely determined by n. The group $\mathbb{Z}^{(I)}$ is thus the direct sum of the infinite cyclic subgroups $\mathbb{Z}e_i$, $i \in I$. We again identify the elements i with the base elements e_i, $i \in I$, and also write briefly $\mathbb{Z}^{(I)} = \sum_{i \in I} \mathbb{Z}i = \sum_{i \in I}^{\oplus} \mathbb{Z}i = \bigoplus_{i \in I} \mathbb{Z}i$. With the canonical injection $\iota_I : I \to \mathbb{Z}^{(I)}$ this group has the following universal property: ◊

Theorem 2.3.27 (Universal Property of Free Abelian Groups) *Let I be an index set and $x = (x_i)_{i \in I} \in G^I$ an arbitrary family of elements of an abelian group G. Then there exists exactly one group homomorphism $\varphi : \mathbb{Z}^{(I)} \to G$ with $\varphi(i) = \varphi(e_i) = x_i$, $i \in I$. In other words: The mapping*

$$\mathrm{Hom}(\mathbb{Z}^{(I)}, G) \xrightarrow{\sim} G^I = \mathrm{Abb}(I, G), \quad \varphi \mapsto \varphi \circ \iota_I = \varphi|I,$$

is an isomorphism of abelian groups. It is

$$\varphi(n) = x^n = \prod_{i \in I} x_i^{n_i}, \quad n = (n_i)_{i \in I} = \sum_{i \in I} n_i e_i \in \mathbb{Z}^{(I)}.$$

The image of φ is the subgroup of G generated by the family $x = (x_i)_{i \in I}$.

In additive notation, which we prefer to use from now on in this example, φ is the mapping $n \mapsto nx = \sum_{i \in I} n_i x_i$. An (additive) abelian group G is called a **free abelian group**, if G has a family $x = (x_i)_{i \in I} \in G^I$ such that the associated homomorphism $\varphi : \mathbb{Z}^{(I)} \to G$ is an isomorphism, i.e., every element $g \in G$ has a representation

$$g = nx = \sum_{i \in I} n_i x_i$$

with coefficients uniquely determined by $g. n_i \in \mathbb{Z}$ possesses (almost all of which disappear). The family $x = (x_i)_{i \in I}$ is then called a **basis** of G. In contrast to the free abelian monoids, the set of elements of a basis of a free abelian group is not uniquely determined. For example, one can replace each basis element x_i

with its negative or also with $x_i + a_j x_j$, where $j \neq i$ and $a_j \in \mathbb{Z}$ is arbitrary. We will go into this in more detail later. The cardinal number $|I|$ of a basis of G is, however, uniquely determined and is called the **rank** of G. If I is finite and H is any *finite* abelian group, then according to Theorem 2.3.27 the number $|H|^{|I|} = |H^I| = |\mathrm{Hom}(G,H)|$ is uniquely determined by G. If I is infinite, then $|I| = |\mathbb{Z}^{(I)}| = |G|$ is also uniquely determined; because it is

$$|I| \leq |\mathbb{Z}^{(I)}| = \left|\bigcup_{J \in \mathfrak{E}(I)} \mathbb{Z}^J\right| \leq \sum_{J \in \mathfrak{E}(I)} |\mathbb{Z}^J| = \aleph_0 \cdot |\mathfrak{E}(I)| = \aleph_0 \cdot |I| = |I|$$

according to Problem 1.8.16b) (and the product rule 1.8.19 for sets).

If x_i, $i \in I$, is a generating system of the abelian group G, then the associated homomorphism $\varphi \colon \mathbb{Z}^{(I)} \to G$, $e_i \mapsto x_i$, $i \in I$, is surjective and consequently $G \cong \mathbb{Z}^{(I)}/\ker \varphi$. In particular, *the quotient groups of the free abelian groups $\mathbb{Z}^{(I)}$ represent all abelian groups up to isomorphism and the quotient groups of the groups $\mathbb{Z}^m$, $m \in \mathbb{N}$, all finitely generated abelian groups.* Thus, the subgroups of free abelian groups are of interest. Here, the following perhaps surprising statement applies:

Theorem 2.3.28 (Subgroups of free abelian groups) *Every subgroup U of a free abelian group G is again a free abelian group, and it holds* rank $U \leq$ rank G.

Proof We can assume $G = \mathbb{Z}^{(I)}$ and consider $\mathbb{Z}^{(J)} = \sum_{i \in J} \mathbb{Z} e_i$ for $J \subseteq I$ as a subgroup of $\mathbb{Z}^{(I)}$. If I is a singleton, i.e., $G = \mathbb{Z}$, then the present theorem is the statement of 2.1.18. If I is infinite, we will use Zorn's Lemma.

We now consider the set $\mathcal{B}$ the triple (J, B_J, f_J), where $J \subseteq I$ is, $B_J \subseteq U$ a basis of $U \cap \mathbb{Z}^{(J)}$ and $f_J \colon B_J \to J$ an injective mapping. We set $(J, B_J, f_J) \leq (K, B_K, f_K)$, if $J \subseteq K$, $B_J \subseteq B_K$ and $f_J = f_K | B_J$ is. This is obviously an order on $\mathcal{B}$. It is even strictly inductive. If (J_r, B_{J_r}, f_{J_r}), $r \in R$, a chain in $\mathcal{B}$, then (J, B_J, f) is an upper bound of this chain, where $J := \bigcup_{r \in R} J_r$ is, $B_J := \bigcup_{r \in R} B_{J_r}$ and f_{B_J} the uniquely determined injective mapping. $f_J \colon B_J \to J$ with $f_J | B_{J_r} = f_{J_r}$, $r \in R$. (For the empty chain, the smallest element $(\emptyset, \emptyset, \emptyset)$ of $\mathcal{B}$ is the upper limit.) Now let (J, B_J, f_J) be a maximum element of $\mathcal{B}$. If I is finite, the existence of such an element is trivial; one chooses an element with maximum $|J|$ ($\leq |I|$). In any case, the existence results from Zorn's Lemma 1.4.15. It suffices to show that $J = I$. Suppose there was a $i_0 \in I - J$. Then let $J' := J \uplus \{i_0\}$ be. It is $\mathbb{Z}^{(J')} = \mathbb{Z}^{(J)} \oplus \mathbb{Z} e_{i_0}$, and we consider the projection $p \colon \mathbb{Z}^{(J')} \to \mathbb{Z} e_{i_0}$ onto $\mathbb{Z} e_{i_0}$ with kernel $\mathbb{Z}^{(J)}$. It is $p(U \cap \mathbb{Z}^{(J')})$ equal to 0 or equal to $\mathbb{Z} a e_{i_0}$ with $a \in \mathbb{N}^*$. In the first case, it is $U \cap \mathbb{Z}^{(J')} = U \cap \mathbb{Z}^{(J)}$ and $(J, B_J, f_J) < (J', B_J, f_J)$. Contradiction! In the second case, let $y \in U \cap \mathbb{Z}^{(J')}$ be an element with $p(y) = a e_{i_0}$. Then $(J, B_J, f_J) < (J', B_J \uplus \{y\}, f_{J'})$ with $f_{J'}|J = f_J$ and $f_{J'}(y) = i_0$ is a genuinely larger element in $\mathcal{B}$, which also results in a contradiction. □

Corollary 2.3.29 *Let G be an abelian group with n generators, $n \in \mathbb{N}$. Then every subgroup $H \subseteq G$ has a generating system of at most n elements.*

2.3 Induced Homomorphisms and Quotient Formation

Proof There is a *surjective* homomorphism $\varphi \colon \mathbb{Z}^n \to G$, and $U := \varphi^{-1}(H) \subseteq \mathbb{Z}^n$ is a subgroup of $\mathbb{Z}^n$, which according to Theorem 2.3.28 is generated by at most n elements. Then also $H = \varphi\bigl(\varphi^{-1}(H)\bigr) = \varphi(U)$ is generated by at most n elements. □

As the proof of Theorem 2.3.28 shows, a subgroup possesses $U \subseteq \mathbb{Z}^n$ always a basis of the form

$$y_1 = a_{11}e_1 + \cdots + a_{1n_1}e_{n_1}, \quad y_2 = a_{21}e_1 + \cdots + a_{2n_2}e_{n_2}, \quad \ldots,$$

$$y_r = a_{r1}e_1 + \cdots + a_{rn_r}e_{n_r}$$

with $1 \leq n_1 < n_2 < \cdots < n_r \leq n$ and $a_{1n_1}, a_{2n_2}, \ldots, a_{rn_r} \in \mathbb{N}^*$, $r = \operatorname{Rang} U$. Apparently, the residue class group $\mathbb{Z}^n/U$ is finite exactly when $r = n$ (and $n_\rho = \rho$, $\rho = 1, \ldots, n$) is. In this case, its order is equal to $a_{11} \cdot a_{22} \cdots a_{nn}$. The more precise structure of the residue class group $\mathbb{Z}^n/U$ is obtained from the following sharper statement: *There is a basis $x_1, \ldots, x_n$ of $\mathbb{Z}^n$ and positive natural numbers $a_1, \ldots, a_r$ such that $a_1 x_1, \ldots, a_r x_r$ a basis of U is. One can also achieve that the divisibility conditions $a_1 \mid a_2 \mid \cdots \mid a_r$ are fulfilled. Then $\mathbb{Z}^n/U \cong \mathbb{Z}_{a_1} \oplus \cdots \oplus \mathbb{Z}_{a_r} \oplus \mathbb{Z}_0^{n-r}$.* We will call this so-called **Elementary divisor theorem** in Vol. 3 prove. From it, of course, also follows the main theorem for finite abelian groups from exercise 2.2.25. It even turns out that every finitely generated abelian group is a direct sum of cyclic groups. We can already prove this here. In preparation, we show the following statement, which is also important in its own right:

Theorem 2.3.30 *Every finitely generated torsion-free abelian group G is a free abelian group.*

Proof Let $G = \mathbb{Z}x_1 + \cdots + \mathbb{Z}x_n$. We choose a maximal subset $J \subseteq \{1, \ldots, n\}$ such that x_j, $j \in J$, is a basis of the subgroup $G_J = \sum_{j \in J} \mathbb{Z}x_j$. In particular, G_J is therefore a free abelian group. After renumbering, we can assume that $J = \{1, \ldots, r\}$ is. Since $x_1, \ldots x_r, x_i$ for $i > r$ is not a basis of $G_J + \mathbb{Z}x_i$, there is an $c_i \in \mathbb{N}^*$ with $c_i x_i \in G_J$, $i = r+1, \ldots, n$. Let $c := c_{r+1} \cdots c_n$. Then $cG \subseteq G_J$. Since G is torsion-free, $x \mapsto cx$ is a group isomorphism $G \xrightarrow{\sim} cG$. Therefore, G is isomorphic to a subgroup of the free abelian group G_J and thus, according to Theorem 2.3.28, itself a free abelian group. (Its rank is of course r.) □

Abelian torsion-free groups that are not finitely generated are, of course, usually not free. The prime example of this is the additive group $(\mathbb{Q}, +)$. We now prove the following already announced theorem:

Theorem 2.3.31 (Main theorem on finitely generated abelian groups) *Every finitely generated abelian group G is a direct sum of cyclic groups, i.e.*

$$G \cong \mathbb{Z}_{m_1} \oplus \cdots \oplus \mathbb{Z}_{m_s} \oplus \mathbb{Z}_0^r \quad \text{with} \quad r, s \in \mathbb{N}, \; m_1, \ldots, m_s > 1.$$

Proof Let TG be the torsion subgroup of G. Then TG like G according to Corollary 2.3.29 is finitely generated and thus finite, hence according to exercise 2.2.25 a finite direct sum of finite cyclic groups. The quotient group $G/$TG is obviously torsion-free and moreover finitely generated, thus free according to Theorem 2.3.30. If then $x_1,\ldots,x_r$ are elements in G, whose residue classes form a basis of $G/$TG, then $x_1,\ldots,x_r$ is a basis of the subgroup $F := \mathbb{Z}x_1 + \cdots + \mathbb{Z}x_r$ of G, which is thus isomorphic to $\mathbb{Z}^r = \mathbb{Z}_0^r$. Moreover, $(TG) + F = G$ and $(TG) \cap F = 0$. Therefore, $G = (TG) \oplus F$ is the direct sum of TG and F. □

The number r in sentence 2.3.31 is the rank of the free part $G/$TG of G and is also called the **rank** of G. The finite cyclic groups $\mathbb{Z}_{m_i}$, $i = 1,\ldots,s$, can be represented as direct sums of cyclic groups of prime power order according to the Chinese Remainder Theorem 2.2.18. Therefore, one can assume that the m_i in 2.3.31 are prime powers. Under this assumption, they are also uniquely determined up to the order, which we have already noted in exercise 2.2.25.

Example 2.3.32 (Free Monoids) Let I be an index set. We are now looking for a monoid W(I) with a canonical mapping $\iota_I : I \to$ W(I), which possesses the analogous universal property of Theorem 2.3.21 for arbitrary (and not just for commutative) monoids. Such can be constructed very simply. One considers I as an **alphabet** and forms the set

$$W(I) := \biguplus_{n \in \mathbb{N}} W_n(I)$$

of all finite **words** over I. The set $W_n(I) = I^n$ is the set of words of **length** n, $n \in \mathbb{N}$, and we identify I with the set $W_1(I)$ of words of length 1. $W_0(I) = \{\emptyset\}$ contains only the empty word (even if I itself is empty). Two words $v = (i_1,\ldots,i_m) \in W_m(I)$ and $w = (j_1,\ldots,j_n) \in W_n(I)$ are multiplied by concatenation (**Concatenation**):

$$v \cdot w := (i_1,\ldots,i_n) \cdot (j_1,\ldots,j_m) := (i_1,\ldots,i_n,j_1,\ldots,j_m) \in W_{n+m}(I).$$

With this multiplication, W(I) is obviously a monoid with the empty word as the neutral element. The word $(i_1,\ldots,i_n)$ is the product of its letters:

$$(i_1,\ldots,i_n) = i_1 \cdots i_n = \prod_{k=1}^n i_k.$$

One calls W(I) with the canonical injection $\iota_I : I \hookrightarrow$ W(I) the **free monoid** over the index set or the alphabet I. It obviously has the desired universal property: ◊

Theorem 2.3.33 (Universal Property of Free Monoids) *Let I be an index set and $x = (x_i)_{i \in I} \in M^I$ an arbitrary family of elements of the monoid M. Then there exists exactly one monoid homomorphism $\varphi : W(I) \to M$ with $\varphi(i) = x_i$, $i \in I$. In other words: The mapping*

2.3 Induced Homomorphisms and Quotient Formation

$$\text{Hom}(W(I), M) \xrightarrow{\sim} M^I = \text{Abb}(I, M), \quad \varphi \mapsto \varphi \circ \iota_I = \varphi | I,$$

is bijective. For $v = (i_1, \ldots, i_n) = i_1 \cdots i_n \in W(I)$ is

$$\varphi(v) = \varphi(i_1 \cdots i_n) = x^v = \prod_{k=1}^{n} x_{i_k} = x_{i_1} \cdots x_{i_n}.$$

The image of φ is the submonoid of M generated by the family $x = (x_i)_{i \in I}$.

A monoid N is called **free**, if there is an isomorphism $W(I) \xrightarrow{\sim} N$. The images of the letters $i \in I$ are referred to as **non-commuting indeterminates** $X_i, i \in I$. The elements of N are the **monomials** $X^v = X_{i_1} \cdots X_{i_n}, v = i_1 \cdots i_n \in W(I)$, which can also be clearly written in the form $X_{j_1}^{n_1} X_{j_2}^{n_2} \cdots X_{j_r}^{n_r}$, where $(j_1, n_1), \ldots, (j_r, n_r)$ is a finite sequence with elements from $I \times \mathbb{N}^*$, for which $j_\rho \neq j_{\rho+1}, \rho < r$, applies. The **length** $\ell(X^v) := n = n_1 + \cdots + n_r$ of a monomial $X^v = X_{i_1} \cdots X_{i_n} = X_{j_1}^{n_1} X_{j_2}^{n_2} \cdots X_{j_r}^{n_r} \in N$ is called the **degree**

$$\deg X^v$$

of X^v. The degree function $\deg: N \to \mathbb{N}$ is a monoid homomorphism. A free monoid is always sharp. The set of indeterminates $X_i, i \in I$, is uniquely determined as the set $N' - (N' \cdot N')$, $N' := N - \{e_N\}$. In particular, the automorphism group of N is isomorphic to the permutation group $\mathfrak{S}(I)$. N is called a free monoid of **rank** $|I|$ with **basis** $X_i, i \in I$.

A monoid M with a generating system $x = (x_i)_{i \in I}$ is isomorphic to the quotient monoid N/R_φ, where $\varphi: N \to M$ is the homomorphism of the free monoid N with basis $X_i, i \in I$, onto M, which maps X_i to $x_i, i \in I$. The image $x^v = \varphi(X^v)$ of the monomial $F = X^v$ we also denote by $F(x)$. *With the quotient monoids of the free monoids, all monoids are given up to isomorphism.* If the compatible equivalence relation $R_\varphi \subseteq N \times N$ is generated by the monomial pairs $(F_j, G_j), j \in J$, one says M has the **representation**

$$\langle x_i, i \in I \mid F_j(x) = G_j(x), j \in J \rangle$$

with **generators $x_i, i \in I$, and relations** $F_j(x) = G_j(x), j \in J$. A monoid M with the above representation has (due to Theorem 2.3.1) the following universal property: If $y = (y_i)_{i \in I}$, a family of elements of any monoid L with $F_j(y) = G_j(y)$ for all $j \in J$, there exists exactly one monoid homomorphism $\varphi: M \to L$ with $\varphi(x_i) = y_i, i \in I$. For example, a free *commutative* monoid with base $x_i, i \in I$, has the representation $\langle x_i, i \in I \mid x_i x_j = x_j x_i, i, j \in I, i \neq j \rangle$. Proof!

A free monoid N with rank $N = 1$ is, up to isomorphism, the monoid $(\mathbb{N}, +)$, which in multiplicative notation contains exactly the powers $X^n, n \in \mathbb{N}$, of an indeterminate X. It coincides with the free *commutative* monoid of rank 1. Its quotient monoids are, up to isomorphism, the **cyclic monoids** M, whose elements are the powers x^n of a single element $x \in M$. If these are pairwise different, then M is free. Otherwise, there is $t \in \mathbb{N}$ and $r \in \mathbb{N}^*$ with $x^{t+r} = x^t$ and thus $x^{n+r} = x^n$ for all $n \geq t$. The sequence of powers is therefore periodic with period length r and pre-period

Fig. 2.9 Finite cyclic monoid of type $(m,k) = (4,8)$

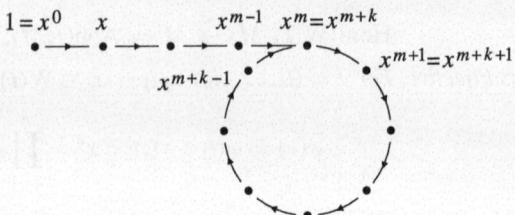

length t. There is therefore a uniquely determined pair $(m,k) \in \mathbb{N} \times \mathbb{N}^*$ with the following property: Every period r of (x^n) is a multiple of k, and if $x^n = x^{n+r}$ holds for $n \geq t$, then $t \geq m$, see exercise 1.7.38. The powers $1 = x^0, \ldots, x^{m-1}$ form the pre-period of length $m \in \mathbb{N}$ of the sequence (x^n) and the powers $x^m, \ldots, x^{m+k-1}$ its period of length $k \in \mathbb{N}^*$, and $x^0, \ldots, x^{m+k-1}$ are the $m+k$ different elements of the monoid M, see Fig. 2.9. We call (m,k) the **type** of the cyclic monoid M. $m+k$ is the smallest exponent $n \in \mathbb{N}^*$ with $x^n \in \{x^0, \ldots, x^{n-1}\}$.

The period members obviously form a k-element *regular* subgroup of M, thus a group G. It is cyclic with neutral element x^{m+i_0}, $0 \leq i_0 < k$, $i_0 \equiv -m$ mod k, and the generating elements $\varphi(k)$ x^{m+i}, $0 \leq i < k$, $\gcd(m+i,k) = 1$. If $m \geq 2$, then x is the only generating element of M. For $m \leq 1$, the generators of the subgroup G are the generators of M. M is a group exactly when $m = 0$, so $M = G$. Overall, the type of the finite cyclic monoid M is uniquely determined regardless of the choice of the generating element x. Two finite cyclic monoids are isomorphic if and only if they have the same type $(m,k) \in \mathbb{N} \times \mathbb{N}^*$.

If M is not a group, so $m \geq 1$, then M has an idempotent element different from the neutral element. It follows: *A finite (not necessarily cyclic) monoid is exactly a group when the neutral element is its only idempotent element.* The cyclic monoid of type $(m,k) \in \mathbb{N} \times \mathbb{N}^*$ is, up to isomorphism, the quotient monoid $\mathbb{N}/\langle(m,m+k)\rangle$ of $\mathbb{N} = (\mathbb{N},+)$ with respect to the compatible equivalence relation generated by $(m,m+k) \in \mathbb{N} \times \mathbb{N}$ on $\mathbb{N}$. It therefore has the representation $(x \mid x^m = x^{m+k})$. An infinite cyclic monoid, by definition, has the type $(\infty, 0)$. If the monoid generated by x has the type (m,k), then the cyclic submonoid generated by the element x^n, $n \in \mathbb{N}^*$, obviously has the type $(\lceil m/n \rceil, k/\gcd(k,n))$.

Example 2.3.34 (Free Groups) Let I be an index set. We are looking for a group $F(I)$ with a canonical mapping $\iota_I: I \to F(I)$, which possesses the analogous universal property from Theorem 2.3.27 for arbitrary (and not only for abelian) groups. For the construction, we double the index set I and consider the set $I \uplus I'$ as well as the free monoid $W(I \uplus I')$ of words over the alphabet $I \uplus I'$, where $I \to I'$, $i \mapsto i'$, is a bijective mapping from I to a set I' that is disjoint from I. Let R_I be the compatible equivalence relation generated by the pairs $(ii', \emptyset)$, $(i'i, \emptyset)$, $i \in I$, on $W(I \uplus I')$. Then

$$F(I) := W(I \uplus I')/R_I$$

2.3 Induced Homomorphisms and Quotient Formation

is a group, since the generating elements $X_i := [i]_{R_I}$ and $X_{i'} := [i']_{R_I}$ in $F(I)$ are invertible with $X_i^{-1} = X_{i'}$, $i \in I$. $F(I)$ with the canonical mapping $\iota_I : I \to F(I)$, $i \mapsto X_i$, has the desired universal property: ◊

Theorem 2.3.35 (Universal Property of Free Groups) *Let I be an index set and $x = (x_i)_{i \in I} \in G^I$ an arbitrary family of elements of the group G. Then there exists exactly one group homomorphism $\varphi : F(I) \to G$ with $\varphi(X_i) = x_i$, $i \in I$. In other words: The mapping*

$$\mathrm{Hom}(F(I), G) \xrightarrow{\sim} G^I = \mathrm{Abb}(I, G), \quad \varphi \mapsto \varphi \circ \iota_I = \varphi | I,$$

is bijective. The image of φ is the subgroup of $x = (x_i)_{i \in I}$ generated by the family G.

Proof According to Theorem 2.3.33, there is a monoid homomorphism $\Phi : W(I \uplus I') \to G$ with $\Phi(i) = x_i$ and $\Phi(i') = x_i^{-1}$. Due to $\Phi(ii') = \Phi(i'i) = e_G$, the compatible equivalence relation R_Φ induced by Φ is coarser than R_I, and according to Theorem 2.3.6, Φ induces a homomorphism $\varphi : F(I) \to G$ with $\Phi = \varphi \circ \pi$, where $\pi : W(I \uplus I') \to F(I)$ is the canonical projection. So it is $\varphi(X_i) = \varphi(\pi(i)) = \Phi(i) = x_i$, $i \in I$. Since the X_i generate the group $F(I)$, φ is also uniquely determined by this condition. □

A group F is called **free**, if there is an isomorphism $F(I) \xrightarrow{\sim} F$ for a suitable set I. The images of the elements $X_i \in F(I)$ are also denoted by X_i, $i \in I$. They form the **basis** $(X_i)_{i \in I}$ of the free group F. Every element of F obviously has a representation as a **monomial** of the form

$$X_{i_1}^{n_1} X_{i_2}^{n_2} \cdots X_{i_r}^{n_r} \quad \text{with} \quad (i_1, n_1), \ldots, (i_r, n_r) \in I \times \mathbb{Z}^* \text{ and } i_\rho \neq i_{\rho+1}, \rho < r.$$

Proving that this representation is unique is not so easy. Of course, the **degree** $n_1 + \cdots + n_r \in \mathbb{Z}$ of such a monomial as an image of the group homomorphism $\deg : F \to \mathbb{Z}$, $X_i \mapsto 1$, is uniquely determined. To *prove* that different such monomials also represent different elements in F, one could proceed as follows: Let $\mathcal{X}$ be the set of sequences $\left(X_{i_1}^{n_1}, \ldots, X_{i_r}^{n_r}\right)$, where the (i_ρ, n_ρ) fulfill the given conditions. For $i \in I$ let $\sigma_i : \mathcal{X} \to \mathcal{X}$ be the mapping with

$$\sigma_i\left(X_{i_1}^{n_1}, X_{i_2}^{n_2}, \ldots, X_{i_r}^{n_r}\right) := \begin{cases} \left(X_i, X_{i_1}^{n_1}, X_{i_2}^{n_2}, \ldots, X_{i_r}^{n_r}\right), & \text{if } i \neq i_1, \\ \left(X_{i_2}^{n_2}, \ldots, X_{i_r}^{n_r}\right), & \text{if } i = i_1, n_1 = -1, \\ \left(X_{i_1}^{n_1+1}, X_{i_2}^{n_2}, \ldots, X_{i_r}^{n_r}\right), & \text{if } i = i_1, n_1 \neq -1. \end{cases}$$

Then σ_i is a permutation of $\mathcal{X}$. σ_i simulates the multiplication with X_i from the left. The inverse of σ_i is the mapping σ_i', which is defined analogously to σ_i, where the effect of X_i is replaced by that of X_i^{-1}, so

$$\sigma_i'\left(X_{i_1}^{n_1}, X_{i_2}^{n_2}, \ldots, X_{i_r}^{n_r}\right) := \begin{cases} \left(X_i^{-1}, X_{i_1}^{n_1}, X_{i_2}^{n_2}, \ldots, X_{i_r}^{n_r}\right), & \text{if } i \neq i_1, \\ \left(X_{i_2}^{n_2}, \ldots, X_{i_r}^{n_r}\right), & \text{if } i = i_1, n_1 = 1, \\ \left(X_{i_1}^{n_1-1}, X_{i_2}^{n_2}, \ldots, X_{i_r}^{n_r}\right), & \text{if } i = i_1, n_1 \neq 1. \end{cases}$$

Due to the universal property of the free group F, there is a group homomorphism $\psi: F \to \mathfrak{S}(\mathcal{X})$ with $X_i \mapsto \sigma_i$ for all $i \in I$. From $\psi\bigl(X_{i_1}^{n_1} \cdots X_{i_r}^{n_r}\bigr) = \bigl(X_{i_1}^{n_1}, \ldots, X_{i_r}^{n_r}\bigr)$ the desired result follows.

As with the free abelian groups, the cardinal number $|I|$ of a basis X_i, $i \in I$, of a free group F is uniquely determined and is called the **rank** of F. This can be proven as in the abelian case with cardinal number arguments. But it can also be traced back to: *The abelization $F_{\mathrm{ab}} = F/[F,F]$ of F with the canonical projection $\pi_{\mathrm{ab}}: F \to F_{\mathrm{ab}}$ is indeed a free abelian group of rank $|I|$.* Proof. The canonical homomorphism $F \to \mathbb{Z}^{(I)}$ with $X_i \mapsto e_i$, $i \in I$, from F into the abelian group $\mathbb{Z}^{(I)}$ induces a homomorphism $F_{\mathrm{ab}} \to \mathbb{Z}^{(I)}$ with $\pi_{\mathrm{ab}}(X_i) \mapsto e_i$. According to the universal property of the free abelian group $\mathbb{Z}^{(I)}$ there is conversely a homomorphism $\mathbb{Z}^{(I)} \to F_{\mathrm{ab}}$ with $e_i \mapsto \pi_{\mathrm{ab}}(X_i)$. Both are inverses of each other and thus isomorphisms.

One must carefully distinguish between free groups and free abelian groups. Only when the rank is ≤ 1 do the two coincide. The isomorphism type of free groups of rank $n \in \mathbb{N}$ is also denoted by

$$\mathbb{F}^n.$$

The isomorphism type of their abelianizations is $\mathbb{Z}^n$.

A group G with a generating system $x = (x_i)_{i \in I}$ is isomorphic to the quotient group $F/\ker \varphi$, where F is a free group with basis $(X_i)_{i \in I}$ and $\varphi: F \to G$ is the homomorphism with $\varphi(X_i) = x_i$, $i \in I$. *Thus, all groups are given up to isomorphism by the quotient groups of the free groups.* If $\ker \varphi$ is the smallest *normal* subgroup of F that contains the elements $F_j G_j^{-1}$, $j \in J$, where F_j and G_j are monomials $X_{i_1}^{n_1} X_{i_2}^{n_2} \cdots X_{i_r}^{n_r} \in F$ of the type described above, then it is said that G has the **representation**

$$\langle x_i, i \in I \mid F_j(x) = G_j(x), j \in J \rangle$$

by generators x_i, $i \in I$, **and relations** $F_j(x) = G_j(x)$, $j \in J$, where again $F_j(x) := \varphi(F_j)$ and $G_j(x) := \varphi(G_j)$, $j \in J$, is. If I and J are finite, we speak of a **finite representation** of the group G. A group G with the above representation has (due to Theorem 2.3.2) the following universal property: *If $y = (y_i)_{i \in I}$ is a family of elements of any group H with $F_j(y) = G_j(y)$ for all $j \in J$, there is exactly one group homomorphism $\varphi: G \to H$ with $\varphi(x_i) = y_i$, $i \in I$.* For example, a free *abelian* group with basis x_i, $i \in I$, has the representation $\langle x_i, i \in I \mid x_i x_j = x_j x_i, i, j \in I, i \neq j \rangle$. An arbitrary group G has the representation $\langle x_g, g \in G \mid x_g x_h = x_{gh}, g, h \in G \rangle$. Proof! If G is finite, then this is a finite representation of G. A cyclic group Z_n, $n \in \mathbb{N}$, has the representation $\langle x \mid x^n = e \rangle$.

If the monoid M has the representation $\langle x_i, i \in I \mid F_j(x) = G_j(x), j \in J \rangle$ by generators and relations and G is the *group* with the same representation, then G has the following universal property with the canonical monoid homomorphism $\iota_M: M \to G$, $x_i \mapsto x_i$, $i \in I$: If $\varphi: M \to L$ is a monoid homomorphism from M into the group L, then there is exactly one group homomorphism $\widetilde{\varphi}: G \to L$ with $\varphi = \widetilde{\varphi} \circ \iota_M$. G is therefore the **Grothendieck group** $G(M)$ of M, cf. Theorem 2.3.19. Construct in a similar way the fraction monoid $S^{-1}M$, where

2.3 Induced Homomorphisms and Quotient Formation

S is a (not necessarily central) submonoid of M, cf. Theorem 2.3.18. Unlike for $S \subseteq Z(M)$, the canonical homomorphism $S \subseteq M^*$ does *not* need to be injective. The first examples of this were constructed by A. I. Mal'cev (1909–1967) in 1937 with $S = M = M^*$. *A regular monoid is therefore not necessarily embeddable into a group.*

In general, it is very difficult to determine the structure of a group from its representation. This set of problems is addressed by **combinatorial group theory**. Important in this context is the **Nielsen-Schreier theorem**: *Every subgroup of a free group F is free.* However, there is no analogue to Theorem 2.3.28 about the rank of subgroups of free abelian groups, if rank F is finite. *Thus, every free group of rank ≥ 2 has free subgroups of infinite rank* (which are not finitely generated in particular). We will prove this theorem and the Nielsen-Schreier theorem in Vol. 5. They follow, for example, from the fact that the fundamental groups of graphs are always free groups.

The last four somewhat more extensive examples are intended to convey the idea of **free objects** and **universal properties.** For a given set I, a free object is a pair $(\mathfrak{F}_I, \iota_I)$ with an object $\mathfrak{F}_I$ of a certain type and a mapping $\iota_I \colon I \to \mathfrak{F}_I$ such that for every object A of the same type as $\mathfrak{F}_I$ the mapping $\varphi \mapsto \varphi \circ \iota_I$ from the set of homomorphisms $\mathrm{Hom}(\mathfrak{F}_I, A)$ to the set of mappings $I \to A$ is bijective. The general framework for this is provided by the language of categories, which will be discussed in more detail in Vol. 3. The reader is also asked to construct free objects with their associated universal properties for semigroups and (somewhat more difficult) for magmas.

Exercises

Exercise 2.3.1 The center $Z(G)$ of a group G is a normal subgroup of G, cf. exercise 2.1.2. The quotient group $G/Z(G)$ is isomorphic to the group $\mathrm{Inn}(G)$ of inner automorphisms of G. If $G/Z(G)$ is cyclic, then $G = Z(G)$, G is therefore abelian. In particular, the index of the center of a group is never a prime number.

Exercise 2.3.2 Let $\varphi \colon G \to G'$ be a homomorphism of groups. N or N' are normal divisors of G or G' with $\varphi(N) \subseteq N'$. Then φ induces a homomorphism $\overline{\varphi} \colon G/N \to G'/N'$ with $\overline{\varphi}(aN) = \varphi(a)N'$.

Exercise 2.3.3 Let G be a group and $[G, G]$ the commutator group of G, cf. Example 2.3.10.

a) Let $\varphi \colon G \to G'$ be a homomorphism of groups. Exactly then is $\mathrm{Im}\,\varphi$ abelian, when $[G, G]$ is contained in $\ker \varphi$.
b) Let $N \subseteq G$ be a normal subgroup in G. Exactly then is G/N abelian, when $[G, G] \subseteq N$is.
c) Every subgroup of G, which includes $[G, G]$, is a normal subgroup in G.

d) Let H and N be normal subgroups in G. Exactly then is $G/(H \cap N)$ abelian, when G/H and G/N are abelian.

Exercise 2.3.4 G is a finite group, N is a normal subgroup of G and H is a subgroup of G, for which the order $\mathrm{Ord}\,H$ and the index $[G:N]$ are coprime. Then $H \subseteq N$. – In particular, a normal subgroup N of G, for which order and index are coprime, is the only subgroup of its order in G. If G is abelian, then this N is a direct factor of G and G is cyclic if and only if N and G/N are cyclic. (Cf. Theorem 2.2.22 and exercise 2.2.11. – In the case of non-abelian groups, this N is still a weak direct factor, i.e., N has a weak complement in G (**Schur-Zassenhaus Theorem**).

Exercise 2.3.5

a) Let G be a group with more than one element. If G has no other subgroups besides the trivial subgroups G and $\{e\}$, then G is cyclic of prime order.
b) An abelian simple group is cyclic of prime order. (A non-abelian finite simple group always has an even order (in fact, an order divisible by 4, cf. Proposition 2.5.21). This is the famous **Feit-Thompson Theorem**.)

Exercise 2.3.6 Let $p \in \mathbb{P}$ be a prime number. An abelian group H is called an **elementary abelian p -group**, if every element of H different from the neutral element has the order p. Let H be such a group, which we write additively. Then $pH = 0$ and for every element $x \neq 0$ in H, $\mathbb{Z}x$ is a cyclic subgroup of H of order p.

a) Show: There exists a subset $B \subseteq H - \{0\}$ such that $H = \sum_{x \in B}^{\oplus} \mathbb{Z}x$ is the direct sum of the subgroups $\mathbb{Z}x$, $x \in B$. In particular, H is isomorphic to the direct sum $\mathbf{Z}_p^{(I)}$ for every index set I with $|I| = |B|$. (Such a B is called a **basis** of H. A subset $F \subseteq H - \{0\}$ is called **free** or **independent**, if the sum $\sum_{x \in F} \mathbb{Z}x$ is direct, cf. Lemma 2.2.21. It is shown that the free subsets of $H - \{0\}$ are strictly inductively ordered with respect to inclusion. Every maximal subset (whose existence follows from Zorn's Lemma 1.4.15 and is trivial for finite H) is a basis of H.) All bases of H have the same cardinality. (It is called the **rank** or the **dimension** of H. – The invariance of the cardinality of B results from $\left|\mathbf{Z}_p^{(I)}\right| = p^{|I|}$ for finite I and from $\left|\mathbf{Z}_p^{(I)}\right| = |I|$ for infinite I.)
b) Every subgroup U of H is a direct summand of H. (If $B \subseteq H$ is a subset that is bijectively mapped onto a basis of H/U with the canonical projection $H \to H/U$, then B generates a complement to U.)

Remark The results of this exercise can be used as a starting point for the proof of the following **generalization of the main theorem on finite abelian groups**, which goes back to H. Prüfer (1896–1934): *Let H be an abelian group with a positive exponent* $\mathrm{Exp}\,H$. *Then H is a direct sum of cyclic subgroups.* For the *proof*, one can assume based on the primary decomposition 2.2.22 that H is a p-group, so that $p^n H = 0$ is for a $n \in \mathbb{N}^*$. The case $n = 1$ is covered by part a) of the present exercise. To conclude from $n \geq 1$ to $n+1$, proceed as follows: Due to $p^n(pH) = 0$,

2.3 Induced Homomorphisms and Quotient Formation

there are elements according to the induction hypothesis $px_i \in pH - \{0\}$, $i \in I$, such that pH is the direct sum $\sum_{i \in I}^{\oplus} \mathbb{Z} p x_i$. Then the sum $\sum_{i \in I}^{\oplus} \mathbb{Z} x_i$ is also direct. (With $\bigoplus_{i \in I} \mathbb{Z} p x_i \to pH$, $\bigoplus_{i \in I} \mathbb{Z} x_i \to H$ is also injective, for example, since $\bigoplus_{i \in I} \mathbb{Z} p x_i$ and $\bigoplus_{i \in I} \mathbb{Z} x_i$ have the same p-base.) If then $U \subseteq {}_pH$ is a complement to ${}_pH \cap \sum_{i \in I}^{\oplus} \mathbb{Z} x_i$ in the p-base ${}_pH$ of H (see part b)), then $H = U \oplus \sum_{i \in I}^{\oplus} \mathbb{Z} x_i$.

Exercise 2.3.7 Let F and G be *non-commutative* simple groups. Then the product $F \times G$ only has the four trivial normal divisors $\{(e_F, e_G)\}$, $\{e_F\} \times G$, $F \times \{e_G\}$ and $F \times G$. Formulate and prove an analogous statement for a product $F_1 \times \cdots \times F_n$ of n simple, non-commutative groups. (The n-fold product $\mathbb{Z}_p \times \cdots \times \mathbb{Z}_p = \mathbb{Z}_p^n$ of the abelian simple group $\mathbb{Z}_p = \mathbb{Z}/\mathbb{Z}p$, p prime, is the additive group of the F_p-vector space $\mathbf{F}_p^n$, and its normal divisors ($=$ subgroups) coincide with the F_p-subspaces of $\mathbf{F}_p^n$, see Example 2.8.3 (4). Their number is $G_{n,p} := \sum_{m=0}^{n} G_m^{[n]}(p)$, cf. Vol. 2, exercise 2.7.9c). The $G_m^{[n]}$ are the so-called **Gauss polynomials**

$$G_m^{[n]} = (T^n - 1)(T^{n-1} - 1) \cdots (T^{n-m+1} - 1)/(T^m - 1)(T^{m-1} - 1) \cdots (T - 1).$$

Explicitly state the $G_{2,p} = p + 3$ subgroups of the group $\mathbb{Z}_p \times \mathbb{Z}_p$.)

Exercise 2.3.8 (Goursat's Lemma) (after É. Goursat (1858–1936)) Let G and H be groups.

a) Through

$$(A, M; B, N; \varphi) \mapsto U := \{(a, b) \in A \times B \mid \varphi(aM) = bN\}$$

a bijective mapping is given from the set of 5-tuples $(A, M; B, N; \varphi)$, in which $A \subseteq G$ and $B \subseteq H$ are subgroups, M is a normal subgroup of A and N is a normal subgroup of B and $\varphi: A/M \xrightarrow{\sim} B/N$ is an isomorphism, to the set of subgroups U of the product group $G \times H$. (If M is a normal divisor of a subgroup $A \subseteq G$, then A/M is called a **subquotient** of G. *The subgroups of $G \times H$ correspond thus reversibly uniquely to the isomorphisms between the subquotients of G and H.*) Under what additional conditions is the subgroup determined by $(A, M; B, N; \varphi)$ of $G \times H$ a normal divisor in $G \times H$?

b) Let G and H be finite. The following conditions are equivalent: (i) Every subgroup of $G \times H$ is of the form $G' \times H'$ with subgroups $G' \subseteq G$ and $H' \subseteq H$. (ii) The orders of G and H are coprime. (Use the Theorem 2.4.8 of Cauchy.)

Exercise 2.3.9 Let G be an abelian group and $m_1, \ldots, m_n \in \mathbb{N}^*$ be pairwise coprime. The mapping $(H_1, \ldots, H_n) \mapsto H_1 \cap \cdots \cap H_n$ is a bijective mapping of the set of n-tuples $(H_1, \ldots, H_n)$ of subgroups $H_i \subseteq G$ with $[G : H_i] = m_i$, $i = 1, \ldots, n$, to the set of subgroups $H \subseteq G$ with $[G : H] = m_1 \cdots m_n$.

Exercise 2.3.10 Let M be a factorial monoid and $\mathbb{P}_M$ a representative system for the associativity classes of the prime elements of M. Furthermore, let S be a submonoid and $\mathbb{P}(S)$ the set of $p \in \mathbb{P}_M$ that divide an element from S. Then the fraction monoid $S^{-1}M$ is also factorial and $\mathbb{P}_{S^{-1}M} := \mathbb{P}_M - \mathbb{P}(S)$ is a representative system for the associativity classes of the prime elements of $S^{-1}M$. Moreover,

$$\left(\varepsilon, (\alpha_p)_{p\in\mathbb{P}(S)}, (\beta_q)_{q\in\mathbb{P}_{S^{-1}M}}\right) \longmapsto \varepsilon \prod_{p\in\mathbb{P}(S)} p^{\alpha_p} \prod_{q\in\mathbb{P}_{S^{-1}M}} q^{\beta_q}$$

is a monoid isomorphism $M^\times \times \mathbb{Z}^{(\mathbb{P}(S))} \times \mathbb{N}^{(\mathbb{P}_{S^{-1}M})} \xrightarrow{\sim} S^{-1}M$ and in particular the group $M^\times \times \mathbb{Z}^{(\mathbb{P}(S))}$ is isomorphic to the unit group of $S^{-1}M$. The Grothendieck group $G(M)$ of M is isomorphic to the group $M^\times \times \mathbb{Z}^{(\mathbb{P}_M)}$, the Grothendieck group $G(M/M^\times)$ of $M/M^\times$ is the free abelian group $\mathbb{Z}^{(\mathbb{P}_M)}$.

Exercise 2.3.11 Let $r \in \mathbb{N}$ and H be a free abelian group of rank $\geq r$. A subgroup $U \subseteq H$ of rank $\leq r$ is exactly maximal in the (ordered by inclusion) set of all subgroups of rank $\leq r$ in H, when U is generated by r elements of a basis of H. (For the non-trivial implication, one can reduce to the case that H itself is of finite rank. Then consider the torsion of the group H/U.)

Exercise 2.3.12
a) $(\alpha_p)_{p\in\mathbb{P}} \mapsto \prod_{p\in\mathbb{P}} p^{\alpha_p}$ is an isomorphism $(\mathbb{Z}^{(\mathbb{P})}, +) \xrightarrow{\sim} \mathbb{Q}_+^\times = G(\mathbb{N}^*)$. $\mathbb{Q}_+^\times = (\mathbb{Q}_+^\times, \cdot)$ is therefore a free abelian group with $\mathbb{P} = \mathbb{P}_{\mathbb{N}^*}$ as a basis.
b) Given are $r + 1$ types of gears each with $n_0, n_1, \ldots, n_r \in \mathbb{N}^* - \{1\}$ teeth, $r \in \mathbb{N}$, potentially any number of each type. The set of gear ratios that can be constructed with these gears is the subgroup G of $\mathbb{Q}_+^\times$ generated by $n_1/n_0, \ldots, n_r/n_0$. G is a free abelian group of rank $\leq r$. Exactly then is rank $G = r$, when there is no $(r + 1)$-tuple of integers $(\alpha_0, \alpha_1, \ldots, \alpha_r) \in \mathbb{Z}^{r+1} - \{0\}$ with $\alpha_0 + \alpha_1 + \cdots + \alpha_r = 0$ and $n_0^{\alpha_0} n_1^{\alpha_1} \cdots n_r^{\alpha_r} = 1$. G is maximal in the set of all subgroups of $\mathbb{Q}_+^\times$ with rank $\leq r$, if $n_1/n_0, \ldots, n_r/n_0$ is part of a basis of $\mathbb{Q}_+^\times$, see exercise 2.3.11. (**Note** If rank $G \geq 2$, then G is dense in the (ordered) group $\mathbb{R}_+^\times$ (see Vol. 2, Lemma 1.4.8) and any translation ratio $d \in \mathbb{R}_+^\times$ can be approximated as closely as desired with a gear made from the given gears. Chr. Huygens (1629–1695) used continued fraction developments to approximate a given ratio $d \in \mathbb{R}_+^\times$ with as few gears as possible, see Example 3.3.11.)
c) With gears of the type triple $(n_0, n_1, n_2) = (9, 15, 21)$ (Lego system), exactly the transmission ratios $3^\alpha 5^\beta 7^\gamma \in \mathbb{Q}_+^\times$ with $\alpha, \beta, \gamma \in \mathbb{Z}$ and $\alpha + \beta + \gamma = 0$ can be realized. They form a free subgroup of $\mathbb{Q}_+^\times$ of rank 2, which is maximal in the set of all subgroups of rank ≤ 2 in $\mathbb{Q}_+^\times$ Examine in a similar way the Fischer-Technik system with the gear types $(10, 15, 20, 30, 40, 58)$.

Exercise 2.3.13 Let $a = (a_1, \ldots, a_n) \in \mathbb{Z}^n$, $a \neq 0$, and $d := \gcd(a) = \gcd(a_1, \ldots, a_n)$.

a) a is part of a basis of the free abelian group $\mathbb{Z}^n$ exactly when $d = 1$. If $1 = r_1 a_1 + \cdots + r_n a_n$ with $r_1, \ldots, r_n \in \mathbb{Z}$, then $U := \ker \varphi$, where $\varphi \colon \mathbb{Z}^n \to \mathbb{Z}$ is the homomorphism given by $e_i \mapsto r_i$, $i = 1, \ldots, n$, a complement of $\mathbb{Z}a$ in $\mathbb{Z}^n$, i.e., it is $\mathbb{Z}^n = \mathbb{Z}a \oplus U$.
b) Let $(a, b) \in \mathbb{Z}^2$. If $1 = ra + sb$, $r, s \in \mathbb{Z}$, then $(a, b), (-s, r)$ is a basis of $\mathbb{Z}^2$.

2.3 Induced Homomorphisms and Quotient Formation

c) Complete $a := (55, 85, 187) \in \mathbb{Z}^3$ to a basis of $\mathbb{Z}^3$. (Construct a basis of a complement $U \subseteq \mathbb{Z}^3$ according to a) using the method of the proof of Theorem 2.3.28.)

d) The residue class group $\mathbb{Z}^n/\mathbb{Z}a$ is isomorphic to $\mathbb{Z}_d \oplus \mathbb{Z}^{n-1}$. (The case $d = 1$ is a).)

(For generalizations see Vol. 3: Linear Algebra 1.)

Exercise 2.3.14 Let M be a submonoid of $\mathbb{Z}^n = (\mathbb{Z}^n, +)$. Then the Grothendieck group $G(M)$ of M is a subgroup of $\mathbb{Z}^n$, thus free. We therefore want to assume that $G(M) = \mathbb{Z}^n$ is.

a) M is factorial if and only if there is a base $a_1, \ldots, a_r, b_1, \ldots, b_s$, $r + s = n$, from $\mathbb{Z}^n$ with $M = \mathbb{Z}a_1 + \cdots + \mathbb{Z}a_r + \mathbb{N}b_1 + \cdots + \mathbb{N}b_s$. (If M is factorial, then the free abelian group $M^\times \subseteq \mathbb{Z}^n$ is a direct summand of $\mathbb{Z}^n$.) If M is factorial, then $\mathbb{P}_M$ is finite.

b) Always is $|\mathbb{P}_M| \leq n$. (The submonoid generated by $\mathbb{P}_M$ is free, see exercise 2.1.18c), and thus its Grothendieck group is a free abelian group.) If $n = 1$, then $\mathbb{I}_M$ is also finite. (In this case, $M = \mathbb{Z}$ or $M \subseteq \mathbb{N}$ or $-M \subseteq \mathbb{N}$. Now use exercise 1.7.27.) The submonoids $M \subseteq \mathbb{N}$ with $G(M) = \mathbb{Z}$ are called **numerical monoids**. They are still studied to this day. For $n = 2$, give an example $M \subseteq \mathbb{N}^2$ such that $\mathbb{I}_M$ is infinite. (E.g. $M := \{(0,0)\} + (\mathbb{N}^*)^2$.)

Exercise 2.3.15 Let M be a commutative monoid. A subset $A \subseteq M$ is called an **ideal** in M, if A includes with every element a also the set Ma of multiples of a, if therefore $A = MA$.[16] The ideals of the form Ma, $a \in A$, are called **principal ideals**. Any unions and intersections of ideals are again ideals. The set $M - M^\times$ of non-units of M is the largest ideal $\neq M$ in M, $\emptyset$ is the smallest ideal in M. $E \subseteq A$ is called a **generating system** of the ideal A, if $A = ME = \bigcup_{a \in E} Ma$ is. M is called **noetherian**, if the set of ideals of M is noetherian ordered.

a) Let $\varphi: M \to N$ be a monoid homomorphism. If B is an ideal in N, then the preimage $\varphi^{-1}(B)$ is an ideal in M. If φ is surjective and A is an ideal in M, then the image $\varphi(A)$ is an ideal in N. If φ is surjective and M is noetherian, then N is also noetherian.

b) M is exactly noetherian when the monoid $M/M||_M$ of divisor classes of M is noetherian.

c) M is Noetherian if and only if every ideal in M is finitely generated. Every ideal is generated in a Noetherian monoid by each representative system of its finitely many minimal divisor classes.

---

[16] In non-commutative monoids M, one has to distinguish between **left-** and **right ideals** $A \subseteq M$ with $A = MA$ or with $A = AM$.

d) If M is Noetherian, then so is the product monoid $M \times (\mathbb{N}^r, +)$, $r \in \mathbb{N}$. (Induction over r. If B is an ideal in $M \times \mathbb{N}$, then $A_n := \{a \in M \mid (a,n) \in B\}$, $n \in \mathbb{N}$, is a monotonically increasing sequence of ideals in M, so $A_n = A_{n_0}$ for all $n \geq n_0$. With finite generating systems of the M ideals $A_0, \ldots, A_{n_0}$, one now obtains a finite generating system of B.)

e) The free abelian monoid $\mathbb{N}^r$ is Noetherian for every $r \in \mathbb{N}$, i.e., every ideal of $\mathbb{N}^r$ is generated by finitely many elements. Since the divisibility relation in $\mathbb{N}^r$ coincides with the product order on $\mathbb{N}^r$, this means: *Every ideal in $\mathbb{N}^r$ (and thus every subset of $\mathbb{N}^r$) has only finitely many minimal elements with respect to the product order* (**Dickson's Lemma**).

f) If M is finitely generated (as a monoid), then M is Noetherian. (Use a) and e).)

g) If M is regular, then M is Noetherian if and only if $M/M^\times (= M/_M||_M)$ is a finitely generated monoid. (Consider a generating system of the ideal $M - M^\times$. – With the combination $(a,b) \mapsto \text{Max}\,(a,b)$, $\mathbb{N}$ is a pointed Noetherian monoid with $\mathbb{N} = \mathbb{N}/_\mathbb{N} ||_\mathbb{N}$, which is not finitely generated. Ideals are only the principal ideals $\mathbb{N}n_0 = \mathbb{N}_{\geq n_0}$, $n_0 \in \mathbb{N}$, as well as $\emptyset$.)

h) A submonoid of $\mathbb{Z}^n$, $n \in \mathbb{N}$, is Noetherian if and only if it is finitely generated. (Note: A finitely generated group is also finitely generated as a monoid.)

i) If M is factorial, then M is Noetherian if and only if $\mathbb{P}_M$ is finite.

Exercise 2.3.16

a) Let M be a commutative monoid, in which the set of principal ideals Ma, $a \in M$, is Noetherian or, equivalently, the set of divisor sets $T(a)$, $a \in M$, is Artinian ordered (which is particularly the case when M is Noetherian). Show that every non-unit in M is a product of irreducible elements. (Otherwise, consider a maximal element in the set of principal ideals $Ma \neq M$, for which a is not a product of irreducible elements.)

b) Construct a pointed regular commutative monoid such that every element is a product of irreducible elements, but the set of principal ideals is not Noetherian ordered.

Exercise 2.3.17 Let M be a pointed commutative regular monoid. Then every element $x \in M$ generates a free submonoid (of rank ≤ 1). Show with examples that the Grothendieck group of such a monoid can have nontrivial torsion elements. (Start with a suitable commutative group with nontrivial torsion.)

2.4 Operation of Monoids and Groups

The theory of groups began concretely with the study of permutation groups, which appeared as symmetry groups of mathematical and especially geometric objects. Only in the course of the 19th century was the abstract group concept developed, with A. Cayley (1821–1895) in particular being seen as a pioneer. The concept of group operation again combines both aspects, an idea that was

2.4 Operation of Monoids and Groups

particularly propagated by F. Klein (1849–1925). In this section, we introduce the relevant elementary concepts.

Let M and X be sets. An **operation** of M on X is a mapping

$$\mu : M \times X \to X.$$

We usually write it in the form $(a, x) \mapsto ax := \mu(a, x)$, $a \in M$, $x \in X$. The mapping ϑ_a defined by $\vartheta_a : x \mapsto ax$ for a fixed $a \in M$ isof X in itself is then the **operation of the element** $a \in M$ on X. The mapping $\vartheta : M \to X^X$, $a \mapsto \vartheta_a$, describes the considered operation of M on X completely. It is called the **action** of M on X. An operation $\mu : M \times X \to X$ of M on X induces an operation of $\mathfrak{P}(M)$ on the power set $\mathfrak{P}(X)$ of X by means of

$$AY := \{ay \mid a \in A, y \in Y\} = \mu(A \times Y) = \bigcup_{a \in A} \vartheta_a(Y), \quad A \subseteq M, \ Y \subseteq X.$$

For $x \in X$ the set $Mx := M\{x\} = \{ax \mid a \in M\}$ is called the **orbit** or the **orbit of** x under the given operation.

Now let M be a monoid. Since the set X^X of mappings from X to itself is also a monoid (with the composition of mappings as the operation), the following concept formation is obvious:

Definition 2.4.1 Let M be a monoid with neutral element $e_M = e$. An operation $\mu : M \times X \to X$, $(a, x) \mapsto ax := \mu(a, x)$, from M on the set X is called a **monoid operation** or an **operation of** M **as a monoid**, if the associated action $\vartheta : M \to X^X$, $a \mapsto (\vartheta_a : x \mapsto ax)$, is a monoid homomorphism, i.e., if the following conditions are fulfilled for all $a, b \in M$ and all $x \in X$:

(1) $(ab)x = a(bx)$, that is $\vartheta_{ab} = \vartheta_a \vartheta_b$. (2) $ex = x$, that is $\vartheta_e = id_X$.

A set $X = (X, \vartheta)$ together with an operation of M on X is also called an **M-set** or an **M-space**.

If M is a monoid, then in the following an operation of M on a set X is always an operation as a monoid, unless explicitly stated otherwise. The operation induced by $M \times X \to X$ $\mathfrak{P}(M) \times \mathfrak{P}(X) \to \mathfrak{P}(X)$ is then also an operation of $\mathfrak{P}(M)$ as a monoid (with complex multiplication as the operation). If the action homomorphism $\vartheta : M \to X^X$ is injective, then the operation is called **faithful** or **effective**.

Let $M \times X \to X$ be an operation of the monoid M. The associated action $\vartheta : M \to X^X$ then maps the unit group $M^\times$ into the unit group $\mathfrak{S}(X) \subseteq X^X$ of X^X. In particular, a **group operation** of a group G on X is given by a group homomorphism $\vartheta : G \to \mathfrak{S}(X)$. Its kernel is called the **kernel** or the **ineffectiveness** of the operation. The operation is faithful exactly when its kernel is trivial.

In a faithful operation of the monoid M on X, M can be considered as a submonoid of X^X with respect to $\vartheta : M \hookrightarrow X^X$. Conversely, every submonoid $N \subseteq X^X$ operates faithfully on X in a canonical way, the action is the canonical embedding $N \hookrightarrow X^X$. This is then referred to as the **natural operation** of N on X and N is called a **transformation monoid** of X. If the action homomorphism $\vartheta: M \to X^X$ is trivial, i.e., every element of M operates like the identity on X, this is referred to as the **trivial operation** of M on X. If X is a mathematical structure with a monoid $\text{End}\, X$ of endomorphisms and a group $\text{Aut}\, X = (\text{End}\, X)^\times$ of automorphisms, e.g., a magma, a semigroup etc., and the image lies $\vartheta(M)$ in the submonoid $\text{End}\, X$ of X^X or in the subgroup $\text{Aut}\, X$ of $\mathfrak{S}(X)$, one says, M **operates as a monoid of endomorphisms** of X or **as a group of automorphisms** of X, when M is a group, $\vartheta(M)$ is thus a subgroup of $\text{Aut}\, X = (\text{End}\, X)^\times$. In particular, $\text{End}\, X$ and $\text{Aut}\, X$ operate in a natural way as a monoid of endomorphisms or as a group of automorphisms on X.

A subset $Y \subseteq X$ is called **invariant** under an operation $\mu: M \times X \to X$, if $\mu(M \times Y) \subseteq Y$ applies. In this case, the restriction $\mu|(M \times Y)$ provides an operation of M on Y. For example, the orbits Mx, $x \in X$, are invariant under any operation. For each subset $A \subseteq X$, $MA := \mu(M \times A) = \{ax \mid a \in M, x \in A\}$ is the smallest A encompassing subset invariant under the operation of M of X. If $\varphi: L \to M$ is a monoid homomorphism, then the composition $\vartheta \circ \varphi: L \to X^X$ is the action homomorphism of the **induced by** φ **induced operation** $L \times X \to X$, $(c, x) \mapsto \varphi(c)x$, from L to X. If L is a submonoid of M, the canonical injection $L \hookrightarrow M$ induces the restriction $L \times X \to X$ on L of the operation $M \times X \to X$ of M on X. Every operation is the operation induced by ϑ of the natural operation of $\vartheta(M)$ on X. If X_i, $i \in I$, is a family of M spaces, then the disjoint union $X := \biguplus_{i \in I} X_i$ is also an M-space in a canonical way, the so-called **sum** (or also the **coproduct**) of the M spaces X_i, $i \in I$, which is also denoted by $\coprod_{i \in I} X_i$.is referred to. The X_i are M-invariant, and the M-operation on X induces the given operations on the X_i, $i \in I$. Similarly, the **product** $\prod_{i \in I} X_i$ of the M spaces X_i, $i \in I$, is an M space with the so-called **diagonal operation** $a(x_i)_{i \in I} := (ax_i)_{i \in I}$, $a \in M$, $(x_i)_{i \in I} \in \prod_{i \in I} X_i$.

In this section, we will primarily focus on operations of groups. Unless explicitly stated otherwise, the following will therefore deal with group operations.

The orbits Gx, $x \in X$, of the operation $G \times X \to X$ of a group G on the set X with the action $\vartheta: G \to \mathfrak{S}(X)$ form a decomposition of X. It is indeed $x = ex \in Gx$, and from $z = gx = hy \in Gx \cap Gy$ follows $x = g^{-1}hy \in Gy$, so $Gx \subseteq Gy$ and analogously $Gy \subseteq Gx$, i.e. $Gx = Gy$. The equivalence relation defined by this decomposition $\sim = \sim_\vartheta$ on X is given by

$$x \sim y \iff y \in Gx \iff \text{there is a } g \in G \text{ with } y = gx.$$

The set of all orbits is called the **orbit space** of X with respect to ϑ. It is denoted by

$$X \backslash G = X \backslash_\vartheta G \quad (\text{often also with } G \backslash X = G \backslash_\vartheta X)$$

It includes the canonical projection $\pi = \pi_\vartheta: X \to X \backslash G$, $x \mapsto Gx$. Their fibers are the orbits of the operation. A full representative system $F \subseteq X$ for the set $X \backslash G$ of orbits is called a **fundamental domain** of the given operation.

2.4 Operation of Monoids and Groups

To describe the orbit Gx we consider the canonical surjective mapping $G \to Gx$, $g \mapsto gx$. Exactly then is $gx = hx$, when $h^{-1}gx = x$ is, i.e. when h and g are the same left coset of the so-called **Isotropy group**

$$G_x := \operatorname{Stab}(x, G) := \operatorname{Stab}(x, G, \vartheta) := \{g \in G \mid gx = \vartheta_g(x) = x\}$$

in G represent. G_x is obviously a subgroup of G and is also called the **stabilizer** or the **stability subgroup** or the **stand group** of x. The mapping $g \mapsto gx$ thus induces the bijection $G/G_x \xrightarrow{\sim} Gx$ with $gG_x \mapsto gx$. Consequently,

$$|Gx| = [G : G_x],$$

i.e. *the orbit through x has exactly as many elements as the index of the isotropy group of x indicates*. For a finite group G this leads to important restrictions about the orbit lengths, namely: *The length $|Gx|$ of each orbit Gx divides the group order.* — It is

$$G_{gx} = gG_x g^{-1};$$

because $h \in G_{gx}$, i.e. $h(gx) = gx$, applies exactly when $g^{-1}hgx = x$, i.e. $g^{-1}hg \in G_x$ is. *The isotropy groups of the elements gx of an orbit Gx are thus exactly the subgroups conjugated to the isotropy group G_x itself*, namely $gG_x g^{-1} = \kappa_g(G_x)$, $g \in G$, see also Example 2.4.5further below. This class of conjugate subgroups of G is called the **isotropy class** of the orbit Gx. It is independent of the specific choice of an element x of the considered orbit. Only then is $G_x = G$, when the orbit Gx of $x$consists only of the element x. It is then said that x is a **fixed point** of the operation of G on X. We denote the set of all these fixed points with[17]

$$\operatorname{Fix}(G, X).$$

The intersection $\bigcap_{x \in X} G_x$ of all isotropy groups is the kernel of the action ϑ. The intersection $\bigcap_{y \in Gx} G_y$ of the isotropy groups from the isotropy class of the orbit Gx is therefore the kernel of the restriction of the given operation to the orbit Gx and in particular a normal subgroup in G. *It is the largest normal subgroup of G, which lies in one of the isotropy groups G_y, $y \in Gx$, from the isotropy class of Gx.*

The decomposition of the set into the orbits of an operation of the group G on X yields the following fundamental equation:

Theorem 2.4.2 (Class Equation) *The group G operates on the set X. If Gx_i, $i \in I$, are the pairwise different orbits with more than one element, then*

$$|X| = |\operatorname{Fix}(G, X)| + \sum_{i \in I} |Gx_i|.$$

In the case of finite G, $|Gx_i| = [G : G_{x_i}]$ is a divisor $\neq 1$ of the order $|G|$ of G, $i \in I$.

We will frequently use the following corollary.

---

[17] If no misunderstandings are to be feared, this set is often also denoted with X^G.

Corollary 2.4.3 *The finite group H of order p^α, p prime, operates on the finite set X. Then the congruence holds*

$$|X| \equiv |\text{Fix}(H, X)| \bmod p.$$

In particular, the operation of H has a fixed point, if p is not a divisor of $|X|$, and $|\text{Fix}(H, X)|$ is divisible by p if $|X|$ itself is divisible by p.

If the isotropy group G_x is trivial for each $x \in X$, then G operates by definition **freely** on X. The operation of G on X is called **transitive**, if it has exactly one orbit, i.e. if $X \neq \emptyset$ is and for every two elements $x, y \in X$ there always exists a $g \in G$ with $gx = y$, if $Gx = X$ is for one and therefore for every $x \in X \neq \emptyset$. It is called **simply transitive**, if it is transitive and free, i.e. if $X \neq \emptyset$ is and for $x, y \in X$ there always exists exactly one $g \in G$ with $gx = y$. This element $g \in G$ is often denoted by $\overrightarrow{xy}$. For every $x \in X$ the mapping $G \to X$ with $g \mapsto gx$ is bijective. Every simply transitive operation is faithful, and every free operation induces a simply transitive operation on each of its orbits. A set X with a simply transitive operation of the group G on X is also called a G-**affine space**. Prime examples of transitive operations are the spaces G/H of left cosets of G with respect to a subgroup $H \subseteq G$ with the natural operation $(g, aH) \mapsto gaH$, $a, g \in G$, see also Example 2.4.14. Because of $\text{Stab}(H, G) = H$ is

$$\text{Stab}(aH, G) = aHa^{-1}, \quad a \in G.$$

Example 2.4.4 The combination of any group G is a simply transitive operation of G on itself. The associated injective group homomorphism $\vartheta: G \to \mathfrak{S}(G)$ is the Cayley representation $L: G \to \mathfrak{S}(G)$, $g \mapsto (x \mapsto gx)$, of G as a transformation group, see Theorem 2.2.9. If one restricts this operation of G on itself to a subgroup H of G, then the orbits are the *right* cosets Hx of G with respect to H, $x \in G$. The class equation is

$$|G| = |G\backslash H| \cdot |H| = [G : H] \cdot |H|,$$

since all isotropy groups are trivial. This is the Theorem 2.1.21 of Lagrange. ◊

The left cosets of G with respect to H can also be obtained similarly to the last example. There are two possibilities for this. In addition to the (left) operations introduced so far, we also consider **right operations** of monoids M on sets X, which are defined as mappings $X \times M \to X$, $(x, a) \mapsto xa$, with

(1) $x(ab) = (xa)b$ and (2) $xe = x$ for all $x \in X, a, b \in M$,

The action homomorphism $\eta: M \to X^X$, $a \mapsto (\eta_a: x \mapsto xa)$, is in this case an antihomomorphism, i.e., a homomorphism $\eta: M^{\text{op}} \to X^X$. The right operations of the monoid M are therefore identical to the (left) operations of the M oppositional monoid M^{op}. For the subgroup H of the group G, the orbits of the right operation $G \times H \to G$ with $(g, h) \mapsto gh$ of H on G are precisely the *left* cosets xH of G with respect to H. Accordingly, we denote the orbit space for any right operation of a group G with action homomorphism $\eta: G^{\text{op}} \to X^X$ with $X/G = X/_\eta G$.

2.4 Operation of Monoids and Groups

On the other hand, every right operation $X \times G \to X$ of a *group* G directly provides a left operation $G \times X \to X$, which has the same orbits as the original right operation. It is defined by $(g,x) \mapsto xg^{-1}$ and uses that the inverse formation $g \mapsto g^{-1}$ is an antihomomorphism $G \to G$, thus an isomorphism $G \xrightarrow{\sim} G^{\mathrm{op}}$. In the case under consideration of a subgroup H of G, this method yields from the given right operation the operation $(h,x) \mapsto xh^{-1}$ of H on G (from the left), whose orbits are thus the left cosets xH, $x \in G$.

Example 2.4.5 (Operation by Conjugation) Let M be a monoid. The automorphism group $\mathrm{Aut}\, M$ of M operates in a natural way as a group of automorphisms. The conjugation homomorphism $\kappa \colon M^\times \to \mathrm{Aut}\, M$, $a \mapsto (\kappa_a \colon x \mapsto axa^{-1})$, cf. Exercise 2.2.3, therefore induces the operation $M^\times \times M \to M$, $(a,x) \mapsto axa^{-1}$, of $M^\times$ on M as a group of automorphisms of M with action homomorphism κ, which is called the **conjugation** of M. The associated orbits $C_{M^\times}(x) := \{axa^{-1} \mid a \in M^\times\}$, $x \in M$, are called the **conjugacy classes** of M. The isotropy group

$$M_x = \{a \in M^\times \mid axa^{-1} = x\} = \{a \in M^\times \mid ax = xa\} = Z_{M^\times}(x)$$

is the subgroup of elements of $M^\times$ that commute with $x \in M$, thus the **centralizer of x in $M^\times$**.

Let $M = G$ be a group. Then the fixed point set of the conjugation is the center $Z(G)$. The center is also the kernel of κ. The general class equation 2.4.2 therefore provides the special **class equation of** G:

$$|G| = |Z(G)| + \sum_{i \in I} |C_i|.$$

Here, C_i, $i \in I$, denote the pairwise different conjugacy classes of G with more than one element. If $x_i \in C_i$, then $|C_i| = [G : Z_G(x_i)]$. Note that for finite G the numbers $|Z(G)|$ and $|C_i|$, $i = 1, \ldots, r$, in this class equation all divide the group order $|G|$. The number of all conjugacy classes, thus $|Z(G)| + |I|$, is called the **class number** of the group G. With corollary 2.4.3 it follows: ◊

Theorem 2.4.6 *A non-trivial finite group, whose order is a power of a prime number, has a non-trivial center.*

Let G again be any group. G is also the unit group of the monoid $\mathfrak{P}(G)$ (with complex multiplication as the operation, cf. Example 2.1.17). Consequently, G also operates on $\mathfrak{P}(G)$ by conjugation. The corresponding paths are again called **conjugacy classes**. The isotropy group of a set $A \in \mathfrak{P}(G)$ is now called the **normalizer of A in G** and is denoted by

$$\mathrm{N}_G(A) := \mathrm{Stab}(A, G, \kappa) = \{g \in G \mid \kappa_g(A) = gAg^{-1} = A\}$$

Its index is the number of subsets of G conjugated to A. A is called **normal** if A is invariant under all conjugations, i.e., if $\mathrm{N}_G(A) = G$ is. Exactly then A is normal

Fig. 2.10 Universal property of the semidirect product

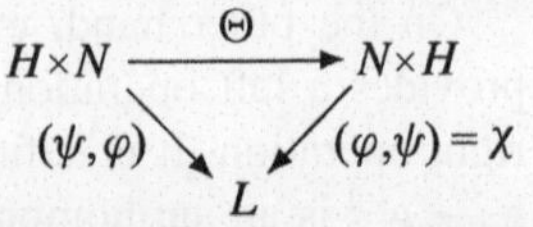

$$H \times N \xrightarrow{\Theta} N \times H$$
$$(\psi,\varphi) \searrow \quad \swarrow (\varphi,\psi) = \chi$$
$$L$$

when A is a union of certain conjugacy classes $C_G(x)$ of G. The kernel of the operation of $N_G(A)$ on A is the **centralizer**

$$Z_G(A) = \bigcap_{a \in A} Z_G(a) = \{g \in G \mid ga = ag \text{ for all } a \in A\}$$

of A in G. In particular, $Z_G(A)$ is a normal subgroup in $N_G(A)$. If $A = H$ is a subgroup of G, then $N_G(H)$ is the largest subgroup of G, in which H is a normal subgroup. The index $[G : N_G(H)]$ is the number of subgroups of G conjugated to H and a divisor of $[G : H]$, if $[G : H]$ is finite. Because of $\text{Stab}(aH, G) = aHa^{-1}$ (where G and thus H operate on G/H in a natural way as above) results in the useful description

$$N_G(H)/H = \text{Fix}(H, G/H) \subseteq G/H.$$

Always, $H \cap Z_G(H) = Z(H)$.

Example 2.4.7 Let G be a finite group of order n and $p \in \mathbb{P}$ a prime number. Following J. McKay, see Amer. Math. Monthly **66**, 119 (1959), we consider on the set G^p of the p-tuples of G the operation of the cyclic group $\mathbb{Z}_p = \mathbb{Z}/\mathbb{Z}p$ of order p by cyclic permutation $\bigl(a,(x_1,\ldots,x_p)\bigr) \mapsto (x_{1+a},\ldots,x_{p+a})$, where with a and the indices $1,\ldots,p$ as residues in $\mathbb{Z}_p$ is to be calculated. The fixed points of this operation are the constant p-tuples $(x,\ldots,x)$. The subset $X \subseteq G^p$ of the p-tuples $(x_1,\ldots,x_p)$ with $x_1 \cdots x_p = e$ is $\mathbb{Z}_p$-invariant; because with $x_1 x_2 \cdots x_p = (x_1 \cdots x_r)(x_{r+1} \cdots x_p) = e$ is also $(x_{r+1} \cdots x_p)(x_1 \cdots x_r) = e$ for $r = 1,\ldots,p-1$. Obviously, $|X| = n^{p-1}$, and according to corollary 2.4.3 is $n^{p-1} \equiv |\text{Fix}(\mathbb{Z}_p, X)| \bmod p$. If now n is divisible by p, this also applies to $|\text{Fix}(\mathbb{Z}_p, X)|$, i.e., *the set of $x \in G$ with $x^p = e$ is divisible by p if $|G|$ is divisible by p.* In particular, we obtain: ◊

Theorem 2.4.8 (Cauchy's Theorem) *A finite group, whose order is divisible by the prime number p, possesses elements of order p.*

As a direct consequence, the following result, which has been mentioned several times, emerges: *The exponent* $\text{Exp}\, G$ *and the order* $\text{Ord}\, G$ *of a finite group G have the same prime divisors. A finite group G is exactly a p-group, $p \in \mathbb{P}$, when the order of G is a p-power.* – By the way, if the prime number p is not a divisor of the group order n, then the fixed point set $\text{Fix}(\mathbb{Z}_p, X)$ of the above operation by $\mathbb{Z}_p$ on X only contains the constant tuple $(e,\ldots,e) \in X \subseteq G^p$ and we get the already known congruence $n^{p-1} \equiv 1 \bmod p$ of the Little Fermat's Theorem (and

2.4 Operation of Monoids and Groups

the Fermat quotient $(n^{p-1} - 1)/p$ is the number $|X\setminus\mathbb{Z}_p|$ of orbits of the operation by $\mathbb{Z}_p$ on X.

Example 2.4.9 (Sylow's Theorems) We take Cauchy's Theorem 2.4.8 as the starting point for the proof of Sylow's theorems about p-subgroups of finite groups, which are of fundamental importance for the theory of finite groups. In the following, let G always be a group. ◊

Definition 2.4.10 Let $p \in \mathbb{P}$ be a prime number. Then a subgroup of G is called a p-**Sylow (sub-)group** of G, if it is maximal in the set of all p-subgroups of G.

G possesses according to Zorn's Lemma 1.4.15 p-Sylow groups for each $p \in \mathbb{P}$, more precisely: *Every p-subgroup of G is contained in a p-Sylow group of G.* Let $S_p \subseteq G$ be a p-Sylow subgroup. Then also $\varphi(S_p)$ for every automorphism φ of G is a p-Sylow group in G. Furthermore, S_p is also a p-Sylow group in every subgroup $H \subseteq G$ with $S_p \subseteq H$. If S_p is normal in G and H is any p-subgroup of G, then $H \subseteq S_p$, because $S_p H = H S_p$ is also a p-group (note $S_p H / S_p \cong H/(S_p \cap H)$), and S_p is consequently the only p-Sylow group in G. More generally, $H \subseteq S_p$ holds for every p-subgroup $H \subseteq N_G(S_p)$. If G is abelian, then the p-primary component $G(p)$ of G is the only p-Sylow group in G (see Theorem 2.2.22). The main theorem about Sylow groups states:

Theorem 2.4.11 (Sylow's Theorem) *Let $p \in \mathbb{P}$ be a prime number, G a finite group of order $n = p^\alpha m$, $p \nmid m$ (i.e. $\alpha = \mathsf{v}_p(|G|))$ and $S_p \subseteq G$ a p-Sylow group in G.*

(1) *S_p has the (maximum possible) order p^α.*
(2) *If $H \subseteq G$ is any p-subgroup of G, then $H \subseteq g S_p g^{-1}$ for some $g \in G$. In particular, all p-Sylow groups in G are conjugate.*
(3) *For the number $s_p = [G : N_G(S_p)]$ of p-Sylow groups in G, $s_p \mid m$ and $s_p \equiv 1 \bmod p$ hold.*

Proof (1) It suffices to show: If $H \subseteq G$ is a subgroup of order p^β, $\beta < \alpha$, then there exists a subgroup $H' \subseteq G$ with $H \subseteq H'$ and $|H'| = p^{\beta+1}$. Since $|G/H|$ is divided by p and

$$|N_G(H)/H| = |\mathrm{Fix}(H, G/H)| \equiv |G/H| \bmod p$$

holds, cf. Corollary 2.4.3, p is also a divisor of $|N_G(H)/H|$. According to Cauchy's Theorem 2.4.8, there exists a subgroup $H'/H \subseteq N_G(H)/H$ of order p. Then $|H'| = |H'/H| \cdot |H| = p \cdot |H| = p^{\beta+1}$.

(2) Since p according to (1) is not a divisor of $|G/S_p|$, the natural operation of H on G/S_p has a fixed point $g S_p$, $g \in G$, cf. Corollary 2.4.3. Then $H \subseteq g S_p g^{-1}$.

(3) After (2), $[G : N_G(S_p)]$ is the number s_p of p-Sylow groups in G, and because of $S_p \subseteq N_G(S_p) \subseteq G$, s_p is a divisor of $[G : S_p] = m$. – We now

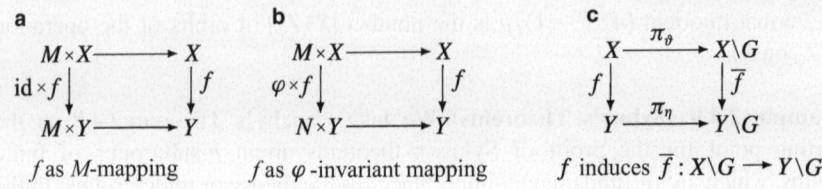

Fig. 2.11 Invariant mappings

consider the natural operation of S_p on $G/N_G(S_p)$ and assert that $N_G(S_p)$ is its only fixed point. If $gN_G(S_p)$, $g \in G$, is a fixed point, i.e., if $S_p \subseteq gN_G(S_p)g^{-1}$ or $g^{-1}S_p g \subseteq N_G(S_p)$, then $g^{-1}S_p g = S_p$, since S_p is the only p-Sylow group in $N_G(S_p)$, thus $g \in N_G(S_p)$. Again with Corollary 2.4.3 we get $s_p = |G/N_G(S_p)| \equiv 1 \bmod p$. □

Example 2.4.12 (Semidirect Products) Let G be a group with neutral element e and N a normal subgroup in G with a weak complement $H \subseteq G$, i.e., H is a subgroup of G with $N \cap H = \{e\}$ and $NH = HN = G$. The product gg' of two elements $g = nh$ and $g' = n'h'$, $n, n' \in N$, $h, h' \in H$, then has the representation

$$gg' = nhn'h' = \bigl(n(hn'h^{-1})\bigr)hh' = \bigl(n\kappa_h(n')\bigr)(hh')$$

with $n\kappa_h(n') \in N$ and $hh' \in H$ (since the normal subgroup N is invariant under the conjugation κ_h with h). *The operation on G is thus determined by the operations of N and H as well as the operation of H on N through conjugation.* This observation leads to the following general construction: ◊

Theorem 2.4.13 *Let N and H be monoids with neutral elements e_N and e_H. H operates on N by means of the action $\vartheta: H \to \operatorname{End} N$ as a monoid of endomorphisms of N. Then on $N \times H$ the operation is defined by*

$$(n, h) \cdot (n', h') := \bigl(n \cdot \vartheta_h(n'), h \cdot h'\bigr)$$

a monoid structure with neutral element (e_N, e_H) defined. $N \times H$ with this operation is called the (abstract) **semidirect product** *of the monoids N and H with respect to the operation ϑ and is denoted by*

$$N \rtimes H = N \rtimes_\vartheta H$$

denoted. It is exactly $(n, h) \in (N \rtimes H)^\times$, when $n \in N^\times$ and $h \in H^\times$ is. In this case, the following applies

$$(n, h)^{-1} = \bigl(\vartheta_{h^{-1}}(n^{-1}), h^{-1}\bigr).$$

In particular, $N \rtimes H$ is exactly a group when N and H are groups. In general,

$$(N \rtimes_\vartheta H)^\times = N^\times \rtimes_{\vartheta^\times} H^\times,$$

where $\vartheta^\times : H^\times \to \operatorname{Aut}(N^\times)$ the group action induced by ϑ is.

2.4 Operation of Monoids and Groups

Proof So it is $\vartheta_h^\times = \vartheta_h|N^\times \in \operatorname{Aut}(N^\times)$, $h \in H$. – The associativity of multiplication in $N \rtimes H$ results from

$$\big((n,h)(n',h')\big)(n'',h'') = (n\vartheta_h(n'), hh')(n'',h'') = (n\vartheta_h(n')\vartheta_{hh'}(n''), hh'h''),$$

$$(n,h)\big((n',h')(n'',h'')\big) = (n,h)(n'\vartheta_{h'}(n''), h'h'') = (n\vartheta_h(n'\vartheta_{h'}(n'')), hh'h'')$$

and $\vartheta_h(n')\vartheta_{hh'}(n'') = \vartheta_h(n'\vartheta_{h'}(n''))$. Apparently, (e_N, e_H) is a neutral element. If (n,h) is invertible, then there exists (n', h') with $n\vartheta_h(n') = e_N = n'\vartheta_{h'}(n)$ and $hh' = e_H = h'h$. Therefore, $h \in H^\times$ with $h^{-1} = h'$ and thus $\vartheta_h \in \operatorname{Aut} N$ with $\vartheta_h^{-1} = \vartheta_{h'}$ as well as $e_N = \vartheta_h(e_N) = \vartheta_h(n')\vartheta_h\vartheta_{h'}(n) = \vartheta_h(n')n$. It follows that n is invertible with inverse $\vartheta_h(n')$ and $n' = \vartheta_{h^{-1}}(n^{-1})$. Conversely, one immediately checks that $(\vartheta_{h^{-1}}(n^{-1}), h^{-1})$ is inverse to (n, h) is. □

In the situation of sentence 2.4.13, $N \hookrightarrow N \rtimes H$, $n \mapsto (n, e_H)$, and $H \hookrightarrow N \rtimes H$, $h \mapsto (e_N, h)$, are embeddings of monoids, with the help of which we consider N and H as submonoids of $N \rtimes H$. The pair (n,h) is then the product $n \cdot h$, and when swapping the factors, one has to apply ϑ_h: $h \cdot n = (e_N, h)(n, e_H) = (\vartheta_h(n), h) = \vartheta_h(n) \cdot h$. If h is invertible, then $n \mapsto \vartheta_h(n) = hnh^{-1}$ is the conjugation $\kappa_h|N$. The mapping $n \cdot h \mapsto h$ is a monoid projection onto H, which in the case of groups is a group homomorphism with N as the kernel. *In this case, therefore, $N \rtimes_\vartheta H$ is the semidirect product of N and H, where $\vartheta_h = \kappa_h|N$ is the conjugation with h on N, $h \in H$.* If the group is G as given at the beginning of this example as a semi-direct product of the normal subgroup N with weak complement H, then $N \rtimes_{\kappa|H} H \xrightarrow{\sim} G$, $n \cdot h \mapsto nh$, is a canonical group isomorphism. H operates trivially on N if and only if $N \rtimes H$ is the product monoid $N \times H$. In the case of groups, this is characterized by H also being a normal subgroup in $N \rtimes H$. $N \rtimes H$ is commutative if and only if N and H are commutative and H operates trivially on N.

The semidirect product $N \rtimes_\vartheta H$ apparently has the following **universal property**: For monoid homomorphisms $\varphi \colon N \to L$ and $\psi \colon H \to L$ there exists exactly one monoid homomorphism $\chi \colon N \rtimes_\vartheta H \to L$ with $\chi|N = \varphi$ and $\chi|H = \psi$, if $\psi(h)\varphi(n) = \varphi(\vartheta_{h(n)})\psi(h)$ for all $n \in N$, $h \in H$ applies, if the diagram from Fig. 2.10 is commutative, in which $\Theta(h,n) = (\vartheta_h(n), h)$ and (ψ, φ) and (φ, ψ) are defined by $(h,n) \mapsto \psi(h)\varphi(n)$ or $(n,h) \mapsto \varphi(n)\psi(h)$.

We mention some concrete examples.

(1) The semidirect product

$$\operatorname{Hol} N := N \rtimes \operatorname{End} N,$$

where $\operatorname{End} N$ operates in a natural way on N, is called the **full holomorph** of N. If $H \subseteq \operatorname{End} N$ is only a submonoid, then the semidirect product $N \rtimes H$, which is a submonoid of the full holomorph $N \rtimes \operatorname{End} N$, is a (restricted) **holomorph** of N. We denote it by $\operatorname{Hol}_H N$. The case $H = \operatorname{Aut} N$ is particularly important. The resulting holomorph is

$$\operatorname{Hol}_{\operatorname{Aut} N} N = N \rtimes \operatorname{Aut} N$$

and is called the **holomorph** of N per se. If $N = G$ is a group, then this holomorph is the unit group $(\mathrm{Hol}\,G)^\times$ of the full holomorph $\mathrm{Hol}\,G$ of G.

The figure $\mathrm{Hol}\,N \to N^N$, $(a,\sigma) \mapsto L_a \circ \sigma$, is a monoid homomorphism. *It is injective if N is regular, especially when N is a group,* cf. exercise 2.4.6. In this case, $\mathrm{Hol}\,N$ and thus every holomorph of N can be viewed in a canonical way as a submonoid of the mapping monoid N^N, where the submonoid $N \subseteq \mathrm{Hol}\,N$ is identified with the monoid of left translations of N according to the Cayley representation 2.2.8 and $\mathrm{End}\,N$ is canonically embedded in N^N. Now let $N = G$ be a group. Then the holomorph $\mathrm{Hol}_{\mathrm{Aut}\,G}\,G = G \rtimes \mathrm{Aut}\,G$ is the subgroup of the permutation group $\mathfrak{S}(G)$, which is generated by the left translations and the automorphisms of G. Because of $R_g = L_g \circ \kappa_{g^{-1}} = \kappa_{g^{-1}} \circ L_g$ for $g \in G$, $\mathrm{Hol}_{\mathrm{Aut}\,G}\,G$ also contains all right translations, and the subgroup generated by the left and right translations of $\mathfrak{S}(G)$ is the holomorph $G \rtimes \mathrm{Inn}\,G \subseteq \mathrm{Hol}_{\mathrm{Aut}\,G}\,G$ to the subgroup $\mathrm{Inn}\,G \subseteq \mathrm{Aut}\,G$ of the inner automorphisms of G.

(2) (**Dihedral groups**) Let H be an (additively written) abelian group. The cyclic group operates on $H.\mathbb{Z}^\times = \{1, -1\}$ of order 2, where -1 acts by forming the negative on H. The associated semidirect product is called the **Dihedral group** to the group H and is denoted by $\mathbf{D}(H)$. The combination on $\mathbf{D}(H) = H \rtimes \mathbb{Z}^\times$ is explicitly given by

$$(n,\varepsilon)(n',\varepsilon') = (n + \varepsilon n', \varepsilon\varepsilon'), \quad n,n' \in N,\ \varepsilon,\varepsilon' \in \mathbb{Z}^\times,$$

Only then is $\mathbf{D}(H)$ the direct product of H and $\mathbb{Z}^\times$, i.e., an abelian group, when the inverse formation on H is the identity of H, H is therefore an elementary (abelian) 2-group.[18] The dihedral group to the cyclic group $\mathbf{Z}_n = \mathbb{Z}/\mathbb{Z}n$, $n \in \mathbb{N}$, is denoted by $\mathbf{D}_n$. Note $\mathrm{Ord}\,\mathbf{D}_n = 2n$ for $n \in \mathbb{N}^*$. The infinite dihedral group $\mathbf{D}_0 := \mathbf{D}(\mathbb{Z})$ is the holomorph $\mathrm{Hol}_{\mathrm{Aut}\,\mathbb{Z}}\,\mathbb{Z} = \mathbb{Z} \rtimes \mathrm{Aut}\,\mathbb{Z}$ of the additive group $\mathbb{Z}$. Furthermore, $\mathbf{D}_1 = \mathbf{Z}_2$ and $\mathbf{D}_2 = \mathbf{Z}_2 \times \mathbb{Z}^\times \cong \mathbf{Z}_2 \times \mathbf{Z}_2$. For $n > 2$, $\mathbf{D}_n = \mathrm{Hol}_{\{\pm id\}}\,\mathbf{Z}_n \subseteq \mathrm{Hol}_{\mathrm{Aut}\,\mathbf{Z}_n}\,\mathbf{Z}_n$ is a holomorph of $\mathbf{Z}_n$, which we, as described in (1), consider as the subgroup of $\mathfrak{S}(\mathbf{Z}_n) = \mathfrak{S}(\{0,\ldots,n-1\})$ generated by the two permutations

$$\begin{pmatrix} 0 & 1 & 2 & \cdots & n-2 & n-1 \\ 1 & 2 & 3 & \cdots & n-1 & 0 \end{pmatrix}, \quad \begin{pmatrix} 0 & 1 & 2 & \cdots & n-2 & n-1 \\ 0 & n-1 & n-2 & \cdots & 2 & 1 \end{pmatrix}$$

If $n = 3$, one obtains $\mathbf{D}_3 = \mathfrak{S}(\{0,1,2\}) \cong \mathfrak{S}_3$. For $n = 3,4,6$ (and only for these) even $\mathbf{D}_n = \mathrm{Hol}_{\mathrm{Aut}\,\mathbf{Z}_n}\,\mathbf{Z}_n$ applies. Often the groups $\mathbf{D}_n$, $n \in \mathbb{N}$, are referred to as the **dihedral groups** par excellence. $\mathbf{D}_n$ is the full symmetry group of a regular n-gon of the Euclidean plane or the group of proper symmetries of an n-dihedron, i.e., a straight double cone whose base is a regular n-gon, cf. Vol. 4. There it is also shown how the dihedral group $\mathbf{D}(\mathbb{R})$ can be identified with the group of movements of an affine Euclidean straight line and the dihedral group $\mathbf{D}(\mathbb{R}/\mathbb{Z})$ with the

---

[18] For their structure see Ex. 2.3.6a).

2.4 Operation of Monoids and Groups

group of isometries of an (oriented) two-dimensional Euclidean vector space. The dihedral group $\mathbf{D}(\mathbb{R}/\mathbb{Z})$ (and occasionally the dihedral group $\mathbf{D}(\mathbf{Z}_\infty) = \mathbf{D}(\mathbb{Q}/\mathbb{Z})$, cf. Ex. 2.2.27) also with $\mathbf{D}_\infty$.

Finally, we briefly discuss those mappings that are compatible with monoid operations, i.e., the homomorphisms of sets with monoid operations. If X and Y are sets on which the monoid M operates, a mapping $f\colon X \to Y$ is called an M-**invariant mapping** or simply an M-**mapping**, if for all $a \in M$ and $x \in X$ applies: $f(ax) = af(x)$, $f \circ \vartheta_a = \eta_a \circ f$ is, where $\vartheta\colon M \to X^X$ and $\eta\colon N \to Y^Y$ are the actions of M on X and Y respectively. This means that the diagram in Fig. 2.11a is commutative, where the horizontal arrows denote the operations. An M-invariant mapping $f\colon X \to Y$ is called an M-**invariant isomorphism** or an M-**isomorphism**, if f is bijective. Then f^{-1} is also M-invariant. Furthermore, the composition of any M-invariant mappings is again M-invariant. In particular, the M-invariant mappings of an M-set X into itself form a submonoid of X^X, whose unit group is the group of M-invariant automorphisms of X. An M-invariant mapping $f\colon X \to Y$ maps the orbit Mx to the orbit $Mf(x)$, $x \in X$. If $M = G$ is a group, it thus induces a mapping $\bar{f}\colon X\backslash G \to Y\backslash G$ of the associated orbit spaces in such a way that the diagram in Fig. 2.11c with the canonical projections $\pi_\vartheta\colon X \to X\backslash G$, $\pi_\eta\colon Y \to Y\backslash G$ is commutative, and also a mapping $f|\mathrm{Fix}(G,X)\colon \mathrm{Fix}(G,X) \to \mathrm{Fix}(G,Y)$ of the fixed point sets. In general, the isotropy group G_x of $x \in X$ is obviously a subgroup of the isotropy group $G_{f(x)}$.

An M-invariant mapping is a special case of an φ-invariant mapping, where $\varphi\colon M \to N$ is a homomorphism of monoids M and N that operate on X and Y respectively. The mapping $f\colon X \to Y$ is called φ-**invariant**, if for all $a \in M$ and $x \in X$ the following holds: $f(ax) = \varphi(a)f(x)$, i.e., if $f \circ \vartheta_a = \eta_{\varphi(a)} \circ f$ is. This means that the diagram in Fig. 2.11b is commutative. Exactly then is $f\colon X \to Y$ φ-invariant, if f is an M-invariant mapping, where the M operation on Y is induced by the N-operation on Y via φ. The M-invariant mappings are identical to the id_M-invariant mappings.

Example 2.4.14 (**Classification of G-Spaces**) Let G be a group. A G-invariant isomorphism $f\colon X \xrightarrow{\sim} Y$ induces for each $x \in X$ a G-invariant isomorphism $f|Gx\colon Gx \xrightarrow{\sim} Gf(x)$. Consequently, the isotropy classes of the orbits Gx and $Gf(x)$ coincide. If $H \subseteq G$ is an element of the isotropy class of Gx, e.g., $H = G_y$, $y \in Gx$, then $G/H \xrightarrow{\sim} Gy = Gx$, $gH \mapsto gy$, is an isomorphism, where G/H is the transitive G-space already introduced above (with the operation $g(g'H) = (gg')H$, $g, g' \in G$). The G-spaces G/H, $H \subseteq G$ subgroup, thus form a complete system of the isomorphism classes of the transitive G-spaces, where G/H and G/H' are exactly G-isomorphic when H and H' are conjugate subgroups of G. If α is a cardinal number, let $\alpha(G/H)$ be the α-fold sum of G-spaces isomorphic to G/H. It follows: ◊

Theorem 2.4.15 *Let H_i, $i \in I$, be a representative system for the conjugation classes $[H]$ of the subgroups H of G, and let X be a G-space. Then X*

G is isomorphic to the sum $\coprod_{i \in I} \alpha_i(G/H_i)$, where α_i is the cardinal number of the G/H_i isomorphic orbits of $X\backslash G$ is, $i \in I$. It is $|X\backslash G| = \sum_{i \in I} \alpha_i$ and $|X| = \sum_{i \in I} \alpha_i \cdot [G : H_i]$.

The function $[H_i] \mapsto \alpha_i$, $i \in I$, is called the **Burnside function** of X. It characterizes X up to G-isomorphism. In particular, it holds: *If G is a finite group and $H_1, \ldots, H_r$ a representative system for the conjugacy classes of the subgroups of G, then the finite G-spaces up to G-isomorphism are characterized by the r-tuples $(n_1, \ldots, n_r) \in \mathbb{N}^r$.* The tuple $(n_1, \ldots, n_r)$ corresponds to the finite G-space $n_1(G/H_1) \sqcup \cdots \sqcup n_r(G/H_r)$ with $n_1[G : H_1] + \cdots + n_r[G : H_r]$ elements and $n_1 + \cdots + n_r$ orbits. The isomorphism classes of finite G-spaces thus form a monoid isomorphic to $\mathbb{N}^r = (\mathbb{N}^r, +)$, when the operation on the isomorphism classes is given by the disjoint union. The Grothendieck group $\mathrm{B}(G)$ of this monoid is therefore a free abelian group of rank r. The product of G-spaces provides an associative multiplication on the set of isomorphism classes of G-spaces. This induces a multiplication on $\mathrm{B}(G)$, with which $\mathrm{B}(G)$ becomes a ring, the so-called **Burnside ring of** G, cf. the construction of the Grothendieck ring in Example 2.6.10.

Example 2.4.16 (Mean Formation – Exact Sequences) The finite group G of order n operates on the (additively written) abelian group H as a group of automorphisms. For each $x \in H$ the sum $\sigma(x) = \sigma_G(x) := \sum_{g \in G} gx$ is a fixed point of the operation due to $h(\sigma(x)) = \sum_{g \in G}(hg)x = \sum_{g \in G} gx = \sigma(x)$ for each $h \in G$, since with g also hg completely runs through G.[19] Now let the multiplication by $x \mapsto nx$ with n on H be bijective. We write the inverse mapping as multiplication with $1/n \colon H \to H$, $x \mapsto x/n$. Both mappings are then G-invariant. The element

$$\mu(x) = \mu(G, H)(x) := \frac{1}{n}\sigma(x) = \frac{1}{n}\sum_{g \in G} gx$$

is called the **mean of** $x \in H$ and is like $\sigma(x)$ a fixed point of the given operation of G. It holds: ◊

Lemma 2.4.17 *The group homomorphism $\mu = \mu(G, H) \colon H \to H$ is a projection from H onto the subgroup $\mathrm{Fix}(G, H)$, i.e. it is $\mu = \mu^2$ and $\mathrm{Im}\,\mu = \mathrm{Fix}(G, H)$. Furthermore, $\ker \mu = \mathrm{Im}(\mathrm{id}_H - \mu)$ is a complement of $\mathrm{Fix}(G, H)$ in H.*

Proof The inclusion $\mu(H) \subseteq \mathrm{Fix}(G, H)$ we have already mentioned. For $x \in \mathrm{Fix}(G, H)$ applies $\mu(x) = \frac{1}{n}\sum_{g \in G} gx = \frac{1}{n}nx = x$. This proves $\mathrm{Fix}(G, H) \subseteq \mu(H)$ and $\mu = \mu^2$. □

---

[19] $\sigma(x)$ is also called the **trace** of x. In multiplicative notation, it is referred to as the **norm**.

2.4 Operation of Monoids and Groups

Lemma 2.4.17 often provides the most convenient method for determining the group of fixed points. It is applicable, for example, whenever H is the additive group of a vector space over a field K with $n \cdot 1_K \neq 0$ (or more generally the additive group of a module over a ring A with $n \cdot 1_A \in A^\times$), see Sect. 2.8.

In the following sentence, we use the term of the **exact sequence** of abelian groups. Let H', H, H'' be abelian groups and $f' \colon H' \to H$ and $f \colon H \to H''$ are group homomorphisms. Then the sequence

$$H' \xrightarrow{f'} H \xrightarrow{f} H'', \quad \text{short} \quad H' \to H \to H'',$$

is called a **complex** (or a **null sequence**), if $f \circ f' = 0$ is, i.e. if $\operatorname{Im} f' \subseteq \ker f$ applies. The quotient group

$$\mathrm{H} = \mathrm{H}\left(H' \to H \to H''\right) := \ker f / \operatorname{Im} f'$$

is then called the **homology (group)** (or – depending on the context – also the **cohomology (group)**) of the given complex. This homology disappears exactly when $\operatorname{Im} f' = \ker f$ is. In this case, the complex (or the sequence) is called **exact**. A homomorphism $H \to H''$ is exactly injective (or surjective), if the sequence $0 \to H \to H''$ (or the sequence $H \to H'' \to 0$) is exact. A longer sequence $\cdots \to H_{i+1} \to H_i \to H_{i-1} \to \cdots$ $(i \in \mathbb{Z})$ of abelian groups and group homomorphisms is called a **complex**(or an **exact sequence**), if all occurring triple sequences $H_{i+1} \to H_i \to H_{i-1}$ are complexes (or exact). For example, a so-called **short triple sequence**

$$0 \to F \xrightarrow{f} G \xrightarrow{g} H \to 0$$

of abelian groups F,G,H is exact only if f is injective and g is surjective, and $\ker g = \operatorname{Im} f$ holds. According to the isomorphism Theorem 2.3.8 for groups, g then induces an isomorphism $\bar{g} \colon G/\operatorname{Im} f \xrightarrow{\sim} H$. Occasionally, a sequence $F \xrightarrow{f} G \xrightarrow{g} H$ of non-abelian groups and group homomorphisms is also called exact if $\ker g = \operatorname{Im} f$ is. $\operatorname{Im} f$ is then necessarily a normal subgroup of G. This also explains short exact sequences for non-abelian groups. A short exact sequence $1 \to F \xrightarrow{f} G \xrightarrow{g} H \to 1$ of arbitrary groups **splits strongly** (or **weakly**) if $\operatorname{Im} f (\cong F)$ is a strong (or a weak) factor of G. In these cases, g induces an isomorphism from any complement of $\operatorname{Im} f$ to H.

An important consequence of the description of the fixed point groups using the mean formation in Lemma 2.4.17 is now:

Theorem 2.4.18 *Let G be a finite group of order n and H', H, H'' abelian groups, on which G each operates as a group of automorphisms. Furthermore, let*

$$H' \xrightarrow{f'} H \xrightarrow{f} H''$$

be an exact sequence with G-invariant group homomorphisms. If the multiplication by non H and H' is bijective,[20] *then the induced sequence*

---

[20] It suffices to assume that it is surjective on H' and injective on $\operatorname{Im} f' = \ker f$.

$$\text{Fix}(G, H') \longrightarrow \text{Fix}(G, H) \longrightarrow \text{Fix}(G, H'')$$

of fixed point groups is exact.

Proof Trivially, with $H' \to H \to H''$ the induced sequence of fixed point groups is also a complex. We still have to find for each $x \in \text{Fix}(G, H)$ with $f(x) = 0$ a $x' \in \text{Fix}(G, H')$ with $f'(x') = x$. μ, μ' denote the mean formations in H or H'. Now let $\widetilde{x} \in H'$ be with $f'(\widetilde{x}) = x$. Then $x' := \mu'(\widetilde{x}) \in \text{Fix}(G, H')$ and

$$f'(x') = f'(\mu'(\widetilde{x})) = \mu(f'(\widetilde{x})) = \mu(x) = x. \qquad \square$$

It should be noted that Theorem 2.4.18 does not generally hold for arbitrary abelian groups. For example, let $G := \mathbb{Z}^\times = \{1, -1\}$ always operate in a natural way, i.e., -1 by forming negatives. Then the canonical projection from $\mathbb{Z}$ to $\mathbb{Z}/\mathbb{Z}2$ is surjective, but the induced homomorphism $0 \to \mathbb{Z}/\mathbb{Z}2$ of the fixed point groups is not.

Exercises

Exercise 2.4.1 A given operation of a group G on a set X induces new operations in various ways. In addition to those already mentioned, we give some further examples. $\vartheta : G \to \mathfrak{S}(X)$ is the associated action. The set Y is arbitrary.

a) If $\varphi : G \to G''$ is a surjective group homomorphism, whose kernel lies in the kernel of the operation of G on X, then G'' operates on X by means of $g''x := gx$, where $g \in \varphi^{-1}(g'')$ is arbitrary. The associated action $\overline{\vartheta} : G'' \to \mathfrak{S}(X)$ is the homomorphism induced by $\vartheta : G \to \mathfrak{S}(X)$.
b) A mapping $f : X \to Y$ is called **compatible** with the operation of G on X, if for all $x, x' \in X$ from $f(x) = f(x')$ always $f(gx) = f(gx')$ for all $g \in G$ follows, if therefore every element $g \in G$ permutes the fibers of f. If f is additionally surjective, then the operation of G on X induces an operation of G on Y by means of $gy := f(gx)$, where $x \in f^{-1}(y)$ is arbitrary. This is then the only operation of G on Y, with respect to which f becomes a G mapping.
c) G operates on the mapping sets X^Y or Y^X by means of $\widetilde{gf} := \vartheta_g \circ \widetilde{f}$ or $gf := f \circ \vartheta_{g^{-1}} = f \circ \vartheta_g^{-1}$, $\widetilde{f} \in X^Y$, $f \in Y^X$, $g \in G$.[21] More generally: If the group H operates on Y by means of the action $\eta : H \to \mathfrak{S}(Y)$, then the group $H \times G$ operates on Y^X by means of $(h, g)f := \eta_h \circ f \circ g^{-1}$, $(h, g) \in H \times G$, $f \in Y^X$. Thus, if G operates on both X and Y, then $G \times G$ and thus also G operate on Y^X,

---

[21] Note the formation of inverses in the operation on Y^X. Without this, one obtains a right operation.

2.4 Operation of Monoids and Groups

where G is identified with the diagonal $\Delta_G \subseteq G \times G$. Then $\mathrm{Fix}(G, Y^X)$ is the set of G-invariant mappings $X \to Y$.

Exercise 2.4.2 An *abelian* group operates simply transitively if and only if it operates transitively and faithfully.

Exercise 2.4.3 Let p be a prime number. Every group of order p^2 is abelian, either cyclic or isomorphic to the product of two cyclic groups of order p. (Use 2.4.6.)

Exercise 2.4.4 Let p be a prime number. Every group of order $2p$ is either cyclic or isomorphic to the dihedral group D_p. ($p = 2$ is a special case.)

Exercise 2.4.5 Let p be a prime number and G a non-abelian group of order p^3. The commutator group and the center of G coincide. (Use exercise 2.3.1.) The class number of G is $p^2 + p - 1$.

Note *There are, up to isomorphism, two non-abelian and three abelian groups of order* p^3. One of the non-abelian groups is the holomorph $\mathrm{Hol}_H \mathbf{Z}_{p^2} = \mathbf{Z}_{p^2} \rtimes H$, where H is the (only) subgroup of order p of the (cyclic) group $\mathrm{Aut}\, \mathbf{Z}_{p^2} \cong (\mathbb{Z}/\mathbb{Z}p^2, \cdot)^\times$. (Note $|\mathrm{Aut}\, \mathbf{Z}_{p^2}| = \varphi(p^2) = p(p-1)$.) The other is for $p > 2$ a holomorph $\mathrm{Hol}_F(\mathbf{Z}_p \times \mathbf{Z}_p) = (\mathbf{Z}_p \times \mathbf{Z}_p) \rtimes F$, where F is one of the subgroups of order p of $\mathrm{Aut}(\mathbf{Z}_p \times \mathbf{Z}_p)$, and for $p = 2$ the so-called quaternion group Q_4 of order 8, see Ex. 2.4.7. It is $|\mathrm{Aut}(\mathbf{Z}_p \times \mathbf{Z}_p)| = (p+1)p(p-1)^2$. The subgroups F of order p in $\mathrm{Aut}(\mathbf{Z}_p \times \mathbf{Z}_p)$ are conjugate according to Theorem 2.7.11 (2). $\mathrm{Hol}_F(\mathbf{Z}_p \times \mathbf{Z}_p)$ has (for $p > 2$) the exponent p. (For $p = 2$ it is the dihedral group of order 8.) The abelian groups of order p^3 are $\mathbf{Z}_p \times \mathbf{Z}_p \times \mathbf{Z}_p$, $\mathbf{Z}_p \times \mathbf{Z}_{p^2}$ and $\mathbf{Z}_{p^3}$, see Ex. 2.2.25.

Exercise 2.4.6 The monoid H operates on the monoid N by means of the action homomorphism $\vartheta : H \to \mathrm{End}\, N$. Then the mapping $N \rtimes_\vartheta H \to N^N$, $(n, h) = nh \mapsto L_n \circ \vartheta_h$, is a monoid homomorphism of the semidirect product $N \rtimes_\vartheta H$ into the mapping monoid N^N. (See Example 2.4.12. Note $\varphi \circ L_n = L_{\varphi(n)} \circ \varphi$ for $\varphi \in \mathrm{End}\, N$ and $n \in N$.) In particular, $(n, \sigma) \mapsto L_n \circ \sigma, n \in N$, $\sigma \in \mathrm{End}\, N$, is a homomorphism from $\mathrm{Hol}\, N = N \rtimes \mathrm{End}\, N$ into N^N. The latter is injective when N is regular. Provide an example that this homomorphism is not always injective.

Exercise 2.4.7 Determine the subgroups of order 8 in the group $\mathrm{Hol}_{\mathrm{Aut}\, \mathbf{Z}_8}\, \mathbf{Z}_8 = \mathbf{Z}_8 \rtimes \mathrm{Aut}\, \mathbf{Z}_8 = (\mathbb{Z}/\mathbb{Z}8) \rtimes (\mathbb{Z}/\mathbb{Z}8)^\times$. They represent all groups of order 8 up to isomorphism. In particular, the subgroup generated by $(2, 1)$ and $(1, 3)$ is the so-called **Quaternion group** Q_4. It is characterized among the groups of order 8 by the fact that it is not commutative and has exactly one element of order 2, and it appears only once. (Catalog the desired

Fig. 2.12 Quaternion group Q_4

	1	-1	i	j	k	-i	-j	-k
1	1	-1	i	j	k	-i	-j	-k
-1	-1	1	-i	-j	-k	i	j	k
i	i	-i	-1	k	-j	1	-k	j
j	j	-j	-k	-1	i	k	1	-i
k	k	-k	j	-i	-1	-j	i	1
-i	-i	i	1	-k	j	-1	k	-j
-j	-j	j	k	1	-i	-k	-1	i
-k	-k	k	-j	i	1	j	-i	-1

subgroups according to their image in $\mathrm{Aut}\, \mathbf{Z}_8 (\cong \mathbf{D}_2)$ with respect to the canonical homomorphism $\mathbf{Z}_8 \rtimes \mathrm{Aut}\, \mathbf{Z}_8 \to \mathrm{Aut}\, \mathbf{Z}_8$. – If we denote the element of order 2 in Q_4 as $-1 (\in Z(\mathbf{Q}_4))$ following W. R. Hamilton (1805–1865), and choose from the 6 elements of order 4 the elements i, j, k such that $j \notin H(i)$ and $k = ij$ are obtained, then for Q_4 the combination table from Fig. 2.12 is obtained. (In it, $-a := (-1)a = a(-1)$ is for $a \in \{i, j, k\}$.) From this table, Hamilton etched the (now weathered) equations $i^2 = j^2 = k^2 = ijk = -1$ into the Broome Bridge of Dublin on October 16, 1843.[22] Consider that the complete table can be deduced from this, see also exercise 2.4.18. – The remaining subgroups of order 8, which are isomorphic to $\mathbf{Z}_8$, $\mathbf{Z}_4 \times \mathbf{Z}_2$, $(\mathbf{Z}_2)^3$ or D_4 occur multiple times. In total, there are 15 subgroups of order 8, namely in addition to the only Q_4 in turn 3, 3, 3 and 5 subgroups each. Note: $\{(0, 1), (4, 1)\} = 4\mathbf{Z}_8$ is the center of $\mathrm{Hol}_{\mathrm{Aut}\, \mathbf{Z}_8} \mathbf{Z}_8$.)

Exercise 2.4.8 For a subgroup H, the set G / H of left cosets of the group G with respect to H is a path with respect to the canonical operation of G on $\mathfrak{P}(G)$ by left translation. The kernel of the induced operation of G on G / H is the normal divisor $N := \bigcap_{g \in G} gHg^{-1} \subseteq G$. In particular, the action of G induces an injective group homomorphism $G/N \to \mathfrak{S}(G/H)$. It follows that: (1) Every subgroup of finite index n in G contains a normal divisor, whose index is finite and divides $n!$. (2) If G is simple and $H \subset G$ is a proper subgroup of G, then G is isomorphic to a subgroup of $\mathfrak{S}(G/H)$ and in particular G is finite and $\mathrm{Ord}\, G$ is a divisor of $n!$, if H has the finite index $n > 1$ in G. (3) If G is finite and H is a subgroup of prime index p, where p is the smallest prime divisor of $\mathrm{Ord}\, G$, then H is a normal divisor in G. In particular, every subgroup of index p is a normal divisor in every finite p-group.

Exercise 2.4.9 If G is a finite group of odd order and $a \in G$, $a \neq e_G$, then a and a^{-1} are in different conjugation classes.

Exercise 2.4.10 Let H be an abelian group. Determine the conjugacy classes of the dihedral group $\mathbf{D}(H)$, and in particular those of $\mathbf{D}_n = \mathbf{D}(\mathbf{Z}_n)$, $n \in \mathbb{N}$.

---

[22] A commemorative plaque reminds of this.

2.4 Operation of Monoids and Groups

Exercise 2.4.11 The finite group G operates on the finite set X. Then the

$$n \in \mathbb{N}$$

applies, i.e., the number of orbits is the arithmetic mean of the numbers of elements of the individual fixed point sets of the group elements. (Count the number of elements of $\{(g,x) \in G \times X \mid gx = x\} \subseteq G \times X$ in two different ways, cf. the remark in exercise 1.6.8. – Note that conjugate elements in G have the same number of fixed points. So it is $n \in \mathbb{N}$, where $g_1, \ldots, g_r$ is a representative system for the different conjugacy classes of G.)

Exercise 2.4.12 G is a group and $H \subset G$ is a subgroup $\neq G$ of finite index. Then $\bigcup_{g \in G} gHg^{-1} \neq G$ is true. (Cf. Example 2.4.5. One can assume that G is finite (consequence (1) in exercise 2.4.18).)

Exercise 2.4.13 Let N be a normal subgroup of the group G, and G (and thus also N) operate on the set X. Furthermore, let $x \in X$.

a) It is $N_x = G_x \cap N$. G also operates on the set $X \backslash N$ of the N orbits of X, and it is $G_{Nx} = NG_x \subseteq G$ as well as $|Gx| = |Nx| \cdot [G : NG_x]$. (It is $[G : NG_x] = [(G/N) : (NG_x/N)]$ and $NG_x/N \cong G_x/N_x$.)

b) The fixed point set $\operatorname{Fix}(N, X) = \bigcap_{g \in N} \operatorname{Fix}(g, X)$ is invariant under G, and N belongs to the kernel of the operation induced by G on $\operatorname{Fix}(N, X)$. The quotient group G/N thus operates in a natural way on $\operatorname{Fix}(N, X)$.

Exercise 2.4.14 Let N be a group. Every semidirect product $N \rtimes_\vartheta H$ with a group H is equal to the direct product $N \times H$, if $|N| \leq 2$ is. (It is to be shown: *Every group N with more than two elements has an automorphism different from the identity.* — Think of conjugations in the non-abelian case, inverses in the abelian case (which yield the dihedral groups) and for elementary abelian 2-groups think of the structure theorem from exercise 2.3.6a). — The result of this exercise can also be formulated as follows: The group N has at most 2 elements if and only if it has the following property: If N is a normal subgroup of a group G and if N is a weak direct factor in G, then N is even a strong direct factor in G. Or expressed with exact sequences: Let N be a group. If $|N| \leq 2$ and only then, is every weakly splitting exact sequence $1 \to N \to G \to H \to 1$ of groups even strongly splitting.)

Exercise 2.4.15 Let H be a subgroup of the group G. The G-space G/H is, up to G-isomorphism, a model for every G-space on which G operates transitively with isotropy class $[H] = \{gHg^{-1} \mid g \in G\}$. For every such space X, therefore, $\operatorname{Aut}_G X \cong \operatorname{Aut}_G(G/H)$.

a) Let $x \in G$. There exists a G-isomorphism $G/H \to G/H$ with $H \mapsto xH$, if $x \in N_G(H)$ is, i.e. if $H = xHx^{-1}$ is. If $x \in N_G(H)$, we denote this G-isomorphism $gH \mapsto gxH$, $g \in G$, with f_x.
b) The mapping $N_G(H) \to \text{Aut}_G(G/H)$, $x \mapsto f_x$, is a surjective antihomomorphism of groups with H as the kernel. In particular, $(N_G(H)/H)^{op} \xrightarrow{\sim} \text{Aut}_G(G/H)$, $\bar{x} \mapsto (f_x : gH \mapsto gxH)$, is a group isomorphism. If N is a normal subgroup of G, then $(G/N)^{op} \xrightarrow{\sim} \text{Aut}_G(G/N)$, $\bar{x} \mapsto f_x$, is an isomorphism. Specifically: If G acts simply transitively on X (so $N = \{e_G\}$), then $\text{Aut}_G X \cong G^{op}$. Specifically: The G automorphisms of $G = G/\{e_G\}$ are the *right* translations R_x, $x \in G$, of G.
c) Describe the group of G automorphisms of any G space X. (See Theorem 2.4.15.)

Exercise 2.4.16 An operation of the free monoid $\mathbb{N} = (\mathbb{N}, +)$ on the set X is given by a mapping $g \colon X \to X$. The associated action $\vartheta \colon \mathbb{N} \to X^X$ is then $n \mapsto g^n$, $n \in \mathbb{N}$. If g is specifically a permutation of X, this operation can be uniquely extended to an operation of the free group $\mathbb{Z} = (\mathbb{Z}, +)$ on the set X, where $n \in \mathbb{Z}$ operates again like g^n. A pair (X, g) with a set X and a mapping $g \colon X \to X$ is called a **discrete dynamic system**. A discrete dynamic system (Y, h) is exactly isomorphic to (X, g) when (Y, h) is (X, g) **conjugated**, i.e., if there is a bijective mapping $f \colon X \xrightarrow{\sim} Y$ with $h = fgf^{-1}$. By transitioning to a conjugated system, considerations for discrete dynamic systems can often be simplified.

Let (X, g) be again an arbitrary discrete dynamic system. For a point $x \in X$ yields the sequence $x_n = g^n(x)$, $n \in \mathbb{N}$ which can be recursively defined by

$$x_0 = x, \quad x_{n+1} = g(x_n), \ n \in \mathbb{N},$$

the path (the orbit) $\mathbb{N}x = \{x_n \mid n \in \mathbb{N}\}$ to the initial value $x_0 = x$.[23] If x is **periodic**, i.e., the sequence (x_n) is periodic with a periodicity type $(m, k) \in \mathbb{N} \times \mathbb{N}^*$, cf. exercise 1.7.38, then the $k + m$ points $x_0 \dots, x_{m+k-1} \in X$ of the path $\mathbb{N}x$ are pairwise different. $x_0, \dots, x_{m-1}$ is the **pre-period** and $x_m, \dots, x_{m+k-1}$ the **period** of x. Furthermore, $x_m = g(x_{m-1}) = x_{m+k}$. If x is **aperiodic**, i.e., the sequence (x_n) is of type $(\infty, 0)$, then all sequence members x_n, $n \in \mathbb{N}$, are pairwise different. If g is injective, then x is obviously aperiodic or **purely periodic**. (This is particularly true if g is a permutation.) The periodicity type of the sequence (x_n) is also the **type** of $x = x_0$ with respect to g. On the additive monoid $\mathbb{N} = (\mathbb{N}, +)$ the relation "$i \sim j \Leftrightarrow x_i = g^i(x) = g^j(x) = x_j$" is a compatible equivalence relation, and the cyclic quotient monoid $\mathbb{N}/\sim$ has the same type as x, see Example 2.3.32 and especially Fig. 2.9. The set of *purely* periodic points $x \in X$ with period length $k \in \mathbb{N}^*$ is denoted by

$$\text{Per}_k(g) = \text{Per}_k(g, X)$$

---

[23] If g is bijective, then $\mathbb{N}x$ is also called the **semi-path** and $\mathbb{Z}x = \{x_n = g^n(x_0) \mid n \in \mathbb{Z}\}$ the (full) **path** to x.

2.4 Operation of Monoids and Groups

For $x \in \text{Per}_k(g)$, $x_0, \ldots, x_{k-1}$ is therefore the orbit of x. It has the length k. If x is periodic of type (m,k), then the k points $x_m, \ldots, x_{m+k-1}$ of the period of x are elements of $\text{Per}_k(g)$. $\text{Per}_1(g,X)$ is the set $\text{Fix}(g,X)$ of fixed points of g in X. $\text{Per}_k(g)$ is a g-invariant subset of X and $g|\text{Per}_k(g)$ is a permutation of $\text{Per}_k(g)$, whose (pairwise disjoint) orbits all have the length k. In particular, $|\text{Per}_k(g)| \equiv 0 \bmod k$, if $\text{Per}_k(g)$ is finite.

a) Let x be periodic of type $(m,k) \in \mathbb{N} \times \mathbb{N}^*$ with respect to g. For $\ell \in \mathbb{N}$, $x_\ell = g^\ell(x)$ is periodic with respect to g of type $(\text{Max}\,(m-\ell,0),k)$. If $q \in \mathbb{N}^*$, then x is periodic with respect to g^q of type $(\lceil m/q \rceil, k/\text{ggT}(k,q))$.[24]

b) For $k \in \mathbb{N}^*$,
$$\text{Fix}(g^k) = \biguplus_{d|k} \text{Per}_d(g).$$
is valid.

c) Let $k \in \mathbb{N}^*$ and $\text{Fix}(g^k)$ be a finite set. Then
$$\left|\text{Fix}(g^k)\right| = \sum_{d|k} |\text{Per}_d(g)| \quad \text{and} \quad \sum_{d|k} \mu(d)\left|\text{Fix}(g^{k/d})\right| = |\text{Per}_k(g)| \equiv 0 \bmod k.$$
applies (Use the Möbius inversion formula from exercise 2.1.23b).

d) To provide an example for the formula derived in c), let $n, k \in \mathbb{N}^* - \{1\}$. As a discrete dynamic system, we choose $(\mathbf{Z}_{n^k-1}, \chi_n)$, where $\chi_n: \mathbf{Z}_{n^k-1} \to \mathbf{Z}_{n^k-1}$ is the multiplication $x \mapsto nx$, which is an automorphism of the (additive) cyclic group $\mathbf{Z}_{n^k-1}$. For a divisor e of k, $\text{Fix}(\chi_n^e = \chi_{n^e}, \mathbf{Z}_{n^k-1})$ is then the (cyclic) subgroup of order $n^e - 1$ of $\mathbf{Z}_{n^k-1}$, so $|\text{Fix}(\chi_n^e, \mathbf{Z}_{n^k-1})| = n^e - 1$. It follows
$$\sum_{H \subseteq \mathbb{P}_k} (-1)^{|H|} n^{k/p^H} = |\text{Per}_k(\chi_n, \mathbf{Z}_{n^k-1})| \equiv 0 \bmod k,$$
where $\mathbb{P}_k$ is the set of prime divisors of k and $p^H = \prod_{p \in H} p$. (Note $\sum_{d|k} \mu(d) = 0$ because $k > 1$.) The given congruence applies to all $n \in \mathbb{Z}$. (Simply replace n with a number $n + ak > 1$.) This is a generalization of Fermat's Little Theorem. For $k = p \in \mathbb{P}$ one obtains $n^p - n \equiv 0 \bmod p$ for all $n \in \mathbb{Z}$. The above formula also provides for $n > 0$ and $p \nmid n$ an interpretation of the so-called **Fermat exponent** $v_p(n^{p-1} - 1) = v_p(n^p - n)$.[25] This is the number of orbits of the restriction of χ_n to the set of purely-p-periodic (=p-periodic) points of $\chi_n: \mathbf{Z}_{n^p-1} \to \mathbf{Z}_{n^p-1}$. – Instead of the initially indicated dynamic systems, one can also consider the dynamic system $(\mathbb{T}, \chi_n)$ on the

---

[24] If $(x_i)_{i \in \mathbb{N}}$ is any periodic sequence of type (m,k), then the type (m',k') of the subsequence $(x_{qi})_{i \in \mathbb{N}}$, $q \in \mathbb{N}^*$, is not necessarily $(\lceil m/q \rceil, k/\gcd(k,q))$. However, $m' \leq \lceil m/q \rceil$ and k' are divisors of $k/\gcd(k,q)$. Consider examples!

[25] The Fermat exponent is rarely >1. For $n = 2$ for example, there are only two prime numbers $p \leq 6{,}7 \cdot 10^{15}$ with $v_p(2^{p-1} - 1) \geq 2$, namely $p = 1093$ (Meißner 1913) and $p = 3511$ (Beeger 1922). Such prime numbers are called **Wieferich primes** (after A. Wieferich (1884–1954)).

additive torus group $\mathbb{T} = \mathbb{R}/\mathbb{Z}$ for $n > 1$, where χ_n again is the n-fold formation $x \mapsto nx$ (or equivalently the power mapping $z \mapsto z^n$ on the circle group $U = \{z \in \mathbb{C} \mid |z| = 1\}$). For $k \in \mathbb{N}^*$ is $\chi_n^k = \chi_{n^k}$ and

$$\text{Fix}(\chi_{n^k}, \mathbb{T}) = T_{n^k-1}(\mathbb{T}) = \frac{1}{n^k - 1}\mathbb{Z}\Big/\mathbb{Z}, \quad \text{e.g.} \quad |\text{Fix}(\chi_{n^k}, \mathbb{T})| = n^k - 1,$$

from which the above formula as a generalization of the Little Fermat's Theorem results again. Also determine the periodic, purely periodic or aperiodic points $x \in \mathbb{T}$ with the help of the representative $\alpha \in [0,1]$ of x and observe the connection with the periodicity behavior of the n-al development of α, see the Examples 3.3.10 and 3.6.18. Exactly then x is aperiodic, when α is irrational.

Exercise 2.4.17 For $n \in \mathbb{N}$ are

$$\langle x, z \mid x^n = z^2 = e, zxz^{-1} = x^{-1} \rangle \quad \text{and} \quad \langle y, z \mid y^2 = z^2 = (yz)^n = e \rangle$$

representations of the dieder group D_n with generators and relations, see example 2.3.34.

Exercise 2.4.18 $\langle \varepsilon, i, j, k \mid \varepsilon^2 = e, i^2 = j^2 = k^2 = ijk = \varepsilon \rangle$ is a representation of the quaternion group Q_4 of order 8. (Hamilton – See exercise 2.4.7.)

Exercise 2.4.19 Let $p \in \mathbb{P}$ and N be a normal subgroup in the finite group G with the canonical projection $\pi: G \to G/N$. If then $S_p \subseteq G$ is a p-Sylow group of G, then $N \cap S_p$ is a p-Sylow group of N and $\pi(S_p) = (S_p N)/N \cong S_p/(N \cap S_p)$ is a p-Sylow group of G/N, (cf. example 2.4.9. — If $H \subseteq N$ is a p-group, then according to theorem 2.4.11(2) there is $x \in G$ with $xHx^{-1} \subseteq N \cap S_p$.)

2.5 Permutation Groups

In this section, we describe the permutation groups of finite sets in more detail and first introduce some general notations.

Let I be a set. We consider the permutation group $\mathfrak{S}(I)$ of I and each of its subgroups always with its natural operation on I. (So it is $\sigma i = \sigma(i)$ for all $\sigma \in \mathfrak{S}(I)$, $i \in I$.) The complement in I of the fixed point set $\text{Fix}\,\sigma = \text{Fix}(\sigma, I)$ of $\sigma \in \mathfrak{S}(I)$ is called the **range of action** of σ and is denoted with $W(\sigma) = W(\sigma, I)$. So it is

$$W(\sigma) := \{i \in I \mid \sigma(i) \neq i\} \subseteq I.$$

As $\text{Fix}\,\sigma$ is also $W(\sigma)$ invariant under σ. Permutations $\sigma, \tau \in \mathfrak{S}(I)$ with disjoint action ranges are obviously interchangeable, i.e. from $W(\sigma) \cap W(\tau) = \emptyset$ follows $\sigma\tau = \tau\sigma$. If I' is a subset of the set I, a canonical embedding of $\mathfrak{S}(I')$ into $\mathfrak{S}(I)$ is obtained by assigning to each $\sigma \in \mathfrak{S}(I')$ the permutation of I that operates on

2.5 Permutation Groups

Fig. 2.13 Cyclic permutation

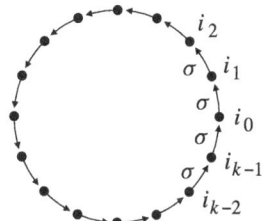

I' like σ and on the complement $I - I'$ like the identity, thus has the same action range as σ. In this way, we identify for $I' \subseteq I$ always $\mathfrak{S}(I')$ with a subgroup of $\mathfrak{S}(I)$ and in particular $\mathfrak{S}_m = \mathfrak{S}(\mathbb{N}^*_{\leq m}) = \mathfrak{S}(\{1,\ldots,m\})$ with a subgroup of $\mathfrak{S}_n$ for $m, n \in \mathbb{N}$, $m \leq n$. Thus, we have the chain of subgroups

$$id = \mathfrak{S}_0 = \mathfrak{S}_1 \subset \mathfrak{S}_2 \subset \cdots \subset \mathfrak{S}_n \subset \mathfrak{S}_{n+1} \subset \cdots \subset \mathfrak{S}(\mathbb{N}^*) \text{ with } \bigcup_{n \in \mathbb{N}} \mathfrak{S}_n = \mathfrak{S}_{\text{fin}}(\mathbb{N}^*),$$

where for a set I in general

$$\mathfrak{S}_{\text{fin}}(I) \subseteq \mathfrak{S}(I)$$

denotes the group of permutations of I with finite action range. $\mathfrak{S}_{\text{fin}}(I)$ is a normal subgroup of $\mathfrak{S}(I)$, because for any $\sigma, \tau \in \mathfrak{S}(I)$ is $\text{Fix}(\tau\sigma\tau^{-1}) = \tau(\text{Fix}(\sigma))$ and thus also $W(\tau\sigma\tau^{-1}) = \tau(W(\sigma))$. Two permutations $\sigma, \tau \in \mathfrak{S}(I)$ are congruent modulo $\mathfrak{S}_{\text{fin}}(I)$ if and only if σ and τ almost everywhere on I coincide. If $I = \biguplus_{j \in J} I_j$ is a decomposition of the set I, then the subgroups $\mathfrak{S}(I_j) \subseteq \mathfrak{S}(I)$, $j \in J$, commute elementwise. Consequently, then $(\sigma_j)_{j \in J} \mapsto \prod_{j \in J} \sigma_j$ is an embedding of the product group $\prod_{j \in J} \mathfrak{S}(I_j)$ into the group $\mathfrak{S}(I)$. Here, the product (even with infinite J) operates $\prod_{j \in J} \sigma_j \in \mathfrak{S}(I)$ on I_j like σ_j, $j \in J$. The product group $\prod_{j \in J} \mathfrak{S}(I_j)$ thus identifies itself with the subgroup of those $\sigma \in \mathfrak{S}(I)$, which leave all I_j, $j \in J$, invariant.

Let I be a finite set. Then $\mathfrak{S}(I)$ is a finite group with $|I|!$ elements, cf. Theorem 1.6.7. For a clear description of an element $\sigma \in \mathfrak{S}(I)$, we consider the orbits of the natural operation of the cyclic subgroup generated by σ, namely $H(\sigma) \subseteq \mathfrak{S}(I)$, on I, which are also called the **orbits** of σ. If $i_0 \in I$, then

$$H(\sigma)i_0 = \{i_0, i_1 := \sigma(i_0), i_2 := \sigma(i_1) = \sigma^2(i_0), \ldots, i_{k-1} := \sigma^{k-1}(i_0)\},$$

applies, where σ^k is the generating element of the isotropy group $H(\sigma)_{i_0} = \{\tau \in H(\sigma) \mid \tau(i_0) = i_0\} \subseteq H(\sigma)$ of i_0 with minimal $k \in \mathbb{N}^*$. The orbit thus has $k = [H(\sigma) : H(\sigma)_{i_0}]$ elements. k divides $\text{Ord}\,\sigma = |H(\sigma)|$ and i_0 is a (purely) periodic point with respect to σ with a period length of k, cf. also exercise 2.4.16, where more general mappings of a set into itself are treated. The orbit through i_0 is exactly the set of elements that are obtained from i_0 by repeatedly applying σ (or also σ^{-1}). So σ operates on this orbit by **cyclic permutation**, cf. Fig. 2.13 or also Fig. 2.5.

We denote the permutation of I, which cyclically swaps the elements $i_0, \ldots, i_{k-1}$ in this way and leaves the remaining elements of I pointwise fixed, with

$$\langle i_0, \ldots, i_{k-1} \rangle$$

and speak of a **cycle** of **length** $L(\langle i_0, \ldots, i_{k-1} \rangle) = k$. For each $j = 0, \ldots, k-1$,

$$\langle i_0, \ldots, i_{k-1} \rangle = \langle i_j, \ldots, i_{k-1}, i_0, \ldots, i_{j-1} \rangle.$$

σ agrees on the orbit $H(\sigma)i_0$ with the cycle $\langle i_0, \ldots, i_{k-1} \rangle$. If $k = 1$, then the cycle is the identity, i_0 is therefore a fixed point of σ. If $k > 1$, then the orbit $H(\sigma)i_0$ belongs to the range of action of σ. Since the orbits of σ form a decomposition of I, we obtain:

Theorem 2.5.1 (Canonical Cycle Representation) *Let I be a finite set. Every permutation σ of I has a representation $\sigma = \sigma_1 \cdots \sigma_r$ with cycles $\sigma_1, \ldots, \sigma_r \in \mathfrak{S}(I)$ of length ≥ 2, whose ranges of action are pairwise disjoint. This representation is unique up to the order of the factors $\sigma_1, \ldots, \sigma_r$.*

To see the uniqueness, note that the action ranges of the cycles $\sigma_1, \ldots, \sigma_r$ necessarily are the orbits of σ with more than one element. An analogous statement also applies to permutations σ of any sets I (possibly with infinitely many nontrivial cycles), provided all orbits of σ are finite, which is certainly the case with Ord $\sigma \in \mathbb{N}^*$.

Example 2.5.2 The permutation

$$\sigma := \begin{pmatrix} 1 & 2 & 3 & 4 & 5 & 6 & 7 & 8 & 9 & 10 & 11 & 12 & 13 & 14 & 15 & 16 & 17 & 18 & 19 & 20 \\ 17 & 3 & 11 & 5 & 4 & 2 & 19 & 12 & 9 & 15 & 7 & 1 & 10 & 18 & 20 & 14 & 8 & 16 & 6 & 13 \end{pmatrix} \in \mathfrak{S}_{20}.$$

has the cycle representation

$$\sigma = \langle 1, 17, 8, 12 \rangle \langle 2, 3, 11, 7, 19, 6 \rangle \langle 4, 5 \rangle \langle 10, 15, 20, 13 \rangle \langle 14, 18, 16 \rangle.$$

It has exactly 5 orbits with more than one point, and its action range is $W(\sigma) = \{1, \ldots, 20\} - \{9\}$. (So it has a total of 6 orbits.)

From the cycle representation $\sigma = \sigma_1 \cdots \sigma_r$ according to 2.5.1, the powers $\sigma^m = \sigma_1^m \cdots \sigma_r^m$, $m \in \mathbb{Z}$, can be easily calculated, since the powers of cycles can be given immediately. *In particular, the formula for the order of σ is*

$$\text{Ord } \sigma = \text{lcm}(\text{Ord } \sigma_1, \ldots, \text{Ord } \sigma_r) = \text{lcm}(L(\sigma_1), \ldots, L(\sigma_r)),$$

since the order of a cycle is equal to its length, see also exercise 2.2.11a). The above permutation $\sigma \in \mathfrak{S}_{20}$ thus has the order Ord $\sigma = \text{gcd}(4, 6, 2, 4, 3) = 12$. The inverse σ^{-1} can also be given directly with the cycle representation $\sigma = \sigma_1 \cdots \sigma_r$. It is indeed $\sigma^{-1} = \sigma_1^{-1} \cdots \sigma_r^{-1}$ (since the σ_i commute pairwise), and for a cycle,

$$\langle i_0, \ldots, i_{k-1} \rangle^{-1} = \langle i_{k-1}, i_{k-2}, \ldots, i_0 \rangle = \langle i_0, i_{k-1}, \ldots, i_1 \rangle.$$

In the above example, the result is

$$\sigma^{-1} = \langle 1, 12, 8, 17\rangle\langle 2, 6, 19, 7, 11, 3\rangle\langle 4, 5\rangle\langle 10, 13, 20, 15\rangle\langle 14, 16, 18\rangle. \quad \diamond$$

The cycles of length 2 are the **transpositions**. The transposition $\langle i,j\rangle \in \mathfrak{S}(I)$, $i \neq j$, swaps the elements i and j and leaves the remaining elements from I fixed. It is therefore involutory with the range of action $\{i,j\}$. A cycle $\langle i_0, i_1, \ldots, i_{k-1}\rangle$ of length k has the following representation as a product of $k - 1$ transpositions:

$$\langle i_0, i_1, \ldots, i_{k-1}\rangle = \langle i_0, i_1\rangle\langle i_1, i_2\rangle \cdots \langle i_{k-2}, i_{k-1}\rangle.$$

The reader should check this carefully. We remind you that permutations, like mappings, are read from right to left. With Theorem 2.5.1 it follows:

Lemma 2.5.3 *Let I be a finite set. Every permutation σ of I has a representation as a product of transpositions. More precisely: If s is the number of orbits of σ (including the single-element ones), then σ has a representation as a product of $|I| - s$ transpositions.*

The existence of a representation of $\sigma \in \mathfrak{S}(I)$ as a product of transpositions can also be easily shown directly by induction over $|W(\sigma)|$: Let $\sigma(i_0) = j_0 \neq i_0$. Then $\sigma' := \langle i_0, j_0\rangle\sigma$ has the fixed point i_0, and because of $|W(\sigma')| < |W(\sigma)|$ there is a representation $\sigma' = \langle i_1, j_1\rangle \cdots \langle i_s, j_s\rangle$. It follows $\sigma = \langle i_0, j_0\rangle\sigma' = \langle i_0, j_0\rangle\langle i_1, j_1\rangle \cdots \langle i_s, j_s\rangle$.

The representation of a permutation as a product of transpositions is (in contrast to the canonical cycle representation) naturally not unique. For example, any such representation can be extended by $\mathrm{id} = \tau\tau$ with any transposition τ. However, as we will show shortly, the parity of the number of transpositions required to represent σ is uniquely determined. The addition in Lemma 2.5.3 thus motivates the following definition:

Definition 2.5.4 Let σ be a permutation of the n-element set I. The number of orbits of σ is s. Then

$$\mathrm{Sign}\,\sigma := (-1)^{n-s}$$

is called the **signum** (or the **sign**) of σ. The permutation σ is called **even**, if $\mathrm{Sign}\,\sigma = 1$, i.e. if $n - s$ is even, otherwise it is called **odd**. (If $\sigma : I \to I$ is not bijective, then set $\mathrm{Sign}\,\sigma = 0$.)

If $B_1, \ldots, B_s$ are the orbits of σ, then

$$n - s = \sum_{k=1}^{s} |B_k| - s = \sum_{k=1}^{s} (|B_k| - 1)$$

and thus

$$\mathrm{Sign}\,\sigma = \prod_{k=1}^{s}(-1)^{|B_k|-1} = (-1)^s,$$

where g is the number of even (numbered) orbits of σ. The permutation σ *is therefore exactly even when the number of even orbits of σ is even.*

Example 2.5.5
(1) The identity is even. A transposition is odd; generally, a cycle of length k has the sign $(-1)^{k-1}$.
(2) The permutation $\sigma \in \mathfrak{S}_{20}$ from Example 2.5.2 is even, as it has exactly 4 even-numbered orbits.
(3) It is $\text{Sign}\,\sigma = \text{Sign}\,(\sigma|W(\sigma))$. With this observation, the signum function for the group $\mathfrak{S}_{\text{fin}}(I)$ of permutations with finite action range on an arbitrary set I can be defined. One sets $\text{Sign}\,\sigma := \text{Sign}\,(\sigma|W(\sigma))$ for $\sigma \in \mathfrak{S}_{\text{fin}}(I)$.
(4) Often, the multiplicative group $\mathbb{Z}^\times$ is replaced by the additive group $\mathbb{Z}_2 = \mathbb{Z}/\mathbb{Z}2$ and then one speaks of the **parity** $\text{Par}\,\sigma$ of the permutation σ. So it is $\text{Par}\,\sigma = 0 \in \mathbb{Z}_2$, if σ is even, and $\text{Par}\,\sigma = 1 \in \mathbb{Z}_2$, if σ is odd as well as $\text{Sign}\,\sigma = (-1)^{\text{Par}\,\sigma}$. ◊

The following theorem is fundamental:

Theorem 2.5.6 *The permutation $\sigma \in \mathfrak{S}(I)$, I finite, is represented as a product $\sigma = \tau_1 \cdots \tau_k$ of k transpositions $\tau_1, \ldots, \tau_k$. Then it is $\text{Sign}\,\sigma = (-1)^k$.*

Proof ((by induction over k)) It suffices to show that σ and $\tau\sigma$ have different signs for any transposition τ. To do this, it suffices to show that the numbers of orbits of $\tau\sigma$ and σ differ by 1. Let $\tau = \langle i,j \rangle$. The orbits of σ that contain neither i nor j are also orbits of $\tau\sigma$. We therefore only consider the orbits of σ that contain i or j.

1. Case: If i and j are in the same orbit of σ, then σ and $\tau\sigma$ have the canonical cycle representations

$$\sigma = \langle i_0, \ldots, i_r, \ldots, i_{s-1} \rangle \cdots \quad \text{resp.} \quad \tau\sigma = \langle i_0, \ldots, i_{r-1} \rangle \langle i_r, \ldots, i_{s-1} \rangle \cdots,$$

with $i_0 = i$ and $i_r = j$, and the number of orbits of $\tau\sigma$ is 1 greater than that of σ.
2. 2. Case: If i and j are in different orbits, then σ and $\tau\sigma$ have the canonical cycle representations

$$\sigma = \langle i_0, \ldots, i_{r-1} \rangle \langle j_0, \ldots, j_{s-1} \rangle \cdots \quad \text{resp.} \quad \tau\sigma = \langle i_0, \ldots, i_{r-1}, j_0, \ldots, j_{s-1} \rangle \cdots$$

with $i_0 = i$ and $j_0 = j$. The number of orbits of $\tau\sigma$ is 1 less than that of σ. □

As already mentioned, according to Theorem 2.5.6, the number of factors in the representation of an even (or odd) permutation as a product of transpositions is always even (or always odd). Furthermore, the following fundamental statement arises:

Theorem 2.5.7 *Let I be a finite set. The mapping*

$$\text{Sign} : \mathfrak{S}(I) \longrightarrow \{1, -1\}$$

is a homomorphism of groups, i.e. for $\sigma, \tau \in \mathfrak{S}(I)$ is

$$\text{Sign}\,\sigma\tau = (\text{Sign}\,\sigma)(\text{Sign}\,\tau).$$

2.5 Permutation Groups

Proof We write $\sigma = \sigma_1 \cdots \sigma_s$ and $\tau = \tau_1 \cdots \tau_t$ as a product of transpositions $\sigma_1, \ldots, \sigma_s$ or $\tau_1, \ldots, \tau_t$ and obtain the representation $\sigma\tau = \sigma_1 \cdots \sigma_s \tau_1 \cdots \tau_t$. According to Theorem 2.5.6, $\operatorname{Sign}\sigma\tau = (-1)^{s+t} = (-1)^s(-1)^t = (\operatorname{Sign}\sigma)(\operatorname{Sign}\tau)$ is true. □

Theorem 2.5.7 also obviously applies to the group $\mathfrak{S}_{\mathrm{fin}}(I)$ of any set I. – A particularly important subgroup of a permutation group is the kernel of the sign function, i.e., the group of even permutations.

Definition 2.5.8 Let I be a finite set. Then the group of even permutations of I is called the **alternating group** of I. It is denoted by

$$\mathfrak{A}(I)$$

$\mathfrak{A}_n \subseteq \mathfrak{S}_n$ denotes the alternating group of the set $\{1, \ldots, n\}$.

Also for an arbitrary set I, $\mathfrak{A}(I) \subseteq \mathfrak{S}_{\mathrm{fin}}(I)$ denotes the kernel of the signum function Sign : $\mathfrak{S}_{\mathrm{fin}}(I) \to \mathbb{Z}^\times$. *As the kernel of a group homomorphism, $\mathfrak{A}(I)$ is a normal subgroup in* $\mathfrak{S}(I)$. It has the index 2, when $|I| \geq 2$ is, because $\operatorname{Sign}\langle i,j\rangle = -1$ for a transposition $\langle i,j\rangle$. The two cosets are then the set $\mathfrak{A}(I)$ of even permutations and the set $\mathfrak{S}(I) - \mathfrak{A}(I) = \tau\mathfrak{A}(I) = \mathfrak{A}(I)\tau$ of odd permutations of I, where $\tau \in \mathfrak{S}(I)$ is any odd permutation (e.g., a transposition). For a finite set I with $|I| \geq 2$ in particular,

$$\operatorname{Ord}\mathfrak{A}(I) = (\operatorname{Ord}\mathfrak{S}(I))/2 = |I|!/2$$

(and the number of odd permutations is also $|I|!/2$).

In the case of $I = \{1, \ldots, n\}$ (or more generally for a completely ordered finite set I), the sign of a permutation $\sigma \in \mathfrak{S}(I)$ can also be described using the so-called inversions. For $\sigma \in \mathfrak{S}(I)$, a pair $(i,j) \in I \times I$ is an **inversion** of σ, if $i<j$, but $\sigma(i) > \sigma(j)$ is. We denote the number of inversions of σ with $F(\sigma)$. It is $F(\sigma) = F(\sigma|W(\sigma))$.[26]

Example 2.5.9
(1) The transposition $\langle i,j\rangle \in \mathfrak{S}_n$, $i<j$, has the inversions

$$(i, i+1), \ldots, (i,j); \quad (i+1,j), \ldots, (j-1,j).$$

Therefore, $F(\langle i,j\rangle) = 2(j-i) - 1$.
(2) In the permutation $\sigma := \begin{pmatrix} 1 & 2 & \cdots & n \\ n & n-1 & \cdots & 1 \end{pmatrix} \in \mathfrak{S}_n$ all (i,j) with $1 \leq i < j \leq n$ are inversions. Therefore, $F(\sigma) = \binom{n}{2}$. Any other permutation from $\mathfrak{S}_n$ has fewer than $\binom{n}{2}$ inversions. The identity is the only permutation in $\mathfrak{S}_n$ with 0 inversions.
(3) The permutation $\sigma := \begin{pmatrix} 1 & 2 & 3 & 4 & 5 \\ 3 & 1 & 5 & 2 & 4 \end{pmatrix} \in \mathfrak{S}_5$ has the inversions $(1,2)$, $(1,4)$, $(3,4)$ and $(3,5)$. Therefore, $F(\sigma) = 4$.

---

[26] Note that the number of inversions of a permutation $\sigma \in \mathfrak{S}(I)$ generally depends on the complete order chosen on I. The minimum number of inversions for $\sigma \in \mathfrak{S}(I)$, which can be achieved by a clever choice of the order of I, is the number $|I| - s$ already appearing in Definition 2.5.4, where s is the number of orbits of σ, cf. Exercise 2.5.10.

Theorem 2.5.10 Let $\sigma \in \mathfrak{S}_n$ be a permutation. Then $\operatorname{Sign} \sigma = (-1)^{F(\sigma)}$ holds. Thus, the parity of the permutation σ coincides with the parity of its inversions.

Proof Since according to Example 2.5.9 (1) a transposition has an odd number of inversions, it suffices to show the following: For $\sigma, \tau \in \mathfrak{S}_n$ is

$$(-1)^{F(\sigma\tau)} = (-1)^{F(\sigma)}(-1)^{F(\tau)}.$$

Obviously, for $\sigma \in \mathfrak{S}_n$

$$(-1)^{F(\sigma)} = \prod_{1 \le i < j \le n} \operatorname{Sign}(\sigma(j) - \sigma(i))$$

(where on the right side of the equation Sign the signum function on $\mathbb{Z}$ is denoted, cf. Example 1.2.2(3)). If in the difference $\sigma(j) - \sigma(i)$ the arguments $i \ne j$ are swapped, the sign changes. This gives:

$$(-1)^{F(\sigma\tau)} = \prod_{1 \le i < j \le n} \operatorname{Sign}(\sigma(\tau(j)) - \sigma(\tau(i)))$$

$$= (-1)^{F(\tau)} \prod_{1 \le r < s \le n} \operatorname{Sign}(\sigma(s) - \sigma(r)) = (-1)^{F(\tau)}(-1)^{F(\sigma)}.$$

The second equality sign results from the fact that the components in exactly $F(\tau)$ of the pairs $(\tau(i), \tau(j))$, $1 \le i < j \le n$, need to be swapped to get back the set of pairs (r, s), $1 \le r < s \le n$, again. □

Example 2.5.11 The permutation σ from Example 2.5.9 (2) has the sign $\operatorname{Sign} \sigma = (-1)^{\binom{n}{2}}$ according to Theorem 2.5.10. This also results from the canonical cycle decomposition

$$\sigma = \langle 1, n \rangle \langle 2, n-1 \rangle \ldots \langle [n/2], n+1 - [n/2] \rangle,$$

which consists of $[n/2]$ transpositions. It is

$$(-1)^{[n/2]} = (-1)^{\binom{n}{2}} = \begin{cases} 1, & \text{if } n \equiv 0, 1 \bmod 4, \\ -1, & \text{if } n \equiv 2, 3 \bmod 4. \end{cases}$$

This is a complicated proof for the simple congruence $\left[\frac{n}{2}\right] \equiv \binom{n}{2} \bmod 2$. ◊

Example 2.5.12 Let I be a set with more than two elements. *Then the center of the permutation group $\mathfrak{S}(I)$ is trivial.* If $\sigma \in \mathfrak{S}(I)$, $\sigma \ne \operatorname{id}$, $\sigma(a) \ne a$ and τ is a transposition $\langle \sigma(a), c \rangle$ with $c \notin \{a, \sigma(a)\}$, then $\tau(a) = \tau^{-1}(a) = a$ and $\tau \sigma \tau^{-1}$ maps the element a to c. Thus, $\tau \sigma \tau^{-1} \ne \sigma$ and σ is not interchangeable with τ. ◊

Fig. 2.14 Young tableaux for the partitions $(1^2, 3, 5)$ or $(1^2, 2^2, 4)$ of 10

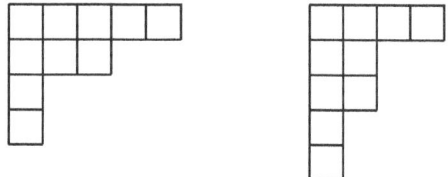

Example 2.5.13 (Conjugate Permutations – Type of a Permutation) A bijective mapping $f: I \to J$ induces the group isomorphism $\mathfrak{S}(I) \to \mathfrak{S}(J)$ with $\sigma \mapsto f \sigma f^{-1}$, which is called the **conjugation** with f. If $I = J$, it is the ordinary conjugation with f in the group $\mathfrak{S}(I)$, cf. Example 2.4.5. f maps the orbits of σ bijectively onto the orbits of $f \sigma f^{-1}$, and the cycle $\langle a_0, \ldots, a_{k-1} \rangle$ transforms into the cycle $\langle f(a_0), \ldots, f(a_{k-1}) \rangle$. For finite I, we thus obtain the cycle decomposition of the conjugated permutation $f \sigma f^{-1} \in \mathfrak{S}(J)$ from that of $\sigma \in \mathfrak{S}(I)$, by replacing the elements in the cycles of σ with their f-images. If $|I| = n$, we refer to the **type** of a permutation $\sigma \in \mathfrak{S}(I)$ as the sequence $\nu = \nu(\sigma) = (\nu_1(\sigma), \ldots, \nu_n(\sigma)) = (\nu_1, \ldots, \nu_n)$, where ν_k is the number of orbits of length k of σ, $k = 1, \ldots, n$. The type $(\nu_1, \ldots, \nu_n) \in \mathbb{N}^n$ defines a **decomposition** or **partition of the natural number** n, i.e., it is $1\nu_1 + \cdots + n\nu_n = n$. Specifically, it is $\nu_1(\sigma) = |\text{Fix}(\sigma, I)|$ is the number of fixed points of σ, and the equation $n = \text{Fix}(\sigma, I) + 2\nu_2 + \cdots + n\nu_n$ is the class Equation 2.4.2 for the operation of the group $H(\sigma)$ on I. The number of orbits of this operation is the number $|\nu| = \nu_1 + \cdots + \nu_n$ of orbits of σ. If we consider (I, σ) as a discrete dynamic system, cf. exercise 2.4.16, then the set $\text{Per}_k(\sigma, I)$ of (purely-)k-periodic points is equal to the union of the k-element orbits of σ. In particular, $|\text{Per}_k(\sigma, I)| = k \, \nu_k(\sigma)$ and

$$\nu_1(\sigma^k) = \left|\text{Fix}\left(\sigma^k, I\right)\right| = \sum_{d \mid k} d \, \nu_d(\sigma),$$

$$\nu_k(\sigma) = k^{-1} \sum_{d \mid k} \mu(d) \left|\text{Fix}(\sigma^{k/d}, I)\right| = k^{-1} \sum_{d \mid k} \mu(d) \, \nu_1(\sigma^{k/d}),$$

cf. exercise 2.4.16c). The partition $\nu(\sigma)$ and the sequence $\nu_1(\sigma^0) = n, \nu_1(\sigma), \ldots, \nu_1(\sigma^{\text{Ord}\,\sigma - 1})$ thus determine each other.

A partition $(\nu_1, \nu_2, \ldots)$ of n is often written as a (finite) monotonically increasing sequence $(1, \ldots, 1, 2, \ldots, 2, \ldots)$ with ν_1 ones, ν_2 twos, etc., or also in the form $(1^{\nu_1}, 2^{\nu_2}, \ldots)$, where terms of the form i^0 are omitted ($(1^2, 3, 4^3)$ thus denotes the partition $(1, 1, 3, 4, 4, 4)$ of the number $n = 17$. Partitions are often described with monotonically decreasing sequences, including infinite stationary monotonically decreasing sequences in $\mathbb{N}^{\mathbb{N}}$, whose limit is 0, thus $(4, 4, 4, 3, 1, 1, 0, 0, \ldots)$ for the above partition of the number 17. Furthermore, so-called **Young tableaux** are used to represent partitions as in Fig. 2.14 for the partitions $(1^2, 3, 5)$ or $(1^2, 2^2, 4)$ of 10. The two tableaux are **associated**: The columns of one are the rows of the other. ◊

We have proven:

n	0	1	2	3	4	5	6	7	8	9	10	11	12	13	14	15
$P(n)$	1	1	2	3	5	7	11	15	22	30	42	56	77	101	135	176

Fig. 2.15 Number $P(n)$ of partitions of n for $n \leq 15$

Theorem 2.5.14 *Let I be a finite set. Two permutations in $\mathfrak{S}(I)$ are conjugate if and only if they have the same type.*

The number of conjugation classes of the group $\mathfrak{S}_n$, that is, the class number of $\mathfrak{S}_n$, is thus equal to the number $P(n)$ of the partitions of n. In exercise 3.7.16 we describe how the numbers $P(n)$ can be easily calculated for not too large n. A list can be found in Fig. 2.15.

The number $|C_{\mathfrak{S}_n}(\sigma)|$ of permutations $\sigma \in \mathfrak{S}_n$ of type $(\nu_1, \ldots, \nu_n)$ conjugated to a permutation is equal to

$$\frac{n!}{\nu_1! \cdots \nu_n! 1^{\nu_1} \cdots n^{\nu_n}},$$

this is the index of the centralizer of this permutation σ, cf. exercise 2.5.17.

If $\sigma \in \mathfrak{A}_n$ is an even permutation, then $C_{\mathfrak{A}_n}(\sigma) = C_{\mathfrak{S}_n}(\sigma)$ holds if and only if the centralizer $Z_{\mathfrak{S}_n}(\sigma)$ of σ contains an odd permutation. Otherwise, $C_{\mathfrak{S}_n}(\sigma)$ decomposes into $\mathfrak{A}_n$ into the two conjugacy classes of the same cardinal number $C_{\mathfrak{A}_n}(\sigma)$ and $C_{\mathfrak{A}_n}(\tau\sigma\tau^{-1})$, where $\tau \in \mathfrak{S}_n$ is any odd permutation, see exercise 2.4.13. *The first case occurs exactly when σ has an even orbit or two odd orbits of the same length.* Proof!

Example 2.5.15 (Structure of the groups $\mathfrak{A}_n$ and $\mathfrak{S}_n$) The group $\mathfrak{A}_n$ includes as the kernel of the homomorphism Sign from $\mathfrak{S}_n$ into the *commutative* group $\mathbb{Z}^\times = \{1, -1\}$ the commutator group of $\mathfrak{S}_n$. It even holds:

Proposition 2.5.16 *For every $n \in \mathbb{N}$ is $\mathfrak{A}_n = [\mathfrak{S}_n, \mathfrak{S}_n]$ the commutator group of $\mathfrak{S}_n$.*

Proof If $a, b, c, d \in \{1, \ldots, n\}$ with $a \neq b$ and $c \neq d$ and is $\tau \in \mathfrak{S}_n$ a permutation with $\tau(a) = c, \tau(b) = d$, then according to Example 2.5.13

$$\langle a, b \rangle \langle c, d \rangle = \langle a, b \rangle \tau \langle a, b \rangle \tau^{-1}$$

is a commutator. Since the permutations $\langle a, b \rangle \langle c, d \rangle$ generate the group $\mathfrak{A}_n$ according to (Lemma 2.5.3 and) Theorem 2.5.6, the assertion follows. □

The sign homomorphism Sign is therefore (for $n \geq 2$) essentially the only non-trivial homomorphism from $\mathfrak{S}_n$ into an abelian group. – To prove the next proposition, we use the following lemma:

Lemma 2.5.17 *For every $n \in \mathbb{N}$ the three cycles generate $\langle a, b, c \rangle$ the group $\mathfrak{A}_n$.*

Proof According to Theorem 2.5.6 the products $\langle a,b\rangle\langle c,d\rangle$ of two transpositions each generate the group $\mathfrak{A}_n$. But if a,b,c,d are pairwise different, then

$$\langle a,b\rangle\langle b,c\rangle = \langle a,b,c\rangle \, , \, \langle a,b\rangle\langle c,d\rangle = \langle a,b\rangle\langle b,c\rangle\langle b,c\rangle\langle c,d\rangle = \langle a,b,c\rangle\langle b,c,d\rangle. \quad \square$$

Proposition 2.5.18 *For every $n \geq 5$ is $\mathfrak{A}_n = [\mathfrak{A}_n, \mathfrak{A}_n]$, the group $\mathfrak{A}_n$ thus perfect.*

Proof Because of $\langle a,b,c\rangle = \langle a,b,d\rangle\langle c,e,a\rangle\langle a,b,d\rangle^{-1}\langle c,e,a\rangle^{-1}$ for pairwise different elements a,b,c,d,e, every three-cycle in $\mathfrak{A}_n$, $n \geq 5$, is a commutator in $\mathfrak{A}_n$. The assertion now follows with Lemma 2.5.17. $\quad \square$

As a significant tightening of Proposition 2.5.18 even applies.

Theorem 2.5.19 *For every $n \geq 5$ the alternating group is $\mathfrak{A}_n$ simple.*

Proof We prove the theorem in the following two steps: (1) The group $\mathfrak{A}_5$ is simple. (2) If the group $\mathfrak{A}_5$ is simple, then every group $\mathfrak{A}_n$, $n \geq 5$, is simple.

Proof of (1): The permutations of $\mathfrak{A}_5$ have the types (1^5), $(1^2, 3)$, $(1, 2^2)$, (5). In the group $\mathfrak{S}_5$ the corresponding conjugacy classes have the element numbers 1, 20, 15, 24. The first three conjugacy classes are also conjugacy classes with respect to $\mathfrak{A}_5$. The last one splits into $\mathfrak{A}_5$ into two conjugacy classes with 12 elements each. (See the end of Example 2.5.13.) In total, $\mathfrak{A}_5$ therefore has 5 conjugacy classes with 1,20,15,12,12 elements. Every normal divisor is a union of conjugacy classes, with the trivial conjugacy class $\{id\}$ always appearing. However, with the given numbers, no divisor of 60 different from 1 and 60 can be realized as a sum with 1 as one of the summands. (Using the same method, it is also easy to prove that the group $\mathfrak{A}_6$ is simple.)

Proof of (2): Let $n \geq 5$ and $N \subseteq \mathfrak{A}_n$ be a normal subgroup $\neq \{id\}$. Furthermore, let $\sigma \in N$, $\sigma \neq id$, and let $i,k \in \{1,\ldots,n\}$ with $j := \sigma(i) \neq i$ and $k \notin \{i,j,\sigma(j)\}$. For the three-cycle $\tau := \langle i,j,k\rangle$, the commutator $\rho := \sigma\tau\sigma^{-1}\tau^{-1} = \langle \sigma(i),\sigma(j),\sigma(k)\rangle\tau^{-1}$ is due to $\langle \sigma(i),\sigma(j),\sigma(k)\rangle \neq \tau$ an element of N different from id. Thus, $N \cap \mathfrak{A}(A)$ is a non-trivial normal subgroup of $\mathfrak{A}(A)$, where $A \subseteq \{1,\ldots,n\}$ is a 5-element subset with $i,j,k,\sigma(j),\sigma(k) \in A$. According to (1), $N \cap \mathfrak{A}(A) = \mathfrak{A}(A)$. In particular, N contains a three-cycle. Since in $\mathfrak{A}_n$ all three-cycles are conjugate — they are first conjugate in $\mathfrak{S}_n$ and then due to $n \geq 5$ also in $\mathfrak{A}_n$ –, contains N all three cycles and is therefore equal to $\mathfrak{A}_n$ according to Lemma 2.5.17. $\quad \square$

From Theorem 2.5.19 also follows the simplicity of the alternating groups $\mathfrak{A}(I)$ for any infinite sets I, see Ex. 2.5.23c). The group $\mathfrak{A}_5$ of order 60 is the smallest non-abelian simple group in the following sense: *Every non-abelian simple group with an order ≤ 60 is isomorphic to the group $\mathfrak{A}_5$.* See 15, Appendix VI.A, Ex. 9. For $n \geq 5$, $\mathfrak{A}_n$ is the only non-trivial normal divisor in the group $\mathfrak{S}_n$, see Ex. 2.5.24a).

The group $\mathfrak{S}_1$ is the trivial group, the group $\mathfrak{S}_2 \cong \mathbf{Z}_2$ is cyclic, the group $\mathfrak{S}_3$ is isomorphic to the Dieder group D_3 and has the group $\mathfrak{A}_3 \cong \mathbf{Z}_3$ as the only non-trivial normal divisor. The group $\mathfrak{S}_4$ has the chain of normal subgroups

$$\{id\} \subset \mathfrak{V}_4 \subset \mathfrak{A}_4 \subset \mathfrak{S}_4$$

with $\mathfrak{S}_4/\mathfrak{V}_4 \cong \mathfrak{S}_3$ and $\mathfrak{A}_4/\mathfrak{V}_4 \cong \mathfrak{A}_3$. In this case, $\mathfrak{V}_4 \subseteq \mathfrak{S}_4$ the standard model for the group $\mathbf{Z}_2 \times \mathbf{Z}_2$ isomorphic so-called **Klein four-groups**. In addition to the identity, $\mathfrak{V}_4$ contains the permutations $\langle 1, 2 \rangle \langle 3, 4 \rangle$, $\langle 1, 3 \rangle \langle 2, 4 \rangle$, $\langle 1, 4 \rangle \langle 2, 3 \rangle$, from which it is immediately apparent that it is a normal subgroup of $\mathfrak{S}_4$, see Example 2.5.13. The subgroup in $\mathfrak{S}_4$ generated by $\langle 1, 2 \rangle$ and $\langle 3, 4 \rangle$ (for example) is also a Klein four-group (but not **the** Klein four-group $\mathfrak{V}_4$). The groups $\mathfrak{V}_4$ and $\mathfrak{A}_4$ are the only nontrivial normal divisors in the group $\mathfrak{S}_4$, and $\mathfrak{V}_4$ is also the only nontrivial normal subgroup in the group $\mathfrak{A}_4$, cf. exercise 2.5.24. ◊

Example 2.5.20 The finite group G operates on itself by left translation with the action $L: G \to \mathfrak{S}(G)$, $a \mapsto (L_a: x \mapsto ax)$. The composition $\mathrm{Sign} \circ L: a \mapsto \mathrm{Sign}(L_a)$ is a group homomorphism. We denote it by $\Lambda = \Lambda_G: G \to \mathbb{Z}^\times = \{\pm 1\}$. ◊

Proposition 2.5.21 *Let G be a finite group. With the above notations, the following applies*

$$\Lambda_G(a) = (-1)^{|G| - |G|/\mathrm{Ord}\, a}, \quad a \in G.$$

In particular, the homomorphism $\Lambda_G: G \to \mathbb{Z}^\times$ is nontrivial, i.e., the image of the Cayley homomorphism $L: G \to \mathfrak{S}(G)$ is not a subgroup of $\mathfrak{A}(G)$, when $|G|$ is even and G contains an element of order $2^{\nu_2(|G|)}$ (i.e., when the 2-Sylow groups of G are nontrivial cyclic groups, cf. Theorem 2.4.11 (1)). If these two conditions are met, then Λ_G is the only nontrivial homomorphism from G to $\mathbb{Z}^\times$. $\ker \Lambda_G$ is therefore the only normal subgroup of index 2 in G. Specifically, a group of order $2k$ with an odd $k \in \mathbb{N}^*$ contains exactly one normal subgroup of index 2 (and is therefore not simple).

Proof The orbits of the permutation $L_a: G \to G$ of G are the right cosets of the subgroup generated by a $H(a) \subseteq G$. The formula for $\Lambda_G(a)$ therefore follows directly from the Definition 2.5.4 of the sign of a permutation. Apparently, $\Lambda_G(a)$ is exactly -1 when $|G|$ is even and $\nu_2(\mathrm{Ord}\, a) = \nu_2(|G|)$ is. This proves the addition about the non-triviality of Λ_G.

Now let $\Lambda = \Lambda_G$ be non-trivial, $a \in G$ an element with $\nu_2(\mathrm{Ord}\, a) = \nu_2(|G|)$ (i.e. $H(a) = S_2$ a 2-Sylow group of G) and $\varphi: G \to \mathbb{Z}^\times$ a surjective homomorphism $\neq \Lambda$. Then $N := \ker \Lambda \cap \ker \varphi \subseteq G$ is a normal subgroup in G of index 4 and $G/N \cong \mathbb{Z}_2 \times \mathbb{Z}_2$ a Klein four-group. This contradicts the result of exercise 2.4.19, according to which G/N as a 2-group like S_2 is cyclic. □

Example 2.5.22 (Jacobi Symbol—Reciprocity Formula) Let $a, b \in \mathbb{Z}^* = \mathbb{Z} - \{0\}$ with $\mathrm{ggT}(a, b) = 1$. Then the multiplication of multiples $x \mapsto ax$ is an automorphism $_b\chi_a: \mathbf{Z}_{|b|} \to \mathbf{Z}_{|b|}$ of the cyclic group $\mathbf{Z}_{|b|} = \mathbb{Z}/\mathbb{Z}b$. If there is no doubt about b, we simply write χ_a for $_b\chi_a$.

Following G. Frobenius (1849–1917) and E. Zolotarev (1847–1878), we define: ◊

2.5 Permutation Groups

Definition 2.5.23 If $a, b \in \mathbb{Z}^*$ with $\mathrm{ggT}(a,b) = 1$ and ${}_b\chi_a: \mathbb{Z}_{|b|} \to \mathbb{Z}_{|b|}$ the automorphism $x \mapsto ax$, then let

$$\left(\frac{a}{b}\right) = (a/b) := \mathrm{Sign}\, {}_b\chi_a.$$

If b is odd, then (a/b) is called the **Jacobi symbol** "a over b" (after C. Jacobi (1804–1851)); if $b \geq 2$, $b \in \mathbb{P}$, an odd prime number, then we also speak of the **Legendre symbol** (after A.-M. Legendre (1752–1833)). (If $a \in \mathbb{Z}$ is not coprime to b, then we set $(a/b) := 0$.)

So it is $(a/b) = (a/|b|)$. Of course, in the definition of (a/b) the cyclic group $\mathbb{Z}/\mathbb{Z}b$ can be replaced by any cyclic group of order $|b|$. In multiplicative notation, χ_a is then exponentiation $x \mapsto x^a$. Because of $\chi_{a_1 a_2} = \chi_{a_1} \circ \chi_{a_2}$ it holds: *The symbol (a/b) is multiplicative in the numerator*, i.e.

$$\left(\frac{a_1 a_2}{b}\right) = \left(\frac{a_1}{b}\right)\left(\frac{a_2}{b}\right).$$

$[a]_b \mapsto (a/b)$ is therefore a homomorphism $(\mathbb{Z}/\mathbb{Z}b)^\times \to \{\pm 1\}$ of multiplicative groups.

It is more surprising that the Jacobi symbol is also multiplicative in the denominator. It holds that: *If $b_1, b_2 \in \mathbb{Z}^*$ both are odd and $a \in \mathbb{Z}^*$ with $\gcd(a, b_1 b_2) = 1$, then it holds*

$$\left(\frac{a}{b_1 b_2}\right) = \left(\frac{a}{b_1}\right)\left(\frac{a}{b_2}\right).$$

But if b_1 is odd and b_2 is even, then $(a/b_1 b_2) = (a/b_2)$. Both statements follow directly from the result of exercise 2.5.13, applied to the canonical exact sequence

$$0 \to \mathbb{Z}/\mathbb{Z}b_1 \to \mathbb{Z}/\mathbb{Z}b_1 b_2 \to \mathbb{Z}/\mathbb{Z}b_2 \to 0$$

with odd b_1 and the automorphisms ${}_{b_1}\chi_a$, ${}_{b_1 b_2}\chi_a$ or ${}_{b_2}\chi_a$. For powers of 2, it holds: *If a is odd, then $(a/2) = 1$ and $(a/2^n) = (a/4) = (-1)^{(a-1)/2}$, if $n \geq 2$*, see exercise 2.7.5. Therefore, in the following, we will only deal with the Jacobi symbol (a/b) with odd b.

Due to the multiplicity in the denominator, it is sufficient to know the Legendre symbols (a/p) for odd prime numbers p to determine all Jacobi symbols. The multiplication $\chi_a: \mathbb{Z}/\mathbb{Z}p \to \mathbb{Z}/\mathbb{Z}p$ has 0 as a fixed point and is the multiplication with the residue class $a = [a]_p$ on the multiplicative group $(\mathbb{Z}/\mathbb{Z}p)^\times = (\mathbb{Z}/\mathbb{Z}p) - \{[0]_p\}$. According to Proposition 2.5.21,

$$\left(\frac{a}{p}\right) = (-1)^{(p-1) - (p-1)/\mathrm{Ord}_p a}.$$

We will later see that the group $(\mathbb{Z}/\mathbb{Z}p)^\times$ is cyclic, see Theorem 2.7.13. Here, however, we only use that its 2-primary component is cyclic, i.e., $[-1]_p$ is the only element of order 2 in $(\mathbb{Z}/\mathbb{Z}p)^\times$. This is however trivial. If $a^2 \equiv 1 \bmod p$, then p is a divisor of $a^2 - 1 = (a-1)(a+1)$ and thus a divisor of $a - 1$ or of $a + 1$, i.e., $a \equiv 1$ or $a \equiv -1 \bmod p$. It follows, again with Proposition 2.5.21, that the Legendre symbol defines a nontrivial homomorphism $(\mathbb{Z}/\mathbb{Z}p)^\times \to \{\pm 1\}$ and that this is the only such homomorphism. On the other hand, $[a]_p \mapsto [a]_p^{(p-1)/2}$ according to Euler's criterion from Exercise 2.2.29 is a homomorphism $(\mathbb{Z}/\mathbb{Z}p)^\times \to \{\pm 1\}(\subseteq (\mathbb{Z}/\mathbb{Z}p)^\times)$, whose kernel is the group of squares in $(\mathbb{Z}/\mathbb{Z}p)^\times$ and which is therefore also not trivial, we have proven:

Theorem 2.5.24 (Euler Criterion for Quadratic Residues) *Let $p > 2$ be an odd prime number and $a \in \mathbb{Z}$, $p \nmid a$. Then the following holds*

$$\left(\frac{a}{p}\right) \equiv a^{(p-1)/2} \bmod p,$$

and the residue class $[a]_p$ is a square in $(\mathbb{Z}/\mathbb{Z}p)^\times$ if and only if $(a/p) = 1$ is.[27]

To calculate the general Jacobi symbol (a/b) for odd $b > 0$, we determine the parity of the number of inversions of the permutation $\chi_a \colon \mathbb{Z}/\mathbb{Z}b \to \mathbb{Z}/\mathbb{Z}b$, where we use the natural order of the absolutely smallest residues $-(b-1)/2, \ldots, -1, 0, 1, \ldots, (b-1)/2$ as the order on the set $\mathbb{Z}/\mathbb{Z}b$. The following important formula by Gauss is fundamental:

Theorem 2.5.25 (Gauss's sign formula) *Let $a, b \in \mathbb{Z}$ with $\gcd(a,b) = 1$, $b > 0$ odd. Furthermore, let v be the number of $r \in \mathbb{N}^*$ with $0 < r \leq \frac{1}{2}(b-1)$, for which the absolutely smallest residue of ar when divided by b is negative. Then*

$$\left(\frac{a}{b}\right) = (-1)^v.$$

Proof In this proof, we fundamentally identify a residue class from $\mathbb{Z}/\mathbb{Z}b$ with its absolutely smallest representative and consider the restriction of χ_a to $(\mathbb{Z}/\mathbb{Z}b) - \{0\}$. According to Theorem 2.5.10, $(a/b) = (-1)^F$, where $F = |\mathcal{F}|$ with

$$\mathcal{F} := \left\{(r,s) \in \mathbb{Z}^2 \,\middle|\, r \neq 0 \neq s, -\frac{1}{2}(b-1) \leq r < s \leq \frac{1}{2}(b-1), [ar]_b > [as]_b\right\}$$

is. With (r,s) also belongs to $(-s,-r)$ of $\mathcal{F}$ (due to $[-ak]_b = -[ak]_b$ for all k). Therefore, $|\mathcal{F}|$ and the number of fixed points of the involution $(r,s) \mapsto (-s,-r)$ of $\mathcal{F}$ have the same parity.[28] A pair $(-r,r)$ with $0 < r \leq \frac{1}{2}(b-1)$ belongs exactly then to $\mathcal{F}$, when $[ar]_b < 0$ is, i.e. it is $F \equiv v \bmod 2$. □

This immediately gives the first formula of the following theorem:

---

[27] For any odd b, $(a/b) = 1$ can hold, without $[a]_b$ being a square in $(\mathbb{Z}/\mathbb{Z}b)^\times$. Example? But if $(a/b) = -1$, then b has at least one prime factor p such that $[a]_p$ is not a square in $(\mathbb{Z}/\mathbb{Z}p)^\times$.

[28] We repeat once again the trivial, but important observation: *If σ is an involution of a finite set X, then $|X|$ and $|\mathrm{Fix}(\sigma,X)|$ have the same parity.*

2.5 Permutation Groups

Theorem 2.5.26 *For odd $b \in \mathbb{N}^*$ the following holds*

$$\left(\frac{-1}{b}\right) = (-1)^{\frac{b-1}{2}}, \quad \left(\frac{2}{b}\right) = (-1)^{\frac{b^2-1}{8}} = \begin{cases} 1, & \text{if } b \equiv \pm 1 \bmod 8, \\ -1, & \text{if } b \equiv \pm 3 \bmod 8. \end{cases}$$

Proof For $a = 2$, in the Gaussian sign Formula 2.5.25, the numbers r with $0 < r \leq \frac{1}{2}(b-1)$ and $[2r]_b < 0$ are exactly the numbers $r \in \mathbb{N}^*$ with $\frac{1}{4}(b-1) < r \leq \frac{1}{2}(b-1)$. If $b \equiv 1 \bmod 4$, then their number is equal to $\frac{1}{2}(b-1) - \frac{1}{4}(b-1) = \frac{1}{4}(b-1) \equiv \frac{1}{4 \cdot 2}(b-1)(b+1) = \frac{1}{8}(b^2 - 1) \bmod 2$. If $b \equiv 3 \bmod 4$, then this number is equal to $\frac{1}{2}(b-1) - \frac{1}{4}(b-3) = \frac{1}{4}(b+1) \equiv \frac{1}{4 \cdot 2}(b+1)(b-1) = \frac{1}{8}(b^2 - 1) \bmod 2$. In total, this gives the second formula. $\square$

Let now $a, b \in \mathbb{N}^*$ both be positive, odd and coprime. The number ν in the above formula of Gauss can also be described as the number of pairs $(r, s) \in \mathbb{N}^* \times \mathbb{N}^*$ with $0 < r \leq \frac{1}{2}(b-1)$, $0 < s \leq \frac{1}{2}(a-1)$ and $-\frac{1}{2}(b-1) \leq ar - bs < 0$. From the condition $-\frac{1}{2}(b-1) \leq ar - bs < 0$ it follows that s is uniquely determined by a given $r \in \{1, \ldots, \frac{1}{2}(b-1)\}$ and $s \in \{1, \ldots, \frac{1}{2}(a-1)\}$ applies. Similarly, $(b/a) = (-1)^\mu$, where μ is the number of pairs $(r, s) \in \mathbb{N}^* \times \mathbb{N}^*$ with $0 < s \leq \frac{1}{2}(a-1)$, $0 < r \leq \frac{1}{2}(b-1)$ and $-\frac{1}{2}(a-1) \leq bs - ar < 0$, i.e. $0 < ar - bs \leq \frac{1}{2}(a-1)$. So it is $(a/b)(b/a) = (-1)^{\nu+\mu}$, where $\nu + \mu = |\mathcal{G}|$ is with

$$\mathcal{G} := \left\{ (r, s) \in (\mathbb{N}^*)^2 \,\middle|\, 0 < r \leq \frac{1}{2}(b-1), \, 0 < s \leq \frac{1}{2}(a-1), \right.$$
$$\left. -\frac{1}{2}(b-1) \leq ar - bs \leq \frac{1}{2}(a-1) \right\}.$$

Note that for the $\mathcal{G}$ appearing in the definition of (r, s) due to the coprimality of the numbers a and b, we never have ar - bs = 0. We now consider the involution

$$\tau: (r, s) \mapsto \left(\frac{1}{2}(b+1) - r, \frac{1}{2}(a+1) - s\right).$$

It maps $\mathcal{G}$ onto itself. For $(r, s) \in \mathcal{G}$ is namely

$$a\left(\frac{1}{2}(b+1) - r\right) - b\left(\frac{1}{2}(a+1) - s\right) = -ar + bs + \frac{1}{2}(a - b)$$

and $-\frac{1}{2}(a-1) \leq -ar + bs \leq \frac{1}{2}(b-1)$, from which $\tau(r, s) \in \mathcal{G}$ results. The parity of $\nu + \mu$ is therefore equal to the parity of $|\text{Fix}(\tau, \mathcal{G})|$. τ has but $\left(\frac{1}{4}(b+1), \frac{1}{4}(a+1)\right)$ as the only fixed point. It lies in $\mathcal{G}$ only when $a, b \equiv 3 \bmod 4$ are, which is exactly the case when $\frac{1}{2}(a-1)$ and $\frac{1}{2}(b-1)$ are both odd. We have thus obtained the main result about the Jacobi symbol:

Theorem 2.5.27 (Reciprocity formula for the Jacobi symbol) *For coprime odd numbers $a, b \in \mathbb{N}^*$ the following holds*

$$\left(\frac{a}{b}\right)\left(\frac{b}{a}\right) = (-1)^{\frac{a-1}{2} \cdot \frac{b-1}{2}}.$$

(a/b) and (b/a) thus coincide exactly when at least one of the numbers a, b is congruent to 1 modulo 4.

For the Legendre symbol, we formulate the proven statements separately. They form the so-called quadratic reciprocity law, which was first proven by Gauss in the "Disquisitiones Arithmeticae" (1801).

Theorem 2.5.28 (Quadratic Reciprocity Law) *For prime numbers p, $q > 2$, $p \neq q$, the following holds*

$$\left(\frac{-1}{q}\right) = (-1)^{\frac{q-1}{2}}, \quad \left(\frac{2}{q}\right) = (-1)^{\frac{q^2-1}{8}} \quad \text{and} \quad \left(\frac{p}{q}\right)\left(\frac{q}{p}\right) = (-1)^{\frac{p-1}{2} \cdot \frac{q-1}{2}}.$$

The first two formulas are also called the **complementary laws** to the quadratic reciprocity law. We illustrate the reciprocity formula with two small examples:

(1) Let $n := 1.234.567.890$. For the two prime numbers $p := n + 1 = 1.234.567.891$ and $q := 10^{10}n + n + 1 = 12.345.678.901.234.567.891$, which are both $\equiv 3 \bmod 4$, it follows that, since 10^{10} is a square,

$$\left(\frac{p}{q}\right) = \left(\frac{n+1}{10^{10}n + n + 1}\right) = -\left(\frac{10^{10}n + n + 1}{n+1}\right) = -\left(\frac{-10^{10}}{n+1}\right) = -\left(\frac{-1}{p}\right) = 1.$$

The residue class of $p = 1.234.567.891$ is therefore a square in the group $(\mathbb{Z}/\mathbb{Z}q)^\times$. For the calculation of a square root, see the remarks on exercise 2.2.29.

(2) Let $b \in \mathbb{N}^*$ be odd and coprime to 3. Because of $(3/b) = (-1)^{(3-1)(b-1)/4}(b/3) = (-1)^{(b-1)/2}(b/3)$, $(3/b) = 1$ is, if $b \equiv \pm 1 \bmod 12$, or $(3/b) = -1$, if $b \equiv \pm 5 \bmod 12$ is. Therefore, if $q > 3$ is prime, then 3 is exactly a square in $(\mathbb{Z}/\mathbb{Z}q)^\times$, if $q \equiv \pm 1 \bmod 12$ is. For a Fermat prime number $q = F_m = 2^{2^m} + 1$, $m \geq 1$, therefore $(3/F_m) = -1 \equiv 3^{(F_m-1)/2}$ modulo F_m. The residue class $[3]_{F_m}$ thus generates the cyclic group $(\mathbb{Z}/\mathbb{Z}F_m)^\times$ of order $F_m - 1 = 2^{2^m}$. If, conversely, $3^{(F_m-1)/2} \equiv -1$ modulo F_m, then $[3]_{F_m}$ is an element of order $F_m - 1$ in $(\mathbb{Z}/\mathbb{Z}F_m)^\times$ and F_m is necessarily prime. We have proven:

Theorem 2.5.29 (Pépin Test) (after Th. Pépin (1826–1904)) *The Fermat number $F_m = 2^{2^m} + 1$, $m \geq 1$, is prime if and only if $3^{(F_m-1)/2} \equiv -1$ modulo F_m is.*

With this test, it is proven that F_m is usually not prime (without having to provide a specific decomposition).

Example 2.5.30 (Cycle Polynomials – Pólya's Enumeration Method) In this example, we calculate (elementarily) in commutative rings, especially in polynomial rings, see Sects. 2.6 and 2.9. The methods are based on the work: Combinatorial Counting for Groups, Graphs and Chemical Compounds, Acta Mathematica **68**, 145–254 (1937), by G. Pólya (1887–1985), which has significantly influenced the development of combinatorics.

The goal is, among other things, to systematically determine the number of orbits in the operation of a (finite) group on a (finite) set. This is usually a classification

2.5 Permutation Groups

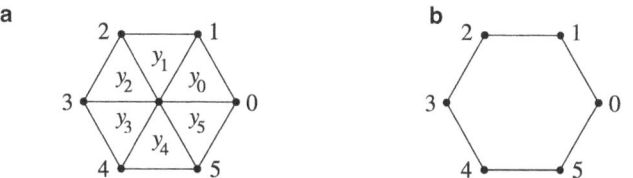

Coulering of the vertices of a 6-gon   Graph of a cycle with 6 vertices

Fig. 2.16 Examples of sets on which C_6 and D_6 operate

problem. The group is interpreted as a symmetry group, and the elements of an orbit are patterns of the same type. To be a bit more concrete, we understand a pattern on a finite set X to be a mapping $f: X \to Y$ into a set Y, where we interpret Y as a set of "colors". The operation $G \times X \to X$ of a finite group G on X induces an operation of G on the set Y^X of patterns with color set Y. It is $gf = f \circ \vartheta_g^{-1}$ for $g \in G$, $f \in Y^X$, cf. exercise 2.4.1c). The orbits then represent the essentially different patterns with respect to the symmetries given by the operation of G ϑ_g, $g \in G$, of X.

For example, a pattern of the sectors of a regular n-gon in the Euclidean plane, $n \geq 3$, is given by a mapping $f: E \to F$ of its set of vertices E, cf. the drawing in Fig. 2.16a. Two such patterns, which arise from a rotation around the center of the n-gon, will be identified. If one is a bit more generous, reflections on the axes of symmetry of the n-gon are also allowed. In the first case, the symmetry group G is a cyclic group Z_n of order n, which we denote in the given geometric realization with C_n. The operation of C_n on the sectors is isomorphic to the Cayley operation of C_n on C_n (by left translation). In the second case, the symmetry group is the dihedral group D_n of order 2n, which we denote in the current geometric situation with D_n. Note that D_n is also the automorphism group of the graph of the cycle with the n vertices $0, 1, \ldots, n-1$ and the n edges $\{0, 1\}, \{1, 2\} \ldots, \{n-1, 0\}$, cf. Fig. 2.16b.[29] One could also interpret the vertices of the graph as beads of a closed chain, which have different colors. In this case, D_n is the appropriate symmetry group. Its operation is isomorphic to the natural operation of the dihedral group D_n as a subgroup of the permutation group $\mathfrak{S}(\mathbf{Z}_n)$, cf. Example 2.4.12 (2). If the chain is open, then the graph is a segment, in which the edge $\{n-1, 0\}$ is missing, and the symmetry group to be considered is a subgroup of order 2 of D_n, which, besides the identity, contains only one reflection.

For the general discussion, let X now be a finite set with n elements, on which the finite group G operates with the action $\vartheta: G \to \mathfrak{S}(X)$. We encode the type $v = v(\sigma) = (v_1(\sigma), \ldots, v_n(\sigma)) = (v_1, \ldots, v_n)$ of a permutation $\sigma \in \mathfrak{S}(X)$, cf. Example 2.5.13, through the monomial

$$Z^v = Z_1^{v_1} \cdots Z_n^{v_n}$$

---

[29] The reader will not find it difficult to define the automorphism group of a graph.

from a multiplicative free commutative monoid M_n in the indeterminates $Z = (Z_1, \ldots, Z_n)$, cf. Example 2.3.20. In order to be able to not only multiply such monomials but also add them, we consider the direct sum $\mathbb{Q}[Z] = \mathbb{Q}[Z_1, \ldots, Z_n] = \mathbb{Q}^{(M_n)}$, where we extend the multiplication of M_n to $\mathbb{Q}[Z]$ in such a way that the distributive laws apply. This results in the polynomial ring over $\mathbb{Q}$ in the indeterminates $Z_1, \ldots, Z_n$ with the multiplication

$$\left(\sum_{\alpha \in \mathbb{N}^n} a_\alpha Z^\alpha \right) \cdot \left(\sum_{\beta \in \mathbb{N}^n} b_\beta Z^\beta \right) = \sum_{\gamma \in \mathbb{N}^n} \left(\sum_{\alpha + \beta = \gamma} a_\alpha b_\beta \right) Z^\gamma,$$

cf. Sect. 2.9. In the following, we use the abbreviation $v(g) = v(g, \vartheta) := v(\vartheta_g)$. ◊

Definition 2.5.31 The polynomial

$$\psi(G) = \psi(G, \vartheta) := |G|^{-1} \sum_{g \in G} Z^{v(g)} = |G|^{-1} \sum_{g \in G} Z_1^{v_1(g)} \cdots Z_n^{v_n(g)}$$

from $\mathbb{Q}[Z] = \mathbb{Q}[Z_1, \ldots, Z_n]$ is called the **cycle polynomial** of $G = (G, \vartheta)$.

For $v \in \mathbb{N}^n$, the coefficient is $a_v \in \mathbb{Q}$ of Z^v in the cycle polynomial $\psi(G)$ the frequency of the elements $g \in G$ with $v(g) = v$. Apparently, $\psi(G, \vartheta) = \psi(\vartheta(G))$, where $\vartheta(G) \subseteq \mathfrak{S}(X)$ naturally operates on X. Furthermore, isomorphic operations have the same cycle polynomial. Since conjugate permutations have the same type, it follows: *If $g_1, \ldots, g_r$ are representatives of the different conjugation classes $C_G(g_i)$, $\rho = 1, \ldots, r$, of G, then it is*

$$\psi(G) = \sum_{\rho=1}^{r} \frac{1}{|Z_G(g_\rho)|} Z^{v(g_\rho)},$$

where $Z_G(g) = \{h \in G \mid hg = gh\}$ is the centralizer of $g \in G$. Note the equation $|G| = |Z_G(g)| \cdot |C_G(g)|$, $g \in G$. Furthermore, remember that according to Example 2.5.13 for the exponents in the monomial $Z^{v(g)} = Z_1^{v_1(g)} \cdots Z_n^{v_n(g)}$ applies:

$$v_i(g) = i^{-1} \sum_{d \mid i} \mu(d) v_1(g^{i/d}).$$

For example, with the concluding remarks in Example 2.5.13 for $n \geq 2$, the cycle polynomials are obtained.

$$\psi(\mathfrak{S}_n) = \sum_{1 \cdot v_1 + \cdots + n v_n = n} \frac{1}{v_1! \cdots v_n!} \left(\frac{Z_1}{1} \right)^{v_1} \cdots \left(\frac{Z_n}{n} \right)^{v_n}$$

and

$$\psi(\mathfrak{A}_n) = 2 \cdot \sum_{\substack{1 \cdot v_1 + \cdots + n v_n = n \\ v_2 + v_4 + \cdots \equiv 0(2)}} \frac{1}{v_1! \cdots v_n!} \left(\frac{Z_1}{1} \right)^{v_1} \cdots \left(\frac{Z_n}{n} \right)^{v_n}$$

2.5 Permutation Groups

for the symmetric group $\mathfrak{S}_n$ or for the alternating group $\mathfrak{A}_n$ with respect to their natural operations on $\{1, \ldots, n\}$. The polynomial

$$\psi(\mathfrak{A}_5) = \frac{1}{60}\left(Z_1^5 + 20Z_1^2 Z_3 + 15Z_1 Z_2^2 + 24Z_5\right)$$

we have already calculated in the proof of Theorem 2.5.19.

If the finite group G of order n operates on itself by left translation (Cayley operation = regular representation of G), it is obviously

$$\psi(G) = \frac{1}{n} \sum_{d|n} \alpha(d) Z_d^{n/d},$$

where $\alpha(d)$ is the number of elements of order d in G, cf. the proof of Proposition 2.5.21. In particular, the cycle polynomial for the Cayley operation of a finite cyclic group Z_n is equal to

$$\psi(\mathbf{Z}_n) = \frac{1}{n} \sum_{d|n} \varphi(d) Z_d^{n/d}.$$

For the natural operation of the dihedral group $\mathbf{D}_n \subseteq \mathfrak{S}(\mathbf{Z}_n)$ the n involutions are added (which are all conjugate for odd n and split into two conjugacy classes for even n). This results in the cycle polynomial

$$\psi(\mathbf{D}_n) = \frac{1}{2n} \sum_{d|n} \varphi(d) Z_d^{n/d} + \begin{cases} \frac{1}{4}(Z_1^2 + Z_2) Z_2^{(n-2)/2}, & \text{if } n \equiv 0(2), \\ \frac{1}{2} Z_1 Z_2^{(n-1)/2}, & \text{if } n \equiv 1(2). \end{cases}$$

With the help of the cycle polynomial, G. Pólya has given a formula for the number $|Y^X \backslash G|$ of equivalence classes of patterns on X with finite color set Y. For this, we generally consider a **weight function** $\gamma : Y \to A$ from Y into a commutative ring A, in which $|G| = |G| \cdot 1_A$ is a unit. We assign the weight

$$\gamma(f) := \prod_{x \in X} \gamma(f(x)) = \prod_{y \in Y} \gamma(y)^{|f^{-1}(y)|} \in A$$

to a pattern $f : X \to Y$. Equivalent patterns obviously have the same weight. Thus, γ induces a weight function $Y^X \backslash G \to A$, which we also denote by γ. If γ is constantly equal to 1, then $\gamma(f) = 1$ is also for all $f \in Y^X$. With these notations, the following applies:

Theorem 2.5.32 (Pólya's Enumeration Theorem) *Let $\psi(G)$ be the cycle polynomial of the operation of the finite group G on the finite set X. Furthermore, let Y be a finite (color) set and $\gamma : Y \to A$ a weight function with values in a commutative ring A with $|G| \cdot 1_A \in A^\times$. If then π_i is the power sum $\sum_{y \in Y} \gamma(y)^i$, $i = 1, \ldots, n$, then it holds*

$$\sum_{[f] \in Y^X \backslash G} \gamma([f]) = \psi(G)(\pi_1, \ldots, \pi_n).$$

In particular, if $|Y| = m$ and are $g_1, \ldots, g_r$ representatives of the different conjugacy classes of G, then it holds

$$|Y^X \backslash G| = \psi(G)(m, \ldots, m) = \frac{1}{|G|} \sum_{g \in G} m^{|v(g)|} = \sum_{\rho=1}^{r} \frac{m^{|v(g_\rho)|}}{|Z_G(g_\rho)|}.$$

Proof Let $M = Y^X$ and $[M] = M \backslash G$. First, we notice that a pattern $f \in M$ is invariant under the operation of $g \in G$ if f has a constant value $y_1, \ldots, y_s$ on the orbits $X_1, \ldots, X_s$ of g. It follows

$$\sum_{f \in \text{Fix}_g M} \gamma(f) = \sum_{(y_1, \ldots, y_s) \in Y^s} \gamma(y_1)^{|X_1|} \cdots \gamma(y_s)^{|X_s|} = \pi_{|X_1|} \cdots \pi_{|X_s|} = \pi_1^{v_1} \cdots \pi_n^{v_n} = \pi^{v(g)},$$

where $v(g) = (v_1, \ldots, v_n)$ is the type of g. Now one obtains – $G_f = \text{Stab}_G(f, M)$ is the isotropy group of a pattern $f \in M$ –

$$|G| \sum_{[f] \in [M]} \gamma([f]) = \sum_{f \in M} |G_f| \gamma(f) = \sum_{(g,f), g \in G_f} \gamma(f) = \sum_{(g,f), f \in \text{Fix}(g,M)} \gamma(f)$$
$$= \sum_{g \in G} \pi^{v(g)} = |G| \cdot \psi(G)(\pi_1, \ldots, \pi_n). \qquad \square$$

As the proof shows, in Theorem 2.5.32 one may also allow an infinite set of colors Y, if for example $A = \mathbb{C}$ and $\gamma(y)$, $y \in Y$, are summable, which can occasionally be very useful.[30] The art in applying Theorem 2.5.32 consists in cleverly choosing the set of colors and their weights according to the problem. The greatest difficulty usually lies in determining the cycle polynomial $\psi(G)$. We illustrate the procedure with some examples.

(1) As a first example, we consider the colorings of the sectors of a regular n-gon, $n \geq 3$, with the set of colors $Y := \{1, \ldots, m\}$, $m \in \mathbb{N}^*$, where we choose as symmetry group G the group $C_n (\cong \mathbb{Z}_n)$ of rotations of the n-gon. As already indicated, the associated cycle polynomial is $\psi(C_n) = \psi(\mathbb{Z}_n) = n^{-1} \sum_{d|n} \varphi(d) Z_d^{n/d}$. If we assign the color $i \in Y$ as weight the indeterminate T_i, so for a pattern $f: E \to Y$ the total weight is equal to $\gamma(f) = T_1^{\alpha_1} \cdots T_m^{\alpha_m}$, where α_i is the number of sectors with color i. According to Theorem 2.5.32, in the polynomial

$$n^{-1} \sum_{d|n} \varphi(d) (T_1^d + \cdots + T_m^d)^{n/d}$$

the coefficient of $T_1^{\alpha_1} \cdots T_m^{\alpha_m}$ is the number of inequivalent patterns in which color i appears exactly α_i times. According to the polynomial Theorem 2.6.3, see also 1.6.16, this is the case for $\alpha_1 + \cdots + \alpha_m = n$ and $\alpha := \gcd(\alpha_1, \ldots, \alpha_m)$ the sum

---

[30] To formulate it as generally as possible: Theorem 2.5.32 *applies to a commutative complete topological ring A, in which $|G| = |G| \cdot 1_A$ is invertible, and weight functions $\gamma: Y \to A$, for which the sums $\sum_{y \in Y^r} \gamma(y_1) \cdots \gamma(y_r)$, $1 \leq r \leq |X|$, exist, see Sect. 4.5.* The last condition is usually already fulfilled when $\pi_1 = \sum_{y \in Y} \gamma(y)$ exists. Often A is a formal power series ring.

2.5 Permutation Groups

$$\frac{1}{n}\sum_{d\mid\alpha}\varphi(d)\frac{(n/d)!}{(\alpha_1/d)!\cdots(\alpha_m/d)!},$$

at $\alpha = 1$ in particular $(n-1)!/\alpha_1!\cdots\alpha_m!$. The total number of inequivalent patterns with (at most) m colors is

$$b_m := n^{-1}\sum_{d\mid n}\varphi(d)m^{n/d}.$$

Thus, a regular hexagon has $(2^6 + 2^3 + 2\cdot 2^2 + 2\cdot 2^1)/6 = 14$ inequivalent colorings with at most 2 colors and 12 colorings in which both colors appear.

To generally compare the numbers $m \in \mathbb{N}$ a_m and $b_m = \sum_{j=0}^{m}\binom{m}{j}a_j$ of inequivalent patterns, in which exactly m colors or at most m colors occur, we consider the exponential generating functions (= power series)

$$P := \sum_{m\in\mathbb{N}}\frac{a_m}{m!}X^m \quad \text{and} \quad Q := \sum_{m\in\mathbb{N}}\frac{b_m}{m!}X^m$$

of the sequences (a_m) and (b_m) in $\mathbb{Q}\,X$. With the exponential series $e^X = \sum_{m\in\mathbb{N}}X^m/m!$, $Q = e^X P$ then applies and consequently $P = e^{-X}Q$ due to $e^X e^{-X} = 1$ (proof!), from which

$$a_m = \sum_{j=0}^{m}(-1)^{m-j}\binom{m}{j}b_j \quad \text{with} \quad b_m = \sum_{j=0}^{m}\binom{m}{j}a_j, \quad m \in \mathbb{N},$$

follows. These formulas are called the **binomial inversion formulas** for the sequences (a_m) and (b_m).

(2) If we identify two patterns even when they are produced by the operation of the dihedral group $\mathbf{D}_n \supseteq \mathbf{C}_n$ as is appropriate for the necklace problem mentioned at the beginning, then we have to consider the natural operation of the dihedral group $\mathbf{D}_n \subseteq \mathfrak{S}(\mathbf{Z}_n)$ on $X = \mathbf{Z}_n$, whose cycle polynomial $\psi(\mathbf{D}_n) = \psi(\mathbf{D}_n)$ we have also given above. The total number of inequivalent patterns with (at most) m colors is thus reduced to

$$\psi(\mathbf{D}_n)(m,\ldots,m) = \frac{1}{2n}\sum_{d\mid n}\varphi(d)\,m^{n/d} + \begin{cases} \frac{1}{4}(m+1)m^{n/2}, & \text{if } n \equiv 0(2), \\ \frac{1}{2}m^{(n+1)/2}, & \text{if } n \equiv 1(2), \end{cases}$$

in the case $n = 6$ and $m = 2$, thus to $7 + (3\cdot 8)/4 = 13$ or to 11, if both colors occur.[31] Also formulate the general inversion formulas for the number of inequivalent patterns with exactly m colors for the operation of $\mathbf{D}_n$.

---

[31] Specify the two patterns that are equivalent with respect to the operation of the group $\mathbf{D}_6$, but not with respect to the operation of the group $\mathbf{C}_6$.

Similar to above for the group C_n, the number of inequivalent patterns in which the color $i \in \{1,\ldots,m\}$ occurs exactly α_i times can also now be explicitly stated. This is the coefficient of $T_1^{\alpha_1} \cdots T_m^{\alpha_m}$ in the polynomial obtained from the cycle polynomial $\psi(D_n)$ by replacing the indeterminate Z_i with $T_1^i + \cdots + T_m^i$, $i = 1,\ldots,n$.

For non-closed chains, the symmetry group is the reflection group S, which, in addition to the identity, only contains the involution $i \mapsto n-1-i$, $i = 0,\ldots,n-1$, for $n \geq 2$. The cycle polynomial is

$$\psi(S) = \frac{1}{2}Z_1^n + \begin{cases} \frac{1}{2}Z_2^{n/2}, & \text{if } n \equiv 0(2), \\ \frac{1}{2}Z_1Z_2^{(n-1)/2}, & \text{if } n \equiv 1(2). \end{cases}$$

The number of inequivalent open chains with exactly α_i beads of color i, $i = 1,\ldots,m$, $\sum_{i=1}^{m} \alpha_i = n$, is the coefficient of $T^\alpha = T_1^{\alpha_1} \cdots T_m^{\alpha_m}$ in the polynomial

$$\frac{1}{2}(T_1 + \cdots + T_m)^n + \begin{cases} \frac{1}{2}(T_1^2 + \cdots + T_m^2)^{n/2}, & \text{if } n \equiv 0(2), \\ \frac{1}{2}(T_1 + \cdots + T_m)(T_1^2 + \cdots + T_m^2)^{(n-1)/2}, & \text{if } n \equiv 1(2). \end{cases}$$

We leave it to the reader to explicitly note this coefficient using the polynomial theorem. The number of all equivalence classes of open chains with n beads of m different types is

$$\psi(S)(m,m) = \frac{1}{2}m^n + \begin{cases} \frac{1}{2}m^{n/2}, & \text{if } n \equiv 0(2), \\ \frac{1}{2}m^{(n+1)/2}, & \text{if } n \equiv 1(2). \end{cases}$$

(3) A graph with the vertex set $\{1,\ldots,n\}$ is defined by a subset of the set $\mathfrak{E}_2 = \mathfrak{E}_2(\mathbb{N}_n^*)$ of 2-element subsets of $\mathbb{N}_n^* = \{1,\ldots,n\}$. Two graphs that arise from a permutation of the vertex set are equivalent (= isomorphic). The permutation group $\mathfrak{S}_n$ operates in the natural way (transitively for $n \geq 2$) on $\mathfrak{E}_2$. The associated cycle polynomial is ψ_n. The isomorphism classes of graphs form the set $\{0,1\}^{\mathfrak{E}_2}\backslash\mathfrak{S}_n$, where we identify a subset in $\mathfrak{E}_2$ with its indicator function. $\{0,1\}^{\mathfrak{E}_2}$ identify. Let's assign the weight 1 to color 0 and the weight T to color 1, then $\pi_d = 1 + T^d$ and the coefficient of T^α in the polynomial

$$\psi_n(1+T, 1+T^2, \ldots)$$

is the number of isomorphism classes of graphs with n vertices and α edges, $\alpha \in \mathbb{N}$. The total number of isomorphism classes of graphs with n vertices is $\psi_n(2,2,\ldots)$. The number of weighted graphs, where an edge can take one of the values $1,\ldots,m$, $m \in \mathbb{N}^*$, is $\psi_n(1+m, 1+m, \ldots)$. To determine ψ_n, for the type $\nu = (\nu_1,\ldots,\nu_n)$ of a permutation $\sigma \in \mathfrak{S}_n$ the type $\tilde{\nu} := \nu(\sigma, \mathfrak{E}_2)$ of the permutation induced by σ on $\mathfrak{E}_2$ has to be determined. With some concentration (exercise!) one shows

$$\tilde{\nu}_i = \frac{1}{2}\nu_i(i\nu_i - e_i) + \nu_{2i} + \sum_{\substack{1 \leq r < s \leq n \\ \text{lcm}(r,s) = i}} \gcd(r,s)\nu_r\nu_s, \quad i = 1,\ldots,\binom{n}{2} = |\mathfrak{E}_2|,$$

2.5 Permutation Groups

where $e_i := 1$ is, if i is odd, and $e_i := 2$, if i is even. The cycle polynomial is then

$$\psi_n = \sum_{1 \cdot \nu_1 + \cdots + n\nu_n = n} \frac{1}{\nu_1! \cdots \nu_n! 1^{\nu_1} \cdots n^{\nu_n}} \prod_{i=1}^{\binom{n}{2}} Z_i^{\bar{\nu}_i}.$$

The number of isomorphism classes of graphs with n vertices is therefore

$$\psi_n(2, 2, \ldots) = \sum_{\nu, 1 \cdot \nu_1 + \cdots + n\nu_n = n} \frac{2^{|\bar{\nu}|}}{\nu_1! \cdots \nu_n! 1^{\nu_1} \cdots n^{\nu_n}}$$

with $|\bar{\nu}| = \bar{\nu}_1 + \bar{\nu}_2 + \cdots = \frac{1}{2} \sum_i i \nu_i^2 - \frac{1}{2} \sum_{i \equiv 1(2)} \nu_i + \sum_{r<s} \gcd(r, s) \nu_r \nu_s$.

For $n = 4$, for example, $\psi_4 = \frac{1}{24}(Z_1^6 + 9Z_1^2 Z_2^2 + 6Z_2 Z_4 + 8Z_3^2)$ results (see also exercise 2.5.25) and $\psi_4(2, 2, 2, 2) = 11$ as the number of isomorphism classes of graphs with 4 corners. A representative should be given for each class. The number of isomorphism classes of graphs with n corners, $1 \le n \le 10$, is in order 1, 2, 4, 11, 34, 156, 1044, 12.346, 274.668, 12.005.168.

(4) The group W of the 24 rotations in Euclidean space that transform a cube into itself operates (transitively) on the 6 sides of this cube. Since the isotropy group of one side consists of the 4 rotations around the center of this side, $|W| = 6 \cdot 4 = 24$. It is $W \cong \mathfrak{S}_4$, because W operates faithfully on the set of 4 cube diagonals. See also exercise 2.5.25. The identity, the 3 half-rotations around the cube axes, the 6 quarter-rotations around these axes, the 6 half-rotations around opposite edge centers and the 8 third-rotations around the diagonals have the cycle polynomials Z_1^6, $Z_1^2 Z_2^2$, $Z_1^2 Z_4$, Z_2^3, Z_3^2. The cycle polynomial for the operation of W on the set of 6 sides of the cube is therefore

$$\psi = \psi(W) = \frac{1}{24}\left(Z_1^6 + 3Z_1^2 Z_2^2 + 6Z_1^2 Z_4 + 6Z_2^3 + 8Z_3^2\right).$$

The number of inequivalent colorings of the cube sides with at most m colors, where color i appears exactly α_i times, is according to Theorem 2.5.32 equal to the coefficient of $T_1^{\alpha_1} \cdots T_m^{\alpha_m}$ in the polynomial that results from $\psi(W)$ is evident from the fact that the indeterminate Z_i is replaced by $T_1^i + \cdots + T_m^i$. In particular, the total number of inequivalent colorings is equal to

$$\frac{1}{24} m^2 (m^4 + 3m^2 + 12m + 8),$$

for $m = 2$, thus 10 (and 8, if both colors occur). If one interprets the (infinitely many) colors $i \in \mathbb{N}$ as numbers and assigns the weight T^i to the color i, $i \in \mathbb{N}$, then

$$\pi_d = \sum_{i=0}^{\infty} T^{id} = \frac{1}{1 - T^d},$$

and the coefficient of T^a in the power series

$$\psi\left(1/(1-T), 1/(1-T^2), 1/(1-T^3), 1/(1-T^4)\right)$$

gives the number of inequivalent numberings of the dice sides with natural numbers, where the total sum of the numbers is equal to α, $\alpha \in \mathbb{N}$.

For extensions of the Pólya counting method, we refer to the recommended book by A. Kerber: Algebraic Combinatorics Via Finite Group Actions, Mannheim 1991, and the literature cited therein.

Exercises

Exercise 2.5.1 For the following permutations, give the canonical cycle representation, a representation as a product of transpositions, the number of inversions, the signum, and the order.

a) $\begin{pmatrix} 1 & 2 & 3 & 4 & 5 & 6 & 7 & 8 & 9 & 10 & 11 & 12 \\ 3 & 2 & 9 & 10 & 8 & 12 & 4 & 6 & 1 & 11 & 7 & 5 \end{pmatrix} \in \mathfrak{S}_{12}.$

b) $\begin{pmatrix} 1 & 2 & 3 & 4 & 5 & 6 & 7 & 8 & 9 & 10 & 11 & 12 & 13 & 14 & 15 \\ 1 & 4 & 10 & 12 & 5 & 7 & 11 & 2 & 15 & 14 & 9 & 8 & 6 & 3 & 13 \end{pmatrix} \in \mathfrak{S}_{15}.$

c) $\begin{pmatrix} 1 & 2 & 3 & 4 & 5 & 6 & 7 & 8 & 9 & 10 & 11 & 12 \\ 7 & 12 & 1 & 10 & 8 & 2 & 11 & 4 & 6 & 5 & 3 & 9 \end{pmatrix} \in \mathfrak{S}_{12}.$

Exercise 2.5.2 For the following permutations, state the number of inversions and the signum.

a) $\begin{pmatrix} 1 & 2 & \ldots & n & n+1 & \ldots & 2n \\ 1 & 3 & \ldots & 2n-1 & 2 & \ldots & 2n \end{pmatrix} \in \mathfrak{S}_{2n}.$

b) $\begin{pmatrix} 1 & 2 & \ldots & n & n+1 & \ldots & 2n \\ 2 & 4 & \ldots & 2n & 1 & \ldots & 2n-1 \end{pmatrix} \in \mathfrak{S}_{2n}.$

c) $\begin{pmatrix} 1 & \ldots & n-r+1 & n-r+2 & \ldots & n \\ r & \ldots & n & 1 & \ldots & r-1 \end{pmatrix} \in \mathfrak{S}_n, 1 \leq r \leq n.$

d) $\begin{pmatrix} 1 & 2 & 3 & 4 & 5 & 6 & \ldots & 2n \\ 1 & 2n & 3 & 2(n-1) & 5 & 2(n-2) & \ldots & 2 \end{pmatrix} \in \mathfrak{S}_{2n}.$

Exercise 2.5.3 For a subset $J \subseteq \{1, \ldots, n\}$ with $J = \{j_1, \ldots, j_m\}$, $j_1 < \cdots < j_m$, let σ_J be the permutation

$$\sigma_J = \begin{pmatrix} 1 & \ldots & m & m+1 & \ldots & n \\ j_1 & \ldots & j_m & i_1 & \ldots & i_{n-m} \end{pmatrix} \in \mathfrak{S}_n,$$

where the numbers $i_1 < \cdots < i_{n-m}$ are the elements of the complement of J in $\{1, \ldots, n\}$. Then the number of inversions of σ_J is equal to

$$F(\sigma_J) = \left(\sum_{k=1}^m j_k\right) - \binom{m+1}{2}$$

and thus Sign$(\sigma_J) = (-1)^{F(\sigma_J)}$. (The permutations σ_J are called **Shuffle-Permutations**. For a fixed $m = |J|$, $0 \leq m \leq n$, they form a canonical representative system for the left cosets of the subgroup $\mathfrak{S}(\{1,\ldots,m\}) \times \mathfrak{S}(\{m+1,\ldots,n\}) \subseteq \mathfrak{S}_n$ in $\mathfrak{S}_n$. This group is the isotropy group of the set $\{1,\ldots,m\}$ under the canonical operation of the group $\mathfrak{S}_n$ on the power set $\mathfrak{P}(\mathbb{N}_n^*)$. The orbit of $\{1,\ldots,m\}$ is the set $\mathfrak{E}_m(\mathbb{N}_n^*)$ of m-element subsets of $\mathbb{N}_n^*$. The inverses σ_J^{-1} of the shuffle permutations σ_J, $|J| = m$, form a representative system for the right cosets of the group $\mathfrak{S}(\{1,\ldots,m\}) \times \mathfrak{S}(\{m+1,\ldots,n\})$ in $\mathfrak{S}_n$.)

Exercise 2.5.4 Let I be a finite set. The inverse σ^{-1} of a permutation $\sigma \in \mathfrak{S}(I)$ has the same orbits and the same sign as σ.

Exercise 2.5.5 A subgroup of the permutation group $\mathfrak{S}_n$, which contains an odd permutation, consists of an equal number of odd and even permutations.

Exercise 2.5.6
a) A permutation $\sigma \in \mathfrak{S}_n$ of odd order is even.
b) The square σ^2 of a permutation $\sigma \in \mathfrak{S}_n$ is an even permutation.

Exercise 2.5.7 The exponent of the group $\mathfrak{S}_n$, $n \in \mathbb{N}^*$, is equal to

$$B(n) := \mathrm{lcm}(1,\ldots,n) = \prod_{p \in \mathbb{P}} p^{\alpha_p} \quad \text{with} \quad \alpha_p := [\log_p n] = [\ln n / \ln p].$$

(We will discuss the function $B(n)$, $n \in \mathbb{N}^*$, in Vol. 5 in connection with the prime number theorem.)

Exercise 2.5.8 Let $m = p_1^{\alpha_1} \cdots p_r^{\alpha_r}$ be the canonical prime factor decomposition of $m \in \mathbb{N}^*$. The group $\mathfrak{S}_n$ contains an element of order m if and only if $n \geq p_1^{\alpha_1} + \cdots + p_r^{\alpha_r}$ is. Determine the maximum $C(n)$ of the orders of the elements of $\mathfrak{S}_n$ for $n = 1,\ldots,20$. (For a discussion of the function $C(n)$, $n \in \mathbb{N}^*$, see W. Miller: The maximum order of an element of a finite symmetric group, Amer. Math. Monthly **94**, 497–506 (1987).)

Exercise 2.5.9
a) What type can a permutation σ of order 40 or 60 have in the group $\mathfrak{S}_{20}$?
b) Let $p \in \mathbb{P}$ be a prime number. For which $n \in \mathbb{N}$ is there an element of order p^3 in the group $\mathfrak{S}_n$ and what type can such a permutation have?

Exercise 2.5.10 Let $n \in \mathbb{N}^*$.

a) Let m_i for $1 \leq i < n$ be the number of inversions (i,j), $i < j \leq n$, in the permutation $\sigma \in \mathfrak{S}_n$, and let $\sigma_i := \langle i + m_i, i + m_i - 1 \rangle \cdots \langle i+1, i \rangle$. Then $\sigma = \sigma_1 \cdots \sigma_{n-1}$. In particular, σ is a product of $F(\sigma)$ transpositions of the form $\langle 1,2 \rangle, \langle 2,3 \rangle, \ldots, \langle n-1, n \rangle$. (This proves again 2.5.10 and reconstructs the permutation σ from its inversions. More precisely: The permutation σ is

Fig. 2.17 Commutative diagram of groups with exact sequences of three as rows

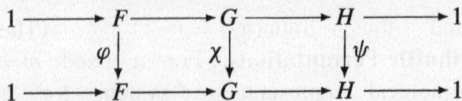

determined by the $(n-1)$-tuple $(m_1,\ldots,m_{n-1})$ with $0 \le m_i \le n-i$ uniquely, and each such tuple determines a $\sigma \in \mathfrak{S}_n$. This encoding of the elements of $\mathfrak{S}_n$ is used quite frequently. – Consider also the analogous problem with the numbers m'_i of inversions $(j,i), j<i, i = 2, \ldots, n$.)
b) Let $\sigma \in \mathfrak{S}_n$ and $\sigma' := \sigma \circ \langle i, i+1 \rangle$ with $i \in \{1,\ldots,n-1\}$. Then $|F(\sigma) - F(\sigma')| = 1$. It follows: σ is not a product of less than $F(\sigma)$ transpositions of adjacent elements.

Exercise 2.5.11 Let $\sigma \in \mathfrak{S}_n$ and $m \in \mathbb{Z}$. Every orbit of σ of length k decomposes into $\mathrm{ggT}(k,m)$ orbits of length $k/\mathrm{ggT}(k,m)$ of σ^m. (See also Example 2.5.13.)

Exercise 2.5.12
a) Let X be a finite set. Determine the sign of the involution $(x_1, x_2) \mapsto (x_2, x_1)$ of $X \times X$.
b) Let $X = \biguplus_{i \in I} X_i$ be a decomposition of the finite set X and $\sigma \in \mathfrak{S}(X)$ a permutation that leaves each X_i invariant, $i \in I$. Then $\mathrm{Sign}\,\sigma = \prod_{i \in I} \mathrm{Sign}\,(\sigma | X_i)$.
c) Let X_i be finite sets with $m_i := \prod_{j \ne i} |X_j|$ and $\sigma_i \in \mathfrak{S}(X_i)$, $i \in I$, I finite. Furthermore, let $\sigma \in X := \prod_{i \in I} X_i \to X$, $(x_i)_{i \in I} \mapsto (\sigma_i(x_i))_{i \in I}$, be the product of the permutations σ_i, $i \in I$. Then $\mathrm{Sign}\,\sigma = \prod_{i \in I}(\mathrm{Sign}\,\sigma_i)^{m_i}$. (One can assume that σ_i is the identity except for one exception.)
d) Let $\sigma \in \mathfrak{S}(X)$ be a permutation of the finite set X with $|X| = m$, and let $\mathfrak{E}_r(X)$ be the set of r-element subsets of X, $0 \le r \le m$. Then the sign of the permutation induced by σ on $\mathfrak{E}_r(X)$ is equal to $(\mathrm{Sign}\,\sigma)^{\binom{m-2}{r-1}}$, where $\binom{m-2}{-1} = 0$ is to be set for all $m \in \mathbb{N}$. (One can assume that σ is a transposition.) Also determine the sign of the permutation induced by σ on $\mathfrak{P}(X) = \mathfrak{E}(X) = \biguplus_{r=0}^{m} \mathfrak{E}_r(X)$.

Exercise 2.5.13 Let $1 \to F \to G \to H \to 1$ an exact sequence of (not necessarily abelian) finite groups and $\varphi: F \to F$, $\chi: G \to G$, $\psi: H \to H$ automorphisms of groups such that the diagram in Fig. 2.17 is commutative. F has odd order or contains no element of order $2^{\nu_2(|F|)}$, cf. Proposition 2.5.21. Then

$$\mathrm{Sign}\,\chi = (\mathrm{Sign}\,\varphi)^{|H|} \cdot (\mathrm{Sign}\,\psi)^{|F|}.$$

applies. In particular: *If the order of G is odd, then* $\mathrm{Sign}\,\chi = \mathrm{Sign}\,\varphi \cdot \mathrm{Sign}\,\psi$. (For the proof, one can assume that F is a normal subgroup of G, H the quotient group G/F and $\varphi = \chi|F$ and ψ the automorphism induced by $G/F \to G/F$. Then let $A \subseteq G$ be a representative system for the elements of G/F and $f: A \times F \to G$ the bijective mapping $(a, x) \mapsto ax$. It holds $\chi(aF) = \chi(a)F = \sigma(a)F$ with a permutation $\sigma \in \mathfrak{S}(A)$. Now let $g_a := \sigma(a)^{-1}\chi(a) \in F$, $a \in A$ and

2.5 Permutation Groups

h, g are the permutations $h(a,x) = (\sigma(a), \chi(x))$ and $g(a,x) = (a, g_a x)$ of $A \times F$. According to the assumption about F, $\text{Sign } g = 1$ is, see exercise 2.5.12b) and Proposition 2.5.21. Furthermore, $\chi = f(gh)f^{-1}$ is. It follows $\text{Sign } \chi = \text{Sign}(gh) = \text{Sign } h = (\text{Sign } \psi)^{|F|} \cdot (\text{Sign } \varphi)^{|H|}$, see exercise 2.5.12c).)

Exercise 2.5.14 Let $n \in \mathbb{N}^*$ and T be a set of transpositions in the group $\mathfrak{S}_n$. We assign to T the graph Γ_T, whose vertices are the numbers $1, \ldots, n$ and in which two vertices i and j, $i \neq j$, are connected by an edge exactly when the transposition $\langle i,j \rangle = \langle j,i \rangle$ belongs to T. $\Gamma_1, \ldots, \Gamma_r$ are the connected components of Γ_T, see Example 1.3.5.

a) The transpositions from T generate the group $\mathfrak{S}_n$ exactly when Γ_T is connected, i.e., when $r = 1$, meaning that any two vertices of Γ_T can be connected by an edge path. Generally, the subgroup generated by T is the product $\mathfrak{S}(\Gamma_1) \times \cdots \times \mathfrak{S}(\Gamma_r) \subseteq \mathfrak{S}_n$.
b) If T is a generating system of the group $\mathfrak{S}_n$, then T contains at least $n-1$ elements. (In general, if $\tau_1, \ldots, \tau_m$ are the elements of T (possibly with repetitions) and if $\tau_1 \cdots \tau_m = id$, then m is even and $m \geq 2 \sum_{\rho=1}^{r} (|\Gamma_\rho| - 1)$.)
c) Every generating system of $\mathfrak{S}_n$ from transpositions contains a (minimal) generating system of $\mathfrak{S}_n$ with $n-1$ elements. (The graphs of such minimal generating systems are the trees with vertex set $\mathbb{N}^*_{\leq n}$. Every connected graph has a **generating tree**, i.e., a subgraph that is a tree and has the same vertex set. *Every generating tree of a connected graph with n vertices has $n-1$ edges. Proof! – By the way, the group $\mathfrak{S}_n$ exactly has n^{n-2} generating systems from $n-1$ transpositions* (**Cayley's theorem**). For the *proof*, show in general: Let $f_{n,k}$ for $1 \leq k \leq n$ denote the number of forests with vertex set $\mathbb{N}^*_{\leq n}$ and exactly k marked trees (so-called **root trees**),[32] then $f_{n,n} = 1$, $(n-k+1)f_{n,k-1} = n(k-1)f_{n,k}$. (By "grafting" one obtains from a forest with $k \geq 2$ root trees $n(k-1)$ forests with $k-1$ root trees and by removing one edge from a forest with $k-1$ root trees $n-k+1$ forests with k root trees.) Thus

$$f_{n,k} = \binom{n-1}{k-1} n^{n-k}, \quad 1 \leq k \leq n.$$

The sought number is $f_{n,1}/n$.)
d) The transpositions $\langle 1,2 \rangle, \langle 2,3 \rangle, \ldots, \langle n-1,n \rangle$ as well as $\langle 1,2 \rangle, \langle 1,3 \rangle, \ldots, \langle 1,n \rangle$ each form a minimal generating system of $\mathfrak{S}_n$. Every permutation $\sigma \in \mathfrak{S}_n$ is a product of transpositions of adjacent elements, see Problem 2.5.10.
e) A statement analogous to a) also applies to the alternating groups. For a "triangle" $\Delta = \{a,b,c\} \in \mathfrak{E}_3(\mathbb{N}^*_{\leq n})$, let $\alpha(\Delta)$ denote the set of the two 3-cycles $\langle a,b,c \rangle$, $\langle a,c,b \rangle = \langle a,b,c \rangle^{-1}$(which are independent of an order or "orientation" on Δ).

---

[32] A **marked set** is a pair (X,P), consisting of a set X and a point $P \in X$. So a set X contains exactly $|X|$ marked sets (X,P). A tree with $m \in \mathbb{N}^*$ corners has exactly m markings.

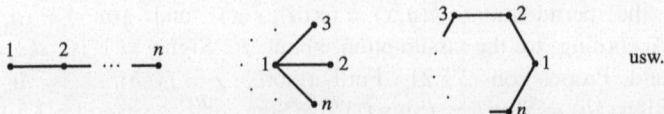

Fig. 2.18 Examples of generating systems of $\mathfrak{S}_n$ from transpositions

Then the following applies: For 3-element sets $\Delta_1, \ldots, \Delta_m \in \mathfrak{E}_3(\mathbb{N}^*_{\leq n})$, the cycles from $\alpha(\Delta_1) \cup \cdots \cup \alpha(\Delta_m)$ generate the group $\mathfrak{A}(\Gamma_1) \times \cdots \times \mathfrak{A}(\Gamma_r) \subseteq \mathfrak{A}_n$. Here, $\Gamma_1, \ldots, \Gamma_r$ are the connected components of the graph with vertex set $\mathbb{N}^*_{\leq n}$, whose edges each belong to one of the triangles $\Delta_1, \ldots, \Delta_m$ belong. (Prove by induction over t: If $\Delta_1, \ldots, \Delta_t$ are 3-element sets with $\Delta_i \cap \Delta_{i+1} \neq \emptyset$ for $i = 1, \ldots, t-1$, then $\alpha(\Delta_1) \cup \cdots \cup \alpha(\Delta_t)$ generates the alternating group $\mathfrak{A}(\Delta_1 \cup \cdots \cup \Delta_t)$.) Conclude: The minimum number of 3-cycles that generate the group $\mathfrak{A}_n$, $n \geq 3$, is $\lceil (n-1)/2 \rceil$. Provide three 3-cycles that generate the group $\mathfrak{A}_5$, but no two of which $(= \lceil (5-1)/2 \rceil)$ generate the group $\mathfrak{A}_5$.

f) For $n \geq 3$, the 3-cycles $\langle 1, 2, 3 \rangle, \langle 2, 3, 4 \rangle, \ldots, \langle n-2, n-1, n \rangle$ and $\langle 1, 2, 3 \rangle, \langle 1, 2, 4 \rangle, \ldots, \langle 1, 2, n \rangle$ each form a generating system of $\mathfrak{A}_n$.

Exercise 2.5.15 A permutation $\sigma \in \mathfrak{S}_n$ with s orbits has a representation as a product of $n - s$ transpositions and no representation as a product of fewer than $n - s$ transpositions. (This exercise has a natural generalization: Let $T \subseteq \mathfrak{S}_n$ be a set of transpositions that generates the group $\mathfrak{S}_n$ (given by the connected graph $\Gamma = \Gamma_T$ on the vertex set $\mathbb{N}^*_{\leq n}$, cf. exercise 2.5.14a). For $\sigma \in \mathfrak{S}_n$ determine the minimum $\ell(\sigma) = \ell_T(\sigma)$ of the $m \in \mathbb{N}$, for which there is a representation $\sigma = \tau_1 \cdots \tau_m$ with $\tau_i \in T$ exists. By the way, $\ell(\sigma) = \ell(\sigma^{-1})$, and $d(\sigma_1, \sigma_2) := \ell(\sigma_2 \sigma_1^{-1})$, $\sigma_1, \sigma_2 \in \mathfrak{S}_n$, is a metric on $\mathfrak{S}_n$ (see Sect. 4.1), for which left and right translations are distance-preserving. Consider for Γ_T in addition to the complete graph also the examples in Fig. 2.18. For the first of these graphs see exercise 2.5.10. For $T \subseteq T'$ naturally $\ell_{T'} \leq \ell_T$.)

Exercise 2.5.16
a) The two cycles $\langle 1, 2 \rangle, \langle 2, \ldots, n \rangle$ generate the group $\mathfrak{S}_n$, $n \geq 2$.
b) The two cycles $\langle 1, 2 \rangle, \langle 1, 2, \ldots, n \rangle$ generate the group $\mathfrak{S}_n$, $n \geq 2$. More generally, if k,n are natural numbers with $1 < k \leq n$, then the two cycles $\langle 1, k \rangle, \langle 1, 2, \ldots, n \rangle$ generate the group $\mathfrak{S}_n$ if $\gcd(k-1, n) = 1$ is.
c) Let $n \geq 3$. The cycles $\langle 1, 2, 3 \rangle, \langle 1, 2, 3, \ldots, n \rangle$, if n is odd, or $\langle 1, 2, 3 \rangle, \langle 2, 3, \ldots, n \rangle$, if n is even, generate the group $\mathfrak{A}_n$.

(Use exercise 2.5.14d) and 2.5.14f).)

Exercise 2.5.17 The number of permutations interchangeable with a permutation $\sigma \in \mathfrak{S}_n$ of type $(\nu_1, \ldots, \nu_n)$ is $\nu_1! \cdots \nu_n! 1^{\nu_1} \cdots n^{\nu_n}$, $n \in \mathbb{N}$. (These permutations form the centralizer $Z_{\mathfrak{S}_n}(\sigma)$. – See Examples 2.5.13 and 2.4.5.)

2.5 Permutation Groups 227

Exercise 2.5.18
a) The number of involutions (= reflections) in $\mathfrak{S}_{2n}$, which have no fixed point, is $1 \cdot 3 \cdots (2n-1) = (2n)!/(n!2^n) \sim \sqrt{2}(2n/e)^n$ for $n \to \infty$. This is also the number of partitions of $\mathbb{N}^*_{\leq 2n}$ into pairs.
b) The number of involutions (= reflections) in $\mathfrak{S}_n$ is $\sum_{k \geq 0} \binom{n}{2k} \frac{(2k)!}{k!2^k}$.

Exercise 2.5.19
a) The number of permutations in $\mathfrak{S}_n$ that have a fixed orbit $B \in \mathfrak{P}(\mathbb{N}^*_{\leq n})$ of length k is $(k-1)!(n-k)!$, $1 \leq k \leq n$.
b) The number of orbits of all permutations in $\mathfrak{S}_n$ is $n!H_n$, where $H_n = \sum_{k=1}^n 1/k$ is the n-th harmonic number. (Determine the number of pairs $(\sigma, B) \in \mathfrak{S}_n \times \mathfrak{P}(\mathbb{N}^*_{\leq n})$ for which B is an orbit of σ, and use a). – The average number of orbits of a permutation $\sigma \in \mathfrak{S}_n$ is therefore H_n. It is $H_n = \ln n + \gamma + \rho_n$ with the Euler number γ and a monotonically decreasing null sequence (ρ_n) cf. Example 3.3.8 (2). The average number of orbits therefore increases very slowly for $n \to \infty$ (like $\ln n$) towards ∞.)

Exercise 2.5.20 The number of permutations in $\mathfrak{S}_n$, in whose canonical cycle decomposition one (and thus exactly one) cycle of length $>n/2$ occurs, is $n! \sum_{n/2 < k \leq n} 1/k = n!(H_n - H_{\lfloor n/2 \rfloor}) \sim n! \ln 2$ for $n \to \infty$, cf. exercise 3.3.27a). So their average number is $H_n - H_{\lfloor n/2 \rfloor} \xrightarrow{n \to \infty} \ln 2 = 0{,}693147 \ldots$

Exercise 2.5.21
a) The number of permutations in $\mathfrak{S}_n$ that have no fixed point is $n! \sum_{k=0}^n (-1)^k/k! \sim n!/e$ for $n \to \infty$. Their average number is $\sum_{k=0}^n (-1)^k/k! \to e^{-1} = 0{,}367879 \ldots$ (The number of permutations in $\mathfrak{S}_n$, $n \geq 1$, that have a given fixed point for $n \to \infty$ is $(n-1)!$. Using the sieve formula from exercise 1.6.23, determine the number of permutations in $\mathfrak{S}_n$ that have at least one fixed point.)
b) The number of permutations in $\mathfrak{S}_n$ with exactly m fixed points is $[n]_{n-m} \sum_{k=0}^{n-m} (-1)^k/k!$, $0 \leq m \leq n$.

Exercise 2.5.22
a) Let $n \geq 2$. The number of permutations $\sigma \in \mathfrak{S}_n$ that have the inversion (i_0, j_0), $1 \leq i_0 < j_0 \leq n$, is $\binom{n}{2}(n-2)! = n!/2$.
b) The number of inversions of all permutations in $\mathfrak{S}_n$ is $\frac{1}{2}\binom{n}{2}n!$. (Determine the number of pairs $(\sigma, (i,j))$ with $\sigma \in \mathfrak{S}_n$ and $1 \leq i < j \leq n$, for which (i,j) is an inversion of σ, and use a). – The average number of inversions is therefore $\frac{1}{2}\binom{n}{2}$. The maximum number of inversions that a permutation from $\mathfrak{S}_n$ can have is $\binom{n}{2}$. It is only reached by the permutation $i \mapsto n+1-i$, $1 \leq i \leq n$, see Example 2.5.9 (2). – The average number of inversions can also be easily calculated with the result of exercise 2.5.10a).)

Exercise 2.5.23
a) Using the simplicity of the groups $\mathfrak{A}_n$, $n \geq 5$, cf. Theorem 2.5.19, prove that the group $\mathfrak{A}_n$ is the only non-trivial normal subgroup in the group $\mathfrak{S}_n$ for $n \geq 5$.
b) Let $n \geq 2$. The group $\mathfrak{S}_n$ is isomorphic to a subgroup of $\mathfrak{A}_{n+2}$, but to no subgroup of $\mathfrak{A}_{n+1}$.
c) For every infinite set I, the alternating group $\mathfrak{A}(I) \subseteq \mathfrak{S}_{\text{fin}}(I)$ is simple.

Exercise 2.5.24
a) The subgroups $\mathfrak{A}_4$ and $\mathfrak{V}_4$ are the only non-trivial normal divisors in $\mathfrak{S}_4$. (See Example 2.5.15. – If X is any set with four elements, then the **Klein four-group** $\mathfrak{V}(X) \subseteq \mathfrak{S}(X)$ as the only normal divisor of order 4 in $\mathfrak{S}(X)$ is well-defined. If $G \cong \mathbf{Z}_2 \times \mathbf{Z}_2$ is an abstract Klein four-group, then the Klein four-group $\mathfrak{V}(G)$ is the image of the regular (= Cayley) representation $L: G \to \mathfrak{S}(G)$.)
b) The group $\mathfrak{V}_4$ is the only non-trivial normal divisor in $\mathfrak{A}_4$.
c) Let G be an arbitrary group of order $12 = 2^2 \cdot 3$. Then one of the nontrivial Sylow groups of G is a normal subgroup in G and G is thus a semidirect product of Sylow groups. (Count the elements of G and use Sylow's Theorem 2.4.11). If both the 2- and 3-Sylow groups of G are normal, then G is abelian, thus isomorphic to $\mathbf{Z}_{12}$ or to $\mathbf{Z}_6 \times \mathbf{Z}_2$. Let G be non-abelian. If the 2-Sylow group of G is normal, then $G \cong \mathfrak{A}_4$. If the 2-Sylow group of G is not normal, then $G \cong \mathbf{Z}_3 \rtimes_\vartheta \mathbf{Z}_4$, where $\vartheta: \mathbf{Z}_4 \to \text{Aut}\,\mathbf{Z}_3$ is the only nontrivial homomorphism, or $G \cong \mathbf{Z}_3 \rtimes_\eta (\mathbf{Z}_2 \times \mathbf{Z}_2) \cong \mathbf{D}_6 \cong \mathbf{Z}_2 \times \mathbf{D}_3$, where $\eta: \mathbf{Z}_2 \times \mathbf{Z}_2 \to \text{Aut}\,\mathbf{Z}_3$ is one of the three nontrivial homomorphisms. Thus, up to isomorphism, there are 5 groups of order 12. (The group $\mathbf{Z}_3 \rtimes_\vartheta \mathbf{Z}_4$ is the so-called quaternion group Q_6.)[33]

Exercise 2.5.25 The cube group W in Example 2.5.30 (4) also operates transitively on the 8 corners, 12 edges, 4 diagonals, and 6 diagonal planes of the cube. Determine the associated cycle polynomials. The last two are identical to the cycle polynomial $\psi(\mathfrak{S}_4)$ and the cycle polynomial ψ_4 from Example 2.5.30 (3). Justify this a priori.

Exercise 2.5.26
a) The square group $\mathbf{D}_4$ operates in a natural way on the set of n^2 squares of an $n \times n$ chessboard, $n \in \mathbb{N}^*$. Determine the associated cycle polynomial. (Distinguish the cases that n is odd or even, and conclude by induction from n to $n+2$.) Also determine the cycle polynomial for the subgroup $\mathbf{C}_4 \subseteq \mathbf{D}_4$ of the four rotations. In how many essentially different ways can the squares of an $n \times n$ chessboard be colored with two colors, where one color is used α times and the other color $\beta \ (= n^2 - \alpha)$ times?

---

[33] The reader should create a list of the 28 groups of order ≤ 15 (up to isomorphism). A group of order 15 is cyclic, its 5-Sylow group is necessarily normal, and any operation of the group $\mathbf{Z}_3$ on $\mathbf{Z}_5$ as a group of automorphisms is trivial. For groups of order 8 see Problem 2.4.7.

2.6 Rings

b) For the natural operation of the dihedral group D_2 on the set of squares of a rectangular $m \times n$ chessboard, $m, n \in \mathbb{N}^*$, answer the questions analogous to a).
c) The cube group W (see Example 2.5.30 (4)) operates in a natural way on the total 54 squares of the 6 sides of a cube, each divided into 9 square fields. (Think of the Rubik's cube.) Determine the associated cycle polynomial. How many essentially different colorings with 6 colors are there, each color appearing exactly 9 times?

Exercise 2.5.27 The finite groups G and H operate on the finite sets X and Y. Then the product group $G \times H$ operates in a natural way on the disjoint union $X \uplus Y$ with $\vartheta_{(g,h)}|X = \vartheta_g$ and $\vartheta_{(g,h)}|Y = \vartheta_h$. For the cycle polynomials, $\psi(G \times H) = \psi(G)\psi(H)$ applies.

Exercise 2.5.28 In the situation of the Pólya's enumeration formula 2.5.32, let the weight $\gamma(y)$ for each color y be a monomial $\prod_{i \in I} T_i^{\beta_i(y)}$ in the (commuting) indeterminates T_i, $i \in I$. Interpret the coefficient of $\prod_{i \in I} T_i^{\alpha_i}$, $(\alpha_i) \in \mathbb{N}^{(I)}$, in the polynomial $\sum_{[f]} \gamma([f]) = \psi(G)(\pi_1, \ldots, \pi_n)$ from Theorem 2.5.32. (G. Pólya discusses this case in detail in the work mentioned at the beginning of Example 2.5.30.)

2.6 Rings

In this section, we study sets with two (compatible) operations. Such structures are familiar to us from the number ranges $\mathbb{Z}, \mathbb{Q}, \mathbb{R}, \mathbb{C}$ or also $\mathbb{Z}/\mathbb{Z}n$, $n \in \mathbb{N}^*$. One of the two operations is addition, usually written additively with the operation symbol +, the other is multiplication, usually written multiplicatively.

Definition 2.6.1 A set $A = (A, +, \cdot)$ with the two operations + (**addition**) and $\cdot$ (**multiplication**) is called a **ring**, if the following conditions are met:

(1) $(A, +)$ is an abelian group.
(2) $(A, \cdot)$ is a monoid.
(3) The so-called **distributive laws** : For all $x, y, z \in A$ is

$$x \cdot (y + z) = (x \cdot y) + (x \cdot z) \quad \text{and} \quad (y + z) \cdot x = (y \cdot x) + (z \cdot x).$$

The neutral element of the **additive group** $(A, +)$ of a ring A is called the **zero** of A and is (as before) denoted by $0 = 0_A$. The neutral element of the **multiplicative monoid** $(A, \cdot)$ of A is called the **one** of A and is (as occasionally before) denoted by $1 = 1_A$. – If $(A, \cdot)$ is a commutative monoid, then A is called a **commutative ring**. – A mapping $\varphi: A \to B$ of rings $A = (A, +, \cdot)$ and $B = (B, +, \cdot)$ is called a **(ring-)homomorphism**, if φ is both a homomorphism of the additive groups $(A, +)$ and $(B, +)$ as well as a homomorphism of the multiplicative monoids $(A, \cdot)$ and $(B, \cdot)$, if for all $x, y \in A$ applies

$$\varphi(x + y) = \varphi(x) + \varphi(y), \quad \varphi(x \cdot y) = \varphi(x) \cdot \varphi(y) \quad \text{as well as} \quad \varphi(1_A) = 1_B.$$

To simplify the notation, it is generally agreed that multiplication takes precedence over addition (*multiplication before addition!*). The distributive laws (3) are then briefly $x(y+z) = xy+xz$, $(y+z)x = yx+zx$. More precisely, multiplication is said to be distributive over addition. In a commutative ring, the second distributive law naturally follows from the first. Elements x,y of a ring are called **interchangeable** or **commuting**, if they are interchangeable with respect to multiplication, i.e., if $xy = yx$. A ring is commutative if and only if any two of its elements commute. We will apply the concepts and results about groups and monoids and their homomorphisms developed mainly in Sects. 2.1, 2.2 and 2.3– sometimes without comment – to the additive groups and the multiplicative monoids of rings. Distinguish the translations $\tau_a \colon x \mapsto x+a = a+x$ in the group $(A, +)$ and the translations $L_a \colon x \mapsto ax$ or $R_a \colon x \mapsto xa$ in the monoid $(A, \cdot)$, $a \in A$. The former are also called **shifts**, the latter **dilations**. The ring $(A, +, \cdot^{\mathrm{op}})$ with the multiplication $a \cdot^{\mathrm{op}} b = ba$ oppositional to $\cdot$ is called the **oppositional ring** A^{op} to A.

The distributive laws describe the compatibility of addition and multiplication in a ring A. They state that the left and right translations L_a and R_a, $a \in A$, of the multiplicative monoid $(A, \cdot)$ are homomorphisms of the additive group $(A, +)$. This already implies the known sign rules, see Proposition 2.2.3:

Proposition 2.6.2 *A is a ring. Then for all $x, y, z \in A$:*

(1) $x \cdot 0 = 0 = 0 \cdot x.$ (2) $x(-y) = (-x)y = -xy.$ (3) $(-x)(-y) = xy.$

(4) $x(y-z) = xy - xz$ *and* $(y-z)x = yx - zx.$

The zero element 0_A is thus an absorbing element for the multiplicative monoid of a ring A. This alone prevents $(A, \cdot)$ from being a group, unless A is the so-called **zero ring**, which contains only 0_A, for which $0_A = 1_A$ is. This last equation already implies that A is the zero ring; because then $x = 1_A \cdot x = 0_A \cdot x = 0_A$ for all $x \in A$. The distributive laws provide by induction the **general distributive law**: If $x_i, i \in I$, and y_j, $j \in J$, are finite families of elements of the ring A (or families whose elements are almost all 0), then

$$\left(\sum_{i \in I} x_i\right)\left(\sum_{j \in J} y_j\right) = \sum_{(i,j) \in I \times J} x_i y_j.$$

applies. The special case

$$\left(\sum_{i=1}^m x_i\right)\left(\sum_{j=1}^n y_j\right) = \sum_{1 \leq i \leq m, 1 \leq j \leq n} x_i y_j = \sum_{i=1}^m \sum_{j=1}^n x_i y_j$$

illustrates the scheme from Fig. 2.19. If the x_i and the y_j match, the general distributive law yields the calculation rules for the multiples

$$(mx)(ny) = (mn)(xy),$$

which, due to the sign rules, apply to all $x, y \in A$ and all $m, n \in \mathbb{Z}$ (and not just for $m, n \in \mathbb{N}$). In particular, $mx = (m1_A)x$ applies to all $x \in A$ and $m \in \mathbb{Z}$. From

2.6 Rings

$$\sum_{i=1}^{m}\sum_{j=1}^{n} x_i y_j = \begin{array}{l} x_1 y_1 + \\ + x_2 y_1 + \\ \\ + x_m y_1 + \end{array} \begin{array}{l} x_1 y_2 + \cdots + \\ x_2 y_2 + \cdots + \\ \cdots \cdots \cdots \cdots \\ x_m y_2 + \cdots + \end{array} \begin{array}{l} x_1 y_n \\ x_2 y_n \\ \\ x_m y_n \end{array}$$

$$= \left(\sum_{i=1}^{m} x_i\right) y_1 + \left(\sum_{i=1}^{m} x_i\right) y_2 + \cdots + \left(\sum_{i=1}^{m} x_i\right) y_n = \left(\sum_{i=1}^{m} x_i\right)\left(\sum_{j=1}^{n} y_j\right)$$

Fig. 2.19 General distributive law

Fig. 2.20 Ring with 2 elements

+	0	1
0	0	1
1	1	0

·	0	1
0	0	0
1	0	1

$m1_A = 0_A$ it follows that $mx = 0_A$ applies to all $x \in A$. In other words: The order *in the additive group* Ord 1_A *is the exponent of* $(A, +)$. This natural number is called the **characteristic** of A and is denoted by

$$\operatorname{Char} A = \operatorname{Ord} 1_A = \operatorname{Exp}(A, +)$$

Char $A = 0$ means that all multiples $n1_A$, $n \in \mathbb{Z}$, are pairwise different, Char $A = n > 0$ means that the elements $1_A = 1 \cdot 1_A, \ldots, (n-1)1_A$ of 0_A are different, but $n1_A = 0_A$ is. After the zero ring, whose characteristic is equal to 1, the ring that contains only the two *different* elements $0 = 0_A$ and $1 = 1_A$ is the smallest ring. The operation tables for its addition and multiplication are necessarily those from Fig. 2.20. It has the characteristic 2 and is even a field, see Example 2.6.16. It is denoted by A_2 or F_2. In a ring A of characteristic 2, $x = -x$ for all $x \in A$, $(A, +)$ is then an elementary 2-group.

Further calculation rules, which result from the distributive laws, are the binomial formula and more generally the polynomial formula, which we have already given in the Theorems 1.6.15 and 1.6.16 for real or complex numbers, whose proofs can be literally transferred to the general situation given here and therefore are not repeated.

Theorem 2.6.3 (Binomial Theorem – Polynomial Theorem) *For commutable elements x, y of a ring A and every $n \in \mathbb{N}$ applies*

$$(x+y)^n = \sum_{m=0}^{n} \binom{n}{m} x^m y^{n-m}.$$

More generally, for pairwise commutable elements $x_1, \ldots, x_r \in A$

$$(x_1 + \cdots + x_r)^n = \sum_{\substack{m \in \mathbb{N}^r \\ |m|=n}} \binom{n}{m} x^m = \sum_{\substack{m=(m_1,\ldots,m_r) \in \mathbb{N}^r \\ |m|=m_1+\cdots+m_r=n}} \frac{n!}{m_1! \cdots m_r!} x_1^{m_1} \cdots x_r^{m_r}.$$

Note that due to $(x+y)^2 = x^2 + xy + yx + y^2$ the classic "first binomial formula" $(x+y)^2 = x^2 + 2xy + y^2$ *only applies* when x and y are commutable. In the case of prime characteristic, the binomial theorem has an important and interesting corollary:

Corollary 2.6.4 *Let A be a ring of prime characteristic $p \in \mathbb{P}$. Then for commuting elements $x, y \in A$ and all $n \in \mathbb{N}$*

$$(x+y)^{p^n} = x^{p^n} + y^{p^n}.$$

In particular, the mapping $x \mapsto x^{p^n}$ for each $n \in \mathbb{N}$ is a ring endomorphism of A, if A is additionally commutative.

Proof It suffices to handle the case $n = 1$ (induction over n). Since the binomial coefficients $\binom{p}{m}$ for $0 < m < p$ are divisible by p according to exercise 1.7.19, the statement directly follows from the binomial Theorem 2.6.3.

For a commutative ring A of prime characteristic $p \in \mathbb{P}$, the endomorphism

$$f: A \to A, \quad x \mapsto x^p,$$

is called the **Frobenius homomorphism** of the ring A. Its iterates are the endomorphisms $f^n: x \mapsto x^{p^n}, x \in A, n \in \mathbb{N}$.

By directly distributing and multiplying out, one also proves the following formula (which is one of the most frequently used in mathematics).

Theorem 2.6.5 (Geometric Series) *For commutative elements x, y of a ring A and any $n \in \mathbb{N}$ the following holds*

$$(x-y)\sum_{m=0}^{n} x^{n-m}y^m = (x-y)(x^n + x^{n-1}y + \cdots + xy^{n-1} + y^n) = x^{n+1} - y^{n+1}.$$

In particular, for any $y \in A$ and any $n \in \mathbb{N}$

$$(1-y)\sum_{m=0}^{n} y^m = (1-y)(1 + y + \cdots + y^{n-1} + y^n) = 1 - y^{n+1}.$$

Note that because of $(x-y)(x+y) = x^2 + xy - yx - y^2$ the "third binomial formula" $(x-y)(x+y) = x^2 - y^2$ *only then* applies when x and y are interchangeable.

Remark 2.6.6 (Bi- and multi-additive mappings) The distributive law in a ring states that the multiplication $A \times A \to A$ in a ring A is bi-additive in the following sense: ◊

Definition 2.6.7 Let F, G, H be additive abelian groups. A mapping $f: F \times G \to H$ is called **biadditive**, if for all $x, x_1, x_2 \in F$ and all $y, y_1, y_2 \in G$ the following distributivities apply:

2.6 Rings

$$f(x, y_1 + y_2) = f(x, y_1) + f(x, y_2) \quad \text{and} \quad f(x_1 + x_2, y) = f(x_1, y) + f(x_2, y),$$

if all partial mappings $f(x, -) \colon G \to H$, $x \in F$, and $f(-, y) \colon F \to H$, $y \in G$, are group homomorphisms. Similarly, for a *finite* family G_i, $i \in I$, of additive abelian groups a **multiadditive** mapping $f \colon \prod_{i \in I} G_i \to H$ is defined: When all arguments except the i-th are fixed, f defines a homomorphism $G_i \to H$, $i \in I$.

If $I = \emptyset$, then f is simply an element of H, with $I = \{i\}$ f is a homomorphism $G_i \to H$, with $|I| = 3$ we speak of triadditive mappings etc. For multiadditive mappings, the statements analogous to the above general distributive laws apply. In particular, the value of a multiadditive mapping is always equal to 0 when at least one component of the argument tuple vanishes. The general distributive laws have, among other things, the following consequence:

Lemma 2.6.8 *If* x_{ij_i}, $j_i \in J_i$, *are generating systems of the additive abelian groups* G_i, $i \in I$, *then two multiadditive mappings* $f, g \colon \prod_{i \in I} G_i \to H$ *already coincide if they agree on all I-tuples* $(x_{ij_i})_{i \in I}$, $(j_i) \in \prod_{i \in I} J_i$.

An additive group A together with a biadditive, i.e., distributive multiplication $A \times A \to A$ is called a **generalized ring**. $(A, \cdot)$ is thus only a magma. To verify that such a generalized ring A is even a ring, one has to show the associative law and the existence of a neutral element for multiplication. But since the two mappings $A^3 \to A$, $(x, y, z) \mapsto (xy)z$ or $(x, y, z) \mapsto x(yz)$ are triadditive, it is sufficient to show for the associativity of multiplication that $(x_{i_1} x_{i_2}) x_{i_3} = x_{i_1}(x_{i_2} x_{i_3})$ for arbitrary elements $x_{i_1}, x_{i_2}, x_{i_3}$ of a generating system x_i, $i \in I$, of the additive group of A is. Furthermore, in this case, 1_A is already a unit element when $1_A x_i = x_i = x_i 1_A$ for all $i \in I$ applies. We will often use this reduction.

A subset A' of a ring A is called a **subring** of A, if A' is both a subgroup of $(A, +)$ and a submonoid of $(A, \cdot)$. Then A' with the induced operations is also a ring. The multiples of one form the smallest subring $\mathbb{Z} 1_A$ of A. It is called the **minimal ring of** A.[34] A ring that coincides with its minimal ring is called a **minimal ring** per se. Every subring of a ring A has the same characteristic $\text{Char } A = \text{Char } \mathbb{Z} 1_A$ as A.

Let A be a ring. If $a \in A$, then the centralizer $Z_A(a)$ is clearly a subring of A. Since the intersection of subrings is again a subring, for any subset $X \subseteq A$ the centralizer $Z_A(X) = \bigcap_{x \in X} Z_A(x)$ is a subring of A. *In particular, the* **center**

$$Z(A) = \{a \in A \mid ax = xa \text{ for all } x \in A\}$$

is a commutative subring of A.

The **unit group**

$$A^\times$$

---

[34] It is common to denote the element $n 1_A \in A$ for $n \in \mathbb{Z}$ briefly with n, if this cannot lead to misunderstandings. — In 15, the minimal rings are called prime rings.

of a ring A is by definition the unit group of its multiplicative monoid $(A, \cdot)$. Occasionally, it is also called the **multiplicative group** of A. Since an element $b \in A$, which commutes with a unit $a \in A^\times$, also commutes with a^{-1}, we have

$$Z(A)^\times = Z(A) \cap A^\times.$$

Similarly, the **regular elements** of the ring A are by definition the regular elements of $(A, \cdot)$. They form a submonoid

$$A^*$$

of $(A, \cdot)$. An element $a \in A$ is regular exactly when the left and right multiplication L_a or R_a with a are both injective, so if a can be shortened on both the left and right side. Since L_a and R_a are homomorphisms of the group $(A, +)$, this is exactly the case when $ab = 0$ or $ba = 0$ for an $b \in A$ only applies when $b = 0$. Therefore, the regular elements in A are also (not entirely consistently) called **non-zero divisors** and the elements of $A - A^*$ **zero divisors** of A. Always, $A^\times = (A^*)^\times \subseteq A^*$. If A is not the zero ring, then $0 \notin A^*$. A ring A with $A^* = A - \{0\}$ is called a **domain**. If A is also commutative, one speaks of an **integral domain**. *A ring A is exactly a domain when it is not the zero ring and 0 is the only zero divisor in A, so that $ab = 0$ for elements $a, b \in A$ holds exactly when $a = 0$ or $b = 0$. Every subring of a domain is also a domain.*

A domain A, in which all elements different from 0 are even units, for which thus $A^\times = A - \{0\}$ holds, is called a division domain or also a skew field. Because of the great importance of this term, we describe it again explicitly:

Definition 2.6.9 A set $K = (K, +, \cdot)$ with an addition $+$ and a multiplication $\cdot$ is called a **division domain** (or a **skew field**), if the following holds:

(1) $(K, +)$ is an abelian group (with neutral element 0).
(2) $(K - \{0\}, \cdot)$ is a group (with neutral element $1 \neq 0$).
(3) The distributive laws apply, i.e., the multiplication $K \times K \to K$ is biadditive.

If the group $(K - \{0\}, \cdot)$ is also commutative, then K is called a **field**.

If K is a division domain as in Definition 2.6.9, then the associativity of multiplication on all of K (and not just on $K^\times = K - \{0\}$) as well as the equations $1 \cdot 0 = 0 \cdot 1 = 0$ follow from $0 \cdot x = x \cdot 0 = 0$ for all $x \in K$, which is a consequence of the biadditivity of multiplication.

The characteristic of a domain A is 0 or a prime number. If indeed $m := \mathrm{Char}\, A > 0$, then $m > 1$ (because $A \neq 0$). If now $m = k\ell$ with $1 < k, \ell < m$, then $0_A = m \cdot 1_A = (k \cdot 1_A)(\ell \cdot 1_A)$, thus $k \cdot 1_A = 0_A$ or $\ell \cdot 1_A = 0_A$. Contradiction! *A finite domain A is a division ring*, because then $(A - \{0\}, \cdot)$ is a finite regular monoid and thus a group, cf. Exercise 2.1.9.[35] *The center* $Z(D)$ *of a division ring* D *is a field* (because of $Z(D)^\times = D^\times \cap Z(D) = Z(D) - \{0\}$).

Let A be any ring again. For a central unit $b \in Z(A)^\times = A^\times \cap Z(A)$ and any element $a \in A$, the **fractional notation**

$$\frac{a}{b} = a/b := ab^{-1} = b^{-1}a$$

is often used. Then $1/b = b^{-1}$, and for all $a \in A$ and $b, b' \in Z(A)^\times$, the **extension rule** (or **reduction rule** – depending on the perspective) $a/b = (ab')/(bb') = (b'a)/(b'b)$ applies because of $ab^{-1} = ab'b'^{-1}b^{-1} = (ab')(bb')^{-1} = (ab')/(bb')$. If $a' \in A$ is another element, then the **rules for fractions** apply:

$$\frac{a}{b} + \frac{a'}{b'} = \frac{ab'}{bb'} + \frac{ba'}{bb'} = (ab')(bb')^{-1} + (ba')(bb')^{-1} = (ab' + ba')(bb')^{-1}$$
$$= \frac{ab' + ba'}{bb'},$$
$$\frac{a}{b} \cdot \frac{a'}{b'} = ab^{-1}a'b'^{-1} = (aa')(b^{-1}b'^{-1}) = \frac{aa'}{bb'}.$$

Furthermore, $(b/b')^{-1} = b'/b$ because of $(b/b')(b'/b) = (bb')/(b'b) = 1/1 = 1$.

Example 2.6.10 (Semirings – Grothendieck Rings) The model example of a ring is the **ring** $\mathbb{Z}$ **of whole numbers** with its natural operations $+$ and $\cdot$. It is an integral domain and the Grothendieck ring to the semi-ring $\mathbb{N} = (\mathbb{N}, +, \cdot)$ of natural numbers, whose addition and multiplication when interpreting the elements of $\mathbb{N}$ as cardinal numbers of finite sets are represented by the set-theoretical operations "disjoint union" and "product formation", see Sects. 1.5 and 1.6 and in particular Remark 1.5.7.

Generally, a **semi-ring** is understood to be a set $A = (A, +, \cdot)$ with an addition $+$ and a multiplication $\cdot$ such that (1) $(A, +)$ is a (additively written) commutative monoid, (2) $(A, \cdot)$ is a monoid and (3) the left and right multiplications are monoid homomorphisms of $(A, +)$, so that the distributive laws $x(y + z) = xy + xz$, $(y + z)x = yx + zx$ apply and moreover $0_A \cdot y = 0_A = y \cdot 0_A$, $x, y, z \in A$. A semi-ring is a ring if its additive monoid is a group. *To every semi-ring A a ring can be assigned in a canonical way.* Its additive group is the Grothendieck or difference group $(G(A), +)$ of the additive monoid $(A, +)$, see Example 2.3.16. This refers to the quotient monoid of the direct sum $A \oplus A$ with respect to the compatible equivalence relation "$(a, b) \sim (a', b')$ only when there is a $u \in A$ with $a + b' + u = b + a' + u$". $A \oplus A$ also carries a canonical multiplication. Thinking of the distributive law $(a - b)(c - d) = (ac + bd) - (bc + ad)$, one defines

---

[35] *A finite division ring is always commutative, thus a field.* This **Wedderburn's theorem** will be proven in one of the later volumes.

$$(a,b) \cdot (c,d) := (ac+bd, bc+ad), \quad a,b,c,d \in A,$$

and easily checks that $A \oplus A$ with the addition and this multiplication is also a semi-ring with the unit element $(1,0)$. The above equivalence relation $\sim$ is also compatible with this multiplication. If one replaces (a,b) with the equivalent element (a',b'), then $a+b'+u = b+a'+u$ and consequently

$$ac+b'c+uc = bc+a'c+uc, \quad ad+b'd+ud = bd+a'd+ud,$$

$$(ac+bd+b'c+a'd) + (uc+ud) = (bc+ad+a'c+b'd) + (uc+ud),$$

that is also $(a,b)(c,d) = (ac+bd, bc+ad) \sim (a'c+b'd, b'c+a'd) = (a',b')(c,d)$. Analogously, one concludes that if (c,d) is replaced by an equivalent element (c',d'). The multiplication on $A \oplus A$ thus induces a multiplication on $G(A)$ such that $(G(A), +, \cdot)$ is a semi-ring. But since $(G(A), +)$ is an abelian group, it is a ring. It is called the **Grothendieck ring** (or **difference ring**) of A. If A (i.e., $(A, \cdot)$) is commutative, so is $G(A)$. The canonical mapping $\iota_A \colon A \to G(A)$, $a \mapsto [a] = [(a,0)]$, is a homomorphism of semi-rings and it is

$$[(a,b)] = \iota_A(a) - \iota_A(b) = [a] - [b], \quad a,b \in A.$$

The pair $(G(A), \iota_A)$ obviously has the following universal property, see Theorem 2.3.19: ◊

Theorem 2.6.11 (Universal Property of the Grothendieck Ring) *Let $A = (A, +, \cdot)$ be a semiring, $B = (B, +, \cdot)$ a ring and $G(A) = (G(A), +, \cdot)$ the Grothendieck ring of A. If $\varphi \colon A \to B$ is a homomorphism of semirings, then there exists exactly one ring homomorphism $\widetilde{\varphi} \colon G(A) \to B$ with $\varphi = \widetilde{\varphi} \circ \iota_A$. It is $\widetilde{\varphi}([x] - [y]) = \varphi(x) - \varphi(y)$, $x, y \in A$.*

Let the additive semigroup $(A, +)$ of the semiring A be regular. Then ι_A is injective and one identifies A with its image in $G(A)$. Furthermore, it is then obviously $A^* \subseteq G(A)^*$. But if A is a domain, i.e. $A^* = A - \{0\}$, then $G(A)$ does not have to be a domain, see exercise 2.6.12. However, this is the case if $G(A) = AG(A)^* = G(A)^* A$ is (as it is for $A = \mathbb{N}$ from the outset). *The Grothendieck ring of the semiring $\mathbb{N}$ is $\mathbb{Z}$.* The Burnside ring $B(G)$ of a finite group G is the Grothendieck ring to the semiring of isomorphism classes of finite G-spaces, see Example 2.4.14.

Example 2.6.12 (Residue Class Rings of $\mathbb{Z}$ – Classification of Minimal Rings) Let A be a ring. Then there is exactly one homomorphism $\chi = {}_A\chi \colon \mathbb{Z} \to A$ of the additive groups of $\mathbb{Z}$ and A with $\chi(1) = 1_A$. According to the calculation rules for multiplication, χ is also multiplicative. *Therefore, ${}_A\chi$ is the only ring homomorphism $\mathbb{Z} \to A$.* It is called the **characteristic homomorphism** of A. Its image is the minimal ring $\mathbb{Z}1_A$ of A, and its kernel is $\mathbb{Z}\operatorname{Ord} 1_A = \mathbb{Z}\operatorname{Char} A$ (by definition, $\operatorname{Ord} 1_A = \operatorname{Char} A$). Thus, χ initially induces an isomorphism

2.6 Rings

$$\overline{\chi}\colon \mathbb{Z}/\mathbb{Z}\operatorname{Char} A \xrightarrow{\sim} \mathbb{Z}1_A$$

of additive groups. This is obviously also compatible with multiplication and *thus an isomorphism of rings*. For the endomorphism rings of abelian groups, we have already noted this in Example 2.2.23. So it follows: ◊

Theorem 2.6.13 *Every minimal ring A of characteristic $m \in \mathbb{N}$ is isomorphic to the residue class ring $\mathbb{Z}/\mathbb{Z}m$, and there is exactly one isomorphism $\mathbb{Z}/\mathbb{Z}m \xrightarrow{\sim} A$.*

The residue class rings $\mathbb{Z}/\mathbb{Z}m$, $m \in \mathbb{N}$, thus represent all minimal rings up to unique isomorphism. We use

$$\mathbf{A}_m := (\mathbb{Z}/\mathbb{Z}m, +, \cdot), \ m \in \mathbb{N}^*, \quad \text{resp.} \quad \mathbf{A}_0 := (\mathbb{Z}, +, \cdot)$$

as standard models for a minimal ring of characteristic $m \in \mathbb{N}^*$ or for a minimal ring of characteristic 0, but also denote any other minimal ring of characteristic m with $\mathbf{A}_m$. The identification with $\mathbb{Z}/\mathbb{Z}m$ is then unique (unlike for cyclic groups). In particular, $\mathbf{A}_m = \mathbb{Z}1_A \subseteq A$ is the minimal ring for every ring A of characteristic m. The additive group of the *ring A_m* is the cyclic *group $\mathbf{Z}_m = (\mathbf{A}_m, +)$*. The unit group $\mathbf{A}_m^\times = (\mathbf{A}_m, \cdot)^\times$ contains exactly the elements $a \cdot 1_{\mathbf{A}_m}$ with $a \in \mathbb{Z}$, $\gcd(a, m) = 1$. For $m > 0$, its order is $\varphi(m)$. In particular, A_m is an (integrity-) domain exactly when $m \in \overline{\mathbb{P}} := \mathbb{P} \uplus \{0\}$ is.[36] and is a field if and only if $m \in \mathbb{P}$ is a prime number. For $m > 0$, $\mathbf{A}_m^\times$ is called the **prime residue class group modulo** m, see Example 2.7.10.

Example 2.6.14 (Rings of Fractions – Total Quotient Rings) The rational numbers form a field with the usual addition and multiplication $\mathbb{Q}$. This is the field of fractions a/b, $a, b \in \mathbb{Z}$, $b \neq 0$. This transition from the integral domain $\mathbb{Z}$ to the field $\mathbb{Q}$ can be significantly generalized, which is of particular importance in commutative algebra.

Let A be a ring and $S \subseteq Z(A)$ a central submonoid of the multiplicative monoid $(A, \cdot)$ of A. Already in Example 2.3.16 we have the monoid $S^{-1}A = A_S$ of fractions a/s, $a \in A$, $s \in S$, with the canonical homomorphism $\iota_S \colon A \to A_S$, $a \mapsto a/1$, constructed. This is the quotient monoid of the product monoid $A \times S$ with respect to the compatible equivalence relation $\equiv_S$, which is defined by

$$(a, s) \equiv_S (b, t) \quad \Longleftrightarrow \quad \text{there is } u \in S \text{ with } atu = bsu.$$

a/s is the equivalence class of (a, s). Thinking of the addition of fractions (see above), one defines on $A \times S$ the addition

$$(a, s) + (b, t) := (at + bs, st), \quad a, b \in A, \ s, t \in S.$$

---

[36] $\overline{\mathbb{P}}$ is the set of *all* prime elements of $(\mathbb{N}, \cdot)$.

Thus, $A \times S$ is clearly a commutative additive monoid with neutral element $(0, 1)$. The equivalence relation $\equiv_S$ is also compatible with this addition. If $(a', s') \equiv_S (a, s)$, i.e. $a'su = as'u$ with a $u \in S$, then

$$(at + bs)s'tu = ats'tu + bss'tu = a'tstu + bs'stu = (a't + bs')stu$$

and thus $(a, s) + (b, t) = (at + bs, st) \equiv_S (a't + bs', st) = (a', s') + (b, t)$. A_S is therefore with the induced addition

$$\frac{a}{s} + \frac{b}{t} = \frac{at + bs}{st}, \quad a, b \in A, \ s, t \in S,$$

a commutative monoid with zero element $0 = 0/1$, cf. Theorem 2.3.4. *It is even a group*; because $(-a)/s$ is due to $a/s + (-a)/s = (as + (-a)s)/s^2 = 0/s^2 = 0$ the negative of a/s. The multiplication on A_S is also distributive over addition, as the reader can easily confirm.[37] Overall, A_S is a ring, and $\iota_S \colon A \to A_S$ is not only multiplicative but due to $a/1 + b/1 = (a \cdot 1 + b \cdot 1)/1 \cdot 1 = (a + b)/1$ also additive, thus a ring homomorphism

$$S^{-1}A = A_S$$

is called the **ring of fractions** or the **fraction ring of A with denominators in S**. Furthermore, the universal property of (A_S, ι_S) from Theorem 2.3.18 transfers to a universal property of rings. We have proven: ◊

Theorem 2.6.15 *Let A be a ring and $S \subseteq Z(A)$ a central submonoid of the multiplicative monoid $(A, \cdot)$. Then the ring of fractions A_S with the canonical ring homomorphism $\iota_S \colon A \to A_S$, $a \mapsto a/1$, has the following universal property: If $\varphi \colon A \to B$ is a homomorphism of rings with $\varphi(S) \subseteq B^\times$, then there exists exactly one ring homomorphism $\varphi_S \colon A_S \to B$ with $\varphi = \varphi_S \circ \iota_S$. It is*

$$\varphi_S(a/s) = \varphi(a)\varphi(s)^{-1} = \varphi(s)^{-1}\varphi(a), \quad a \in A, \ s \in S.$$

Note that in the situation of Theorem 2.6.15 $\varphi(S)$ lies in the center of the image ring $\varphi(A)$. Exactly then ι_S is injective, if $S \subseteq Z(A) \cap A^*$ holds. In this case, one usually identifies the elements $a \in A$ with the elements $a/1 \in A_S$. Then a/s is simply the quotient $as^{-1} = s^{-1}a$, $a \in A$, $s \in S$. If $T \subseteq A$ is an arbitrary subset of $Z(A)$, then by definition $A_T = T^{-1}A := A_{\langle T \rangle}$, where $\langle T \rangle$ is the central submonoid of $(A, \cdot)$ generated by T. Exactly then is A_T the zero ring, when $0 \in \langle T \rangle$ (but not necessarily $0 \in T$) applies. In particular, for $s \in A$ the ring A_s is the fraction ring with the denominators s^n, $n \in \mathbb{N}$.

A special case is particularly important, namely that A is commutative and S is the monoid A^* of the regular elements of A. In this case,

$$Q(A) := A_{A^*}$$

---

[37] The distributive laws generally do *not* apply in $A \times S$, i.e., $A \times S$ is not a semi-ring.

is called the **total ring of quotients** of A. Every fraction ring A_S of A with denominators $S \subseteq A^*$ can be identified with the subring $\{a/s \mid a \in A, s \in S\} \subseteq Q(A)$ of the total ring of quotients of A. This is exactly the case when $A = Q(A)$, if $A^\times = A^*$ is. If A is an integral domain, then $A^* = A - \{0\}$ and $Q(A)$ is a field, the so-called **field of quotients** of A. The inverse to $a/b \neq 0 = 0/1$ is the reciprocal b/a. *The field $\mathbb{Q}$ of rational numbers is the field of quotients of $\mathbb{Z}$.*

Example 2.6.16 (Classification of Prime Fields) Let D be a division ring. Then the average of all sub-division rings of D is also a division ring and thus the smallest sub-division ring. Since the center $Z(D)$ of D is a field, this smallest sub-division ring of D is also a field. It is called the **prime field** of D and naturally contains the minimal ring $\mathbf{A}_{\mathrm{Char}\, D}$ of D. The characteristic $\mathrm{Char}\, D$ of D is 0 or a prime number $p \in \mathbb{P}$.

If $\mathrm{Char}\, D = p \in \mathbb{P}$, then the minimal ring A_p is already a field and thus the prime field of D. (That $\mathbf{A}_p = \mathbb{Z}/\mathbb{Z}p$ is a field, is also evident from the following simple argument: If $[a]_p \neq [0]_p$, then the subgroup generated by $[a]_p$ of $\mathbb{Z}/\mathbb{Z}p$ is equal to $\mathbb{Z}/\mathbb{Z}p$, since $\mathbb{Z}/\mathbb{Z}p$ has no non-trivial subgroups. Thus, there is an $b \in \mathbb{Z}$ with $[1]_p = b[a]_p = b[1]_p[a]_p = [b]_p[a]_p$.)

Let $\mathrm{Char}\, D = 0$ be. Then the minimal ring $A_0 = \mathbb{Z}$ of D is not a field. However, the canonical embedding $_D\chi : \mathbb{Z} \to D$ can be uniquely extended due to the universal property of the quotient field $\mathbb{Q} = Q(\mathbb{Z})$ to the also injective homomorphism $\mathbb{Q} \to D$ with $a/b \mapsto a/b = a1_D/b1_D$. Its image is the prime field of D. We have proven: ◊

Theorem 2.6.17 *For every $p \in \overline{\mathbb{P}} = \mathbb{P} \uplus \{0\}$ there is exactly one prime field of characteristic p up to isomorphism. This is for $p \in \mathbb{P}$ isomorphic to the residue class field $\mathbb{Z}/\mathbb{Z}p$ and for $p = 0$ isomorphic to $\mathbb{Q}$.*

The residue class fields $\mathbf{A}_p = \mathbb{Z}/\mathbb{Z}p$, $p \in \mathbb{P}$, as well as the field $\mathbb{Q}$ of rational numbers thus represent all prime fields up to isomorphism. We use

$$\mathbf{F}_p := (\mathbb{Z}/\mathbb{Z}p, +, \cdot),\ p \in \mathbb{P}, \quad \text{resp.} \quad \mathbf{F}_0 := (\mathbb{Q}, +, \cdot)$$

as standard models for a prime field of characteristic $p \in \mathbb{P}$ or for a prime field of characteristic 0, but also denote the prime field of any division area of characteristic $p \in \overline{\mathbb{P}}$ with F_p. The identification with $\mathbb{Z}/\mathbb{Z}p$ or $\mathbb{Q}$ is then unique.

Example 2.6.18 (Product Rings) Let B_i, $i \in I$, be a family of rings with the zero and one elements 0_i or 1_i, $i \in I$. Then the product

$$B := \prod_{i \in I} B_i$$

is also a ring, where $(B, +)$ is the product of the additive group $(B_i, +)$ and $(B, \cdot)$ is the product of the multiplicative monoids $(B_i, \cdot)$. In the **product ring** B is therefore added and multiplied component-wise. Its zero element is $0 = 0_B = (0_i)_{i \in I}$

and its unit element is $1 = 1_B = (1_i)_{i \in I}$. Together with the canonical projections $p_i : B \to B_i$ it has the following **universal property**: If A is any ring and $\varphi_i : A \to B_i$ is a family of ring homomorphisms, then there is exactly one ring homomorphism $\varphi = (\varphi_i)_{i \in I} : A \to B$ with $\varphi_i = p_i \varphi$, $i \in I$, i.e., *the canonical mapping*

$$\operatorname{Hom}\left(A, \prod_{i \in I} B_i\right) \xrightarrow{\sim} \prod_{i \in I} \operatorname{Hom}(A, B_i), \quad \varphi \mapsto (p_i \varphi)_{i \in I},$$

is bijective. It is $\varphi(a) = \bigl(\varphi_i(a)\bigr)_{i \in I}$. If the family $\psi_i : C \to B_i$, $i \in I$, defines an isomorphism $\psi = (\psi_i)_{i \in I} : C \xrightarrow{\sim} \prod_{i \in I} B_i$, then one says that C together with the homomorphisms ψ_i, $i \in I$, represents the product of the B_i, $i \in I$. For the product ring $B = \prod_i B_i$ it holds that $B^* = \prod_i B_i^*$, $B^\times = \prod_i B_i^\times$ and $Z(B) = \prod_i Z(B_i)$. Also important are the elements

$$e_J := (e_{J,k})_{k \in I} \quad \text{with} \quad \begin{cases} e_{J,k} = 1_k, & \text{if } k \in J, \\ e_{J,k} = 0_k, & \text{if } k \in I - J, \end{cases} \quad J \subseteq I.$$

These elements e_J are evidently central idempotent elements of B, i.e., $e_J \in Z(B)$ and $e_J^2 = e_J$ apply. In general, the **idempotent elements** of a ring are by definition the idempotent elements of its multiplicative monoid (cf. also exercise 2.6.6). It is $e_{J \cap K} = e_J e_K$ and $e_{J \triangle K} = (e_J - e_K)^2$ for $J, K \subseteq I$. The family $e_i := e_{\{i\}} = (\delta_{ik})_{k \in I}$, $i \in I$, is **orthogonal**, i.e.,

$$e_i e_j = \delta_{ij} e_i, \quad i, j, \in I.$$

applies. Furthermore, $Be_i = e_i B$ with the operations induced by B is a ring with unit element e_i (but not a subring of B, if $B_j \neq \{0_j\}$ is for at least one $j \neq i$), and the mapping $p_i | Be_i : Be_i \xrightarrow{\sim} B_i$ is a ring isomorphism, $i \in I$. If I is finite, then the family e_i, $i \in I$, is also **complete**, i.e., the completeness relation

$$\sum_{i \in I} e_i = 1 = 1_B.$$

additionally applies. For finite I, e_i, $i \in I$, is therefore a complete orthogonal family of central idempotent elements of the product ring B. Conversely, one can immediately see: If B is any ring and e_i, $i \in I$, a finite complete orthogonal family of central idempotent elements of B, then $B_i := Be_i = e_i B$, $i \in I$, with the operations induced by B are rings and $\prod_i B_i \xrightarrow{\sim} B$, $(b_i)_{i \in I} \mapsto \sum_i b_i$, is a ring isomorphism. The finite product decompositions of a ring B thus correspond reversibly uniquely to the finite complete orthogonal families of central idempotent elements of B. In particular, every central idempotent element e of a ring B provides the central orthogonal family $e, 1 - e$ and thus the product decomposition $Be \times B(1 - e) \xrightarrow{\sim} B = Be \oplus B(1 - e)$. A ring B is called **indecomposable** or **connected** if it is not the zero ring and is not isomorphic to a product $B_1 \times B_2$ of two rings $B_1, B_2 \neq 0$. *A ring that is different from the zero ring is*

2.6 Rings

thus indecomposable exactly when it has no central idempotent elements other than 0 and 1. A domain is always indecomposable.—Another method to describe finite product representations of rings is given at the end of Example 2.7.8.—If $B_i = A$ for all $i \in I$ with a fixed ring A, then the product ring $\prod_i B_i$ is the ring A^I of the A-valued functions $I \to A$, which are added and multiplied pointwise: $(f+g)(i) = f(i) + g(i)$, $(fg)(i) = f(i)g(i)$, $f, g \in A^I$, $i \in I$. The elements $e_J \in A^I$, $J \subseteq L$, are the indicator functions with $e_J(i) = 1$ for $i \in J$ and $e_J(i) = 0$ for $i \in I - J$. Such **rings of functions** and their subrings are important objects in mathematics, especially in analysis and topology. ◊

Example 2.6.19 (Cayley Representation of a Ring – Affine Group of a Ring) The most natural examples of rings are the endomorphism rings of abelian groups, which we have already introduced following Proposition 2.2.4.

Let H be an abelian group written additively. Then $\operatorname{End} H \subseteq H^H$ is both a subgroup of the additive product group H^H and a submonoid of the multiplicative mapping monoid H^H with the composition ∘ of endomorphisms as multiplication. Since this is distributive over addition, cf. loc. cit., $\operatorname{End} H = (\operatorname{End} H, +, \circ)$ is a ring, the so-called **endomorphism ring** of H.[38]

Now let A be a ring. Then the left translations $L_a \colon x \mapsto ax$ are endomorphisms of the additive group of A due to the distributive laws, and the Cayley mapping $a \mapsto L_a$ of A into $\operatorname{End}(A, +)$ is not only multiplicative but also additive: $L_{a+b}(x) = (a+b)x = ax + bx = L_a(x) + L_b(x)$. In total: ◊

Theorem 2.6.20 (Cayley's Representation Theorem) *Let A be a ring. Then the Cayley mapping*

$$L \colon A \longrightarrow \operatorname{End}(A, +), \quad a \mapsto L_a,$$

is an injective ring homomorphism. In particular, a ring can always be identified with a subring of the endomorphism ring of its additive group.

The Cayley homomorphism $L \colon A \to \operatorname{End} A$ is also called the **regular representation** of A. It is (of course) not generally surjective. However, it is surjective and thus a ring isomorphism when A is a minimal ring, i.e., *the minimal ring A_m, $m \in \mathbb{N}$, can be identified with the endomorphism ring of the cyclic group $\mathbb{Z}_m = (\mathbf{A}_m, +)$*. Specifically, the prime residue class group $\mathbf{A}_m^\times$ is canonically isomorphic to the automorphism group of the cyclic group $\mathbb{Z}_m$. We have already described this in Example 2.2.23.

Similarly, for the prime field $\mathbf{F}_0 = \mathbb{Q}$ of characteristic 0, the Cayley homomorphism $\mathbb{Q} \to \operatorname{End}(\mathbb{Q}, +)$ is an isomorphism, see Exercise 2.2.7. *For the field $\mathbb{R}$ the Cayley homomorphism is not surjective.* For a general discussion of this so-called **Cauchy problem**, see Example 2.8.19.

---

[38] H^H is not a ring with + and ∘ if H is not the zero group. Proof!

The right translations of A provide the Cayley representation $R\colon A^{\mathrm{op}} \to \mathrm{End}(A^{\mathrm{op}}, +) = \mathrm{End}(A, +)$, $a \mapsto R_a$, of the ring opposed to A. In particular, the multiplicative monoid $(A^{\mathrm{op}}, \cdot)$ operates as a monoid of endomorphisms of $(A, +)$. The associated semidirect product $A \rtimes A^{\mathrm{op}} = (A, +) \rtimes (A^{\mathrm{op}}, \cdot)$ is a submonoid of the full holomorph $A \rtimes \mathrm{End}(A, +)$. The canonical homomorphism $A \rtimes A^{\mathrm{op}} \to A^A$ identifies $A \rtimes A^{\mathrm{op}}$ with the monoid of **affine transformations** $x \mapsto xa + b$, $(b, a) \in A \rtimes A^{\mathrm{op}}$, of A.[39] Its unit group is the group of **affinities** $x \mapsto xa + b$, $(b, a) \in A \rtimes (A^{\times})^{\mathrm{op}}$, or the **affine group** of A and is denoted by

$$\mathrm{A}_1(A).$$

It is $A \rtimes A^{\times} = \mathrm{A}_1(A^{\mathrm{op}})$.

At the end of this section, we already prove an important theorem about integral domains, which we will revisit in a more general context in Sect. 2.9 during the discussion of polynomial functions.

Theorem 2.6.21 *Let A be an integral domain and $a_0, \ldots, a_n \in A$, $n \in \mathbb{N}$, with $a_n \neq 0$. Then there are at most n elements $x \in A$ with*

$$P(x) := \sum_{k=0}^{n} a_k x^k = a_0 + a_1 x + \cdots + a_n x^n = 0.$$

Proof We use induction over n. For $n = 0$ (and $n = 1$) the statement is trivial. Let now $n > 0$ and $x_1 \in A$ be an element with $P(x_1) = 0$. If then $x \neq x_1$ is another element in A with $P(x) = 0$, then

$$0 = P(x) - P(x_1) = (x - x_1)\bigl(b_1 + b_2 x + \cdots + b_n x^{n-1}\bigr)$$

applies with $b_\ell := a_\ell + a_{\ell+1} x_1 + \cdots + a_n x_1^{n-\ell}$, $\ell = 1, \cdots, n$, cf. Theorem 2.6.5. Since A is an integral domain, it follows $b_1 + b_2 x + \cdots + b_n x^{n-1} = 0$ due to $x - x_1 \neq 0$. Moreover, $b_n = a_n \neq 0$. By induction hypothesis, there are thus at most $n-1$ such elements $x \in A$. □

An important consequence of the above statement is:

Theorem 2.6.22 *Let A be an integral domain. Then every finite subgroup G of the unit group $A^{\times}$ of A is cyclic. – In particular, the multiplicative group of a finite field is cyclic.*

Proof G fulfills according to Theorem 2.6.21 the prerequisites of the cyclicity criterion 2.2.15.

---

[39] The right translations R_a, $a \in A$, are the A-linear endomorphisms of the A-(left-)module A, cf. Sect. 2.8. Therefore, we speak of the affine transformations of A and not of the affine transformations of A ^{op}. In the case of commutative A, this distinction is omitted.

Exercises

Exercise 2.6.1
a) Justify that the power set $\mathfrak{P}(I)$ of a non-empty set I, equipped with the union as addition and the intersection as multiplication, is never a ring.
b) Show that $\mathfrak{P}(I)$ however, with the symmetric difference $\triangle$ as addition and the intersection $\cap$ as multiplication, is a commutative ring with $\emptyset$ as zero and A as the unit element. This ring is called the **set ring** of I. If $|I| = 1$, it is the field with two elements, otherwise the set ring of I is not a field. The mapping $\mathfrak{P}(I) \xrightarrow{\sim} \mathbf{F}_2^I$, $J \mapsto e_J$, is a ring isomorphism.[40] (This bijection provides the most convenient method to verify the ring axioms for $(\mathfrak{P}(I), \triangle, \cap)$.) Determine, for example, $\sum_{k \in \mathbb{N}^*} e_{k\mathbb{N}^*}$ in $\mathbf{F}_2^{\mathbb{N}^*}$. (The sum is indeed infinite, due to $e_{k\mathbb{N}^*}(n) = 0$ for $k > n$ but well-defined. See Example 4.5.43 (1).)

Exercise 2.6.2 Show that the commutativity of addition in a ring follows from the other ring axioms.

Exercise 2.6.3 Let A be a ring. An element $x \in A$ with $x^m = 0$ for a $m \in \mathbb{N}$ is called **nilpotent**. The smallest m with $x^m = 0$ is then called the **nilpotency degree** of x. If A is not the zero ring, then this is >0. An element of the form $1 - x$ with nilpotent $x \in A$ is called **unipotent**. A ring without non-zero nilpotent elements is called **reduced**.

a) Let $x \in A$ be nilpotent with nilpotency degree m. Then $1 - x$ is a unit in A with the inverse $1 + x + \cdots + x^{m-1}$. *Unipotent elements are therefore units.*
b) Let x and y be *commutative* nilpotent elements of A with the nilpotency degrees m and n respectively. Then $x + y$ is also nilpotent, with a nilpotency degree $\leq \text{Max}(0, m + n - 1)$. (Use the binomial Theorem 2.6.3.)
c) *If the elements $x, y \in A$ are commutative* and one of them is nilpotent with a nilpotency degree m, then their product xy is nilpotent with a nilpotency degree $\leq m$.
d) Let A be commutative. The results from b) and c) then imply that the set $\mathfrak{n}_A$ of nilpotent elements of A is an ideal of A (in the sense of Definition 2.7.1). It is called the **nilradical** of A and is exactly the zero ideal 0 when A is reduced. In particular, $\mathfrak{n}_A$ is a subgroup of $(A, +)$. Show that the set $1 + \mathfrak{n}_A$ of unipotent elements of A is a subgroup of the unit group $A^\times$.

Exercise 2.6.4 Let A be a ring $\neq \{0\}$, in which every equation of the form $ax + b = 0$ with $a, b \in A$, $a \neq 0$, has a solution $x \in A$. Then A is a division ring.

---

[40] Do not confuse the difference $J - K$ in the set ring $\mathfrak{P}(I)$, which is identical with the addition, i.e., the symmetric difference $J \triangle K$, $J, K \subseteq I$, with the set difference $J - K$. They coincide exactly when $J \supseteq K$ is.

Exercise 2.6.5 In generalization of Theorem 2.6.5, show: For not necessarily commutative elements x,y of a ring, the following holds

$$\sum_{m=0}^{n} x^{n-m}(x-y)y^m = x^{n+1} - y^{n+1}.$$

Exercise 2.6.6 Let A be a ring and let $x, y \in \mathrm{Idp}(A)\, (= \mathrm{Idp}(A, \cdot))$.

a) If $xy = yx$, then $xy, x + y - xy$ and $(x-y)^2$ are also idempotent. If A is commutative, then the set $\mathrm{Idp}(A)$ of idempotent elements of A with the addition $x \triangle y := (x-y)^2$ and the multiplication induced by A is a commutative ring of characteristic 2 for $A \neq 0$ (and 1 for $A = 0$), in which every element is idempotent.
(**Remark** A ring B is called a **Boolean ring** (after G. Boole (1815–1864)), when every element of B is idempotent. Let B be such a ring. Then $4 = 2 \cdot 2 = 2$, so $2 = 0$ in B and $\mathrm{Char}\, B = 2$ for $B \neq 0$. Furthermore, $x + y = (x+y)^2 = x^2 + xy + yx + y^2 = x - xy + yx + y$ and thus $xy = yx$ for all $x, y \in B$. A Boolean ring is necessarily commutative. Classic examples of Boolean rings are the set rings $\mathfrak{P}(I) \cong \mathbf{F}_2^I$ and their subrings, cf. Ex. 2.6.1. According to the Stone representation theorem from Ex. 4.4.28a), every Boolean ring is isomorphic to a subring of a suitable set ring. If B is a Boolean ring, then the ring structure defined above on $\mathrm{Idp}(B) = B$ coincides with the given one. *A finite Boolean ring B is always isomorphic to one of the set rings* $\mathbf{F}_2^n$, $n \in \mathbb{N}$. For the *proof* by induction over $|B|$, let B have more than two elements. If then $e \in B$, $0 \neq e \neq 1$, then $B \cong Be \times B(1-e)$, cf. example 2.6.18, and by induction hypothesis $Be \cong \mathbf{F}_2^k$, $B(1-e) \cong \mathbf{F}_2^\ell$ and consequently $B \cong \mathbf{F}_2^{k+\ell}$. – So, if the Boolean ring $\mathrm{Idp}(Z(A))$ is finite, then $|\mathrm{Idp}(Z(A))|$ is a power of 2.)

b) $x + y$ is idempotent exactly when $xy = yx$ and $2xy = 0$.

c) $x - y$ is idempotent exactly when $xy = yx$ and $2(1-x)y = 0$ holds. (This can easily be traced back to b).)

d) $1 - 2x$ is **involutory**, i.e., self-inverse (with respect to multiplication in A). If A is commutative, then $x \mapsto 1 - 2x$ is a homomorphism of the group $(\mathrm{Idp}(A), \triangle)$ (see a)) into the group $\mathrm{Inv}(A) \subseteq A^\times$ of involutory elements of A, which is an isomorphism when $2 = 2 \cdot 1_A \in A^\times$.

Exercise 2.6.7 Prove the following so-called **polarization formulas** for pairwise commuting elements $x_1, \ldots, x_r$, $r \in \mathbb{N}^*$, of a ring A.

a) $2^{r-1} r!\, x_1 \cdots x_r = \sum_{\varepsilon} \varepsilon_2 \cdots \varepsilon_r (x_1 + \varepsilon_2 x_2 + \cdots + \varepsilon_r x_r)^r$,

where on the right side all sign tuples $\varepsilon = (\varepsilon_2, \ldots, \varepsilon_r) \in \{1, -1\}^{r-1}$ are to be summed. For $r = 2$ this is the formula $4x_1 x_2 = (x_1 + x_2)^2 - (x_1 - x_2)^2$, which reduces multiplication to squaring twice when $2 \in A^\times$.

2.6 Rings

b) $(-1)^r r! x_1 \cdots x_r = \sum_{H \subseteq \{1,\ldots,r\}} (-1)^{|H|} x_H^r = \sum_e (-1)^{e_1+\cdots+e_r} (e_1 x_1 + \cdots + e_r x_r)^r$

(with $x_H := \sum_{i \in H} x_i$ for $H \subseteq \{1,\ldots,r\}$). Here, e runs through all tuples $(e_1,\ldots,e_r) \in \{0,1\}^r$. This formula generalizes the equation $2 x_1 x_2 = (x_1 + x_2)^2 - x_1^2 - x_2^2$.

Exercise 2.6.8 The additive group $(K, +)$ and the multiplicative group $(K^\times, \cdot)$ of a field K are never isomorphic.

Exercise 2.6.9 Let A be a ring.

a) A is exactly a minimal ring when its additive group is cyclic. If A is finite with a square-free number of elements, then A is a minimal ring. Determine up to isomorphism all rings with a number of elements $n \leq 7$. (For $n = 4$ there are four, including a field.)

b) If A is finite, then $|A|$ and Char A have the same prime divisors. In particular, the number of elements of a finite field K is a prime power. (The additive group of K is even an elementary abelian p-group, $p := $ Char $K = \mathrm{Exp}(K, +)$.)

Exercise 2.6.10 Provide an example of a ring B with a subring $A \subseteq B$ such that $A^\times \subset B^\times \cap A$ and $A^* \supset B^* \cap A$ apply. (The inclusions $A^\times \subseteq B^\times \cap A$ and $A^* \supseteq B^* \cap A$ are trivial.)

Exercise 2.6.11 Let $\varphi: A \to B$ be a homomorphism of rings and $A' \subseteq A$ and $B' \subseteq B$ subrings of A and B respectively. Then $\varphi(A')$ is a subring of B, $\varphi^{-1}(B')$ is a subring of A, which includes $\ker \varphi = \varphi^{-1}(0)$, and $\varphi^{-1}(\varphi(A')) = A' + \ker \varphi$. (However, for $B \neq 0$, $\ker \varphi$ is not a subring of A.) If φ is surjective, then $A' \mapsto \varphi(A')$ and $B' \mapsto \varphi^{-1}(B')$ are bijective and mutually inverse mappings between the set of subrings of A that include $\ker \varphi$, and the set of all subrings of B.

Exercise 2.6.12 Let H be an additive abelian group and A_H the direct sum $A_H := \mathbb{Z} \oplus H$. We identify $\mathbb{Z}$ and H with the summands $\mathbb{Z} \oplus \{0\}$ respectively $\{0\} \oplus H$ of A_H.

a) A_H is a commutative ring with the given addition and multiplication

$$(a + x)(b + y) := ab + (ay + bx), \quad a, b \in \mathbb{Z},\; x, y \in H,$$

The mapping $(\varepsilon, x) \mapsto \varepsilon(1 + x)$ is a group isomorphism of the product group $\mathbb{Z}^\times \times H$ onto the group $A_H^\times = \{\varepsilon + h \mid \varepsilon \in \mathbb{Z}^\times, h \in H\}$ of units of A_H. The monoid A_H^* of regular elements of A_H contains exactly the elements $n + h \in A_H$ with $n \in \mathbb{Z}^*$, $h \in H$ and $_nH = \{x \in H \mid nx = 0\} = 0$.

b) $A := (\mathbb{N}^* \oplus H) \uplus \{0\}$ is a subfield of A_H (i.e., A is a subring with $A^* = A - \{0\}$). Its Grothendieck ring is $G(A) = A_H$. There are semi-fields whose Grothendieck ring is not a field and therefore(!) cannot be embedded into any field.

2.7 Ideals and Quotient Rings

Let A,B be rings and $\varphi \colon A \to B$ a ring homomorphism. Then $\mathfrak{a} := \ker \varphi = \varphi^{-1}(0_B)$ is a subgroup of the additive group of A, and the congruence relation defined by $\mathfrak{a} \equiv_\mathfrak{a}$ on $A = (A, +)$ is not only compatible with the addition of A, but also with the multiplication of A. From $a \equiv_\mathfrak{a} 0$ it follows for any $x \in A$ in particular $xa \equiv_\mathfrak{a} x \cdot 0 = 0$ and $ax \equiv_\mathfrak{a} 0 \cdot x = 0$, i.e. $xa \in \mathfrak{a}$ and $ax \in \mathfrak{a}$. $\mathfrak{a}$ is thus a two-sided ideal in the sense of the following definition.

Definition 2.7.1 Let A be a ring and $\mathfrak{a}$ a subgroup of the additive group of A.

(1) $\mathfrak{a}$ is called a **two-sided ideal** of A, if $\mathfrak{a}$ with every element a also contains all left and right multiples xa and ax, $x \in A$, of a.
(2) $\mathfrak{a}$ is called a **left ideal** (or a **right ideal**) of A, if $\mathfrak{a}$ with every element a also contains all left multiples xa (or all right multiples ax), $x \in A$, of a.

A subset $\mathfrak{a} \subseteq A$ is exactly a two-sided ideal when $\mathfrak{a}$ is both a left and a right ideal of A. The right ideals in A are the left ideals in the opposite ring A^{op}. If A is commutative, then two-sided ideals, left ideals, and right ideals of A coincide, and we simply speak of **ideals** in A. In arbitrary rings, we also use "ideal" as a collective term for the different types of ideals.

Lemma 2.7.2 *Let $\mathfrak{a} \subseteq A$ be a two-sided ideal in the ring A. Then the congruence relation $\equiv_\mathfrak{a}$ is compatible (not only with addition, but also) with the multiplication of A.*

Proof Let $x \equiv_\mathfrak{a} x'$ and $y \equiv_\mathfrak{a} y'$, i.e. $x = x' + a$ and $y = y' + b$ with $a, b \in \mathfrak{a}$. Then
$$xy = (x' + a)(y' + b) = x'y' + (x'b + ay' + ab),$$
is therefore $xy \equiv_\mathfrak{a} x'y'$ due to $x'b + ay' + ab \in \mathfrak{a}$. □

Let $\mathfrak{a} \subseteq A$ be a two-sided ideal in the ring A. Then the addition and multiplication on A induce corresponding operations on $A/\mathfrak{a}$ such that $A/\mathfrak{a}$ is an abelian group, $(A/\mathfrak{a}, \cdot)$ a monoid and moreover the canonical projection $\pi = \pi_\mathfrak{a} \colon A \to A/\mathfrak{a}$, $x \mapsto \bar{x} = [x] = [x]_\mathfrak{a} = x + \mathfrak{a}$, a surjective group and monoid homomorphism (cf. Theorem 2.3.4). Thus, the distributive laws of A also transfer to $A/\mathfrak{a}$. In total, we obtain:

2.7 Ideals and Quotient Rings

Theorem 2.7.3 *Let $\mathfrak{a} \subseteq A$ be a two-sided ideal in the ring A. Then $A/\mathfrak{a}$ with the operations*

$$(x+\mathfrak{a})+(y+\mathfrak{a}) = (x+y)+\mathfrak{a}, \quad (x+\mathfrak{a})\cdot(y+\mathfrak{a}) = xy+\mathfrak{a}$$

is a ring, and the canonical projection $\pi_\mathfrak{a}: A \to A/\mathfrak{a}$ is a surjective ring homomorphism with $\ker \pi_\mathfrak{a} = \mathfrak{a}$ and the following universal property: If $\varphi: A \to B$ is an arbitrary homomorphism of rings with $\mathfrak{a} \subseteq \ker \varphi$, then there exists exactly one ring homomorphism $\overline{\varphi}: A/\mathfrak{a} \to B$ with $\varphi = \overline{\varphi} \circ \pi_\mathfrak{a}$. It is $\overline{\varphi}(x+\mathfrak{a}) = \varphi(x)$, $\operatorname{Im} \overline{\varphi} = \operatorname{Im} \varphi$ and $\ker \overline{\varphi} = (\ker \varphi)/\mathfrak{a}$. In particular, $\overline{\varphi}$ is an isomorphism if and only if $\ker \varphi = \mathfrak{a}$ is and φ is surjective. So it is

$$A/\ker \varphi \xrightarrow{\sim} \operatorname{Im} \varphi.$$

The **residue class rings** $A/\mathfrak{a}$, $\mathfrak{a}$ two-sided ideal in A, thus represent up to isomorphism all homomorphic images of the ring A. This motivates the following definition in analogy to the definition of simple groups:

Definition 2.7.4 A ring A is called **simple**, if it is not the zero ring and 0 and A are the only two-sided ideals in A.

$A \neq 0$ is therefore simple exactly when every homomorphism $A \to B$ into a ring $B \neq 0$ is injective. Division domains are obviously simple, but not every simple ring is a division domain, see exercise 2.8.8e). *Commutative* simple rings are however fields. See Proposition 2.7.16 below.

We also note the general Theorem of the induced homomorphism, see 2.3.1 and 2.3.2, as it presents itself for rings.

Theorem 2.7.5 (Theorem of the induced homomorphism for rings) *Let $\psi: A \to C$ and $\varphi: A \to B$ be homomorphisms of rings. Let ψ be surjective, and let $\ker \psi \subseteq \ker \varphi$ hold. Then there exists exactly one ring homomorphism $\overline{\varphi}: C \to B$ with $\varphi = \overline{\varphi} \circ \psi$. The following holds $\operatorname{Im} \overline{\varphi} = \operatorname{Im} \varphi$ and $\ker \overline{\varphi} = \psi(\ker \varphi)$. (See Fig. 2.21.)*

Exactly then is $\overline{\varphi}$ surjective, when φ is surjective. – Exactly then is $\overline{\varphi}$ injective, when $\ker \psi = \ker \varphi$ is. – Exactly then is $\overline{\varphi}$ an isomorphism, when φ is surjective and $\ker \psi = \ker \varphi$ applies (**Isomorphism theorem for rings**) .

Let $\mathfrak{a}_i, i \in I$, be a family of left ideals in the ring A. Then the sum $\sum_{i \in I} \mathfrak{a}_i$ and the intersection $\bigcap_{i \in I} \mathfrak{a}_i$ are also left ideals in A. The left ideals thus form a complete sublattice of the lattice of subgroups of $(A, +)$. The same applies to right ideals and

Fig. 2.21 The ring homomorphism $\overline{\varphi}$ is induced by φ

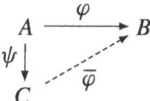

two-sided ideals. If $\mathfrak{a}, \mathfrak{b} \subseteq A$ are left ideals, then the complex product of $\mathfrak{a}$ and $\mathfrak{b}$ is usually not a subgroup of $(A, +)$. *Therefore, in ring theory, the product*

$$\mathfrak{a}\mathfrak{b}$$

always refers to the subgroup generated by the complex product $\{ab \mid a \in \mathfrak{a}, b \in \mathfrak{b}\}$ *of* $(A, +)$.[41] It contains the finite sums of products ab, $a \in \mathfrak{a}$, $b \in \mathfrak{b}$. The product thus defined $\mathfrak{a}\mathfrak{b}$ is then obviously again a left ideal, and the same applies to right ideals and two-sided ideals. Clearly, the associative and distributive laws

$$(\mathfrak{a} + \mathfrak{b}) + \mathfrak{c} = \mathfrak{a} + (\mathfrak{b} + \mathfrak{c}), (\mathfrak{a}\mathfrak{b})\mathfrak{c} = \mathfrak{a}(\mathfrak{b}\mathfrak{c}),$$
$$\mathfrak{a}(\mathfrak{b} + \mathfrak{c}) = \mathfrak{a}\mathfrak{b} + \mathfrak{a}\mathfrak{c}, (\mathfrak{b} + \mathfrak{c})\mathfrak{a} = \mathfrak{b}\mathfrak{a} + \mathfrak{c}\mathfrak{a}$$

apply to ideals $\mathfrak{a}, \mathfrak{b}, \mathfrak{c}$. The set of respective ideals is therefore a semiring with the **zero ideal** $0 = \{0\}$ as the zero element and the **unit ideal** A as the unit element.[42]

The smallest left ideal that contains a given element $a \in A$ is the set $Aa = \{xa \mid x \in A\}$ of left multiples of $\mathfrak{a}$, and an element $b \in A$ belongs to Aa if and only if a is a right divisor of b. Then $Ab \subseteq Aa$. Aa is called the **principal left ideal** generated by a of A. If every left ideal of A is a principal left ideal, it is called a **principal left ideal ring** or a **principal left ideal domain** if A is also a domain. Similarly, **principal right ideals** and **principal right ideal rings** or — **domains** are defined. In the commutative case, we simply speak of **principal ideals** and **principal ideal rings** or **principal ideal domains**. The smallest left ideal that contains a family $a_i, i \in I$, is the sum $\sum_{i \in I} Aa_i$. The smallest two-sided ideal that contains the element a is the subgroup generated by the products xay, $x, y \in A$, of $(A, +)$ and in the non-commutative case often difficult to oversee. According to the above convention, it is also denoted by AaA. For any subset $M \subseteq A$, AM, MA and AMA are the ones generated by M generated left, right, or two-sided ideal. If $a_i, i \in I$ is a family of elements from A, we denote the smallest left, right, or two-sided ideal that contains the $a_i, i \in I$, often simply with $(a_i, i \in I)$, at $I = \{1, \ldots, n\}$ thus with $(a_1, \ldots, a_n)$, if both the ring A and the type of ideal are apparent from the context.

Remark 2.7.6 (Divisibility in Rings) Let A be a commutative ring. The concepts of divisibility in A always refer to the multiplicative monoid of A. $b \in A$ is therefore exactly a divisor of $a \in A$, if a lies in the principal ideal Ab or if $Aa \subseteq Ab$ is. Following a terminology of R. Dedekind (1831–1916), for a long time any ideal $\mathfrak{b} \subseteq A$ was called a divisor of the ideal $\mathfrak{a} \subseteq A$, if the inclusion $\mathfrak{a} \subseteq \mathfrak{b}$ holds. This is obviously the divisibility in the monoid of ideals of A with the intersection $\mathfrak{a} \cap \mathfrak{b}$ of ideals as operation. This divisibility relation is to be distinguished from the divisibility of ideals in the (also commutative) multiplicative monoid of ideals with the

---

[41] We also note that for a two-sided ideal $\mathfrak{a} \subseteq A$ the product $(x + \mathfrak{a})(y + \mathfrak{a}) = xy + \mathfrak{a}$, $x, y \in A$, in the ring $A/\mathfrak{a}$ is usually not the complex product of $x + \mathfrak{a}$ and $y + \mathfrak{a}$ in $(\mathfrak{P}(A), \cdot)$.

[42] The associated Grothendieck ring is trivial due to $\mathfrak{a} + \mathfrak{a} = \mathfrak{a}$ for all ideals $\mathfrak{a}$.

2.7 Ideals and Quotient Rings

product introduced above as operation. The ideal $\mathfrak{b}$ is a divisor of $\mathfrak{a}$, if there is an ideal $\mathfrak{c}$ with $\mathfrak{a} = \mathfrak{b}\mathfrak{c}$ given. (Note that then also $\mathfrak{a} \subseteq \mathfrak{b}$ applies.) Dedekind was the first, after preliminary work by E. Kummer (1810–1893), to notice that it is very useful for the divisibility theory in rings not only to consider principal ideals, but any ideals. He also introduced the term "ideal". — Furthermore, one has to distinguish between the ideals of the ring A and the ideals of the multiplicative monoid $(A, \cdot)$, cf. exercise 2.3.15. Every ring ideal is of course an ideal of the monoid, but the reverse is usually wrong. In the ring $\mathbb{Z}$ for example, every subgroup is cyclic and in particular every ideal is a principal ideal, in the multiplicative monoid $(\mathbb{Z}, \cdot)$ however, for example, $\mathbb{Z} - \mathbb{Z}^\times$ is an ideal that is not even finitely generated. ◊

We also mention the following statement analogous to Proposition 2.2.7, the simple proof of which is left to the reader.

Proposition 2.7.7 *Let $\varphi\colon A \to B$ be a homomorphism of rings.*

(1) *If $\mathfrak{b} \subseteq B$ is a two-sided ideal in B, then $\varphi^{-1}(\mathfrak{b})$ is a two-sided ideal in A that includes $\ker \varphi$, and φ induces an injective ring homomorphism $\overline{\varphi}\colon A/\varphi^{-1}(\mathfrak{b}) \to B/\mathfrak{b}$.*
(2) *Let φ be surjective. If $\mathfrak{a} \subseteq A$ is a two-sided ideal in A, then $\varphi(\mathfrak{a})$ is a two-sided ideal in B and $\varphi^{-1}(\varphi(\mathfrak{a})) = \mathfrak{a} + \ker \varphi$. Furthermore, $\mathfrak{a} \mapsto \varphi(\mathfrak{a})$ and $\mathfrak{b} \mapsto \varphi^{-1}(\mathfrak{b})$ are bijective and mutually inverse mappings between the set of two-sided ideals in A, which include $\ker \varphi$, and the set of all two-sided ideals in B. For $\mathfrak{b} \subseteq B$, the induced homomorphism $\overline{\varphi}\colon A/\varphi^{-1}(\mathfrak{b}) \xrightarrow{\sim} B/\mathfrak{b}$ is a ring isomorphism.*
The correspondence of two-sided ideals in (1) and (2) specified in A and B applies in an analogous way also for left and right ideals.

If φ is not surjective in the situation of 2.7.7, then the image $\varphi(\mathfrak{a})$ of an ideal $\mathfrak{a} \subseteq A$ is not necessarily a corresponding ideal of B. Instead of $\varphi(\mathfrak{a})$, one therefore considers the ideal $B\varphi(\mathfrak{a})$ generated by $\varphi(\mathfrak{a})$, $\varphi(\mathfrak{a})B$ or $B\varphi(\mathfrak{a})B$ in B.

Example 2.7.8 (Chinese Remainder Theorem) Here we consider the Chinese Remainder Theorem from Example 2.2.17 for rings. Let A be a ring and $\varphi_i\colon A \to A_i$ be surjective homomorphisms of rings with the kernels $\mathfrak{a}_i$, $i = 1, \ldots, n$. The φ_i induce the ring homomorphism

$$\varphi = (\varphi_1, \ldots, \varphi_n)\colon A \to A_1 \times \cdots \times A_n, \quad x \mapsto (\varphi_1(x), \ldots, \varphi_n(x)).$$

It is $\mathfrak{a} := \ker \varphi = \mathfrak{a}_1 \cap \cdots \cap \mathfrak{a}_n$. According to Theorem 2.7.3, φ induces an injective ring homomorphism $\overline{\varphi}\colon A/\mathfrak{a} \to A_1 \times \cdots \times A_n$. The question of the surjectivity of φ and thus from $\overline{\varphi}$ is answered by Theorem 2.2.18. Namely: φ is surjective if and only if for every $i = 1, \ldots, n$ the following holds:

$$\mathfrak{a}_i + \mathfrak{a}'_i = A \quad \text{with} \quad \mathfrak{a}'_i := \bigcap_{j \neq i} \mathfrak{a}_j.$$

This condition for the two-sided ideals $\mathfrak{a}_1,\ldots,\mathfrak{a}_n$ can be formulated more clearly. First, it implies that $\mathfrak{a}_i + \mathfrak{a}_j = A$ is for all $i \neq j$. Conversely, the original condition also follows from this. For the *proof*, let i be fixed, say $i = 1$. Because of $\mathfrak{a}_1 + \mathfrak{a}_j = A$ for $j = 2,\ldots,n$ there are elements $a_j \in \mathfrak{a}_1$ and $a'_j \in \mathfrak{a}_j$ with $1 = a_j + a'_j$. Then

$$1 = (a_2 + a'_2) \cdots (a_n + a'_n) = b + b' \in \mathfrak{a}_1 + \mathfrak{a}'_1$$

with $b \in \mathfrak{a}_1$ and $b' = a'_2 \cdots a'_n \in \mathfrak{a}_2 \cap \cdots \cap \mathfrak{a}_n = \mathfrak{a}'_1$. Since $\mathfrak{a}_1$ and $\mathfrak{a}'_1$ are (two-sided) ideals, then $\mathfrak{a}_1 + \mathfrak{a}'_1 = A$.

Ideal $\mathfrak{a}, \mathfrak{b}$ of a ring A with $\mathfrak{a} + \mathfrak{b} = A$ are called **comaximal**, see also Problem 2.7.10.[43] We have thus proven: ◊

Theorem 2.7.9 (Chinese Remainder Theorem for Rings) *If $\varphi_i : A \to A_i$, $i = 1,\ldots,n$, are surjective ring homomorphisms with pairwise comaximal kernels $\mathfrak{a}_1 := \ker \varphi_1, \ldots, \mathfrak{a}_n := \ker \varphi_n$, then the canonical ring homomorphism is*

$$\overline{\varphi}: A/(\mathfrak{a}_1 \cap \cdots \cap \mathfrak{a}_n) \xrightarrow{\sim} A_1 \times \cdots \times A_n, \quad \text{with} \quad \overline{\varphi}(\overline{a}) = (\varphi_1(a),\ldots,\varphi_n(a))$$

an isomorphism of rings.

The inverse mapping of $\overline{\varphi}$ can be explicitly given with a representation $1 = a'_1 + \cdots + a'_n$, $a'_i \in \mathfrak{a}'_i = \bigcap_{j \neq i} \mathfrak{a}_j$, as it exists according to Problem 2.2.15 (see also exercise 2.7.10c)). For

$$\overline{\varphi}^{-1}(\varphi_1(a_1),\ldots,\varphi_n(a_n)) = \overline{a_1 a'_1 + \cdots + a_n a'_n} = \overline{a'_1 a_1 + \cdots + a'_n a_n}, \quad a_1,\ldots,a_n \in A.$$

The Chinese Remainder Theorem 2.7.9 is often applied for pairwise comaximal two-sided ideals $\mathfrak{a}_1,\ldots,\mathfrak{a}_n \subseteq A$ and the canonical projections $\pi_i : A \to A/\mathfrak{a}_i$, $i = 1,\ldots,n$, for example. If $e_1,\ldots,e_n$ is a complete orthogonal family of central idempotent elements in the ring B, then the principal ideals $\mathfrak{b}_i := B(1 - e_i) = (1 - e_i)B = \sum_{j \neq i} Be_j$, $i = 1,\ldots,n$, are pairwise comaximal with $\mathfrak{b}_1 \cap \cdots \cap \mathfrak{b}_n = 0$ and we thus obtain the product representation $B \xrightarrow{\sim} B_1 \times \cdots \times B_n$ with $B_i = B/\mathfrak{b}_i \xrightarrow{\sim} Be_i$, $i = 1,\ldots,n$, from Example 2.6.18. So, if in the ring B there are pairwise comaximal two-sided ideals $\mathfrak{b}_1,\ldots,\mathfrak{b}_n$ with intersection $\mathfrak{b}_1 \cap \cdots \cap \mathfrak{b}_n = 0$, then $B \xrightarrow{\sim} (B/\mathfrak{b}_1) \times \cdots \times (B/\mathfrak{b}_n)$ and there is a (uniquely determined) complete orthogonal family $e_1,\ldots,e_n$ of central idempotent elements of B with $\mathfrak{b}_i = B(1 - e_i)$, $i = 1,\ldots,n$. In particular, a ring $B \neq 0$ is decomposable if and only if there are two-sided comaximal ideals $\mathfrak{b}$ and $\mathfrak{c}$ in B with $\mathfrak{b} \neq B \neq \mathfrak{c}$ and $\mathfrak{b} \cap \mathfrak{c} = 0$.

Example 2.7.10 (Prime Residue Class Groups) In the integrity domain $\mathbb{Z}$ of the whole numbers, the subgroups of the additive group $(\mathbb{Z}, +)$ coincide with the ideals. For each ideal $\mathfrak{a}$ in $\mathbb{Z}$, according to Theorem 2.1.18, there is (exactly)

---

[43] According to Remark 2.7.6 some authors also call comaximal ideals **coprime** ideals.

one $a \in \mathbb{N}$ with $\mathfrak{a} = \mathbb{Z}a$. In particular, $\mathfrak{a}$ is a principal ideal, and the residue class rings of $\mathbb{Z}$ are exactly the minimal rings $\mathbf{A}_m = \mathbb{Z}/\mathbb{Z}m$, $m \in \mathbb{N}$. Because of $\mathbb{Z}a + \mathbb{Z}b = \mathbb{Z}\gcd(a,b)$, see 2.1.19, the ideals $\mathfrak{a} = \mathbb{Z}a$ and $\mathfrak{b} = \mathbb{Z}b$ are co-maximal exactly when a and b are coprime. Then $\mathbb{Z}a \cap \mathbb{Z}b = \mathbb{Z}ab$ is. The Chinese Remainder Theorem 2.7.9 thus provides for the ring $\mathbb{Z}$: ◊

Theorem 2.7.11 (Chinese Remainder Theorem for $\mathbb{Z}$) *If $m_1, \ldots, m_n \in \mathbb{N}^*$ are pairwise coprime, then the canonical homomorphism*

$$\overline{\varphi}: \mathbf{A}_m = \mathbb{Z}/\mathbb{Z}m \xrightarrow{\sim} \mathbb{Z}/\mathbb{Z}m_1 \times \cdots \times \mathbb{Z}/\mathbb{Z}m_n = \mathbf{A}_{m_1} \times \cdots \times \mathbf{A}_{m_n}, \quad m := m_1 \cdots m_n,$$

is a ring isomorphism.

With a representation $1 = c_1 m_1' + \cdots + c_n m_n'$, $m_i' := m/m_i$, $i = 1, \ldots, n$, (see also Proposition 2.1.19) the inverse mapping of $\overline{\varphi}$ is explicitly given by

$$(r_1 + \mathbb{Z}m_1, \ldots, r_n + \mathbb{Z}m_n) \longmapsto (r_1 c_1 m_1' + \cdots + r_n c_n m_n') + \mathbb{Z}m.$$

The Chinese Remainder Theorem 2.7.11 is used for calculating with large integers. If, for example, the product of two integers b,c is $\leq r$ in absolute value, it can be determined in the following way: One chooses (relatively small) pairwise coprime positive natural numbers (e.g., different prime numbers) $m_1, \ldots, m_n$ with $m := m_1 \cdots m_n \geq 2r + 1$ and calculates the product bc modulo the individual numbers $m_1, \ldots, m_n$. With 2.7.11 one obtains from this the product bc modulo m, whereby bc is already uniquely determined due to $|bc| \leq r$. — The isomorphism from Theorem 2.7.11 we have used several times for the additive groups $\mathbb{Z}_m$ or $\mathbb{Z}_{m_1}, \ldots \mathbb{Z}_{m_n}$ of the minimal rings $\mathbf{A}_m = \mathbb{Z}/\mathbb{Z}m$ or $\mathbf{A}_{m_1} = \mathbb{Z}/\mathbb{Z}m_1, \ldots, \mathbf{A}_{m_n} = \mathbb{Z}/\mathbb{Z}m_n$, see Corollary 2.2.19.

We now choose the finest possible decomposition for $m = m_1 \cdots m_n = p_1^{\alpha_1} \cdots p_n^{\alpha_n}$ with coprime prime powers $m_i = p_i^{\alpha_i} > 1$ and obtain the decomposition

$$\mathbf{A}_m \xrightarrow{\sim} \mathbf{A}_{p_1^{\alpha_1}} \times \cdots \times \mathbf{A}_{p_n^{\alpha_n}}.$$

Let $e_1, \ldots, e_n$ be the associated complete orthogonal family of idempotent elements of A_m and A an arbitrary ring of characteristic m. According to the last remark in the previous example, the ring A then has the product decomposition

$$A \xrightarrow{\sim} Ae_1 \times \cdots \times Ae_n \quad \text{with} \quad A/A(1 - e_i) \xrightarrow{\sim} Ae_i, \, i = 1, \ldots, n.$$

The Ae_i are obviously the p_i-primary components of the additive group of A, and it is $\operatorname{Char} Ae_i = p_i^{\alpha_i}$. The ring A (with $\operatorname{Char} A > 0$) *is therefore isomorphic as a ring to the direct product of the primary components of its additive group.*

The unit groups of the minimal rings are the prime residue class groups. The Chinese Remainder Theorem 2.7.11 implies:

Corollary 2.7.12 *If $m_1, \ldots, m_n \in \mathbb{N}^*$ are pairwise coprime, then the canonical homomorphism is*

$$\overline{\varphi}^\times : \mathbf{A}_m^\times \xrightarrow{\sim} \mathbf{A}_{m_1}^\times \times \cdots \times \mathbf{A}_{m_n}^\times, \quad m := m_1 \cdots m_n,$$

an isomorphism of groups. Specifically for the canonical prime factor decomposition $m = p_1^{\alpha_1} \cdots p_n^{\alpha_n}$ of a number $m \in \mathbb{N}^*$ we get

$$\mathbf{A}_m^\times \xrightarrow{\sim} \left(\mathbf{A}_{p_1^{\alpha_1}}\right)^\times \times \cdots \times \left(\mathbf{A}_{p_n^{\alpha_n}}\right)^\times.$$

The structure of the primary residue class groups $\mathbf{A}_m^\times$ is reduced with this corollary to the structure of the primary residue class groups $(\mathbf{A}_{p^\alpha})^\times$, $p \in \mathbb{P}$. First of all, the following applies:

Theorem 2.7.13 *Let $p \in \mathbb{P}$ be a prime number >2 and $\alpha \in \mathbb{N}^*$. Then the primary residue class group $(\mathbf{A}_{p^\alpha})^\times$ is cyclic.*

Proof For $\alpha = 1$, $\mathbf{A}_p = \mathbf{F}_p$ is a field, and the statement follows from Theorem 2.6.22. – Now let $\alpha > 1$. We consider the surjective homomorphism $(\mathbf{A}_{p^\alpha})^\times \to \mathbf{A}_p^\times$, $[a]_{p^\alpha} \mapsto [a]_p$, with the kernel $1 + p\mathbf{A}_{p^\alpha} \subseteq (\mathbf{A}_{p^\alpha})^\times$ of order $p^{\alpha-1}$. Due to the cyclicity of $\mathbf{A}_p^\times$ and $\gcd\left(|1 + p\mathbf{A}_{p^\alpha}|, |\mathbf{A}_p^\times|\right) = 1$ it is sufficient to show that the p-socle $1 + p\mathbf{A}_{p^\alpha}$ is also cyclic. For this, it is sufficient to show that its p-base only contains p elements, see Problem 2.2.22a). So let $r \in \mathbb{Z}$ and $1 = (1 + rp)^p = 1 + rp^2 + \binom{p}{2}r^2p^2 + \cdots + r^p p^p$ in $\mathbf{A}_{p^\alpha}$, i.e. p^α divides $rp^2\left(1 + \binom{p}{2}r + \cdots + r^{p-1}p^{p-2}\right)$. Since $p > 2$, p is a divisor of $\binom{p}{2}r + \cdots + r^{p-1}p^{p-2}$ and therefore not a divisor of $1 + \binom{p}{2}r + \cdots + r^{p-1}p^{p-2}$. Thus, $p^{\alpha-2}$ is a divisor of r and therefore $1 + rp \in 1 + p^{\alpha-1}\mathbf{A}_{p^\alpha}$, i.e. $1 + p^{\alpha-1}\mathbf{A}_{p^\alpha}$ is the socle of $1 + p\mathbf{A}_{p^\alpha}$. □

It is also easy to show directly that the element $1 + p \in 1 + p\mathbf{A}_{p^\alpha}$ has the order $p^{\alpha-1}$. More generally: Exactly the elements $1 + pr, r \in \mathbb{Z}, p \nmid r$, are generating elements of the cyclic subgroup $1 + p\mathbf{A}_{p^\alpha} \subseteq \mathbf{A}_{p^\alpha}^\times$, $\alpha \in \mathbb{N}^*$. – In the case of powers of two, the following applies:

Theorem 2.7.14 *Let $\alpha \in \mathbb{N}^*$. For $\alpha \leq 2$, $(\mathbf{A}_{2^\alpha})^\times$ is cyclic, and for $\alpha \geq 3$, $(\mathbf{A}_{2^\alpha})^\times$ is the direct product of the cyclic subgroup $\{\pm 1\}$ of order 2 and of the by 3 (or of 5) generated cyclic subgroup of order $2^{\alpha-2}$.*

Proof Let $\alpha \geq 3$. Then $(\mathbf{A}_{2^\alpha})^\times = 1 + 2\mathbf{A}_{2^\alpha}$ is a 2-group of order $2^{\alpha-1}$. We consider the surjective homomorphism $(\mathbf{A}_{2^\alpha})^\times \to \mathbf{A}_8^\times \; (\cong \mathbf{D}_2 \cong \mathbf{Z}_2 \times \mathbf{Z}_2)$ with the kernel $1 + 8\mathbf{A}_{2^\alpha}$ of index 4 and show below that this kernel is equal to the group ${}^2(\mathbf{A}_{2^\alpha})^\times$ of squares in $(\mathbf{A}_{2^\alpha})^\times$. Then, any two elements of $(\mathbf{A}_{2^\alpha})^\times$, whose images in $\mathbf{A}_8^\times$ generate this group, also generate the group $(\mathbf{A}_{2^\alpha})^\times$, cf. Lemma 2.3.13. Thus, $(\mathbf{A}_{2^\alpha})^\times$ is the product of the cyclic subgroups generated by -1 and 3 (or also by 5), and the

2.7 Ideals and Quotient Rings

intersection of these two cyclic subgroups is necessarily trivial, so their product is direct.

We now show $^2(\mathbf{A}_{2^\alpha})^\times = 1 + 8\mathbf{A}_{2^\alpha}$. Since all squares in $\mathbf{A}_8^\times$ are trivial, $^2(\mathbf{A}_{2^\alpha})^\times \subseteq 1 + 8\mathbf{A}_{2^\alpha}$ is certain, and it suffices to show that the 2-socle of $(\mathbf{A}_{2^\alpha})^\times$ contains only 4 elements. So let $r \in \mathbb{Z}$ and $1 = (1+2r)^2 = 1 + 4r + 4r^2$ be in $\mathbf{A}_{2^\alpha}$, i.e., 2^α divides $4r(1+r)$ or $2^{\alpha-2}$ divides $r(1+r)$. If r is even, then $2^{\alpha-2} | r$ and $1 + 2r \in \{1, 1 + 2^{\alpha-1}\} \subseteq (\mathbf{A}_{2^\alpha})^\times$ apply. If r is odd, then $2^{\alpha-2}|(1+r)$ applies and there is $1 + 2r = 1 - 2 + 2^{\alpha-1}s \in \{-1, -1 + 2^{\alpha-1}\} \subseteq (\mathbf{A}_{2^\alpha})^\times$ with $s \in \mathbb{Z}$. □

We recall the designation

$$\mathrm{Ord}_m\, a = \mathrm{Ord}[a]_m, \ a \in \mathbb{Z}, \ m \in \mathbb{N}^*, \ \gcd(a, m) = 1.$$

The last proof shows that -1 and a generate the group $(\mathbf{A}_{2^\alpha})^\times$ if only $a \equiv \pm 3 \bmod 8$ is. *These are exactly the elements a with $\mathrm{Ord}_{2^\alpha} a = 2^{\alpha-2}$ for $\alpha > 3$.* One can also infer $\mathrm{Ord}_{2^\alpha} 7 = \mathrm{Ord}_{2^\alpha} 9 = 2^{\alpha-3}$ for $\alpha \geq 4$ (perhaps by induction over α).

Let now $m = 2^\alpha p_1^{\alpha_1} \cdots p_r^{\alpha_r}$ with prime numbers $2 < p_1 < \cdots < p_r$ and $\alpha \geq 0$, $\alpha_1, \ldots, \alpha_r \geq 1$, the canonical prime factor decomposition of m. Then the Euler formula $a^{\varphi(m)} \equiv 1 \bmod m$, $\gcd(a, m) = 1$, with the Euler φ-function

$$\varphi(m) = 2^{\mathrm{Max}(\alpha-1, 0)} p_1^{\alpha_1-1}(p_1 - 1) \cdots p_r^{\alpha_r-1}(p_r - 1)$$

can be improved to

$$a^{\varepsilon(m)} \equiv 1 \bmod m, \quad a \in \mathbb{Z}, \ m \in \mathbb{N}^*, \ \gcd(a, m) = 1, \quad \text{where}$$

$$\varepsilon(m) := \mathrm{Exp}\, \mathbf{A}_m^\times = \begin{cases} \mathrm{lcm}\left(2^{\alpha-2}, p_1 - 1, \ldots, p_r - 1, p_1^{\alpha_1-1} \cdots p_r^{\alpha_r-1}\right), & \alpha \geq 3 \\ \mathrm{lcm}\left(2, p_1 - 1, \ldots, p_r - 1, p_1^{\alpha_1-1} \cdots p_r^{\alpha_r-1}\right), & \alpha = 2 \\ \mathrm{lcm}\left(p_1 - 1, \ldots, p_r - 1, p_1^{\alpha_1-1} \cdots p_r^{\alpha_r-1}\right), & \alpha = 0, 1, \end{cases}$$

the exponent of the group $\mathbf{A}_m^\times$ is, see Theorem 2.7.13 and Theorem 2.7.14. The exponent $\varepsilon(m)$ is a divisor of $\varphi(m)$ and *is equal only when $\varphi(m)$, if $\mathbf{A}_m^\times$ is cyclic, i.e. if m is one of the numbers 1, 2, 4, p^α, $2p^\alpha$ with a prime number $p \in \mathbb{P}$, $p \geq 3$, and $\alpha \geq 1$.* If $\mathbf{A}_m$ is cyclic, then every integer $a \in \mathbb{Z}$, whose residue class $[a]_m$ generates the group $\mathbf{A}_m^\times$, is called a **primitive residue modulo m**. There are then exactly $\varphi(\varphi(m))$ classes of primitive residues modulo m. Particularly important are the primitive residues modulo $p \in \mathbb{P}$, $p \geq 3$. Their proportion of all residues $\neq 0$ is $\varphi(p-1)/(p-1) = \prod_{q \in \mathbb{P}, q|(p-1)}(1 - q^{-1})$. For the powers of ten $m = 10, 100, 1000, 10^\alpha$, $\alpha \geq 4$, is $\varphi(m) = 4, 40, 400, 4 \cdot 10^{\alpha-1}$ but $\varepsilon(m) = 4, 20, 100, 10^{\alpha-1}/2$. The primitive residues modulo 10 are $\equiv 3, 7 \bmod 10$.

The multiplicative monoids $(\mathbf{A}_m, \cdot)$ and their unit groups $\mathbf{A}_m^\times$ are used for public key cryptosystems according to remark (1) in Example 2.2.23. We adopt the considerations and designations (slightly modified) from there. Let $m = p_1 \cdots p_n \in \mathbb{N}^*$ be square-free with the pairwise different (very large) prime factors $p_1, \ldots, p_n$.

Then the exponentiation $x \mapsto x^a$ in $\mathbf{A}_m^\times$ is bijective, i.e., an automorphism of $\mathbf{A}_m^\times$, if a is coprime to $\varphi(m)$ or to the exponent $\varepsilon(m)$. In this case, $x \mapsto x^a$ is even an automorphism of the monoid $(\mathbf{A}_m, \cdot)$, cf. exercise 2.7.6, and the inverse mapping is the exponentiation $x \mapsto x^b$, where $b \in \mathbb{N}^*$ is arbitrary with $ab \equiv 1 \bmod \varphi(m)$ or with $ab \equiv 1 \bmod \varepsilon(m)$ is. To obtain the original message x from the encrypted message x^a, you need this b, which is difficult to determine if the prime factor decomposition of m is not known. With this prime factor decomposition, however, you know $\varphi(m)$ and $\varepsilon(m)$ and can then quickly calculate b (with the help of the Euclidean algorithm).

In the so-called **RSA codes** (named after R. Rivest, A. Shamir, and L. Adleman), everyone who wants to receive an encrypted message publishes two numbers m and a as a public key (public key), where $m = pq$ is the product of $n = 2$ very large and significantly different prime numbers $p, q \in \mathbb{P}$ and $a \in \mathbb{N}^*$ is coprime to $\varphi(m) = (p-1)(q-1) = m - p - q + 1$. The sender of the message sends the encrypted message $y = x^a$ modulo m. The recipient obtains the message $x = y^b$ modulo m back from it. The private key (private key) b is determined by $ab \equiv 1 \bmod (p-1)(q-1)$ or by $\varepsilon(m) = \mathrm{lcm}(p-1, q-1)$ and is only known to the recipient. Note that knowing the product $(p-1)(q-1)$ is equivalent to knowing the factors p and q of $m = pq$. Currently (without quantum computers), there is no known method that calculates the prime factor decomposition of such a number m in a reasonable time if p and q have a few 1000 decimal places. However, it is not a big problem to generate such prime numbers, for example with the prime number test from exercise 2.7.8.

Primitive residues modulo prime numbers are used in the following way for **sender identification**: One chooses a large prime number $p \in \mathbb{P}$ and a fixed primitive residue a modulo p. Each participant of a communication network publishes the residue class $x_s \in \mathbf{A}_p^\times = \mathbf{F}_p^\times$ of a^s, where s is a natural number chosen by the participant and kept secret by him, which is coprime to $p - 1$ (so that x_s is also a primitive residue modulo p). It is ensured that different participants choose different numbers s, i.e., different residue classes x_s are published. Now, if the participant T, who has chosen the number t, sends a message to the participant S with the number s, he gives the power $x_s^t \in \mathbf{F}_p^\times$ for sender identification (also publicly). Since this is equal to the power $a^{st} = (a^t)^s = x_t^s$, the recipient can check the sender's information. To pretend to be T as the sender, one would have to determine the exponent t for the known power $x_t = a^t$. This is a discrete logarithm problem, see remark (3) in Example 2.2.23, and only very time-consuming methods are known for this. Unpublished, $x_s^t = x_t^s$ from S and T can also be used as a secret key known only to them. — Here too, as with the RSA codes above, one-way functions are used, namely the exponential functions $t \mapsto a^t$ of $\mathbf{F}_p^\times$, whose values are easy to calculate, while the values of the inverse functions $y \mapsto \mathrm{Log}_a\, y$ are difficult to determine. However, this does not apply if the order $p - 1$ of the element a has only small prime factors (which can also be the case with large p), see loc. cit. — Instead of the prime fields F_p, one often also chooses other finite fields, see Example 2.10.33.

2.7 Ideals and Quotient Rings

Example 2.7.15 (Maximal Ideals) We start with a characterization of division domains. ◊

Proposition 2.7.16 *For a ring A the following conditions are equivalent:*

(i) *A is a division ring.*
(ii) *The trivial ideals 0 and A are the only left ideals in A, and $0 \neq A$.*
(ii') *The trivial ideals 0 and A are the only right ideals in A, and $0 \neq A$.*

Proof It is sufficient to show the equivalence of (i) and (ii). If A is a division ring and $\mathfrak{a} \subseteq A$ is a left ideal $\neq 0$, then $\mathfrak{a}$ contains an element $a \neq 0$ and thus also $1 = a^{-1}a$, i.e., it is $A = A \cdot 1 \subseteq \mathfrak{a}$. Conversely, if $0 \neq A$ and A are the only left ideals in A and $a \in A$, $a \neq 0$. Then the principal left ideal Aa is equal to A, and there is an $a' \in A$, $a' \neq 0$, with $a'a = 1$. Every element in $A - \{0\}$ therefore has a left inverse in $(A - \{0\}, \cdot)$, and $(A - \{0\}, \cdot)$ is a group. □

We explicitly note once again that A is not necessarily a division ring when 0 and $A \neq 0$ are the only two-sided ideals in A, i.e., when A is a simple ring, cf. Exercise 2.8.8e). Proposition 2.7.16 rather states that A is a division ring exactly when 0 is a maximal left ideal or a maximal right ideal in A in the sense of the following definition.

Definition 2.7.17 Let A be a ring. A left, right, or two-sided ideal $\mathfrak{a} \subseteq A$ is called a **maximal** left, right, or two-sided ideal in A, if $\mathfrak{a}$ (with respect to inclusion) is maximal in the set of all left, right, or two-sided ideals of A that are different from A (i.e., if $\mathfrak{a}$ is an anti-atom in the set of the respective ideals of A).

If $\mathfrak{a} \subseteq A$ is a two-sided ideal $\neq \mathfrak{a}$, then according to Proposition 2.7.7 (2), the respective maximal ideals in the quotient ring $A/\mathfrak{a}$ correspond to the respective maximal ideals in A that include $\mathfrak{a}$. In particular, according to Proposition 2.7.16, *$A/\mathfrak{a}$ is a division ring exactly when $\mathfrak{a}$ is even a maximal left ideal or a maximal right ideal in A itself.*

If the ring A is not the zero ring, then the set $\mathcal{I}$ of left, right, or two-sided ideals different from A is inductively ordered with respect to inclusion (even strictly). It is $0 \in \mathcal{I}$ and if $\mathcal{K} \subseteq \mathcal{I}$ is a non-empty chain, then $\mathfrak{b} := \bigcup_{\mathfrak{a} \in \mathcal{K}} \mathfrak{a}$ is also in $\mathcal{I}$ due to $1 \notin \mathfrak{b}$ and thus an upper bound of $\mathcal{K}$ in $\mathcal{I}$. The Zorn's Lemma 1.4.15 thus implies the following theorem:

Theorem 2.7.18 (Krull's Theorem) *Let A be a ring and $\mathfrak{a}$ a left, right, or two-sided ideal $\neq A$ in A. Then there exists a maximal left, right, or two-sided ideal $\mathfrak{m} \subset A$ with $\mathfrak{a} \subseteq \mathfrak{m}$.*

In the commutative case, we obtain the following important corollary:

Corollary 2.7.19 *Let A be a commutative ring and $\mathfrak{a}$ an ideal $\neq A$ in A. Then there exists a maximal ideal $\mathfrak{m}$ in A with $\mathfrak{a} \subseteq \mathfrak{m} \subset A$, for which $A/\mathfrak{m}$ is a field.* – *In particular, every commutative ring $\neq 0$ has (at least) one maximal ideal.*

Atoms, i.e., minimal elements of the set of ideals $\neq 0$, are usually not found in the set of ideals of a commutative ring. Consider, for example, the ring $\mathbb{Z}$. – The set of maximal ideals of a commutative ring A is called the **maximal spectrum** of A and is denoted by

$$\operatorname{Spm} A$$

A commutative ring A is exactly the zero ring when $\operatorname{Spm} A = \emptyset$ is. The maximal spectrum of $\mathbb{Z}$ is $\operatorname{Spm} \mathbb{Z} = \{\mathbb{Z}p \mid p \in \mathbb{P}\}$. The maximal spectrum of a field contains only the zero ideal.

Theorem 2.7.20 *For a ring A the following are equivalent:*

(i) *A has exactly one maximal left ideal $\mathfrak{m}$.*
(ii) *A has exactly one maximal right ideal $\mathfrak{n}$. If these conditions are met, then $\mathfrak{m} = \mathfrak{n}$ is a two-sided ideal $\mathfrak{m}_A$ in A, $A/\mathfrak{m}_A$ a division ring and $A^\times = A - \mathfrak{m}_A$.*

Proof Let $\mathfrak{m}$ be the only maximal left ideal in A. We show that $\mathfrak{m}$ is also a right ideal of A, i.e., that with $a \in A$ the left ideal $\mathfrak{m}a$ in $\mathfrak{m}$ is included. For this, we can immediately assume $a \notin \mathfrak{m}$, thus $Aa + \mathfrak{m} = A$. We consider the surjective homomorphism of additive groups $\varphi \colon A \to A/\mathfrak{m}$, $x \mapsto [xa]_\mathfrak{m}$, for which

$$\mathfrak{a} := \ker \varphi = \{x \in A \mid xa \in \mathfrak{m}\}$$

is also a left ideal different from A. Since $\mathfrak{m}$ is the only maximal left ideal of A, $\mathfrak{a} \subseteq \mathfrak{m}$ applies according to Theorem 2.7.18. If now $\mathfrak{a}$ were strictly contained in $\mathfrak{m}$, then $\varphi(\mathfrak{m}) = (\mathfrak{m}a + \mathfrak{m})/\mathfrak{m}$ would be a non-trivial subgroup of $A/\mathfrak{m}$, i.e., $\mathfrak{m}a + \mathfrak{m}$ a left ideal strictly between $\mathfrak{m}$ and A. Contradiction! Therefore, as desired, $\mathfrak{a} = \mathfrak{m}$. In particular, $A/\mathfrak{m}$ is a division ring (cf. Proposition 2.7.16), $\mathfrak{m}$ a maximal right ideal of A and $A - \mathfrak{m}$ a submonoid of $(A, \cdot)$. It follows that $A - \mathfrak{m}$ is a group and therefore equal to $A^\times$; because: if $x \notin \mathfrak{m}$ is arbitrary, then $Ax = A$ due to $Ax \not\subseteq \mathfrak{m}$ and x has a left inverse, which also does not lie in $\mathfrak{m}$. Now it also follows that $\mathfrak{m}$ is the only maximal right ideal in A. If $\mathfrak{b}$ is a right ideal with $\mathfrak{b} \not\subseteq \mathfrak{m}$, then there is an $b \in \mathfrak{b} - \mathfrak{m}$. According to what has been proven, b is a unit in A and $\mathfrak{b} = A$. With this, the theorem is fully proven. □

Definition 2.7.21 A ring A that fulfills the equivalent conditions of theorem 2.7.20 is called a **local ring**. The only maximal left ideal $\mathfrak{m}_A$ of a local ring A is called the **Jacobson radical** of A.

The Jacobson radical $\mathfrak{m}_A$ of a local ring is also the only maximal right ideal of A and thus also the only maximal two-sided ideal of A. A commutative ring is local

2.7 Ideals and Quotient Rings

if and only if its maximal spectrum $\operatorname{Spm} A = \{\mathfrak{m}_A\}$ is single-element. For any ring A, the **Jacobson radical** $\mathfrak{m}_A$ is defined as the intersection of all maximal left ideals of A. $\mathfrak{m}_A$ is then also the intersection of all maximal right ideals of A and in particular a two-sided ideal, see exercise 2.8.7.

Example 2.7.22 (Prime Ideals) We only define prime ideals for commutative rings. ◊

Definition 2.7.23 Let A be a *commutative* ring. An ideal $\mathfrak{p} \subseteq A$ of A is called **prime** or a **prime ideal**, if the quotient ring $A/\mathfrak{p}$ is an integral domain. The set of all prime ideals of A is called the **(prime) spectrum** of A. We denote it with

$$\operatorname{Spec} A.$$

Let A continue to be a commutative ring until the end of this example. An ideal $\mathfrak{p}$ is a prime ideal if and only if $\mathfrak{p} \neq A$ is and if for elements $a, b \in A$ from $[0]_\mathfrak{p} = [a]_\mathfrak{p} [b]_\mathfrak{p} = [ab]_\mathfrak{p}$ always $[a]_\mathfrak{p} = [0]_\mathfrak{p}$ or $[b]_\mathfrak{p} = [0]_\mathfrak{p}$ follows, in other words, if for all $a, b \in A$ applies:

From $ab \in \mathfrak{p}$ follows $a \in \mathfrak{p}$ or $b \in \mathfrak{p}$.

Often the following reformulation is useful. An ideal $\mathfrak{p} \neq A$ is prime if and only if for any ideals $\mathfrak{a}, \mathfrak{b} \subseteq A$ the following holds: If $\mathfrak{ab} \subseteq \mathfrak{p}$, then $\mathfrak{a} \subseteq \mathfrak{p}$ or $\mathfrak{b} \subseteq \mathfrak{p}$. An ideal $\mathfrak{c} \neq A$ is therefore *not* prime if there are ideals $\mathfrak{a}, \mathfrak{b} \subseteq A$ with $\mathfrak{c} \subset \mathfrak{a}, \mathfrak{c} \subset \mathfrak{b}$, but $\mathfrak{ab} \subseteq \mathfrak{c}$. Since for a principal ideal $\mathfrak{p} = Ap$ and an element $c \in A$ the condition $c \in Ap$ is equivalent to the divisibility condition $p \mid c$, *a principal ideal is $\mathfrak{p} = Ap$ prime if and only if p is not a unit in A and if for all $a, b \in A$ the following holds*:

From $p \mid ab$ follows $p \mid a$ or $p \mid b$,

i.e. if p is a prime element in the multiplicative monoid $(A, \cdot)$ in the sense of Sect. 2.1. The (commutative) ring A is an integral domain exactly when its zero ideal is prime, i.e., when 0 is a prime element in A. Maximal ideals in A are always prime, i.e., it is $\operatorname{Spm} A \subseteq \operatorname{Spek} A$. In particular, $\operatorname{Spek} A = \emptyset$ is the case exactly when A is the zero ring. The spectrum of $\mathbb{Z}$ is $\operatorname{Spek} \mathbb{Z} = \{0\} \uplus \operatorname{Spm} \mathbb{Z} = \{\mathbb{Z}p \mid p \in \overline{\mathbb{P}} = \{0\} \uplus \mathbb{P}\}$. The prime ideal property of an ideal in A can also be formulated as follows: *The ideal $\mathfrak{p} \subseteq A$ is prime exactly when the complement $S_\mathfrak{p} := A - \mathfrak{p}$ is a submonoid of $(A, \cdot)$.*

This leads to considering the ring of fractions $A_{S_\mathfrak{p}}$, which is always (somewhat misleadingly) denoted by $A_\mathfrak{p}$. So it is

$$A_\mathfrak{p} = \{a/s \mid a, s \in A,\ s \notin \mathfrak{p}\},$$

where two fractions $a/s, b/t \in A_\mathfrak{p}$ are equal exactly when there is an $u \in A - \mathfrak{p}$ with $atu = bsu$, see Example 2.6.14. $A_\mathfrak{p}$ *is a local ring with a unique maximal ideal* $\mathfrak{m}_{A_\mathfrak{p}} = \mathfrak{p}A_\mathfrak{p} := \iota_\mathfrak{p}(\mathfrak{p})A_\mathfrak{p}$, where $\iota_\mathfrak{p} : A \to A_\mathfrak{p}$ is the canonical homomorphism $a \mapsto a/1$, cf. exercise 2.7.15. $A_\mathfrak{p}$ is called the **localization** of A with respect to the prime ideal $\mathfrak{p}$. These localizations play a fundamental role in algebraic number theory, commutative algebra, and algebraic geometry. Thus, we have the following

null test: *An element $a \in A$ is exactly 0, when $a/1 = 0 = 0/1$ is in all localizations $A_\mathfrak{m}$, $\mathfrak{m} \in \operatorname{Spm} A$.* If $a \neq 0$ and $\mathfrak{m} \in \operatorname{Spm} A$ with $\{x \in A \mid xa = 0\} \subseteq \mathfrak{m}$, then $a/1 \neq 0$ in $A_\mathfrak{m}$. An ideal $\mathfrak{a} \subseteq A$ is therefore exactly the zero ideal, when $\mathfrak{a} A_\mathfrak{m} = 0$ is for all $\mathfrak{m} \in \operatorname{Spm} A$. More generally, it follows: *If $\mathfrak{a}, \mathfrak{b} \subseteq A$ are ideals in A, then is $\mathfrak{a} \subseteq \mathfrak{b}$ exactly when $\mathfrak{a} A_\mathfrak{m} \subseteq \mathfrak{b} A_\mathfrak{m}$ applies to all $\mathfrak{m} \in \operatorname{Spm} A$.* For proof, consider the ideal $(\mathfrak{a} + \mathfrak{b})/\mathfrak{b} \subseteq A/\mathfrak{b}$. – For a prime element $p \in A$ one must carefully distinguish between the localization A_{Ap} and the ring of fractions $A_p = \{a/p^n \mid a \in A, n \in \mathbb{N}\}$ with respect to the denominator monoid generated by p, namely $\langle p \rangle = \{p^n \mid n \in \mathbb{N}\}$.

Exercises

Exercise 2.7.1 Calculate the inverse of [40] in $A_{91}^\times$ and in $F_{97}^\times$.

Exercise 2.7.2 Let $m \in \mathbb{N}^*$. The number of idempotent elements in A_m is $2^{\omega(m)}$, where $\omega(m)$ is the number of *different* prime divisors of m. Determine the idempotent elements in A_{60} and $A_{10,000}$.

Exercise 2.7.3 For two coprime numbers $a, b \in \mathbb{N}^*$ is $a^{\varphi(b)} + b^{\varphi(a)} \equiv 1 \bmod ab$.

Exercise 2.7.4 Let $p \in \mathbb{P}$ be a prime number >2 and $a \in \mathbb{Z}$ with $p \nmid a$ and $\alpha \in \mathbb{N}^*$.

a) It is $\operatorname{Ord}_{p^\alpha} a = p^\beta \operatorname{Ord}_p a$ with $\beta := \operatorname{Max}(0, \alpha - v_p(a^{p-1} - 1))$.
b) The following conditions are equivalent: (i) a is a primitive residue modulo p^α for all $\alpha \in \mathbb{N}^*$. (ii) a is a primitive residue modulo p^2. (iii) a is a primitive residue modulo p and the Fermat exponent $v_p(a^{p-1} - 1)$ is 1. (See Theorem 2.7.13 and the remarks on it. 14 is a primitive residue modulo 29, but not modulo 29^2. However, 14 is not the smallest primitive residue modulo 29. This is rather 2, and 2 is also a primitive residue modulo 29^2 and thus a primitive residue modulo all powers 29^α, $\alpha \geq 1$. To date, no p is known for which 2 is a primitive residue modulo p, but not modulo p^2. For the Wieferich primes 1093 and 3511 with Fermat exponents $v_p(2^{p-1} - 1) > 1$, see exercise 2.4.16, 2 is not a primitive residue. For $p = 3511$ this already follows from the fact that $(2/3511) = 1$ is, see Theorem 2.5.28, so 2 is a square in F_{3511}. Also determine $\operatorname{Ord}_{1093} 2$ (< 1092). The smallest prime number p, for which the smallest positive primitive residue is not a primitive residue modulo p^2, is $p = 40.487$ with smallest positive primitive residue 5.)
c) If a is a primitive residue modulo p, then a or $a + p$ (or a or $a(1 + p)$) is a primitive residue modulo p^α for all $\alpha \geq 1$.

Exercise 2.7.5 Let $\alpha \in \mathbb{N}^*$, $\alpha \geq 2$, and $a \in \mathbb{Z}$ be an odd integer. Then $(a/2^\alpha) = (a/4) = (-1)^{(a-1)/2}$ applies. (For the symbol $(a/2^\alpha)$ see Example 2.5.22. – For $\beta = 0, \ldots, \alpha$ the sets of elements of order 2^β in $\mathbb{Z}_{2^\alpha}$ are each invariant under multiplication with a. It follows $(a/2^\alpha) = \prod_{\beta=0}^\alpha \operatorname{Sign} \lambda_\beta$, where λ_β is the multiplication with a in $\mathbf{A}_{2^\beta}^\times$. Now use Proposition 2.5.21 and the (trivial) statement that $\mathbf{A}_{2^\beta}^\times$ is not cyclic for $\beta \geq 3$.)

Exercise 2.7.6 Let $m = p_1 \cdots p_n \in \mathbb{N}^*$ be a square-free number with pairwise different prime factors $p_1, \ldots, p_n$. For $a \in \mathbb{N}^*$, the exponentiation $x \mapsto x^a$ in A_m is bijective, i.e., an automorphism of the multiplicative monoid $(\mathbf{A}_m, \cdot)$, if a is coprime to $\varphi(m)$. (It is $\mathbf{A}_m \xrightarrow{\sim} \mathbf{F}_{p_1} \times \cdots \times \mathbf{F}_{p_n}$.) If $m \in \mathbb{N}^*$ is not square-free, there is no $a \in \mathbb{N}^*$, $a \geq 2$, such that $\mathbf{A}_m \to \mathbf{A}_m$, $x \mapsto x^a$, is bijective.

Exercise 2.7.7 Let m be an odd natural number >1. The following statements are equivalent: (i) The exponent $\varepsilon(m)$ ($= \operatorname{Exp} \mathbf{A}_m^\times$) is a proper divisor of $m - 1$. (ii) m is *not* a prime number, but $a^{m-1} \equiv 1 \bmod m$ holds for all numbers $a \in \mathbb{Z}$ coprime to m. (iii) m is *not* a prime number, but $a^m \equiv a \bmod m$ holds for all $a \in \mathbb{Z}$. (iv) m is square-free with at least three different prime factors, and $p - 1$ divides $m - 1$ for every prime factor p of m. (Regarding $\varepsilon(m)$ see the remarks following the theorems 2.7.13 and 2.7.14, – numbers that meet the specified conditions are called **Carmichael numbers**. $561 = 3 \cdot 11 \cdot 17$ is the smallest Carmichael number. If the numbers $6t + 1$, $12t + 1$ and $18t + 1$ are prime and m their product, then $\varepsilon(m) = \operatorname{kgV}(6t, 12t, 18t) = 36t$ is a proper divisor of $m - 1$ and m a Carmichael number. This yields, for example, the Carmichael number $7 \cdot 13 \cdot 19 = 1729$. By the way, there are infinitely many Carmichael numbers.)

Exercise 2.7.8 Let $m \in \mathbb{N}^*$ and $p_1, \ldots, p_r$ be the different prime factors of $m - 1$. Are there numbers $a_1, \ldots, a_r \in \mathbb{Z}$ with $a_i^{m-1} \equiv 1 \bmod m$ and $a_i^{(m-1)/p_i} \not\equiv 1 \bmod m$, then m is a prime number. (**Fermat's primality test** – Show that $\mathbf{A}_m^\times$ has at least $m - 1$ elements. – If m is prime, there are always such numbers $a_1, \ldots, a_r$. They can even all be chosen the same, because $\mathbf{A}_m^\times$ is then cyclic. – The Fermat primality test is particularly suitable in the case that the prime factors of $m - 1$ are easy to determine.) More generally, prove: Let $m - 1 = ab$ with coprime natural numbers a,b, and let $p_1, \ldots, p_r$ be the different prime factors of a. If there are numbers $a_1, \ldots, a_r \in \mathbb{Z}$ with $a_i^{m-1} \equiv 1 \bmod m$ and $\operatorname{ggT}(a_i^{(m-1)/p_i} - 1, m) = 1$, then for every prime divisor q of m the congruence $q \equiv 1 \bmod a$ applies. In particular, m is prime if $a > b$. (Show: a divides $|\mathbf{F}_q^\times| = q - 1$.)

Exercise 2.7.9 Let $p \in \mathbb{P}$ be a prime number ≥ 3 such that $2p + 1$ is also a prime number.[44]

a) If $p \equiv 3 \bmod 4$, then $2p + 1$ is a divisor of $M(p) = 2^p - 1$. (According to Theorem 2.5.28, $(2/(2p + 1)) = 1$.)
b) If $p \equiv 1 \bmod 4$, then 2 is a primitive residue modulo $2p + 1$. $((2/(2p + 1)) = -1$.)

---

[44] In general, prime number pairs of the form $(p, 2p + 1)$ are called **Sophie Germain pairs** (after S. Germain (1776–1831)).

Exercise 2.7.10 Let $\mathfrak{a}, \mathfrak{b}, \mathfrak{c}, \mathfrak{a}_1, \ldots, \mathfrak{a}_n$ be two-sided ideals in the ring A and $\mathfrak{a}'_i := \bigcap_{j \neq i} \mathfrak{a}_j, i = 1, \ldots, n$.

a) If $\mathfrak{a}, \mathfrak{b}$ are comaximal, then $\mathfrak{a} \cap \mathfrak{b} = \mathfrak{a}\mathfrak{b} + \mathfrak{b}\mathfrak{a}$ applies. In particular, $\mathfrak{a} \cap \mathfrak{b} = \mathfrak{a}\mathfrak{b}$ is true when A is commutative.
b) If both $\mathfrak{a}, \mathfrak{b}$ and $\mathfrak{a}, \mathfrak{c}$ are comaximal, then $\mathfrak{a}$ and $\mathfrak{b}\mathfrak{c}$ are also comaximal. In particular, $\mathfrak{a}^m, \mathfrak{b}^n$ are comaximal for all $m, n \in \mathbb{N}$, if $\mathfrak{a}, \mathfrak{b}$ are comaximal.
c) The following statements are equivalent: (i) $\mathfrak{a}_1, \ldots, \mathfrak{a}_n$ are pairwise comaximal. (ii) $\mathfrak{a}_i$ and $\mathfrak{a}'_i$ are comaximal for all $i = 1, \ldots, n$. (iii) It is $\mathfrak{a}'_1 + \cdots + \mathfrak{a}'_n = A$. – If these conditions are met, then $\mathfrak{a}_1 \cap \cdots \cap \mathfrak{a}_n = \sum_{\sigma \in \mathfrak{S}_n} \mathfrak{a}_{\sigma 1} \cdots \mathfrak{a}_{\sigma n}$ applies.

Exercise 2.7.11 Let A be a finite commutative ring. Then is $\operatorname{Spm} A = \operatorname{Spec} A$. Determine $\operatorname{Spm} \mathbf{A}_m = \operatorname{Spec} \mathbf{A}_m$ for $m \in \mathbb{N}^*$.

Exercise 2.7.12 If B is a Boolean ring, then show $\operatorname{Spm} B = \operatorname{Spec} B$. ($F_2$ is the only Boolean ring that is also an integral domain.)

Exercise 2.7.13 Let $\mathfrak{a}_i$ be left, right, or two-sided ideals in the rings $A_i, i \in I$, and $A := \prod_{i \in I} A_i$ the Cartesian product of the A_i and $\mathfrak{a} := \prod_{i \in I} \mathfrak{a}_i \subseteq A$.

a) $\mathfrak{a}$ is a left, right, or two-sided ideal in A. If I is finite, then all ideals in A are of this form. For two-sided ideals $\mathfrak{a}_i$, $A/\mathfrak{a} \xrightarrow{\sim} \prod_{i \in I} A_i/\mathfrak{a}_i$ applies. In the case that for infinitely many $i \in I$ the ring A_i is not the zero ring, provide an ideal of A that is *not* a product ideal.
b) Let the rings A_i all be commutative. Only then is $\mathfrak{a}$ a maximal ideal (or a prime ideal) in A, if there is an index $i_0 \in I$ such that $\mathfrak{a}_{i_0}$ is maximal (or prime) in A_{i_0} and $\mathfrak{a}_i = A_i$ for all $i \neq i_0$. If I is finite, then one can identify $\operatorname{Spm} A$ (or $\operatorname{Spec} A$) with the disjoint union $\biguplus_{i \in I} \operatorname{Spm} A_i$ (or $\biguplus_{i \in I} \operatorname{Spec} A_i$). If $A_i \neq 0$ for infinitely many $i \in I$, then there are maximal ideals in A that do not belong to $\biguplus_{i \in I} \operatorname{Spm} A_i$.

Exercise 2.7.14 Let $\varphi : A \to B$ be a homomorphism of commutative rings. If $\mathfrak{q} \in \operatorname{Spek} B$, then $\varphi^{-1}(\mathfrak{q}) \in \operatorname{Spek} A$. φ thus induces in a canonical way a mapping

$$\operatorname{Spec} \varphi : \operatorname{Spec} B \to \operatorname{Spec} A.$$

If φ is surjective with $\mathfrak{a} := \ker \varphi$, then $\operatorname{Spec} \varphi$ is injective with the image

$$V(\mathfrak{a}) = V_A(\mathfrak{a}) := \{\mathfrak{p} \in \operatorname{Spec} A \mid \mathfrak{a} \subseteq \mathfrak{p}\} \subseteq \operatorname{Spec} A.$$

In particular, one can identify the spectrum $\operatorname{Spec}(A/\mathfrak{a})$ of $A/\mathfrak{a}$ with $V(\mathfrak{a})$.

Note Note that φ does not generally induce a mapping from $\operatorname{Spm} B$ to $\operatorname{Spm} A$ in the same way. If φ is surjective, however, the restriction $(\operatorname{Spek} \varphi)|\operatorname{Spm} B$ is a bijective mapping $\operatorname{Spm} B \xrightarrow{\sim} V(\mathfrak{a}) \cap \operatorname{Spm} A$. In particular, $\operatorname{Spm}(A/\mathfrak{a})$ identifies itself with $V(\mathfrak{a}) \cap \operatorname{Spm} A$.

2.7 Ideals and Quotient Rings

Exercise 2.7.15 A is a commutative ring and S is a submonoid of $(A, \cdot)$. Furthermore, $\iota_S \colon A \to A_S$ is the canonical homomorphism $a \mapsto a/1$ from A to the ring of fractions A_S. Then $\operatorname{Spec} \iota_S \colon \operatorname{Spec} A_S \to \operatorname{Spec} A$ (see exercise 2.7.14) is an injective mapping, whose image is $\operatorname{Spec}_S A := \{\mathfrak{p} \in \operatorname{Spec} A \mid \mathfrak{p} \cap S = \emptyset\} \subseteq \operatorname{Spec} A$. The inverse mapping $\operatorname{Spek}_S A \xrightarrow{\sim} \operatorname{Spek} A_S$ is given by

$$\mathfrak{p} \mapsto \mathfrak{p} A_S := \iota_S(\mathfrak{p}) A_S = \{a/s \mid a \in \mathfrak{p}, s \in S\} \subseteq A_S.$$

In particular, $\operatorname{Spec}_S A \neq \emptyset$, when $A_S \neq 0$, i.e. $0 \notin S$ is. In the case of $S = A - \mathfrak{p}$ with a prime ideal $\mathfrak{p} \in \operatorname{Spec} A$ one obtains the following: The spectrum of the localization $A_\mathfrak{p} = A_S$ consists exactly of the prime ideals $\mathfrak{q} A_\mathfrak{p}$, $\mathfrak{q} \in \operatorname{Spec} A$, $\mathfrak{q} \subseteq \mathfrak{p}$. Specifically, $A_\mathfrak{p}$ is a local ring with the only maximal ideal (= Jacobson radical) $\mathfrak{m}_{A_\mathfrak{p}} = \mathfrak{p} A_\mathfrak{p}$, cf. Definition 2.7.21, and the canonical injection $A/\mathfrak{p} \to A_\mathfrak{p}/\mathfrak{m}_{A_\mathfrak{p}}$ induces a canonical isomorphism $Q(A/\mathfrak{p}) \xrightarrow{\sim} A_\mathfrak{p}/\mathfrak{m}_{A_\mathfrak{p}}$, cf. Example 2.6.14.

Exercise 2.7.16 Let A be a commutative ring. The intersection $\bigcap_{\mathfrak{p} \in \operatorname{Spek} A} \mathfrak{p}$ of all prime ideals of A is equal to the nilradical $\mathfrak{n}_A$ of A (which contains exactly the nilpotent elements of A, cf. exercise 2.6.3). (The inclusion $\mathfrak{n}_A \subseteq \mathfrak{p}$ for each prime ideal $\mathfrak{p}$ is trivial. Let $x \in A$ be non-nilpotent. Then $S := \{x^n \mid n \in \mathbb{N}\} \subseteq (A, \cdot)$ does not contain the zero element, and according to exercise 2.7.15 there exists a prime ideal $\mathfrak{p} \subseteq A$ with $\mathfrak{p} \cap S = \emptyset$.) If S is a multiplicative submonoid of A, then $\mathfrak{n}_{A_S} = \iota_S(\mathfrak{n}_A) A_S$. A is reduced (i.e., $\mathfrak{n}_A = 0$) if and only if $A_\mathfrak{m}$ is reduced for all $\mathfrak{m} \in \operatorname{Spm} A$.

Exercise 2.7.17 Let $\varphi \colon A \to B$ be a surjective homomorphism of commutative rings, whose kernel contains only nilpotent elements (for which $\ker \varphi \subseteq \mathfrak{n}_A$ applies). Then the one from φ induced homomorphism $\operatorname{Idp}(\varphi) \colon \operatorname{Idp}(A) \to \operatorname{Idp}(B)$ an isomorphism of Boolean rings. (See Ex. 2.6.6a).) In particular, the ring A is connected exactly when the ring B is connected. (The injectivity of $\operatorname{Idp}(\varphi)$ is trivial. To prove surjectivity, let $e \in A$ be an element with $\varphi(e) \in \operatorname{Idp}(B)$, i.e. with $e - e^2 = e(1-e) \in \ker \varphi$ and $e^n(1-e)^n = 0$ for a $n \in \mathbb{N}^*$. Then $Ae^n + A(1-e)^n = A$ and $Ae^n \cap A(1-e)^n = Ae^n(1-e)^n = 0$. If c, d are elements in A with $ce^n + d(1-e)^n = 1$, then ce^n is idempotent and $\varphi(e) = \varphi(ce^{n+1}) = \varphi(ce^n)$. – Note that this lifting of idempotent elements is constructive. If furthermore $\varphi \colon A \to B$ is a surjective homomorphism of *arbitrary* rings, whose kernel contains only nilpotent elements, then $\operatorname{Idp}(\varphi) \colon \operatorname{Idp}(A) \to \operatorname{Idp}(B)$ is also surjective (but generally not injective). If $b \in \operatorname{Idp}(B)$ and $a \in A$ with $\varphi(a) = b$, consider for proof the surjective restriction $\varphi|\mathbb{Z}[a] \colon \mathbb{Z}[a] \to \mathbb{Z}[b]$.)

Exercise 2.7.18 If $\varphi, \psi \colon A \to B$ are surjective ring homomorphisms with $\ker \varphi = \ker \psi$, then there is a ring automorphism $\chi \colon B \xrightarrow{\sim} B$ with $\psi = \chi \circ \varphi$. (A similar theorem naturally also applies to groups.)

2.8 Modules and Vector Spaces

Let A be a ring and V an additively written abelian group. If

$$A \times V \to V, \quad (a,x) \mapsto ax,$$

is an operation of the multiplicative monoid $(A, \cdot)$ of A on V as a monoid of group homomorphisms, then the action homomorphism

$$\vartheta : A \to \mathrm{End}\, V, \quad a \mapsto (\vartheta_a : x \mapsto ax),$$

is a homomorphism from $(A, \cdot)$ into the multiplicative monoid $(\mathrm{End}\, V, \circ)$ of the endomorphism ring $\mathrm{End}\, V = (\mathrm{End}\, V, +, \circ)$, cf. Example 2.6.19. Since $\mathrm{End}\, V$ is a ring, it is natural to consider such operations of A for which ϑ is even a ring homomorphism.

Definition 2.8.1 Let A be a ring. An additively written abelian group V together with an operation $A \times V \to V$ is called an A-**module** or also a **module over** A, if this operation is given by a ring homomorphism $\vartheta = \vartheta_V : A \to \mathrm{End}\, V$, if therefore for all $a, b \in A$ and all $x, y \in V$ applies:

(1) $(ab)x = a(bx)$, (2) $a(x+y) = ax + by$, (3) $(a+b)x = ax + bx$, (4) $1 \cdot x = x$.

A mapping $f : V \to W$ of the A-module V into the A-module W is called an A-**homomorphism** or a **homomorphism of** A-**modules** or an A-**linear mapping**, if f is a homomorphism of the additive groups of V and W, which is compatible with the operations ϑ_V and ϑ_W of A on V and W, if therefore $f \circ \vartheta_{V,a} = \vartheta_{W,a} \circ f$ for all $a \in A$ is, i.e. if for all $a \in A$ and all $x, y \in V$ applies:

$$f(x+y) = f(x) + f(y), \quad f(ax) = af(x).$$

The modules over a division domain K are called K-**vector spaces** or **vector spaces over** K.

The operation $A \times V \to V$ of an A-module V is called the **scalar multiplication** of V. It is biadditive, cf. Definition 2.6.7. In particular, $0 \cdot x = 0 = a \cdot 0$ for all $x \in V$ and all so-called **scalars** $a \in A$. The operation $\vartheta_a : V \to V$, $x \mapsto ax$, of $a \in A$ on V is called the **homothety** or **dilation** with a. If this is injective, then $a \in A$ is called **regular** for the A-module V. The additive translation $\tau_{x_0} : x \mapsto x_0 + x$ of V with a $x_0 \in V$ is also called the **translation** by x_0. The composition of A-homomorphisms is again an A-homomorphism. The set of all A-homomorphisms $f : V \to W$ of an A-module V into an A-module W we denote by

$$\mathrm{Hom}_A(V, W).$$

This is obviously a subgroup of the group $\mathrm{Hom}(V, W)$ of homomorphisms of the additive groups of V and W. Accordingly, $\mathrm{Iso}_A(V, W) \subseteq \mathrm{Hom}_A(V, W)$ denotes the

2.8 Modules and Vector Spaces

set of A-**isomorphisms** from V to W, i.e., the set of bijective A-homomorphisms $V \to W$. The set $\mathrm{End}_A V := \mathrm{Hom}_A(V, V)$ of A-**endomorphisms** of V is a subring of $\mathrm{End}\, V$, whose unit group $(\mathrm{End}_A V)^\times$ is the group

$$\mathrm{GL}_A V = \mathrm{Aut}_A V = \mathrm{Iso}_A(V, V)$$

of A-**automorphisms** of V is.[45] If A is commutative, then the homotheties $\vartheta_{V,a}$, $a \in A$, of an A-module V are A-linear. It follows that then the a-multiple $af = \vartheta_{W,a} \circ f = f \circ \vartheta_{V,a}$ of an A-homomorphism $V \to W$ is also an A-homomorphism, and it is immediately checked that with this scalar multiplication $\mathrm{Hom}_A(V, W)$ is an A-module. We summarize again:

Proposition 2.8.2 *If V and W are modules over the ring A, then $\mathrm{Hom}_A(V, W)$ is a subgroup of $\mathrm{Hom}(V, W)$. If A is commutative, then $\mathrm{Hom}_A(V, W)$ is an A-module with the scalar multiplication $af : x \mapsto af(x) = f(ax)$, $a \in A$, $f \in \mathrm{Hom}_A(V, W)$.*

An A-**submodule** U of V is a subgroup of $(V, +)$, which is invariant under scalar multiplication, for which therefore $ax \in U$ is for all $a \in A$ and $x \in U$. If $U_i \subseteq V$, $i \in I$, are submodules of V, then their sum $\sum_{i \in I} U_i \subseteq V$ is not only a subgroup, but even an A-submodule of V and thus the smallest submodule of V that includes all U_i, $i \in I$. Images and pre-images of submodules with respect to an A-linear mapping $f : V \to W$ of A-modules are obviously again submodules. In particular, $\mathrm{Im}\, f$ is a submodule of W and $\ker f = f^{-1}(0)$ a submodule of V.

Example 2.8.3 (1) Let W be an abelian group. The characteristic homomorphism $\chi : \mathbb{Z} \to \mathrm{End}\, W$, $a \mapsto \chi_a = a\,\mathrm{id}_W$, is the only ring homomorphism from $\mathbb{Z}$ to $\mathrm{End}\, W$. Consequently, W has exactly one $\mathbb{Z}$-module structure, and this is given by the multiplication of multiples $(a, x) \mapsto ax$, $a \in \mathbb{Z}$, $x \in W$. Abelian groups and $\mathbb{Z}$-modules are thus one and the same, and it is $\mathrm{Hom}(V, W) = \mathrm{Hom}_\mathbb{Z}(V, W)$ *for abelian groups V, W.*

(2) Let A be a ring. The Cayley homomorphism $A \to \mathrm{End}(A, +)$, $a \mapsto L_a$, defines an A-module structure on A, whose homotheties are the left translations L_a, $a \in A$. The associated operation $A \times A \to A$ is the multiplication of A. A is always considered with this A-module structure, unless otherwise stated. *The A-submodules of A are then exactly the left ideals of A.*

If $f : A \to V$ is an A-homomorphism, then $f(a) = f(a \cdot 1) = af(1)$ holds for all $a \in A$. f is therefore uniquely determined by the value $f(1)$. Conversely, for any $v \in V$, the mapping $a \mapsto av$ is an A-homomorphism. *The mapping*

---

[45] Often the index A for sets of A-homomorphisms is suppressed when there is no doubt about the scalar ring A. In general, the scalar ring A is also often called the **base ring**, when it is chosen once and for all.

$$\mathrm{Hom}_A(A,V) \xrightarrow{\sim} V, \quad f \mapsto f(1),$$

is therefore bijective and even a group isomorphism or even an A -module homomorphism, if A is commutative. For $V = A$, $(f \circ g)(1) = f(g(1)) = g(1)f(1)$ holds for the composition of two A-endomorphisms $f, g \colon A \to A$. The mapping $f \mapsto f(1)$ *is therefore a ring isomorphism* $\mathrm{End}_A A \xrightarrow{\sim} A^{\mathrm{op}}$.

The right translations $R_a \colon x \mapsto xa$, $a \in A$, define a ring homomorphism $R \colon A^{\mathrm{op}} \to \mathrm{End}(A, +)$ and thus an A^{op}-module structure on A, for which the right translations R_a, $a \in A$, are homotheties and the right ideals are the A^{op}-submodules. In general, an A^{op}-module is called an A-**right module** and the previously considered A-modules, if clarity requires, A-**left modules**. If one writes an A-right module structure on V as a right operation $V \times A \to V$, $(x,a) \mapsto xa$, then the clear calculation rules

(1) $x(ab) = (xa)b$, (2) $(x + y)a = xa + ya$, (3) $x(a + b) = xa + xb$, (4) $x \cdot 1 = x$

apply for all $a, b \in A$, $x, y \in V$. The ring A itself thus carries two module structures, which are moreover compatible in the following sense: The homotheties of one structure commute with the homotheties of the other structure: $L_a \circ R_b = R_b \circ L_a$ for all $a, b \in A$. In general, two A or B-(left-)module structures on the same abelian group V with the action homomorphisms $\vartheta \colon A \to \mathrm{End}\, V$ or $\eta \colon B \to \mathrm{End}\, V$ are called **compatible**, if the homotheties ϑ_a, $a \in A$, and η_b, $b \in B$, commute, i.e., if $a(bx) = b(ax)$ is for all $a \in A$, $b \in B$, $x \in V$, i.e., when the homotheties of one structure are linear with respect to the other structure. This is then referred to as an A-B-**Bimodule** V. Every ring A is thus an A-A^{op}-Bimodule, and every module over a *commutative* ring A with the same A-module structure is an A-A-Bimodule. If V is an A-B-Bimodule and W is an A-module, then $\mathrm{Hom}_A(V,W)$ with the right operation $\mathrm{Hom}_A(V,W) \times B \to \mathrm{Hom}_A(V,W)$, $(f,b) \mapsto fb := f \circ \eta_b$, is a B-right module, i.e., a B^{op}-(left-)module. In a similar way, $\mathrm{Hom}_A(V,W)$ with the left operation $bf := \eta_b \circ f$ is a B-left module, if W carries an A-B-Bimodule structure. In this way, the A-module structure on $\mathrm{Hom}_A(V,W)$ was obtained from the canonical A-A-Bimodule structure of V (or of W) when A is commutative. The above isomorphism $\mathrm{Hom}_A(A,V) \xrightarrow{\sim} V$ is an A-module isomorphism for every ring A and every A-module V, if $\mathrm{Hom}_A(A,V)$ carries the A-module structure that is induced by the A-A^{op}-Bimodule structure on A. – In contrast to the elements of $\mathrm{Hom}_A(A,V)$, the so-called A-**Linear forms** $f \in V^* := \mathrm{Hom}_A(V,A)$ on V is harder to describe. However, according to the last remark, the A-A^{op}-bimodule structure on A induces an A-*right* module structure on

$$V^* = \mathrm{Hom}_A(V,A)$$

with the scalar multiplication $fa \colon x \mapsto f(x)a$, $a \in A$, $f \in V^*$. V^* is called the **dual module** of V.

(3) Let V be an A-module with action homomorphism $\vartheta \colon A \to \mathrm{End}\, V$. If $\varphi \colon A' \to A$ is a ring homomorphism, then the composition $\vartheta \circ \varphi \colon A' \to \mathrm{End}\, V$ defines an A'-module structure on V with the operation $(a',x) \mapsto a'x = \varphi(a')x$. It

2.8 Modules and Vector Spaces

is called the by φ **induced** A'-**module structure** on V. Particularly important is the case where A' is a subring of A and φ is the canonical inclusion $A' \hookrightarrow A$. Then the A'-operation on V is simply the restriction of the A-operation. Without comment, we will consider an A-module also as an A'-module in this way. For example, every complex (i.e., $\mathbb{C}$-) vector space is also a real (i.e., $\mathbb{R}$-) vector space and every $\mathbb{R}$-vector space is also an $\mathbb{Q}$-vector space.

(4) Let V be an A-module with action homomorphism $\vartheta: A \to \text{End}\, V$. Then the two-sided ideal

$$\text{Ann}_A V := \ker \vartheta = \{a \in A \mid ax = 0 \text{ for all } x \in V\} = \{a \in A \mid aV = 0\}$$

is the **annihilator** *of the module* V. It is $\text{Ann}_A V = \bigcap_{x \in V} \text{Ann}_A x$, where

$$\text{Ann}_A x := \{a \in A \mid ax = 0\}$$

is the **annihilator** *of the element* $x \in V$. $\text{Ann}_A x$ is the kernel of the A-homomorphism $A \to V$, $a \mapsto ax$, and therefore (only) a left ideal. V is called a **faithful** A-**module**, if $\text{Ann}_A V = 0$ is. If $\mathfrak{a} \subseteq \text{Ann}_A V$ is a two-sided ideal in the annihilator of V, then the action homomorphism $\vartheta: A \to \text{End}\, V$ induces a homomorphism $\overline{\vartheta}: A/\mathfrak{a} \to \text{End}_A V$ of rings and thus a $(A/\mathfrak{a})$-module structure on V with scalar multiplication $[a]_\mathfrak{a} x = ax$, $a \in A$, and $\text{Ann}_{A/\mathfrak{a}} V = (\text{Ann}_A V)/\mathfrak{a}$. Conversely, an $(A/\mathfrak{a})$-module structure on V using the canonical projection $A \to A/\mathfrak{a}$ induces an A-module structure on V with $\mathfrak{a} \subseteq \text{Ann}_A V$. *For a two-sided ideal* $\mathfrak{a} \subseteq A$ *are thus* $(A/\mathfrak{a})$-*modules and* A-*modules, whose annihilator* $\mathfrak{a}$ *includes, one and the same.* Thus, the annihilator of an abelian group W (considered as a $\mathbb{Z}$-module) is the ideal $\mathbb{Z}\, \text{Exp}\, W \subseteq \mathbb{Z}$. For $m \in \mathbb{N}$, abelian groups with $(\text{Exp}\, W)|m$ and A_m-modules are the same. Specifically, for a prime number $p \in \mathbb{P}$, elementary abelian p-groups and F_p-vector spaces are identical objects.

(5) Let V be a module over the *commutative* ring A. An element $x \in V$ is called a **torsion element** of V, if there is a non-zero divisor $a \in A^*$ with $ax = 0$, so if $\text{Ann}_A x$ contains a non-zero divisor. The set of all torsion elements of V is denoted by

$$T_A V$$

$T_A V$ is obviously an A-submodule of V. For a non-zero divisor $a \in A^*$, $T_a V := \ker \vartheta_a = \{x \in V \mid ax = 0\}$ is called the a-**torsion** of V. Then $Aa \subseteq \text{Ann}_A T_a V$, $T_a V$ is therefore an (A/Aa)-module (see (4)), and $T_A V = \bigcup_{a \in A} T_a V$. V is called a **torsion module**, if $T_A V = V$ is, and **torsion-free**, if $T_A V = 0$ is.

(6) Let W be an additive abelian group. The identity $\text{End}\, W \to \text{End}\, W$ defines the so-called **tautological** $(\text{End}\, W)$-**module structure** on W with scalar multiplication

$$fx := f(x), \quad f \in \text{End}\, W, \ x \in W.$$

In particular, W is a module over any subring of $\text{End}\, W$. Any A-module structure on W is induced by the action homomorphism $\vartheta: A \to \text{End}\, W$. ◊

Example 2.8.4 (Direct Sums and Direct Products) (1) Let W_i, $i \in I$, be a family of A-modules. Then the **direct product** $\prod_{i \in I} W_i$ with (component-wise addition and) component-wise scalar multiplication is an A-module. Analogous to abelian groups, it has the following universal property with the canonical A-linear projections $p_i \colon \prod_{i \in I} W_i \to W_i$, $i \in I$, see Example 2.2.17: *For every A-module V, the canonical mapping*

$$\mathrm{Hom}_A\Big(V, \prod_{i \in I} W_i\Big) \xrightarrow{\sim} \prod_{i \in I} \mathrm{Hom}_A(V, W_i), \quad f \mapsto (p_i f)_{i \in I},$$

is a group isomorphism and for commutative A an A-module isomorphism. The I-tuple $(f_i)_{i \in I} \in \prod_{i \in I} \mathrm{Hom}_A(V, W_i)$ is the image of the A-homomorphism $V \to \prod_{i \in I} W_i$, $x \mapsto (f_i(x))_{i \in I}$, which is also denoted with $(f_i)_{i \in I}$.

(2) Let V_j, $j \in J$, be a family of A-modules. The restricted direct product or the **direct sum** $\bigoplus_{j \in J} V_j$ of the V_j is the submodule of those elements $(x_j)_{j \in J}$ of the direct product $\prod_{j \in J} V_j$, where almost all components vanish. In addition to the canonical projections $(v_j)_{j \in J} \mapsto v_j$, the canonical injections $\iota_j \colon V_j \to \bigoplus_{j \in J} V_j$, $j \in J$, play a special role now. For $x_j \in V_j$, $\iota_j(x_j) = (\delta_{ij} x_j)_{i \in J}$ is the J-tuple, whose j-th component is equal to x_j and whose remaining components vanish. Analogous to abelian groups, the direct sum with the canonical A-linear injections $\iota_j \colon V_j \to \bigoplus_{j \in J} V_j$ has the following universal property, cf. Example 2.2.20: *For every A-module W, the canonical mapping*

$$\mathrm{Hom}_A\Big(\bigoplus_{j \in J} V_j, W\Big) \xrightarrow{\sim} \prod_{j \in J} \mathrm{Hom}_A(V_j, W), \quad g \mapsto (g \iota_j)_{j \in J},$$

is a group isomorphism and for commutative A an A-module isomorphism. The J-tuple $(f_j)_{j \in J} \in \prod_{j \in J} \mathrm{Hom}_A(V_j, W)$ is the image of the A-homomorphism

$$\sum_{j \in J} f_j \colon \bigoplus_{j \in J} V_j \to W, \quad (x_j)_{j \in J} \mapsto \sum_{j \in J} f_j(x_j).$$

The combination of the universal properties of direct product and direct sum yields the following important theorem: ◊

Theorem 2.8.5 *Let V_j, $j \in J$, and W_i, $i \in I$, be families of A-modules. Then the canonical mapping*

$$\mathrm{Hom}_A\Big(\bigoplus_{j \in J} V_j, \prod_{i \in I} W_i\Big) \xrightarrow{\sim} \prod_{(i,j) \in I \times J} \mathrm{Hom}_A(V_j, W_i),$$

$$f \mapsto (f_{ij})_{(i,j) \in I \times J}, \quad f_{ij} := p_i f \iota_j, \quad i \in I, j \in J,$$

is a group isomorphism and for commutative A an A-module isomorphism. The matrix $(f_{ij})_{(i,j) \in I \times J} \in \prod_{(i,j) \in I \times J} \mathrm{Hom}_A(V_j, W_i)$ is the image of the homomorphism

$$f \colon \bigoplus_{j \in J} V_j \to \prod_{i \in I} W_i, \quad (x_j)_{j \in J} \mapsto (y_i)_{i \in I} \quad \text{with} \quad y_i := \sum_{j \in J} f_{ij}(x_j), i \in I.$$

2.8 Modules and Vector Spaces

Note that for finite index sets, direct sums and direct products coincide. Let I,J,K be *finite* index sets and U_k, $k \in K$, another family of A-modules. Then the matrices $\mathsf{B} = (g_{jk}) \in \prod_{j,k} \operatorname{Hom}_A(U_k, V_j)$ and $\mathsf{A} = (f_{ij}) \in \prod_{1,j} \operatorname{Hom}_A(V_j, W_i)$ describe the homomorphisms $g\colon \bigoplus_{k \in K} U_k \to \bigoplus_{j \in J} V_j$ and $f\colon \bigoplus_{j \in J} V_j \to \bigoplus_{i \in I} W_i$, respectively. The composition $f \circ g\colon \bigoplus_{k \in K} U_k \to \bigoplus_{i \in I} W_i$ is given by the **product matrix**

$$\mathsf{AB} := (f_{ij})_{i,j}(g_{jk})_{j,k} = (h_{ik})_{i,k} \in \prod_{(i,k) \in I \times K} \operatorname{Hom}_A(U_k, W_i)$$

with $\quad h_{ik} := \sum_{j \in J} f_{ij} \circ g_{jk}, \ (i,k) \in I \times K.$

We leave it to the reader to determine the restrictions on the matrices A or B to formulate, when the index sets I, J, K are not necessarily finite. In the frequently used case of direct sums A^n and A^m, Theorem 2.8.5 reads, taking into account the identification of $\operatorname{End}_A A$ with A^{op}, cf. Example 2.8.3 (2): *Every A-module homomorphism $f\colon A^n \to A^m$ is given by an $m \times n$-matrix* $\mathsf{A} = (a_{ij}) \in \mathrm{M}_{m,n}(A^{\mathrm{op}}) = (A^{\mathrm{op}})^{\{1,\ldots,m\} \times \{1,\ldots,n\}}$. If one writes – as is customary – the elements $x \in A^n$ or $y \in A^m$ as single-column matrices with n or m rows, then

$$f(\mathsf{x}) = \mathsf{Ax} = \begin{pmatrix} a_{11} & a_{12} & \cdots & a_{1n} \\ a_{21} & a_{22} & \cdots & a_{2n} \\ \vdots & \vdots & \ddots & \vdots \\ a_{m1} & a_{m2} & \cdots & a_{mn} \end{pmatrix} \begin{pmatrix} x_1 \\ x_2 \\ \vdots \\ x_n \end{pmatrix} = \begin{pmatrix} y_1 \\ y_2 \\ \vdots \\ y_m \end{pmatrix} = \mathsf{y}$$

with $\quad y_i = \sum_{j=1}^{n} x_j a_{ij}, \ 1 \le i \le m.$

Note that the matrices have coefficients in the opposite ring A^{op} and are to be multiplied there! This provides the summands $x_j a_{ij}$ instead of $a_{ij} x_j$ and is also to be considered when multiplying matrices. *The endomorphism ring of the A-module A^n is thus the ring $\mathrm{M}_n(A^{\mathrm{op}})$ of the square $n \times n$-matrices with coefficients in A^{op}.* The identity of A^n is represented by the **unit matrix**

$$\mathsf{E}_n := (\delta_{ij}) = \begin{pmatrix} 1 & 0 & \cdots & 0 \\ 0 & 1 & \cdots & 0 \\ \vdots & \vdots & \ddots & \vdots \\ 0 & 0 & \cdots & 1 \end{pmatrix} \in \mathrm{M}_n(A)$$

In the important case that A is commutative, there is of course no need to distinguish between A and A^{op}.

As a rule, it is easier to specify direct sum representations of a module than direct product representations. For example, Lemma 2.2.21 implies:

Lemma 2.8.6 (Direct sums of submodules) *Let U_i, $i \in I$, be a family of submodules of the A-module V and $h\colon \bigoplus_{i \in I} U_i \to V$, $(u_i)_{i \in I} \mapsto \sum_{i \in I} u_i$, the canonical A-homomorphism with $\sum_{i \in I} U_i$ as image. Exactly then is h injective, the sum of the U_i thus direct, when the following condition is fulfilled: For each $i \in I$ applies*

$$U_i \cap \sum_{j \neq i} U_j = \{0\}.$$

If I is completely ordered, then this condition is also equivalent to the following: It applies $U_i \cap \sum_{j < i} U_j = \{0\}$ for all $i \in I$.

If the sum $\sum_{i \in I} U_i \subseteq V$ is direct, then this sum is also denoted by

$$\overset{\oplus}{\underset{l \in I}{\sum}} U_i.$$

For a submodule U of an A-module V, the quotient group

$$V/U$$

is also an A-module. Its scalar multiplication is $a[x]_U = [ax]_U$, $a \in A$, $x \in V$. The operation $\bar{\vartheta}_a$ of $a \in A$ on V/U is thus induced by the operation ϑ_a of a on V (which is well-defined due to $\vartheta_a(U) \subseteq U$). The canonical projection $\pi_U\colon V \to V/U$ is A-linear and has the following universal property, see Theorem 2.3.7 and Theorem 2.3.8:

Theorem 2.8.7 *Let U be a submodule of the A-module V and $f\colon V \to W$ an A-linear mapping into an A-module W with $U \subseteq \ker f$. Then there exists exactly one A-linear mapping $\bar{f}\colon V/U \to W$ with $f = \bar{f} \circ \pi_U$. Here, $\bar{f}([x]_U) = f(x)$, $x \in V$, $\mathrm{Im}\,\bar{f} = \mathrm{Im}\,f$ and $\ker \bar{f} = (\ker f)/U$ $(\subseteq V/U)$. – Exactly then is $\bar{f}$ an isomorphism, when f is surjective and $U = \ker f$. In particular, is*

$$V/\ker f \overset{\sim}{\to} \mathrm{Im}\,f \quad \textbf{(Theorem of ismorphisms for modules)}.$$

The cosets $x + U$, $x \in V$, of a K-subvector space U of a K-vector space V (K skew field) are also called (parallel to U) **affine subspaces** of V.[46]

The general theorem of the induced homomorphism, of which Theorem 2.8.7 is a special case, has the following form for modules (cf. Theorem 2.3.2):

Theorem 2.8.8 (Theorem of the induced homomorphism for modules) *Let $g\colon V \to W$ and $f\colon V \to X$ be homomorphisms of A-modules. g is surjective, and*

---

[46] Sometimes a similar terminology is also used for modules.

2.8 Modules and Vector Spaces

let $\ker g \subseteq \ker f$ *hold. Then there exists exactly one homomorphism* $\bar{f}: W \to X$ *with* $f = \bar{f} \circ g$. – *Exactly then is* $\bar{f}$ *surjective, iff f is surjective.* – *Exactly then is* $\bar{f}$ *injective, if* $\ker g = \ker f$ *is.* – *Exactly then is* $\bar{f}$ *an isomorphism, iff f is surjective and* $\ker g = \ker f$ *holds* (**Theorem of isomorphisms for modules**).

Let V be an A-module. The elements of V can often be clearly represented with the help of the scalars $a \in A$ (which are usually more or less well known). To this end, we first mention the **general distributive law**: Let a_i, $i \in I$, and x_j, $j \in J$, families of elements in A or V, of which almost all vanish, then

$$\left(\sum_{i \in I} a_i\right)\left(\sum_{j \in J} x_j\right) = \sum_{(i,j) \in I \times J} a_i x_j.$$

applies. For any family v_i, $i \in I$, of elements of V, the elements

$$\sum_{i \in I} a_i v_i, \quad (a_i)_{i \in I} \in A^{(I)},$$

are called the **linear combinations** of the v_i, $i \in I$, (with coefficients in A). They form the submodule of V generated by v_i, $i \in I$, i.e., the smallest A-submodule of V that contains all v_i, $i \in I$. If this submodule is equal to V, then v_i, $i \in I$, is called a **generating system** of V. The linear combinations of the family $v_1, \ldots, v_n$ in V are the elements

$$a_1 v_1 + \cdots + a_n v_n, \quad (a_1, \ldots, a_n) \in A^n.$$

For $v \in V$ is $Av = \{av \mid a \in A\}$ the submodule generated by v and thus $\sum_{i \in I} Av_i$ is the one generated by the v_i, $i \in I$, generated submodule. If $M \subseteq V$ is a subset of V, then AM is the submodule of V generated by M. Here we use the following convention already introduced for rings: For a subset R of A and a subset M of V, RM denotes the subgroup of $(V, +)$ generated by the complex product $\{ax \mid a \in R, x \in M\}$. The infimum of the cardinal numbers of the generating systems of V, which exists according to Theorem 1.8.20, we denote by

$$\mu_A(V).$$

If $\mu_A(V) \in \mathbb{N}$ is finite, then V is called a **finite A-module**.[47] If $\mu_A(V) \leq 1$, i.e. V is generated by (at most) one element, then V is called **cyclic**. It is $\mu_A(0) = 0$. We mention the following simple lemma:

Lemma 2.8.9 *Let V be an A-module.*

(1) *If $\mu_A(V)$ is finite, then every generating system of V contains a finite generating subsystem.*

---

[47] If A has infinitely many elements, a finite A-module usually also has infinitely many elements, e.g. $A = A \cdot 1_A$ itself.

(2) Let $\mu_A(V)$ be infinite. Then every generating system of V contains a generating subsystem with $\mu_A(V)$ elements. In particular, every minimal generating system of V $\mu_A(V)$ elements.
(3) If $0 \to U \xrightarrow{f} V \xrightarrow{g} W \to 0$ is an exact sequence of A-modules and A-homomorphisms, then $\mu_A(V) \leq \mu_A(U) + \mu_A(W)$ applies. – In particular, V is a finite A-module, if this is the case for U and W.

Proof (1) Let $v_1, \ldots, v_n$ be a (finite) generating system of V and w_j, $j \in J$, an arbitrary generating system. Then there exist finite subsets $J_1, \ldots, J_n \subseteq J$ with $v_i \in \sum_{j \in J_i} Aw_j$, $i = 1, \ldots, n$. For the finite subset $J' := J_1 \cup \cdots \cup J_n \subseteq J$ then $x_i \in \sum_{j \in J'} Aw_j$ holds for all $i = 1, \ldots, n$. Consequently, $\sum_{j \in J'} Aw_j = V$.

(2) Let v_i, $i \in I$, be a generating system of V with $|I| = \mu_A(V)$ and w_j, $j \in J$, be any generating system. Then there are finite subsets $J_i \subseteq J$ with $v_i \in \sum_{j \in J_i} Aw_j$, $i \in I$, and it is w_j, $j \in J' := \bigcup_{i \in I} J_i$, also a generating system of V. According to exercise 1.8.17, $|J'| \leq |I| = \mu_A(V)$ is true. On the other hand, since $\mu_A(V) \leq |J'|$ by definition of $\mu_A(V)$, $|J'| = \mu_A(V)$ holds according to the Bernstein equivalence Theorem 1.8.16.

(3) Exact sequences of A-modules are defined in exactly the same way as for abelian groups, see Example 2.4.16. To prove (3), let u_i, $i \in I$, be a generating system of U with $|I| = \mu_A(U)$ and v_j, $j \in J$, be a family of elements of V such that $g(v_j)$, $j \in J$, is a generating system of W with $|J| = \mu_A(W)$. Then the elements $f(u_i)$, $i \in I$, and v_j, $j \in J$, together form a generating system of V with $|I| + |J| = \mu_A(U) + \mu_A(W)$ elements. If $x \in V$ and $g(x) = \sum_{j \in J} a_j g(v_j) = g(\sum_{j \in J} a_j v_j)$, thus $g(x - \sum_{j \in J} a_j v_j) = 0$, then $x - \sum_{j \in J} a_j v_j = f(u) = f(\sum_{i \in I} b_i u_i)$ with $u \in U$ and consequently $x = \sum_{j \in J} a_j v_j + \sum_{i \in I} b_i f(u_i)$. □

We explicitly note that a minimal generating system of a finite A-module V can contain more than $\mu_A(V)$ elements. Thus, 2.3 is a minimal generating system of the cyclic $\mathbb{Z}$-module $\mathbb{Z}$. Furthermore, an A-module does not need to have a minimal generating system, cf. Ex. 2.8.12. (Then, of course, $\mu_A(V)$ is infinite.)

Definition 2.8.10 An A-module V is called **Artinian** (or **Noetherian**), if the set of A-submodules of V (with respect to inclusion) is Artinian (or Noetherian), i.e., if the following equivalent conditions are met, cf. Lemma 1.4.19:

(i) Every non-empty set of submodules of V has a minimal (or a maximal) element.
(ii) There is no infinite strictly monotonically decreasing (or strictly monotonically increasing) sequence of submodules of V.
(iii) Every infinite monotonically decreasing (or monotonically increasing) sequence of submodules of V is stationary.

2.8 Modules and Vector Spaces

A ring A is called **left Artinian** (or **left Noetherian**), if it is Artinian (or Noetherian) as an A-(left-)module, i.e., if the left ideals in A satisfy the above equivalent conditions. It is called **right Artinian** (or **right Noetherian**), if it is Artinian (or Noetherian) as an A-right module, i.e., if the right ideals in A satisfy the above equivalent conditions. If A is commutative, we simply speak of an **Artinian** (or **Noetherian**) **ring**.

We mention without proof the following **Hopkins' theorem**: *Every left Artinian ring is left Noetherian.* Submodules and homomorphic images of Artinian (or Noetherian) modules are obviously again Artinian (or Noetherian). In the important Noetherian case, the following interesting characterization is added (cf. Ex. 2.3.15c)):

Proposition 2.8.11 *For an A-module V the following conditions are equivalent:*

(i) *V is Noetherian.*
(ii) *Every submodule of V is finitely (generated).*
(iii) *Every infinite monotonically increasing sequence of finite submodules of V is stationary.*

Proof

- (i) $\Rightarrow$ (ii): Let U be a submodule of V and U' a maximal element in the set of finite submodules of U. If then $x \in U$, then $U' + Ax \supseteq U'$ is also finite and thus equal to U', so $x \in U'$ and consequently $U' = U$ is finite.
- (ii) $\Rightarrow$ (iii): Let $U_0 \subseteq U_1 \subseteq \cdots$ be an infinite sequence of (finite) submodules of V. Then $U := \bigcup_{n \in \mathbb{N}} U_n$ is a submodule of V and thus has a finite generating system $x_1, \ldots, x_r$. There is an $n_0 \in \mathbb{N}$ with $x_1, \ldots, x_r \in U_{n_0}$, so $U \subseteq U_{n_0} \subseteq U_{n_0+1} \subseteq \cdots \subseteq U$ and $U = U_n$ for all $n \geq n_0$.
- (iii) $\Rightarrow$ (i): Suppose there is an infinite strictly monotonically increasing sequence $U_0 \subset U_1 \subset U_2 \subset \cdots$ of submodules of V. If then $x_n \in U_{n+1} - U_n$ for all n, then $0 \subset Ax_0 \subset Ax_0 + Ax_1 \subset Ax_0 + Ax_1 + Ax_2 \subset \cdots$ is an infinite strictly monotonically increasing sequence of *finite* submodules of V. Contradiction.
□

Proposition 2.8.12 *Let $0 \to U \xrightarrow{f} V \xrightarrow{g} W \to 0$ be an exact sequence of A-modules and A-homomorphisms. Exactly then is V Artinian (or Noetherian), if U and W are Artinian (or Noetherian).*

Proof We only deal with the Noetherian case. The proof in the Artinian case proceeds completely analogously. With V, U and W are also Noetherian. To prove the converse, let $V_0 \subseteq V_1 \subseteq V_2 \subseteq \cdots$ be a monotonically increasing sequence of submodules of V. Then $f^{-1}(V_0) \subseteq f^{-1}(V_1) \subseteq f^{-1}(V_2) \subseteq \cdots$ and $g(V_0) \subseteq g(V_1) \subseteq g(V_2) \subseteq \cdots$ are monotonically increasing sequences of submodules in U and W respectively. By assumption, these sequences are stationary. But if $V' \subseteq V''$ are submodules of V with $f^{-1}(V') = f^{-1}(V'')$ and $g(V') = g(V'')$, then obviously $V' = V''$. Thus, the original sequence V_n, $n \in \mathbb{N}$, also becomes stationary. □

In the Noetherian case, Proposition 2.8.12 can also be proven as follows: If V' is a submodule of V, then the induced sequence $0 \to f^{-1}(V') \to V' \to g(V') \to 0$ is exact and with $f^{-1}(V')$ and $g(V')$, V' is finite according to Lemma 2.8.9 (3).

According to the last proposition, a finite (direct) sum of Artinian (or Noetherian modules) is again Artinian (or Noetherian). In particular, every finite A-module is Artinian (or Noetherian) if the ring A itself is left Artinian (or left Noetherian). In the Noetherian case, this results with Proposition 2.8.11:

Proposition 2.8.13 *Let A be a left Noetherian ring. An A-module V is Noetherian if and only if V is a finite A-module. If V is finite and all left ideals of A are generated by $\leq m$ elements, $m \in \mathbb{N}$, then all submodules of V are generated by $\leq m\mu_A(V)$ elements. In particular, all submodules of a finite A-module V are generated by $\leq \mu_A(V)$ elements, if all left ideals of A are principal left ideals.*

The additions result from induction over $\mu_A(V)$ with lemma 2.8.9 (3).

Let A again be any ring and I an index set. In the direct sum $A^{(I)}$ one identifies the element

$$e_i = (\delta_{ij})_{j \in I},$$

that has 1 as the i-th component and 0 at the other positions, often identified with i itself when $A \neq 0$: $i = e_i$. Then every element $(a_i)_{i \in I} \in A^{(I)}$ has the representation

$$(a_i)_{i \in I} = \sum_{i \in I} a_i e_i = \sum_{i \in I} a_i i.$$

One calls $A^{(I)}$ the **free** A **-module** to the (index-)set I. Because of $\mathrm{Hom}_A(A, V) = V$ for every A-module V, see example 2.8.3 (2), the module $A^{(I)}$ together with the mapping $\iota_I : I \to A^{(I)}$, $i \mapsto e_i$, according to example 2.8.4 (2) has the following universal property, which indeed characterizes it as a free object:

Theorem 2.8.14 *Let A be a ring and I a set. Then for every A-module V the mapping*

$$\mathrm{Hom}(A^{(I)}, V) \xrightarrow{\sim} V^I, \quad f \mapsto f \circ \iota_I = (f(e_i))_{i \in I},$$

is a group isomorphism and for commutative A even an A-module isomorphism. For an I-tuple $(v_i)_{i \in I} \in V^I$ the preimage of the homomorphism is

$$f : A^{(I)} \to V, \quad (a_i)_{i \in I} \mapsto \sum_{i \in I} a_i v_i,$$

2.8 Modules and Vector Spaces

whose image is the submodule of V generated by the v_i, $i \in I$.

The kernel of the homomorphism $f: A^{(I)} \to V$, $(a_i)_{i \in I} \mapsto \sum_{i \in I} a_i v_i$, is the submodule

$$\mathrm{Rel}_A(v_i, i \in I) = \mathrm{Syz}_A(v_i, i \in I) := \left\{ (a_i) \in A^{(I)} \,\middle|\, \sum_{i \in I} a_i v_i = 0 \right\}.$$

It is called the **module of relations** or the **module of syzygies** of the family $(v_i)_{i \in I} \in V^I$. Its elements are the so-called **relations** or **syzygies** of the v_i, $i \in I$.[48] So it is

$$A^{(I)}/\mathrm{Syz}_A(v_i, i \in I) \xrightarrow{\sim} \mathrm{Im} f = \sum_{i \in I} A v_i.$$

In particular, it is $A^{(I)}/\mathrm{Syz}_A(v_i, i \in I) \xrightarrow{\sim} V$, if v_i, $i \in I$, a generating system of V is. *Every A-module, which has a generating system with $|I|$ elements, is therefore isomorphic to a quotient module of $A^{(I)}$.* Specifically, the quotient modules of A^n represent up to isomorphism all finite modules with n generators, $n \in \mathbb{N}$. A cyclic A-module $V = Ax$ is isomorphic to a quotient module of A, more precisely to $Ax \cong A/\mathrm{Syz}_A x = A/\mathrm{Ann}_A x$. If one wants to specify an A-module V, one often only gives a submodule $U \subseteq A^{(I)}$ which is the syzygy module of a generating system of V, and again restricts itself to a generating system of U. If I is finite and A is left Noetherian, then U will always be generated by finitely many elements according to Proposition 2.8.13. The module V is then itself Noetherian.

Definition 2.8.15 Let v_i, $i \in I$, be a family of elements of the A-module V.

(1) The family v_i, $i \in I$, is called **linearly independent** (over A), if $\mathrm{Syz}_A(v_i, i \in I) = 0$ is, if therefore a linear combination $\sum_{i \in I} a_i v_i$ of the v_i, $i \in I$, over A is only 0 if *all* (and not just almost all) coefficients a_i, $i \in I$, vanish.
(2) The family v_i, $i \in I$, is called an **A-basis** of V, if it is a linearly independent generating system of V. V is called a **free A-module**, if V has a basis over A.

The family v_i, $i \in I$, is exactly linearly independent when the homomorphism defined by the v_i $f: A^{(I)} \to V$, $(a_i)_{i \in I} \mapsto \sum_{i \in I} a_i v_i$, is injective, if therefore the coefficients a_i of a linear combination $x = \sum_{i \in I} a_i v_i$ through x is uniquely determined. Precisely then, v_i, $i \in I$, is a basis of V, when f is bijective, i.e., when every element $x \in V$ has a representation $x = \sum_{i \in I} a_i v_i$ with coefficients $a_i \in A$, $i \in I$, uniquely determined by x. The v_i, $i \in I$, are linearly independent exactly when the sum of the cyclic submodules $A v_i$, $i \in I$, is direct and moreover $\mathrm{Ann}_A v_i = 0$ is for every $i \in I$. If $f : V \to W$ is an A-linear mapping and the image family $f(v_i), i \in I$, is linearly independent, then the family v_i, $i \in I$, itself is also. A family v_i, $i \in I$,

---

[48] The use of the word "syzygy" in this context goes back to D. Hilbert (1862–1943).

is linearly independent exactly when every *finite* subfamily is linearly independent. *With $A \neq 0$ every basis v_i, $i \in I$, from V a minimal generating system of V* (it is $v_i \notin \sum_{j \neq i} Av_j$ for every $i \in I$). Knowledge of a basis of V provides a complete overview of the elements of V. Often such a basis is given by the construction, for example for $V = A^{(I)}$ the so-called **standard basis** e_i, $i \in I$. A free A-module V with basis v_i, $i \in I$, has the following **universal property**, which characterizes it as a free object, analogous to $A^{(I)}$: *For every A-module W the mapping*

$$\mathrm{Hom}_A(V, W) \xrightarrow{\sim} W^I, \quad f \mapsto (f(v_i))_{i \in I},$$

is a group isomorphism and for commutative A an A-module isomorphism. The preimage $f: V \to W$ of the I-tuple $(w_i) \in W^I$ forms the linear combination $\sum_i a_i v_i \in V$ onto the linear combination $\sum_i a_i w_i \in W$. Exactly then is f surjective, when the w_i, $i \in I$, generate the module W, and exactly then injective, when the $w_i, i \in I$, are linearly independent.

The following simple lemma, the proof of which is left to the reader, is useful.

Lemma 2.8.16 *Let v_i, $i \in I$, be a family in the A-module V and $I = I' \uplus I''$ a decomposition of I. Then v_i, $i \in I$, is linearly independent (or a basis), if the subfamily v_i, $i \in I'$, is linearly independent and if the family of residue classes $[v_i]$, $i \in I''$, is linearly independent in (or a basis of) V/U, $U := \sum_{i \in I'} Av_i$.*

Vector spaces are free. The proof is based on the following trivial lemma:

Lemma 2.8.17 *Let V be a vector space over the division ring K. Furthermore, let v_i, $i \in I$, be a linearly independent family in V. Then for every vector $v \in V$ with $v \notin U := \sum_{i \in I} Kv_i$ the family extended by v, v_i, $i \in I$, is also linearly independent and thus a basis of $U' := Kv + U$. If furthermore $w \in U' - U$ is arbitrary, then w, v_i, $i \in I$, is also a basis of U'.*

Proof Let $0 = av + \sum_{i \in I} a_i v_i$ with $a \in A$, $(a_i) \in A^{(I)}$. If $a \neq 0$, then $v = -\sum_{i \in I} a^{-1} a_i v_i$ would be $\in \sum_{i \in I} Kv_i$. Contradiction. Therefore, $a = 0$ and then also $(a_i) = 0$, since the v_i are linearly independent. We leave the proof of the addition to the reader. □

We now prove the following fundamental theorem.

Theorem 2.8.18 *Let V be a vector space over the division ring K, and let v_i, $i \in I$, be a generating system of V. Furthermore, let v_i, $i \in I' \subseteq I$, be a linearly independent subsystem. Then there exists a subset $I'' \subseteq I$ with $I' \subseteq I''$ such that v_i, $i \in I''$, forms a basis of V. In particular, V always has a basis.*

Proof The addition results for $I' = \emptyset$. – To prove the existence of I'' let's consider the set $\mathcal{M}$ of those subsets J of I with $I' \subseteq J \subseteq I$, for which v_i, $i \in J$, is linearly independent. $\mathcal{M}$ is inductively ordered with respect to inclusion (even strictly):

2.8 Modules and Vector Spaces

Due to $I' \in \mathcal{M}$ is $\mathcal{M} \neq \emptyset$, and for a non-empty chain $\mathcal{K} \subseteq \mathcal{M}$ is $\bigcup_{J \in \mathcal{K}} J \in \mathcal{M}$ an upper bound of $\mathcal{K}$. According to Zorn's Lemma 1.4.15 $\mathcal{M}$ has maximum elements, and each such maximum element $I'' \in \mathcal{M}$ provides a basis v_i, $i \in I''$, of V. If $U := \sum_{i \in I''} K v_i \subset V$ were the case, there would be an $i_0 \in I$ with $v_{i_0} \notin U$ (since the v_i, $i \in I$, which generates the K-vector space V), and $I'' \uplus \{i_0\}$ would be a strictly larger element than I'' of $\mathcal{M}$ according to lemma 2.8.17. Contradiction. □

Note that in the proof of Theorem 2.8.18 the existence of a maximum $I'' \in \mathcal{M}$ is trivial when I is finite, so V is a finite K-vector space.

Example 2.8.19 (Hamel Bases) The existence of bases for arbitrary vector spaces was first shown in 1905 for the special case of $\mathbb{R}$ as a $\mathbb{Q}$-vector space (see Example 2.8.3 (3)) by G. Hamel (1877–1954). Today, the $\mathbb{Q}$ bases of $\mathbb{R}$ are still called **Hamel bases**. Hamel proved Theorem 2.8.18 using a well-ordering of I with $i' < i$ for all $i' \in I'$, $i \in I - I'$ – the well-ordering Theorem 1.4.17 had just been proven by Zermelo – and defined I'' in the following way: $i \in I''$ exactly when $x_i \notin \sum_{j < i} K x_j$. Then $I' \subseteq I''$ and v_i, $i \in I''$, is a basis of V. Proof! – Hamel used a $\mathbb{Q}$-basis v_i, $i \in I$, of $\mathbb{R}$ (whose cardinality must be $\aleph$), to solve the following **Cauchy problem**: Are there additive mappings $\mathbb{R} \to \mathbb{R}$, which are not stretches $L_a: x \mapsto ax$ with a $a \in \mathbb{R}$? Since according to the universal property of a basis, the canonical isomorphism $\text{End}\,\mathbb{R} = \text{End}_\mathbb{Q}\,\mathbb{R} \xrightarrow{\sim} \mathbb{R}^I$ applies, in which the homothety L_a, $a \in \mathbb{R}$, corresponds to the I-tuple $(av_i)_{i \in I}$, most additive endomorphisms of $\mathbb{R}$ are not stretches L_a, $a \in \mathbb{R}$. The cardinality of $\text{End}\,\mathbb{R}$ is equal to $\aleph^\aleph = 2^\aleph > \aleph$, since every $\mathbb{Q}$-basis of $\mathbb{R}$ has the power $\aleph$. See also Example 2.6.19 and exercise 2.8.17 as well as exercise 3.10.6. ◊

It is by no means self-evident that two bases of a free module over a ring $\neq 0$ have the same cardinal number. In fact, this is also wrong. There are rings $A \neq 0$, for which the A-module A has a basis of two (and thus also from $n \in \mathbb{N}^*$) elements, so $A \cong A^n$ for all $n \in \mathbb{N}^*$ is, see exercise 2.8.15.

Definition 2.8.20 A free module V over the ring A by definition has a rank if all bases of V have the same cardinal number. This common cardinal number is then called the **rank** of V (over A) and is denoted by

$$\text{rank}\,V = \text{rank}_A V$$

For vector spaces over a division domain K, one generally speaks of the **dimension** of V instead of the rank and writes for it

$$\dim V = \dim_K V.$$

Non-finite free modules are not a problem for the rank.

Theorem 2.8.21 *Every free module V with an infinite basis v_i, $i \in I$, over a ring $A \neq 0$ has the rank* $\text{rank}_A V = |I|$.

Proof Since v_i, $i \in I$, is a minimal generating system of V, there is no finite generating system of V according to Lemma 2.8.9 (1) and therefore no finite basis. Let w_j, $j \in J$, be any basis of V. According to 2.8.9 (2), $|J| = |I| = \mu_A(V)$ then applies. □

For vector spaces, the following generally applies:

Theorem 2.8.22 *Every vector space V over a division ring K has a dimension, i.e., all bases of V have the same cardinal number.*

Proof According to theorem 2.8.21, we can assume that V is a finite K-vector space. In this case, the statement follows from the subsequent lemma. □

Lemma 2.8.23 *Let V be a K-vector space with basis $v_1, \ldots, v_n$. Then any $n+1$ vectors $w_1, \ldots, w_{n+1} \in V$ are linearly dependent.*

Proof We use induction over n. The statement is trivial for $n = 0$ (and $n = 1$). When concluding from n to $n + 1$, we assume that $v_1, \ldots, v_{n+1}$ is a basis of V and that $w_1, \ldots, w_{n+2} \in V$ are linearly independent. Then, by the induction hypothesis, not all w_i are in the subspace $U := Kv_1 + \cdots + Kv_n \subseteq V$. Let's say $w_{n+2} \notin U$. According to Lemma 2.8.17, $v_1, \ldots, v_n, w_{n+2}$ is a basis of V, and the residue classes $[v_1], \ldots, [v_n] \in V/Kw_{n+2}$ form a basis of V/Kw_{n+2} according to Lemma 2.8.16. Also according to Lemma 2.8.16, $[w_1], \ldots, [w_{n+1}]$ are linearly independent in V/Kw_{n+2}, which is not possible according to the induction hypothesis. □

So, up to isomorphism, the vector spaces K^n, $n \in \mathbb{N}$, represent all finite-dimensional vector spaces over the division ring K. However, this should not lead to considering only these spaces as finite-dimensional vector spaces. The identification of an n-dimensional K-vector space V with K^n, i.e., a **calibration** of V, means the selection of a K-basis $v_1, \ldots, v_n$ of V, which is a nontrivial process (even for $n = 1$ and $|K| > 2$). – From lemma 2.8.16 it follows immediately:

Theorem 2.8.24 (Rank Theorem) *Let $0 \to U \xrightarrow{f} V \xrightarrow{g} W \to 0$ be an exact sequence of K-vector spaces and K-homomorphisms. Then the following holds*

$$\dim_K V = \dim_K U + \dim_K W.$$

In particular, $\dim_K V = \dim_K U + \dim_K(V/U)$ for a K-vector space V and a K-subspace $U \subseteq V$.

Free modules over a commutative ring $A \neq 0$ also have a rank. This can be traced back to the field case with the following construction: Let V be any module over the arbitrary ring A and $\mathfrak{a} \subseteq A$ a two-sided ideal in A. Then $\mathfrak{a}V$ is a submodule of V with $\mathfrak{a}V = \sum_{i \in I} \mathfrak{a}v_i$ for each generating system v_i, $i \in I$, of V. Moreover, $\mathfrak{a} \subseteq \operatorname{Ann}_A(V/\mathfrak{a}V)$ and thus $V/\mathfrak{a}V$ is a $A/\mathfrak{a}$-module, see Example 2.8.3

2.8 Modules and Vector Spaces

(4). If now v_i, $i \in I$, is a basis of V, then $\mathfrak{a}V = \sum_{i \in I}^{\oplus} \mathfrak{a}v_i \subseteq V = \sum_{i \in I}^{\oplus} Av_i$ and $(A/\mathfrak{a})^{(I)} \cong \bigoplus_{i \in I}(Av_i/\mathfrak{a}v_i) = V/\mathfrak{a}V$. Consequently, the residue classes $[v_i]$, $i \in I$, form a $(A/\mathfrak{a})$-basis of $V/\mathfrak{a}V$. It follows: If all free $(A/\mathfrak{a})$-modules have a rank, then all free A-modules do as well. Since a commutative ring $A \neq 0$ has maximal ideals $\mathfrak{m}$ for which $A/\mathfrak{m}$ is a body, see Theorem 2.7.18, it follows specifically with theorem 2.8.22:

Theorem 2.8.25 *Every free module over a commutative ring $\neq 0$ has a rank.*

Similarly, Lemma 2.8.23 has an analogue for commutative rings $\neq 0$, see exercise 2.8.16c). More general than theorem 2.8.25 is the following statement:

Theorem 2.8.26 *Let $\varphi: A \to B$ be a homomorphism of rings. If every free B-module has a rank, then every free A-module does as well.*

Proof Using Theorem 2.8.21, we have to show: If $m, n \in \mathbb{N}$ and $A^m \cong A^n$ holds, then $m = n$. Let $f: A^n \to A^m$ and $g: A^m \to A^n$ be inverse A-isomorphisms, represented by the matrices $A = (a_{ij}) \in M_{m,n}(A^{op})$ and $B = (b_{jk}) \in M_{n,m}(A^{op})$ are described, see Example 2.8.4 (2). Then the product matrices $BA \in M_n(A^{op})$ and $AB \in M_m(A^{op})$ describe the compositions $g \circ f = \mathrm{id}_{A^n}$ or $f \circ g = \mathrm{id}_{A^m}$, so they are the unit matrices E_n or E_m. The φ images $\varphi(A) = (\varphi(a_{ij})) \in M_{m,n}(B^{op})$ and $\varphi(B) = (\varphi(b_{jk})) \in M_{n,m}(B^{op})$ then describe inverse B-isomorphisms $B^n \to B^m$ or $B^m \to B^n$. By assumption about B, $m = n$ is therefore as desired. □

The theory of rings is essentially identical to the theory of modules over rings, with commutative algebra mainly considering modules over Noetherian commutative rings. Linear algebra, on the other hand, largely deals with linear mappings of free modules and in particular with the structure of linear mappings between vector spaces (which are readily free according to Theorem 2.8.18). In the case of fields, the homomorphism groups $\mathrm{Hom}_K(V, W)$ are even K-vector spaces. Linear algebra is discussed in detail in volumes 3 and 4.

Exercises

Exercise 2.8.1 For submodules U, W of an A-module V, the following canonical isomorphisms apply: (1) $U/(U \cap W) \xrightarrow{\sim} (U + W)/W$. (2) If $U \subseteq W$, then $(V/U)/(W/U) \xrightarrow{\sim} V/W$ applies.

Exercise 2.8.2 Let U, W be submodules of the A module V. Then the two so-called **Meyer-Vietoris sequences**

$$0 \to U \cap W \to U \oplus W \to U + W \to 0,$$

$$0 \to V/(U \cap W) \to (V/U) \oplus (V/W) \to V/(U + W) \to 0$$

are exact, where the nontrivial homomorphisms in the first sequence are given by $x \mapsto (x, -x)$ and $(x, y) \mapsto x + y$ and in the second sequence analogously by $[x] \mapsto ([x], -[x])$ and $([x], [y]) \mapsto [x + y]$. From the first exact sequence, infer in the case that $A = K$ is a division ring, the so-called **dimension formula**

$$\dim_K U + \dim_K W = \dim_K(U \cap W) + \dim_K(U + W)$$

and from the second the so-called **codimension formula**

$$\text{codim}_K(U, V) + \text{codim}_K(W, V) = \text{codim}_K(U \cap W, V) + \text{codim}_K(U + W, V).$$

Here,

$$\text{codim}_K(U, V) := \dim_K(V/U)$$

for any subspace U of a K vector space V is the (K-)**codimension** of U in V. In particular, in the case of a finite-dimensional K vector space V, the two inequalities $\dim_K(U \cap W) \geq \dim_K U + \dim_K W - \dim_K V$ and $\text{codim}_K(U \cap W, V) \leq \text{codim}_K(U, V) + \text{codim}_K(W, V)$ apply. If in particular $\dim_K U + \dim_K W > \dim_K V$, then $U \cap W \neq 0$ is true.

Exercise 2.8.3 For a family $v_i, i \in I$, of vectors of a K-vector space V are equivalent: (i) $v_i, i \in I$, is a basis of V. (ii) $v_i, i \in I$, is a minimal generating system of V. (iii) $v_i, i \in I$, is a maximum linearly independent family in V.

Exercise 2.8.4 Let $A \neq 0$ and x be a basis of the cyclic A-module $V := Ax$. Exactly then is $y = ax \in V$, $a \in A$, also a basis of V, when a is a unit in A. (Note that Ax can be a free A-module $\neq 0$ without x being a basis of Ax, cf. exercise 2.8.15c).) □

Exercise 2.8.5 Let $A \subseteq B$ be an extension of rings. Let B be free as an A-(left-)module with basis $b_i, i \in I$ (cf. Example 2.8.3 (3)). If then W is a free B-module with basis $w_j, j \in J$, then W is also a free A-module, specifically with $b_i w_j$, $(i, j) \in I \times J$, as a basis. If all free A and B modules each have a rank, then

$$\text{rank}_A W = \text{rank}_A B \cdot \text{rank}_B W \quad \textbf{(Formula for ranks)}.$$

applies. In particular, for division rings K,L with $K \subseteq L$ and every L-vector space W, the **dimension formula** $\dim_K W = \dim_K L \cdot \dim_L W$ applies. Specifically, every $\mathbb{C}$-vector space is an $\mathbb{R}$-vector space of double dimension. The dimension $\text{Dim}_K L$ of the extension $K \subseteq L$ of division rings is also called the **degree** of the extension and is denoted by

$$[L : K] := \dim_K L.$$

Exercise 2.8.6 Let V be an A-module, $S \subseteq Z(A)$ a central submonoid of $(A, \cdot)$ and A_S is the ring of fractions of A with respect to S, see Example 2.6.14. Then $V \times S$

2.8 Modules and Vector Spaces

with the addition $(x,s) + (y,t) = (tx + sy, st)$ is a commutative monoid and through

$$(x,s) \equiv_S (y,t) \iff \text{there is } u \in S \text{ with } utx = usy,$$

a compatible equivalence relation on $V \times S$ is defined. $V_S = S^{-1}V$ denotes the quotient monoid and x/s, $x \in V$, $s \in S$, the equivalence class of (x,s). V_S is even an abelian group with the addition $x/s + y/t = (tx + sy)/st$, and through

$$\frac{a}{s} \cdot \frac{x}{t} := \frac{ax}{st}, \quad a \in A, \ x \in V, \ s,t \in S,$$

a scalar multiplication $A_S \times V_S \to V_S$ on V_S is well-defined, with respect to which V_S is an A_S-module. V_S has the following universal property with the canonical A-module homomorphism $\iota_S: V \to V_S$, $x \mapsto x/1$: If $f: V \to W$ is an A-module homomorphism from V into an A_S-module W, then there is exactly one A_S-module homomorphism $f_S: V_S \to W$ with $f = f_S \circ \iota_S$. It is $f_S(x/s) = (1/s)f(x) = f(x)/s$.

Remark V_S is called the **module of fractions** of V with respect to S. – If, for example, A is *commutative* and $Q(A) = A_{A^*}$ is the total quotient ring of A, then V_{A^*} is a module over $Q(A)$ and $\ker \iota_{A^*}$ is the torsion submodule $T_A V$ of V. It is said that V has a **rank over** A if $A \neq 0$ and V_{A^*} is a free $Q(A)$-module. Then $\operatorname{rank}_A V := \operatorname{rank}_{Q(A)} V_{A^*}$ is set. If V itself is free over A with basis v_i, $i \in I$, then $v_i/1, i \in I$, a $Q(A)$-basis of V_{A^*}, i.e., the new rank definition is compatible with the one from Definition 2.8.20. In particular, for an integrity domain A, the module of fractions V_{A^*} is a vector space over the quotient field $K := Q(A)$ of A and V is an A-module with $\operatorname{rank}_A V = \dim_K V_{A^*}$.

Exercise 2.8.7 Let A be a ring $\neq 0$. An A-module V is called **simple**, if $V \neq 0$ and V only has the trivial submodules 0 and V.

a) For an A-module V the following are equivalent: (i) V is simple. (ii) $V \neq 0$ and every homomorphism $V \to W$ of A-modules is either the zero homomorphism or injective. (iii) $V \neq 0$ and $V = Ax$ for every $x \in V - \{0\}$. (iv) V is isomorphic to a quotient module $A/\mathfrak{a}$, where $\mathfrak{a}$ is a maximal left ideal in A.
b) Let V be a simple A-module. Then the two-sided ideal $\operatorname{Ann}_A V$ is the intersection of the maximal left ideals $\operatorname{Ann}_A x$, $x \in V - \{0\}$. (This provides a simple proof that the intersection of all maximal left ideals of A is a two-sided ideal in A. This is the **Jacobson radical** of A, cf. the remarks following Definition 2.7.21.)

Exeercise 2.8.8 Let $f: V \to W$ be a homomorphism of A-modules.

a) For a submodule $U \subseteq V$, $f^{-1}(f(U)) = U + \ker f$ and $U/(U \cap \ker f) \xrightarrow{\sim} (U + \ker f)/\ker f \xrightarrow{\sim} f(U)$.

b) If f is surjective, then $U \mapsto f(U)$ and $X \mapsto f^{-1}(X)$ are inverse mappings between the set of submodules U of V, which include $\ker f$, and the set of all submodules X of W.

c) Let V and W be simple A-modules, cf. Ex. 2.8.7. Then every A-homomorphism $V \to W$ is either the zero homomorphism or an isomorphism. In particular, $\mathrm{End}_A V$ is a division ring (**Lemma of (Issai) Schur**).

d) If A is commutative, then the modules $A/\mathfrak{m}$, $\mathfrak{m} \in \mathrm{Spm}\, A$, up to isomorphism, are the only simple A-modules, and different maximal ideals of A define non-isomorphic simple A-modules. (Note $\mathrm{Ann}_A(A/\mathfrak{m}) = \mathfrak{m}$. – The classification of simple modules over non-commutative rings is more complicated. A local ring A with Jacobson radical $\mathfrak{m}_A$ has the residue class division ring $A/\mathfrak{m}_A$ (considered as an A-module) up to isomorphism as the only simple A-module.)

e) Let V be a vector space $\neq 0$ over the division ring K. Then V is a simple ($\mathrm{End}_K V$) module, see Example 2.8.3 (5). The endomorphisms of V as a ($\mathrm{End}_K V$) module are the homotheties ϑ_a, $a \in K$, of V. Thus, $\mathrm{End}_{\mathrm{End}_K V} V \cong K$ is the image of the action homomorphism $\vartheta : K \to \mathrm{End}\, V$. The Jacobson radical of $\mathrm{End}_K V$ is 0. (It is $\mathrm{Ann}_{\mathrm{End}_K V} V = \bigcap_{x \in V} \mathrm{Ann}_{\mathrm{End}_K V} x = 0$.)

f) Let V be a finite-dimensional K-vector space of dimension $n > 0$. Then $\mathrm{End}_K V$ is a simple ring. (If $f \in \mathrm{End}_K V$ and $v_1, \ldots, v_n$ is a K-basis of V with $f(v_1) \neq 0$, then the two-sided ideal generated by f in $\mathrm{End}_K V$ contains an element f_1 with $f_1(v_j) = \delta_{1j}(v_j)$, $j = 1, \ldots, n$, and then also id_V.) For each K-basis $v_1, \ldots, v_n$ of V, the mapping $\mathrm{End}_K V \to V^n$, $f \mapsto (f(v_1), \ldots, f(v_n))$, is an isomorphism of $\mathrm{End}_K V$-modules (see the Chinese Remainder Theorem 2.2.18). V is, up to isomorphism, the only simple ($\mathrm{End}_K V$)-module.

g) If $\alpha := \dim_K V \geq \aleph_0$, that is V is infinite-dimensional, then

$$\mathfrak{a}_\alpha := \{f \in \mathrm{End}_K V \mid \dim_K \mathrm{Im}\, f < \alpha\}$$

is the only maximal two-sided ideal in $\mathrm{End}_K V$. The ring $(\mathrm{End}_K V)/\mathfrak{a}_\alpha$ is simple (but due to exercise 2.8.15c) not a division domain because of theorem 2.8.26). If $\alpha > \aleph_0$, then $\mathfrak{a}_{\aleph_0} := \{f \in \mathrm{End}_K V \mid \dim_K \mathrm{Im}\, f < \aleph_0\}$ is another two-sided ideal in $\mathrm{End}_K V$.

Exercises 2.8.9 Let V be a module over the ring A and $U \subseteq V$ a submodule of V. We recall that U is by definition a **direct summand** of V if there is a module complement $W \subseteq V$ to U, for which $V = U \oplus W$ is.

a) U is a direct summand of V if and only if there is a projection $p \in \mathrm{End}_A V$ with $\mathrm{Im}\, p = U$. In this case, $V = U \oplus W$ with $W := \ker p$, $p = p_{U,W}$ is the **projection onto** U **along** W, and the complementary projection $q = q_{U,W} = \mathrm{id}_V - p_{U,W} = p_{W,U}$ is the **projection along** U **onto** W. (See also exercise 2.2.13.)

b) If $A = K$ is a division ring, then every subspace $U \subseteq V$ has a complement.

c) Let W be a complement of U. Then $f \mapsto \Gamma_f = \{f(y) + y \mid y \in W\} \subseteq V$ is a bijective mapping from $\mathrm{Hom}_A(W, U)$ to the set of all complements of U in V.

2.8 Modules and Vector Spaces

Exercise 2.8.10 Let V be an A-module over the ring $A \neq 0$. V is called **indecomposable** or **irreducible**, if $V \neq 0$ and there is no direct sum decomposition $V = U \oplus W$ with submodules $U \neq 0 \neq W$ of V.

a) V is indecomposable exactly when $V \neq 0$ and the endomorphism ring $\operatorname{End}_A V$ contains no nontrivial idempotent elements. Every simple A-module is indecomposable. Provide an example of an indecomposable module that is not simple. A is indecomposable as an A-(left or right) module exactly when the ring A contains no nontrivial idempotent elements. (This should be clearly distinguished from the indecomposability of A as a ring. This is equivalent to A containing no nontrivial *central* idempotent elements.)
b) The only indecomposable vector spaces over a division ring K are the one-dimensional vector spaces. (In general, it is difficult—if not impossible—to classify the indecomposable modules over a given ring A. The finitely generated indecomposable abelian groups ($= \mathbb{Z}$-modules) are exactly the cyclic groups $\mathbb{Z} = \mathbf{Z}_0$ and $\mathbf{Z}_{p^\alpha}$, $p \in \mathbb{P}$, $\alpha \in \mathbb{N}^*$. This is essentially the main Theorem 2.3.31 on finitely generated abelian groups. But there are many more indecomposable abelian groups, e.g., all subgroups $\neq 0$ of $\mathbb{Q} = (\mathbb{Q}, +)$ are indecomposable, as are all Prüfer p-groups $\mathrm{I}(p)$, $p \in \mathbb{P}$, cf. exercise 2.2.26c). Every abelian p-group with a one-dimensional (i.e., non-zero cyclic) p-base is indecomposable. Up to isomorphism, these are exactly the groups $\mathbf{Z}_{p^\alpha}$, $\alpha \in \mathbb{N}^*$, and $\mathrm{I}(p)$. Why?)

Exercise 2.8.11 A ring $A \neq 0$ is exactly a division ring when all A-(left-)modules (or when all A-right modules) are free.

Exercise 2.8.12 Let A be an integral domain, but not a field. Then the quotient field $\mathrm{Q}(A)$ of A has no minimal generating system as an A-module. In particular, $\mathrm{Q}(A)$ is not finitely generated over A.

Exercise 2.8.13 Let $m \in \mathbb{N}^*$. Provide a minimal generating system of the abelian group $\mathbb{Z}$ with exactly m elements.

Exercise 2.8.14 Let V be a module over the local ring A with Jacobson radical $\mathfrak{m}_A$ and v_i, $i \in I$, a family of elements of V.

a) If v_i, $i \in I$, is a generating system of V, then v_i, $i \in I$, is minimal if and only if $\operatorname{Syz}_A(v_i, i \in I) \subseteq \mathfrak{m}_A A^{(I)}$ is. (It is $A^\times = A - \mathfrak{m}_A$.) In this case, the residue classes $[v_i] \in V/\mathfrak{m}_A V$, $i \in I$, form an $(A/\mathfrak{m}_A)$-basis of $V/\mathfrak{m}_A V$, and we get
$$\mu_A(V) = |I| = \dim_{A/\mathfrak{m}_A}(V/\mathfrak{m}_A V).$$
In particular, for every finite A-module V: It is $\mu_A(V) = \dim_{A/\mathfrak{m}_A}(V/\mathfrak{m}_A V)$ and $V = 0$ if and only if $V = \mathfrak{m}_A V$ is.

b) Is $U \subseteq V$ a submodule of V such that the quotient module V/U is finite, and if $V = U + \mathfrak{m}_A V$ holds, then $V = U$ (**Nakayama's Lemma**). (It is $V/U = \mathfrak{m}_A(V/U)$ and consequently $V/U = 0$.) If V is finite, then the elements v_i, $i \in I$, generate V when their residue classes generate the vector space $V/\mathfrak{m}_A V$.

Exercise 2.8.15 Let A be a ring $\neq 0$.

a) If $A^m \cong A^{m+1}$ (as A-modules) for a $m \in \mathbb{N}$, then $A^m \cong A^n$ for all $n \geq m$.
b) Exactly then is $x, y \in A$ a basis of the A-module A, if there are elements $a, b \in A$ with (1) $ax + by = 1$, (2) $xa = 1$, (3) $xb = 0$, (4) $ya = 0$ and (5) $yb = 1$. (So it is

$$(x, y)\begin{pmatrix} a \\ b \end{pmatrix} = (1), \quad \begin{pmatrix} a \\ b \end{pmatrix}(x, y) = \begin{pmatrix} 1 & 0 \\ 0 & 1 \end{pmatrix},$$

where all matrices are to be understood over the oppositional ring A^{op}, see Example 2.8.4 (2).)
c) Let B be a ring $\neq 0$ and V a B-module $\neq 0$ with $V \cong V \oplus V$ (e.g., a free B-module with infinite basis). Then there are elements a,b,x,y in the endomorphism ring $A := \text{End}_B V$ that satisfy equations (1) to (5) from b). In particular, the finite free A-modules have no rank. (Describe mutually inverse isomorphisms $V \xrightarrow{\sim} V \oplus V$ and $V \oplus V \xrightarrow{\sim} V$ with matrices, whose coefficients lie in $\text{End}_A V$, according to theorem 2.8.5.)

Exercise 2.8.16 Let A be a ring $\neq 0$.

a) $w_1, \ldots, w_{n+1}$ are linearly independent elements in the free A-module V with the basis $v_1, \ldots, v_n$. Then V has a free submodule with a countably infinite basis. (One constructs recursively an infinite sequence $u_0, u_1, u_2, \ldots$ of linearly independent elements in V and free submodules $U_0, U_1, U_2, \ldots \subseteq V$ each with a basis of n elements such that for each $k \in \mathbb{N}$ the direct sum decomposition $Au_1 \oplus \cdots \oplus Au_k \oplus U_k$ holds. Then $u_0, u_1, u_2, \ldots$ generate the desired submodule. One starts with $u_0 := w_1$, $U_0 := Aw_2 \oplus \cdots \oplus Aw_{n+1}$.)
b) If A is left Noetherian and $n \in \mathbb{N}$, then every $n + 1$ elements in an A-module V with $\mu_A(V) \leq n$ are linearly dependent. (One can assume that $V \cong A^n$ is, and note that A^n is a Noetherian A-module. – From a) also follows the analogous statement for left Artinian rings. However, this also results from the already cited theorem of Hopkins.)
c) With a trick, the result from b) can be transferred to any *commutative* rings $A \neq 0$: If A is commutative and $n \in \mathbb{N}$, then each $n + 1$ elements in an A-module V are $\mu_A(V) \leq n$ linearly dependent. (Assuming that A^n $n + 1$ linearly independent elements $w_j = \sum_{i=1}^{n} a_{ij} e_i$, $j = 1, \ldots, n+1$, contains. According to Hilbert's Basis Theorem 2.9.21 the smallest subring $B := \mathbb{Z}[a_{ij}, 1 \leq i \leq n, 1 \leq j \leq n+1] \subseteq A$ of A, which contains all coefficients a_{ij}, is Noetherian, and the elements $w_1, \ldots, w_{n+1} \in B^n$ are also

linearly independent over B in B^n in contradiction to b). – The method used here to reduce a problem to the Noetherian case is called **Noetherization** of a problem.)

Exercise 2.8.17 Let V be an additively written abelian group. Only then is V the additive group of an $\mathbb{Q}$-vector space, if V is torsion-free and divisible. In this case, the $\mathbb{Q}$-vector space structure of V is uniquely determined. For $a, b \in \mathbb{Z}$, $b \neq 0$, is $\vartheta_{a/b} = \vartheta_a \vartheta_b^{-1} = \vartheta_b^{-1} \vartheta_a$. If V,W are any torsion-free and divisible abelian groups, then $\text{Hom}(V, W) = \text{Hom}_{\mathbb{Z}}(V, W) = \text{Hom}_{\mathbb{Q}}(V, W)$. (If V is torsion-free and divisible, then the characteristic homomorphism forms $\mathbb{Z} \to \text{End}\, V$ the monoid $\mathbb{Z}^*$ in Aut V and can therefore be uniquely extended to a ring homomorphism $\mathbb{Q} \to \text{End}\, V$.) Up to isomorphism, the torsion-free and divisible abelian groups are exactly the direct sums $\mathbb{Q}^{(I)}$, I any set. Because of $|\mathbb{Q}^{(I)}| = |I|$ for infinite sets I, in particular, two uncountable torsion-free and divisible abelian groups are isomorphic if and only if they have the same cardinal number. For example, the additive groups of the $\mathbb{R}$-vector spaces $\mathbb{R}^n$, $n \in \mathbb{N}^*$, as well as $\mathbb{R}^{(\mathbb{N})}$ and $\mathbb{R}^{\mathbb{N}}$ are all isomorphic to each other (which usually surprises the beginner). The endomorphism ring $\text{End}(\mathbb{R}, +) = \text{End}_{\mathbb{Q}}(\mathbb{R}, +)$ has exactly two non-trivial two-sided ideals if and only if the continuum hypothesis from remark 1.8.21 holds, see exercise. 2.8.8g).

2.9 Algebras

In the concept of an algebra, ring and module structures merge. Let S be a ring and A an additive abelian group. A is an S-algebra if A is both a ring and an S-module, where the addition for both structures is the given addition on A. The only reasonable compatibility condition for both structures is that the left and right multiplications L_x and R_y, $x, y \in A$, of the ring $(A, +, \cdot)$ are S-linear, so that $L_x \circ \vartheta_b = \vartheta_b \circ L_x$ and $R_y \circ \vartheta_a = \vartheta_a \circ R_y$ hold or – explicitly – that $x(by) = b(xy)$ and $(ax)y = a(xy)$ for all $a, b \in S$ and $x, y \in A$ is. In particular, then $\vartheta_a = L_{a \cdot 1_A} = R_{a \cdot 1_A}$ for all $a \in S$, i.e. $S \cdot 1_A$ lies in the center $Z(A)$ of A. For this reason, we want to assume from the outset that the base ring S is commutative, and define:

Definition 2.9.1 Let S be a *commutative* ring. An S-**Algebra** A is an S-module with a multiplication $\cdot : A \times A \to A$ such that $(A, +, \cdot)$ is a ring, for which all left and right translations L_x and R_y, $x, y \in A$, are S-linear, i.e., for all $a, b \in S$ and all $x, y \in A$ applies:

$$(1)\ x(by) = b(xy), \quad (2)\ (ax)y = a(xy).$$

A mapping $f: A \to B$ of S-algebras is called an S-**Algebra homomorphism**, if f is both a ring and an S-module homomorphism.

We adopt all relevant statements and constructions that concern rings and modules and that can be transferred without further ado, usually without comment for algebras. The set of S-algebra homomorphisms $A \to B$ we denote with

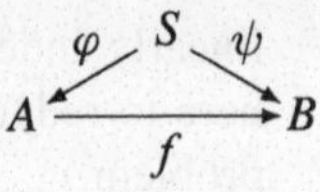

Fig. 2.22 S-Algebra homomorphism f

$$\begin{array}{c} & S & \\ \varphi \swarrow & & \searrow \psi \\ A & \xrightarrow{\ \ f\ \ } & B \end{array}$$

$$\mathrm{Hom}_{S\text{-Alg}}(A,B).$$

It is a subset of $\mathrm{Hom}_S(A,B)$. The set $\mathrm{End}_{S\text{-Alg}}(A)$ of S-algebra endomorphisms of A is a monoid with respect to composition with unit group $\mathrm{Aut}_{S\text{-Alg}}A$.

Let A be an algebra over the commutative ring S. The conditions (1) and (2) in Definition 2.9.1 can be summarized into one condition

$$(ax)(by) = (ab)(xy) \quad \text{for all } a,b \in S,\ x,y \in A$$

The mapping $\varphi\colon S \to A$, $a \mapsto a \cdot 1_A$, is a ring homomorphism, *whose image lies in the center of A*, and is called the **structure homomorphism** of A. Through it, the scalar multiplication on A is uniquely determined: It is $ax = (a \cdot 1_A)x = \varphi(a)x$ for all $a \in S$, $x \in A$. If A,B are S-algebras with structure homomorphisms φ or ψ, then a ring homomorphism $f\colon A \to B$ is an S-algebra homomorphism if and only if $f \circ \varphi = \psi$ is, i.e., if the diagram in Fig. 2.22 is commutative.

If, conversely, $\varphi\colon S \to A$ is a homomorphism of the commutative ring S into the (arbitrary) ring A with $\mathrm{Im}\,\varphi \subseteq Z(A)$, then A with the scalar multiplication $ax := \varphi(a)x$, $a \in S$, $x \in A$, is an S-algebra.

For a commutative ring S, therefore, S-algebras and pairs (A,φ), A ring, $\varphi\colon S \to A$ ring homomorphism with $\varphi(S) \subseteq Z(A)$ are the same. In particular, every homomorphism $S \to A$ of commutative rings defines an S-algebra structure on A. Every ring is an algebra over its center and its subrings. Furthermore, the only $\mathbb{Z}$-module structure of its additive group always provides an $\mathbb{Z}$-algebra structure. Rings and $\mathbb{Z}$-algebras are therefore the same. A subring A' of an S-algebra A is exactly an S-subalgebra when A' includes the smallest S-subalgebra $S \cdot 1_A \cong S/\mathrm{Ann}_S\,1_A$ of A. The quotient algebras $S/\mathfrak{s}$ of S, $\mathfrak{s}$ ideal in S, are therefore, up to isomorphism, the minimal S-algebras (which by definition do not have any proper S-subalgebras).

Remark 2.9.2 (*S-bilinear and S-multilinear mappings*) Let S be a commutative ring. If A is a ring that is also an S-module (with the same addition), then A is by definition an S-algebra if and only if the multiplication $A \times A \to A$ is S-bilinear in the following sense, cf. Remark 2.6.6: ◊

Definition 2.9.3 Let V, W, X be modules over the commutative ring S. A mapping $f\colon V \times W \to X$ is called (S-)**bilinear**, if for all $a \in S$, $x,x_1,x_2 \in V$ and all $y,y_1,y_s \in W$ the following applies:

$$f(x,y_1+y_2) = f(x,y_1)+f(x,y_2),\quad f(x_1+x_2,y) = f(x_1,y)+f(x_2,y),$$

$$f(ax,y) = af(x,y) = f(x,ay),$$

2.9 Algebras

that is, all partial mappings $f(x,-)\colon W \to X$, $x \in V$, and $f(-,y)\colon V \to X$, $y \in W$, are S-linear. Similarly, for a *finite* family V_i, $i \in I$, of S-modules, a (S-)**multilinear** mapping $f\colon \prod_{i\in I} V_i \to X$ is defined: When all arguments except the i-th are fixed, f defines an S-linear mapping $V_i \to X$, $i \in I$.

If $I = \emptyset$ is true, then f is an element of X, for $I = \{i\}$, f is an S-linear mapping $V_i \to X$, for $|I| = 3$ we speak of S-trilinear mappings etc. For multilinear mappings, analogous statements apply as for multiadditive mappings, in particular:

Lemma 2.9.4 *If v_{ij_i}, $j_i \in J_i$, are generating systems of the S-modules V_i, $i \in I$, then two S-multilinear mappings $f, g \colon \prod_{i\in I} V_i \to X$ already coincide if they agree on all I-tuples $(v_{ij_i})_{i\in I}$, $(j_i) \in \prod_{i\in I} J_i$, coincide.*

An S-module A together with an S-bilinear multiplication $A \times A \to A$ is called a **generalized S-algebra**. The multiplication on a generalized S-algebra A is uniquely determined by the products $v_i v_j$, $i,j \in I$, where v_i, $i \in I$, is an S-module generating system of A. To then verify that A is even an S-algebra (in the narrower sense) with unit element 1_A, one only has to check the equations $(v_i v_j) v_k = v_i (v_j v_k)$ and $1_A \cdot v_i = v_i = v_i \cdot 1_A$ for all $i, j, k \in I$.

If A is an S-algebra and x_i, $i \in I$, is a family of elements in A, then

$$S\langle x_i, i \in I \rangle$$

denotes the smallest S-subalgebra of A that contains the elements x_i, $i \in I$. It is generated as an S-module by the submonoid $M(x_i, i \in I)$ of the multiplicative monoid of A, which is generated by the x_i is generated. $M(x_i, i \in I)$ consists of the monoms $x^\nu = x_{i_1} \cdots x_{i_n}$, where $\nu = i_1 \cdots i_n \in W(I)$ is a word over the alphabet I. If the x_i commute pairwise, then the S-subalgebra generated by the x_i, $i \in I$, is also commutative and one then writes

$$S[x_i, i \in I]$$

for this subalgebra. The monomers in the x_i can in this case be written in the form $x^\nu = \prod_{i \in I} x_i^{\nu_i}$, $\nu = (\nu_i) \in \mathbb{N}^{(I)}$. An S-algebra A is called **finite** (or **free**), if A is finite (or free) as an S-module.[49] If $v = (v_i)_{i\in I}$ is an S-module basis of the free S-algebra A, then the multiplication on A is already determined by the products

$$v_i \cdot v_j = \sum_{k \in I} a_{ij}^{(k)} v_k, \quad i,j \in I,$$

see Lemma 2.9.4. The $a_{ij}^{(k)} \in S$, $i,j,k \in I$, are called the **structure constants** of A with respect to the S-basis v. An S-algebra A is called **of finite type** if it has a finite S-algebra generating system, i.e., if there is a finite family x_i, $i \in I$, gives with $A = S\langle x_i, i \in I \rangle$. Finite algebras and algebras of finite type must not be confused. Of course, every finite algebra is also of finite type.

---

[49] One should carefully distinguish between free algebras in the sense given here and the freely generated algebras, see Example 2.9.12.

Example 2.9.5 In the following, let S always denote a commutative ring $\neq 0$.

(1) Prime examples for S-algebras are the endomorphism algebras $\mathrm{End}_S V$ of S-modules V. According to Proposition 2.8.2 is $\mathrm{End}_S V$ both a ring and an S-module with the composition as ring multiplication or the scalar multiplication $af = \vartheta_a \circ f = f \circ \vartheta_a$, $a \in S$, $f \in \mathrm{End}_S V$. The defining property as an S-algebra is fulfilled due to

$$(af)(bg) = \vartheta_a f \vartheta_b g = \vartheta_a \vartheta_b f g = \vartheta_{ab}(fg) = (ab)(fg), \quad a,b \in S, f,g \in \mathrm{End}_S V,$$

In the case of $V = S^I = S^{(I)}$ with a *finite* index set I we obtain the finite free S-algebra of rank $|I|^2$ of the **square $I \times I$ -matrices** over S

$$\mathrm{End}_S S^I = \mathrm{M}_I(S^{\mathrm{op}}) = \mathrm{M}_I(S) = S^{I \times I},$$

cf. sentence 2.8.5. It is

$$\mathsf{AB} = \mathsf{C} = (c_{ik}) \quad \text{with} \quad c_{ik} = \sum_{j \in I} a_{ij} b_{jk} \quad \text{for} \quad \mathsf{A} = (a_{ij}), \mathsf{B} = (b_{jk}) \in \mathrm{M}_I(S),$$

cf. loc. cit.[50] The automorphism group $\mathrm{Aut}_S S^I = \mathrm{GL}_S S^I$ we accordingly denote with

$$\mathrm{GL}_I(S) = \mathrm{M}_I(S)^\times.$$

After distinguishing a basis $v_1, \ldots, v_n$ the endomorphism algebra of a free S -module of rank $n \in \mathbb{N}$ can be identified with the matrix algebra $\mathrm{M}_n(S)$ of rank n^2 and its unit group with the group $\mathrm{GL}_n(S) \subseteq \mathrm{M}_n(S)$ of invertible $n \times n$ matrices. This applies in particular to vector spaces of dimension n over a field.

Let A be a ring that is simultaneously an S-module, and $L : A \to \mathrm{End}\, A = \mathrm{End}(A, +)$ the regular representation of A, cf. Example 2.6.19. If A is an S-algebra, then the left translations $L_a : A \to A$, $a \in A$, are S-linear and the image of L lies in the S-algebra $\mathrm{End}_S A$. Since the right translations $R_b : A \to A$, $b \in A$, are S-linear, L is even S-linear and thus an S-algebra homomorphism. In other words: *A is an S -algebra exactly when the image of the regular representation $L : A \to \mathrm{End}\, A$ lies in the S -algebra $\mathrm{End}_S A \subseteq \mathrm{End}\, A$ and L is furthermore S-linear.* In particular, an S-algebra A can always be identified with an S-subalgebra of $\mathrm{End}_S A$ and a free S-algebra of rank $n \in \mathbb{N}$ after designating an S-basis $v_1, \ldots, v_n$ with an S-subalgebra of the matrix algebra $\mathrm{M}_n(S)$. The latter is completely analogous to the corresponding Theorem 2.2.9 about groups, which states that every finite group of order n can be identified with a subgroup of the permutation group $\mathfrak{S}_n$.

---

[50] For an arbitrary (not necessarily commutative) ring R, $\mathrm{M}_I(R)$ with this multiplication is also a ring, namely the R^{op}-endomorphism ring of $(R^{\mathrm{op}})^I$. So it is $\mathrm{End}_R R^I = \mathrm{M}_I(R^{\mathrm{op}})$. Pay attention to this (already emphasized several times) transition to the oppositional ring.

2.9 Algebras

Let A be an S-algebra. An A-module V is then also an S-module, and the homotheties $\vartheta_a\colon V \to V$, $a \in A$, are S-linear. The image of the action homomorphism ϑ thus lies in the S-algebra $\operatorname{End}_S V \subseteq \operatorname{End} V$. In other words: *An A-module is an S-module V, together with an S-algebra homomorphism $\vartheta\colon A \to \operatorname{End}_S V$.* For example, if $S = K$ is a field, then V is a K-vector space, thus free, and $\operatorname{End}_K V$ is easy to oversee, especially when V is finite-dimensional.

(2) Due to their great importance, we explicitly mention the commutative functional algebras S^I as special cases of general product algebras $\prod_{i \in I} A_i$ of families A_i, $i \in I$, of S-algebras. A subring C of S^I is exactly an S-subalgebra when C contains the constant functions $a\colon I \to S$, $i \mapsto a$. Such special **functional algebras** often occur, especially in analysis and topology. For example, the polynomial functions $S \to S$ form an S-subalgebra of S^S. This is generated by the identity $x \mapsto x$ generated by S, which is often simply referred to again as x. The polynomial functions are thus the S-linear combinations $a_0 + a_1 x + \cdots + a_n x^n$, $a_\nu \in S$, of the power functions x^ν, $\nu \in \mathbb{N}$. Similarly, the functions $I \to S$ with finitely many values or with at most countably many values each form an S-subalgebra of S^I. More generally, this applies to the functions with $(< \alpha)$—or with $(\leq \alpha)$ -many values, where α is any infinite cardinal number, cf. exercise 1.8.15a). If X is a topological space, then the continuous real or continuous complex functions on X each form an $\mathbb{R}$ — or $\mathbb{C}$ -algebra $C_\mathbb{R}(X) \subseteq \mathbb{R}^X$ or $C_\mathbb{C}(X) \subseteq \mathbb{C}^X$, cf. Theorem 4.2.31, as do the differentiable functions on an interval $I \subseteq \mathbb{R}$ etc., cf. vol. 2. ◊

Example 2.9.6 (Monoidal Algebras) A crucial method for constructing new algebras is provided by the so-called monoidal algebras. In the following, S is always a commutative ring $\neq 0$. A is an S-algebra and M is a (multiplicative) monoid with neutral element ι. The elements $e_\sigma, \sigma \in M$, the standard basis of the A-module $A^{(M)}$ we briefly denote by σ. Then the multiplication

$$\left(\sum_{\sigma \in M} a_\sigma \sigma\right)\left(\sum_{\tau \in M} b_\tau \tau\right) := \sum_{\rho \in M} c_\rho \rho \quad \text{with} \quad c_\rho := \sum_{\sigma \tau = \rho} a_\sigma b_\tau,$$

on $A^{(M)}$ is obviously S-bilinear, and on the products $a\sigma$, $a \in A$, $\sigma \in M$, it agrees with the multiplication on the product monoid $A \times M$, is therefore particularly associative there and then even associative on all $A^{(M)}$, cf. Lemma 2.9.4. Because of $\iota(a\sigma) = a\sigma = (a\sigma)\iota$ for all $(a, \sigma) \in A \times M$, ι is also a unit element for the multiplication on $A^{(M)}$. Overall, $A^{(M)}$ is an S-algebra. It is called the **monoidal algebra** over A to the monoid M and is denoted by

$$A[M].$$

If M is a group, we also speak of a **group algebra**. We always identify A with the S-subalgebra $A\iota$ of $A[M]$. For $A \neq 0$, the canonical monoid homomorphism $\iota_M\colon M \to A[M]$, $\sigma \mapsto \sigma$, is injective. It then allows us to also identify M with its image in $A[M]$. $A[M]$ has the following universal property: ◊

Theorem 2.9.7 *Let A be an S -Algebra and M a monoid. Furthermore, let B be another S-Algebra. Then $f \mapsto (f|A, f \circ \iota_M)$ is a bijective mapping of the set $\mathrm{Hom}_{S\text{-Alg}}(A[M], B)$ onto the set of pairs $(g, h) \in \mathrm{Hom}_{S\text{-Alg}}(A, B) \times \mathrm{Hom}(M, B)$ with $g(a)h(\sigma) = h(\sigma)g(a)$ for all $a \in A$ and all $\sigma \in M$ (where $\mathrm{Hom}(M, B)$ is the set of monoid homomorphisms $M \to (B, \cdot)$). Such a pair (g, h) defines the S-Algebra homomorphism $f : A[M] \to B$, $\sum_{\sigma \in M} a_\sigma \sigma \mapsto \sum_{\sigma \in M} g(a_\sigma) h(\sigma)$. In the case $A = S$ we particularly obtain: $f \mapsto f \circ \iota_M$ is a bijective mapping $\mathrm{Hom}_{S\text{-Alg}}(S[M], B) \xrightarrow{\sim} \mathrm{Hom}(M, B)$ with inverse mapping*

$$h \mapsto \left(\sum_{\sigma \in M} a_\sigma \sigma \mapsto \sum_{\sigma \in M} a_\sigma h(\sigma) \right).$$

Proof The reader immediately checks that the mappings $f \mapsto (f|A, f \circ \iota_M)$ and $(g, h) \mapsto f$ are inverses of each other. The addition results from the fact that $\mathrm{Hom}_{S\text{-Alg}}(S, B)$ contains only one element, namely the structure homomorphism $S \to B$. □

According to the addition in Theorem 2.9.7, a $S[M]$-module is given by an S-module V and a monoid homomorphism $M \to (\mathrm{End}_S V, \circ)$, i.e., by an operation of M on V as a monoid of S-endomorphisms. This is also referred to as a **representation of M in the S-module** V. *The theory of representations of M in S-modules is therefore equivalent to the module theory of $S[M]$.* The trivial representation of M in V provides the $S[M]$-module structure on V with $\sigma x = x$ for all $\sigma \in M$ and all $\in V$. $S[M]$-modules are also called *M-S-modules* and for $S = \mathbb{Z}$ simply *M-modules*. An M-module is therefore an abelian group on which M operates as a monoid of group endomorphisms.

The S-algebra homomorphism defined by the trivial homomorphism $M \to A$, $\sigma \to 1_A$, $A[M] \to A$, $\sum_\sigma a_\sigma \sigma \mapsto \sum_\sigma a_\sigma$, is called the **augmentation** of $A[M]$, its kernel the **augmentation ideal**. It is generated by the elements $\sigma - 1 = \sigma - \iota$, $\sigma \in M$. With Lemma 2.8.26 it follows: *If all free A -modules have a rank, then all free $A[M]$-modules do as well.*

Calculations in $A[M]$ are made particularly clear by the direct sum decomposition $A[M] = \sum_{\sigma \in M}^{\oplus} A\sigma$ with $(A\sigma)(A\tau) \subseteq A(\sigma\tau)$. In general, one defines:

Definition 2.9.8 Let B be an S-algebra and M a monoid with neutral element ι. An *M*-grading on B is a direct sum decomposition $B = \sum_{\sigma \in M}^{\oplus} B_\sigma$ with S-submodules $B_\sigma \subseteq B$, for which $1_B \in B_\iota$ is and $B_\sigma B_\tau \subseteq B_{\sigma\tau}$ holds for all $\sigma, \tau \in M$. B together with an M-grading is called an **M-graded S-algebra**. If $S = \mathbb{Z}$, one simply speaks of an **M-graded ring**.

For an element $b = \sum_{\sigma \in M} b_\sigma \in B$, $b_\sigma \in B_\sigma$, b_σ is called the σ**-th homogeneous component**. Because of $1_B \in B_\iota$ is $S1_B \subseteq B_\iota$. An element $b \in B$ is called **homogeneous of degree** σ, if $b \in B_\sigma$ is. The degree of a homogeneous element $\neq 0$ is uniquely determined. The zero element $0 \in B$ has by definition every degree

2.9 Algebras

$\sigma \in M$. The condition $1_B \in B_\iota$ is automatically fulfilled, if M is a regular monoid. Proof! If B is commutative, then $B_\sigma B_\tau = B_\tau B_\sigma \subseteq B_{\sigma\tau} \cap B_{\tau\sigma}$ applies. In commutative M-graded algebras, the monoid M itself will usually also be commutative. In additive notation of M, $B_\sigma B_\tau \subseteq B_{\sigma+\tau}$ and $1_B \in B_0$.

Particularly important is the case that $M = (M, \leq)$ is a *totally* ordered monoid (see exercise 2.1.16) is. If then $b = \sum_{\sigma \in M} b_\sigma \in B$ with $b_\sigma \in B_\sigma$ is an element $\neq 0$, then

$$\omega = \deg b := \mathrm{Max}\,\{\sigma \in M \mid b_\sigma \neq 0\}$$

is called the **(upper) degree** and $\mathrm{LF}(b) := b_\omega$ the **leading form** of b. Obviously, the inequality $\deg(bc) \leq (\deg b)(\deg c)$ holds when b,c,bc $\neq 0$. In a monoidal algebra $A[M]$ with $A \neq 0$, the leading form of $b \in A[M]$ has the shape $a\sigma$, $a \in A - \{0\}$, $\sigma \in M$. σ is then called the **leading monomial** $\mathrm{LM}(b)$ and a the **leading coefficient** $\mathrm{LK}(b)$ of b. The element $b \neq 0$ is called **normalized**, if its leading coefficient is equal to 1_A. **Subdegree**, **initial form** $\mathrm{AF}(b)$ as well as **initial monomial** $\mathrm{AM}(b)$ and **initial coefficient** $\mathrm{AK}(b)$ of b are by definition the upper degree, the leading form, the leading monomial and the leading coefficient of b with respect to the opposite order $\leq^{\mathrm{op}} = \geq$ on M.

Lemma 2.9.9 *Let M be a regular totally ordered monoid. In the M-graded algebra B, one of the leading forms $\mathrm{LF}(b)$, $\mathrm{LF}(c)$ of the elements $b, c \in B - \{0\}$ is a non-zero divisor in B. Then*

$$\mathrm{LF}(bc) = \mathrm{LF}(b)\,\mathrm{LF}(c) \text{ and } \deg(bc) = (\deg b)(\deg c) \quad \textbf{(Formula for Grades)}.$$

In particular, b is a non-zero divisor in B, if $\mathrm{LF}(b)$ is a non-zero divisor, and B is a domain if all homogeneous elements $\neq 0$ in B are non-zero divisors. If the S-algebra A is a domain, so is the monoid algebra $A[M]$.

For the *proof*, it is only necessary to note that $\rho\sigma < \rho\tau$ and $\sigma\rho < \tau\rho$ hold for arbitrary elements ρ, σ, τ with $\sigma < \tau$ of the regular totally ordered monoid M.

Example 2.9.10 *The Mal'cev regular monoid M mentioned at the end of Example 2.3.34, which cannot be embedded into a group, can be totally ordered so that M becomes a totally ordered monoid. So if S is an integral domain, then the monoid algebra $S[M]$ is a domain that cannot be embedded into a division domain.* ◊

Let $B = \sum_{\sigma \in M}^{\oplus} B_\sigma$ continue to be an M-graded ring. A (left, right, or two-sided) ideal $\mathfrak{b} \subseteq B$ is called **homogeneous** or **graded**, if $\mathfrak{b} = \sum_{\sigma \in M}^{\oplus} (\mathfrak{b} \cap B_\sigma)$ is, that is, an element $b = \sum_\sigma b_\sigma \in B$ belongs to $\mathfrak{b}$ exactly when all its homogeneous components b_σ belong to $\mathfrak{b}$. Obviously, $\mathfrak{b}$ is graded exactly when $\mathfrak{b}$ is generated by homogeneous elements. Then also $B/\mathfrak{b} = \sum_{\sigma \in M}^{\oplus} B_\sigma/\mathfrak{b}_\sigma$ is M-graded. If M is totally ordered, then to any arbitrary ideal $\mathfrak{b} \subseteq B$ its so-called **leading form ideal** $\mathrm{LF}(\mathfrak{b})$ can be naturally assigned. It is the homogeneous (left, right, or two-sided) ideal, which is generated by the leading forms of the elements $b \in \mathfrak{b} - \{0\}$. To illustrate the usefulness of leading form ideals, we prove the following lemma as an example:

Lemma 2.9.11 $B = \sum_{\sigma \in M}^{\oplus} B_{\sigma}$ *let be a M-graded ring, $\mathfrak{b}$ a left or right ideal in B, and M is a regular ordered monoid with respect to a well-ordering. If the leading forms $c_{j,\mu_j} := \mathrm{LF}(c_j) \in B_{\mu_j}$, $j \in J$, of the elements $c_j \in \mathfrak{b} - \{0\}$ generate the leading form ideal $\mathrm{LF}(\mathfrak{b})$ of $\mathfrak{b}$, then the c_j, $j \in J$, generate the ideal $\mathfrak{b}$.*

Proof We consider the case of a left ideal $\mathfrak{b}$. Let $\mathfrak{b}' \subseteq \mathfrak{b}$ be the ideal generated by the c_j, $j \in J$. Assume that $\mathfrak{b}' \neq \mathfrak{b}$. Then let b be an element of smallest degree in $\mathfrak{b} - \mathfrak{b}'$ with leading form $b_\mu \in \mathrm{LF}(\mathfrak{b}) \cap B_\mu$. There is a finite subset $J' \subseteq J$ and – since M is regular – homogeneous elements a_j, $j \in J'$, in B with $b_\mu = \sum_{j \in J'} a_j c_{j,\mu_j}$. Then $b - \sum_{j \in J'} a_j c_j$ is in $\mathfrak{b} - \mathfrak{b}'$ and has a degree $<\mu$. Contradiction! □

A system of elements $c_j \in \mathfrak{b}$, $j \in J$, whose leading forms, as in the situation of Lemma 2.9.11, generate the leading form ideal of $\mathfrak{b}$, is called a **Gröbner basis** of $\mathfrak{b}$ (after W. Gröbner (1899–1980)). With the help of Gröbner bases, many operations with ideals can be computationally mastered. We refer to the relevant literature, such as T. Becker, B. V. Weispfenning: Gröbner Bases: A Computational Approach to Commutative Algebra. Graduate Texts in Mathematics 141, New York 1993.

Often, a given M-gradation is coarsened with a monoid homomorphism $\varphi: M \to N$: If $B = \sum_{\sigma}^{\oplus} B_\sigma$ is an M-gradation of B, then $B = \sum_{\tau \in N}^{\oplus} B_\tau$ with $B_\tau := \sum_{\sigma \in \varphi^{-1}(\tau)}^{\oplus} B_\sigma$ is an N-gradation of B. For each submonoid $M' \subseteq M$, $\sum_{\sigma \in M'}^{\oplus} B_\sigma$ is a M'-graded subalgebra of B. If D is a left ideal in M (i.e., $MD \subseteq D$), then $\sum_{\sigma \in D}^{\oplus} B_\sigma$ is a homogeneous left ideal in B.

We mention two extensions of the concept of monoidal algebra.

(1) The monoid M operates on the S-algebra A as a monoid of S-algebra endomorphisms with action homomorphism $\vartheta : M \to \mathrm{End}_{S\text{-Alg}} A$. Then the semidirect product $A \rtimes_\vartheta M$ with the multiplication $(a\sigma)(b\tau) = (a\vartheta_\sigma(b))(\sigma\tau)$ is a monoid, see Theorem 2.4.13, and $A^{(M)}$ is made an S-algebra with unit element $\iota = 1_A \iota$ by the multiplication

$$\left(\sum_{\sigma \in M} a_\sigma \sigma\right)\left(\sum_{\tau \in M} b_\tau \tau\right) = \sum_{\rho \in M} c_\rho \rho, \quad c_\rho := \sum_{\sigma\tau = \rho} a_\sigma \vartheta_\sigma(b_\tau),$$

It is called the **braided monoidal algebra** over A with respect to the monoid M associated with ϑ and is denoted by

$$A[M, \vartheta].$$

The above ordinary monoidal algebra is obtained with the trivial operation of M on A (where $\vartheta_\sigma = \mathrm{id}_A$ is for all $\sigma \in M$). Through $A \rtimes_\vartheta M \to \mathrm{End}_S A$, $a\sigma \mapsto L_a \vartheta_\sigma$, an S-algebra homomorphism $A[M, \vartheta] \to \mathrm{End}_S A$ is defined, see Problem 2.4.6. Also, the braided monoidal algebras $A[M, \vartheta]$ are M-graded algebras with the homogeneous components $A\sigma$, $\sigma \in M$. For example, if M is the free monoid of monomials X^ν, $\nu \in \mathbb{N}$, then $A[M]$ is the **polynomial algebra** $A[X]$ and an operation of M on the S-algebra A is simply given by an S-algebra endomorphism φ of A. We then also write $A[X, \varphi]$ for the associated so-called **braided polynomial algebra**. For $b \in A$ and $\nu \in \mathbb{N}$, $X^\nu b = \varphi^\nu(b) X^\nu$ is given.

(2) Let M be a monoid for which the fibers of the multiplication $M \times M \to M$ are all finite, so that every element of M has only finitely many left and right divisors. Then, on the product $\prod_{\sigma \in M} A\sigma$, as for the monoidal algebra $A[M]$, a multiplication can be defined as follows:

$$(a_\sigma \sigma)_{\sigma \in M} (b_\tau \tau)_{\tau \in M} := (c_\rho \rho)_{\rho \in M}, \quad c_\rho := \sum_{\sigma\tau = \rho} a_\sigma b_\tau.$$

This again provides an S-algebra structure on A^M. The resulting S-algebra is called the **formal monoidal algebra** over A to M. It is denoted by

$$A[[M]].$$

Also, the element $(a_\sigma \sigma)_{\sigma \in M}$ is written as (now generally infinite) sum $\sum_{\sigma \in M} a_\sigma \sigma$. This can be justified if one equips $A[[M]] = A^M$ with the product topology, where A itself carries the discrete topology, cf. Example 4.5.43 (1). The ordinary monoidal algebra $A[M]$ is a subalgebra of the formal monoidal algebra $A[[M]]$. Analogous to (1), twisted formal monoidal algebras $A[[M, \vartheta]]$ can also be defined as extensions of the twisted monoidal algebras $A[M, \vartheta]$. For example, the twisted polynomial algebra $A[X, \varphi]$ is extended to the so-called **twisted power series algebra** $A[[X, \varphi]]$, whose elements are the so-called **power series** $\sum_{\nu \in \mathbb{N}} a_\nu X^\nu$, $a_\nu \in A$. If $\varphi = \mathrm{id}_A$, one obtains the ordinary **power series algebra** $A[[X]]$ for the polynomial algebra $A[X]$.

Example 2.9.12 (Freely generated algebras) Let S be a commutative ring $\neq 0$ and M a free monoid in the (non-commuting) indeterminates X_i, $i \in I$, see Example 2.3.32. The elements of N are thus the monomials $X^\nu = X_{i_1} \cdots X_{i_n}$, $\nu = i_1 \cdots i_n \in W(I)$, which can also be uniquely written in the form $X_{i_1}^{n_1} X_{i_2}^{n_2} \cdots X_{i_r}^{n_r}$, where $(i_1, n_1), \ldots, (i_r, n_r)$ is a finite sequence with elements from $I \times \mathbb{N}^*$, for which $i_\rho \neq i_{\rho+1}$, $\rho < r$, is, see Example 2.3.32. The monoidal algebra $S[M]$ is denoted by

$$S\langle X_i, i \in I \rangle.$$

It is called the **freely generated** S-**algebra** in the (non-commuting) variables X_i, $i \in I$. Its elements are called **polynomials in the non-commuting variables** X_i, $i \in I$. The monomials X^ν, $\nu \in W(I)$, form an S-module basis of $S\langle X_i, i \in I\rangle$. In particular, the X_i, $i \in I$, an S-algebra generating system of $S\langle X_i, i \in I\rangle$. The universal property of the free monoid M, cf. 2.3.33, together with the universal property of the monoidal algebra $S[M]$ according to Theorem 2.9.7 result in the following universal property of the freely generated algebra $S\langle X_i, i \in I\rangle$: ◊

Theorem 2.9.13 *For every S-algebra A, the mapping*

$$\text{Hom}_{S\text{-Alg}}(S\langle X_i, i \in I\rangle, A) \xrightarrow{\sim} A^I, \quad f \mapsto (f(X_i))_{i \in I},$$

is a bijective mapping. The $x := (x_i)_{i \in I} \in A^I$ *defined S-algebra homomorphism* $S\langle X_i, i \in I\rangle \to A$ *is called the so-called* **substitution homomorphism**. *It is denoted with*

$$\varphi_x : S\langle X_i, i \in I\rangle \to A, \quad X_i \mapsto x_i, i \in I,$$

and maps the monomial $X^\nu = X_{i_1} \cdots X_{i_n}$, $\nu \in W(I)$, *in the indeterminates* X_i *to the monomial* $x^\nu = x_{i_1} \cdots x_{i_n}$ *in the elements* x_i, $i \in I$, *of A. The image of* φ_x *is the by the* x_i, $i \in I$, *generated subalgebra* $S\langle x_i, i \in I\rangle$ *of A. In particular,* φ_x *is surjective exactly when the* x_i, $i \in I$, *form an S-algebra generating system of A. We denote the image of any element* $F \in S\langle X_i, i \in I\rangle$ *with*

$$\varphi_x(F) = F\langle x\rangle = F\langle x_i, i \in I\rangle.$$

The kernel of φ_x is called the **relations ideal** of the x_i, $i \in I$.[51] According to the isomorphism theorem,

$$S\langle X_i, i \in I\rangle / \ker \varphi_x \xrightarrow{\sim} S\langle x_i, i \in I\rangle.$$

For $x \in S^I$ is $S\langle x_i, i \in I\rangle = S$, and $\ker \varphi_x$ is generated by the polynomials $X_i - x_i$, $i \in I$, generated, see exercise 2.9.3. In particular, the indeterminates X_i, $i \in I$, generate the kernel of φ_0. The image $\varphi_0(F) = F(0)$ is the so-called **constant term** of F. The quotient algebras of the freely generated S-algebras thus represent up to isomorphism all S-algebras (and for $S = \mathbb{Z}$ all rings). If the polynomials $G_j \in S\langle X_i, i \in I\rangle$, $j \in J$, generate the two-sided ideal $\ker \varphi_x$, then one says

$$\langle x_i, i \in I \mid G_j(x) = 0, j \in J\rangle$$

is a **representation of the algebra** $S\langle x_i, i \in I\rangle$ **by generators and relations**. Every such represented S-algebra $S\langle x_i, i \in I\rangle$ has the following universal property: *If A is any S-algebra, then the mapping*

$$\text{Hom}_{S\text{-Alg}}(S\langle x_i, i \in I\rangle, A) \xrightarrow{\sim} \text{NS}_A(G_j, j \in J) := \{a = (a_i) \in A^I \mid G_j(a) = 0, j \in J\}$$

is bijective,. Therefore, if one wants to understand the zero set $\text{NS}_A(G_j, j \in J) = \bigcap_{j \in J} \text{NS}_A(G_j)$ for arbitrary S-algebras A, one has to study the algebra $S\langle x_i, i \in I\rangle$. If the index sets I and J are finite, the representation is called **finite**. As with groups, it is generally difficult to describe the associated algebra clearly, even for simple relational systems.

---

[51] Note that the relations module $\text{Rel}_S(x_i, i \in I) = \text{Syz}_S(x_i, i \in I) = \left(\sum_{i \in I} SX_i\right) \cap \ker \varphi_x k$, cf. the definitions following Theorem 2.8.14, only contains the *linear* relations of the $x_i, i \in I$.

2.9 Algebras

The free monoid M in the indeterminates X_i can easily be provided with a total order so that it becomes a totally ordered monoid. A common method is as follows: One provides the index set I with a total order, arranges for each $n \in \mathbb{N}$ the set $W_n(I)$ of words of length n lexicographically and provides the disjoint union $W(I) = \biguplus_{n \in \mathbb{N}} W_n(I)$ of all words with the total sum order according to Example 1.4.6, where the index set $\mathbb{N}$ carries the natural order. We also call the order thus obtained the (induced by the given order on I) **homogeneous lexicographic order** $\leq_{\text{hlex}}$ on $W(I)$.[52] It is transferred to M, making M clearly a regular sharp totally ordered monoid. If I is well-ordered, then so is $W(I)$. The direct sum $S\langle X_i, i \in I \rangle_n := \sum_{\nu \in W_n(I)}^{\oplus} SX^{\nu}$ is the n-th homogeneous component of $S\langle X_i, i \in I \rangle$ with respect to the $\mathbb{N}$-gradation, which arises from the M-gradation by means of the homomorphism $M \to (\mathbb{N}, +)$, $X_i \mapsto 1$. The degree of a polynomial $F = \sum_{n \in \mathbb{N}} F_n$ with respect to this gradation is called the **degree** ($\deg F$) of F per se. Occasionally, it is clever to assign arbitrary weights $\gamma_i = \gamma(X_i) \in \mathbb{Z}$, $i \in I$, to the indeterminates X_i instead of the weight 1. One then speaks of the γ-**degree**

$$\deg_{\gamma} F$$

of F. – With Lemma 2.9.9 we get:

Proposition 2.9.14 *Let S be an integral domain. Then every freely generated algebra $Q := S\langle X_i, i \in I \rangle$ is a domain. Furthermore, $Q^{\times} = S^{\times}$.*

Obviously, for every S-algebra A that is a domain, $A\langle X_i, i \in I \rangle = A[M]$ is also a domain. Perhaps surprising is the following theorem:

Theorem 2.9.15 *Let K be a field. Then every left ideal in the freely generated algebra $A := K\langle X_i, i \in I \rangle$ is a free A-module. More generally, every submodule of a free A-module is also free.*

Proof Let $\mathfrak{a} \subseteq A$ be a left ideal $\neq 0$. In addition to the homogeneous lexicographic well-ordering of M with respect to a *well-ordering* on I, we also use the order of M determined by the right divisibility $\preceq$ (which is not total at $|I| \geq 2$): Exactly then is $X^{\mu} \preceq X^{\nu}$, when there is a $X^{\lambda} \in M$ with $X^{\lambda} X^{\mu} = X^{\nu}$, when thus the word μ is a suffix of ν. Like the well-ordering, $\preceq$ is also an Artinian order on M. Let G_j, $j \in J$, be the family of those normalized polynomials in $\mathfrak{a}$, whose leading monomials are minimal with respect to the order $\preceq$ in the set of all leading monomials $\text{LM}(F)$, $F \in \mathfrak{a} - \{0\}$.

The G_j, $j \in J$, *form an A-basis of* $\mathfrak{a}$. Since the leading monomials of the G_j, $j \in J$, generate the leading form ideal $\text{LF}(\mathfrak{a})$ of $\mathfrak{a}$ by construction, the G_j, $j \in J$, generate the ideal $\mathfrak{a}$ according to Lemma 2.9.11. The G_j, $j \in J$, are also linearly

---

[52] It does not quite match the order of words in a dictionary. There, "four" comes before "two", but here it comes after. The term "homogeneous" is meant to remind us that $\leq_{\text{hlex}}$ refines the quasi-order on $W(I)$ given by the length of the words.

independent over A. For proof by contradiction, let $J' \subseteq J$ be a finite subset $\neq \emptyset$ and F_j, $j \in J'$, polynomials $\neq 0$ in A with $\sum_{j \in J'} F_j G_j = 0$. According to Lemma 2.9.9, $\operatorname{LM}(F_j G_j) = \operatorname{LM}(F_j)\operatorname{LM}(G_j)$ holds for all $j \in J'$. Since the monomials $\operatorname{LM}(G_j)$ are pairwise incomparable with respect to $\preceq$, the elements $\operatorname{LM}(F_j G_j)$ are pairwise different. The largest among them with respect to the well-ordering on M is then the leading monomial of the given linear combination, which therefore cannot be 0.

The additional result follows from lemma 2.9.16, whose proof runs completely analogous to the proof of Theorem 2.3.28 (without paying attention to the cardinal numbers of the occurring bases) and is left to the reader. □

Lemma 2.9.16 *Let A be a ring, all of whose left ideals are free A-modules. Then A-submodules of any free A-modules are free.*

As the proof of Theorem 2.9.15 shows, every antichain with respect to $\preceq$ of monomials in M forms a basis of a left ideal of $K\langle X_i, i \in I\rangle$. In $|I| \geq 2$, A therefore has free submodules of rank $\operatorname{Max}(\aleph_0, |I|)$ (but none of larger rank). In the case of $|I| = 1$, $M = \{X^\nu \mid \nu \in \mathbb{N}\}$ and $K\langle X_i, i \in I\rangle = K\langle X\rangle = K[X]$ is the (commutative) polynomial ring over K in one indeterminate X.

Corollary 2.9.17 *If K is a field, then the polynomial ring $K[X]$ in one indeterminate over K is a principal ideal domain. If $\mathfrak{a} \subseteq K[X]$ is an ideal $\neq 0$, then $\mathfrak{a} = K[X]G$, where G is a polynomial of minimal degree in $\mathfrak{a} - \{0\}$.*

Let S is again a commutative ring $\neq 0$ and I is an index set. Furthermore, let M ($\cong \mathbb{N}^{(I)}$) be a free commutative monoid in the (commuting) indeterminates X_i, $i \in I$. The elements of M are the monomials $X^\nu = \prod_{i \in I} X_i^{\nu_i}$, $\nu = (\nu_i)_{i \in I} \in \mathbb{N}^{(I)}$, cf. Example 2.3.20. The (commutative) monoidal algebra $S[M]$ over S is called the **polynomial algebra over** S **in the (commuting) indeterminates** X_i, $i \in I$. It is denoted by

$$S[X_i, i \in I].$$

The monomials X^ν, $\nu \in \mathbb{N}^{|I|}$, form an S-module basis of the polynomial algebra $S[X_i, i \in I]$, and the multiplication is determined by $(aX^\mu)(bX^\nu) = (ab)X^{\mu+\nu}$, $a, b \in S$, $\mu, \nu \in \mathbb{N}^{(I)}$. The formal monoidal algebra $S\!\!\mid\!\! M = \prod_{\nu \in \mathbb{N}^{(I)}} SX^\nu$ is called the **(formal) power series algebra over** S **in the (commuting) indeterminates** X_i, $i \in I$. It is denoted by

$$S[[X_i]], i \in I$$

The universal property of the monoid M, cf. Theorem 2.3.21, together with the universal property of monoidal algebras, cf. Theorem 2.9.7, yields the following universal property of commutative polynomial algebras:

2.9 Algebras

Theorem 2.9.18 *For every S-algebra A, the mapping*

$$\mathrm{Hom}_{S\text{-Alg}}(S[X_i, i \in I], A) \longrightarrow A^I, \quad f \longmapsto (f(X_i))_{i \in I},$$

*is an injective mapping onto the set of I-tuples $x = (x_i)_{i \in I} \in A^I$ with $x_i x_j = x_j x_i$ for all $i, j \in I$. The S-algebra homomorphism defined by such an I-tuple is the **substitution homomorphism** $\varphi_x : S[X_i, i \in I] \to A$, $X_i \mapsto x_i, i \in I$, with*

$$\varphi_x(F) = F(x) = F(x_i, i \in I) = \sum_{\nu \in \mathbb{N}^{(I)}} a_\nu x^\nu, \quad F = \sum_{\nu \in \mathbb{N}^{(I)}} a_\nu X^\nu \in S[X_i, i \in I].$$

Specifically: If A is commutative, then the mapping $\mathrm{Hom}_{S\text{-Alg}}(S[X_i, i \in I], A) \xrightarrow{\sim} A^I$ with $f \mapsto (f(X_i))_{i \in I}$ is bijective. – The image of the substitution homomorphism φ_x is the commutative S-subalgebra $S[x_i, i \in I]$ of A. In particular, φ_x is surjective exactly when the $x_i, i \in I$, form an S-algebra generating system of A.

The kernel of φ_x is called the **relation ideal** of the pairwise commuting elements $x_i, i \in I$. According to the isomorphism theorem,

$$S[X_i, i \in I]/\mathrm{ker}\, \varphi_x \xrightarrow{\sim} S[x_i, i \in I].$$

If the relation ideal $\ker \varphi_x$ is the zero ideal, i.e., the substitution homomorphism φ_x induces an isomorphism $S[X_i, i \in I] \xrightarrow{\sim} S[x_i, i \in I]$, then the family $x = (x_i)_{i \in I}$ is called **algebraically independent** or **transcendental**, otherwise **algebraically dependent** (over S). For $|I| = 1$ we simply speak of **transcendental** or **algebraic elements**. An element $x \in A$ is therefore algebraic over S exactly when there is a polynomial $F \neq 0$ in $S[X]$ with $F(x) = 0$. If there is a (with respect to the $\mathbb{N}$-graduation of $S[X]$) normalized polynomial F in $S[X]$ with $F(x) = 0$, then x is called **entire (algebraic)** over S. The S-algebra A is called **algebraic** or **entire (algebraic)** over S, if every element of A is algebraic or entire (algebraic) over S. Every finite S-algebra is entire, cf. Ex. 2.9.17a). This is trivial for a field $S = K$, over which the terms "algebraic" and "entire (algebraic)" naturally coincide: If $\dim_K A$ is finite, then for $x \in A$ the substitution homomorphism $\varphi_x : K[X] \to A$ cannot be injective.

The residue class algebras of the commutative polynomial algebras over S represent up to isomorphism all commutative S-algebras (and for $S = \mathbb{Z}$ all commutative rings). If the polynomials $G_j, j \in J$, generate the relations ideal $\ker \varphi_x$, then one says

$$\langle x_i, i \in I \mid G_j(x) = 0, j \in J \rangle$$

is a **representation of the commutative S-Algebra** $S[x_i, i \in I]$ **by generators and relations**. Every such represented commutative S-Algebra $S[x_i, i \in I]$ has the following universal property: *If A is any commutative S-Algebra, then the mapping*

$$\mathrm{Hom}_{S\text{-Alg}}(S[x_i, i \in I], A) \xrightarrow{\sim} \mathrm{NS}_A(G_j, j \in J) := \{a = (a_i) \in A^I \mid G_j(a) = 0, j \in J\}$$

is bijective. So, if one wants to understand the zero sets $\mathrm{NS}_A(G_j, j \in J) = \bigcap_{j \in J} \mathrm{NS}_A(G_j)$ for arbitrary *commutative S-Algebras A*, one has to study the Algebra $S[X_i, i \in I]$. If the index sets I and J are finite, the representation is called **finite**. *In particular, the quotient algebras of the polynomial algebras* $S[X_1, \ldots, X_n]$, $n \in \mathbb{N}$, *up to isomorphism, represent all commutative S-Algebras of finite type.*

The substitution homomorphisms $\varphi_x \colon S[X_i, i \in I] \to S$, $x \in S^I$, together yield a homomorphism

$$\varphi \colon S[X_i, i \in I] \longrightarrow S^{S^I} = \mathrm{Abb}(S^I, S), \quad F \longmapsto (x \mapsto \varphi_x(F) = F(x)),$$

whose images are the so-called **polynomial functions** on S^I. Since $\varphi(X_i)$ is the i-th projection (= i-th coordinate function) $S^I \to S$, $x \mapsto x_i$, the S-algebra $\mathrm{Im}\,\varphi$ of polynomial functions on S^I is generated by these coordinate functions. In general, φ is not injective: *Different polynomials can define the same polynomial functions.* For example, over a finite commutative ring S, the normalized polynomial $\prod_{a \in S}(X - a) \in S[X]$ of degree $|S|$ defines the zero function on S, cf. however Corollary 2.9.33.

Example 2.9.19 (Horner's Scheme) To calculate the values of a polynomial in one variable

$$F = a_0 + a_1 X + \cdots + a_n X^n \in S[X]$$

at a point $a \in S$, it is advisable to use the so-called **Horner's Scheme**. For this purpose, a sequence of polynomials

$$F_0 = a_n, \quad F_1 = a_{n-1} + F_0 X = a_{n-1} + a_n X, \quad \ldots,$$

$$F_{k+1} = a_{n-k-1} + F_k X = a_{n-k-1} + \cdots + a_{n-1} X^k + a_n X^{k+1}, \quad \ldots,$$

$$F_n = a_0 + F_{n-1} X = F,$$

is defined recursively, from which the value $F(a) = F_n(a)$ results in the recursion scheme

$$F_0(a) = a_n, \quad F_{k+1}(a) = a_{n-k-1} + F_k(a)a, \; k = 0, \ldots, n-1,$$

It can also be used to calculate the value $F(a)$ for an element a of any S-algebra. A supplement to this can be found in exercise 2.9.21b). ◊

One should carefully distinguish between the freely generated S-algebra $S\langle X_i, i \in I\rangle$ and the commutative polynomial algebra $S[X_i, i \in I]$. Only in $|I| \leq 1$ do both coincide. (Note that $S \neq 0$.) The kernel of the surjective substitution homomorphism $\varphi_X \colon S\langle X_i, i \in I\rangle \longrightarrow S[X_i, i \in I]$, $X_i \mapsto X_i$, $i \in I$, is generated by the so-called **commutators** $[X_i, X_j] := X_i X_j - X_j X_i$, $i, j \in I$. Proof! If the index set $I = I' \uplus I''$ is decomposed into disjoint subsets I', I'', then the substitution homomorphism

$$S[X_i, i \in I] \xrightarrow{\sim} \left(S[X_i, i \in I']\right)[X_i, i \in I''], \quad X_i \mapsto X_i, \; i \in I,$$

is obviously an isomorphism of S-algebras, with the help of which one identifies both algebras. For example,

$$S[X_1, \ldots, X_{n+1}] = (S[X_1, \ldots, X_n])[X_{n+1}], \quad n \in \mathbb{N}.$$

This allows for induction proofs to be often conducted for commutative polynomial algebras in finitely many indeterminates; analogous isomorphisms do not apply for freely generated algebras.

Like a free monoid, a commutative free monoid $\mathbb{N}^{(I)}$ also possesses canonical total orders, which make it an ordered monoid, so that one can then also speak of leading terms, leading coefficients and in particular of **normalized polynomials** (i.e., those with leading coefficient 1). Let $\mathbb{N}^{(I)} \to \mathbb{N}$, $\nu = (\nu_i) \mapsto |\nu| = \sum_i \nu_i$, the standard weight homomorphism, with the help of which the (standard) **degree** of a polynomial is defined. Then one chooses a total order on I and extends it to a total order on all of $\mathbb{N}^{(I)}$. Two such extensions are particularly popular. In the case of the **homogeneous lexicographic order** $\leq = \leq_{\text{hlex}}$ by definition $\mu = (\mu_i) < \nu = (\nu_i)$ holds exactly when $|\mu| < |\nu|$ is or when $|\mu| = |\nu|$ is, but $\mu \neq \nu$, and for the *smallest* index i_0 with $\mu_{i_0} \neq \nu_{i_0}$ holds $\mu_{i_0} > \nu_{i_0}$.[53] In the case of the **reverse homogeneous lexicographic order** $\leq = \leq_{\text{rev hlex}}$ holds $\mu = (\mu_i) < \nu = (\nu_i)$ exactly when $|\mu| < |\nu|$ is or when $|\mu| = |\nu|$ is, but $\mu \neq \nu$, and for the *largest* index i_0 with $\mu_{i_0} \neq \nu_{i_0}$ holds $\mu_{i_0} < \nu_{i_0}$. In both cases, the null tuple is the smallest element $e_i < e_j$ for $i < j$ and $\mathbb{N}^{(I)}$ a (regular) totally ordered monoid. We leave it to the reader to show that these are well-orders if the given order on I is a well-order. (This is trivial for finite I.) From $|I| \geq 3$ both orders are different. (Examples?) These orders are transferred to the monomials X^ν, $\nu \in \mathbb{N}^{(I)}$, and make $S[X_i, i \in I]$ each a $\mathbb{N}^{(I)}$-graded S-algebra with the free homogeneous components SX^ν, $\nu \in \mathbb{N}^{(I)}$, of rank 1. From lemma 2.9.9 it follows:

Proposition 2.9.20 *Let S be an integral domain. Then every polynomial algebra $P := S[X_i, i \in I]$ is an integral domain. Furthermore, $P^\times = S^\times$.*

Let S be a commutative ring $\neq 0$ with total quotient ring $Q(S) = S_{S^*}$, cf. Example 2.6.14, and P the polynomial algebra $S[X_i, i \in I]$. Then the canonical homomorphism $P_{S^*} \to Q(S)[X_i, i \in I]$, which extends the inclusion $P \hookrightarrow Q(S)[X_i, i \in I]$, is obviously an isomorphism. The total quotient rings of P and $Q(S)[X_i, i \in I]$ therefore coincide. We denote this common quotient ring with

$$Q(S)(X_i, i \in I).$$

---

[53] Note the reversal of the order symbol.

If S is an integral domain, then so is P. Then $Q(S)$ is the quotient field K of S, and the quotient field of P is the so-called field $K(X_i, i \in I)$ of the **rational functions** in the indeterminates X_i, $i \in I$, over K.

Let S be any commutative ring $\neq 0$ again. The graduation $P = \sum_{n\in\mathbb{N}}^{\oplus} P_n$ enlarged with the degree homomorphism $v \to |v|$ on the polynomial algebra $P := S[X_i, i \in I]$ has the free S-modules $P_n := \sum_{|v|=n} SX^v$, $n \in \mathbb{N}$, as homogeneous components. For finite I and $S \neq 0$, P_n has the rank $\binom{n+|I|-1}{|I|-1}$, cf. Example 1.6.13. We repeat the **degree formula**

$$\deg(FG) \leq \deg F + \deg G, \quad F, G \in P, \ FG \neq 0,$$

where the equality sign applies when the base ring S is an integral domain, cf. Lemma 2.9.9.

The following Hilbert's Basis Theorem is of utmost importance.

Theorem 2.9.21 (Hilbert's Basis Theorem) *If S is a Noetherian commutative ring and $n \in \mathbb{N}$, then the polynomial algebra $S[X_1, \ldots, X_n]$ is also Noetherian.*

Proof We conclude by induction over n and can therefore limit ourselves to the case $n = 1$. According to Lemma 2.9.11, it is sufficient to show that every *homogeneous* ideal $\mathfrak{a} \subseteq S[X]$ is finitely generated. Obviously, then $\mathfrak{a} = \sum_{n\in\mathbb{N}}^{\oplus} \mathfrak{a}_n X^n$ with a monotonically increasing sequence $\mathfrak{a}_0 \subseteq \mathfrak{a}_1 \subseteq \mathfrak{a}_2 \subseteq \cdots \subseteq S$ of ideals in S. Since S is Noetherian, all $\mathfrak{a}_n$ are finitely generated and there is an n_0 with $\mathfrak{a}_n = \mathfrak{a}_{n_0}$ for all $n \geq n_0$. If a_{i,j_i}, $j_i \in J_i$, are finite generating systems of $\mathfrak{a}_i$, $i = 0, \ldots, n_0$, then the finitely many elements $a_{i,j_i} X^i$, $0 \leq i \leq n_0$, $j_i \in J_i$, generate the homogeneous ideal $\mathfrak{a}$ in $S[X]$. □

Remark 2.9.22 (1) More general than Theorem 2.9.21 is (with the same proof): *If A is a left-Noetherian S-algebra, then so is $A[X_1, \ldots, X_n] (= A[\mathbb{N}^n])$.*

(2) Compare Hilbert's Basis Theorem 2.9.21 with the analogous statement in Problem 2.3.15d) about Noetherian monoids. Dickson's lemma, that the monoids $(\mathbb{N}^n, +)$, $n \in \mathbb{N}$, are Noetherian, also follows directly from Hilbert's Basis Theorem. If $A \subseteq \mathbb{N}^n$ is an ideal in the monoid $\mathbb{N}^n$, then $\sum_{v \in A} KX^v \subseteq K[X_1, \ldots, X_n]$ is an ideal in the Noetherian polynomial ring $K[X_1, \ldots, X_n]$ over an (arbitrary) field K and thus finitely generated. If X^v, $v \in N \subseteq A$, are finitely many monomials that generate this ideal, then the v, $v \in N$, generate the monoid ideal A.

(3) If one wants to avoid referring to the leading form ideals in the proof of Hilbert's Basis Theorem, one can conclude as follows: Assume that $\mathfrak{a} \subseteq S[X]$ is an ideal (not necessarily homogeneous) that is not finitely generated. Then one defines recursively polynomials $F_m \in \mathfrak{a}$, $m \in \mathbb{N}$, with the following properties: F_0 is a polynomial $\neq 0$ of smallest degree in $\mathfrak{a}$, and F_{m+1} is a polynomial of smallest degree in $\mathfrak{a} - \sum_{j=0}^{m} S[X] F_j$. Then the sequence $\deg F_m$, $m \in \mathbb{N}$, is monotonically increasing, and it holds $\mathrm{LK}(F_{m+1}) \notin \sum_{j=0}^{m} S \cdot \mathrm{LK}(F_j)$, $m \in \mathbb{N}$, i.e., the sequence $0 \subset S \cdot \mathrm{LK}(F_0) \subset S \cdot \mathrm{LK}(F_0) + S \cdot \mathrm{LK}(F_1) \subset \cdots$ of S-ideals is strictly monotonically increasing. Contradiction!

2.9 Algebras

(4) A statement analogous to Hilbert's Basis Theorem also holds for power series algebras: *If S is a Noetherian commutative ring and $n \in \mathbb{N}$, then the power series algebra $S \, X_1, \ldots, X_n$ is also Noetherian.* In the *proof* by induction, one can again assume $n = 1$. Instead of the leading forms, one now uses the **initial forms** $\mathrm{AF}(F)$ of a non-zero power series $F = \sum_{n \in \mathbb{N}} a_n X^n \in S \, X$. This is the form $a_{n_0} X^{n_0}$, where n_0 is the *smallest* index with $a_{n_0} \neq 0$. a_{n_0} is then called the **initial coefficient** $\mathrm{AK}(F)$ and n_0 is the **order** of the power series F. If now $\mathfrak{a} \subseteq S \, X$ is an ideal, then the initial forms of degree $n \in \mathbb{N}$ of the elements $F \neq 0$ from $\mathfrak{a}$ together with 0 form a set of the form $\mathfrak{a}_n X^n$ with an ideal $\mathfrak{a}_n \subseteq S$. It holds $\mathfrak{a}_0 \subseteq \mathfrak{a}_1 \subseteq \mathfrak{a}_2 \subseteq \cdots \subseteq S$. If $\mathfrak{a}_n = \mathfrak{a}_{n_0}$ for all $n \geq n_0$ and the $a_{i,j_i} \neq 0$, $j_i \in J_i$, are a finite generating systems of $\mathfrak{a}_i$, $i = 0, \ldots, n_0$, then the power series $F_{i,j_i} \in \mathfrak{a}$, $0 \leq i \leq n_0$, $j_i \in J_i$, with initial forms $a_{i,j_i} X^i$ generate the ideal $\mathfrak{a}$. We leave it to the reader to verify this. ◊

Corollary 2.9.23 *Every commutative algebra A of finite type over a Noetherian commutative ring S is Noetherian and has a finite representation $A \cong S[X_1, \ldots, X_n]/(G_1, \ldots, G_r)$ with $n, r \in \mathbb{N}$. In particular, this applies to commutative algebras of finite type over a field or over the ring $\mathbb{Z}$ of integers.*

The commutative algebras of finite type over commutative Noetherian rings are classic objects of Commutative Algebra and Algebraic Geometry.

A useful tool for calculating with polynomials is formal differentiation, whose calculus is taken from analysis and whose essence is the product rule. Let S be a commutative ring and A a (not necessarily commutative) S-algebra. A mapping $\delta \colon A \to A$ is called an S-**derivation**, if δ is S-linear and satisfies the **product rule**

$$\delta(xy) = \delta(x)y + x\delta(y), \quad x, y \in A,$$

Because of $\delta(1_A) = \delta(1_A \cdot 1_A) = 2\delta(1_A)$, $\delta(1_A) = 0$ and thus $\delta(s) = \delta(s \cdot 1_A) = s \cdot \delta(1_A) = 0$. The kernel of an S-derivation $\delta \colon A \to A$ is therefore due to the product rule an S-subalgebra of A. By induction, the general **power rule**

$$\delta(x^n) = nx^{n-1}\delta(x), \quad x \in A, \, n \in \mathbb{N}^*.$$

is obtained. The set

$$\mathrm{Der}_S \, A$$

of S-derivations of A is obviously an S-submodule and even an $Z(A)$-submodule of $\mathrm{End}_S \, A$.

Let $A = S[X]$ be a polynomial algebra in an indeterminate X over $S \neq 0$ and δ a S-derivation of $S[X]$. For a polynomial $F = \sum_n a_n X^n \in S[X]$, the power rule

$$\delta(F) = \sum_{n \in \mathbb{N}^*} n a_n X^{n-1} \delta(X) = F'\delta(X) \quad \text{with} \quad F' := \sum_{n \in \mathbb{N}^*} n a_n X^{n-1}.$$

applies. It is immediately shown that $F \mapsto F'$ is a S-derivation of $S[X]$. It is called the **derivative** of $S[X]$ and is also denoted by

$$\partial_X \quad \text{or with} \quad D = D_X$$

It follows:

Proposition 2.9.24 *The mappings* $S[X] \to S[X]$, $F \mapsto F'H$, *with fixed* $H \in S[X]$ *are the only S-derivations of* $S[X]$. *The module* $\mathrm{Der}_S S[X]$ *of the S-derivations of* $S[X]$ *is therefore a free* $S[X]$*-module of rank* 1 *with basis* D_X. *For* $\delta \in \mathrm{Der}_S S[X]$ *is* $\delta = \delta(X) D_X$.

Let's generalize $P := S[X_i, i \in I]$ to be any polynomial algebra in the (commuting) indeterminates X_i, $i \in I$. For fixed $i \in I$ we then consider $P = P_i[X_i]$ as a polynomial algebra in the indeterminate X_i over the polynomial algebra $P_i := S[X_j, j \neq i]$. The P_i-derivation

$$\partial_{X_i} = D_{X_i} : P \to P,$$

which is naturally also an S-derivation, is called the i-th **partial derivative** of P. Apparently, the partial derivatives are interchangeable: $D_{X_i} D_{X_j} = D_{X_j} D_{X_i}$ for all $i, j \in I$. The mapping $\delta \mapsto (\delta(X_i))_{i \in I}$ is a P-module isomorphism $\mathrm{Der}_S P \to P^I$ with the inverse isomorphism

$$P^I \xrightarrow{\sim} \mathrm{Der}_S P, \quad (G_i)_{i \in I} \mapsto \left(F \mapsto \sum_{i \in I} (\partial_{X_i} F) G_i \right),$$

cf. Ex. 2.9.5. For $\nu \in \mathbb{N}^{(I)}$ the composition $D_X^\nu = \prod_{i \in I} D_{X_i}^{\nu_i}$ is a so-called **higher partial derivative**. The ν-th iterated D_X^ν, $\nu \in \mathbb{N}$, of the ordinary derivative $F \mapsto F'$ of the polynomial algebra $S[X]$ in one variable is usually denoted by $F \mapsto F^{(\nu)}$. – The partial derivatives $\partial_{X_i} = D_{X_i}$, $i \in I$, are defined in an analogous way also for power series algebras $S\ X_i, i \in I$. They are also derivations.

Example 2.9.25 (Kronecker's Method of Indeterminates) Polynomial rings $S[X_i, i \in I]$ are used, among other things, for proving universally valid identities. The basis for this are the substitution homomorphisms. If one wants to prove an equation for the family $x = (x_i)_{i \in I}$, of any commutative S-algebra A, it is often sufficient to show the corresponding identity for the indeterminates X_i, $i \in I$, and then transport these with the substitution homomorphism $\varphi_x : X_i \mapsto x_i$, $i \in I$, to A. This procedure is called the **(Kronecker's) method of indeterminates**.

The binomial equation

$$(1+x)^n = \sum_{k=0}^n \binom{n}{k} x^k, \quad x \in A,$$

for example, cf. Theorem 2.6.3, only needs to be proven for the indeterminate $X \in \mathbb{Z}[X]$. In this case, one can replace $\mathbb{Z}$ with $\mathbb{Q}$ due to $X \in \mathbb{Z}[X] \subseteq \mathbb{Q}[X]$. The equation

$$(1+X)^n = \sum_{k=0}^{n} \binom{n}{k} X^k$$

is in $\mathbb{Q}[X]$ but according to the Taylor formula, cf. Exercise 2.9.4b), it is obvious, since the k-th derivative of the polynomial $(1+X)^n$ is equal to $n(n-1)\cdots(n-k+1)(1+X)^{n-k}$, $0 \le k \le n$. In a similar way, the polynomial theorem follows from the corresponding representation of $(X_1 + \cdots + X_r)^n$ in $\mathbb{Z}[X_1, \ldots, X_r]$, which is also obtained with the Taylor formula. ◊

Let S be a commutative ring $\ne 0$ again. As already mentioned, the study of the roots of a polynomial $F \in S[X]$ in an indeterminate X in S-algebras A requires a good overview of the quotient algebras $S[X]/(F)$ or $A[X]/(F)$ (where $F \in A[X]$ denotes the canonical image of F in $A[X]$): The canonical mappings

$$\underset{S\text{-Alg}}{\mathrm{Hom}}(S[X]/(F), A) \xrightarrow{\sim} \underset{A}{Z}(F) = \{x \in A \mid F(x) = 0\}$$

are bijective. If A is commutative, then of course the canonical mapping $\mathrm{Hom}_{A-\mathrm{Alg}}(A[X]/(F), A) \xrightarrow{\sim} \mathrm{NS}_A(F)$ is also bijective. The discussion of these root sets $Z_A(F)$ was the origin and has been a central subject of algebra for thousands of years. In this context, division with remainder is an important tool for investigating polynomials in an indeterminate.

Theorem 2.9.26 (Division with Remainder) *Let S be a commutative ring $\ne 0$ and F, G be polynomials in $S[X]$. G is normalized. Then there exist uniquely determined polynomials Q and R in $S[X]$ with*

$$F = QG + R \quad \text{and} \quad R = 0 \text{ or } \deg R < \deg G.$$

G is a divisor of F in $S[X]$, if $R = 0$.

Proof The existence of Q and R is trivial at $F = 0$ and otherwise proven by induction over degree F. In the case of $n := \deg F < m := \deg G$ one sets $Q := 0$ and $R := F$. Let now $n \ge m$ and $F = a_n X^n + \cdots + a_0$ as well as $G = X^m + \cdots + b_0$ with $a_n \ne 0$. Then $F_0 := F - a_n X^{n-m} G$ is a polynomial of lower degree than F. According to the induction hypothesis, there are polynomials Q_0 and R_0 with $F_0 = Q_0 G + R_0$ and $R_0 = 0$ or $\deg R_0 < \deg G$. From this follows

$$F = (a_n X^{n-m} + Q_0)G + R_0 = QG + R \quad \text{with} \quad Q := a_n X^{n-m} + Q_0, \ R := R_0.$$

To prove uniqueness, let $F = Q_1 G + R_1$ also be a representation as in the theorem. Then $0 = F - F = (Q - Q_1)G + (R - R_1)$. At $Q \ne Q_1$ we would get

$$\deg G \le \deg(Q - Q_1) + \deg G = \deg((Q - Q_1)G) = \deg(R_1 - R) < \deg G.$$

Contradiction! Therefore, $Q = Q_1$ and then also $R = R_1$. □

The proof of existence for Theorem 2.9.26 is constructive and provides the well-known method for obtaining the quotient Q and the remainder R in the division

$$(3X^4+\tfrac{3}{2}X^3+\tfrac{7}{2}X^2+2X+2) : (2X^2+X+1) = \tfrac{3}{2}X^2+1 \quad \text{Rest} \quad X+1$$
$$\underline{-(3X^4+\tfrac{3}{2}X^3+\tfrac{3}{2}X^2)}$$
$$2X^2+2X+2$$
$$\underline{-(2X^2+X+1)}$$
$$X+1$$

Fig. 2.23 Division with remainder

of two polynomials F and Q. Q is called the **quotient** and R the **remainder** in the division of F by G. Instead of the prerequisite that G is normalized, it is sufficient that $G \neq 0$ and the leading coefficient $\text{LK}(G)$ is a unit in S. Theorem 2.9.26 is then applied to F and the normalized polynomial $\widetilde{G} := \text{LK}(G)^{-1}G$. In particular, division with remainder is executable for any polynomials $G \neq 0$ over fields. The method is effective. We illustrate it with an example.

Example 2.9.27 For the polynomials $F := 3X^4 + \tfrac{3}{2}X^3 + \tfrac{7}{2}X^2 + 2X + 2$ and $G := 2X^2 + X + 1$, the division of F by G in $\mathbb{Q}[X]$ using the calculation scheme from Fig. 2.23 results in the representation $F = (\tfrac{3}{2}X^2 + 1)G + (X + 1)$, i.e., the quotient $\tfrac{3}{2}X^2 + 1$ and the remainder $X + 1$. ◊

Let S be a subring of the commutative ring T. If F and G are polynomials in $S[X]$ as in Theorem 2.9.26, it is irrelevant whether the division with remainder is performed in $S[X]$ or in $T[X]$. Specifically:

Corollary 2.9.28 *Let S be a subring of the commutative ring T and let $F, G \in S[X]$ be polynomials as in Theorem 2.9.26. Exactly then is G a divisor of F in $S[X]$, if G is a divisor of F in $T[X]$.*

For $G = X - a \in S[X]$, the division with remainder of $F \in S[X]$ by G gives the equation $F = Q \cdot (X - a) + R$ with $R = 0$ or $\deg R = 0$. In any case, $R \in S$ holds. The substitution homomorphism $\varphi_a \colon S[X] \to S$ yields $R = F(a)$. It follows:

Corollary 2.9.29 *Let $a \in S$ and $F \in S[X]$. Then there exists exactly one polynomial $Q \in S[X]$ with $F = Q \cdot (X - a) + F(a)$. In particular, $X - a$ is a divisor of F if and only if a is a root of F, i.e. $a \in Z_S(F)$.*

If one applies Corollary 2.9.29 to $Q \neq 0$, one obtains a representation $F = Q_2(X - a)^2 + Q(a)(X - a) + F(a)$. Continuing in this way, one eventually obtains the so-called **Taylor expansion** of f around a: If $F \in S[X] - \{0\}$ and $\deg F = n \in \mathbb{N}$, then

$$F = a_n(X - a)^n + a_{n-1}(X - a)^{n-1} + \cdots + a_1(X - a) + a_0$$

with uniquely determined coefficients $a_0, \ldots, a_n \in S$, $a_n \neq 0$, see also exercise 2.9.4b) and exercise 2.9.21a). – Corollary 2.9.29 has important consequences for polynomial functions. First, we prove once again the result from Theorem 2.6.21 in a more general form.

Theorem 2.9.30 *Let $F \in S[X]^*$ be a polynomial $\neq 0$ over the integral domain S. Then there exist uniquely determined, pairwise different elements $a_1, \ldots, a_r \in S$, $r \in \mathbb{N}$, positive natural numbers $\alpha_1, \ldots, \alpha_r \in \mathbb{N}^*$ and a polynomial $G \in S[X]^*$ without roots in S with*

$$F = (X - a_1)^{\alpha_1} \cdots (X - a_r)^{\alpha_r} G.$$

The factors G and $(X - a_i)^{\alpha_i}$, $i = 1, \ldots, r$, are uniquely determined up to the order. In particular, $r = |Z_S(F)| \leq \deg F$, and the polynomial function to F is certainly not the zero function, if $|S| > \deg F$ is.

Proof ((Induction over $\deg F$)) First, we note that $S[X]$ like S is an integral domain, see Proposition 2.9.20. If F has no zeros in S, then necessarily $r = 0$ and $G = F$. Otherwise, there is an $a_1 \in S$ with $F(a_1) = 0$ and according to Corollary 2.9.28 an $Q \in S[X]$ with $F = Q \cdot (X - a_1)$. Then $\deg Q = \deg F - 1$, and the induction hypothesis provides the existence of the indicated representation of F. For uniqueness, it is only necessary to note that $Z_S(F) = \{a_1, \ldots, a_r\}$ is. □

Corollary 2.9.31 *Let S be an integrity domain, $F \in S[X]$ and $n \in \mathbb{N}$. If the polynomial function $S \to S$ of F has more than n zeros, then $F = 0$ or $\deg F > n$.*

The linear polynomials $X - a$, $a \in S$, are due to $S[X]/(X - a) \xrightarrow{\sim} S$ prime elements in the polynomial algebra $S[X]$ over the integral domain S. The exponents α_i in Theorem 2.9.30 are thus the $(X - a_i)$-exponents of F. In general, $v_a(F)$ denotes the $(X - a)$-exponent of the polynomial $F \neq 0$. It is also called the **multiplicity** of the root a in F and is exactly 0 when a is not a root of F. a is called a **simple root** of F when $v_a(F) = 1$ is. We will pursue this aspect further in the next section. Here we extend Theorem 2.9.30 partially to polynomial algebras in several indeterminates. If $F \in S[X_i, i \in I]$ is a polynomial $\neq 0$ in the indeterminates $X_i, i \in I$, we denote for $k \in I$ as **partial degree** $\deg_{X_k} F$ of F with respect to X_k the degree of $F \in S[X_i, i \in I]$, considered as a polynomial in $(S[X_i, i \neq k])[X_k]$. Then the following applies:

Theorem 2.9.32 (Identity theorem for polynomials) *Let $F \in S[X_i, i \in I]^*$ be a polynomial $\neq 0$ in the indeterminates X_i, $i \in I$, over the integrity domain S. If then $N_i \subseteq S$ are subsets of S with $|N_i| > \deg_{X_i} F$, $i \in I$, then $N := \prod_i N_i \not\subseteq Z_S(F)$, i.e. there exists an $x = (x_i)_{i \in I} \in N$ with $F(x) \neq 0$. – In particular, the polynomial function defined by F is not the zero function when S is infinite.*

Proof Since only finitely many variables occur in F, we can assume that I is finite, say $I = \{1,\ldots,n\}$, and then conclude by induction over n. For $n = 0$ the statement is trivial. When concluding from $n-1$ to $n \geq 1$, let

$$F = \sum_{k=0}^{d} F_k(X_1,\ldots,X_{n-1})X_n^k, \quad d := \deg_{X_n} F \geq 0.$$

Then $F_d(X_1,\ldots,X_{n-1}) \in S[X_1,\ldots,X_{n-1}]$ is not the zero polynomial. Because of $\deg_{X_i} F_d \leq \deg_{X_i} F < |N_i|$ for $i = 1,\ldots,n-1$, there is, by induction hypothesis, a $(n-1)$-tuple $(x_1,\ldots,x_{n-1}) \in N_1 \times \cdots \times N_{n-1}$ with $F_d(x_1,\ldots,x_{n-1}) \neq 0$. Consequently, $F(x_1,\ldots,x_{n-1},X_n) = \sum_{k=0}^{d} F_k(x_1,\ldots,x_{n-1})X_n^k \in S[X_n]$ is a polynomial of degree $d < |N_n|$. According to Theorem 2.9.30, there is an $x_n \in N_n$ with $F(x_1,\ldots,x_{n-1},x_n) \neq 0$. □

Corollary 2.9.33 *Let S be an integral domain with infinitely many elements. Then for each index set I the S-algebra homomorphism $\varphi \colon S[X_i, i \in I] \to \mathrm{Abb}(S^I, S)$, which maps a polynomial $F \in S[X_i, i \in I]$ to the corresponding polynomial function $x \mapsto F(x)$, is injective.*

Example 2.9.34 Let furthermore S be a commutative ring $\neq 0$. Also let

$$G = c_0 + c_1 X + \cdots + c_{n-1} X^{n-1} + X^n \in S[X]$$

be a *normalized* polynomial of degree $n \in \mathbb{N}$ and $(G) = S[X]G$ the principal ideal generated by G in $S[X]$. According to Theorem 2.9.26 every element in the quotient algebra

$$S[x] = S[X]/(G), \quad x := [X]_G = X + (G),$$

has a uniquely determined representative $R \in S[X]$ with $R = 0$ or $\mathrm{Grad}\, R < n$. In other words: $S[x]$ *is a finite free S-algebra of rank n with S-module basis* $1 = x^0, x, \ldots, x^{n-1}$. For an arbitrary polynomial $F \in S[X]$ is

$$F(x) = R(x) = a_0 + a_1 x + \cdots + a_{n-1} x^{n-1},$$

where $R = a_0 + a_1 X + \cdots + a_{n-1} X^{n-1}$ is the remainder of the division of F by G. Note that the multiplication in $S[x]$ is determined by the operations in S and the equation

$$0 = G(x) = c_0 + c_1 x + \cdots + c_{n-1} x^{n-1} + x^n, \quad \text{i.e.} \quad x^n = -c_0 - c_1 x - \cdots - c_{n-1} x^{n-1},$$

In $S[x]$ the polynomial G thus has the root x, and in $(S[x])[X]$ the equation

$$G = H \cdot (X - x) \quad \text{with} \quad H = b_0 + b_1 X + \cdots + b_{n-1} X^{n-1} \in (S[x])[X],$$

holds, where the coefficients $b_{n-1},\ldots,b_0$ can conveniently be calculated recursively with the **Horner scheme**

$$b_{n-1} = 1, \quad b_{n-(i+1)} = c_{n-i} + b_{n-i} x, \quad i = 1,\ldots,n-1,$$

see Example 2.9.19. If $S = K$ is a field, $K[x]$ a cyclic K-algebra and x *algebraic over K*, then the relation ideal $\ker \varphi_x = \{F \in K[X] \mid F(x) = 0\}$ is the principal ideal (μ_x), where the so-called **minimal polynomial of** x

$$\mu_x$$

is the (uniquely determined) normalized polynomial of smallest degree in $\ker \varphi_x$, see corollary 2.9.17. The K-algebra $K[x] \xleftarrow{\sim} K[X]/(\mu_x)$ is therefore finite and has the dimension $n := \deg \mu_x$ with K-(vector space-)basis $1, x, \ldots, x^{n-1}$. n is then also called the **degree of** x (over K). We will deal with such finite cyclic algebras in more detail in the next section, see Example 2.10.20. ◊

Example 2.9.35 (Free quadratic algebras) Let S again be a commutative ring $\neq 0$. Continuing from the previous example, we consider the simplest non-trivial case $n = 2$. We then write G in the form[54]

$$G = X^2 - pX + q, \quad p, q \in S.$$

The so-called **free quadratic** S **-algebra** $S[x] = S[X]/(G)$ has the S-basis $1, x$, and because of $x^2 = px - q$ for $a, b, c, d \in S$

$$(a + bx)(c + dx) = ac + (ad + bc)x + bdx^2 = (ac - bdq) + (ad + bc + bdp)x.$$

In $(S[x])[X]$ $G = X^2 - pX + q = (X - x)(X - (p - x))$ applies. Since $p - x$ is also a root of G, (due to the universal property of $S[x]$) an S-algebra homomorphism $x \mapsto p - x$ is defined by $\kappa : S[x] \to S[x]$. Because of $\kappa^2(x) = \kappa(p - x) = p - \kappa(x) = x$ κ is an involutory S-algebra automorphism of $S[x]$. It is called the **conjugation** of $S[x]$. For

$$z = a + bx \in S[x], \quad a, b \in S, \quad \text{is} \quad \bar{z} := \kappa(z) = a + b\kappa(x) = (a + bp) - bx.$$

It follows $\mathrm{Sp}(z) := z + \bar{z} = 2a + bp \in S$ and $\mathrm{N}(z) := z\bar{z} = a^2 + pab + qb^2 \in S$ as well as

$$(X - z)(X - \bar{z}) = X^2 - \mathrm{Sp}(z)X + \mathrm{N}(z) \in S[X].$$

The **trace(mapping)** $\mathrm{Sp} = \mathrm{Sp}_S^{S[x]} : S[x] \to S$, $z \mapsto \mathrm{Sp}(z) = z + \bar{z}$, is S-linear, and the **norm(mapping)** $\mathrm{N} = \mathrm{N}_S^{S[x]} : S[x] \to S$, $z \mapsto \mathrm{N}(z) := z\bar{z}$, is a homomorphism of the multiplicative monoids of $S[x]$ or S. As an important application of the norm, we note: *An element $z = a + bx \in S[x]$ is exactly a unit in $S[x]$, when $\mathrm{N}(z)$ is a unit in S.* In this case,

$$z^{-1} = \frac{\bar{z}}{\mathrm{N}(z)} = \frac{a + bp}{a^2 + pab + qb^2} - \frac{b}{a^2 + pab + qb^2} x.$$

---

[54] Note the convention about the sign of the coefficient of X, which may differ from school teaching.

In the case $p = 0$, we speak of a **purely quadratic** S-algebra. Then $x = \sqrt{-q}$ is a square root of $-q$ and
$$S[\sqrt{-q}] \xleftarrow{\sim} S[X]/(X^2 + q).$$
The multiplication in $S[\sqrt{-q}]$ is
$$(a + b\sqrt{-q})(c + d\sqrt{-q}) = (ac - bdq) + (ad + bc)\sqrt{-q},$$
and for $z = a + b\sqrt{-q}$ simply applies $\bar{z} = a - b\sqrt{-q}$ and
$$\mathrm{Sp}(z) = 2a, \quad \mathrm{N}(z) = a^2 + qb^2,$$
$$z^{-1} = \frac{a}{a^2 + qb^2} - \frac{b}{a^2 + qb^2}\sqrt{-q}, \quad \text{if} \quad a^2 + qb^2 \in S^\times. \quad \diamondsuit$$

If 2 is a non-zero divisor in S, then $z = \bar{z}$ is equivalent to $z \in S$, i.e., it is Fix$(\kappa, S[\sqrt{-q}]) = S$, and $z = -\bar{z}$, i.e., $\mathrm{Sp}(z) = 0$, is equivalent to $z \in S\sqrt{-q}$. For an integral domain S with $2 \neq 0$ and an element $z = a + b\sqrt{-q} \in S[\sqrt{-q}]$, $z \neq 0$, $\mathrm{Sp}(z) = 2a = 0$ is also equivalent to $z \notin S$ and $z^2 \in S$. This follows from $z^2 = a^2 - qb^2 + 2ab\sqrt{-q}$.

If S is arbitrary and $-q$ is a square already in S, for example $-q = s^2$, $s \in S$, so one must not confuse the quadratic algebra $S[\sqrt{-q}]$ with the algebra $S[s] = S$. Due to the universal property of $S[\sqrt{-q}]$ there is the surjective substitution homomorphism $S[\sqrt{-q}] \to S$, $\sqrt{-q} \mapsto s$, whose kernel is already generated as an S-module by $s - \sqrt{-q}$. For $q = 0$, one usually denotes $\sqrt{0}$ with ε and calls the algebra $S[\varepsilon] \xleftarrow{\sim} S[X]/(X^2)$ the **algebra of dual numbers** over S.[55]

The case $q = 1$ provides the historically first examples of purely quadratic algebras. For the base ring S, this is the algebra
$$\mathbb{C}_S := S[\sqrt{-1}] \xleftarrow{\sim} S[X]/(X^2 + 1)$$
of the **complex numbers over** S. Following Euler, we set
$$\mathrm{i} := \sqrt{-1}$$
and speak of the **imaginary unit** (since for a long time it was not known what kind of number this should be). It is $\mathrm{i}^2 = -1$ and $\mathrm{i}^{-1} = -\mathrm{i}$. For a complex number $z = a + b\mathrm{i} \in S[\mathrm{i}]$, $a, b \in S$, accordingly $\Re z := a$ is the **real part** and $\Im z := b$ is the **imaginary part** of z. It applies
$$\bar{z} = a - b\mathrm{i}, \quad 2a = 2\Re z = z + \bar{z} = \mathrm{Sp}(z), \quad 2b = 2\Im z = \mathrm{i}^{-1}(z - \bar{z}) = \mathrm{i}(\bar{z} - z),$$
$$\mathrm{N}(z) = a^2 + b^2, \quad z^{-1} = \frac{a}{a^2 + b^2} - \frac{b}{a^2 + b^2}\mathrm{i}, \quad \text{if} \quad a^2 + b^2 \in S^\times.$$
The multiplication in $S[\mathrm{i}]$ is (due to $\mathrm{i}^2 = -1$) given by

---

[55] In choosing the letter "ε", one certainly thought of the ε in analysis, especially in differential calculus. There, ε^2 is (often) "negligibly small".

$$(a+bi)(c+di) = (ac-bd) + (ad+bc)i, \quad a,b,c,d \in S.$$

If 2 is a non-zero divisor in S, then the elements $z \in S$ are characterized by $z = \bar{z}$ and the elements $z \in S$i of the "imaginary axis" Si by $z = -\bar{z}$, i.e. $\mathrm{Sp}(z) = 0$. $S[i]$ is exactly a field when S is a field in which $a^2 + b^2 = 0$ only holds for $a = b = 0$, $a,b \in S$. This is obviously equivalent to S being a field in which -1 is not a square. For the prime field F_p, $p \in \mathbb{P}$, this is exactly the case when $p \equiv 1 \bmod 4$ is. Because for $p > 2$, $\mathbf{F}_p^\times$ is a group of order $p - 1$, in which -1 is the only element of order 2. -1 is therefore exactly a square when $\mathbf{F}_p^\times$ has an element of order 4. Since the 2-primary component of $\mathbf{F}_p^\times$ is cyclic (even $\mathbf{F}_p^\times$ is cyclic!) this is equivalent to $4 \mid (p-1)$, see also exercise 2.2.29. One can directly specify a square root of -1 for $p \equiv 1 \bmod 4$. This is $m! \in \mathbf{F}_p^\times$ with $m := \frac{1}{2}(p-1)$. Because of $\frac{1}{2}(p-1) \equiv 0 \bmod 2$, $(m!)^2 = (-1)^{(p-1)/2}(p-1)! = (p-1)! = -1$ is. For $S = \mathbb{R}$ is the field

$$\mathbb{C} := \mathbb{C}_\mathbb{R} = \mathbb{R}[i]$$

the **complex number field** par excellence. It contains the quadratic algebras $\mathbb{C}_S = S[i] = S + S$i for every subring $S \subseteq \mathbb{R}$ of $\mathbb{R}$. For a description of the multiplicative group $\mathbb{C}^\times$ of $\mathbb{C}$, refer to Example 2.2.16 (2) and also to Sect. 3.5.

In conclusion, we note that *every quadratic S-algebra is purely quadratic if $2 \in S^\times$ is a unit in S*. If $S[x]$ is a quadratic S-algebra with defining equation $x^2 - px + q = 0$, then by **completing the square**

$$0 = x^2 - px + q = \left(x - \frac{1}{2}p\right)^2 + \frac{1}{4}(4q - p^2) \quad \text{resp.} \quad \left(x - \frac{1}{2}p\right)^2 = \frac{1}{4}(p^2 - 4q).$$

we have $S[x] = S[\tilde{x}]$ for $\tilde{x} := 2(x - \frac{1}{2}p)$, and $\tilde{x}$ satisfies the purely quadratic equation $\tilde{x}^2 = p^2 - 4q$. Thus,

$$S[x] \cong S\left[\sqrt{p^2 - 4q}\right].$$

$p^2 - 4q \in S$ is called the **discriminant** of the quadratic equation $x^2 - px + q = 0$ or also of the polynomial $X^2 - pX + q \in S[X]$. In particular, we have:

Proposition 2.9.36 (Babylonian Solution Formula = p-q-Formel) *Let S be a commutative ring with $2 \in S^\times$ and $p, q \in S$. The roots of the polynomial $X^2 - pX + q \in S[X]$ in any S-algebra A are the elements $\frac{1}{2}(p + u)$, where $u \in A$ is the set of square roots from the discriminant $p^2 - 4q$ of the polynomial $X^2 - pX + q$ in A.*

Exercises

Exercise 2.9.1 Let A be an S-algebra over the commutative ring $S \neq 0$ and V be an A-left-right-bimodule with $sv = vs$ for all $s \in S$, $v \in V$. Then the direct sum $A \oplus V$ with the multiplication

$$(x, v) \cdot (y, w) = (xy, xw + vy), \quad x, y \in A, \ v, w \in V,$$

is an S-algebra, in which $V = \{0\} \oplus V$ is a two-sided ideal with $(A \oplus V)/V \xrightarrow{\sim} A$ and $V^2 = 0$. Furthermore, $(A \oplus V)^\times = A^\times \oplus V$. (It is $(x, v)^{-1} = (x^{-1}, -x^{-1}vx^{-1})$ for $x \in A^\times$, $v \in V$. – The S-algebra $A \oplus V$ is called the **idealization** of V. If A is commutative, then the idealization for any A-module V is defined and again commutative.)

Exercise 2.9.2 Let M be a regular totally ordered monoid with neutral element ι and $B = \sum_{\sigma \in M}^{\oplus} B_\sigma$ a M-graded domain.

a) Every left or right divisor of a homogeneous element $\neq 0$ of B is also homogeneous. In particular, all units of B are homogeneous.
b) Let B be also commutative (thus an integral domain), $\mathfrak{p} \subseteq B$ a homogeneous prime ideal in B and $b = \sum_{\sigma \in M} b_\sigma \in B$ an element $\neq 0$ with leading form b_ω and initial form b_α, $\alpha \leq \omega$. If $b_\sigma \in \mathfrak{p}$ holds for all $\sigma \neq \omega$, $b_\omega \notin \mathfrak{p}$ and $b_\alpha \notin \mathfrak{p}^2$, then every divisor of b in B is homogeneous. Specifically, if p is a homogeneous prime element $\neq 0$ in B and $p \mid b_\sigma$ holds for $\sigma \neq \omega$ and $p \nmid b_\omega$ and $p^2 \nmid b_\alpha$, then all divisors of b in B are homogeneous. (This so-called **Eisenstein's Lemma** (after G. Eisenstein (1823–1852)) can be varied in many ways.)

Exercise 2.9.3 Let S be a commutative ring $\neq 0$ and P the freely generated S-algebra $S\langle X_i, i \in I\rangle$ or the polynomial algebra $S[X_i, i \in I]$ in the indeterminates X_i, $i \in I$. For $x = (x_i) \in S^I$, the kernel of the substitution homomorphism $\varphi_a : P \to S$, $X_i \mapsto x_i$, $i \in I$, (as a two-sided ideal) is generated by the linear polynomials $X_i - x_i$, $i \in I$.

Exercise 2.9.4 Let S be a commutative ring $\neq 0$ and $P := S[X_i, i \in I]$ the polynomial algebra over S in the indeterminates X_i, $i \in I$. Further, let $F = \sum_{\nu \in \mathbb{N}^{(I)}} a_\nu X^\nu \in P$.

a) It is $\nu! a_\nu = \left(D_X^\nu F\right)(0)$, $\nu \in \mathbb{N}^{(I)}$. (We recall the definition $\nu! := \prod_i \nu_i!$. – It suffices to verify the formula for monomials $F = X^\mu$, $\mu \in \mathbb{N}^{(I)}$.)
b) Let $c = (c_i) \in S^I$. The polynomials $(X - c)^\nu = \prod_i (X_i - c_i)^{\nu_i}$, $\nu \in \mathbb{N}^{(I)}$, form an S-module basis of P. (The substitution homomorphism $P \to P$, $X_i \mapsto X_i - c_i$, $i \in I$, is an S-algebra automorphism of P, a so-called **translation automorphism** of P.) If $F = \sum_{\nu \in \mathbb{N}^{(I)}} b_\nu (X - c)^\nu$, then the **Taylor formula**

$$\nu! b_\nu = \left(D_X^\nu F\right)(c).$$

applies (If for example $\mathbb{Q} \subseteq S$, then $b_\nu = (1/\nu!)\left(D_X^\nu F\right)(c)$, $\nu \in \mathbb{N}^{(I)}$ applies.)
c) Let $r \in \mathbb{N}^*$ and $n \in \mathbb{N}$. Prove using the Taylor formula in $\mathbb{Z}[X_1, \ldots, X_r]$ the universal polynomial formula

$$(X_1 + \cdots + X_r)^n = \sum_{\nu \in \mathbb{N}^r, |\nu| = n} \binom{n}{\nu} X^\nu$$

2.9 Algebras

as well as the two universal polarization formulas (see Ex. 2.6.7)

$$2^{r-1} r! X_1 \cdots X_r = \sum_{\varepsilon} \varepsilon_2 \cdots \varepsilon_r (X_1 + \varepsilon_2 X_2 + \cdots + \varepsilon_r X_r)^r$$

(on the right side, sum over all sign tuples $\varepsilon = (\varepsilon_2, \ldots, \varepsilon_r) \in \{1, -1\}^{r-1}$) and

$$(-1)^r r! X_1 \cdots X_r = \sum_{H \subseteq \{1,\ldots,r\}} (-1)^{|H|} X_H^r = \sum_e (-1)^{e_1 + \cdots + e_r} (e_1 X_1 + \cdots + e_r X_r)^r$$

(in the last sum, e runs through all tuples $(e_1, \ldots, e_r) \in \{0, 1\}^r$).

Exercise 2.9.5 Let S be a commutative ring $\neq 0$ and $P := S[X_i, i \in I]$ the polynomial algebra over S in the indeterminates X_i, $i \in I$. The mapping $\mathrm{Der}_S P \xrightarrow{\sim} P^I$, $\delta \mapsto \bigl(\delta(X_i)\bigr)_{i \in I}$, is a P-module isomorphism. The I-tuple $(G_i) \in P^I$ corresponds to the S-derivation $\delta: F \mapsto \sum_{i \in I} (\mathrm{D}_{X_i} F) G_i$. In particular, $\mathrm{Der}_S P$ is a free P-module of rank $|I|$ with basis D_{X_i}, $i \in I$, when I is finite.

Note It follows in particular that for sets I,J with $|I| \neq |J|$ the S-algebras P and $Q := S[X_j, j \in J]$ are not isomorphic if one of the two sets is finite. This also applies if both sets I and J are infinite, and then follows from $\mathrm{Rang}_S P = |\mathbb{N}^{(I)}| = |I|$ for infinite I. Note that the algebras P and Q can be isomorphic as rings, even if $|I| \neq |J|$ is. For infinite I, for example, P and $P[Y_k, k \in K]$ are isomorphic rings when $|K| \leq |I|$ is.

Exercise 2.9.6 Let S be a commutative ring $\neq 0$. The polynomial algebra $S[X_i, i \in I]$ is Noetherian exactly when S is Noetherian and I is finite.

Exercise 2.9.7 Let S be a commutative ring $\neq 0$ and $P := S\langle X_i, i \in I\rangle$ the freely generated algebra in the indeterminates X_i, $i \in I$, over S.

a) A P-module is the same as a pair $V = (V, (f_i)_{i \in I})$, consisting of an S-module V and a family f_i, $i \in I$, of S-endomorphisms of V. (f_i is the homothety ϑ_{X_i}, $i \in I$.) If $W = (W, (g_i)_{i \in I})$ another such pair, then a P-module homomorphism $V \to W$ is the same as an S-module homomorphism $h: V \to W$ with $h \circ f_i = g_i \circ h$, $i \in I$. In particular, V and W are exactly P-isomorphic when there is an S-isomorphism $h: V \to W$ with $g_i = h \circ f_i \circ h^{-1}$, $i \in I$. (It is an interesting problem, for a field $S = K$ to classify the P-modules whose K-dimension is a fixed number $m \in \mathbb{N}$. This involves describing the orbit space $(\mathrm{End}_K K^m)^I \backslash \mathrm{Aut}_K K^m = \mathrm{M}_m(K)^I \backslash \mathrm{GL}_m(K)$ of the operation of $\mathrm{Aut}_K K^m = \mathrm{GL}_m(K)$ on $(\mathrm{End}_K K^m)^I = \mathrm{M}_m(K)^I$ by conjugation: $(\mathrm{B}, (\mathrm{A}_i)_{i \in I}) \mapsto (\mathrm{B A}_i \mathrm{B}^{-1})_{i \in I}$, $\mathrm{B} \in \mathrm{GL}_m(K)$, $(\mathrm{A}_i)_{i \in I} \in \mathrm{M}_m(K)^I$, in a clear way. The case $|I| = 1$ is one of the main subjects of Linear Algebra and is essentially dealt with in Theorem 2.10.17, see volumes 3 and 4 on Linear Algebra.)

b) Formulate and prove the corresponding statements as in a) for the polynomial algebra $S[X_i, i \in I]$. (The $f_i: V \to V$, $i \in I$, must commute pairwise!)

Exercise 2.9.8 Let K be a field, I a well-ordered index set and P the freely generated K-algebra $K\langle X_i, i \in I\rangle$ or the polynomial algebra $K[X_i, i \in I]$ each with the homogeneous lexicographic order of the monomials. Furthermore, let $\mathfrak{a}$ be a two-sided ideal in P and $\mathrm{LF}(\mathfrak{a})$ the leading form ideal of $\mathfrak{a}$. Then the residue classes of those monomials in P that do *not* belong to $\mathrm{LF}(\mathfrak{a})$ form a K-vector space basis of the residue class algebra $P/\mathfrak{a}$.

Exercise 2.9.9 Let A be an S-algebra. For $x, y \in A$ the term $[x,y] := xy - yx$ is called the **commutator** of x and y. The connection $A \times A \to A$, $(x,y) \mapsto [x,y]$, on A is called the **Lie bracket**. It is S-bilinear, and $[x,x] = 0$ holds for all $x \in A$. (Such a bilinear mapping is called **alternating**. Because of $0 = [x+y, x+y] = [x,x] + [x,y] + [y,x] + [y,y] = [x,y] + [y,x]$, it follows that $[y,x] = -[x,y]$.) For $x, y, z \in A$, $[z, xy] = [z,x]y + x[z,y]$ applies, i.e. $\delta_z : A \to A$, $x \mapsto [z,x]$, is an S-derivation of A. These derivations δ_z, $z \in A$, are called the **inner derivations** of A.

a) Verify the so-called **Jacobi identity**: For all $x, y, z \in A$ the following holds
$$[x,[y,z]] + [y,[z,x]] + [z,[x,y]] = 0.$$
(In the Jacobi identity, the arguments x,y,z are cyclically swapped! – Any arbitrary S-module L with an alternating S-bilinear operation $L \times L \to L$, $(x,y) \mapsto [x,y]$, that fulfills the Jacobi identity, is called an **S-Lie algebra** (after S. Lie (1842–1899)). The Jacobi identity for the operation $[-,-]$ on the Lie algebra L can also be interpreted as follows: The S-linear left translation $\delta_z : L \to L$, $x \mapsto [z,x]$, (and thus also the right translation $x \mapsto [x,z] = -\delta_z(x)$) is for each $z \in L$ an S-derivation on L. Every (ordinary) S-algebra A with the above Lie bracket $[-,-]$ is an S-Lie algebra. It is called the **associated Lie algebra** to A, and is denoted by $[A]$. A is commutative exactly when the Lie bracket $[-,-]$ on A is identically 0. In general, a Lie algebra L is called **commutative**, when its Lie product is identically 0. – For the general algebra concept, see the definition following Lemma 2.9.4.)

b) The derivation module $\mathrm{Der}_S A$ is an S-Lie subalgebra of the $\mathrm{End}_S A$ associated S-Lie algebra $[\mathrm{End}_S A]$, and the mapping $[A] \to \mathrm{Der}_S A$, $z \mapsto \delta_z$, is an S-Lie algebra homomorphism, whose kernel is the center $Z(A)$ of A.

Exercise 2.9.10 Let A be an algebra over the commutative ring $S \neq 0$ and $\delta : A \to A$ an S-derivation.

a) **(Quotient rules)** For $x \in A$ and $y \in A^\times$ applies $\delta(y^{-1}) = -y^{-1}\delta(y)y^{-1}$ as well as
$$\delta(y^{-1}x) = y^{-1}(\delta(x) - \delta(y)y^{-1}x) \quad \text{resp.} \quad \delta(xy^{-1}) = (\delta(x) - xy^{-1}\delta(y))y^{-1}.$$
If $y \in Z(A) \cap A^\times = Z(A)^\times$, then $\delta(x/y) = (\delta(x)y - x\delta(y))/y^2$.

2.9 Algebras

b) Let $T \subseteq Z(A)$ be a central submonoid of the multiplicative monoid of A and $\iota_T: A \to A_T$ the canonical S-algebra homomorphism from A into the S-algebra $A_T = \{x/t \mid x \in A, t \in T\}$ of fractions of A with respect to T. Then there exists exactly one S-derivation $\delta_T: A_T \to A_T$ with $\iota_T \circ \delta = \delta_T \circ \iota_T$. It is

$$\delta_T\left(\frac{x}{t}\right) = \frac{\delta(x)t - x\delta(t)}{t^2}, \quad x \in A, \ t \in T.$$

In particular, the partial derivatives D_{X_i}, $i \in I$, of the polynomial algebra $P := S[X_i, i \in I]$ can be uniquely extended to S-derivations on the total quotient ring $Q(P) = Q(S)(X_i, i \in I)$ of P. These extensions are also denoted by D_{X_i}.

Exercise 2.9.11 Let K be a field. The K-algebra automorphisms of $K[X]$ are exactly the substitution homomorphisms $X \mapsto aX + b$, $a, b \in K$, $a \neq 0$. The group $\mathrm{Aut}_{K\text{-Alg}} K[X]$ of K-algebra automorphisms of $K[X]$ is therefore anti-isomorphic and thus isomorphic to the affine group $\mathrm{A}_1(K) = K \rtimes K^\times$ of K, cf. Example 2.6.19. (The K-automorphism group of a polynomial algebra $K[X_1, \ldots, X_n]$, $n \geq 2$, in more than one variable is much harder to describe and still a current research topic. For an important subgroup see the next Exercise.)

Exercise 2.9.12 Let K be a field and L_i, $i \in I$, a family of homogeneous polynomials of degree 1 in the polynomial algebra $P := K[Y_j]_{j \in J}$. The substitution homomorphism $Y_i \mapsto L_i$, $i \in I$, from $K[Y_i]_{i \in I}$ to P is exactly injective or surjective or bijective when the L_i, $i \in I$, are linearly independent or a generating system or a basis in the K-vector space P_1 of all homogeneous polynomials of degree 1 in $K[Y_j]_{j \in J}$. In particular, in the case $I = J$ the substitution endomorphism $P \to P$, $Y_j \mapsto L_j$, $j \in J$, is exactly a K-algebra automorphism when its restriction to P_1 is a K-vector space automorphism of P_1. (In this way, the linear group $\mathrm{GL}_K(P_1) = \mathrm{Aut}_K(P_1)$ is always considered as a subgroup of $\mathrm{Aut}_{K\text{-Alg}} P$. Together with the translation automorphisms from exercise 2.9.4b) they generate the so-called group of **affine** K-algebra automorphisms of P.)

Exercise 2.9.13 For $m \in \mathbb{N}$ let $P_m := K[X_1 \ldots, X_m]$ be the polynomial algebra in m indeterminates over the field K. If $\varphi: P_m \to P_n$ is an injective (or surjective) K-algebra homomorphism, then $m \leq n$ (or $m \geq n$). In particular, $m = n$ when φ is an isomorphism. (If $\mathrm{Grad}\, \varphi(X_i) \leq d$, $i = 1, \ldots, m$, then $\mathrm{Grad}\, \varphi(F) \leq d \cdot \mathrm{Grad}\, F$ for all $F \in P_m$. Furthermore, use that the polynomials in P_m of degree $\leq r \in \mathbb{N}$ form a K-vector space of dimension $\binom{r+m}{m}$. – The case that φ is surjective is reduced to the case that φ is injective. – Another proof for $m = n$, if φ is a K-algebra isomorphism, can be found in Exercise 2.9.5. If $m \neq n$, P_m and P_n are also not isomorphic as rings; because every ring isomorphism induces an automorphism of K due to $K^\times = P_m^\times = P_n^\times$.)

Exercise 2.9.14 Let K be an infinite field and F,G be polynomials in $K[X_i, i \in I]$. If $F \neq 0$ and G vanishes on $K^I - Z_K(F)$, then $G = 0$.

Exercise 2.9.15 Let K be an infinite field and V a K-vector space. Every linearly independent family $f_i \in V^*$, $i \in I$, of K-linear forms $V \to K$ is algebraically independent in the K-algebra K^V of the K-valued functions on V. (One can reduce to the case that V is finite-dimensional. – Often, the K-subalgebra of the K-algebra K^V of all K-valued functions on V generated by the K-linear forms $V \to K$ is called the algebra of **polynomial functions** on V. For $V = K^I$ and *finite I*, this agrees with the usual definition.)

Exercise 2.9.16 Let K be a field and A a K-algebra. Furthermore, let $x \in A^*$ be a non-zero divisor and integral (i.e., algebraic) over K. Then $x \in A$ is even a unit in A and $x^{-1} \in K[x]$. (The multiplication with x is injective and thus bijective on the finite K-algebra $K[x]$.) Also determine the minimal polynomial $\mu_{x^{-1}}$ of x^{-1} using the minimal polynomial μ_x of x. (The constant term of μ_x is $\neq 0$.) In particular, A is a division ring when A is a ring and integral over K.

Exercise 2.9.17 Let A be an algebra over the commutative ring $S \neq 0$.

a) For $x \in A$ the following statements are equivalent: (i) x is integral over S. (ii) $S[x]$ is a finite S-algebra. (iii) $S[x]$ is contained in a finite S-subalgebra of A. – In particular, A is integral over S when A is a finite S-algebra. (The implication (iii) $\Rightarrow$ (ii) is trivial when S is Noetherian. The general case should be reduced to this, see the method used in exercise 2.8.16 c). But one could also conclude directly. – If S is Noetherian, then (iii) is of course equivalent to the following condition: (iv) $S[x]$ is contained in a finite S-sub*module* of A. In general, however, conditions (iii) and (iv) are *not* equivalent. Example?)

b) If A is commutative, then the elements of A over S form an S-subalgebra of A. (If $x, y \in A$ is integral over S, then $S[x, y]$ is a finite S-subalgebra of A.)

c) If S itself is an integral algebra over the commutative ring T and if A is integral over S, then A is also integral over T (**Transitivity of integrality**) . If A is finite over S and S is finite over T, then A is also finite over T. (A subalgebra of a finite S-algebra does not need to be finite over S. For an example, one can use Exercise 2.9.1.)

d) Let A be integral over S. If $x \in A^\times$, then $x^{-1} \in S[x]$. If $A' \subseteq A$ is an S-subalgebra of A, then $A'^\times = A' \cap A^\times$. In particular, A' is a division ring if A is a division ring. Conversely, if A' is a division ring and A is free of zero divisors, then A is also a division ring. (According to the first statement in d), the kernel of the structure homomorphism $S \to A' \subseteq A$ is a maximal ideal in S. Now use exercise 2.9.16.)

Exercise 2.9.18 Let K be a field. Provide a K-algebra A with algebraic elements $x, y \in A$ for which both $x + y$ and xy are transcendental over K. (Take, for example, $A := \text{End}_K V$, where V is an infinite-dimensional K-vector space, such as $V = K^{(\mathbb{N})}$, or a suitable group algebra, such as $K[\mathbf{D}_0]$.)

2.9 Algebras

Exercise 2.9.19 Let $K \subseteq L$ be an extension of fields. Then the elements of L that are algebraic over K form a subfield of L, which includes K. It is called the **algebraic closure** or the **algebraic hull** of K in L. If K is finite, then the algebraic hull of K in L is at most countable. If K is infinite, then K and the algebraic hull of K in L have the same cardinality. (The polynomial ring $K[X]$ has the same cardinality as K for infinite K.)

Note The elements over the prime field of L are called **algebraic** per se or **absolutely algebraic**. The absolutely algebraic elements of L thus form a countable subfield of L. In particular, the subfield $\overline{\mathbb{Q}}$ of the (absolutely) algebraic numbers in $\mathbb{C}$ is countable, and the set $\mathbb{C} - \overline{\mathbb{Q}}$ of the (over $\mathbb{Q}$) transcendental complex numbers has the power $\aleph$ of the continuum. This is Cantor's proof from 1874 for the existence of transcendental complex numbers, cf. Cantor, G.: On a property of the set of all real algebraic numbers, J. for pure and applied Math. **74**, 258–262 (1874). He does not provide a single complex transcendental number explicitly. Such numbers were first given in 1844 by J. Liouville (1809–1882). (Since $\mathbb{C}$ is algebraically closed according to the Fundamental Theorem of Algebra 3.9.7, $\overline{\mathbb{Q}}$ is also algebraically closed.)

Exercise 2.9.20 Let $K \subseteq L \subseteq M$ be extensions of fields. If M is algebraic over L and L is algebraic over K, then M is also algebraic over K with $\mathrm{Dim}_K M = \mathrm{Dim}_K L \cdot \mathrm{Dim}_L M$. (The formula holds for any field extensions $K \subseteq L \subseteq M$.)

Exercise 2.9.21 Let S be a commutative ring $\neq 0$.

a) $G \in S[X]$ is a normalized polynomial of degree $m \geq 1$. For every polynomial $F \neq 0$, there are uniquely determined polynomials $P_0, \ldots, P_r$ with $P_r \neq 0$ and
$$F = P_0 + P_1 G + \cdots + P_r G^r, \quad P_i = 0 \text{ or } \deg P_i < m, \ i = 0, \ldots r.$$
(This development corresponds to the g-al development of natural numbers, cf. Example 1.7.6, and is called the **G-al development** of F. If $G = X - c$ is of degree 1, it is the Taylor development of F at c.)

b) Let $F \in S[X]$ be a polynomial of degree n and $c \in S$. The coefficients $b_0, \ldots, b_{n-1}$ of the quotient $Q = b_{n-1} + b_{n-2} X + \cdots + b_0 X^{n-1}$ in $F = F(c) + Q \cdot (X - c)$ are the values $F_0(c), \ldots, F_{n-1}(c)$ in the Horner scheme for the calculation of $F(c) = F_n(c)$ according to Example 2.9.19. If you continue this process with the polynomial Q instead of F, you successively obtain the coefficients of the Taylor development of F at c. – Develop the polynomial $X^4 - 3X^3 + 5X^2 - X + 2 \in \mathbb{Z}[X]$ with the Horner scheme around $c = 2$ and around $c = -1$.

Exercise 2.9.22 Let S be a commutative ring $\neq 0$ and $G = c_0 + c_1 X + \cdots + c_{n-1} X^{n-1} + X^n \in S[X]$ be a normalized polynomial of degree $n \in \mathbb{N}^*$. Furthermore, let n be a unit in S. In the free residue class algebra $S[x] = S[X]/(G)$ of rank n, the element $\widetilde{x} := x + \frac{1}{n}c_{n-1}$ satisfies an equation $\widetilde{c}_0 + \cdots + \widetilde{c}_{n-2}\widetilde{x}^{n-2} + \widetilde{x}^n = 0$ with coefficients $\widetilde{c}_i$, $i = n-2, \ldots, 0$, in S. Thus, it is $S[x] = S[\widetilde{x}] \xleftarrow{\sim} S[X]/(\widetilde{G})$, where the coefficient at X^{n-1} in the normalized polynomial $\widetilde{G} := \widetilde{c}_0 + \cdots + \widetilde{c}_{n-2}X^{n-2} + X^n \in S[X]$ vanishes. The transition from G to $\widetilde{G}$ is called the (linear) **Tschirnhaus(en) transformation** (after W. Tschirnhaus(en) (1651–1708)).

Exercise 2.9.23 Let S be a non-zero Noetherian commutative ring and G_j, $j \in J$, an arbitrary family of polynomials in the polynomial algebra $P := S[X_1, \ldots, X_n]$. Then there exists a *finite* subset $J' \subseteq J$ with the following property: For every commutative S-algebra A, $Z_A(G_j, j \in J) = \{x \in A^n \mid G_j(x) = 0, j \in J\} = Z_A(G_j, j \in J')$ is true. (P is a Noetherian ring and the ideal $\sum_j PG_j \subseteq P$ is finitely generated.)

Exercise 2.9.24 Determine the quotient and the remainder when dividing $X^m - 1$ by $X^n - 1$ in $\mathbb{Z}[X]$ for $m, n \in \mathbb{N}^*$. (Cf. exercise 1.7.11.)

2.10 Principal Ideal Domains and Factorial Integral Domains

The ring $\mathbb{Z}$ of integers was one of the first rings to be consciously studied as a ring. Accordingly, many terms of ring theory are derived from properties of the ring $\mathbb{Z}$. For example, in $\mathbb{Z}$ every ideal is a principal ideal, cf. Theorem 2.1.18. This leads to the following terms, which were already mentioned earlier:

Definition 2.10.1 A ring A is called a **principal ideal ring**, if A is commutative and every ideal in A is a principal ideal Ab with $b \in A$. If A is even an integral domain, then A is called a **principal ideal domain**.

$\mathbb{Z}$ is therefore a principal ideal domain. In proving that every ideal in $\mathbb{Z}$ is a principal ideal, the division with remainder in $\mathbb{Z}$ was used. This can also be axiomatized:

Definition 2.10.2 Let A be an integral domain. A function $\varphi: A^* \to \mathbb{N}$ is called a **Euclidean degree function** on A, if for all $a \in A$ and $b \in A^* = A - \{0\}$ elements $q, r \in A$ exist with

$$a = qb + r \quad \text{and} \quad r = 0 \text{ or } \varphi(r) < \varphi(b).$$

A is called a **Euclidean domain**, if A has a Euclidean degree function.

Note that the uniqueness of the quotient q and the remainder r is not required. $\mathbb{Z}$ is a Euclidean domain with the absolute value $\mathbb{Z}^* \to \mathbb{N}$, $a \mapsto |a|$, as a Euclidean degree function. As for $A = \mathbb{Z}$ one proves:

2.10 Principal Ideal Domains and Factorial Integral Domains

Theorem 2.10.3 *Every Euclidean domain A is a principal ideal domain.*

Proof Let $\varphi: A^* \to \mathbb{N}$ be a Euclidean degree function on A and $\mathfrak{a} \subseteq A$ an ideal $\neq 0$ in A. Every element $c \in \mathfrak{a} - \{0\}$, for which $\varphi(c)$ is minimal, then generates $\mathfrak{a}$. If $a \in \mathfrak{a}$ and $a = qc + r$ with $r = 0$ or $\varphi(r) < \varphi(c)$, the second case cannot occur here because of $r = a - qc \in \mathfrak{a}$. So $r = 0$ and thus $\mathfrak{a} \subseteq Ac \subseteq \mathfrak{a}$. □

A large class of Euclidean domains is provided by the polynomial algebras $K[X]$ in an indeterminate X over fields K. The degree deg: $K[X]^* \to \mathbb{N}$, $F \mapsto \deg F$, is a Euclidean degree function. This is stated (among other things) in Theorem 2.9.26. *The polynomial algebras $K[X]$, K being fields, are therefore principal ideal domains*, which was already noted earlier as Corollary 2.9.17.

The main theorem of elementary number theory 1.7.11 states that the (regular commutative) multiplicative monoid $\mathbb{Z}^*$ is factorial, i.e., that $\mathbb{Z}^*/\mathbb{Z}^\times$ is a free commutative monoid, cf. Definition 2.3.23. This leads to the following general definition:

Definition 2.10.4 Let A be an integral domain. A is called **factorial**, if the multiplicative monoid $A^* = A - \{0\}$ is factorial.

Note that the property of an integral domain A being factorial only concerns the multiplicative structure of A. For factorial monoids, we refer to the discussion in Example 2.3.20, in particular, greatest common divisors (gcd) and least common multiples (lcm) are defined in factorial integral domains A. For $a, b \in A^*$, $A \gcd(a, b)$ is the smallest *principal* ideal that includes Aa and Ab, and $A \operatorname{lcm}(a, b) = Aa \cap Ab$ is the largest ideal that is contained in both Aa and Ab. In particular, $Aa + Ab = A \gcd(a, b)$, if $Aa + Ab$ is a principal ideal.[56] Furthermore, let $\mathbb{P} = \mathbb{P}_{A^*}$ as before denote a representative system for the associativity classes of the prime elements of the multiplicative regular monoid A^* of an integral domain A. Now let A be factorial. Then every element $a \in A^*$ has a unique prime factor decomposition

$$a = \varepsilon \prod_{p \in \mathbb{P}} p^{v_p(a)}$$

with a unit $\varepsilon \in A^\times$ and the tuple $(v_p(a)) \in \mathbb{N}^{(\mathbb{P})}$ of the p-**exponents** of a. If you let for $(v_p(a))$ all tuples from $\mathbb{Z}^{(\mathbb{P})}$ to, you get a unique representation of the elements of the multiplicative group $Q(A)^\times$ of the quotient field $Q(A)$ of A. In particular, every element $x \in Q(A)^\times$ has a (unique up to units of A) **reduced representation** $x = a/b$ with $a, b \in A^*$, $\gcd(a, b) = 1$. $Q(A)^\times / A^\times$ is isomorphic to the free abelian group $\mathbb{Z}^{(\mathbb{P})}$. The elements of the free commutative monoid $M(\mathbb{P})$ generated by $\mathbb{P}$ form a representative system for the associativity classes of the elements of A^*. As

---

[56] Occasionally, it is convenient to also include the zero element $0 \in A$ as an absorbing element, i.e., as the largest element with respect to the divisibility relation, in divisibility considerations.

a rule, we prefer these representatives (which of course depend on the choice of $\mathbb{P}$ depends; for $A = \mathbb{Z}$ is $M(\mathbb{P}) = \mathbb{N}^*$). Every principal ideal $\neq 0$ in A is generated by exactly one of these representatives. Among them are the ideals Ap, $p \in \mathbb{P}$, exactly the prime principal ideals $\neq 0$. We remind that a principal ideal $\mathfrak{p}$ is exactly a prime ideal, i.e. that $A/\mathfrak{p}$ is an integral domain, when one and thus every generating element of $\mathfrak{p}$ is a prime element $p \in A$. With Lemma 2.3.24 one can now easily prove the following theorem, which generalizes the main theorem of elementary number theory:

Theorem 2.10.5 *Every principal ideal domain A is factorial.*

Proof According to Lemma 2.3.24 we show the following:

(1) *Every non-unit $a \in A^*$ is a product of irreducible elements.* But since A is Noetherian, the set of principal ideals $\neq 0$ (i.e., the set of all ideals $\neq 0$) is Noetherian ordered, and the assertion follows from Exercise 2.3.16a).
(2) *Every irreducible element $p \in A^*$ is prime.* By definition, p generates a principal ideal $Ap \neq A$, which is maximal in the set of all principal ideals $\neq A$. But since all ideals in A are principal ideals, Ap is a maximal ideal in A and therefore prime, since A/Ap is even a field. □

As the last proof shows, the following applies:

Corollary 2.10.6 *Let A be a principal ideal domain. For an element $p \in A^* - A^\times$ the following statements are equivalent:* (i) *p is prime.* (ii) *p is irreducible.* (iii) *The ideal Ap is a prime ideal.* (iv) *The ideal Ap is a maximal ideal.* (v) *A/Ap is an integral domain.* (vi) *A/Ap is a field.*

Note that in the situation of Corollary 2.10.6 also 0 is a prime element in A. It generates a maximal ideal exactly when A is a field. Also, fields K are factorial integral domains (with $\mathbb{P}_{K^*} = \emptyset$). Since in a principal ideal domain A for $a, b \in A^*$ the ideals $Aa + Ab$ and $Aa \cap Ab$ are principal ideals per se, so

$$Aa + Ab = A \gcd(a,b) \quad \text{resp.} \quad Aa \cap Ab = A \operatorname{lcm}(a,b)$$

applies, the factoriality of A can also be proven with conditions (iv) or (v) from Lemma 2.3.24 (after verifying that every element $a \in A^* - A^\times$ is a product of irreducible elements). As the first example class to Theorem 2.10.5 we mention:

Corollary 2.10.7 *If K is a field, then the polynomial algebra $K[X]$ in one indeterminate over K is factorial.*

As a standard representative system $\mathbb{P} = \mathbb{P}_{K[X]^*}$ for the prime elements in $K[X]^*$ we choose the set of *normalized* irreducible polynomials in $K[X]$. The simplest elements in $\mathbb{P}$ are the normalized polynomials $X - c$, $c \in K$, of degree 1. It is $M(\mathbb{P}_{K[X]^*}) \subseteq K[X]^*$ the multiplicative monoid of normalized polynomials in $K[X]$.

2.10 Principal Ideal Domains and Factorial Integral Domains

In Euclidean integral domains A and especially in the polynomial algebras $K[X]$ one calculates the greatest common divisor of two elements $a, b \in A^*$ (and thus of finitely many elements) conveniently with the **Euclidean Algorithm**, which runs quite analogously to the Euclidean algorithm for $\mathbb{Z}$, where one only has to replace the absolute value function by a Euclidean degree function φ on A. It also explicitly provides a representation

$$\gcd(a, b) = sa + tb \quad \text{with } s, t \in A$$

of $\gcd(a.b)$ as a linear combination of a and b, cf. Theorem 1.7.7 and the comments on it.

Let A be again an arbitrary factorial integrity domain. If $S \subseteq A^*$ is a submonoid, then for the ring of fractions $A_S = \{a/s \mid a \in A, s \in S\} \subseteq Q(A)$ the equation $(A_S)^* = (A^*)_S$ holds and from Exercise 2.3.10 it follows:

Theorem 2.10.8 *If A is a factorial integrity domain and $S \subseteq A^*$ is a submonoid of A^*, then the fraction ring A_S is also factorial. If $\mathbb{P}(S) \subseteq \mathbb{P}_{A^*}$ is the set of prime elements in $\mathbb{P}_{A^*}$, which divide an element from S, then $\mathbb{P}_{A^*} - \mathbb{P}(S)$ is a representative system for the associativity classes of prime elements $\neq 0$ in A_S.*

The Nagata Lemma 2.3.25 provides the following partial reversal of the last theorem:

Theorem 2.10.9 *Let A be an integral domain and $S \subseteq A^*$ a submonoid with the following properties: (1) Every element from $A^* - A^\times$ is a product of irreducible elements of A^*. (2) S is generated by prime elements from A^*. – If the fraction ring $S^{-1}A$ is factorial, then so is A.*

As an application of the last theorem, we show Gauss's theorem that (commutative) polynomial algebras over factorial integral domains are again factorial and mention in advance the following lemma (which is often also referred to as Gauss's lemma):

Lemma 2.10.10 *Let $\mathfrak{p} \subseteq S$ be a prime ideal in the commutative ring S. Then the extension ideal $\mathfrak{p}S[X_i, i \in I] = \sum_{\nu \in \mathbb{N}^{(I)}}^{\oplus} \mathfrak{p}X^\nu$ is a prime ideal in the polynomial algebra $S[X_i, i \in I]$. In particular, a prime element in S is also a prime element in $S[X_i, i \in I]$.*

Proof The homomorphism $S[X_i, i \in I] \to (S/\mathfrak{p})[X_i, i \in I]$, which is induced by the canonical projection $S \mapsto S/\mathfrak{p}$, is surjective and has the kernel $\mathfrak{p}S[X_i, i \in I]$. Therefore[57]

$$S[X_i, i \in I]/\mathfrak{p}S[X_i, i \in I] \xrightarrow{\sim} (S/\mathfrak{p})[X_i, i \in I].$$

---

[57] The analogous isomorphism holds for any ideals $\mathfrak{a} \subseteq S$.

Since $(S/\mathfrak{p})[X_i, i \in I]$ is an integral domain, the assertion follows. □

Theorem 2.10.11 (Gauss) *Let A be a factorial integral domain. Then every polynomial algebra $P := A[X_i, i \in I]$ over A is factorial.*

Proof Every polynomial $F \in P$ is in a polynomial algebra $P' := A[X_i, i \in I']$ with a finite subset $I' \subseteq I$. Since the prime elements in P' are also prime in P according to Lemma 2.10.10, we can assume that I is finite and then (induction over $|I|$), that $|I| = 1$ is.

So let $P = A[X]$ be the polynomial algebra in an indeterminate X over A. Furthermore, let S be the free commutative submonoid generated by the prime elements $p \in \mathbb{P}_{A^*}$ of $A^* \subseteq P^*$. According to Lemma 2.10.10, the $p \in \mathbb{P}_{A^*}$ are also prime in P. Moreover, $P_S = Q(A)[X]$ is factorial according to Corollary 2.10.7. According to Theorem 2.10.9, it is now sufficient to show that every polynomial $F = a_0 + a_1 X + \cdots + a_n X^n \in P^* - P^\times = P^* - A^\times$, $a_n \neq 0$, is a product of irreducible polynomials. To this end, we conclude by induction over $n = \deg F$. For $n = 0$, this follows from the fact that every element from $A^* - A^\times$ is a product of prime elements and thus of irreducible elements in A.

Finally, let $n > 0$. If F is a product of polynomials of degree smaller than n, the assertion follows from the induction hypothesis. Otherwise, let $d := \mathrm{ggT}(a_0, \ldots, a_n)$. Every constant divisor of F is then a divisor of d, and F/d is irreducible. Together with a prime factor decomposition of d, this results in a decomposition of F as a product of irreducible polynomials. □

We still want to describe a canonical representative system $\mathbb{P}_{P^*}$ for the associativity classes of the prime elements $\neq 0$ in the polynomial algebra $P := A[X_i, i \in I]$ over a factorial integrity domain A with quotient field $K := Q(A)$ and use Theorem 2.10.8 for $S := A^* \subseteq P^*$ with $Q := S^{-1}P = K[X_i, i \in I]$. We give ourselves the representative system $\mathbb{P}_{A^*}$ for the associativity classes of prime elements in A^* and thus $\mathrm{M}(\mathbb{P}_{A^*}) \subseteq A^*$ as a representative system for the associativity classes of all elements of A^*. Furthermore, we use the homogeneous lexicographic order (or the reverse homogeneous lexicographic order) of the monomials X^ν, $\nu \in \mathbb{N}^{(I)}$, with respect to a total order of I, to be able to speak of leading forms and leading coefficients. If $A = K$ is a field, we choose for $\mathbb{P}_{P^*}$ the normalized prime polynomials in P^* (whose leading coefficients are therefore 1). In the general case, we first introduce the following terminology: If $F = \sum_\nu a_\nu X^\nu \in P^*$, then

$$\mathrm{I}(F) := \gcd\left(a_\nu, \nu \in \mathbb{N}^{(I)}\right),$$

is called the **content** of F. For $\mathrm{I}(F)$ we choose the representative of the GCD for which the leading coefficient of $(\mathrm{I}(F))^{-1} F$ is in $\mathrm{M}(\mathbb{P}_{A^*})$ is given. Then

$$F = \mathrm{I}(F) F^* \quad \text{with} \quad \mathrm{I}(F^*) = 1.$$

holds. Exactly then is $F = F^*$, i.e. $\mathrm{I}(F) = 1$, when the coefficients of F are coprime and the leading coefficient belongs to $\mathrm{M}(\mathbb{P}_{A^*})$. F *is prime in P, if and*

2.10 Principal Ideal Domains and Factorial Integral Domains

only if F is prime in Q and $I(F) \in A^\times$. Apparently, for every polynomial $G \in Q^*$ there is exactly one $c \in K^\times$ such that $G^* := c^{-1}G$ is in P and $I(G^*) = 1$ is. For $G = \sum_\nu a_\nu X^\nu \in Q^*$ as above, $c = \gcd_{A^*}(a_\nu, \nu \in \mathbb{N}^{(I)})$ is (up to a unit $\varepsilon \in A^\times$). For example,

$$G := 2 + \frac{1}{3}X + 2Y - \frac{5}{7}X^2 - 3XY^2$$
$$= -\frac{1}{21}(-42 - 7X - 42Y + 15X^2 + 63XY^2) \in \mathbb{Q}[X, Y],$$

is thus $c = -1/21$ and $G^* = -42 - 7X - 42Y + 15X^2 + 63XY^2$. c is again called the **content** $I(G) \in K^\times$ and $G^* \in P$ the **primitive part** of G. G is called **primitive**, if $G = G^*$ is or $I(G) = 1$. For arbitrary polynomials $G, H \in Q^*$ applies

$$G = I(G)G^*, \quad I(GH) = I(G)I(H), \quad (GH)^* = G^*H^*.$$

To prove the product formulas, note that $LK(GH) = LK(G)LK(H)$ is and G^*H^* is not divisible by any prime element $p \in \mathbb{P}_{A^*}$, since neither G^* nor H^* are divisible by p. We now take

$$\mathbb{P}_{P^*} := \mathbb{P}_{A^*} \uplus \mathbb{P}^*_{Q^*} \quad \text{with} \quad \mathbb{P}^*_{Q^*} := \{\pi^* \mid \pi \in \mathbb{P}_{Q^*}\}.$$

$\mathbb{P}_{P^*}$ is thus the set of prime elements in P^*, whose leading coefficient belongs to $M(\mathbb{P}_{A^*})$. As a corollary, we note:

Corollary 2.10.12 (Gauss's Lemma) *With the previous notations, the following applies: If the polynomial* $F \in P^*$ *has a decomposition* $F = GH$ *with non-constant polynomials* $G, H \in Q^*$*, then F also has such a decomposition in* P^*. *– If* $G, H \in Q^*$ *are normalized polynomials, whose product* $F := GH$ *is in* P^**, then G and H are already in* P^*. *In particular, if* $G \in A[X]$ *is a normalized polynomial (in one variable) with a root* $c \in K = Q(A)$*, then* $c \in A$ *applies.*

Proof It is $F = I(F)F^* = I(F)G^*H^*$. – To prove the addition, let $G, H \in Q^*$ be normalized. Then $F = GH \in P^*$ is also normalized, and $F = F^* = G^*H^*$ holds. Therefore, G^* and H^* are normalized and therefore equal to G or H. □

Let A now specifically be a principal ideal domain and $\mathbb{P} = \mathbb{P}_{A^*}$ a representative system for the associativity classes of its prime elements $\neq 0$. Because of $Aa + Ab = A \ggT(a, b)$, two (main) ideals $Aa, Ab \subseteq A$ are comaximal exactly when a and b are coprime. The Chinese Remainder Theorem now reads as follows, see Theorem 2.7.11 and the remarks there:

Theorem 2.10.13 (Chinese Remainder Theorem for Principal Ideal Domains) *If* $a_1, \ldots, a_n \in A^*$ *are pairwise coprime elements of the principal ideal domain A, then the canonical homomorphism is*

$$\overline{\varphi} \colon A/Aa \xrightarrow{\sim} A/Aa_1 \times \cdots \times A/Aa_n, \quad a := a_1 \cdots a_n,$$

an isomorphism of both rings and A-modules.

Let V be an A-module. For an $p \in \mathbb{P}$ we define in analogy to $\mathbb{Z}$-modules (= abelian groups) the p-**primary component** of V as the submodule

$$V(p) := \bigcup_{n \in \mathbb{N}} \mathrm{T}_{p^n} V = \{x \in V \mid \text{there is } n \in \mathbb{N} \text{ with } p^n x = 0\} \subseteq \mathrm{T}_A V \subseteq V$$

of those elements of V that are annihilated by a power of p. As for abelian groups, one proves, see Theorem 2.2.22:

Theorem 2.10.14 (Primary decomposition of torsion modules over principal ideal domains) *Let V be a module over the principal ideal domain A. Then the torsion submodule $\mathrm{T}_A V$ of V is the direct sum of the primary components $V(p)$, $p \in \mathbb{P}$, of V, i.e it is*

$$\mathrm{T}_A V = \sum_{p \in \mathbb{P}}^{\oplus} V(p).$$

The remaining results about abelian groups can be generalized to modules over A with completely analogous proofs. We leave it to the reader to carry this out. *In the following theorems, A is as before a principal ideal domain.*

Theorem 2.10.15 (Submodules of free A-modules) *Every submodule U of a free A-module V is free, and it holds $\mathrm{rank}_A U \leq \mathrm{rank}_A V$. (See Theorem 2.3.28.)*

Theorem 2.10.16 *Every finite torsion-free A-module is free. (See Theorem 2.3.30.)*

Theorem 2.10.17 (Main theorem for finite A-modules) *Every finite A-module V is a direct sum of cyclic A-modules, i.e.*

$$V \cong A/Aa_1 \oplus \cdots \oplus A/Aa_s \oplus A^r \quad \text{with} \quad r, s \in \mathbb{N}, a_1, \ldots, a_s \in A^*. \quad (Cf. \text{ Theorem 2.3.31.})$$

Theorem 2.10.18 (Prüfer's Theorem) *Every A-module V with $\mathrm{Ann}_A V \neq 0$ is a direct sum of cyclic A-modules. (See the remark on Exercise 2.3.6.)*

For the next statement, we recall the following notations: For $p \in \mathbb{P}_{A^*}$ is

$$A_p = \{a/p^n \mid a \in A, n \in \mathbb{N}\} \subseteq \mathrm{Q}(A) \quad \text{and} \quad A_{(p)} = \{a/b \mid a \in A, p \nmid b\} \subseteq \mathrm{Q}(A).$$

$A_{(p)}$ is the localization of A with respect to the prime ideal $(p) = Ap \subseteq A$.

Theorem 2.10.19 *Let $K := \mathrm{Q}(A)$ be the quotient field of A. Then the following holds (see Exercise 2.2.26):*

(1) *A_p/A is the p-primary component of K/A. Therefore, $K/A = \sum_{p \in \mathbb{P}}^{\oplus} A_p/A$.*

(2) *The A-module $K/A_{(p)}$ is equal to its p-primary component. The canonical homomorphism $\pi: K/A \to \prod_{p \in \mathbb{P}} K/A_{(p)}$ is injective and its image is the direct sum $\bigoplus_{p \in \mathbb{P}} K/A_{(p)} \subseteq \prod_{p \in \mathbb{P}} K/A_{(p)}$. Therefore, $\pi: K/A \xrightarrow{\sim} \bigoplus_{p \in \mathbb{P}} K/A_{(p)}$.*

(3) *π induces an isomorphism $A_p/A \xrightarrow{\sim} K/A_{(p)}$.*

2.10 Principal Ideal Domains and Factorial Integral Domains

Every A_p/A isomorphic A-module is called a **Prüfer A-p-module** and is denoted by $I(p)$. It naturally depends only on the prime ideal (p) and not on its generating element p.

Example 2.10.20 (Polynomials in $K[X]$) In this example, we discuss in more detail the polynomial rings in one indeterminate over fields. Let K be a field and $P := K[X]$ the polynomial ring over K. The standard representatives for the associativity classes of the polynomials $F \in P^*$ are the normalized polynomials. In particular, $\mathbb{P} = \mathbb{P}_{P^*}$ is the set of normalized prime polynomials (= set of normalized irreducible polynomials) in P. Obviously, $\mathbb{P}$ contains infinitely many elements. (If K is finite, one concludes as in the proof of Theorem 1.7.2.) A polynomial $F \in P^*$ has the **canonical prime factor decomposition**

$$F = \varepsilon \prod_{\pi \in \mathbb{P}} \pi^{v_\pi(F)} = \varepsilon \prod_{c \in K} (X-c)^{v_c(F)} \prod_{\pi \in \mathbb{P}, \text{Grad } \pi \geq 2} \pi^{v_\pi(F)}$$

with $\varepsilon \in K^\times$, $v_\pi(F) \in \mathbb{N}$, $\pi \in \mathbb{P}$, $\deg \pi \geq 2$, and the multiplicities $v_c(F) := v_{X-c}(F)$, $c \in K$. So it is $(v_c(F)) \in \mathbb{N}^{(K)}$, and $\text{NS}_K(F) = \{c \in K \mid v_c(F) > 0\}$ is the set of zeros of F in K. In particular, $|Z_K(F)| \leq \deg F$ is, more precisely: $\sum_{c \in K} v_c(F) \leq \deg F$. Exactly then is $|Z_K(F)| = \deg F$, when F decomposes into linear factors, i.e., has no prime divisor of degree ≥ 2, and all zeros $c \in Z_K(F)$ are simple. A polynomial $F \in P^*$ of degree 2 or 3 is irreducible and thus exactly prime when it has no zero in K. If $F \in P^*$, then the residue class algebra $A := P/(F)$ is a finite K-algebra of dimension $\dim_K A = n := \deg F$ with K-basis $1, x, \ldots, x^{n-1}$, $x = [X]_F$, see Example 2.9.34. According to corollary 2.10.6 specifically applies: $\diamond$

Theorem 2.10.21 *Let $F \in P^*$. Then F is a prime polynomial if and only if $K[x] := P/(F)$ is a field. In this case, $K[x]$ is a finite field extension of K with $\dim_K K[x] = \deg F$.*

For the general investigation of the algebras $P/(F)$, the Chinese Remainder Theorem 2.10.13 is useful:

Theorem 2.10.22 *If $F = F_1 \cdots F_n$ with pairwise coprime polynomials $F_1, \ldots, F_n \in P^*$, then the canonical homomorphism*

$$\widetilde{\varphi} \colon P/(F) \xrightarrow{\sim} P/(F_1) \times \cdots \times P/(F_n)$$

is an isomorphism of K-algebras of dimension $\deg F = \deg F_1 + \cdots + \deg F_n$. In particular, if $F = \varepsilon(X - c_1)^{\alpha_1} \cdots (X - c_r)^{\alpha_r} \pi_1^{\beta_1} \cdots \pi_s^{\beta_s}$ with $\varepsilon \in K^\times$, $r, s \in \mathbb{N}$, pairwise different $c_1, \ldots, c_r \in K$, $\alpha_1, \ldots, \alpha_r \in \mathbb{N}^*$ and pairwise different $\pi_1, \ldots, \pi_s \in \mathbb{P}$, $\deg \pi_1 \geq 2, \ldots, \deg \pi_s \geq 2$, and $\beta_1, \ldots, \beta_s \in \mathbb{N}^*$ the canonical prime factor decomposition of F, then the canonical K-algebra homomorphism

$$P/(F) \xrightarrow{\sim} P/((X-c_1)^{\alpha_1}) \times \cdots \times P/((X-c_r)^{\alpha_r}) \times P/\left(\pi_1^{\beta_1}\right) \times \cdots \times P/\left(\pi_s^{\beta_s}\right)$$

is an isomorphism.

Note that the surjectivity of $\overline{\varphi}$ already follows from the injectivity of $\overline{\varphi}$, since the pre-image and image range of $\overline{\varphi}$ have the same finite K-dimension $\deg F$. Furthermore, note that the algebras $P/((X-c)^\alpha)$, $c \in K$, $\alpha \in \mathbb{N}$, are all isomorphic to $P/(X^\alpha)$.

The K-derivation $F \mapsto F' = D_X F$ of $P = K[X]$ provides an important tool to distinguish whether the prime factors $X - c_\rho$ or π_σ in Theorem 2.10.22 are simple or multiple. The basis for this is the following simple lemma:

Lemma 2.10.23 *Let $F, H \in P^*$ and $\gcd(H, H') = 1$. Exactly then is H^2 a divisor of F in P, when H is a common divisor of F and F'.*

Proof We can assume $\deg H \geq 1$. If $F = GH^\alpha$ with $\alpha \in \mathbb{N}^*$ and $H \nmid G$, then $F' = G'H^\alpha + \alpha GH^{\alpha-1}H' = (G'H + \alpha GH')H^{\alpha-1}$. For $\alpha \geq 2$, it follows that F' is divided by H. Conversely, let $\alpha \geq 1$ and H also be a divisor of F'. Then H is a divisor of $GH^{\alpha-1}$ due to the coprimality of H and H'. For $\alpha = 1$, H would be a divisor of G. Contradiction! □

The lemma 2.10.23 motivates the following definition:

Definition 2.10.24 A polynomial $H \in P^*$ is called **separable**, if $\gcd(H, H') = 1$ is.

If $F \in P^*$, then $F' = 0$ or $\deg F' < \deg F$. Exactly then is $F' = 0$ for a non-constant polynomial $F \in P = K[X]$, when $\operatorname{Char} K = p \in \mathbb{P}$ is and $F \in K[X^p]$. A prime polynomial $\pi \in P^*$ is separable exactly when $\pi' \neq 0$ is. In particular, every polynomial of degree 1 is separable and at $\operatorname{Char} K = 0$ every prime polynomial. A separable prime divisor π of F in P^* is simple exactly when π is not a divisor of F'. In particular, $c \in K$ is a simple root of $F \in P^*$, when $F(c) = 0$ and $F'(c) \neq 0$ is. The product $H = H_1 H_2$ of two polynomials from P^* is separable exactly when H_1 and H_2 are separable and $\gcd(H_1, H_2) = 1$ is. Every divisor of a separable polynomial is separable. A polynomial of the form $X^n - c \in K[X]$ with $n \in \mathbb{N}^*$ and $c \in K^\times$ is separable exactly when n is not a multiple of $\operatorname{Char} K$.

Proposition 2.10.25 *Let $F = \varepsilon(X - c_1)^{\alpha_1} \cdots (X - c_r)^{\alpha_r} \pi_1^{\beta_1} \cdots \pi_s^{\beta_s}$ be with $\varepsilon \in K^\times$, $r, s \in \mathbb{N}$, pairwise different $c_1, \ldots, c_r \in K$, $\alpha_1, \ldots, \alpha_r \in \mathbb{N}^*$ as well as pairwise different $\pi_1, \ldots, \pi_s \in \mathbb{P}$, $\deg \pi_1 \geq 2, \ldots, \deg \pi_s \geq 2$, and $\beta_1, \ldots, \beta_s \in \mathbb{N}^*$ the canonical prime factor decomposition of $F \in P^*$. Exactly then is F separable, when all prime factors of F are simple and the nonlinear prime factors $\pi_1, \ldots, \pi_s$ of F are separable. In particular, over a field of characteristic 0, a polynomial $\neq 0$ is separable exactly when all its prime factors are simple.*

The set $\mathbb{P}$ of normalized prime polynomials in $K[X]$ immediately provides a K-vector space basis of the rational function field $K(X)$.

2.10 Principal Ideal Domains and Factorial Integral Domains

Theorem 2.10.26 (Partial Fraction Decomposition) *The monomials X^ν, $\nu \in \mathbb{N}$, together with the rational functions*

$$X^\mu/\pi^\kappa, \quad 0 \leq \mu < \deg \pi, \; \kappa \in \mathbb{N}^*, \; \pi \in \mathbb{P},$$

form a K-vector space basis of $K(X)$. Therefore,

$$\dim_K K(X) = |K(X)| = \operatorname{Max}(\aleph_0, |K|)$$

*(in contrast to $\dim_K K[X] = \aleph_0$). The representation of a rational function $F/G \in K(X)$ as a K-linear combination of this basis is called the **partial fraction decomposition** of F/G (over K).*

Proof We first show that the given functions are linearly independent. For this, it is sufficient to exploit the π-al-developments of polynomials (cf. exercise 2.9.22a)), to show the following: If

$$F + \frac{F_1}{\pi_1^{k_1}} + \cdots + \frac{F_s}{\pi_s^{k_s}} = 0$$

holds with pairwise different $\pi_1, \ldots, \pi_s \in \mathbb{P}$, $k_1, \ldots, k_s \in \mathbb{N}^*$ and polynomials $F, F_1, \ldots, F_s \in K[X]$, $F_\sigma = 0$ or $\deg F_\sigma < k_\sigma \cdot \deg \pi_\sigma = \deg \pi_\sigma^{k_\sigma}$, then $F = F_1 = \cdots = F_s = 0$. Let this be $\pi := \pi_1^{k_1} \cdots \pi_s^{k_s}$ and $\widetilde{\pi}_\sigma := \pi/\pi_\sigma^{k_\sigma}$, $\sigma = 1, \ldots, s$. Multiplication with π then yields $\pi F + \widetilde{\pi}_1 F_1 + \cdots + \widetilde{\pi}_s F_s = 0$, i.e. $\pi_1^{k_1} \cdots \pi_s^{k_s} F = -\widetilde{\pi}_1 F_1 - \cdots - \widetilde{\pi}_s F_s$. For each $\sigma = 1, \ldots, s$ thus divides $\pi_\sigma^{k_\sigma}$ the polynomial F_σ, which would lead to a contradiction at $F_\sigma \neq 0$ $\deg F_\sigma \geq \deg \pi_\sigma^{k_\sigma}$. Therefore, all $F_\sigma = 0$ and then also $F = 0$.

To obtain a representation of the desired type, let $F/G \in K(X)^\times$ be a rational function $\neq 0$ with $F, G \in K[X]^*$ and $G = \pi_1^{k_1} \cdots \pi_s^{k_s}$, $\pi_1, \ldots, \pi_s \in \mathbb{P}$ pairwise different and $k_1, \ldots, k_s \in \mathbb{N}^*$. We divide F by G with remainder and can thus assume $\operatorname{Grad} F < \operatorname{Grad} G$. Now let $\widetilde{G}_\sigma = G/\pi_\sigma^{k_\sigma}$, $\sigma = 1, \ldots, s$. Since the $\widetilde{G}_1, \ldots, \widetilde{G}_s$ are coprime, there is a representation $H_1 \widetilde{G}_1 + \cdots + H_s \widetilde{G}_s = 1$ with $H_\sigma \in K[X]$. Let $H_\sigma F = Q_\sigma \pi_\sigma^{k_\sigma} + F_\sigma$, $\sigma = 1, \ldots, s$, be the division with remainder of $H_\sigma F$ by $\pi_\sigma^{k_\sigma}$ with $Q_\sigma, F_\sigma \in K[X]$ and $\deg F_\sigma < \deg \pi_\sigma^{k_\sigma}$. Then $F = H_1 F \widetilde{G}_1 + \cdots + H_s F \widetilde{G}_s = F_1 \widetilde{G}_1 + \cdots + F_s \widetilde{G}_s + QG$, $Q := Q_1 + \cdots + Q_s \in K[X]$. For $Q \neq 0$, $\deg G \leq \deg QG = \deg(F - (F_1 \widetilde{G}_1 + \cdots + F_s \widetilde{G}_s)) < \deg G$ would be a contradiction. Therefore, $Q = 0$ and

$$\frac{F}{G} = \frac{F_1}{\pi_1^{k_1}} + \cdots + \frac{F_s}{\pi_s^{k_s}}.$$

With the π_σ-al development of the F_σ, we now obtain the desired partial fraction decomposition of F/G. □

As the proof shows, the partial fraction decomposition of $F/G \in K[X]^\times$ can be explicitly obtained if the prime factor decomposition of the denominator G is known and the division with remainder in $K[X]$ is constructively executable. We also note that *the above representation $F/G = F_1/\pi_1^{k_1} + \cdots + F_s/\pi_s^{k_s}$ already from the proven linear independence of the $X^{\mu_\sigma}/\pi_\sigma^{k_\sigma}$, $0 \leq \mu_\sigma < \deg \pi_\sigma$,*

$1 \leq \kappa_\sigma \leq k_\sigma$, $\sigma = 1, \ldots, s$, follows. These $k_1 \deg \pi_1 + \cdots + k_s \deg \pi_s$ linearly independent rational functions all belong to the vector space

$$V_G := \{F/G \mid F \in K[X], F = 0 \text{ or } \deg F < \deg G\}$$

of dimension $\deg G = k_1 \deg \pi_1 + \cdots + k_s \deg \pi_s$ and therefore generate it. – If $F, G \in K[X]^*$ are coprime with $\deg F < \deg G$ and the denominator G has at least two non-associated prime divisors, it is recommended to perform the partial fraction decomposition of F/G in such a way that one $G = G_1 G_2$ nontrivially decomposes into two coprime factors G_1, G_2, a representation $F/G = F_1/G_1 + F_2/G_2$, $\deg F_i < \deg G_i$, $i = 1,2$, is determined with the help of division with remainder and then continues in an analogous way with the two summands F_1/G_1 and F_2/G_2. If the denominator G has a zero c of order $v_c > 0$, then in the partial fraction decomposition of F/G the coefficient $a_{c,v(c)}$ of the summand $a_{c,v(c)}/(X-c)^{v(c)}$ is equal to

$$a_{c,v(c)} = F(c)/\widetilde{G}(c) \quad \text{with} \quad \widetilde{G} := G/(X-c)^{v_c},$$

as is immediately apparent by multiplying this partial fraction decomposition with $(X-c)^{v_c}$. Apparently, $v_c! \widetilde{G}(c) = G^{(v_c)}(c)$ and in particular $\widetilde{G}(c) = G'(c)$ at $v_c = 1$, cf. exercise 2.9.4b). – If A is any principal ideal domain with quotient field $L = Q(A)$, then one can interpret the direct sum decomposition $L/A = \sum_{p \in \mathbb{P}_{A^*}}^\oplus A_p/A$ from Theorem 2.10.19 (1) as a **partial fraction decomposition for** A.

The prime factor decomposition is most clear in $K[X]^*$ (and thus also in $K(X)^\times$), if there are no prime polynomials in $K[X]^*$ of degree ≥ 2.

Definition 2.10.27 The field K is called **algebraically closed**, if every prime polynomial $\neq 0$ in $K[X]$ has degree 1.

Theorem 2.10.28 *For a field K the following are equivalent:* (i) *K is algebraically closed.* (ii) *Every prime polynomial in $K[X]^*$ has degree 1.* (iii) *Every polynomial $F \in K[X]^*$ has a unique representation*

$$F = \varepsilon \prod_{c \in K} (X-c)^{\alpha_c} \quad \text{with} \quad \varepsilon \in K^\times, (\alpha_c)_{c \in K} \in \mathbb{N}^{(K)}.$$

(iv) *Every non-constant polynomial in $K[X]$ has a root in K.* (v) *Every prime polynomial in $K[X]$ has a root in K.* (vi) *If $K \subseteq L$ is an algebraic field extension, then $L = K$.* (vii) *If $K \subseteq L$ is a finite field extension, then $L = K$.*

Proof The equivalences from (i) to (v) result from the definitions or are trivial. To prove (ii) $\Rightarrow$ (vi) let $x \in L$. Then x is algebraic and consequently its minimal polynomial μ_x is different from 0. Because of $K[X]/(\mu_x) \xrightarrow{\sim} K[x] \subseteq L$, $K[X]/(\mu_x)$ is an integral domain and therefore $\mu_x \neq 0$ is prime, thus of degree 1 according to (ii). From $\text{Dim}_K K[x] = \text{Grad}\,\mu_x = 1$ follows $K[x] = K$ and thus $x \in K$. The implication (vi) $\Rightarrow$ (vii) is again trivial, as every finite field extension is

2.10 Principal Ideal Domains and Factorial Integral Domains

algebraic. Finally, (ii) follows from (vii). If $\pi \in K[X]^*$ is a prime polynomial, then $L := K[X]/(\pi)$ is a finite field extension of dimension $\deg \pi$ over K according to Theorem 2.10.21. Assuming the validity of (vii), $L = K$, thus $\deg \pi = 1$. □

Let K be an algebraically closed field. Then the K-vector space basis of $K(X)$ simplifies according to the partial fraction decomposition 2.10.26 to

$$X^\nu, \ \nu \in \mathbb{N}, \quad 1/(X-c)^\kappa, \ \kappa \in \mathbb{N}^*, \ c \in K.$$

As Theorem 3.9.7 we will later prove the so-called **Fundamental theorem of algebra**: *Every non-constant polynomial over $\mathbb{C}$ has a root in $\mathbb{C}$.* In other words:

Theorem 2.10.29 *The field $\mathbb{C}$ of complex numbers is algebraically closed.*

This results in:

Theorem 2.10.30 *Every normalized prime polynomial of degree ≥ 2 over the field $\mathbb{R}$ of real numbers is quadratic of the form $X^2 - pX + q$ with $p, q \in \mathbb{R}$, $p^2 - 4q < 0$.*

Proof The normalized quadratic polynomial $X^2 - pX + q \in \mathbb{R}[X]$ has a root in $\mathbb{R}$ exactly when its discriminant $p^2 - 4q$ is a square in $\mathbb{R}$ (see the end of Example 2.9.35), and this is exactly the case when $p^2 - 4q \geq 0$ is (see Example 3.3.7), the given quadratic polynomials are exactly the normalized prime polynomials over $\mathbb{R}$ of degree 2. It remains to show that there are no prime polynomials over $\mathbb{R}$ of degree >2. Let $\pi = a_0 + \cdots + a_{n-1}X^{n-1} + X^n \in \mathbb{R}[X]$ be normalized and prime of degree $n \geq 3$. In particular, then π has no root in $\mathbb{R}$. Let $z = a + ib \in \mathbb{C} - \mathbb{R}$, $a, b \in \mathbb{R}$, $b \neq 0$, be a non-real complex root of π, i.e. let it be $\pi(z) = 0$. Then it follows $\pi(\bar{z}) = a_0 + \cdots + a_{n-1}\bar{z}^{n-1} + \bar{z}^n = \overline{\pi(z)} = 0$, and thus $\bar{z} = a - ib \neq z$ is also a root of π. Therefore, the polynomial

$$G := (X - z)(X - \bar{z}) = X^2 - \mathrm{Sp}(z)X + \mathrm{N}(z) = X^2 - 2aX + (a^2 + b^2) \in \mathbb{R}[X]$$

of degree 2 is a divisor of π in $\mathbb{C}[X]$. Then G is also a divisor of π in $\mathbb{R}[X]$, see Corollary 2.9.28. Contradiction! □

The canonical prime factor decomposition of a real polynomial $F \in \mathbb{R}[X]^*$ therefore has the following form:

$$F = \varepsilon(X - c_1)^{\alpha_1} \cdots (X - c_r)^{\alpha_r}(X^2 - p_1X + q_1)^{\beta_1} \cdots (X^2 - p_sX + q_s)^{\beta_s}$$

with $\varepsilon \in \mathbb{R}^\times$, $r, s \in \mathbb{N}$, pairwise different $c_1, \ldots, c_r \in \mathbb{R}$, $\alpha_1, \ldots, \alpha_r \in \mathbb{N}^*$ and the pairwise different irreducible quadratic polynomials $X^2 - p_1X + q_1, \ldots$, $X^2 - p_sX + q_s \in \mathbb{R}[X]$, $p_1^2 - 4q_1 < 0, \ldots, p_s^2 - 4q_s < 0$, $\beta_1, \ldots, \beta_s \in \mathbb{N}^*$. *In particular, a real polynomial of odd degree has a real root.* A normalized quadratic polynomial $X^2 - pX + q \in \mathbb{R}[X]$ has at $p^2 - 4q = 0$ the double root $\frac{1}{2}p$ and at $p^2 - 4q > 0$ the two real roots $\frac{1}{2}(p \pm \sqrt{p^2 - 4q})$.

Example 2.10.31 We determine the partial fraction decomposition of the real rational function
$$R := \frac{X^6 + 1}{X^4 - X^2 - 2X + 2}.$$
Division with remainder yields
$$R = (X^2 + 1) + \frac{F}{G}, \quad F := 2X^3 - X^2 + 2X - 1, \ G := X^4 - X^2 - 2X + 2.$$
The denominator G has over $\mathbb{R}$ or $\mathbb{C}$ the prime factor decompositions
$$G = (X - 1)^2(X^2 + 2X + 2) = (X - 1)^2(X - (-1 + i))(X - (-1 - i)).$$
The complex partial fraction decomposition of F/G therefore has the form
$$\frac{F}{G} = \frac{\alpha}{X - 1} + \frac{\beta}{(X - 1)^2} + \frac{\gamma}{X - (-1 + i)} + \frac{\overline{\gamma}}{X - (-1 - i)}.$$
Following the remark after the proof of Theorem 2.10.26, the coefficients $\gamma, \overline{\gamma}, \beta, \alpha$ result in
$$\gamma = \frac{F(-1 + i)}{G'(-1 + i)} = \frac{1 + 8i}{8 + 6i} = \frac{1}{50}(28 + 29i), \quad \overline{\gamma} = \frac{1}{50}(28 - 29i).$$

$$\beta = \frac{F(1)}{\widetilde{G}(1)} = \frac{2}{5}, \quad \alpha = \frac{\widetilde{\widetilde{F}}(1)}{\widetilde{\widetilde{G}}(1)} = \frac{22}{25},$$

with
$$\widetilde{G} := \frac{G}{(X - 1)^2}, \quad \frac{\widetilde{\widetilde{F}}}{(X - 1)(X^2 + 2X + 2)} := \frac{F}{G} - \frac{\beta}{(X - 1)^2}.$$

As a real partial fraction decomposition, one finally obtains
$$\frac{F}{G} = \frac{\alpha}{X - 1} + \frac{\beta}{(X - 1)^2} + \frac{\gamma(X - (-1 - i)) + \overline{\gamma}(X - (-1 + i))}{X^2 + 2X + 2}$$
$$= \frac{\alpha}{X - 1} + \frac{\beta}{(X - 1)^2} + \frac{\delta X + \varepsilon}{X^2 + 2X + 2}, \quad \alpha = \frac{22}{25}, \beta = \frac{2}{5}, \delta = \frac{28}{25}, \varepsilon = -\frac{1}{25}.$$

The coefficients $\alpha, \beta, \gamma, \overline{\gamma}$ could also have been determined by multiplying both sides of the equation defining these coefficients by G, comparing the coefficients of the powers of X, and solving the resulting linear system of equations. This method is particularly useful when the degree of the denominator G is not too large, i.e., the linear system of equations does not have too many unknowns. ◊

A **theorem by Steinitz** (1871–1928) from 1910 states: *Every field K can be embedded in an algebraically closed field.* A first step towards this is Theorem 2.10.21: If $F \in K[X]^*$ and $\pi \in K[X]$ is a prime factor of F, then

2.10 Principal Ideal Domains and Factorial Integral Domains 327

$L := K[X]/(\pi) = K[x] \supseteq K$ is a field extension of K with $\dim_K L = \deg \pi$, in which F has the zero $x := [X]_\pi$. In L thus $F = (X - x)G$ with $G \in L[X]$. If G over L does not yet decompose into linear factors, we can adjoin to L a zero of a prime factor of G and obtain — continuing in this way — the following theorem:

Theorem 2.10.32 (Theorem of Kronecker) *Let K be a field and $F \in K[X]^*$ a polynomial of degree $n \in \mathbb{N}^*$. Then there is an extension field L of K, over which F decomposes into linear factors. L can be chosen so that L is finite over K with $\dim_K L \le n!$.*

A field L as in Theorem 2.10.32 is called a **decomposition field** of F.

Example 2.10.33 (Finite Fields) With Theorem 2.10.21 or more generally the Theorem 2.10.32 by Kronecker also yields all finite fields. These play a central role in modern cryptography. Let K be such a field of characteristic $p \in \mathbb{P} = \mathbb{P}_{\mathbb{Z}^*}$, $(\mathbb{Z}/\mathbb{Z}p =)\mathbf{F}_p \subseteq K$ its prime field and $m := \text{Dim}_{\mathbf{F}_p} K$. Then K has exactly p^m elements. Since the multiplicative group $K^\times$ has the order $p^m - 1$, $x^{p^m-1} = 1$ is for all $x \in K^\times$ and thus $x^{p^m} - x = 0$ for all $x \in K$. K is therefore the set of roots of $X^{p^m} - X \in \mathbf{F}_p[X]$. Consequently,

$$X^{p^m} - X = \prod_{x \in K}(X - x).$$

K is thus a splitting field of the polynomial $X^{p^m} - X \in \mathbf{F}_p[X]$. Coefficient comparison at X yields $-1 = \prod_{x \in K^\times} x$. (Note that this also applies when $p = 2$.) For $K = \mathbf{F}_p$ we get once again the **Theorem of Wilson**: *It holds $-1 \equiv (p - 1)! \mod p$ for every $p \in \mathbb{P}$.* – We now prove:

Theorem 2.10.34 *For every prime power p^m, $p \in \mathbb{P}$ prime, $m \in \mathbb{N}^*$, there exists exactly one field K with p^m elements, up to isomorphism. It is denoted by*[58]

$$\mathbf{F}_{p^m} \quad or \quad \mathbf{GF}_{p^m} = \mathbf{GF}(p^m).$$

Proof According to the preliminary remark, we consider an upper field L of F_p, over which the polynomial $F := X^{p^m} - X$ decomposes into linear factors (see Theorem 2.10.32), and in L the set K of the roots of F. Because of $(x + y)^{p^m} = x^{p^m} + y^{p^m}$ and $(xy)^{p^m} = x^{p^m} y^{p^m}$ for all $x, y \in L$ (see corollary 2.6.4), K is a subfield of L. It has p^m elements: To show this, it is only necessary to prove that all roots of F are simple. This follows from $F' = -1$ and $F'(a) = -1 \ne 0$ for all roots a of F (see lemma 2.10.23) or directly from $X^{p^m} - X = (X^{p^m} - a^{p^m}) - (X - a) = (X - a)^{p^m} - (X - a) = (X - a)((X - a)^{p^m - 1} - 1)$. K is therefore a field with p^m elements.

To prove the uniqueness of K up to isomorphism, let $x \in K$ be a generating element of the cyclic multiplicative group $K^\times$ of K (cf. Theorem 2.6.22)

---

[58] *GF* is the abbreviation for **Galois Field** and recalls E. Galois (1811–1832).

and $\mu_x \in \mathbf{F}_p[X]$ the minimal polynomial of x over $\mathbf{F}_p \subseteq K$. Then $K = \mathbf{F}_p[x] \cong \mathbf{F}_p[X]/(\mu_x)$ is, and due to $X^{p^m-1} - 1 = \prod_{a \in K^\times}(X - a)$ in $K[X]$, μ_x is a prime divisor of $X^{p^m-1} - 1$. If K' is another field with p^m elements, then $X^{p^m-1} - 1$ also decomposes in $K'[X]$ into linear factors, in particular μ_x has a zero $x' \in K'$. Thus, $K \cong \mathbf{F}_p[X]/(\mu_x) \cong \mathbf{F}_p[x'] \subseteq K'$ is, and for reasons of number, $\mathbf{F}_p[x'] = K'$ holds. □

The proof shows that the normalized prime divisors of the polynomial $X^{p^m} - X \in \mathbf{F}_p[X]$ are exactly the normalized prime polynomials in $\mathbf{F}_p[X]$, whose degree is a divisor of m. Apparently, it can be generalized to a proof of the following result: *If $\mathbf{F}_q$ is a finite field with q elements, then the normalized prime divisors of the polynomial $X^{q^m} - X \in \mathbf{F}_q[X]$ are exactly the normalized prime polynomials in $\mathbf{F}_q[X]$, whose degree is a divisor of m.* Each of these prime polynomials appears with multiplicity 1 in $X^{q^m} - X$.

Over F_2 are $X^2 + X + 1$ and $X^3 + X + 1$ as well as $X^3 + X^2 + 1 = (X + 1)^3 + (X + 1) + 1$ the only prime polynomials of degree 2 or 3. They provide the fields F_4 or F_8. Now let p be a prime number > 2. Then the quadratic polynomial $X^2 - aX + b \in \mathbf{F}_p[X]$ is prime and $\mathbf{F}_p[X]/(X^2 - aX + b)$ a field with p^2 elements, if the discriminant $a^2 - 4b$ is not a square in F_p, i.e. if $((a^2 - 4b)/p) = -1$ applies for the Legendre symbol, see Theorem 2.5.24. *In particular, the complex numbers $\mathbb{C}_{\mathbf{F}_p} = \mathbf{F}_p[i]$, $i^2 = -1$, over F_p are exactly a field, when $p \equiv 3$ mod 4 is.*

Example 2.10.35 (Lucas Test) For the Mersenne number $M(q) = 2^q - 1$ to the prime number $q > 2$ applies $(3/M(q)) = -(M(q)/3) = -1$ (see the reciprocity formula 2.5.27). If $p := M(q)$ is also a prime number, then $L := \mathbf{F}_p[\sqrt{3}] = \mathbf{F}_p[X]/(X^2 - 3)$ is a field with p^2 elements and $L^\times$ a cyclic group with $p^2 - 1 = (p-1)(p+1) = (2^q - 2)2^q$ elements, and consequently $L^\times/\mathbf{F}_p^\times$ a cyclic group of order 2^q. If $x \in L$ is not a square in L, then the residue class $[x] \in L^\times/\mathbf{F}_p^\times$ generates this group. In other words: It is $y := x^{2^{q-1}} \notin \mathbf{F}_p^\times$, but $y^2 = x^{2^q} \in \mathbf{F}_p^\times$. Then $\mathrm{Sp}(y) = 0$ is, see Example 2.9.35. For x one can choose the element $x := 1 + \sqrt{3} \in L$; because $\mathrm{N}(x) = x\bar{x} = (1 + \sqrt{3})(1 - \sqrt{3}) = -2$ is due to $(-2/p) = (-1/p)(2/p) = (-1) \cdot (+1) = -1$, see Theorem 2.5.26, not a square in F_p. Therefore, for the quadratic algebra $\mathbf{A}_{M(q)}[\sqrt{3}]$: *If $M(q)$ is a prime number, then* $\mathrm{Sp}(x^{2^{q-1}}) = 0$ *for* $x := 1 + \sqrt{3}$.

The converse also holds: If $\mathrm{Sp}(x^{2^{q-1}}) = 0$ for $x := 1 + \sqrt{3} \in \mathbf{A}_{M(q)}[\sqrt{3}]$, then $M(q)$ is prime. Because of $(3/M(q)) = -1$ there exists a prime divisor p of $M(q)$ with $(3/p) = -1$. It is $M(q) = 2^q - 1 = pr$ with a $r \in \mathbb{N}^*$, and $L := \mathbf{F}_p[\sqrt{3}]$ is again a field. For $x := 1 + \sqrt{3} \in L$ is $\mathrm{Sp}((x^{2^{q-1}})) = 0$ in F_p. Therefore, $y := x^{2^{q-1}} \in \mathbf{F}_p^\times\sqrt{3}$, see Example 2.9.35, and $y^2 \in \mathbf{F}_p^\times$. Thus, $L^\times/\mathbf{F}_p^\times$ and therefore also $L^\times$ has an element of order 2^q, and $2^q = pr + 1$ is a divisor of $|L^\times| = p^2 - 1 = (p-1)(p+1)$. This is, as one can easily see, only possible for $r = 1$. (For this conclusion, see also the addition to exercise 2.7.8.) In summary: *The Mersenne number $M(q)$, $q \geq 3$, is exactly a prime number when*

2.10 Principal Ideal Domains and Factorial Integral Domains

$\text{Sp}(x^{2^{q-1}}) = 0$ *is for* $x = 1 + \sqrt{3} \in \mathbf{A}_{M(q)}[\sqrt{3}]$. Because of $N(x) = -2 \in \mathbf{A}_{M(q)}^\times$ this is equivalent to $r_{q-1} := \text{Sp}(x^{2^{q-1}})/N(x^{2^{q-1}}) = 0$. With $\text{Sp}\, x^2 = \text{Sp}(4 + 2\sqrt{3}) = 8$ and $\text{Sp}(y^2) = y^2 + \bar{y}^2 = (y + \bar{y})^2 - 2y\bar{y} = (\text{Sp}(y))^2 - 2N(y)$ the recursion

$$r_1 = \frac{\text{Sp}(x^2)}{N(x)} = -4, \quad r_{n+1} = \frac{\text{Sp}(x^{2^{n+1}})}{N(x^{2^n})} = \left(\frac{\text{Sp}(x^{2^n})}{N(x^{2^{n-1}})}\right)^2 - 2 = r_n^2 - 2,\ n \geq 1,$$

results and overall: ◊

Theorem 2.10.36 (Lucas Test) *Let q be a prime number ≥ 3. The Mersenne number $M(q) = 2^q - 1$ is a prime number if and only if for the sequence defined recursively by $r_1 = -4$, $r_{n+1} = r_n^2 - 2$, $n \geq 1$, in $\mathbf{A}_{M(q)} = \mathbb{Z}/\mathbb{Z}M(q)$ the following holds: $r_{q-1} = 0$.*

With the quadratic algebras over the minimal rings A_m, $m \in \mathbb{N}^*$ odd, prime number tests can generally be constructed for such odd numbers $p \in \mathbb{N}^*$ for which the prime factor decomposition of $p + 1$ is manageable (similar to the Fermat prime number tests described in exercise 2.7.9 for numbers p, where the prime factor decomposition of $p - 1$ is manageable, see also the Pépin test in example (2) following Theorem 2.5.28).

Example 2.10.37 (Examples of Euclidean quadratic $\mathbb{Z}$ algebras) We consider some further examples of Euclidean domains from number theory. These are quadratic algebras A over $\mathbb{Z}$. These can be easily classified: Let $A = \mathbb{Z}[y] \xleftarrow{\sim} \mathbb{Z}[Y]/(G)$ be with the normalized quadratic polynomial $G = Y^2 - pY + q \in \mathbb{Z}[Y]$ and the discriminant $\Delta = p^2 - 4q$. Then $\Delta \equiv 0, 1 \bmod 4$, *and all numbers $\equiv 0, 1 \bmod 4$ appear as discriminants*. For convenience (so that 2 becomes a unit), we embed $A = \mathbb{Z}[y]$ into the corresponding quadratic $\mathbb{Q}$ algebra $\mathbb{Q}[y] \xleftarrow{\sim} \mathbb{Q}[Y]/(G)$. $\mathbb{Q}[y]$ is then the pure quadratic $\mathbb{Q}$ algebra $\mathbb{Q}[\sqrt{\Delta}]$, and it is $y = \tfrac{1}{2}(p + \sqrt{\Delta})$.

Let initially $\Delta = p^2 - 4q = -4q' \equiv 0 \bmod 4$. Then p is even, and $A = \mathbb{Z}[x]$ with $x := y - \tfrac{1}{2}p$ is the pure quadratic algebra

$$\mathbb{Z}[x] = \mathbb{Z}\left[\sqrt{\Delta/4}\right] \xleftarrow{\sim} \mathbb{Z}[X]/\left(X^2 - \frac{1}{4}\Delta\right) \quad \text{with} \quad x^2 - \frac{1}{4}\Delta = 0.$$

Now let $\Delta = p^2 - 4q = 1 - 4q' \equiv 1 \bmod 4$. Then $p - 1$ is even and $A = \mathbb{Z}[x]$ with $x := y - \tfrac{1}{2}(p - 1)$ is the quadratic algebra

$$\mathbb{Z}[x] = \mathbb{Z}\left[\frac{1}{2}(1 + \sqrt{\Delta})\right] \xleftarrow{\sim} \mathbb{Z}[X]/\left(X^2 - X + \frac{1}{4}(1 - \Delta)\right)$$

$$\text{with} \quad x^2 - x + \frac{1}{4}(1 - \Delta) = 0.$$

The discriminant Δ thus determines the quadratic $\mathbb{Z}$-algebra A up to isomorphism.[59] $\mathbb{Z}[x]$ is an integral domain exactly when G is a prime polynomial. Since $\mathbb{Z}$ is factorial, this is equivalent to G being prime in $\mathbb{Q}[X]$, i.e., it has no root in $\mathbb{Q}$. It follows: *The quadratic $\mathbb{Z}$-algebra A with discriminant Δ is an integral domain exactly when Δ is not a square in $\mathbb{Q}$ is.* If then $\Delta > 0$, we choose the positive square root $\sqrt{\Delta}$ from Δ, identify A with the subalgebra $\mathbb{Z}[\frac{1}{2}\sqrt{\Delta}]$ or $\mathbb{Z}[\frac{1}{2}(1+\sqrt{\Delta})]$ of $\mathbb{R}$ and speak of a **real-quadratic** $\mathbb{Z}$-algebra. If $\Delta < 0$, we choose the complex root $\sqrt{\Delta} = i\sqrt{|\Delta|} \in \mathbb{C}$, identify A with the subalgebra $\mathbb{Z}[\frac{1}{2}\sqrt{\Delta}]$ or $\mathbb{Z}[\frac{1}{2}(1+\sqrt{\Delta})]$ of $\mathbb{C}$ and speak of an **imaginary-quadratic** $\mathbb{Z}$-algebra. Note that in the last case, the conjugation of A is the restriction of the conjugation of $\mathbb{C}$ to A. The following formula is useful, which is a special case of a much more general formula, see Vol. 3:

Lemma 2.10.38 *Let z be a non-zero divisor in the quadratic $\mathbb{Z}$-Algebra A. Then $|A/Az| (= \operatorname{Kard} A/Az) = |N(z)|$.*

Proof With z, $m := N(z) = z\bar{z} \in \mathbb{Z}^*$ is also a non-zero divisor. Furthermore, $|A/Am| = |\mathbb{Z}^2/m\mathbb{Z}^2| = |m|^2 \in \mathbb{N}^*$. The conjugation $x \mapsto \bar{x}$ induces an isomorphism $A/Az \xrightarrow{\sim} A/A\bar{z}$ and the multiplication with z an isomorphism $A/A\bar{z} \xrightarrow{\sim} Az/Az\bar{z} = Az/Am$. Consequently, $|m|^2 = |A/Am| = |A/Az| \cdot |Az/Am| = |A/Az|^2$, thus $|A/Az| = |m|$. $\square$

The discriminant $\Delta = -4$ corresponds to the imaginary-quadratic $\mathbb{Z}$-algebra

$$\mathbb{C}_\mathbb{Z} = \mathbb{Z}[i] := \{a + ib \mid a, b \in \mathbb{Z}\} \subseteq \mathbb{C}$$

of complex numbers over $\mathbb{Z}$, which is also called the **ring of entire Gaussian numbers**. Its field of fractions is $\mathbb{Q}[i] \subseteq \mathbb{C}$. For $z = a + ib \in \mathbb{C} = \mathbb{R}[i]$ is $N(z) = z\bar{z} = a^2 + b^2$.

Theorem 2.10.39 *The norm is a Euclidean degree function on $\mathbb{Z}[i]$.*

Proof Let $x = a + ib$ and $y = c + id \neq 0$ be from $\mathbb{Z}[i]$ with $a, b, c, d \in \mathbb{Z}$. Then $z = x/y = u + iv$ is with $u, v \in \mathbb{Q}$. There exists $s, t \in \mathbb{Z}$ with $|u - s| \leq \frac{1}{2}$ and $|v - t| \leq \frac{1}{2}$. For $q := s + it \in \mathbb{Z}[i]$ and $r := x - qy = (z - q)y \in \mathbb{Z}[i]$ then $x = qy + r$ and

$$N(r) = N(z-q)N(y) = \left(|u-s|^2 + |v-t|^2\right)N(y) \leq \frac{1}{2}N(y) < N(y). \quad \square$$

$\mathbb{Z}[i]$ is therefore a principal ideal domain, see Theorem 2.10.3. The ideals in $\mathbb{Z}[i]$ are exactly the subgroups $\mathfrak{a} \subseteq \mathbb{Z}^2 = \mathbb{Z} \oplus i\mathbb{Z}$, which are mapped onto themselves by multiplication with i (i.e., by rotating the complex number plane by $\pi/2$. The

---

[59] Show that normalized quadratic polynomials F, G from $\mathbb{Z}[X]$ with different discriminants define non-isomorphic quadratic $\mathbb{Z}$-algebras $\mathbb{Z}[X]/(F)$ or $\mathbb{Z}[X]/(G)$.

2.10 Principal Ideal Domains and Factorial Integral Domains

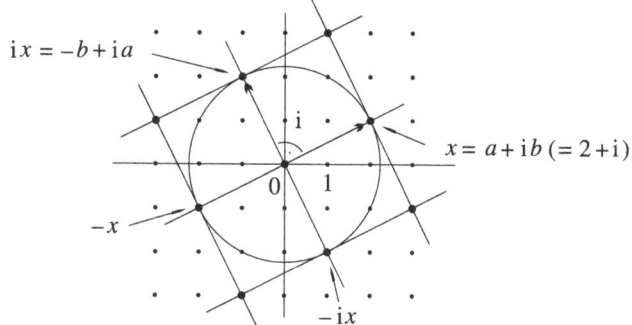

Fig. 2.24 The four generators of the $\mathbb{Z}[i]$-ideal $\mathbb{Z}[i]x$

non-zero ones among them are, according to the proven, exactly the **square lattices**

$$\mathbb{Z}x + \mathbb{Z}ix = \mathbb{Z}(a+ib) \oplus \mathbb{Z}(-b+ia), \quad x = a+ib \neq 0,$$

with the orthogonal $\mathbb{Z}$-bases $x = a+ib$, $ix = -b+ia$, whose elements also each have the same norm $a^2 + b^2$, see Fig. 2.24.

Exactly then is $z \in \mathbb{Z}[i]$ a unit in $\mathbb{Z}[i]$, if $N(z) = z\bar{z} \in \mathbb{Z}^\times$ is, i.e. if z is one of the fourth roots of unity 1, i, −1, −i. For $x \in \mathbb{Z}[i]^*$ the four elements x, ix, $-x$, $-ix$ are exactly the generating elements of the principal ideal $\mathfrak{a} = \mathbb{Z}[i]x$.[60] Among the four generators of an ideal $\mathfrak{a} \neq 0$ in $\mathbb{Z}[i]$, one often chooses the element in the quadrant $\Re z > 0$, $\Im z \geq 0$ or better in the sector $-\Re z < \Im z \leq \Re z$ as a natural representative.

To determine the multiplicative structure of the factorial monoid $\mathbb{Z}[i]^*$ and its Grothendieck group $\mathbb{Q}[i]^\times$, it is sufficient to know the set of prime elements of $\mathbb{Z}[i]^*$ in one of these areas. We denote by $\mathbb{P}_{\mathbb{Z}[i]^*}$ the set of prime elements in the second specified area. Note that with x also $\bar{x}$ is prime in $\mathbb{Z}[i]$, since the conjugation is an automorphism of $\mathbb{Z}[i]$. Furthermore, the prime elements of $\mathbb{Z}[i]^*$ are exactly the irreducible elements of $\mathbb{Z}[i]^*$. We first prove the following lemma:

Lemma 2.10.40 *Let $x \in \mathbb{Z}[i]^*$ be.*

(1) *If x is prime in $\mathbb{Z}[i]$, then x is a divisor of $N(x) = x\bar{x} \in \mathbb{N}^*$ and thus a divisor of a prime divisor $p \in \mathbb{P} \subseteq \mathbb{N}^*$ of $N(x)$. If $N(x) \in \mathbb{P}$ is prime in $\mathbb{Z}$, then x is prime in $\mathbb{Z}[i]$.*
(2) *If $p \in \mathbb{P}$ is a prime number $\equiv 3 \bmod 4$, then p is prime in $\mathbb{Z}[i]$.*
(3) *If $p \in \mathbb{P}$ is a prime number $\equiv 1 \bmod 4$, then $p = N(y) = y\bar{y}$ with $y \in \mathbb{P}_{\mathbb{Z}[i]^*}$.*
(4) *It is $1+i \in \mathbb{P}_{\mathbb{Z}[i]^*}$, and it holds $2 = N(1+i) = (1+i)(1-i) = -i(1+i)^2$.*

---

[60] These are exactly the points $\neq 0$ in $\mathfrak{a}$, which have minimal distance from 0. By considering these points for a given ideal $\mathfrak{a} \neq 0$, it can easily be shown elementarily geometrically that $\mathbb{Z}[i]$ is a principal ideal domain.

Proof (1) The first part of (1) is trivial. Now let $N(x) \in \mathbb{P}$. If $x = yz$ with $y,z \in \mathbb{Z}[i]^*$, then $N(x) = N(y)N(z)$ and thus $N(y) = 1$ or $N(z) = 1$, i.e. $y \in \mathbb{Z}[i]^\times$ or $z \in \mathbb{Z}[i]^\times$. Thus, x is irreducible and therefore prime. Because of $|\mathbb{Z}[i]/\mathbb{Z}[i]x| = N(x)$, according to Lemma 2.10.38, the residue class ring $\mathbb{Z}[i]/\mathbb{Z}[i]x \cong \mathbf{F}_{N(x)}$ is a field.

(2) For each $m \in \mathbb{N}^*$, the canonical homomorphism $\mathbb{Z}[i] \to \mathbf{A}_m[i]$ induces the canonical isomorphism $\mathbb{Z}[i]/\mathbb{Z}[i]m \xrightarrow{\sim} \mathbf{A}_m[i]$. Because of $p \equiv 3 \bmod 4$, -1 is not a square in $\mathbf{A}_p = \mathbf{F}_p$ and $\mathbf{F}_p[i]$ is a field, see Example 2.10.33. Consequently, $\mathbb{Z}[i]p$ is a prime ideal.

(3) If $p \equiv 1 \bmod 4$, then -1 is a square in F_p, see Example 2.9.35. Then $\mathbf{F}_p[i] \xrightarrow{\sim} \mathbf{F}_p \times \mathbf{F}_p$ is not a field and p is not prime in $\mathbb{Z}[i]$. Therefore, $p = uv$ with non-units $u, v \in \mathbb{Z}[i]$. Because of $p^2 = N(p) = N(u)N(v)$, $N(u) = N(v) = p$ is true, and u,v are prime according to (1). Then $p = u\bar{u}$ is true (and necessarily $v = \bar{u}$), see Example 2.10.34. One of the four elements associated with u then belongs to $\mathbb{P}_{\mathbb{Z}[i]^*}$.

(4) According to (1), $1 + i$ is prime in $\mathbb{Z}[i]$. □

It follows:

Theorem 2.10.41 *The prime elements in $\mathbb{P}_{\mathbb{Z}[i]^*}$, i.e. the prime elements $x = a + ib$ in $\mathbb{Z}[i]^*$ with $-a < b \le a$, are*

(1) $1 + i$. (2) $p \in \mathbb{P}$, $p \equiv 3 \bmod 4$.

(3) $a + ib$ and $a - ib$ with $0 < b < a$ and $a^2 + b^2 = p \in \mathbb{P}$, $p \equiv 1 \bmod 4$.

For $p \in \mathbb{P}$, $p \equiv 1 \bmod 4$, there is exactly one $a + ib \in \mathbb{Z}[i]$ with $0 < b < a$ and $a^2 + b^2 = p$.

In Fig. 2.25, the first prime elements in $\mathbb{P}_{\mathbb{Z}[i]^*}$ are marked according to theorem 2.10.41. Except for $1 + i$, they are symmetric to the real axis $\mathbb{R}$. By the way, it is unknown whether there are infinitely many prime elements $a + i$ in $\mathbb{Z}[i]$, i.e., whether there are infinitely many $a \in \mathbb{N}^*$ for which $a^2 + 1 \in \mathbb{P}$ is. Since according

Fig. 2.25 Prime elements in $\mathbb{Z}[i]$

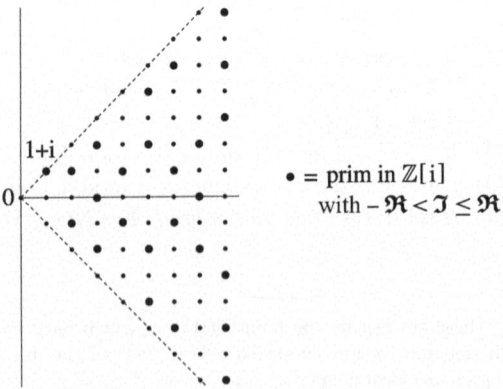

to exercise 1.7.6 there are infinitely many prime numbers $\equiv 3 \mod 4$, there are infinitely many prime elements of $\mathbb{Z}[i]$ on the real axis.

Corollary 2.10.42 (Fermat-Euler Two-Squares Theorem) *The number $R(n)$ of representations of the positive natural number $n \in \mathbb{N}^*$ as the sum of two squares of whole numbers, i.e., the number of pairs $(a,b) \in \mathbb{Z}^2$ with $a^2 + b^2 = n$, is*

$$R(n) := 4 \cdot \prod_{p \in \mathbb{P},\, p \equiv 1\ (4)} (\nu_p(n) + 1),$$

if the multiplicities $\nu_p(n)$ are even for all $p \in \mathbb{P}$ with $p \equiv 3 \mod 4$, and 0 otherwise. – In particular, n is exactly the sum of two squares if all prime divisors p of n with $p \equiv 3 \mod 4$ have an even multiplicity. Exactly then is n the sum of two coprime squares if n is of the form $n = m$ or $n = 2m$, where the prime divisors of m are all $\equiv 1 \mod 4$. If this condition is met, there are exactly $4 \cdot 2^r = 2^{r+2}$ different representations of n as the sum of two coprime squares, where r is the number of different prime divisors $\equiv 1 \mod 4$ of n.

Proof $R(n)$ is the number of $z \in A := \mathbb{Z}[i]$ with $N(z) = n = |A/Az|$. Because of $|A^\times| = 4$, $\frac{1}{4}R(n)$ is the number of (principal) ideals $\mathfrak{a} \subseteq A$ with $|A/\mathfrak{a}| = n$. We now use the representation of prime elements from $\mathbb{P}_{A^*}$ according to Theorem 2.10.41 and thus the unique representation of ideals $\neq 0$ in A by exactly one generating element from $M(\mathbb{P}_{A^*})$. Since moreover $|A/A(1+i)^\alpha| = 2^\alpha$, $|A/Ap^\alpha| = p^{2\alpha}$ for $p \in \mathbb{P}$, $p \equiv 3 \mod 4$ and $|A/A(a+ib)^\alpha| = |A/A(a-ib)^\alpha| = p^\alpha$ for $0 < b < a$, $a^2 + b^2 = p \equiv 1 \mod 4$, $p \in \mathbb{P}$, the assertion follows. – The first addition follows directly from the description of $R(n)$. The ideals Am, $m \in \mathbb{N}^*$, are exactly the products of the principal ideals $A(1+i)^2$, Ap, $p \equiv 3 \mod 4$ and $A(a+ib)(a-ib)$, where $a + ib$ is chosen as in theorem 2.10.41 (3). We have found those ideals $\mathfrak{a}$ in A with $|A/\mathfrak{a}| = n$ that can manage without a factor of the form $A(1+i)^2$ and of the form Ap, $p \equiv 3 \mod 4$. This is exactly possible for the specified n. The number of possible representations of n as the sum of two coprime squares also results from this. □

In a similar way, it is shown that as for the algebra $\mathbb{Z}[i]$ with the discriminant -4, the norm is also a Euclidean degree function for the imaginary quadratic $\mathbb{Z}$ algebras with the discriminants $-3, -7, -8, -11$ (see Ex. 2.10.25 and 2.10.26) and likewise for the real quadratic $\mathbb{Z}$ algebras with the discriminants 5, 8, 12, 13, 24. (There are further real quadratic $\mathbb{Z}$ algebras, whose norm function is a Euclidean degree function, however, there are no further Euclidean imaginary quadratic $\mathbb{Z}$ algebras, see Ex. 2.10.27.) In general, the real quadratic $\mathbb{Z}$ algebras are more difficult to handle, among other things because their unit groups are infinite. The imaginary quadratic $\mathbb{Z}$ algebras, which are still principal ideal domains, are those with the discriminants $-19, -43, -67, -163$ in addition to the already mentioned Euclidean domains. *So there are a total of nine imaginary quadratic $\mathbb{Z}$ algebras that are principal ideal domains.* This **theorem of Stark** was already suspected by

Gauss. Furthermore, it is suspected that there are infinitely many real quadratic $\mathbb{Z}$ algebras that are principal ideal domains. ◊

Example 2.10.43 (Quaternions) We consider another historically significant example of non-commutative algebras. Let R be any *commutative* ring $\neq 0$. As already noted, every free R algebra A of rank $n \in \mathbb{N}$ is isomorphic to a subalgebra of the matrix algebra $M_n(R)$, which itself is free of rank n^2 with the unit matrix E_n as a unit element. The Cayley representation, after distinguishing an R-basis of A, provides an embedding $A \to \mathrm{End}_R A \xrightarrow{\sim} M_n(R)$. The study of such matrix algebras is a significant subject of linear algebra, which is covered in volumes 3 and 4. Here we only consider 2×2 matrices. Of fundamental importance is the **determinant mapping**

$$\mathrm{Det}: M_2(R) \to R, \quad A = \begin{pmatrix} a & b \\ c & d \end{pmatrix} \mapsto \mathrm{Det}\, A := ad - bc.$$

As the reader immediately confirms, the mapping Det is multiplicative, i.e., it is $\mathrm{Det}\, E_2 = 1$ and $\mathrm{Det}(AB) = (\mathrm{Det}\, A) \cdot (\mathrm{Det}\, B)$. In particular, the determinant $ad - bc$ of an invertible matrix $A = \begin{pmatrix} a & b \\ c & d \end{pmatrix} \in M_2(R)^\times = \mathrm{GL}_2(R)$ is a unit in R. In this case, A is also invertible with

$$A^{-1} = \frac{1}{ad - bc} \begin{pmatrix} d & -b \\ -c & a \end{pmatrix},$$

as confirmed again by direct verification. We now choose for R the S-algebra $\mathbb{C}_S = S[i] = S[\sqrt{-1}]$ of complex numbers over the commutative ring $S \neq 0$ with the conjugation $z = a + bi \mapsto \bar{z} = a - bi$, $a, b \in S$, see Example 2.9.35. Then $M_2(\mathbb{C}_S)$ is a free $\mathbb{C}_S$ algebra of rank 4 and thus a free S-algebra of rank 8. In it,

$$\mathbb{H}_S := \left\{ \begin{pmatrix} z & w \\ -\bar{w} & \bar{z} \end{pmatrix} \,\bigg|\, z, w \in \mathbb{C}_S \right\}$$

is obviously a free S-subalgebra of rank 4 with the S-basis

$$1 = E_2 = \begin{pmatrix} 1 & 0 \\ 0 & 1 \end{pmatrix}, \quad i := \begin{pmatrix} i & 0 \\ 0 & -i \end{pmatrix}, \quad j := \begin{pmatrix} 0 & 1 \\ -1 & 0 \end{pmatrix}, \quad k := \begin{pmatrix} 0 & i \\ i & 0 \end{pmatrix}. \quad ◊$$

The multiplication in $\mathbb{H}_S$ is determined by the products

$$i^2 = j^2 = k^2 = -1, \quad ij = -ji = k, \quad jk = -kj = i, \quad ki = -ik = j$$

$\mathbb{H}_S$ is called the **quaternion algebra** over S. The algebra

$$\mathbb{H} := \mathbb{H}_\mathbb{R}$$

2.10 Principal Ideal Domains and Factorial Integral Domains

over $\mathbb{R}$ is the quaternion algebra par excellence.[61] Note that $\mathbb{H}_S$ is invariant both under the conjugation of $\mathbb{C}_S$ and under the transposition of matrices. The conjugation provides an S-automorphism of $M_2(\mathbb{C}_S)$ and the transposition a $\mathbb{C}_S$-anti-automorphism of $M_2(\mathbb{C}_S)$. In particular, the composition of both mappings induces an involutory S-anti-automorphism

$$x = a + b\mathrm{i} + c\mathrm{j} + d\mathrm{k} = \begin{pmatrix} z & w \\ -\overline{w} & \overline{z} \end{pmatrix} \mapsto \begin{pmatrix} \overline{z} & -w \\ \overline{w} & z \end{pmatrix} =: \overline{x} = a - b\mathrm{i} - c\mathrm{j} - d\mathrm{k},$$

$z := a + b\mathrm{i}$, $w := c + d\mathrm{i}$, of $\mathbb{H}_S$. It is called the **conjugation** of $\mathbb{H}_S$. The two S-algebras $\mathbb{H}_S$ and $\mathbb{H}_S^{\mathrm{op}}$ are therefore isomorphic. The determinant $\mathrm{N}(x) := \mathrm{Det}\, x = z\overline{z} + w\overline{w} = a^2 + b^2 + c^2 + d^2 = x\overline{x} \in S$ is called the **norm** and $\mathrm{Sp}(x) := x + \overline{x} = 2a \in S$ the **trace** of x. Obviously, $x^2 - \mathrm{Sp}(x)x + \mathrm{N}(x) = 0$. So if $S = K$ is a field and $x \notin S$, then $X^2 - \mathrm{Sp}(x)X + \mathrm{N}(x)$ is the minimal polynomial of x. The norm is like Det multiplicative and the trace S-linear. Moreover, it follows: *Exactly when is $x \in \mathbb{H}_S^\times$, when $\mathrm{N}(x) \in S^\times$ is. In this case, $x^{-1} = \overline{x}/\mathrm{N}(x)$ applies.* It follows: *Exactly when is $\mathbb{H}_S$ a division ring, when S is a field, in which the equation $a^2 + b^2 + c^2 + d^2 = 0$ only has the trivial solution $a = b = c = d = 0$.* Thus, for every ordered field K *(see Definition 3.1.1)* the quaternion algebra $\mathbb{H}_K$ and specifically $\mathbb{H} = \mathbb{H}_\mathbb{R}$ a division ring. The unit group $\mathbb{H}_\mathbb{Z}^\times$ of $\mathbb{H}_\mathbb{Z}$ contains exactly the quaternions $x = a + b\mathrm{i} + c\mathrm{j} + d\mathrm{k} \in \mathbb{H}_\mathbb{Z}$ with $\mathrm{N}(x) = a^2 + b^2 + c^2 + d^2 \in \mathbb{Z}^\times = \{\pm 1\}$, i.e. the eight elements ± 1, $\pm\mathrm{i}$, $\pm\mathrm{j}$, $\pm\mathrm{k}$, and is the already discussed in exercise 2.4.7 **Quaternion group** Q_4. The Theorem 2.6.22 cannot therefore be extended to division rings. However, it can be shown that: *A finite abelian subgroup of the multiplicative group $L^\times$ of a division ring L is always cyclic.*[62] According to a **theorem of Frobenius**, which we will prove in one of the later volumes, $\mathbb{R}, \mathbb{C}$ and $\mathbb{H}$ *up to isomorphism are the only finite $\mathbb{R}$-algebras that are also division rings.*

Every homomorphism $\varphi: S \to T$ of commutative rings $\neq 0$ induces a homomorphism $\mathbb{H}(\varphi): \mathbb{H}_S \to \mathbb{H}_T$ of quaternion algebras with $\mathbb{H}(\varphi)(a + b\mathrm{i} + c\mathrm{j} + d\mathrm{k}) = \varphi(a) + \varphi(b)\mathrm{i} + \varphi(c)\mathrm{j} + \varphi(d)\mathrm{k}$. It is surjective if and only if φ is surjective. Its kernel is $(\ker \varphi)\mathbb{H}_S$. In particular, the canonical isomorphism $\mathbb{H}_S/\mathfrak{a}\mathbb{H}_S \xrightarrow{\sim} \mathbb{H}_{S/\mathfrak{a}}$ holds for every ideal $\mathfrak{a} \neq S$ in S.

Similar to the proof of the Two-Squares Theorem 2.10.42 using the Gaussian integers, the quaternions can be used to prove the so-called **Four-Squares Theorem** of Lagrange: *Every natural number $m \in \mathbb{N}$ is the sum of four square numbers.* (See also exercise 2.10.28. – Three squares are not enough, as the numbers $m \equiv 7 \bmod 8$ show.) One considers the slightly larger $\mathbb{Z}$ subalgebra $\widetilde{\mathbb{H}}_\mathbb{Z} \subseteq \mathbb{H}_\mathbb{Q}$

---

[61] $\mathbb{H}$ reminds of W. Hamilton, who introduced the quaternions for the first time after preliminary work by Euler.

[62] With the theorem of Wedderburn, it can also be shown that: Every finite subgroup of the multiplicative group of a division ring with *positive* characteristic is cyclic.

of the so-called **Hurwitz Quaternions** (after A. Hurwitz (1859–1919)). It contains exactly the elements

$$x = \frac{1}{2}(a + bi + cj + dk) \in \mathbb{H}_\mathbb{Q} \quad \text{with} \quad a, b, c, d \in \mathbb{Z}, a \equiv b \equiv c \equiv d \bmod 2,$$

and is the largest $\mathbb{Z}$ subalgebra of $\mathbb{H}_\mathbb{Q}$, which includes $\mathbb{H}_\mathbb{Z}$ and all of its elements have integer norm and trace. $\widetilde{\mathbb{H}}_\mathbb{Z}$ has the $\mathbb{Z}$ basis $1, i, j, \frac{1}{2}(1 + i + j + k)$, and it is $[\widetilde{\mathbb{H}}_\mathbb{Z} : \mathbb{H}_\mathbb{Z}] = 2$. The unit group $\widetilde{\mathbb{H}}_\mathbb{Z}^\times$ contains, in addition to the eight elements of $\mathbb{H}_\mathbb{Z}^\times$, the sixteen half-integral quaternions $\frac{1}{2}(\varepsilon_0 + \varepsilon_1 i + \varepsilon_2 j + \varepsilon_3 k)$ with $\varepsilon_0, \ldots, \varepsilon_3 \in \{\pm 1\}$, and it is $\widetilde{\mathbb{H}}_\mathbb{Z} = \widetilde{\mathbb{H}}_\mathbb{Z}^\times \cdot \mathbb{H}_\mathbb{Z}$.

In conclusion, it should be noted that the above construction of the quaternion algebra $\mathbb{H}_S$ can be carried out in a similar way for any free quadratic S-algebra and also results in a free S-algebra of rank 4. When is this commutative, when is it a division ring?

Exercise

Exercise 2.10.1 Using the Euclidean algorithm, determine the greatest common divisor of the polynomials $F, G \in \mathbb{Q}[X]$ and represent it as a linear combination $SF + TG$, $S, T \in \mathbb{Q}[X]$, as follows:

$$F := X^4 + X^2 + 1, \quad G := X^2 + X + 1;$$

$$F := 3X^4 + \frac{3}{2}X^3 + \frac{7}{2}X^2 + 2X + 2, \quad G := 2X^2 + X + 1;$$

$$F := 2X^3 - 4X^2 + X - 2, \quad G := X^3 - X^2 - X - 2;$$

$$F := X^5 + X - 1, \quad G := X^5 - X^4 + 2X^3 + 1.$$

Exercise 2.10.2 A is an integrity domain with Euclidean degree function $\varphi : A^* \to \mathbb{N}$.

a) Every element $a \in A^*$, for which $\varphi(a)$ is minimal in $\varphi(A^*)$, is a unit in A.
b) Every element $p \in A^* - A^\times$, for which $\varphi(p)$ is minimal in $\varphi(A^* - A^\times)$, is a prime element, for which the canonical group homomorphism $A^\times \to (A/Ap)^\times = (A/Ap) - \{0\}$ is surjective.
c) If φ is a monoid homomorphism $(A^*, \cdot) \to (\mathbb{N}, +)$ with $\varphi(a+b) \leq \text{Max}(\varphi(a), \varphi(b))$ for $a, b, a+b \in A^*$, then A is a field or isomorphic to a polynomial ring in one indeterminate over a field. (For $A^* \neq A^\times$ choose a p as in b) as indeterminate.)

2.10 Principal Ideal Domains and Factorial Integral Domains

Exercise 2.10.3 A finite product of principal ideal rings is a principal ideal ring.

Exercise 2.10.4 Let S be a commutative ring $\neq 0$ and $S[X]$ the polynomial algebra over S in one variable.

a) X is prime in $S[X]$ if and only if S is an integral domain.
b) X is irreducible in $S[X]$ if and only if S (as a ring) is indecomposable.
c) $S[X]$ is a principal ideal domain if and only if S is a field. In particular, the polynomial rings $K[X_i]_{i\in I}$, K field, $|I| \geq 2$, and $\mathbb{Z}[X_i]_{i\in I}$, $|I| \geq 1$, are not principal ideal domains. (When is $S[X]$ a principal ideal ring?)

Exercise 2.10.5 Let A be a principal ideal domain and B be a ring with $A \subseteq B \subseteq K := Q(A)$.

a) B is the ring of fractions A_S with $S := A \cap B^\times$ and also a principal ideal domain.
b) The mapping $S \mapsto A_S = A_{M(S)}$ is a bijective mapping of the set $\mathfrak{P}(\mathbb{P}_{A^*})$ onto the set of intermediate rings C with $A \subseteq C \subseteq K$.
c) If φ is a Euclidean degree function on A, then the mapping $\psi : B^* \to \mathbb{N}$, $x \mapsto \operatorname{Min} \varphi(A^* \cap Bx)$, is a Euclidean degree function on B. (If $B = A$, then ψ is a Euclidean degree function on A, which is monotonic with respect to divisibility: If $x \mid y$ holds, then $\psi(x) \leq \psi(y)$.)

Exercise 2.10.6 Let A be a factorial integrity domain with $|\mathbb{P}_{A^*}| = r \in \mathbb{N}$ finite and $\mathbb{P}_{A^*} = \{p_1, \ldots, p_r\}$. We set $v_\rho := v_{p_\rho}$ for the multiplicities v_{p_ρ} of the p_ρ, $\rho = 1, \ldots, r$. Then

$$\varphi : A^* \to \mathbb{N}, \quad \varphi(a) := \sum_{\rho=1}^{r} v_\rho(a),$$

is a Euclidean degree function on A. In particular, A is a principal ideal domain. (Let $a, b \in A^*$. For the existence of an element $q \in A^*$ with $\varphi(a - qb) < \varphi(b)$ (for $a \neq qb$), one can assume $a \neq qb$ for all $q \in A$. Then $v_{p_0}(b) > v_{p_0}(a)$ for at least one $\rho_0 \in \{1, \ldots, r\}$, and for q the element $\prod_{\rho=1}^{r} p_\rho^{\alpha_\rho}$ with $\alpha_\rho := 0$, if $v_\rho(a) \neq v_\rho(b)$, and $\alpha_\rho := 1$, if $v_\rho(a) = v_\rho(b)$, can be chosen. Note: $v_\rho(x + y) = \operatorname{Min}(v_\rho(x), v_\rho(y))$, if $v_\rho(x) \neq v_\rho(y)$.)

Note If $r = 1$, i.e., A has exactly one prime element p apart from associativity, then the principal ideal domain A is called a **discrete valuation ring** and p a **(local) uniformizer** of A. The elements of A^* then have the representation $x = \varepsilon p^{v_p(x)}$ with $\varepsilon \in A^*$. The p-exponent $v_p : A \to \overline{\mathbb{N}} = \mathbb{N} \cup \{\infty\}$ (with $v_p(0) = \infty$) is called the **(discrete) standard valuation** of A. According to the proven, its restriction to A^* is a Euclidean degree function on A. The discrete valuation rings are therefore exactly the local principal ideal domains that are not fields. If p is a prime element $\neq 0$ in any factorial integrity domain A, then the localization A_{Ap} is a discrete valuation ring with local uniformizer p. If K is a field, then the formal power series ring $K \llbracket X \rrbracket$ in an indeterminate X is a discrete valuation ring with local uniformizer X. The standard valuation agrees with the order, see Note 2.9.22(4).

Exercise 2.10.7
a) A factorial integrity domain A is exactly a principal ideal domain when every prime element $p \in A^*$ generates (not just a prime ideal, but even) a maximal ideal.
b) A finite algebra A over a principal ideal domain S, which is a factorial integrity domain, is a principal ideal domain. In particular, every finite factorial $\mathbb{Z}$-algebra or even every finite factorial $K[X]$-algebra, K field, is a principal ideal domain. (One can assume $S \subseteq A$. If $p \in A^*$ is a prime element, then $\mathfrak{m} := S \cap Ap$ is a prime ideal $\neq 0$, thus maximal; because the constant term of an integrality equation of p over S, which lies in Ap, can be assumed to be different from 0 after (possible) shortening of a power of p. Then, however, A/Ap is also a field like $S/\mathfrak{m} \subseteq A/Ap$, cf. exercise 2.9.17d).

Exercise 2.10.8 Let $F, G \in S[X]^*$ be non-zero polynomials over the integrity domain S with $\deg G \geq 1$. Then $\deg F(G) = \deg F \cdot \deg G$.

Exercise 2.10.9 The field K is algebraically closed and $n \in \mathbb{N}$, $n \geq 2$. Then every non-constant polynomial $F \in K[X_1, \ldots, X_n]$ has infinitely many zeros in K^n.

Exercise 2.10.10 Let K be a field. Then every non-constant rational function $F/G \in K(X_i, i \in I)^\times$ is transcendental over K. (Without loss of generality, let $|I| = 1$.)

Exercise 2.10.11 (Hermite Interpolation) (after Ch. Hermite (1822–1901)) Let $c_1, \ldots, c_r$ be pairwise different elements in the field K and $\alpha_1, \ldots, \alpha_r \in \mathbb{N}^*$. For any given $a_1^{(0)}, \ldots, a_1^{(\alpha_1-1)}, \ldots, a_r^{(0)}, \ldots, a_r^{(\alpha_r-1)} \in K$, there is a uniquely determined polynomial $F \in K[X]$ of degree $< \alpha_1 + \cdots + \alpha_r$, whose Taylor expansion around c_ρ has the form

$$F = a_\rho^{(0)} + a_\rho^{(1)}(X - c_\rho) + \cdots + a_\rho^{(\alpha_\rho-1)}(X - c_\rho)^{\alpha_\rho-1} + \cdots, \quad \rho = 1, \ldots, r.$$

. (Theorem 2.10.22)

Note If $\operatorname{Char} K = 0$ or $\operatorname{Char} K \geq \operatorname{Max}(\alpha_1, \ldots, \alpha_r)$, the coefficients $a_\rho^{(\nu_\rho)}$ can be determined using the Taylor formula $F^{(\nu_\rho)}(c_\rho) = \nu_\rho! a_\rho^{(\nu_\rho)}$ also by the values of the derivatives $F^{(\nu_\rho)}(c_\rho)$, see exercise 2.9.4b). – In Vol. 2, Sect. 2.9, we will describe an algorithm for determining the interpolating polynomial F. In the case of $\alpha_1 = \cdots = \alpha_r = 1$, F can be directly specified using the so-called **Lagrange interpolation formula**:

$$F = \sum_{\rho=1}^r a_\rho^{(0)} G_\rho, \quad G_\rho := \prod_{\sigma \neq \rho} \frac{X - c_\sigma}{c_\rho - c_\sigma}, \quad \rho = 1, \ldots, r.$$

2.10 Principal Ideal Domains and Factorial Integral Domains

However, in this case, the so-called **Newton interpolation** (after I. Newton (1643–1727)) is often more convenient for calculations: Here, one successively determines polynomials F_i of degree $<i$, which at the points $c_1, \ldots, c_i$ already take the required values $a_1^{(0)}, \ldots, a_i^{(0)}$, $i = 1, \ldots, r$. Apparently, the following recursively defined polynomials do this:

$$F_1 = a_1^{(0)}, \quad F_{i+1} = F_i + \left(a_{i+1}^{(0)} - F_i(c_{i+1})\right) \frac{(X - c_1) \cdots (X - c_i)}{(c_{i+1} - c_1) \cdots (c_{i+1} - c_i)}, \quad i = 1, \ldots, r-1.$$

Exercise 2.10.12 Let K be a field and A a K-algebra.

a) If $x \in A$ is algebraic over K of degree n, then every element $y \in K[x]$ is algebraic of degree $\leq n$ and exactly of degree n when $K[y] = K[x]$ is.
b) Let $x \in A$ be nilpotent. Then x is algebraic over K, and the degree of x is equal to the nilpotency degree of x. What degree does $(a + x)^m$ have for $a \in K$ and $m \in \mathbb{N}^*$? (Distinguish the cases $a = 0$ and $a \neq 0$ and consider in particular the case always present at $\operatorname{Char} K = 0$ that $m = m \cdot 1_K \neq 0$ is.)

Exercise 2.10.13 Let $\varphi \colon A \to B$ be a surjective K-algebra homomorphism of algebraic algebras over the field K. Then the induced homomorphism $\varphi^\times \colon A^\times \to B^\times$ is also surjective. (Let $y \in B^\times$ and $\varphi(x) = y$. Then $y \in K[y]^\times$. Now consider the restriction $\varphi | K[x] \colon K[x] \to K[y]$ and use Theorem 2.10.22. – Similarly, if $\varphi \colon A \to B$ is a surjective homomorphism of finite rings, then $\varphi^\times \colon A^\times \to B^\times$ is also surjective.)

Exercise 2.10.14 Let $p \in \mathbb{P}$ be a prime number ≥ 3. There exists a ring with exactly p units if and only if $p = M(q) = 2^q - 1$ (q prime) is a Mersenne prime number.

Exercise 2.10.15 Let $p \in \mathbb{P}$ be a prime number. Up to isomorphism, there are exactly 4 rings A with p^2 elements, namely $\mathbf{A}_{p^2}, \mathbf{F}_{p^2}, \mathbf{F}_p \times \mathbf{F}_p, \mathbf{F}_p[\varepsilon] = \mathbf{F}_p[X]/(X^2)$. (They are all commutative. If $\operatorname{Char} A = p$, then A is a square F_p-algebra. For $p = 2$ see Exercise 2.6.9a). – With the remark following the Chinese Remainder Theorem 2.7.11, the structure of all finite rings of an order m with $v_p(m) \leq 2$ for all $p \in \mathbb{P}$ is clarified.)

Exercise 2.10.16 Let $K = \mathbf{F}_q$ be a finite field with q elements (see Example 2.10.33) and $F \in K[X]$ a normalized polynomial with the prime factor decomposition $F = \pi_1^{\alpha_1} \cdots \pi_r^{\alpha_r}$, $\pi_1, \ldots, \pi_r \in \mathbb{P}_{K[X]^*}$ pairwise different, $\alpha_1, \ldots, \alpha_r \in \mathbb{N}^*$, as well as the residue class algebra $A = K[x] := K[X]/(F)$. What order does the unit group $A^\times$ have? When is $A^\times$ cyclic?

Exercise 2.10.17 Let $K = \mathbf{F}_q$ be a finite field with q elements (see Example 2.10.33).

a) There are $\binom{q}{2}$ normalized prime polynomials of degree 2 over K and exactly $2\binom{q+1}{3}$ normalized prime polynomials of degree 3 over K.

b) Let $s_q(m)$ be the number of normalized prime polynomials of degree m in $K[X]$. Then $q^m = \sum_{d|m} s_q(d)d$ applies (see the remark at the end of Example 2.10.33). With the Möbius inversion formula from Exercise 2.1.23b) one can deduce the Gaussian formula

$$s_q(m) = \frac{1}{m} \sum_{d|m} \mu\left(\frac{m}{d}\right) q^d.$$

(By the way, this directly implies $s_q(m) > 0$ due to

$$\sum_{d|m, d \neq m} q^d < \sum_{k=1}^{m-1} q^k = (q^m - q)/(q-1) < q^m$$

and specifically $s_q(p) = q(q^{p-1} - 1)/p$ for any prime number p.)

Exercise 2.10.18 Let A be an (not necessarily factorial) integrity domain with quotient field K.

a) Two polynomials $F, G \in A[X]^*$ are exactly co-prime in $K[X]$, if there are polynomials $S, T \in A[X]$ with $SF + TG \in A^*$.

b) A *normalized* polynomial $F \in A[X]^*$ is prime in $A[X]$ if and only if it is prime in $K[X]$. (It is $A[X]/(F) \hookrightarrow K[X]/(F)$. – Provide an example of an integral domain A and a (non-normalized) polynomial $F \in A[X]^*$ without a divisor $a \in A^* - A^\times$, which is prime in $K[X]$, but not prime in $A[X]$.)

c) If $a, b \in A^*$ with $\mathrm{kgV}(a,b) = ab$ (i.e. $Aab = Aa \cap Ab$), then $bX - a$ is prime in $A[X]$ and the substitution homomorphism $X \mapsto a/b \in K$ induces an isomorphism $A[X]/(bX - a) \xrightarrow{\sim} A[a/b] (\subseteq K)$.

d) Let $F = a_0 + a_1 X + \cdots + a_n X^n \in A[X]^*$ be a non-constant polynomial and $p \in A^*$ a prime element with $v_p(a) < \infty$ for all $a \in A^*$ (i.e. $\bigcap_{k \in \mathbb{N}} Ap^k = 0$) and $p \mid a_i$ for $i = 0, \ldots, n-1$, $p \nmid a_n$ and $p^2 \nmid a_0$. Then F is a prime polynomial in $K[X]$. (This is a classic formulation of the **Eisenstein's Lemma**, cf. Ex. 2.9.2b). – Thus, for example, all polynomials $X^n - a$ are prime, $n \in \mathbb{N}^*$, if A is factorial and a prime element $p \in A^*$ exists with $v_p(a) = 1$.)

Exercise 2.10.19 Let $p \in \mathbb{P}$ be a prime number. For a polynomial $F \in \mathbb{Z}[X]$, denote $\overline{F}$ the image of F in $\mathbb{F}_p[X]$ under the canonical homomorphism $\mathbb{Z}[X] \to \mathbb{F}_p[X]$. Show: If $\mathrm{Grad}\, F = \mathrm{Grad}\, \overline{F}$ and $\overline{F}$ are prime in $\mathbb{F}_p[X]$, then F is prime in $\mathbb{Q}[X]$. – With this result, the irreducibility of a polynomial in $\mathbb{Q}[X]$ can often be proven very simply. Since, for example, the polynomial $X^4 + X^3 + 1$ is prime in $\mathbb{F}_2[X]$, for even integers a_1, a_2 and odd integers a_0, a_3, a_4 all polynomials $a_4 X^4 + a_3 X^3 + a_2 X^2 + a_1 X + a_0 \in \mathbb{Z}[X]$ are prime in $\mathbb{Q}[X]$. (However, there are normalized irreducible polynomials F in $\mathbb{Z}[X]$ such that $\overline{F}$ is decomposable for *all* prime numbers $p \in \mathbb{P}$, e.g. $F := X^4 + 1 = \mu_{\zeta_8}$, $\zeta_8 = e^{2\pi i/8} = (1+i)/\sqrt{2}$, (Proof!)

2.10 Principal Ideal Domains and Factorial Integral Domains

— One should formulate and prove analogous statements for a prime element $p \neq 0$ in a factorial ring A and polynomials $F \in A[X]$ or $F \in Q(A)[X]$.

Exercise 2.10.20
a) Determine the minimal polynomials of the following real algebraic numbers over $\mathbb{Q}$: $\sqrt[n]{p}$, $n \in \mathbb{N}^*$, p prime ; $\sqrt{2} + \sqrt{3}$; $\sqrt{2} + \sqrt[3]{3}$; $\sqrt[3]{2} + \sqrt[3]{3}$.
b) For a prime number $p \in \mathbb{P}$, the numbers $p^r \in \mathbb{R}$, $r \in \mathbb{Q}$, $0 \leq r < 1$, are linearly independent over $\mathbb{Q}$. (The polynomials $X^n - p$, $n \in \mathbb{N}^*$, are all prime in $\mathbb{Q}[X]$, cf. Exercise 2.10.18d).

Exercise 2.10.21 (Polynomials with coefficients in $k(Z)$) Let k be a field. We consider the polynomial algebra $K[X] = k(Z)[X]$ with the rational function field $K := k(Z)$ as the base field. Then every polynomial possesses $F \in K[X]$, $F \neq 0$, has a unique representation $F = I(F)F^*$ with the content $I(F) \in k(Z)^\times$ of F and the primitive part $F^* \in k[Z][X] =: k[Z, X]$ of F (whose coefficients are coprime polynomials in $k[Z]$ and whose leading coefficient is also normalized). When examining $F^* \in k[Z, X]$, the roles of Z and X can be swapped, i.e., F^* can also be considered as a polynomial in $k[X][Z] \subseteq k(X)[Z]$. For example, the (in X normalized) polynomial $F = F^* = (X-1)^2(X^2 + Z^2) - X^2$ with $\text{Grad}_X F = 4$ from $k(Z)[X]$ is prime for Char $k \neq 2$ due to $F = (X-1)^2 Z^2 + X^3(X-2)$ and since $Z^2 + X^3(X-2)/(X-1)^2 \in k(X)[Z]$ is prime. – Another simple example is the following: If $R \in k(Z)^\times$, $R \notin k$, $R = F/G$ with coprime polynomials $F, G \in k[Z]^*$, then R is transcendental over k (see exercise 2.10.10) and $k(Z)$ is algebraic over $k(R)$. The minimal polynomial of Z over $k(R)$ is, up to a normalization factor from $k(R)^\times$, the (in R even linear) polynomial $F(X) - G(X)R \in k(R)[X]$. In particular, $[k(Z):k(R)] = \dim_{k(R)} k(Z) = \text{Max}(\deg F, \deg G)$ applies, and only then is $k(R) = k(Z)$, $\varphi_R : k(Z) \to k(Z)$, $Q \mapsto Q(R)$, thus a k-algebra automorphism of $k(Z)$, when R is a piecewise linear function:

$$R = \frac{a + bZ}{c + dZ} \quad \text{with} \quad a, b, c, d \in k, \ ad - bc \neq 0.$$

(The condition $ad - bc \neq 0$ ensures that R does not lie in k.) The group of these **piecewise linear functions**, i.e., the group $\text{Aut}_{k\text{-Alg}}\, k(Z)$, is called the group of **Möbius transformations** over k or simply the **Möbius group** of k. (Note the contravariance $\varphi_R \circ \varphi_S = \varphi_{S(R)}$ for two piecewise linear functions R,S. In Vol. 3, the Möbius group is identified with the projective group $\text{PGL}_2(k)$.) — More generally, one shows: If $\pi \in k[X]$ is prime and the polynomials $F_0, F_1 \in k[Z]$, $F_1 \neq 0$, are not both constant, then π and $\pi(F_1 X + F_0)$ is prime in $k(Z)[X]$ and it holds $I(\pi(F_1 X + F_0)) = 1$, i.e. exactly then is $\pi(F_1 X + F_0) \notin k(X)$ also prime in $k[Z, X]$ and $k(X)[Z]$, if $\gcd(\pi(F_0), F_1) = 1$ is.

Exercise 2.10.22 Let A be a factorial integrity domain with quotient field K. Furthermore, let $F, G \in K[X]$ be non-constant coprime polynomials over K. Then the set of $\gcd\bigl(F(a), G(a)\bigr)$, $a \in A$, is finite (where the greatest common divisors are to be chosen as elements of $M(\mathbb{P}_{A^*})$, i.e., as divisor classes). If $A^\times$ is finite (i.e., each $a \in A^*$ has only a finite number of divisors in A, e.g., $A = \mathbb{Z}$), then there are only a finite number of $a \in A$, for which the values $F(a)/G(a)$ of the rational function $F/G \in K(X) - K[X]$ (are defined and) also lie in A. Provide an algorithm to determine these numbers a.

Exercise 2.10.23 Let $A \subseteq B$ be an extension of factorial integral domains. Furthermore, let $P_A \subseteq P_B$ be the polynomial algebras $P_A := A[X_i, i \in I]$ and $P_B := B[X_i, i \in I]$. For any two elements $a, b \in A^*$, let the $\gcd_A(a,b)$ of a,b in A also be a greatest common divisor $\gcd_B(a,b)$ of a,b in B (this condition is equivalent to $\operatorname{lcm}_A(a,b) = \operatorname{lcm}_B(a,b)$ for all $a, b \in A^*$). Then $\gcd_{P_A}(F,G) = \gcd_{P_B}(F,G)$ also applies to $F, G \in P_A^*$. (Without restriction, let $|I| = 1$.) In particular, the last equation always holds when A is a principal ideal domain. Provide an example of factorial integral domains $A \subseteq B$, for which the condition about the greatest common divisors is *not* fulfilled.

Exercise 2.10.24 Let $A = \sum_{n\in\mathbb{Z}}^{\oplus} A_n$ be a $\mathbb{Z}$-graded factorial integral domain. If $a_k \in A_k$ and $a_{k+1} \in A_{k+1}$ are two coprime homogeneous elements $\neq 0$ of adjacent degrees $k, k+1$, then $a_k + a_{k+1}$ is prime in A.

Exercise 2.10.25 Let A be the imaginary quadratic $\mathbb{Z}$-algebra $\mathbb{Z}[\sqrt{-2}] = \{a + ib\sqrt{2} \mid a, b \in \mathbb{Z}\} = \mathbb{Z}[i\sqrt{2}] \subseteq \mathbb{C}$ with the discriminant -8. It is $A^\times = \{\pm 1\}$. In this exercise treat the algebra A similarly to the algebra $\mathbb{Z}[i]$ of the entire Gaussian numbers at the end of Example 2.10.37.

a) A is a Euclidean integral domain with the norm $N(a + ib\sqrt{2}) = a^2 + 2b^2$ as a Euclidean degree function. In particular, A is a principal ideal domain. For $\mathbb{P}_{A^*}$ we choose the prime elements $a + ib \in A^*$ with $a > 0$ or with $a = 0$ and $b > 0$.

b) The elements of $\mathbb{P}_{A^*}$ are: (1) $i\sqrt{2}$. (2) $p \in \mathbb{P} = \mathbb{P}_{\mathbb{Z}^*}$, $p \equiv 5$ or $p \equiv 7 \bmod 8$. (3) $a + ib\sqrt{2}$ and $a - ib\sqrt{2}$ with a,b > 0 and $a^2 + 2b^2 = p \in \mathbb{P}$, $p \equiv 1$ or $p \equiv 3 \bmod 8$. For the prime number $p \equiv 1$ or $p \equiv 3 \bmod 8$ there is exactly one pair $(a,b) \in \mathbb{N}^* \times \mathbb{N}^*$ with $a^2 + 2b^2 = p$. (It $(-2/p) = (-1/p)(2/p) = 1$, if $p \equiv 1$ or $p \equiv 3 \bmod 8$, and $(-2/p) = -1$, if $p \equiv 5$ or $p \equiv 7 \bmod 8$.)

c) Let $n \in \mathbb{N}^*$. The number $S(n)$ of pairs $(a,b) \in \mathbb{Z}^2$ with $a^2 + 2b^2 = n$ is

$$S(n) = 2 \cdot \prod_{p \in \mathbb{P}, p \equiv 1, 3 \ (8)} (v_p(n) + 1),$$

if $v_p(n)$ is even for all $p \in \mathbb{P}$ with $p \equiv 5$ or $p \equiv 7 \bmod 8$, and 0 otherwise. – In particular, n has a representation $n = a^2 + 2b^2$ with $a, b \in \mathbb{N}$ exactly when all prime divisors p of n with $p \equiv 5$ or $p \equiv 7 \bmod 8$ have an even multiplicity. When can a and b be chosen to be coprime?

Exercise 2.10.26
a) Let $D \in \mathbb{N}^*$, $D > 2$. In the imaginary quadratic $\mathbb{Z}$-algebra $A := \mathbb{Z}[\sqrt{-D}] = \mathbb{Z}[i\sqrt{D}]$ (with the discriminant $-4D$), the ideal $\mathfrak{m}$, generated by 2 and $i\sqrt{D}$, if D is

2.10 Principal Ideal Domains and Factorial Integral Domains 343

even, or by 2 and $1 + i\sqrt{D}$, if D is odd, is generated, at most with $A/\mathfrak{m} = \mathbb{F}_2$. The ideal $\mathfrak{m}$ is not a principal ideal (since there is no element of norm 2 in A). In particular, A is not a principal ideal domain and therefore not factorial, see Problem 2.10.7b). (**Remark** The algebra $\mathbb{Z}[\sqrt{-5}]$ is R. Dedekind's prime example of a finite free $\mathbb{Z}$-algebra, which is a non-factorial integrity domain. For example, in $\mathbb{Z}[\sqrt{-5}]$ the equation $2 \cdot 3 = (1 + \sqrt{-5})(1 - \sqrt{-5}) = N(1 + \sqrt{-5})$ holds, and all four factors $2, 3, (1 + \sqrt{-5}), (1 - \sqrt{-5})$ are indecomposable and not associated in pairs. – In $\mathbb{Z}[\sqrt{-3}]$ the normalized polynomial $X^2 - X + 1$ has no root, but in the quotient field $\mathbb{Q}[\sqrt{-3}]$ the roots $\frac{1}{2}(1 \pm \sqrt{-3}) = \zeta_6^{\pm 1}$. This polynomial is therefore irreducible in $\mathbb{Z}[\sqrt{-3}][X]$, but not prime, see Lemma 2.10.12. There are no such examples over $\mathbb{Z}[\sqrt{-5}]$.)

b) The quadratic $\mathbb{Z}$ algebra $\mathbb{Z}[\frac{1}{2}(1 + \sqrt{-3})] = \mathbb{Z}[\zeta_6]$ with the discriminant -3 is, like the algebras with the discriminants -7 and -11, Euclidean with the norm as the Euclidean degree function.

Exercise 2.10.27 An imaginary quadratic $\mathbb{Z}$ algebra A with discriminant < -11 cannot be a Euclidean domain. (*A* only has the two units ±1 and does not possess an element with norm 2 or 3. Now use exercise 2.10.2b).)

Exercise 2.10.28 We consider the free $\mathbb{Z}$ algebras $\mathbb{H}_\mathbb{Z} \subseteq \widetilde{\mathbb{H}}_\mathbb{Z}$ of the quaternions or the Hurwitz quaternions over $\mathbb{Z}$ and want to propose a proof of the **Four-Squares Theorem**, see Example 2.10.43. N: $\widetilde{\mathbb{H}}_\mathbb{Z} \to \mathbb{N} \subseteq \mathbb{Z}$ denotes the multiplicative norm function.

a) The natural numbers $m \in \mathbb{N}$, which can be represented as the sum of four square numbers, form a submonoid of $(\mathbb{N}, \cdot)$. To prove the Four-Squares Theorem, it is therefore sufficient to show that every prime number $p \in \mathbb{P}$ is the sum of four squares.

b) For each element $x \neq 0$ from $\mathbb{H}_\mathbb{Z}$ or from $\widetilde{\mathbb{H}}_\mathbb{Z}$, $|\mathbb{H}_\mathbb{Z}/\mathbb{H}_\mathbb{Z}x| = N(x)^2$ or $|\widetilde{\mathbb{H}}_\mathbb{Z}/\widetilde{\mathbb{H}}_\mathbb{Z}x| = N(x)^2$ applies. (One concludes as in the proof of Lemma 2.10.38.)

c) Let $p \in \mathbb{P}$, $p > 2$. Then $\mathbb{H}_{\mathbb{F}_p} = \mathbb{H}_\mathbb{Z}/\mathbb{H}_\mathbb{Z}p \xrightarrow{\sim} \widetilde{\mathbb{H}}_\mathbb{Z}/\widetilde{\mathbb{H}}_\mathbb{Z}p$ is not a division domain. (The equation $a^2 + b^2 + c^2 + d^2 = 0$ (and even the equation $a^2 + b^2 + c^2 = 0$) has nontrivial solutions in F_p. If ^2F_p is the set of $(p+1)/2$ squares in F_p, then $^2F_p + ^2F_p = F_p$ applies, see exercise 2.1.14c).) In $\widetilde{\mathbb{H}}_\mathbb{Z}$ there is therefore (according to Proposition 2.7.16) a left ideal, whose index is a nontrivial divisor of p^4.

d) In $\widetilde{\mathbb{H}}_\mathbb{Z}$ the following division with remainder can be performed: If $x, y \in \widetilde{\mathbb{H}}_\mathbb{Z}$ with $x \neq 0$, then there exists $q, r \in \widetilde{\mathbb{H}}_\mathbb{Z}$ with $y = qx + r$ and $N(r) < N(x)$. (See the proof of Theorem 2.10.39.) $\widetilde{\mathbb{H}}_\mathbb{Z}$ is a left principal ideal domain (and a right principal ideal domain). ($\mathbb{H}_\mathbb{Z}$ is not a left principal ideal domain!)

e) If $p \in \mathbb{P}$, $p > 2$, then there exists an element $\widetilde{x} \in \widetilde{\mathbb{H}}_\mathbb{Z}$ with $N(\widetilde{x}) = p$. Because of $\widetilde{\mathbb{H}}_\mathbb{Z} = \widetilde{\mathbb{H}}_\mathbb{Z}^\times \cdot \mathbb{H}_\mathbb{Z}$ there also exists an element $x \in \mathbb{H}_\mathbb{Z}$ with $N(x) = p$, which proves the Four-Squares Theorem. (Use b), c) and d).)

Exercise 2.10.29 Let A be an algebra $\neq 0$ over the field K and $x \in A$.

a) For pairwise different elements $c_1, \ldots, c_n \in K$ with $x - c_1, \ldots, x - c_n \in A^\times$ the following statements are equivalent: (i) $(x - c_1)^{-1}, \ldots, (x - c_n)^{-1}$ are linearly independent over K. (ii) $1, x, \ldots, x^{n-1}$ are linearly independent over K. (iii) There is $\text{Dim}_K K[x] \geq n$.

b) Let A be a division ring. If $\text{Dim}_K A < |K|$, then A is algebraic over K. In particular, $A = K$, when K is algebraically closed and $\text{Dim}_K A < |K|$.

c) Let K be an uncountable algebraically closed field (e.g. $K = \mathbb{C}$). Then K is, up to isomorphism, the only division ring that is a countably generated K-algebra. If $\mathfrak{m}$ is a maximal ideal in the polynomial algebra $P := K[X_i, i \in I]$ and I is countable, then $\mathfrak{m}$ is a point ideal $\mathfrak{m}_c = \sum_{i \in I} P \cdot (X_i - c_i)$, $c := (c_i) \in K^I$, see exercise 2.10.3. (**Remark** If I is finite, the last statement also holds for countable algebraically closed fields K. This is the so-called **Hilbert's Nullstellensatz**.) □

Exercise 2.10.30 Let $E \subseteq \mathbb{Z}$ is a finite subset of $\mathbb{Z}$ with $m := |E| \geq 2$ and $\sigma \in \mathfrak{S}(E)$. Furthermore, let $P \in \mathbb{Q}[X]$ be the uniquely determined interpolating polynomial of degree $< m$ with $P(x) = \sigma x (= \sigma(x))$ for all $x \in E$.

a) Exactly then is $P \in \mathbb{Z}[X]$, when $|\sigma y - \sigma x| = |y - x|$ is for all $x, y \in E$. This is exactly the case when $P = X$ or $P = -X + a_0$ with a $a_0 \in \mathbb{Z}$ is. (Let $x, y \in E$ be arbitrary with $x \neq y$. Furthermore, let $x_0 = x$, $x_1 = y, x_2, \ldots, x_{m-1}$ is a count of E. Then

$$P = a_0 + a_1(X - x_0) + \cdots + a_{m-1}(X - x_0) \cdots (X - x_{m-2})$$

with $a_0, a_1, \ldots, a_{m-1} \in \mathbb{Q}$, cf. the Newton interpolation in exercise 2.10.11. Only then is $P \in \mathbb{Z}[X]$, when all a_i, $i = 0, \ldots, m-1$, are in $\mathbb{Z}$. For $P \in \mathbb{Z}[X]$ in particular, $a_1 = (\sigma x_1 - \sigma x_0)/(x_1 - x_0) \in \mathbb{Z}$ applies. Because of

$$\prod_{i<j}(\sigma x_j - \sigma x_i) = (\text{Sign}\, \sigma) \prod_{i<j}(x_j - x_i) \quad \text{follows} \quad \prod_{i<j}(\sigma x_j - \sigma x_i)/(x_j - x_i) = \text{Sign}\, \sigma$$

and $(\sigma y - \sigma x)/(y - x) = (\sigma x_1 - \sigma x_0)/(x_1 - x_0) = \pm 1$.)

b) One can deduce from a): If $Q \in \mathbb{Z}[X]$ is a polynomial of degree $n > 1$, then the set E_Q of purely periodic points of the discrete dynamic system defined by Q $Q: \mathbb{Z} \to \mathbb{Z}$ has at most n elements. Either E_Q is the set of fixed points of Q or E_Q contains at least two elements and $Q(x) = -x + a_0$ is true for all $x \in E_Q$ (with a $a_0 \in \mathbb{Z}$). Given Q, one should decide which case applies, and provide an algorithm for determining E_Q. (Consider, for example, Q and $Q \circ Q$.– Of course, the dynamic system $(\mathbb{Z}, Q)$ can have periodic points that are not purely periodic.)

Real and Complex Numbers 3

3.1 Ordered Fields

The aim of a theory of real numbers is, among other things, to mathematically structure the set of points on a line $\mathfrak{g}$ of our perceptual space in an appropriate way. The first steps in this direction were taken over 2300 years ago in the "Elements" of Euclid (around 300 BC). They go back to Eudoxus (408–355 BC), were then expanded by Archimedes (ca. 287–212 BC) and only found their preliminary conclusion in the second half of the 19th century through B. Bolzano (1781–1848), A. Cauchy (1789–1857), K. Weierstraß (1815–1897), R. Dedekind, G. Cantor and others. Whether the concept of a complete ordered field introduced further below is an adequate description of the number line is ultimately a problem of physics. Without a doubt, however, the real numbers are the foundation of mathematics, especially for analysis, topology, and geometry.

After calibrating the line $\mathfrak{g}$ by selecting a pair of points $(O, E) \in \mathfrak{g}^2$ with $O \neq E$, the elementary geometric operations of (parallel) shifts along $\mathfrak{g}$ and stretches with center O provide an addition or a multiplication on $\mathfrak{g}$ with $0 = O$ and $1 = E$ as neutral elements according to the known constructions in Fig. 3.1.

For the addition, $O \in \mathfrak{g}$ is sufficient as a calibration point. Construct the sum $x + y$ and the product $x \cdot y$ also in the case that O and E exchange their roles, and construct points $x \in \mathfrak{g}$ for $-x \in \mathfrak{g}$ with $x + (-x) = 0$ and, if $x \neq 0$, $x^{-1} \in \mathfrak{g}$ with $x \cdot x^{-1} = 1$. The simplest geometric axioms imply that $\mathfrak{g}$ with the thus defined addition and multiplication is a field, cf. Vol. 3. $\mathfrak{g}$ is furthermore totally ordered by the half-ray originating from O through E, as indicated by the arrow in the drawings. Thus, $(\mathfrak{g}, +)$ becomes an ordered group, whose positivity area $\mathfrak{g}_+ = \{y \in \mathfrak{g} \mid y \geq 0 = O\}$ is this half-ray, cf. example 2.1.16. Thus, the law of monotonicity of addition applies: The shift $\tau_x \colon y \mapsto x + y$ is monotonically increasing for every $x \in \mathfrak{g}$. Moreover, $\mathfrak{g}_+$ is also closed with respect to multiplication. This is equivalent to the law of monotonicity of multiplication: The stretch

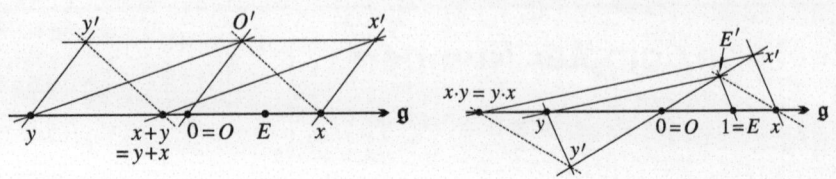

Fig. 3.1 Addition and multiplication on a line $\mathfrak{g}$

$\vartheta_x: y \mapsto x \cdot y$ is monotonically increasing for every $x \in \mathfrak{g}_+$. Overall, $\mathfrak{g}$ is thus an ordered field.[1] We will not generally deal with ordered rings, but only consider ordered fields, whose order is assumed to be total from the outset. So we define:

Definition 3.1.1 $K = (K, +, \cdot, \leq)$ is an **ordered field**, when $(K, +, \cdot)$ is a field and $\leq$ is a total order on K such that $(K, +)$ is an ordered group, whose **positivity area** $K_+ := \{y \in K \mid y \geq 0\}$ is also a submonoid of $(K, \cdot)$, when thus the following **monotonicity laws** for all $x, y, z \in K$ apply:

(1) **Monotonicity of addition:** From $x \leq y$ follows $x + z \leq y + z$.
(2) **Monotonicity of multiplication:** From $x \leq y$ and $0 \leq z$ follows $xz \leq yz$.

If K' and K are ordered fields, then a field homomorphism $\varphi: K' \to K$ is called a **homomorphism of ordered fields**, when φ is monotonic (increasing).

Note that $x \leq y$ is equivalent to $0 (= x - x) \leq y - x$. From the monotonicity laws, the corresponding strict inequalities for all $x, y, z \in K$: (1) From $x < y$ follows $x + z < y + z$. (2) From $x < y$ and $0 < z$ it follows that $xz < yz$. In particular, the set $K_+^\times := \{x \in K \mid x > 0\}$ of the **positive** elements of K with respect to multiplication is also a totally ordered group $((K^\times, \cdot)$ but not!). The completeness of the order on K is equivalent to $K^\times = K_+^\times \uplus K_-^\times$, where $K_-^\times := \{y \in K \mid y < 0\} = -K_+^\times$ is the set of **negative** elements of K. The inequality $0 \leq 1$ already follows from the fact that 0 and 1 are comparable and K_+ is multiplicatively closed; because from $1 \leq 0$ it follows that $-1 \geq 0$ and $1 = (-1)(-1) \geq 0$. This conclusion generally shows $y^2 \geq 0$ for all $y \in K$. From $0 < 1$ it follows by induction that $0 \leq n (= n \cdot 1_K)$ for all $n \in \mathbb{N}$. In particular, an ordered field always has characteristic 0. It contains the model $\mathbb{N} \cdot 1_K = \{0_K < 1_K < 1_K + 1_K = 2 \cdot 1_K < 2 \cdot 1_K + 1_K = 3 \cdot 1_K < \cdots\}$ of the natural numbers as the smallest inductive subset of K_+ with respect to the (injective) successor function. $f: y \mapsto y + 1$ and the element $0 \notin f(K_+)$, cf. remark 1.5.7. In every ordered group, the formation of inverses is monotonically decreasing. For the groups $(K, +)$ and $(K_+^\times, \cdot)$ this means: From $x \leq$

---

[1] It makes no sense to demand the monotonicity of multiplication with x for all x. If $y < z$, then $-z = -z - y + y < -z - y + z = -y$ applies, so multiplication by -1 is not monotonically increasing, if there are at all two different comparable elements.

3.1 Ordered Fields

y follows $-x \geq -y$, and from $0 < x \leq y$ follows $x^{-1} \geq y^{-1} > 0$ — A homomorphism $\varphi: K' \to K$ of ordered fields is injective and due to $0 = \varphi(0) < \varphi(1) = 1$ necessarily strictly monotonically increasing. A field homomorphism $\varphi: K' \to K$ is already monotonically increasing when $\varphi(K'_+) \subseteq K_+$ is. If then $x', y' \in K'$ with $x' \leq y'$ then $y' - x' \in K'_+$ and $\varphi(y') = \varphi(y' - x') + \varphi(x') \geq \varphi(x')$.

Example 3.1.2 (1) The ring $\mathbb{Z}$ is an ordered ring with the natural order. This is the only way to arrange $\mathbb{Z}$ so that $\mathbb{Z}$ becomes an ordered ring; because from $0 < 1$ it follows by induction that $0 \leq n$ for all $n \in \mathbb{N}$ and thus $\mathbb{Z}_+ = \mathbb{N}$. From the uniqueness of the order on $\mathbb{Z}$ also follows the uniqueness of the order on the quotient field $\mathbb{Q}$ of $\mathbb{Z}$, with respect to which $\mathbb{Q}$ becomes an ordered field. In general: ◊

Lemma 3.1.3 *Let A be an integrity domain with a total order $\leq$ such that A is an ordered ring, i.e. that $A_+ = \{a \in A \mid a \geq 0\}$ is a submonoid of both $(A, +)$ and $(A, \cdot)$. Then there is exactly one total order $\leq$ on the quotient field $K = Q(A)$, which continues the order given on A, so that K becomes an ordered field.*

Proof Because of $b^2 > 0$ for all $b \in A^*$, a fraction $a/b \in K$ is exactly ≥ 0 when $(a/b)b^2 = ab \geq 0$ is. This shows the uniqueness of the order on K. Conversely, it is easy to see that K with

$$K_+ := \{a/b \mid a \in A, b \in A^*, ab \in A_+\} = \{a/b \mid a \in A_+, b \in A^*_+\}$$

as a positivity area is an ordered field. □

In addition to $\mathbb{Q}$, the lemma yields further ordered fields: Let K be any ordered field and $P := K[X_i, i \in I]$ be the polynomial algebra in the indeterminates X_i, $i \in I$. We choose a total order on I and extend this to the monomials X^ν, $\nu \in \mathbb{N}^{(I)}$, using the homogeneous lexicographic order (with respect to which the monomials form a totally ordered regular monoid, cf. Sect. 2.9). Apparently, P with the positivity area $P_+ := \{F \in P^* \mid 0 < \text{LK}(F)\} \uplus \{0\}$ is a totally ordered integrity area. Consequently, the field $K(X_i, i \in I) = Q(P)$ of rational functions in the indeterminates X_i, $i \in I$, is an ordered field. The positive rational functions are exactly the fractions F/G, where $F, G \in P^*$ have positive leading coefficients. For each indeterminate X_i and all $a \in K$, $a < X_i$ therefore applies. In particular, $n < X_i$ or $0 < 1/X_i < 1/n$ applies for all $i \in I$ and all $n \in \mathbb{N}^*$.

(2) Every sub-field of an ordered field is (with the induced order) also an ordered field. $\mathbb{R}$ and thus all sub-fields of $\mathbb{R}$ are ordered fields (which also includes $\mathbb{Q}$). In Sect. 3.4 we will construct the ordered field $\mathbb{R}$ from $\mathbb{Q}$. ◊

Let K continue to be an ordered field. As with real numbers,

$$\text{Sign}\, x := \begin{cases} 1, & \text{if } x > 0, \\ 0, & \text{if } x = 0, \\ -1, & \text{if } x < 0, \end{cases}$$

is called the **sign** or **signum** of $x \in K$. The sign is a surjective monoid homomorphism $(K, \cdot) \to \{-1, 0, 1\}$, i.e., it is $\text{Sign}\, xy = \text{Sign}\, x \cdot \text{Sign}\, y$ for all $x, y \in K$ and $\text{Sign}\, 1 = 1$. Similarly, the **absolute value function** can be defined as in $\mathbb{R}$:

Definition 3.1.4 K is an ordered field. For $x \in K$ is called

$$|x| := \begin{cases} x, & \text{if } x \geq 0, \\ -x, & \text{if } x < 0, \end{cases}$$

the **value** of x.

Apparently, $|x| = x \cdot \text{Sign}\, x$. Furthermore, for all $x, y \in K$: (1) $|x| = \text{Max}\,(x, -x)$. (2) $|x| = |-x|$. (3) It is $|x| \geq 0$, and $|x| = 0$ holds exactly when $x = 0$. (4) $|xy| = |x||y|$, and $|x/y| = |x|/|y|$, if $y \neq 0$, i.e. $x \mapsto |x|$ is a surjective monoid homomorphism $(K, \cdot) \to (K_+, \cdot)$ or a surjective group homomorphism $(K^\times, \cdot) \to (K_+^\times, \cdot)$. The group $K^\times$ itself is the direct product of the subgroups $\{\pm 1\}$ and $K_+^\times$. The inverse isomorphism of $\{\pm 1\} \times K_+^\times \xrightarrow{\sim} K_+^\times$, $(\varepsilon, x) \mapsto \varepsilon x$, is $y \mapsto (\text{Sign}\, y, |y|)$. — Of particular importance is the triangle inequality:

Theorem 3.1.5 (Triangle Inequality) *For elements x, y of an ordered field, the following applies*

$$|x + y| \leq |x| + |y|, \quad |x - y| \geq ||x| - |y||.$$

Proof Because of $x \leq |x|$ and $y \leq |y|$, $x + y \leq |x| + |y|$ is true, and likewise $-(x + y) \leq |x| + |y|$ is true because of $-x \leq |x|$ and $-y \leq |y|$. This results in $|x + y| = \text{Max}\,(x + y, -(x + y)) \leq |x| + |y|$. With the already proven first inequality, $|x| = |(x - y) + y| \leq |x - y| + |y|$ follows, so $|x| - |y| \leq |x - y|$. If you swap x and y, you get $|y| - |x| \leq |y - x| = |x - y|$. Overall, you get $|x - y| \geq \text{Max}\,(|x| - |y|, |y| - |x|) = ||x| - |y||$. □

For $x, y \in K$, $d(x, y) := |y - x| \in K_+$ is called the **distance** between x and y. We recall the following designations for intervals, see Sect. 1.4: Let a and b be elements of an ordered field K with $a \leq b$. Then there are

$$[a, b] = \{x \in K \mid a \leq x \leq b\} \quad \text{the \textbf{closed interval}},$$
$$]a, b[= \{x \in K \mid a < x < b\} \quad \text{the \textbf{open interval}},$$
$$[a, b[= \{x \in K \mid a \leq x < b\},$$
$$]a, b] = \{x \in K \mid a < x \leq b\} \quad \text{the \textbf{halfopen intervals}},$$

which are each determined by a and b. The difference $b - a$ of the **interval boundaries** a, b is called the **length** of these intervals. An interval always contains all elements that lie between any two elements within it. For $a, \varepsilon \in K$ with $\varepsilon > 0$ let

$$B(a; \varepsilon) := {]a - \varepsilon, a + \varepsilon[} = \{x \in K \mid |x - a| < \varepsilon\} \quad \text{and}$$

3.1 Ordered Fields

Fig. 3.2 Open and closed ε-surrounding of a

$$\overline{B}(a;\varepsilon) := [a-\varepsilon, a+\varepsilon] = \{x \in K \mid |x-a| \leq \varepsilon\}.$$

We call these intervals the **open** and the **closed ε-surrounding** of a. They both have the length 2ε, see Fig. 3.2.

A **neighborhood** of $a \in K$ is a subset of K, which contains such an (open or closed) ε-neighborhood of a with $\varepsilon > 0$. This ε-neighborhood also contains all ε'-neighborhoods of a with $0 < \varepsilon' < \varepsilon$. *The order of an ordered field is always dense*, i.e., every open interval $]a,b[$, $a<b$, is not empty. For example, the **center** $m := \frac{1}{2}(a+b)$ lies in $]a,b[$, and $]a,b[$ is the open ε-neighborhood of m with $\varepsilon := \frac{1}{2}(b-a)$. More generally, $]a,b[$ contains exactly the points $(1-t)a + tb$, $0 < t < 1$.

Occasionally, it is convenient to extend K by adding two different elements ∞ and $-\infty$ (which are not already in K) to an ordered set.

$$\overline{K} := K \uplus \{\pm\infty\}$$

with the largest element $\infty = +\infty$ and smallest element $-\infty$ to expand. So we set

$$-\infty \leq x \leq \infty$$

for all $x \in \overline{K}$. Then the above intervals are also defined if the interval boundaries a and b lie in $\overline{K}$. If $-\infty$ or ∞ is the boundary of an interval with more than one point, we speak of an **infinite** or **unbounded** interval, otherwise of a **finite** or **bounded**. For calculating with ∞ and $-\infty$ we generally agree on the following rules:

(1) $x + \infty := \infty + x := \infty$ for all $x \in \overline{K}$, $x \neq -\infty$.
(2) $x + (-\infty) := (-\infty) + x := -\infty$ for all $x \in \overline{K}$, $x \neq \infty$.
(3) $x \cdot \infty := \infty \cdot x := \infty$, $x \cdot (-\infty) := (-\infty) \cdot x := -\infty$ for all $x \in \overline{K}$, $x > 0$.
(4) $x \cdot \infty := \infty \cdot x := -\infty$, $x \cdot (-\infty) := (-\infty) \cdot x := \infty$ for all $x \in \overline{K}$, $x < 0$.
(5) $0 \cdot \infty := \infty \cdot 0 := 0 \cdot (-\infty) := (-\infty) \cdot 0 := 0$.

So only the sums of $-\infty$ and ∞ or of ∞ and $-\infty$ are not defined.[2] Furthermore, let $x - y$ be defined for $x, y \in \overline{K}$ exactly when $x + (-y)$ is defined, and then let $x - y := x + (-y)$.[3]— We repeat the definitions for bounded sets, see Sect. 1.4:

Definition 3.1.6 Let K be an ordered field. A subset A of K is called **bounded above** if there is an $S \in K$ with $x \leq S$ for all $x \in A$. It is called **bounded below**

---

[2] We note that often the products of 0 with $\pm\infty$ are not determined.
[3] In Chap. 4 we will occasionally also allow the differences $K = \mathbb{R}$ and $\infty - \infty = \infty + (-\infty) = 0$ and $(-\infty) - (-\infty) = (-\infty) + \infty = 0$.

if there is an $s \in K$ with $x \geq s$ for all $x \in A$. It is called **bounded**, if it is bounded above and below.

A number S (or s) as in the preceding definition is an **upper** (or **lower**) **bound** of A in K. A set $A \subseteq K$ is exactly upper or lower bounded when it is a subset of an interval $]-\infty, S]$ or $[s, \infty[$ with $S, s \in K$. It is bounded when it is a subset of a finite interval $[s, S] \subseteq K$. Then it is also a subset of an interval of the form $[-R, R]$ with $R \in K_+$, i.e., it is $|x| \leq R$ for all $x \in A$, e.g., for $R := \mathrm{Max}\,(|s|, |S|)$.

Exercises

Unless otherwise stated, in the following exercises the elements are in an ordered field K. Occasionally we use for $K = \mathbb{R}$ the existence of an n-th root $\sqrt[n]{x} \in \mathbb{R}_+^\times$ with $(\sqrt[n]{x})^n = x$ for $x \in \mathbb{R}_+^\times$ and $n \in \mathbb{N}^*$, see example 3.3.7. Furthermore, we recommend orienting yourself on the case $K = \mathbb{R}$ (possibly with a suitable sketch).

Exercise 3.1.1 Determine the set of $x \in K$ or the $(x, y) \in K^2$, for which the given inequalities hold.

a) $1/|x - 2| > 1/(1 + |x - 1|)$, $x \neq 2$.
b) $(2 - |x - 1|)/|x - 4| \geq 1/2$, $x \neq 4$.
c) $(|x| - 1)/(x^2 - 1) \geq 1/2$, $x \neq \pm 1$.
d) $2(x + y)^2 \leq y(3x + 2y)^2$.

Exercise 3.1.2

a) If $x \geq 0$, then $x \geq \left(3x/(3 + x)\right)^2$ follows.
b) If $x \geq 1$, then $x \geq \left((3x + 1)/(3 + x)\right)^2$ follows.

Exercise 3.1.3 For all $m, n \in \mathbb{N}$ the following holds:

a) From $0 \leq x < y$ and $n > 0$, $0 \leq x^n < y^n$ follows.
b) From $1 \leq x$ and $m \leq n$, $x^m \leq x^n$ follows.
c) From $0 \leq x \leq 1$ and $m \leq n$, $x^m \geq x^n$ follows.

Exercise 3.1.4
a) $x/y + y/x \geq 2$, if $x, y > 0$, and $x/y + y/x > 2$, if furthermore $x \neq y$.
b) $2xy \leq \frac{1}{2}(x + y)^2 \leq x^2 + y^2$.
c) $xy + xz + yz \leq x^2 + y^2 + z^2$.
d) $(x + y)(y + z)(z + x) \geq 8xyz$, if $x, y, z \geq 0$.

Exercise 3.1.5 For all $n \in \mathbb{N}$ and all $x, y > 0$, applies

$$\left(1 + \frac{x}{y}\right)^n + \left(1 + \frac{y}{x}\right)^n \geq 2^{n+1}.$$

3.1 Ordered Fields

Exercise 3.1.6 For all $n \in \mathbb{N}^*$ $\left(\frac{1}{2}(x+y)\right)^n \leq \frac{1}{2}(x^n + y^n) \leq \frac{1}{2}(x+y)^n$ applies, if $x, y \geq 0$.

Exercise 3.1.7 For all x, y with $x + y \geq 0$, the following applies:

a) $x^3 + y^3 \geq xy(x+y)$.
b) $x/y^2 + y/x^2 \geq 1/x + 1/y$, if $x, y \neq 0$.

Exercise 3.1.8 For all x, y, z with $x + y \geq 0$, $x + z \geq 0$, $y + z \geq 0$, the following applies:

$$x^3 + y^3 + z^3 \geq \frac{1}{3}(x^2 + y^2 + z^2)(x+y+z).$$

Exercise 3.1.9 It is $\text{Max}(x, y) = \frac{1}{2}(x+y+|x-y|)$ and $\text{Min}(x, y) = \frac{1}{2}(x+y-|x-y|)$.

Exercise 3.1.10 Let $n \in \mathbb{N}^*$. For all $x_1, \ldots, x_n$, $y_1, \ldots, y_n$ with $y_1, \ldots, y_n > 0$ applies:

$$\text{Min}\left(\frac{x_1}{y_1}, \ldots, \frac{x_n}{y_n}\right) \leq \frac{x_1 + \cdots + x_n}{y_1 + \cdots + y_n} \leq \text{Max}\left(\frac{x_1}{y_1}, \ldots, \frac{x_n}{y_n}\right).$$

Exercise 3.1.11
a) $|x| \leq |x+y| + |y|$.
b) From $|x| \leq 1, |y| \leq 1$ follows $|x+y| \leq 1 + xy$.
c) $|x+y|/(1+|x+y|) \leq |x|/(1+|x|) + |y|/(1+|y|)$.

Exercise 3.1.12 From $x = y + z$ and $xz \leq 0$ follows $x = \theta y$ with $0 \leq \theta \leq 1$. (This trivial statement often provides important error estimates.)

Exercise 3.1.13 $\prod_{i=1}^{n}(1+x_i) \geq 1 + x_1 + \cdots + x_n$, if all $x_i \geq 0$ are or if $0 \geq x_i \geq -1$ applies for all i. In particular, one obtains $(1+x)^n \geq 1 + nx$ for all x with $x \geq -1$ and all $n \in \mathbb{N}$ (**Bernoulli inequalities**).

Exercise 3.1.14
a) $\prod_{i=1}^{n}(1-x_i) \leq 1/(1+x_1 + \cdots + x_n)$, if $0 \leq x_i \leq 1$ applies for all $i = 1, \ldots, n$. If $0 < x_i$ is for at least one i, then the inequality is strict. For all $n \in \mathbb{N}^*$ in particular $(1-x)^n < 1/(1+nx)$ applies, if $0 < x \leq 1$.
b) $\prod_{i=1}^{n}(1+x_i) \leq 1/\left(1 - \sum_{i=1}^{n} x_i\right)$, if all $x_i \geq 0$ are and $\sum_{i=1}^{n} x_i < 1$ is. When is the inequality strict?

Exercise 3.1.15 For $x_1, \ldots, x_n \geq 1$ applies

$$\prod_{i=1}^{n}(1+x_i) \geq \frac{2^n}{n+1}(1 + x_1 + \cdots + x_n).$$

In particular, one obtains $(1+x)^n \geq \frac{2^n}{n+1}(1+nx)$ for $x \geq 1$ and $n \in \mathbb{N}$.

Exercise 3.1.16 For all x with $0 \leq x$ and all $n \geq 2$ applies $(1+x)^n \geq \frac{1}{4}n^2x^2$.

Exercise 3.1.17 For all x, y with $(x, y) \neq (0, 0)$ and all positive even natural numbers n applies $x^n + x^{n-1}y + \cdots + xy^{n-1} + y^n > 0$.

Exercise 3.1.18 In case of $x_1, \ldots, x_n > 0$ applies $(x_1 + \cdots + x_n)(x_1^{-1} + \cdots + x_n^{-1}) \geq n^2$. (See exercise 3.1.23.)

Exercise 3.1.19 For all $x_1, \ldots, x_n > 0$ with $x_1 \cdots x_n = 1$ applies $(1+x_1)\cdots(1+x_n) \geq 2^n$. The equality sign applies exactly when $x_1 = \cdots = x_n = 1$ is. (When concluding from $n \geq 1$ to $n+1$, let x_n be the smallest and x_{n+1} the largest of the numbers $x_1, \ldots, x_{n+1}$. Then apply the induction assumption to $x_1, \ldots, x_{n-1}, x_n x_{n+1}$.)

Exercise 3.1.20 For all $x_1, \ldots, x_n$ with $x_1, \ldots, x_n > 0$ and $x_1 \cdots x_n = 1$ applies $x_1 + \cdots + x_n \geq n$. The equality sign applies exactly when $x_1 = \cdots = x_n = 1$ is. (Use the hint to exercise 3.1.19.)

Exercise 3.1.21 For all $x_1, \ldots, x_n$ with $x_1, \ldots, x_n > 0$ and $x_1 + \cdots + x_n = n$, $x_1 \cdots x_n \leq 1$ applies. The equality sign applies exactly when $x_1 = \cdots = x_n = 1$ is. (This can be traced back to exercise 3.1.20 or a similar induction proof as in exercise 3.1.19 can be conducted, whereby the induction assumption is then applied to $x_1, \ldots, x_{n-1}, x_n + x_{n+1} - 1$.)

Exercise 3.1.22 Let $n \in \mathbb{N}^*$. For all $x_1, \ldots, x_n$ with $x_1, \ldots, x_n > 0$ applies

$$\left(\frac{x_1 + \cdots + x_n}{n}\right)^n \geq x_1 \cdots x_n \geq \left(\frac{n}{x_1^{-1} + \cdots + x_n^{-1}}\right)^n.$$

The equality sign applies exactly when $x_1 = \cdots = x_n$ is. The second inequality follows from the first. To prove this, one can apply exercise 3.1.21 to $x_1/a, \ldots, x_n/a$, $a := (x_1 + \cdots + x_n)/n$, or, if in K all positive numbers have an n-th root, for example at $K = \mathbb{R}$, exercise 3.1.20 to $x_1/g, \ldots, x_n/g$ apply, $g := \sqrt[n]{x_1 \cdots x_n}$, or generally also conclude as follows by induction: It is sufficient for a and $b := (x_1 + \cdots + x_{n+1})/(n+1)$ the inequality $b^{n+1} \geq a^n x_{n+1} = a^n((n+1)b - na)$ or $(b/a)^{n+1} \geq (n+1)(b/a) - n$ to verify, i.e. for $x > 0$ the inequality

$$0 \leq x^{n+1} - (n+1)x + n = (x-1)((x^n - 1) + \cdots + (x-1))$$
$$= (x-1)^2(x^{n-1} + 2x^{n-2} + \cdots + n).$$

Remark For positive real numbers $x_1, \ldots, x_n$, the term

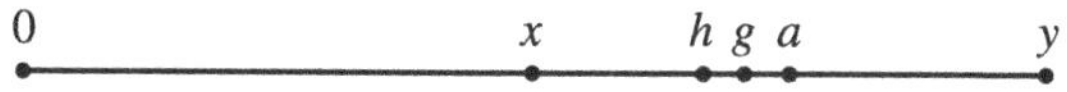

3.1 Ordered Fields

```
0                    x      h g a           y
•────────────────────•──────••••────────────•
```

Fig. 3.3 Harmonic (h), geometric (g) and arithmetic (a) mean of x and y

$$a = \frac{x_1 + \cdots + x_n}{n}, \quad g = \sqrt[n]{x_1 \cdots x_n} \quad \text{and} \quad h = \frac{n}{x_1^{-1} + \cdots + x_n^{-1}}$$

refers to the **arithmetic**, **geometric**, or **harmonic mean** of the numbers $x_1, \ldots, x_n$. So, $a \geq g \geq h$. For $n = 2$, the inequality $a \geq g$ is trivial due to $(x_1 + x_2)^2 - 4x_1x_2 = (x_1 - x_2)^2 \geq 0$.

A sequence of positive real numbers is called **arithmetic** or **geometric** or **harmonic**, if each term of the sequence (excluding the first term) is the arithmetic, geometric, or harmonic mean of the two adjacent terms. A sequence is exactly arithmetic when the sequence of reciprocals is harmonic. From the arithmetic sequence $1, 2, 3, \ldots$ of positive natural numbers, the harmonic sequence $1, \frac{1}{2}, \frac{1}{3}, \ldots$ of proper fractions is derived.

The term "harmonic mean" has the following origin: At constant tension, the frequency of the sound of a string is inversely proportional to the length of its vibrating part. Thus, if the string lengths x and y, $x < y$, produce tones with the frequencies ν and μ, $\nu > \mu$, then the mean of these tones, i.e., the tone with the arithmetic mean $(\nu + \mu)/2$ as frequency, is produced by a string length that is the harmonic mean h of x and y, cf. Fig. 3.3. This mean h can also be characterized by the proportion $(y - h) : (h - x) = y : x$, as has already been done in antiquity. For example, the (pure) fifth is the (arithmetic) mean of the fundamental and the octave. Therefore, one plucks $2/(\frac{1}{1} + \frac{1}{1/2}) = 2/3$ of the string to generate the fifth. The (pure) major third is the mean of the fundamental and the fifth. How do you pluck it on the string?—If the tone interval $\nu : \mu$ is to be divided by the frequency γ in such a way that the two resulting subintervals $\nu : \gamma$ and $\gamma : \mu$ coincide, then γ is the geometric mean $\sqrt{\mu\nu}$ of the frequencies ν and μ to choose and for the string length the geometric mean $g = \sqrt{xy}$ of the string lengths x, y. With the octave $\nu : \mu = 2 : 1$, this results in the interval $\sqrt{2} : 1$. This is the tritone (= three whole tone steps = diminished fifth = augmented fourth) in tempered tuning, which in classical harmony theory is counted among the dissonances as "diabolus in musica". Examples: siren or (downwards) the Hagen motif in the first two acts of "Götterdämmerung". The semitone "reine Quinte" : "temperierter Tritonus" = $\frac{3}{2} : \sqrt{2} = \frac{3}{4}\sqrt{2} = 1{,}06066\ldots$ is a good approximation of the tempered semitone $\sqrt[12]{2} : 1 = 1{,}05946\ldots$

Exercise 3.1.23 For all $x_1, \ldots, x_n, y_1, \ldots, y_n$ the **Cauchy-Schwarz inequality**

$$\left(\sum_{i=1}^{n} x_i y_i\right)^2 \leq \left(\sum_{i=1}^{n} x_i^2\right)\left(\sum_{i=1}^{n} y_i^2\right).$$

applies (One can assume $x_1, \ldots, x_n, y_1, \ldots, y_n \geq 0$ and use $\left(\sum_{i=1}^n x_i^2\right)\left(\sum_{i=1}^n y_i^2\right) = \left(\sum_{i=1}^n x_i y_i\right)^2 + \sum_{1 \leq i < j \leq n}(x_i y_j - x_j y_i)^2$ or add in the case of $K = \mathbb{R}$ and $x := \sum_{i=1}^n x_i^2 > 0, y := \sum_{i=1}^n y_i^2 > 0$ the n inequalities

$$\frac{x_i}{\sqrt{x}} \cdot \frac{y_i}{\sqrt{y}} \leq \frac{1}{2}\left(\frac{x_i^2}{x} + \frac{y_i^2}{y}\right), \quad i = 1, \ldots, n.)$$

Exercise 3.1.24 For all $x_1, \ldots, x_n$ applies $\left(\sum_{i=1}^n x_i\right)^2 \leq n \sum_{i=1}^n x_i^2$.

Exercise 3.1.25 For $n \in \mathbb{N}^*$ applies:

$$\left(\sum_{k=1}^n \frac{1}{k}\right)^2 < 2n \quad \text{and} \quad \left(\sum_{k=n+1}^{2n} \frac{1}{k}\right)^2 < \frac{1}{2}.$$

Exercise 3.1.26 For all $x_1, \ldots, x_n, y_1, \ldots, y_n \in \mathbb{R}$ applies

$$\sqrt{\sum_{i=1}^n (x_i + y_i)^2} \leq \sqrt{\sum_{i=1}^n x_i^2} + \sqrt{\sum_{i=1}^n y_i^2} \quad \text{(\textbf{Minkowski's inequality})}$$

(To prove this, square both sides and use the Cauchy-Schwarz inequality from exercise 3.1.23.)

Exercise 3.1.27 Let $k, n \in \mathbb{N}^*$, $k \leq n$. For each k of the positive numbers $x_1, \ldots, x_n$ the product is ≥ 1. Then also $x_1 \cdots x_n \geq 1$ is.

Exercise 3.1.28 The union of finitely many bounded subsets of K is again bounded.

Exercise 3.1.29 Let $f: \mathbb{R} \to \mathbb{R}$ be the function $x \mapsto (x-1)x(x+1)$. Sketch on the number line the point sets $\{x \in \mathbb{R} \mid f(x) \geq 0\}$ and $\{x \in \mathbb{R} \mid f(x) \leq 0\}$.

Exercise 3.1.30 Sketch for the following functions $f: \mathbb{R}^2 \to \mathbb{R}$ in the number plane $\mathbb{R}^2$ each the set of points (x, y) with $f(x, y) > 1$ or $= 1$ or < 1:

a) $f(x, y) = |x - y|$.
b) $f(x, y) = x^2 y^2$.
c) $f(x, y) = x^2 + xy + 1$.

Exercise 3.1.31 Sketch the set $\{(x, y) \in \mathbb{R}^2 \mid x^2 \leq y \leq x^4\} \subseteq \mathbb{R}^2$.

Exercise 3.1.32 Sketch the set of pairs $(x, y) \in \mathbb{R}^2$, for which $xy > x + y$ or $xy = x + y$ or $xy < x + y$ is.

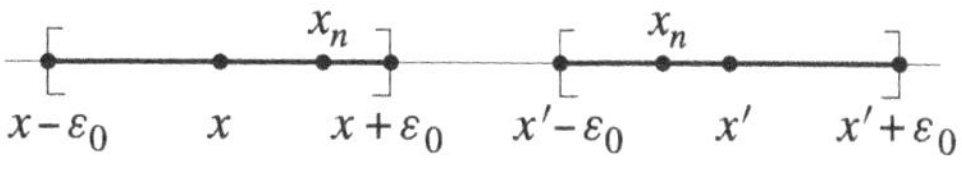

Fig. 3.4 Uniqueness of the limit

3.2 Convergent Sequences

To distinguish the real number fields among the ordered fields K, we use the properties of sequences in K. The concept of the convergent sequence is of fundamental importance. In the following, K denotes an ordered field in the sense of Definition 3.1.1 (unless otherwise stated).

Definition 3.2.1 A sequence $(x_n) = (x_n)_{n\in\mathbb{N}}$ of elements from K is called **convergent** (in K), if there is an $x \in K$ with the following property: For every (no matter how small) positive $\varepsilon \in K_+^\times$ there is an $n_0 \in \mathbb{N}$ with $|x_n - x| \leq \varepsilon$ for all natural numbers $n \geq n_0$.

This element x is uniquely determined by the sequence (x_n). If $x' \in K$ were another different element in K with the corresponding property, then $\varepsilon_0 := \frac{1}{3}|x - x'| > 0$ and there would be natural numbers n_0 and n_0' with $|x_n - x| \leq \varepsilon_0$ for all $n \geq n_0$ and $|x_n - x'| \leq \varepsilon_0$ for all $n \geq n_0'$, see Fig. 3.4. Then, with any arbitrary $n \geq \mathrm{Max}\,(n_0, n_0')$ we obtain the contradiction

$$|x - x'| = |x - x_n + x_n - x'| \leq |x_n - x| + |x_n - x'| \leq \varepsilon_0 + \varepsilon_0 = \frac{2}{3}|x - x'| < |x - x'|.$$

The element x uniquely determined by the convergent sequence (x_n) according to Definition 3.2.1 $\lim x_n = \lim_{n\to\infty} x_n$ is called the **limit** or the **limes** of the sequence (x_n). We denote it with

$$\lim x_n = \lim_{n\to\infty} x_n.$$

If x is the limes of (x_n), we also briefly describe this situation with

$$x_n \longrightarrow x \quad \text{or} \quad x_n \xrightarrow{n\to\infty} x$$

and say, (x_n) **converges to** x. Apparently, the sequence (x_n) converges to x exactly when the sequence $(x_n - x)$ converges to 0. A convergent sequence with the limit 0 is called a **null sequence**. A sequence that does not converge is called **divergent**. A constant sequence (x_n) with the value $x_n = x$ for all n or more generally a stationary sequence with limit x converges to x.

We say that a number x is **approximated by the number** y **up to an error** $\leq \varepsilon$ approximated $\leq \varepsilon$ if $|y - x| \leq \varepsilon$ is. Given $\varepsilon > 0$, the members of a sequence converging to x from a point n_0 approximate the number x up to an error $\leq \varepsilon$. For practical applications, of course, the **quality of the approximation** is important, i.e., from which n_0 the members of the sequence differ from x by at most ε. We will occasionally address such problems later. For the concept of convergence itself, this rate of convergence does not matter. *By definition, a sequence (x_n) has the limit x if in every neighborhood of x almost all members of the sequence, i.e.,* all members with at most finitely many exceptions, lie. The above proof for the uniqueness of the limit is based on the fact that in the disjoint ε_0-neighborhoods of x and x' not simultaneously almost all members of the sequence (x_n) can lie. Furthermore, for each permutation $\sigma \in \mathfrak{S}(\mathbb{N})$: *The sequence converges exactly when* $(x_n)_{n \in \mathbb{N}}$, *if the sequence* $(x_{\sigma n})_{n \in \mathbb{N}}$ *converges. Both sequences then have the same limit.* The convergence of a sequence has nothing to do with the order of $\mathbb{N}$. It is generally defined: An *infinite* family $(x_i)_{i \in I}$ from K converges to $x \in K$, if in every neighborhood of x almost all members of the family lie. The limit value x is then again uniquely determined.[4]— From Definition 3.2.1 it follows immediately:

Proposition 3.2.2 *Let* (x_n) *be a sequence converging in K.*

(1) *Every subsequence* $(x_{n_k})_{k \in \mathbb{N}}$ *converges with the same limit as* (x_n).
(2) *If you change finitely many members of the sequence, the sequence remains convergent with the same limit.*

A sequence $(x_{n_k})_{k \in \mathbb{N}}$ is a **subsequence** of (x_n), if the sequence $(n_k)_{k \in \mathbb{N}}$ of indices is *strictly* monotonically increasing. From 3.2.2 (1) it follows for example, that a sequence, which has a non-convergent subsequence or two subsequences with different limits, cannot be convergent.

Definition 3.2.3 A sequence (x_n) in K is **bounded above** or **bounded below** or **bounded**, if the same applies to the set $\{x_n \mid n \in \mathbb{N}\} \subseteq K$ of sequence members.

Since in every ε-neighborhood of the limit of a convergent sequence almost all members of this sequence lie, it follows:

Proposition 3.2.4 *Every convergent sequence is bounded.*

Of course, not every bounded sequence is convergent in return.

Theorem 3.2.5 (Calculation rules for limits) *Let* (x_n) *and* (y_n) *be K convergent sequences with the limit values x and y. Then the following applies:*

(1) *The sum sequence* $(x_n + y_n)$ *converges, and it is*
$$\lim(x_n + y_n) = \lim x_n + \lim y_n = x + y.$$
(2) *The product sequence* $(x_n y_n)$ *converges, and it is*
$$\lim(x_n y_n) = (\lim x_n)(\lim y_n) = xy.$$
In particular, $\lim(\lambda x_n) = \lambda \cdot \lim x_n = \lambda x$ *applies to all* $\lambda \in K$.
(3) *If* $y_n \neq 0$ *applies to all* $n \in \mathbb{N}$ *and* $y \neq 0$, *then the quotient sequence* (x_n/y_n) *also converges and it is*

---

[4] A finite family would converge to any $x \in K$.

3.2 Convergent Sequences

$$\lim \frac{x_n}{y_n} = \frac{\lim x_n}{\lim y_n} = \frac{x}{y}.$$

In particular, then $\lim 1/y_n = 1/y$.

Proof Let $\varepsilon > 0$ given. For $\varepsilon' := \varepsilon/2$ there is $n_1, n_2 \in \mathbb{N}$ with $|x_n - x| \leq \varepsilon'$ or $|y_n - y| \leq \varepsilon'$ for all $n \geq n_1$ or $n \geq n_2$. For all $n \geq n_0 := \mathrm{Max}\,(n_1, n_2)$ then applies

$$|(x_n + y_n) - (x + y)| = |(x_n - x) + (y_n - y)| \leq |x_n - x| + |y_n - y| \leq \varepsilon' + \varepsilon' = \varepsilon.$$

(2) Let $\varepsilon > 0$ be given. It is

$$|x_n y_n - xy| = |x_n y_n - x_n y + x_n y - xy| \leq |x_n y_n - x_n y| + |x_n y - xy|$$
$$= |x_n||y_n - y| + |x_n - x||y|.$$

According to Proposition 3.2.4 there is an $R > 0$ with $|x_n| \leq R$ for all $n \in \mathbb{N}$. We choose for $\varepsilon' := \varepsilon/2\mathrm{Max}\,(R, |y|)$ an n_0 with $|x_n - x| \leq \varepsilon'$ and $|y_n - y| \leq \varepsilon'$ for all $n \geq n_0$, so it follows for these n:

$$|x_n y_n - xy| \leq |x_n||y_n - y| + |x_n - x||y| \leq R\varepsilon' + \varepsilon'|y| \leq \varepsilon/2 + \varepsilon/2 = \varepsilon.$$

(3) Because of (2), it is sufficient to show the special case indicated in (3). For this purpose, let $\varepsilon > 0$ given. It is

$$\left|\frac{1}{y_n} - \frac{1}{y}\right| = \left|\frac{y - y_n}{y y_n}\right| = \frac{|y_n - y|}{|y|} \cdot \frac{1}{|y_n|}.$$

Since $\lim y_n = y \neq 0$ is, only finitely many terms of the sequence (y_n) lie in the $(|y|/2)$ environment of 0. Because of $y_n \neq 0$ for all n there is therefore an $r > 0$ with $|y_n| \geq r$, thus with $1/|y_n| \leq 1/r$ for all n. Now we choose for $\varepsilon' := \varepsilon r|y|$ an n_0 with $|y_n - y| \leq \varepsilon'$ for all $n \geq n_0$, so for these n the following estimate applies:

$$\left|\frac{1}{y_n} - \frac{1}{y}\right| = \frac{|y_n - y|}{|y|} \cdot \frac{1}{|y_n|} \leq \frac{\varepsilon'}{|y|} \cdot \frac{1}{r} = \varepsilon. \quad \square$$

Theorem 3.2.5 can be summarized as follows: The convergent sequences in K form a K-subalgebra $K_{\mathrm{kon}}^{\mathbb{N}}$ of the algebra $K^{\mathbb{N}}$ of all K-valued sequences, whose unit group $\left(K_{\mathrm{kon}}^{\mathbb{N}}\right)^{\times}$ contains exactly those sequences from $K_{\mathrm{kon}}^{\mathbb{N}} \cap \left(K^{\mathbb{N}}\right)^{\times}$ whose limit $\neq 0$ is. *The mapping* $\lim: K_{\mathrm{kon}}^{\mathbb{N}} \to K$ *is a surjective K-algebra homomorphism.* Its kernel is the ideal of null sequences. Thus, they form a maximal ideal $\mathfrak{n}$ in $K_{\mathrm{kon}}^{\mathbb{N}}$, which includes the ideal $K^{(\mathbb{N})}$ of stationary null sequences. It is $K_{\mathrm{kon}}^{\mathbb{N}} = K \oplus \mathfrak{n}$ and $K_{\mathrm{kon}}^{\mathbb{N}}/\mathfrak{n} = K$.[5] — Another useful calculation rule is:

---

[5] By the way, no maximal ideal in $K^{\mathbb{N}}$, which includes $K^{(\mathbb{N})}$, is explicitly known, although these ideals $\mathfrak{m}$ with their residue fields $K^{\mathbb{N}}/\mathfrak{m}$ play an extraordinary role in the so-called **Non-Standard Analysis**, especially for $K = \mathbb{R}$, see also the remark on exercise 4.2.31. — Furthermore, we note that there are ordered fields K whose only null sequences are the stationary null sequences, see exercise 3.1.12. In these fields, the stationary sequences are the only convergent sequences, and it is $K_{\mathrm{kon}}^{\mathbb{N}} = K \oplus K^{(\mathbb{N})}$.

Proposition 3.2.6 *If (x_n) is a sequence convergent in K with limit x, then $(|x_n|)$ is also convergent and it holds $\lim |x_n| = |\lim x_n| = |x|$.*

Proof The assertion follows from $||x_n| - |x|| \leq |x_n - x|$, cf. theorem 3.1.5. □

The following criterion is often used to determine limit values:

Proposition 3.2.7 (Enclosure Criterion) *Let there be $(x_n), (y_n)$ and (z_n) sequences. For (almost) all $n \in \mathbb{N}$ apply $x_n \leq y_n \leq z_n$. If the sequences (x_n) and (z_n) are convergent with the same limit value y, then also (y_n) is convergent with limit value y.*

Proof Let $\varepsilon > 0$ be given. In the ε neighborhood around y, according to the assumption, almost all members of the sequence (x_n) as well as almost all members of the sequence (z_n) are located, thus also almost all members of the sequence (y_n). □

In addition to the sequences considered so far (in the proper sense) convergent, sequences are often considered that converge in the improper sense towards ∞ or $-\infty$.

Definition 3.2.8 A sequence (x_n) in K **converges (improperly)** to ∞ (or $-\infty$), if for every $s \in K$ there is an $n_0 \in \mathbb{N}$ with $x_n \geq s$ (or $x_n \leq s$) for all $n \geq n_0$. It is said that (x_n) **converges in absolute value to ∞ converges**, if the sequence $(|x_n|)$ of absolute values converges to ∞.

The sequence (x_n) converges to ∞ if $(-x_n)$ converges to $-\infty$. Of course, sequences that converge to ∞ or $-\infty$ are unbounded above or below. When we talk about convergent sequences in the following, we usually mean only sequences that are convergent in the proper sense. If improper convergence is also allowed, we will usually mention this explicitly. With the conventions (1)–(4) for calculating with $\pm\infty$ established at the end of Sect. 3.1, the limit calculation rules continue to apply. However, the convention $0 \cdot (\pm\infty) = 0$ has *no* correspondence in the limit calculation rules. In $\mathbb{R}$ for example, $\lim 1/n = 0$ is, but $\lim b_n/n$ depends significantly on the sequence (b_n). Consider for (b_n) the sequences $(\sqrt{n}), (n), (n^2)$, cf. Corollary 3.3.3.

Exercises

Unless otherwise stated, in the following exercises the elements are in a fixed ordered field K.

Exercise 3.2.1 Let (x_n) be a convergent sequence with $s \geq x_n$ (or $s \leq x_n$) for almost all $n \in \mathbb{N}$). Then also $s \geq \lim x_n$ (or $s \leq \lim x_n$) is.

3.2 Convergent Sequences

Exercise 3.2.2 Let (x_n) be a convergent sequence with a positive (or negative) limit. Then there is an $s > 0$ with $s \leq x_n$ (or $x_n \leq -s$) for almost all members of the sequence.

Exercise 3.2.3 Let (x_n) be a sequence with the subsequences (x_{n_k}) and (x_{m_k}) such that each term x_n appears in at least one of the two subsequences (i.e., each index n appears in one of the two index sequences (n_k) and (m_k)). The sequence (x_n) converges exactly when each of the two subsequences converges to the same limit (which is then of course equal to $\lim x_n$).

Exercise 3.2.4 (x_n) is a null sequence if and only if $(|x_n|)$ is a null sequence.

Exercise 3.2.5 Let (x_n) be a null sequence and (y_n) be a bounded sequence. Then $(x_n y_n)$ is also a null sequence.

Exercise 3.2.6
a) A sequence with exclusively positive (or exclusively negative) terms converges exactly when it approaches ∞ (or $-\infty$), if the sequence of reciprocals is a null sequence. A sequence, whose terms are all different from 0, is a null sequence exactly when the sequence of reciprocals converges in absolute value to ∞.
b) Let $\lim x_n = \infty$ and $\lim y_n = a \in \overline{K} - \{0\}$. Then $\lim x_n y_n = \infty$ is true, if $a > 0$, and $\lim x_n = -\infty$ is true, if $a < 0$.
c) The sequence (x_n) converges to ∞, and the sequence (y_n) is bounded from below. Then $(x_n + y_n)$ also converges to ∞.

Exercise 3.2.7 The sequences $(x_n + y_n)$ and $(x_n - y_n)$ are convergent with the limit values α and β, respectively. Then (x_n), (y_n) and $(x_n y_n)$ also converge, and the following holds

$$\lim x_n = (\alpha + \beta)/2, \quad \lim y_n = (\alpha - \beta)/2, \quad \lim x_n y_n = (\alpha^2 - \beta^2)/4.$$

Exercise 3.2.8 For which convergent sequences can the position $n_0 \in \mathbb{N}$ in Definition 3.2.1 be chosen independently of $\varepsilon (> 0)$?

Exercise 3.2.9 Let (x_n) and (y_n) be sequences in K. Let $y_n \neq 0$ for almost all n. Then (x_n) and (y_n) are called **asymptotically equal**, if the (defined for almost all n) sequence (x_n/y_n) converges to 1, i.e., if the sequence $((x_n - y_n)/y_n)$ of the **relative errors** is a null sequence. We then write $x_n \sim y_n$ (for $n \to \infty$).

a) The asymptotic equality is an equivalence relation on the set of sequences in K, whose members are almost all non-zero.
b) Let there be an $s > 0$ with $|y_n| \geq s$ for almost all n. If the sequence $(y_n - x_n)$ of the **absolute errors** is a null sequence, then (x_n) and (y_n) are asymptotically equal.
c) If the sequence (y_n) is bounded and (x_n) and (y_n) are asymptotically equal, then the sequence of absolute errors is a null sequence.

Exercise 3.2.10 The following statements about the ordered field K are equivalent: (i) There is a non-stationary null sequence in K. (ii) There is a strictly monotonically decreasing null sequence in K. (iii) There is a countable subset of $K_+^\times$ with 0 as the lower limit. (iv) There is a countable unbounded subset in K.

Exercise 3.2.11 Let $I \neq \emptyset$ and $K(X) := K(X_i, i \in I)$ be the rational function field in the indeterminates X_i, $i \in I$, with an arrangement according to example 3.1.2. We continue the degree function of $K[X]^* = K[X_i, i \in I]^*$ by $\deg(F/G) := \deg F - \deg G$, $F, G \in K[X]^*$, according to $K(X)^\times$. The sequence $(R^n)_{n\in\mathbb{N}}$, $R \in K(X)^\times$, is a null sequence if and only if $\deg R < 0$ is. In particular, there are non-stationary null sequences in $K(X)$.

Exercise 3.2.12 Let $I \neq \emptyset$ and $K(X) := K(X_i, i \in I)$ be the rational function field in the indeterminates X_i, $i \in I$, where I is well-ordered. On the monomials X^ν, $\nu \in \mathbb{N}^{(I)}$, we define the reverse lexicographic order $\leq_{\text{rev lex}}$ by "$X^\mu <_{\text{rev lex}} X^\nu$ if and only if $\mu \neq \nu$ and $\mu_i < \nu_i$ for the *largest* index i with $\mu_i \neq \nu_i$". (The degrees $|\mu|$ or $|\nu|$ of the monomials therefore play a different role than in the homogeneous reverse lexicographic order $<_{\text{rev hlex}}$. If, for example, $i_1 < i_2$, then $X_{i_1}^n <_{\text{rev lex}} X_{i_2}$ for all $n \in \mathbb{N}$.)

a) $\leq_{\text{rev lex}}$ is a well-ordering, with respect to which the monomials form a regular ordered monoid with 1 as the smallest element, and $K(X)$ is an ordered field with

$$K(X)_+^\times := \{F/G \mid F, G \in K[X]^*, \text{LK}(F), \text{LK}(G) > 0\}$$

b) $K(X)$ has a non-stationary zero sequence if and only if I contains a countable cofinal subset. In particular, there are examples such that $K(X)$ only has stationary zero sequences. (See Example 1.8.12e).)

3.3 Real Number Fields

To characterize real numbers, we formulate the so-called completeness axiom using monotone sequences, following C. Carathéodory (1873–1950):

Definition 3.3.1 (Completeness Axiom) An ordered field K is called a **complete ordered field** or a **real number field**, if in K every bounded monotone sequence converges.

A sequence (x_n) is monotonically increasing if and only if the sequence $(-x_n)$ is monotonically decreasing, so it would have been sufficient to demand the convergence of bounded monotonically increasing sequences. We will soon see, cf. Corollary 3.4.10, that a real number field is uniquely determined in the following sense: If K_1 and K_2 are complete ordered fields, then there is *exactly* one isomorphism (of ordered fields) $K_1 \xrightarrow{\sim} K_2$. We therefore already speak (as before) of *the* field of real numbers and denote it by

$$\mathbb{R}.$$

In the following, let $\mathbb{R}$ always be a real number field. A first important consequence of the completeness axiom is the following statement:

Theorem 3.3.2 *The sequence of natural numbers in $\mathbb{R}$ is unbounded in $\mathbb{R}$.*

Proof The sequence $(n)_{n \in \mathbb{N}}$ is monotonically increasing. If it were bounded, it would have a limit x according to Definition 3.3.1. In the ε neighborhood of x with $\varepsilon := 1/3$, however, there is at most one natural number. Contradiction! □

An ordered field K, in which the set $\mathbb{N}$ of natural numbers is unlimited, is called an **Archimedean ordered field**. Since for an element $S \in K_+^\times$ the condition $S \leq n$ is equivalent to $0 < 1/n \leq 1/S$, $n \in \mathbb{N}^*$, it follows: *Exactly then is K an Archimedean ordered field, when $(1/n)_{n \in \mathbb{N}^*}$ is a null sequence in K.*

Lemma 3.3.3 *Let K be an Archimedean ordered field and $\varepsilon, x, y \in K_+^\times$. Then the following applies:*

(1) *There exists an $n \in \mathbb{N}$ with $n\varepsilon \geq x$.*
(2) *If $y > 1$, there exists a $n \in \mathbb{N}$ with $y^n \geq x$, i.e. (y^n) converges (strictly monotonically increasing) towards ∞.*
(3) *If $y < 1$, there exists a $n \in \mathbb{N}$ with $y^n \leq \varepsilon$, i.e. (y^n) is a (strictly monotonically decreasing) null sequence.*

Proof (1) Since $\mathbb{N}$ is not bounded above by assumption, there exists an $n \in \mathbb{N}$ for x/ε with $n \geq x/\varepsilon$, i.e. with $n\varepsilon \geq x$.

(2) Let $y = 1 + h$ for some $h > 0$. Then $y^n = (1+h)^n = 1 + nh + \cdots + h^n \geq 1 + nh \geq x$ for all $n \in \mathbb{N}^*$ with $n \geq (x-1)/h$.

(3) Since $0 < y < 1$, we have $1/y > 1$. According to (2), there exists an n with $(1/y)^n \geq 1/\varepsilon$, i.e. with $y^n \leq \varepsilon$. □

As the examples in 3.1.2 show, there are ordered fields that are not Archimedean ordered. For example, the rational function field $K(X)$ in an indeterminate over any ordered field K is not Archimedean ordered. The indeterminate X is an upper bound for all elements from $K \supseteq \mathbb{N}$. We recall that the positive elements in $K(X)$ are exactly the rational functions of the form F/G, $F, G \in K[X]^*$, with positive leading coefficients $LK(F)$, $LK(G)$. The significance of the Archimedean axiom was fully recognized by Archimedes first.

From Lemma 3.3.3 (3) it follows that in an Archimedean ordered field K for every $x \in K$ with $|x| < 1$ the sequence (x^n) is a null sequence, since $|x^n| = |x|^n$, $n \in \mathbb{N}$, is a null sequence.

If the ordered field K has a null sequence (ε_m) in $K_+^\times$ (like an Archimedean ordered field the sequence $1/m$, $m \in \mathbb{N}^*$), then a sequence (x_n) from K already converges to $x \in K$, if for every m in the ε_m-neighborhood of x almost all members of the sequence (x_n) lie.

Since in an Archimedean ordered field K the set $\mathbb{Z} = \mathbb{N} \cup (-\mathbb{N})$ is neither bounded above nor below, there is for every $x \in K$ exactly one $m \in \mathbb{Z}$ with $m \leq x < m+1$. This number m is called the **whole part** or also the **Gauss bracket** of x and is denoted by $[x]$ (or with $\lfloor x \rfloor =$ floor of x, cf. Example 1.2.2 (4)). By definition,

$$[x] \leq x < [x] + 1, \quad [x] \in \mathbb{Z}.$$

Furthermore, the notation

$$\{x\} = x - [x]$$

is common. We note the following consequence of the Archimedean axiom.

Lemma 3.3.4 *Let K be an Archimedean ordered field. Then $\mathbb{Q} = \mathbf{F}_0 \subseteq K$ is dense in K, i.e. in every interval $]a, b[\subseteq K$, $a < b$, there are one and therefore infinitely many rational numbers. In particular, every $x \in K$ is the limit of a sequence of rational numbers.*

Proof Let $n \in \mathbb{N}^*$ be chosen such that $1/n < b - a$ is. Then $([na] + 1)/n \in]a, b[$ applies, as the reader can easily confirm. □

We now turn again to a real number field $\mathbb{R}$. The following characterization of a real number dates back to Archimedes.

Theorem 3.3.5 (Interval Nesting) *Let (a_n) be a monotonically increasing and (b_n) a monotonically decreasing sequence in $\mathbb{R}$ with the following properties:*

(1) *It is $a_n \leq b_n$ for all $n \in \mathbb{N}$.* (2) *It is $\lim(b_n - a_n) = 0$.*

Then there is exactly one $x \in \mathbb{R}$ with $a_n \leq x \leq b_n$ for all $n \in \mathbb{N}$. It is $x = \lim a_n = \lim b_n$.

Proof The sequences (a_n) and (b_n) are also bounded, so they are convergent according to 3.3.1. Due to $0 = \lim(b_n - a_n) = \lim b_n - \lim a_n$, (a_n) and (b_n) have the same limit x. From the monotony of the sequences (a_n) and (b_n), $a_n \leq x \leq b_n$ results for all n. If x' is another such number, then $|x - x'| \leq b_n - a_n$ holds for all n and therefore necessarily $|x - x'| = 0$, i.e., $x = x'$. □

Let (a_n) and (b_n) be sequences of real numbers as in 3.3.5. $I_n := [a_n, b_n]$, $n \in \mathbb{N}$, is then a sequence of closed intervals, for which

$$I_0 \supseteq I_1 \supseteq I_2 \supseteq \cdots \supseteq I_n \supseteq I_{n+1} \supseteq \cdots$$

3.3 Real Number Fields

applies and whose lengths form a null sequence. Such a sequence of closed intervals is called an **interval nesting**. According to theorem 3.3.5, there is exactly one number x that lies in each of the intervals I_n. This number x is called **the number defined by the interval nesting**. It is

$$a_0 \leq a_1 \leq a_2 \leq \cdots \leq x \leq \cdots \leq b_2 \leq b_1 \leq b_0$$

and $x = \lim a_n = \lim b_n$. The midpoint $(a_n + b_n)/2$ of the n-th interval I_n approximates the number x up to an error that is at most equal to half the interval length $(b_n - a_n)/2$. According to the enclosure criterion 3.2.7, every sequence (c_n) with $c_n \in I_n$, $n \in \mathbb{N}$, converges to x. Conversely, for a convergent sequence (x_n) in $\mathbb{R}$, there is an interval nesting I_n, $n \in \mathbb{N}$, such that for each $n \in \mathbb{N}$ outside I_n, only finitely many members of the sequence lie. However, one has to use the existence of null sequences in $\mathbb{R}$. Since $\mathbb{Q}$ is dense in $\mathbb{R}$, one even finds such interval nestings with interval boundaries in $\mathbb{Q}$.

Example 3.3.6 (Uncountability of $\mathbb{R}$) We want to present the first proof of the uncountability of $\mathbb{R}$ from the work "On a Property of the Collection of All Real Algebraic Numbers" by G. Cantor in Journal for Pure and Applied Math. **74**, 258–262 (1874) (see also exercise 1.8.13d) and exercise 2.9.19): Assume the sequence $r_0, r_1, r_2, \ldots$ contains all real numbers. In contradiction to this, we construct step by step an interval nesting $[a_n, b_n]$, $n \in \mathbb{N}$, which defines a real number that certainly does not appear in the given sequence. We choose $[a_0, b_0]$ so that $r_0 \notin [a_0, b_0]$ is, and $[a_{n+1}, b_{n+1}] \subseteq [a_n, b_n]$ so that $r_{n+1} \notin [a_{n+1}, b_{n+1}]$ lies. It is important to ensure that the interval lengths converge to 0. For example, divide each interval into thirds $[a_n, b_n]$. In at least one of the outer thirds, r_{n+1} does not lie, and such a third provides the next interval $[a_{n+1}, b_{n+1}]$. (According to exercise 3.3.21, it is not necessary for the interval lengths to converge to 0. — For a generalization, see exercise 3.4.15.) ◊

Example 3.3.7 (Babylonian or Heronian Root Extraction) Let $a \in \mathbb{R}_+^\times$. Then the sequence defined by

$$x_{n+1} = \frac{1}{2}\left(x_n + \frac{a}{x_n}\right), \quad x_0 > 0 \text{ arbitrary, for instance } x_0 = a \text{ or } x_0 = 1,$$

is recursively defined (x_n) converges with $\lim x_n = \sqrt{a}$, i.e. for $x := \lim x_n$ is $x^2 = a$ and $x > 0$. *Proof.* Apparently, all $x_n > 0$. Furthermore, $x_{n+1}^2 \geq a$ holds for all $n \geq 0$, since the square of the arithmetic mean $\frac{1}{2}(x_n + a/x_n)$ of x_n and a/x_n is at least as large as the square $x_n \cdot a/x_n = a$ of their geometric mean. Explicitly: It is

$$x_{n+1}^2 - a = \left(\frac{1}{2}\left(x_n + \frac{a}{x_n}\right)\right)^2 - a = \frac{1}{4}\left(x_n - \frac{a}{x_n}\right)^2 \geq 0.$$

The sequence x_n, $n \geq 1$, is due to

$$x_n - x_{n+1} = x_n - \frac{1}{2}\left(x_n + \frac{a}{x_n}\right) = \frac{1}{2}\left(x_n - \frac{a}{x_n}\right) = \frac{1}{2x_n}(x_n^2 - a) \geq 0$$

monotonically decreasing. Since it is bounded below by 0, it converges according to definition 3.3.1 to an $x \in \mathbb{R}$ with $x \geq 0$. Because of $x_{n+1}^2 \geq a$ is $x^2 \geq a > 0$ and consequently even $x > 0$. With the recursion equation and the calculation rules for limits, we now get $x^2 = a$ due to ·

$$x = \lim x_{n+1} = \frac{1}{2}\left(\lim x_n + \frac{a}{\lim x_n}\right) = \frac{1}{2}\left(x + \frac{a}{x}\right). \quad \square$$

To get an impression of the quality of the approximation of $\sqrt{a}$ by the x_n, we estimate for $n \geq 1$ in the following way:

$$x_{n+1} - \sqrt{a} = \frac{1}{2}\left(x_n + \frac{a}{x_n}\right) - \sqrt{a} = \frac{1}{2x_n}\left(x_n - \sqrt{a}\right)^2 \leq \frac{1}{2\sqrt{a}}\left(x_n - \sqrt{a}\right)^2.$$

The error $x_{n+1} - \sqrt{a}$ in the $(n+1)$-th step, $n \geq 1$, is therefore up to the factor $1/2\sqrt{a}$ at most as large as the square of the error in the n-th step, i.e., the number of correct places after the decimal point approximately doubles with each step. Furthermore,

$$0 \leq x_n - \sqrt{a} = \frac{x_n^2 - a}{x_n + \sqrt{a}} \leq \frac{1}{2\sqrt{a}}(x_n^2 - a), \quad n \geq 1.$$

This is referred to as **quadratic convergence**.

Generally, we say that the convergent sequence (x_n) with $x := \lim x_n$ of order $k > 1$ converges to x if there is a constant $C \geq 0$ with $|x_{n+1} - x| \leq C|x_n - x|^k$ for all $n \geq 0$. If $|x_{n+1} - x| \leq c|x_n - x|$ holds for sufficiently large n with a constant $c \in \mathbb{R}$, $0 \leq c < 1$, we speak of **linear convergence**. Accordingly, we say that the interval nesting $I_n = [a_n, b_n]$, $n \in \mathbb{N}$, for the number x has a **convergence order** $k > 1$, if $b_{n+1} - a_{n+1} \leq C(b_n - a_n)^k$ with a constant $C \geq 0$ holds for all $n \geq 0$. The linear convergence of interval nestings is defined analogously to the linear convergence of sequences.

The Babylonian method is the standard method for root extraction and is usually also used in computers. Similarly, it can be shown that for any $k \in \mathbb{N}^*$ the recursively defined sequence (x_n) with $x_0 > 0$ arbitrary and

$$x_{n+1} = \frac{1}{k}\left((k-1)x_n + \frac{a}{x_n^{k-1}}\right), \quad n \in \mathbb{N},$$

for $n \geq 1$ monotonically decreasing towards $\sqrt[k]{a}$ converges, cf. Example 3.3.19. *This also proves the existence of the k-th root for positive real numbers.* In the last recursion, the approximation x_{n+1} is determined as the arithmetic mean of the edge lengths $x_n, \ldots, x_n, a/x_n^{k-1}$ of a k-dimensional cuboid with volume a. ◊

Example 3.3.8 (Interval nesting for the Euler number e and for the Euler constant γ) (1) The sequences

3.3 Real Number Fields

$$a_n := \left(1 + \frac{1}{n}\right)^n \quad \text{and} \quad b_n := \left(1 + \frac{1}{n}\right)^{n+1}, \quad n \in \mathbb{N}^*,$$

define an interval nesting. *Proof.* For all $n \geq 1$, $a_n < b_n$. Furthermore, a_n is strictly monotonically increasing: The inequality

$$\left(\frac{n+1}{n}\right)^n = \left(1 + \frac{1}{n}\right)^n < \left(1 + \frac{1}{n+1}\right)^{n+1} = \left(\frac{n+2}{n+1}\right)^{n+1}$$

is equivalent to

$$1 - \frac{1}{n+2} = \frac{n+1}{n+2} < \left(\frac{n(n+2)}{(n+1)^2}\right)^n = \left(1 - \frac{1}{(n+1)^2}\right)^n.$$

With the Bernoulli inequality from exercise 3.1.13, applied to $x := -1/(n+1)^2$, we obtain

$$\left(1 - \frac{1}{(n+1)^2}\right)^n \geq 1 - \frac{n}{(n+1)^2} > 1 - \frac{1}{n+2}.$$

The sequence (b_n) is strictly monotonically decreasing, since $\left(1 + \frac{1}{n}\right)^{n+1} > \left(1 + \frac{1}{n+1}\right)^{n+2}$ is equivalent to the inequality

$$\frac{n+1}{n+2} > \left(\frac{n^2 + 2n}{n^2 + 2n + 1}\right)^{n+1} = \left(1 - \frac{1}{(n+1)^2}\right)^{n+1},$$

which is obtained with exercise 3.1.14a) (for $x := 1/(n+1)^2$). Finally, $\lim(b_n - a_n) = \lim a_n/n = 0$. □

The number defined by this interval nesting is called the **Euler's number** e after L. Euler (1707–1783). So it is

$$e = \lim_{n \to \infty} \left(1 + \frac{1}{n}\right)^n = 2{,}71828182845904523536\ldots$$

This interval nesting is hardly suitable for quickly calculating e, as the length a_n/n of the nth interval $[a_n, b_n]$ is still $\geq a_1/n = 2/n$. However, the averages $\frac{1}{2}(a_n + b_n)$, $n \in \mathbb{N}^*$ are quite favorable, see Vol. 2 for an error estimate. A nice interpretation of the interval nesting for e described here can be found in Exercise 1.7.33.

(2) When using the natural logarithm ln to the base e (see Sect. 3.10) the inequalities

$$\left(1 + \frac{1}{n}\right)^n < e < \left(1 + \frac{1}{n-1}\right)^n \quad \text{and} \quad \ln\left(1 + \frac{1}{n}\right) < \frac{1}{n} < \ln\left(1 + \frac{1}{n-1}\right)$$

for $n \geq 2$ are equivalent. With the so-called **harmonic numbers**

$$H_n := \sum_{\nu=1}^{n} \frac{1}{\nu}, \quad n \in \mathbb{N},$$

Fig. 3.5 Circle with inscribed and circumscribed $2^n m$-gon

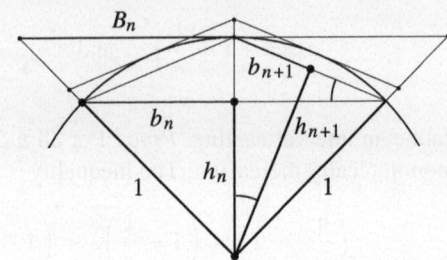

the sequence

$$c_n := \sum_{\nu=1}^{n} \left(\frac{1}{\nu} - \ln\left(1 + \frac{1}{\nu}\right) \right) = \sum_{\nu=1}^{n} \left(\frac{1}{\nu} - \ln(\nu+1) + \ln \nu \right)$$
$$= H_n - \ln(n+1), \quad n \in \mathbb{N}^*,$$

is therefore monotonically increasing and the sequence

$$d_n := 1 + \sum_{\nu=2}^{n} \left(\frac{1}{\nu} - \ln\left(1 + \frac{1}{\nu-1}\right) \right) = H_n - \ln n, \quad n \in \mathbb{N}^*,$$

is monotonically decreasing. Because $d_n - c_n = \ln(n+1) - \ln n = \ln(1 + \frac{1}{n}) < \frac{1}{n}$ they also form an interval nesting. The number defined by this

$$\gamma := \lim_{n \to \infty} (H_n - \ln n) = 0{,}577215664901532860060\ldots$$

is called the **Euler** (or **Mascheroni**) **constant**. So it is

$$H_n = \ln n + \gamma + \rho_n$$

with a monotonically decreasing null sequence $(\rho_n)_{n \in \mathbb{N}^*}$ and in particular $H_n \sim \ln n$ for $n \to \infty$, i.e. $\lim_{n \to \infty}(H_n/\ln n) = 1$.—The number e is irrational, see exercise 3.6.15c), even transcendental (over $\mathbb{Q}$), i.e. not a root of a polynomial function $\neq 0$ with coefficients in $\mathbb{Z}$, see for example 22, 16.A, remark to exercise 3.3.4. It is not known whether γ is irrational. ◊

Example 3.3.9 (Interval nesting for the circle number π) The area of a circle is well-defined in the number plane $\mathbb{R}^2$, see for example Vol. 6 on measure and integration theory. The **circle number** π can therefore be introduced as the area of the unit circle $K := \{(x, y) \in \mathbb{R}^2 \mid x^2 + y^2 \leq 1\}$, see also Vol. 2. In the following, we use some elementary geometric facts (e.g. the Pythagorean theorem).

Let $m \in \mathbb{N}$ with $m \geq 3$ fixed. For $n \in \mathbb{N}$ let f_n be the area of the inscribed regular $2^n m$-corner and F_n the area of the circumscribed regular $2^n m$-corner of K. Then $f_n < \pi < F_n$. Finally, let h_n be the length of the height of one of the isosceles sub-triangles, which is formed by the center of the circle and two adjacent corners of the inscribed $2^n m$-corner, and b_n be the half base length of this triangle, cf. Fig. 3.5. Applying the Pythagorean theorem three times then yields

3.3 Real Number Fields

$$4(1^2 - h_{n+1}^2) = 4b_{n+1}^2 = (1-h_n)^2 + b_n^2 = 1 - 2h_n + (h_n^2 + b_n^2) = 1 - 2h_n + 1^2$$

and consequently $2h_{n+1}^2 = h_n + 1$ or $h_{n+1}^2 = \frac{1}{2}(h_n + 1)$, $n \in \mathbb{N}$. Furthermore, the two angles drawn in Fig. 3.5 are obviously equal and thus base angles of similar right-angled triangles. It follows $b_n : 2b_{n+1} = h_{n+1} : 1$, i.e. $b_n = 2b_{n+1}h_{n+1}$ and

$$f_n = 2^n m b_n h_n = 2^{n+1} m b_{n+1} h_{n+1} h_n = f_{n+1} h_n.$$

If B_n denotes the half base length of the corresponding sub-triangle for the circumscribed $2^n m$-corner, then the intercept theorem yields $B_n : b_n = 1 : h_n$, i.e. $b_n = B_n h_n$ and therefore

$$F_n h_n^2 = 2^n m B_n h_n^2 = 2^n m b_n h_n = f_n = f_{n+1} h_n,$$

thus $F_n h_n = f_{n+1}$ and $h_n = \sqrt{f_n/F_n}$. Because $F_n/f_{n+1} = h_n^{-1} = f_{n+1}/f_n$ results in

$$f_{n+1} = \sqrt{f_n F_n}, \quad F_{n+1} = \frac{f_{n+2}}{h_{n+1}} = \frac{f_{n+1}}{h_{n+1}^2} = \frac{2f_{n+1}}{h_n + 1} = \frac{2f_{n+1}F_n}{h_n F_n + F_n} = \frac{2f_{n+1}F_n}{f_{n+1} + F_n},$$

$n \in \mathbb{N}$. f_{n+1} is thus the geometric mean of f_n and F_n and F_{n+1} the harmonic mean of f_{n+1} and F_n. (Compare this with the recursion for the harmonic-geometric mean in exercise 3.3.18.) *The sequences (f_n) and (F_n) form an interval nesting for π.* Always applies $f_n < f_{n+1} < F_{n+1} < F_n$. In addition, one has

$$\frac{F_{n+1} - f_{n+1}}{F_n - f_n} = \frac{2f_{n+1}F_n - (f_{n+1} + F_n)f_{n+1}}{(f_{n+1} + F_n)(F_n - f_{n+1}^2/F_n)} = \frac{1 - f_{n+1}/F_n}{(1 + F_n/f_{n+1})(1 - f_{n+1}^2/F_n^2)}$$

$$= \frac{1}{(1 + F_n/f_{n+1})(1 + f_{n+1}/F_n)} < \frac{1}{4}$$

due to $(1 + F_n/f_{n+1})(1 + f_{n+1}/F_n) = 1 + 1 + F_n/f_{n+1} + f_{n+1}/F_n > 2 + 2 = 4$. (According to exercise 3.1.4a) applies $a + a^{-1} > 2$ for all $a \in \mathbb{R}_+^\times$, $a \neq 1$.) It follows

$$F_{n+1} - f_{n+1} < (F_n - f_n)/4 < \cdots < (F_0 - f_0)/4^{n+1} \xrightarrow{n \to \infty} 0.$$

The convergence order of the interval nesting is linear. In the n-th step, the center $\frac{1}{2}(f_n + F_n)$ approximates the number π with an error $\leq (F_n - f_n)/2 < (F_0 - f_0)/2 \cdot 4^n = 1/4^n$. (However, the weighted mean is $(f_n + 2F_n)/3$ significantly cheaper. For this and for further acceleration of convergence see Vol. 2.)

In the case of $m = 4$, $h_0 = \frac{1}{2}\sqrt{2}$ and $f_0 = 2$, $F_0 = 4$. This results in

$$f_{11} = 3{,}1415923\ldots < \pi < F_{11} = 3{,}1415928\ldots$$

If we set $c_n := 2h_n$, then $c_0 = \sqrt{2}$ and $c_{n+1} = \sqrt{2 + c_n}$ for $n \in \mathbb{N}$, thus

$$c_n = \sqrt{2 + \sqrt{2 + \cdots + \sqrt{2}}},$$

where in c_n a total of $n+1$ root signs appear, and we obtain the following product representation for π, already given by Vieta (1540–1603):

$$\frac{2}{\pi} = \lim_{n\to\infty} \frac{2}{f_{n+1}} = \lim_{n\to\infty} \frac{2}{f_n} h_n = \cdots = \lim_{n\to\infty} \frac{2}{f_0} h_0 \cdots h_n =: \prod_{n=0}^{\infty} \frac{c_n}{2}.$$

If trigonometric functions are allowed, then $h_n = c_n/2 = \cos(2\pi/2^{n+2})$ (from which, with $\cos(\alpha/2) = \sqrt{(1+\cos\alpha)/2}$, $0 \le \alpha \le \pi$, also the recursion equation $c_{n+1} = \sqrt{2 + c_n}$, $n \in \mathbb{N}^*$, results).

The circle number

$$\pi = 3{,}141592653589793238462\ldots$$

is occasionally also called **Ludolph's number** after Ludolph van Ceulen (1540–1610), who published the first 20 decimal places of π in 1596 and used the inscribed and circumscribed $2^{35} \cdot 15$-corners for calculation, however their perimeters, see below for more on this. In the case of $m = 15$, $f_0 = 15 b_0 h_0 = 15 \sin\frac{\pi}{15} \cos\frac{\pi}{15} = \frac{15}{2} \sin\frac{2\pi}{15}$ and $F_0 = f_0/h_0^2 = \frac{15}{2} \sin\frac{2\pi}{15} / \cos^2\frac{\pi}{15} = 15 \sin\frac{2\pi}{15} / (1 + \cos\frac{2\pi}{15})$. To determine $\sin\frac{2\pi}{15}$ and $\cos\frac{2\pi}{15}$ one (nowadays) conveniently calculates complexly with $\zeta_{15} := \cos\frac{2\pi}{15} + i \sin\frac{2\pi}{15}$, cf. example 2.2.16 (2) and exercise 3.5.28. It is

$$\zeta_{15} = \zeta_{15}^{16} = (\zeta_{15}^5 \zeta_{15}^3)^2 = \zeta_3^2 \zeta_5^2 = \frac{1}{4}(-1+i\sqrt{3})^2 \cdot \frac{1}{16}\left(\sqrt{5}-1+i\sqrt{10+2\sqrt{5}}\right)^2$$

$$= \frac{1}{2}(-1 - i\sqrt{3}) \cdot \frac{1}{4}\left(-1 - \sqrt{5} + i\sqrt{10 - 2\sqrt{5}}\right)$$

$$= \frac{1}{8}\left(1 + \sqrt{5} + \sqrt{30 - 6\sqrt{5}}\right) + \frac{i}{8}\left(\sqrt{3}(1+\sqrt{5}) - \sqrt{10 - 2\sqrt{5}}\right)$$

and thus $F_{35} - f_{35} < (F_0 - f_0)/4^{35} < (3{,}19 - 3{,}05)/4^{35} < 1{,}2 \cdot 10^{-22}$. (Alternatively, one can also use $2\pi/15 = (2\pi/5 + 2\pi/5) - 2\pi/3$ and the addition theorems of the trigonometric functions.) Later, van Ceulen calculated π up to 35 decimal places using the regular 2^{62}-corners.

If one generally starts the recursion $f_{n+1} = \sqrt{f_n F_n}$, $F_{n+1} = 2f_{n+1} F_n/(f_{n+1} + F_n)$, $n \in \mathbb{N}$, with the areas $f_0 = \sin\alpha \cos\alpha$, $F_0 = f_0/h_0^2 = \tan\alpha$ of the inscribed or circumscribed triangle of a circle sector of the unit circle with central angle 2α, $0 < \alpha < \pi/2$, the sequences $(f_n), (F_n)$ form an interval nesting for the area α of this sector and the sequences $(f_n), (F_n)$ with the initial conditions $f_0 := \sin\alpha \cos\alpha / \sin\alpha \cos\alpha = 1$, $F_0 := \tan\alpha / \sin\alpha \cos\alpha = 1/\cos^2\alpha = 1 + \tan^2\alpha > 1$ form an interval nesting for $\alpha/\sin\alpha \cos\alpha$. If one generally replaces the initial condition f_0, F_0 with $0 < f_0 < F_0$ by af_0, aF_0 with an $a \in \mathbb{R}_+^\times$, one obtains for the interval nesting the sequences (af_n) and (aF_n). Therefore, it holds: *If f_0, F_0 are any real numbers with $0 < f_0 < F_0$, the two sequences $(f_n), (F_n)$ provide an interval nesting for the number*

3.3 Real Number Fields

$$\frac{f_0 \arctan \sqrt{F_0/f_0 - 1}}{\sqrt{1 - f_0/F_0}\sqrt{f_0/F_0}} = \frac{F_0 \arctan \sqrt{F_0/f_0 - 1}}{\sqrt{F_0/f_0 - 1}}.$$

$f_0 = 1$, $F_0 = 2$ thus provides (as above) an interval nesting for $\pi/2$, and $f_0 = 1$, $F_0 = 4$ one for $4\pi/3\sqrt{3}$.

By the way, the presented method for nesting the number π goes back to Archimedes, who, like van Ceulen, used the circumferences U_n and u_n instead of the areas F_n and f_n of the circumscribed and inscribed regular $2^n m$-gon of the unit circle. For this, one obtains the formulas $U_n = 2^{n+1} m B_n = 2F_n$, $u_n = 2^{n+1} m b_n = 2^{n+2} m b_{n+1} h_{n+1} = 2f_{n+1}$ with the recursions

$$U_{n+1} = 2F_{n+1} = \frac{4f_{n+1}F_n}{f_{n+1} + F_n} = \frac{2u_n U_n}{u_n + U_n},$$

$$u_{n+1} = 2f_{n+2} = 2\sqrt{f_{n+1}F_{n+1}} = \sqrt{u_n U_{n+1}}.$$

Archimedes started with $m = 6$, i.e. with $u_0 = 6$, $U_0 = 4\sqrt{3}$, calculated up to the 96-gon, so up to $n = 4$, and thus was able to obtain the known estimates $6\frac{20}{71} < u_4 < U_4 < 6\frac{2}{7}$, i.e. $3\frac{10}{71} < \pi < 3\frac{1}{7}$. Also note the value $f_1 = u_0/2 = 3$ for the area of the inscribed regular 12-gon. Incidentally, besides the inscribed or circumscribed square, the inscribed regular 12-gon is the only inscribed or circumscribed regular m-gon of the unit circle ($m \geq 3$), whose area is a rational number. Proof? *For the circumference U of the unit circle, one still obtains* $U = \lim_{n\to\infty} U_n = 2\lim_{n\to\infty} F_n = 2\pi = 2\lim_{n\to\infty} f_{n+1} = \lim_{n\to\infty} u_n$.—The number π is transcendental (over $\mathbb{Q}$), see for example the appendix in 14. ◊

Example 3.3.10 (g-al development of real numbers) Let g be a natural number ≥ 2. Furthermore, let x be a non-negative real number. It is $x = [x] + r$ with $0 \leq r < 1$. For the remainder $r = \{x\}$ we provide a canonical r defining interval nesting. To do this, we divide the half-open interval $[0, 1[$ into g equally long half-open intervals $[i/g, (i+1)/g[$, $i = 0, \ldots, g-1$, and define the digit z_1 by $r \in [z_1/g, (z_1 + 1)/g[$. Then we divide this last interval again into g equally long intervals and define the digit z_2 by

$$r \in \left[\frac{z_1}{g} + \frac{z_2}{g^2}, \frac{z_1}{g} + \frac{z_2 + 1}{g^2}\right[.$$

Continuing in this way, we obtain a sequence of digits $z_n \in \{0, 1, \ldots, g-1\}$ such that for all $n \in \mathbb{N}^*$ the following holds:

$$r \in \left[\sum_{i=1}^n \frac{z_i}{g^i}, \left(\sum_{i=1}^n \frac{z_i}{g^i}\right) + \frac{1}{g^n}\right[.$$

The sequences of these interval ends result in an r defining interval nesting. This representation is called the g-al development of r. In particular,

$$r = \sum_{i=1}^\infty \frac{z_i}{g^i} := \lim_{n\to\infty} \sum_{i=1}^n \frac{z_i}{g^i}.$$

If you use the g-al development $[x] = \sum_{i=0}^{m} a_i g^i$ according to example 1.7.6 for the whole part $[x]$ of x, you get the g-**al-development**

$$x = (a_m \ldots a_0, z_1 z_2 z_3 \ldots)_g$$

of x. The **digits** appearing in it are uniquely determined by x according to the construction. It is $(a_m \ldots a_0, z_1 \ldots z_k)_g = [xg^k]/g^k$ for $k \in \mathbb{N}$, cf. exercise 3.3.24. The digits z_i, $i \geq 1$, after the decimal point are obtained recursively with the scheme $z_0 = 0, r_0 = r$,

$$z_i = [r_{i-1} g], \quad r_i = r_{i-1} g - z_i, \; i \geq 1.$$

If r is a rational number $r = a/b$, then all remainders r_i belong to the b different fractions $0/b, 1/b, \ldots, (b-1)/b$ and must repeat. *The g-al-developments of rational numbers are therefore periodic.* Conversely, every number with a periodic g-al-development is rational, cf. example 3.6.18. Usually, the **decimal development** ($g = 10$) is chosen. In the proof of theorem 1.8.12 the **dual development** ($g = 2$) was used, where however a finite dual fraction $x = (0, z_1 \ldots z_{m-1} 100 \ldots)_2$ with $z_m = 1$ and $z_i = 0$ for all $i > m$ is replaced by the infinite dual fraction $(0, z_1 \ldots z_{m-1} 011 \ldots)_2$ with $z_m = 0$ and $z_i = 1$ for all $i > m$, which has the same value x.

In the above algorithm, it is not possible that almost all digits z_n are equal to $g - 1$. If, from a certain point $n_0 + 1$ all z_n are equal to $g - 1$, then r would from then on be in the last of the g subintervals at each step and would therefore already be at the n_0-th step in the $(z_{n_0} + 1)$-th subinterval as a boundary point. Conversely, each digit sequence z_n, $n \in \mathbb{N}^*$, in which not almost all digits are equal to $g - 1$, describes an interval nesting of the above form, which defines exactly one number $r \in [0, 1[$. From the g-al development it follows in particular that every real number is the limit of a sequence of rational numbers, see Lemma 3.3.4. — For a generalization of the g-al developments see exercise 3.6.15. ◊

Example 3.3.11 (Continued Fractions) The so-called **continued fraction expansion** of x provides very good rational approximations of a real number x. This is a generalization of the Euclidean algorithm from Sect. 1.7 and is essentially also described by Euclid, in the X. Book of the "Elements" as **alternating subtraction**. See in particular example 1.7.12.

Let $a, b \in \mathbb{R}$ with $b > 0$. We set $r_{-1} := a$, $r_0 := b$ and define the quotients $q_i \in \mathbb{N}$ and the remainders r_i for $i \geq 0$ recursively by

$$r_{i-1} = q_i r_i + r_{i+1}, \quad 0 < r_{i+1} < r_i.$$

The procedure stops if the remainder becomes 0, otherwise two infinite sequences $(q_i), (r_i)$ are obtained. If we set

$$x_i := r_{i-1}/r_i,$$

then $x_0 = x := a/b$ and the x_i and q_i are recursively determined by

$$x_0 = x, q_0 = [x_0]; \quad x_i = q_i + \frac{1}{x_{i+1}}, \quad q_{i+1} = [x_{i+1}]$$

3.3 Real Number Fields

The q_i thus depend solely on the quotient $x = a/b$. For each i, for which x_i is still defined, x is the **continued fraction**

$$x = q_0 + \cfrac{1}{q_1 + \cfrac{1}{\ddots + \cfrac{1}{q_{i-1} + \cfrac{1}{x_i}}}}.$$

One writes briefly

$$x = [q_0, q_1, \ldots, q_{i-1}, x_i].$$

If x_{i+1} is no longer defined, then $x_i = [x_i] = q_i \geq 2$ and $x = [q_0, \ldots, q_i]$ are rational. Conversely, if x is a rational number, the continued fraction development stops. *The continued fraction development of $x = a/b$ is therefore finite exactly when x is rational.* With Euclid, the numbers a,b are then called **commensurable**.

In the general case, if q_i is still defined, the fraction $[q_0, \ldots, q_i]$ is the *i*-th **approximate fraction** of x and q_i is the *i*-th **partial denominator**. The *i*-th approximate fraction can be easily calculated recursively. It is indeed

$$[q_0, \ldots, q_i] = \frac{a_i}{b_i},$$

where the *i*-th **approximate numerator** a_i and the *i*-th **approximate denominator** b_i can be determined according to the following recursion scheme:

$$a_{-2} = 0, \; a_{-1} = 1; \quad b_{-2} = 1, \; b_{-1} = 0;$$

$$a_i = q_i a_{i-1} + a_{i-2}; \quad b_i = q_i b_{i-1} + b_{i-2}, \; i \geq 0.$$

(See the algorithm for representing the gcd of two numbers $a, b \in \mathbb{N}^*$ before Lemma 1.7.8 by Bezout.) Note that because of $b_{-1} = 0$ the 0-th partial denominator q_0 is not used for the calculation of the approximate denominators b_i, $i \geq 1$. The recursion scheme is easily proven by induction: For $i = 0$, $a_0 = q_0$ and $b_0 = 1$, so $[q_0] = a_0/b_0$. In the conclusion from i to $i + 1$,

$$[q_0, \ldots q_{i+1}] = \left[q_0, \ldots, q_{i-1}, q_i + q_{i+1}^{-1}\right]$$

has a representation a_i'/b_i' according to the induction hypothesis, where the $a_0', \ldots, a_i'$ or $b_0', \ldots, b_i'$ can be calculated according to the above scheme to the sequence $q_0, \ldots, q_{i-1}, q_i + q_{i+1}^{-1}$. It is obviously $a_j = a_j'$ and $b_j = b_j'$ for $j \leq i - 1$, furthermore

$$q_{i+1} a_i' = q_{i+1}((q_i + 1/q_{i+1})a_{i-1} + a_{i-2}) = (q_{i+1}q_i + 1)a_{i-1} + q_{i+1}a_{i-2}$$
$$= q_{i+1}(q_i a_{i-1} + a_{i-2}) + a_{i-1} = q_{i+1}a_i + a_{i-1} = a_{i+1}$$

and analogously $q_{i+1}b_i' = b_{i+1}$. Therefore, $[q_0, \ldots, q_{i+1}] = a_{i+1}/b_{i+1}$ holds, as claimed.

From the recursion scheme, a trivial induction yields

$$a_{i+1}b_i - a_i b_{i+1} = (-1)^i \quad \text{bzw.} \quad \frac{a_{i+1}}{b_{i+1}} - \frac{a_i}{b_i} = \frac{(-1)^i}{b_i b_{i+1}}, \; i \geq 0.$$

If the continued fraction expansion x_{i+1} is still defined, one specifically obtains from $x = [q_0, \ldots, q_i, x_{i+1}]$ the following **approximation formula**:

$$x - \frac{a_i}{b_i} = \frac{(-1)^i}{b_i(x_{i+1}b_i + b_{i-1})}.$$

With $x_{i+1} > 1$ follows

$$\left| x - \frac{a_i}{b_i} \right| = \frac{1}{b_i(x_{i+1}b_i + b_{i-1})} < \frac{1}{b_i(b_i + b_{i-1})}.$$

Because of $q_i \geq 1$ for $i \geq 0$, for an infinite continued fraction, thus an irrational number x, the sequence of approximation denominators b_i is monotonic and unbounded. Therefore, in particular,

$$x = \lim_{i \to \infty} \frac{a_i}{b_i} = \lim_{i \to \infty} [q_0, \ldots, q_i].$$

Moreover, the approximation fractions a_i/b_i are reduced due to $a_i b_{i-1} - a_{i-1} b_i = (-1)^{i-1}$. More precisely, the approximation formula provides: *The sequences*

$$\frac{a_{2j}}{b_{2j}}, \; j \in \mathbb{N}, \quad \text{and} \quad \frac{a_{2j+1}}{b_{2j+1}}, \; j \in \mathbb{N},$$

of approximation fractions of the number x define an interval nesting for the number x, and the length of the j-th interval is

$$\frac{a_{2j+1}}{b_{2j+1}} - \frac{a_{2j}}{b_{2j}} = \frac{1}{b_{2j}b_{2j+1}}.$$

The approximation fractions a_i/b_i of the continued fraction expansion of x are even **best approximations** of x in the following sense: *Is c/d a rational number with $c \in \mathbb{Z}$, $d \in \mathbb{N}^*$ and $|x - c/d| < |x - a_i/b_i|$, then is $d > b_i$*, i.e. every rational number that approximates x better than a_i/b_i, has a larger denominator than a_i/b_i. *Proof.* At $|x - c/d| < |x - a_i/b_i|$ with $d \leq b_i$ is a_{i+1}/b_{i+1} still defined and $c/d \neq a_{i+1}/b_{i+1}$, from which a contradiction follows:

$$\frac{1}{b_i b_{i+1}} \leq \frac{1}{d b_{i+1}} \leq \left| \frac{c}{d} - \frac{a_{i+1}}{b_{i+1}} \right| \leq \left| \frac{c}{d} - x \right| + \left| x - \frac{a_{i+1}}{b_{i+1}} \right|$$
$$< \left| \frac{a_i}{b_i} - x \right| + \left| x - \frac{a_{i+1}}{b_{i+1}} \right| = \left| \frac{a_i}{b_i} - \frac{a_{i+1}}{b_{i+1}} \right| = \frac{1}{b_i b_{i+1}}. \quad \square$$

Good approximations of x often also provide the so-called **secondary approximation fractions**

$$[q_0, \ldots, q_{i-1}, q] = \frac{q a_{i-1} + a_{i-2}}{q b_{i-1} + b_{i-2}}, \quad 1 \leq q < q_i \; (i \geq 1).$$

3.3 Real Number Fields

The continued fraction development of π, for example, begins with

$$\pi = [3, 7, 15, 1, 292, 1, 1, 1, 2, 1, 3, 1, 14, 2, 1, 1, 2, 2, 2, 2, 1, 84, \ldots].$$

This results in the nesting

$$\frac{3}{1} = [3, 7, 15] = \frac{333}{106} < \cdots < \pi < \cdots < [3, 7, 15, 1] = \frac{355}{113} < [3, 7] = \frac{22}{7},$$

of which the fractions $[3, 7] = 22/7 > \pi$ and $[3, 7, 15, 1] = 355/113 > \pi$ are particularly favorable approximations in relation to the denominator size 7 or 113, as the next partial denominators 15 or 292 are relatively large. No regularity is known in the continued fraction development of π. For e, on the other hand, according to Euler, see about 15, Appendix 2.C, the partial denominators are $q_0 = 2, q_{3j-1} = 2j$ for $j \geq$ and $q_i = 1$ otherwise, i.e. it is:

$$e = [2, 1, 2, 1, 1, 4, 1, 1, 6, 1, 1, 8, 1, 1, 10, \ldots].$$

Conversely, for any finite sequence of integers $q_0, \ldots, q_n, n \in \mathbb{N}^*, q_1, \ldots, q_{n-1} \geq 1$ and $q_n \geq 2$ or for any infinite sequence $q_0, q_1, \ldots$ of integers with $q_i \geq 1$ for $i \geq 1$, there is a $x \in \mathbb{R}$, whose continued fraction development has just these q_i as partial denominators. In the first case, x is the rational number $[q_0, \ldots, q_n]$, otherwise it is $x = \lim_{i \to \infty} [q_0, \ldots, q_i]$. ◊

Exercises

Unless otherwise stated, in the following exercises the elements are in a real number field $\mathbb{R}$.

Exercise 3.3.1 Examine the sequences

$$\frac{(n+1)(n^2-1)}{(2n+1)(3n^2+1)}; \quad \frac{n+1}{n^2+1}; \quad \frac{4^n+1}{5^n}; \quad \frac{1}{n^2} + (-1)^n \frac{n^2}{n^2+1}, \quad n \in \mathbb{N}^*,$$

for convergence and determine the limit value if applicable.

Exercise 3.3.2 Let $t \mapsto F(t) := \sum_{i=0}^{k} a_i t^i$ and $t \mapsto G(t) := \sum_{j=0}^{m} b_j t^j$ be polynomial functions $\mathbb{R} \to \mathbb{R}$ with $a_i, b_j \in \mathbb{R}$, $a_k \neq 0$, $b_m \neq 0$. For all $n \geq n_0$ let $G(n) \neq 0$. Then the sequence $F(n)/G(n)$, $n \geq n_0$, is defined and it holds:

$$\lim_{n \to \infty} \frac{F(n)}{G(n)} = \begin{cases} 0, & \text{if } k < m, \\ a_m/b_m, & \text{if } k = m, \\ \infty, & \text{if } k > m \text{ and } a_k/b_m > 0, \\ -\infty, & \text{if } k > m \text{ and } a_k/b_m > 0. \end{cases}$$

The rational function $F/G \in \mathbb{R}(T)$ is exactly > 0 in the sense of example 3.1.3, when $F(n)/G(n) > 0$ is for sufficiently large n.

Exercise 3.3.3 Calculate

$$\lim_{n\to\infty} \frac{n-\sqrt{n}}{n+\sqrt{n+1}} \quad \text{and} \quad \lim_{n\to\infty} \frac{\sqrt{n}+1}{n+1}.$$

Exercise 3.3.4 Calculate the following limits:

$$\lim_{n\to\infty}\left(\sqrt{n+1}-\sqrt{n}\right); \quad \lim_{n\to\infty}\sqrt{n}(\sqrt{n+a}-\sqrt{n}), a\in\mathbb{R}, n\geq |a|;$$

$$\lim_{n\to\infty}\left(\sqrt{n+\sqrt{n}}-\sqrt{n-\sqrt{n}}\right); \quad \lim_{n\to\infty} n\left(\frac{1}{\sqrt{n+1}}-\frac{1}{\sqrt{n}}\right);$$

$$\lim_{n\to\infty} n\left(\sqrt{1+\frac{1}{n}}-1\right), n\geq 1.$$

Exercise 3.3.5 Let $a > 0$. Then $\lim_{n\to\infty}\sqrt[n]{a}=1$ applies. (For $a \geq 1$ write $\sqrt[n]{a}=1+h_n$ and use the Bernoulli inequality exercise 3.1.13) or take advantage of the fact that for $a \geq 1$ the sequence is monotonically decreasing, thus converging to an $x \geq 1$, on the other hand $x = \lim \sqrt[n]{a} = \lim(\sqrt[2n]{a})^2 = x^2$ is.

Exercise 3.3.6 Show $\lim \sqrt[n]{n} = 1$. (You can proceed similarly as in exercise 3.3.5. For $n \geq 3$ the sequence is monotonically decreasing.)

Exercise 3.3.7 Let (x_n) be a sequence of non-zero real numbers.

a) If there is a q with $0<q<1$ and $|x_{n+1}/x_n| \leq q$ for almost all n, then $\lim x_n = 0$.
b) If there is a q with $q > 1$ and $|x_{n+1}/x_n| \geq q$ for almost all n, then $\lim |x_n| = \infty$.
c) Show for each $k \in \mathbb{N}$

$$\lim_{n\to\infty}\frac{1}{2^n}\binom{n}{k}=0.$$

Exercise 3.3.8 Let $a_1,\ldots,a_m \in \mathbb{R}_+$, $m \geq 1$. Then applies

$$\lim_{n\to\infty}\sqrt[n]{a_1^n+\cdots+a_m^n} = \text{Max}\,(a_1,\ldots,a_m).$$

Exercise 3.3.9 Let (x_n) be a possibly improperly convergent sequence of real numbers with $\lim x_n = x \in \overline{\mathbb{R}}$.

a) The sequence $a_n := \frac{1}{n}(x_1 + \cdots + x_n)$, $n \geq 1$, of arithmetic means also converges to x.
b) Let $x_n > 0$ for all n. The sequence

$$h_n := \frac{n}{\frac{1}{x_1}+\cdots+\frac{1}{x_n}}, \quad n \geq 1,$$

of harmonic means also converges to x. (This follows from a).)

3.3 Real Number Fields

c) Let $x_n > 0$ for all n. Then the sequence $g_n := \sqrt[n]{x_1 \cdots x_n}$ of geometric means also converges to x. (Use exercise 3.1.22. By transitioning to logarithms, the statement follows directly from a) due to the continuity of ln and exp, see Sect. 3.10.)

d) Using c), solve exercise 3.3.5 and 3.3.6 again and also prove

$$\lim_{n \to \infty} \sqrt[n]{n!} = \infty \quad \text{as well as} \quad \lim_{n \to \infty} \frac{\sqrt[n]{n!}}{n} = \frac{1}{e}.$$

(The last two limit statements can also be derived from Stirling's formula, see remark 1.6.8. This even provides the asymptotic equality $n! \sim \sqrt{2\pi n}(n/e)^n$. The sequence of absolute errors $n! - \sqrt{2\pi n}(n/e)^n$ indeed converges to ∞.)

e) Show by counterexamples that the reversals of the statements in a), b), c) are not generally correct.

f) Let (y_n) be a sequence in $\mathbb{R}_+^\times$, for which the sequence (y_{n+1}/y_n) converges to $y \in \mathbb{R}_+ \uplus \{\infty\}$. Then the sequence $(\sqrt[n]{y_n})$ also converges to y.

Exercise 3.3.10 Let (x_n) be a possibly improperly convergent sequence of real numbers with $\lim x_n = x$. Then the sequence

$$\frac{1}{2^n} \sum_{m=0}^{n} \binom{n}{m} x_m, \quad n \in \mathbb{N},$$

is also convergent with limit value x.

Exercise 3.3.11 Let (x_n) and (y_n) be sequences in $\mathbb{R}$ with $y_n > 0$ and $\lim_{n \to \infty}(y_0 + \cdots + y_n) = \infty$. If the sequence (x_n/y_n) converges to a, then the sequence

$$\left(\frac{x_0 + \cdots + x_n}{y_0 + \cdots + y_n} \right), \quad n \in \mathbb{N}.$$

also converges to a.

Exercise 3.3.12

$$\lim_{n \to \infty} \left(1 - \frac{1}{n^2}\right)^n = 1, \quad \lim_{n \to \infty} \left(1 - \frac{1}{n}\right)^n = e^{-1}, \quad \lim_{n \to \infty} \left(1 + \frac{1}{n^2}\right)^n = 1.$$

(For the first formula, use the Bernoulli inequality. The second sequence is, by the way, monotonic.)

Exercise 3.3.13 Show that the sequence F_{n+1}/F_n, $n \geq 1$, of the quotients of successive Fibonacci numbers converges to the number $\Phi = \frac{1}{2}(1 + \sqrt{5})$ of the Golden Ratio. (See example 1.5.6.—By the way, F_{n+1}/F_n is the $(n-1)$-th approximation fraction of the continued fraction development of $\Phi = [1, 1, 1, \ldots]$, see exercise 3.3.26a).)

Exercise 3.3.14 Examine the following recursively defined sequences (x_n) for convergence and calculate their limits if possible.

a) $x_{n+1} = x_n^2 + \frac{1}{4}, n \in \mathbb{N}$, with $0 \le x_0 \le \frac{1}{2}$.
b) $x_0 = 0, x_{n+1} = \frac{1}{2}(a + x_n^2), n \in \mathbb{N}$, with $0 \le a \le 1$.
c) $x_0 = 0, x_{n+1} = \frac{1}{2}(a - x_n^2), n \in \mathbb{N}$, with $0 \le a \le 1$.
d) $x_0 = 2, x_{n+1} = 2 - x_n^{-1}, n \in \mathbb{N}$.
e) $x_0 = 0, x_{n+1} = \sqrt{a + x_n}, n \in \mathbb{N}$, with $a > 0$. (For $a = 2$, the sequence (c_n) from example 3.3.9 is obtained.)
f) $x_{n+1} = 2x_n - ax_n^2, n \in \mathbb{N}$, with $a \in \mathbb{R}, a > 0$ and $0 < x_0 < 2/a$.
g) $x_{n+1} = (x_n + 2)/(x_n + 1)$ with $x_0 \ge 0$.
h) $x_{n+1} = \frac{1}{3}(x_n^2 + 2)$ with arbitrary x_0.

Exercise 3.3.15
a) For $a, b \in \mathbb{R}$, the sequence (x_n) is recursively defined by $x_0 = a, x_1 = b$ and $x_{n+2} = \frac{1}{2}(x_n + x_{n+1})$. Then $\lim x_n = \frac{1}{3}(a + 2b)$ is true. (Examine the subsequences (x_{2n}) and (x_{2n+1}) separately.)
b) The sequence (x_n) is recursively defined by $x_0 = a, x_1 = 1, x_{n+2} = \sqrt{x_n x_{n+1}}$ with $a \in \mathbb{R}_+^\times$. Then $\lim x_n = \sqrt[3]{a}$ applies.

Exercise 3.3.16 Let $a \ge 1$. For the recursively defined sequence (x_n) with $x_0 = a$, $x_{n+1} = a + x_n^{-1}, n \in \mathbb{N}$, show the convergence and calculate the limit. (Also compare with example 3.3.11 about continued fractions.)

Exercise 3.3.17 Let $a, b > 0$. The recursively defined sequences (a_n) and (b_n) with $a_0 = a, b_0 = b$ and

$$a_{n+1} = \frac{2a_n b_n}{a_n + b_n} = \text{harmonic mean of } a_n, b_n,$$

$$b_{n+1} = \frac{a_n + b_n}{2} = \text{arithmetic mean von } a_n, b_n$$

form from $n = 1$ an interval nesting for the geometric mean $\sqrt{ab}$ of a and b. (Note that $a_n b_n = ab$ is for all $n \in \mathbb{N}$. The sequence (x_n) from example 3.3.7 results as the sequence (b_n), if we set $a_0 = a/x_0$ and $b_0 = x_0$. We have here due to $0 \le b_{n+1} - a_{n+1} = (b_n - a_n)^2/2(a_n + b_n)$ quadratic convergence of the interval nesting.)

Exercise 3.3.18 Let $a, b > 0$. The recursively defined sequences (a_n) and (b_n) with $a_0 = a, b_0 = b$ and

$$a_{n+1} = \frac{a_n + b_n}{2} = \text{arithmetic mean of} a_n, b_n,$$

$$b_{n+1} = \sqrt{a_n b_n} = \text{geometric mean of } a_n, b_n$$

3.3 Real Number Fields

form from $n = 1$ an interval nesting $[b_n, a_n]$, $n \in \mathbb{N}^*$. (The number defined by this, $M(a,b)$, is called the **arithmetic-geometric mean** of a and b. Because of

$$0 \leq a_{n+1} - b_{n+1} = (a_n - b_n)^2 / 2\left(a_n + b_n + 2\sqrt{a_n b_n}\right)$$

there is also quadratic convergence of the interval nesting. — By the way, the interval nesting constructed analogously with the harmonic and geometric means defines the so-called **harmonic-geometric mean** of a and b. This is equal to $1/M(a^{-1}, b^{-1})$. Proof!

Exercise 3.3.19 Prove that the recursively defined sequence at the end of example 3.3.7 (x_n) converges to $\sqrt[k]{a}$, where for $n \geq 1$ the following error estimate applies:

$$0 \leq x_n - \sqrt[k]{a} \leq \frac{1}{k(\sqrt[k]{a})^{k-1}} \left(x_n^k - a\right).$$

Exercise 3.3.20 Let $a \in \mathbb{R}_+^\times$. The recursively defined sequence (x_n) with $x_0 > 0$ arbitrary and

$$x_{n+1} = \frac{x_n^2 + 3a}{3x_n^2 + a} x_n$$

converges monotonically towards $\sqrt{a}$, where even cubic convergence exists (due to $x_{n+1} - \sqrt{a} = (x_n - \sqrt{a})^3 / (3x_n^2 + a)$).

Exercise 3.3.21 Let (a_n) be a monotonically increasing and (b_n) a monotonically decreasing sequence of real numbers with $a_n \leq b_n$ for all $n \in \mathbb{N}$ and $a := \lim a_n$, $b := \lim b_n$. Then $\bigcap_{n=0}^{\infty} [a_n, b_n] = [a, b]$ is true.

Exercise 3.3.22
a) Determine the dual and trial developments of $1/7$, $1/8$, $1/9$, $1/10$.
b) Determine the g-al developments of $a/(g-1)$ and $a/(g+1)$. Also confirm $1/(g-1)^2 = (0, \overline{012 \ldots g - 3g - 1})_g$, where the crossed-out digits indicate the period.

Exercise 3.3.23
a) Every interval in $\mathbb{R}$ with more than one point has the power of $\mathbb{R}$, i.e., the power $\aleph$ of the continuum.
b) If $A \subseteq \mathbb{R}$ is countable, then $\mathbb{R} - A$ is dense in $\mathbb{R}$. In particular, the set $\mathbb{R} - \mathbb{Q}$ of irrational numbers is dense in $\mathbb{R}$.

Exercise 3.3.24 Let K be an Archimedean ordered field. For $x \in K$ the sequence $[nx]/n \in \mathbb{Q}$, $n \geq 1$, converges to x. (This proves once again that $\mathbb{Q}$ is dense in K, cf. Lemma 3.3.4.)

Exercise 3.3.25 $c_n = \sqrt{2 + \sqrt{2 + \cdots + \sqrt{2}}} \in \mathbb{R}_+^\times$, $n \in \mathbb{N}^*$, are the numbers from Example 3.3.9.

a) The product representation $2/\pi = \prod_{n=1}^\infty (c_n/2)$ of Vieta can be interpreted geometrically as follows: A square is circumscribed around a circle with radius $r_0 = 1$ and a circle with radius $r_1 = \sqrt{2}$ is circumscribed around this square, then a regular octagon around this circle, then a circle with radius r_2 around this octagon, then a hexadecagon, etc. Then the sequence (r_n) of the radii of the circles obtained in this way converges to $\pi/2$. (It is $r_n = 2^n/c_1 \cdots c_n$, $n \in \mathbb{N}$.)
b) The c_n are algebraic over $\mathbb{Q}$ of degree 2^n, $n \in \mathbb{N}^*$. Their minimal polynomials $\mu_n := \mu_{c_n}$ satisfy the recursion $\mu_1 = X^2 - 2$, $\mu_{n+1} = \mu_n(X^2 - 2)$, $n \in \mathbb{N}^*$. (In μ_n, $X^2 - 2$ is substituted for X. — It is to be shown that the polynomials defined by this recursion are prime in $\mathbb{Q}[X]$ or $\mathbb{Z}[X]$. This can be seen, for example, with the Eisenstein criterion from exercise 2.10.18d). — Because of $c_n = 2\cos(2\pi/2^{n+2})$, by the way, $\mu_n = 2^{2^n} T_{2^n}(X/2)$, where T_k, $k \in \mathbb{N}$, are the Chebyshev polynomials from example 3.5.8. In particular, $\mu_n(0) = 2$ for all $n > 1$.)

Exercise 3.3.26 Prove the following continued fraction developments.

a) For $a, b \in \mathbb{N}^*$ is $[a, b, a, b, a, b, \ldots] = (ab + \sqrt{a^2b^2 + 4ab})/2b$.
b) For $n \in \mathbb{N}^*$ applies $\sqrt{n^2 + 1} = [n, 2n, 2n, 2n, \ldots]$, $\sqrt{n^2 + 2} = [n, n, 2n, n, 2n, \ldots]$ as well as $\sqrt{(n+1)^2 - 1} = [n, 1, 2n, 1, 2n, \ldots]$.

Exercise 3.3.27 For $x \in \mathbb{R}_+$ let $H_x := H_{[x]} = \sum_{n \in \mathbb{N}^*, n \leq x} 1/n$. For $x \geq 1$ applies according to example 3.3.8 $H_x = \ln x + \gamma + c_x/x$ with $|c_x| < 2$.

a) Let $(y_k)_{k \in \mathbb{N}^*}$, $(x_k)_{k \in \mathbb{N}^*}$ be sequences in $\mathbb{R}_+^\times$ with $\lim_k y_k = \lim_k x_k = \infty$ such that $x := \lim_k (y_k/x_k)$ exists in $\mathbb{R}_+^\times$. Then $\lim_{k \to \infty}(H_{y_k} - H_{x_k}) = \ln x$ is. (Use the continuity of the logarithm: If the sequence $x_k \in \mathbb{R}_+^\times$ converges to $x \in \mathbb{R}_+^\times$, then $\ln x_k$ converges to $\ln x$, cf. Sect. 3.10.) — For $x_k := k$, $y_k := 2k$, $k \in \mathbb{N}^*$, for example,

$$\ln 2 = \lim_{k \to \infty} \sum_{n=k+1}^{2k} \frac{1}{n} = \lim_{k \to \infty} \sum_{n=1}^{2k} \frac{(-1)^{n-1}}{n} = \lim_{k \to \infty} \sum_{n=1}^{k} \frac{(-1)^{n-1}}{n} = \sum_{n=1}^{\infty} \frac{(-1)^{n-1}}{n}.$$

As an application of this equation, which we will prove several times, we determine the average value of the differences $U(n) - G(n)$ of the number $U(n)$ of odd and the number $G(n)$ of even divisors of $n \in \mathbb{N}^*$. First, the sum $\sum_{m=1}^{n}(U(m) - G(m))$ is equal to

$$\sum_{k,\ell \in \mathbb{N}^*, k\ell \leq n} (-1)^{k-1} = \sum_{k \leq \sqrt{n}} (-1)^{k-1} \left[\frac{n}{k}\right] + \sum_{\ell < \sqrt{n}} \left(\sum_{\sqrt{n} < k \leq n/\ell} (-1)^{k-1}\right).$$

3.3 Real Number Fields

So

$$\left|\sum_{m=1}^{n}(U(m)-G(m))-\sum_{k=1}^{[\sqrt{n}]}(-1)^{k-1}\frac{n}{k}\right| \leq \left|\sum_{k\leq\sqrt{n}}(-1)^k\left\{\frac{n}{k}\right\}\right|+\left|\sum_{\ell<\sqrt{n}}\left(\sum_{\sqrt{n}<k\leq n/\ell}(-1)^{k-1}\right)\right|$$

$$\leq \sqrt{n}+\sqrt{n}$$

applies (for $t \in \mathbb{R}$ is $\{t\}=t-[t]$) and consequently

$$\lim_{n\to\infty}\frac{1}{n}\sum_{m=1}^{n}(U(m)-G(m))=\sum_{k=1}^{\infty}(-1)^{k-1}\frac{1}{k}=\ln 2.$$

The number of odd divisors of a natural number exceeds the number of even divisors on average by $\ln 2$. — It is $U(n)-G(n)=(1-\nu_2(n))\prod_{p\in\mathbb{P},p>2}(1+\nu_p(n))$. Proof!

b) Let $x \in \mathbb{R}_+^\times$. Then $\lim_{n\to\infty}\left(H_{n^x}-xH_n\right)=(1-x)\gamma$. For $x:=1/2$, for example, we get $\lim_{n\to\infty}(2H_{\sqrt{n}}-H_n)=\gamma$. — To give an application of this equation, we use for the harmonic numbers the representation

$$H_n=\sum_{k=1}^{n}\frac{1}{k}=\frac{1}{n}\sum_{k=1}^{n}\left[\frac{n}{k}\right]+\frac{1}{n}\sum_{k=1}^{n}\left\{\frac{n}{k}\right\}, \quad n\in\mathbb{N}^*.$$

$\sum_{k=1}^{n}[n/k]$ is the number of pairs $(k,\ell)\in(\mathbb{N}^*)^2$ with $k\ell \leq n$. For each such pair, $k \leq \sqrt{n}$ or $\ell \leq \sqrt{n}$, and consequently

$$\sum_{k=1}^{n}\left[\frac{n}{k}\right]=2\sum_{k=1}^{[\sqrt{n}]}\left[\frac{n}{k}\right]-[\sqrt{n}]^2=2nH_{\sqrt{n}}-2C_n[\sqrt{n}]-[\sqrt{n}]^2 \quad \text{with } 0\leq C_n<1.$$

With $\lim_{n\to\infty}\left(2H_{\sqrt{n}}-H_n\right)=\gamma$ we finally get

$$\lim_{n\to\infty}\frac{1}{n}\sum_{k=1}^{n}\left\{\frac{n}{k}\right\}=\lim_{n\to\infty}\left(H_n-2H_{\sqrt{n}}+2C_n\frac{[\sqrt{n}]}{n}+\frac{[\sqrt{n}]^2}{n}\right)=-\gamma+1.$$

At first glance, one might think that this limit should be $1/2$. The reader should also note that $\sum_{k=1}^{n}[n/k]$ is equal to the sum $\sum_{m=1}^{n}\tau(m)$ of the numbers $\tau(m)$ of the divisors of the numbers $m=1,\ldots,n$ is. It follows

$$\frac{1}{n}\sum_{m=1}^{n}\tau(m)=\sum_{k=1}^{n}\frac{1}{k}-\frac{1}{n}\sum_{k=1}^{n}\left\{\frac{n}{k}\right\}=\ln n+2\gamma-1+\varepsilon_n$$

with a null sequence (ε_n). *In particular, the average number of divisors of a number* $n \in \mathbb{N}^*$ *for* $n\to\infty$ *is asymptotically equal* $\ln n$. (However, there are always numbers n with only two divisors.) See also part a), which incidentally results in the following equation:

$$\lim_{n\to\infty}\frac{1}{n}\sum_{k=1}^{n}(-1)^{k-1}\left\{\frac{n}{k}\right\}=0.$$

3.4 Consequences of Completeness

In the following, K denotes an ordered field.

Definition 3.4.1 Let (x_n) be a sequence in K. A point $x \in K$ is called a **cluster point** of (x_n), if in every (no matter how small) neighborhood of x there are infinitely many members of the sequence.

A convergent sequence obviously has its limit as the only accumulation point. The sequence $(-1)^n$, $n \in \mathbb{N}$, has the two accumulation points 1 and -1. The sequence $n + (-1)^n n$, $n \in \mathbb{N}$, has the only accumulation point 0, but is not convergent. If K is Archimedean ordered and (x_n) is a sequence in which every rational number occurs, then the set of accumulation points of (x_n) is entirely K, cf. Lemma 3.3.4.

Lemma 3.4.2 *K possesses a non-stationary null sequence and thus a strictly monotonically decreasing null sequence (ε_k). A sequence (x_n) in K has exactly $x \in K$ as an accumulation point, if (x_n) has a subsequence converging to x.*

Proof If (x_n) has a subsequence converging to x, then almost all members of the subsequence, and thus certainly infinitely many members of (x_n), lie in any neighborhood of x. (This implication holds for any ordered field.) Conversely, let x be an accumulation point of (x_n). We recursively construct a subsequence (x_{n_k}) with $|x - x_{n_k}| \leq \varepsilon_k$. This subsequence then converges to x. Since x is an accumulation point of (x_n), there is an $n_0 \in \mathbb{N}$ with $|x - x_{n_0}| \leq \varepsilon_0$. If $n_0, \ldots, n_k$ is defined, then choose n_{k+1} such that $n_{k+1} > n_k$ and $|x_{n_{k+1}} - x| \leq \varepsilon_{k+1}$ is. This is possible because in the ε_{k+1} neighborhood of x, infinitely many members of the sequence (x_n) lie. □

The most important existence statement about accumulation points is the following theorem:

Theorem 3.4.3 (Bolzano-Weierstrass Theorem) *Every bounded sequence of real numbers has an accumulation point, i.e., a convergent subsequence.*

Proof Let (x_n) be a bounded sequence. Then all terms lie within a bounded interval $[a, b] \subseteq \mathbb{R}$, $a < b$. To determine a limit point of (x_n) we construct an interval nesting $[a_n, b_n]$, $n \in \mathbb{N}$, such that in each interval of this sequence there are infinitely many terms of the sequence (x_n). Then the number x defined by this interval nesting is a limit point of (x_n).

We specify such an interval nesting with the so-called **interval halving method**. For this, we set $a_0 = a$, $b_0 = b$. If a_n and b_n are chosen, then in the interval $[a_n, b_n]$ by construction, there are infinitely many members of the sequence (x_n). This then also applies to at least one of the two subintervals $[a_n, \frac{1}{2}(a_n + b_n)]$ and $[\frac{1}{2}(a_n + b_n), b_n]$, whose endpoints we then take for a_{n+1} and b_{n+1}. If this applies to

3.4 Consequences of Completeness

both subintervals, we take the "left" half $[a_n, \frac{1}{2}(a_n + b_n)]$ for reasons of uniqueness. Because of $b_n - a_n = (b-a)/2^n$, $\lim(b_n - a_n) = 0$ is, see Lemma 3.3.3 (3), and it is an interval nesting. □

We use the Bolzano-Weierstrass theorem to prove the Cauchy convergence criterion.

Definition 3.4.4 A sequence (x_n) in K is called a **Cauchy sequence**, if for every $\varepsilon \in K_+^\times$ there is an $n_0 \in \mathbb{N}$ with $|x_m - x_n| \leq \varepsilon$ for all $m, n \geq n_0$.

(x_n) is apparently a Cauchy sequence exactly when for every $\varepsilon > 0$ there is a $n_0 \in \mathbb{N}$ with $|x_n - x_{n_0}| \leq \varepsilon$ for all $n \geq n_0$. This characterization of a Cauchy sequence avoids the additional index variable m. *Cauchy sequences are bounded.* Furthermore, the following applies:

Lemma 3.4.5 *The Cauchy sequences in K form a K-subalgebra $K_{CF}^\mathbb{N}$ of the algebra $K^\mathbb{N}$ of all sequences in K, which includes the algebra $K_{con}^\mathbb{N}$ of the convergent sequences.*

Proof The constant sequences are in $K_{CF}^\mathbb{N}$. The proof that the sum and product of two Cauchy sequences in K are again Cauchy sequences proceeds completely analogously to the proof of the corresponding statement about convergent sequences, cf. theorem 3.2.5. — Now let the sequence be $(x_n) \in K_{kon}^\mathbb{N}$ and $x = \lim x_n$. For $\varepsilon > 0$ there is a $n_0 \in \mathbb{N}$ with $|x_n - x| \leq \frac{1}{2}\varepsilon$ for all $n \geq n_0$. For all $m, n \geq n_0$ then

$$|x_m - x_n| = |x_m - x + x - x_n| \leq |x_m - x| + |x_n - x| \leq \frac{1}{2}\varepsilon + \frac{1}{2}\varepsilon = \varepsilon.$$

applies. Consequently, (x_n) is a Cauchy sequence. □

Lemma 3.4.6 *Let $\varphi: K' \to K$ be a homomorphism of ordered fields, thus a monotonic field homomorphism. Furthermore, let (x_n) be a sequence in K' and $x \in K'$. Then the following applies:*

(1) *If $\varphi(x_n)$ is convergent in K with $\lim \varphi(x_n) = \varphi(x)$, then (x_n) is convergent in K' with $\lim x_n = x$. If $\varphi(K')$ is dense in K, then the reverse also applies.*
(2) *If $\varphi(x_n)$ is a Cauchy sequence in K, then (x_n) is a Cauchy sequence in K'. If $\varphi(K')$ is dense in K, then the reverse also applies.*

Proof Since φ is injective, φ is strictly monotonically increasing, and we can assume that K' is a subfield of K and φ is the canonical embedding.

(1) Let $\lim x_n = x$ apply in K. Trivially, $\lim x_n = x$ also applies in K'. Now let K' be dense in K and let $\lim x_n = x$ apply in K'. Then there exists a $\varepsilon \in K_+^\times$ corresponding to $\varepsilon' \in K_+'^\times$ with $\varepsilon' \leq \varepsilon$ and consequently a $n_0 \in \mathbb{N}$ with $|x_n - x| \leq \varepsilon' \leq \varepsilon$ for all $n \geq n_0$, and (x_n) is also convergent in K with limit x.

(2) The proof of (2) proceeds completely analogously to that of (1). $\square$

We explicitly note that in the situation of Lemma 3.4.6 the sequence (x_n) can converge in K' towards x without $(\varphi(x_n))$ in K converging towards $\varphi(x)$. For example, $1/n$, $n \in \mathbb{N}^*$, is a null sequence in $\mathbb{Q}$, but not a null sequence in the rational function field $\mathbb{Q}(X)$ with the order according to Example 3.1.2. (Are there also orders on $\mathbb{Q}(X)$, with respect to which $(1/n)$ also converges in $\mathbb{Q}(X)$ towards 0? If $x \in \mathbb{R}$ is transcendental (over $\mathbb{Q}$), then $\mathbb{Q}(X) \cong \mathbb{Q}(x) \subseteq \mathbb{R}$.)

We prepare the proof of the fundamental Cauchy convergence criterion for the field $\mathbb{R}$ of real numbers with the following lemma:

Lemma 3.4.7 *Every Cauchy sequence in K with a limit point is convergent.*

Proof Let x be an accumulation point of the Cauchy sequence (x_n). We show that (x_n) converges to x: For $\varepsilon > 0$ there is an n_0 with $|x_m - x_n| \leq \frac{1}{2}\varepsilon$ for all $m, n \geq n_0$. Furthermore, there is an x_{m_0} with $m_0 \geq n_0$ and $|x_{m_0} - x| \leq \frac{1}{2}\varepsilon$, since x is an accumulation point of the sequence (x_n). For all $n \geq n_0$ follows .

$$|x_n - x| = |x_n - x_{m_0} + x_{m_0} - x| \leq |x_n - x_{m_0}| + |x_{m_0} - x| \leq \frac{1}{2}\varepsilon + \frac{1}{2}\varepsilon = \varepsilon. \quad \square$$

From the theorem 3.4.3 of Bolzano-Weierstraß and the preceding lemmas follows:

Theorem 3.4.8 (Cauchy's Convergence Criterion) *A sequence of real numbers converges if and only if it is a Cauchy sequence.*

To prove that a Cauchy sequence (x_n) of real numbers converges, the interval nesting criterion 3.3.5 can also be used directly: Let $[a_0, b_0]$, $a_0 < b_0$, be a closed interval that contains almost all members of the sequence (x_n). For the recursive construction of the interval $[a_{n+1}, b_{n+1}]$, the interval $[a_n, b_n]$ is divided into thirds. In (at least) one of the two edge thirds, only finitely many members of the sequence lie. (Why?) The other two thirds together form the interval $[a_{n+1}, b_{n+1}]$ of length $2(b_n - a_n)/3$.

We now also have the means to conveniently prove the uniqueness of a real number field (up to isomorphism). More precisely, we show:

Theorem 3.4.9 *Let K be an archimedically ordered field and $\mathbb{R}$ a real number field. Then there is exactly one homomorphism of ordered fields $\varphi: K \to \mathbb{R}$.*

Proof Let $\varphi: K \to \mathbb{R}$ be a monotonic field homomorphism. φ is necessarily the identity on the prime field $\mathbb{Q}$ of K and $\mathbb{R}$. Now let $x \in K$. Then there exists a

3.4 Consequences of Completeness

sequence $(x_n) \in \mathbb{Q}^{\mathbb{N}}$ with $\lim x_n = x$, and it follows $\varphi(x) = \lim \varphi(x_n) = \lim x_n$ in $\mathbb{R}$, cf. Lemma 3.4.6 (1). Thus, there is at most one φ as in the theorem.

The proof of uniqueness also shows how φ is to be defined. Let $x \in K$ and $(x_n) \in \mathbb{Q}^{\mathbb{N}}$ be a sequence of rational numbers with $\lim x_n = x$. According to Lemma 3.4.6 (2), (x_n) is a Cauchy sequence in $\mathbb{R}$ and thus convergent according to Theorem 3.4.8. We set $\varphi(x) := \lim \varphi(x_n)$ and first have to show that $\varphi(x)$ is independent of the choice of the sequence (x_n). But if $(x'_n) \in \mathbb{Q}^{\mathbb{N}}$ with $\lim x'_n = x$ is also the case, then $(x_n - x'_n)$ is a null sequence in K and thus also in $\mathbb{R}$ according to Lemma 3.4.6 (1). Therefore, in $\mathbb{R}$: $\lim x'_n = \lim x_n$, as desired.

φ is a field homomorphism due to the calculation rules 3.2.5 for limits. φ is also monotonically increasing: If $x \in K_+$ exists, then there is a sequence $(x_n) \in \mathbb{Q}_+^{\mathbb{N}}$ with $\lim x_n = x$, and $\varphi(x) = \lim x_n \geq 0$ is also in $\mathbb{R}$. □

Corollary 3.4.10 *Let $\mathbb{R}_1$ and $\mathbb{R}_2$ be real number fields. Then there is exactly one homomorphism $\varphi : \mathbb{R}_1 \to \mathbb{R}_2$ of ordered fields. This is an isomorphism. Every field homomorphism $\mathbb{R}_1 \to \mathbb{R}_2$ is monotonic and thus an isomorphism of ordered fields. — The identity is the only field endomorphism of a real number field.*

Proof Let $\varphi : \mathbb{R}_1 \to \mathbb{R}_2$ be a field homomorphism. Since every element $x_1 \in (\mathbb{R}_1)_+$ is a square, $x_1 = y_1^2$ (cf. Example 3.3.7), $\varphi(x_1) = \varphi(y_1)^2 \in (\mathbb{R}_2)_+$ holds, thus $\varphi((\mathbb{R}_1)_+) \subseteq (\mathbb{R}_2)_+$. According to Theorem 3.4.9, there is exactly one field homomorphism $\varphi : \mathbb{R}_1 \to \mathbb{R}_2$ and then exactly one field homomorphism $\psi : \mathbb{R}_2 \to \mathbb{R}_1$. $\psi \circ \varphi : \mathbb{R}_1 \to \mathbb{R}_1$ and $\varphi \circ \psi : \mathbb{R}_2 \to \mathbb{R}_2$ are field endomorphisms of $\mathbb{R}_1$ and $\mathbb{R}_2$ respectively, and therefore each is the identity. φ and ψ are therefore inverse isomorphisms to each other. □

We explicitly note that an Archimedean ordered field can have field automorphisms that are not monotonic and thus not the identity. For example, the square $\mathbb{Q}$-algebra $\mathbb{Q}[\sqrt{2}] \subseteq \mathbb{R}$ has the conjugation defined by $\sqrt{2} \mapsto -\sqrt{2}$ $a + b\sqrt{2} \mapsto a - b\sqrt{2}$, $a, b \in \mathbb{Q}$, as the (only) non-trivial automorphism. So there are two total orders on $\mathbb{Q}[\sqrt{2}]$, with respect to which $\mathbb{Q}[\sqrt{2}]$ is an Archimedean ordered field. — Another important consequence of the interval nesting principle is the following theorem:

Theorem 3.4.11 (Theorem of the upper and lower bound) *Every non-empty set A of real numbers that is bounded above (or below) has an upper bound (or lower bound) in $\mathbb{R}$.*

Proof We recall that an upper (or lower) bound of A is the smallest upper bound (or the largest lower bound) of A in $\mathbb{R}$.

Let S now be an upper bound of the upper-bounded subset $A \subseteq \mathbb{R}$, and let $a \in A$. To construct the upper limit of A, we again use the interval halving method. We construct the intervals $[a_n, b_n]$ such that at least one element of A lies in $[a_n, b_n]$ and that b_n is an upper bound of A. Then it is clear that the number defined by this interval nesting is the upper limit Sup A of A. We set $a_0 = a$, $b_0 = S$. If a_n and b_n are already defined, then let

$$[a_{n+1}, b_{n+1}]$$

So $[a_{n+1}, b_{n+1}]$ is the "left" or "right" half of $[a_n, b_n]$ depending on whether the midpoint of $[a_n, b_n]$ is an upper bound of A or not. The existence of the lower limit Inf A for a lower-bounded non-empty set A is treated analogously or is derived by considering the set $-A = \{-x \mid x \in A\}$ in the already treated case of the upper limit. □

The validity of sentence 3.4.11 implies the validity of the Carathéodory completeness axiom: If (x_n) is an upwardly bounded, monotonically increasing sequence and x is the upper limit of $\{x_n \mid n \in \mathbb{N}\}$, then $x = \lim x_n$. *In $\overline{\mathbb{R}} = \mathbb{R} \cup \{\infty, -\infty\}$ every subset of $\overline{\mathbb{R}}$ has a supremum and an infimum.* For the empty set, $-\infty$ is the supremum and ∞ is the infimum. The supremum (or the infimum) of a set in $\mathbb{R}$ that is unbounded above (or below) is ∞ (or $-\infty$).

The existence of a real number field can now be easily proven following Cantor. Let $\mathbb{R}$ be a real number field. Then the mapping

$$\lim : \mathbb{Q}_{CF}^{\mathbb{N}} \to \mathbb{R}, \quad (x_n) \mapsto \lim x_n,$$

is a surjective homomorphism of $\mathbb{Q}$-algebras with the maximal ideal $\mathfrak{n}_\mathbb{Q} \subseteq \mathbb{Q}_{CF}^{\mathbb{N}}$ of null sequences in $\mathbb{Q}$ as the core. According to the isomorphism theorem 2.7.3,

$$\mathbb{Q}_{CF}^{\mathbb{N}}/\mathfrak{n}_\mathbb{Q} \xrightarrow{\sim} \mathbb{R}.$$

applies. Conversely, it is easy to show that $\mathfrak{n}_\mathbb{Q}$ is a maximal ideal in $\mathbb{Q}_{CF}^{\mathbb{N}}$. If $(x_n) \in \mathbb{Q}_{CF}^{\mathbb{N}}$ is not a null sequence, there is an $s \in \mathbb{Q}_+^{\times}$ and an n_0 with $|x_n| \geq s$ for all $n \geq n_0$ and every sequence (y_n) with $y_n = x_n^{-1}$ for all $n \geq n_0$ is also a Cauchy sequence and represents an inverse to (x_n) modulo $\mathfrak{n}_\mathbb{Q}$. The residue field $\mathbb{Q}_{CF}^{\mathbb{N}}/\mathfrak{n}_\mathbb{Q}$ can also be canonically ordered: A rational Cauchy sequence (x_n) represents an element > 0 by definition, if there is an $s \in \mathbb{Q}_+^{\times}$ with $x_n \geq s$ for almost all n. It is immediately shown that $\mathbb{Q}_{CF}^{\mathbb{N}}/\mathfrak{n}_\mathbb{Q}$ so that it becomes an Archimedean ordered field. It is also complete. For the *proof* of completeness let x_n, $n \in \mathbb{N}$, be a Cauchy sequence in $\mathbb{Q}_{CF}^{\mathbb{N}}/\mathfrak{n}_\mathbb{Q}$.[6] Since $\mathbb{Q}$ is dense in $\mathbb{Q}_{CF}^{\mathbb{N}}/\mathfrak{n}_\mathbb{Q}$, there are elements $y_n \in \mathbb{Q}$ with $|x_n - y_n| \leq 1/(n+1)$, $n \in \mathbb{N}$. Then y_n, $n \in \mathbb{N}$, is also a Cauchy sequence, and its residue class in $\mathbb{Q}_{CF}^{\mathbb{N}}/\mathfrak{n}_\mathbb{Q}$ is the limit of the given sequence (x_n), $n \in \mathbb{N}$. □

Overall, we get:

Theorem 3.4.12 (Existence of Real Number Fields) *The null sequences form a maximal ideal $\mathfrak{n}_\mathbb{Q}$ in the $\mathbb{Q}$-algebra $\mathbb{Q}_{CF}^{\mathbb{N}}$ of the Cauchy sequences in $\mathbb{Q}$, and the quotient field $\mathbb{Q}_{CF}^{\mathbb{N}}/\mathfrak{n}_\mathbb{Q}$ is naturally a real number field $\mathbb{R}$.*

---

[6] Note that in an *Archimedean* ordered field, a monotone bounded sequence is obviously a Cauchy sequence. So it is sufficient to consider Cauchy sequences.

3.4 Consequences of Completeness

The subfields of $\mathbb{R}$ are according to theorem 3.4.9 (up to isomorphism) exactly the Archimedean ordered fields. Two different such subfields of $\mathbb{R}$ are not isomorphic as *ordered* fields.

Remark 3.4.13 Let K be an ordered field. The quotient field $K_{\mathrm{CF}}^{\mathbb{N}}/\mathfrak{n}_K$ of the Cauchy sequences in K modulo the null sequences in K is always naturally an ordered field and in the case that K is Archimedean ordered, a real number field. If every Cauchy sequence in K is convergent — then we call K **sequentially complete** -, i.e. if $K_{\mathrm{CF}}^{\mathbb{N}} = K_{\mathrm{kon}}^{\mathbb{N}} = K \oplus \mathfrak{n}_K$, then $K_{\mathrm{CF}}^{\mathbb{N}}/\mathfrak{n}_K = K$. This is particularly the case when K only has stationary sequences as convergent sequences, i.e. when no null sequence with purely positive elements exists in K. The sequential completeness is then uninteresting.

The Dedekind approach to the construction of real numbers also provides interesting field extensions for these cases. However, we only introduce them for $K = \mathbb{Q}$, where we will again obtain a real number field $\mathbb{R}$.[7] Dedekind follows the idea of Eudoxus, according to which every real number α is determined by the section $A_\alpha := \{x \in \mathbb{Q} \mid x < \alpha\}$ in $\mathbb{Q}$. Generally, we understand a **Dedekind cut** (in $\mathbb{Q}$) to be a non-empty, upper-bounded subset of $\mathbb{Q}$ without a greatest element, which contains all smaller elements of $\mathbb{Q}$ with each element. If $\mathbb{R}$ is a real number field, then every such Dedekind cut A in $\mathbb{Q}$ is the cut A_α to a real number, namely $A = A_\alpha$, $\alpha := \operatorname{Sup} A$, cf. theorem 3.4.11. The set $\mathbb{R}$ of real numbers can therefore be identified with the set of Dedekind cuts in $\mathbb{Q}$.

One can now conversely define the set $\mathbb{R}$ as the set of Dedekind cuts in $\mathbb{Q}$ *define*, which can be done solely with the help of $\mathbb{Q}$ is possible without recourse to $\mathbb{R}$. Then $\mathbb{R} \subseteq \mathfrak{P}(\mathbb{Q})$ and the order on $\mathbb{R}$ is the inclusion induced by $\mathfrak{P}(\mathbb{Q})$. Positive are the Dedekind sections that contain a positive number. The validity of the Carathéodory completeness axiom 3.3.1 is trivial: If $A_0 \subseteq A_1 \subseteq A_2 \subseteq \cdots$ is an upwardly bounded, monotonically increasing sequence in $\mathbb{R}$, then $\lim A_n = \bigcup_{n \in \mathbb{N}} A_n$. In general, for any non-empty upwardly bounded set $\mathfrak{A} \subseteq \mathbb{R} \subseteq \mathfrak{P}(\mathbb{Q})$, the union $\bigcup_{A \in \mathfrak{A}} A \in \mathbb{R}$ is the upper bound of $\mathfrak{A}$. The addition in $\mathbb{R}$ is simply the Minkowski sum

$$A + B = \{x + y \mid x \in A, y \in B\}, \quad A, B \in \mathbb{R}.$$

Only the multiplication is somewhat more tedious to describe, which is due to the fact that multiplication with negative numbers is monotonically decreasing. Perhaps it is best to initially define multiplication only for *positive* sections, namely by

$$A \cdot B = \mathbb{Q}_- \uplus \{xy \mid x \in A \cap \mathbb{Q}_+^\times, y \in B \cap \mathbb{Q}_+^\times\}, \quad A, B \in \mathbb{R}_+^\times.$$

---

[7] The general construction is somewhat more complicated for non-archimedean ordered fields than is outlined below.

Then $\mathbb{R}_+ = \mathbb{R}_+^\times \uplus \{0\}$ with $0 \cdot A = A \cdot 0 = 0$ for all $A \in \mathbb{R}_+$ a semi-ring and $\mathbb{R}$ the associated Grothendieck ring, see example 2.6.10. ◊

At the end of this section, we introduce some topological terms that often simplify the language and are discussed in a much more general context in Chapter 4. The terms introduced here always refer to K as the base space.

Definition 3.4.14 Let K be an ordered field and $A \subseteq K$ and $x \in K$.

(1) The point x is called a **contact point** of A, if there is at least one element of A in every neighborhood of x.
(2) The point x is called a **accumulation point** of A, if there is a point different from x (and thus infinitely many points) of A in every neighborhood of x.
(3) The point x is called an **interior point** of A, if A is a neighborhood of x.
(4) The set A is called **closed**, if every contact point of A belongs to A.
(5) The set A is called **open**, if every point of A is an interior point of A.

The set of contact points and the set of inner points of A in K we denote with

$$\overline{A} \quad \text{resp.} \quad \mathring{A}.$$

If K has a non-stationary null sequence, then a point $x \in K$ is obviously a contact point of A if and only if there is a sequence $(x_n) \in K_{\text{kon}}^{\mathbb{N}}$ with $x_n \in A$ and $x = \lim x_n$. x is exactly a cluster point if the sequence members x_n can also be chosen differently from x. From the definitions it follows directly:

Proposition 3.4.15 *For every subset A of the ordered field K, the following applies:*

(1) *Every accumulation point of A is a contact point of A, and every contact point of A that does not belong to A is an accumulation point of A.*
(2) *It is $\mathring{A} \subseteq A \subseteq \overline{A}$. Exactly then is $A = \overline{A}$, when A is closed, and exactly then is $A = \mathring{A}$, when A is open.*
(3) *Exactly then is A open (or closed) in K, when the complement $K - A$ of A in K is closed (or open) in K.*

Apparently, $\overline{A}$ is the smallest closed subset of K that includes A, and $\mathring{A}$ is the largest open subset of K that is contained in A. $\overline{A}$ is called the **closed hull** and $\mathring{A}$ the **open core** or the **interior of** A. Their difference set $\overline{A} - \mathring{A}$ is called the **boundary of** A and is denoted by

$$\operatorname{Rd} A \ (= \overline{A} - \mathring{A})$$

Exactly then is $\operatorname{Rd} A = \emptyset$—$A$ is then called **boundaryless** –, when $A = \overline{A} = \mathring{A}$ is both open and closed, cf. Example 3.4.19. Points inside the complement of A are called **outer points** of A. They form the complement of $\overline{A}$. Examples of closed (or open) subsets of K are the closed (or open) intervals. The half-open intervals $]a, b]$ and $[a, b[$, $a, b \in K$, $a < b$, are neither open nor closed.

3.4 Consequences of Completeness

The supremum (or infimum) of a non-empty upwardly (or downwardly) bounded subset of $\mathbb{R}$, cf. Theorem 3.4.11, is a contact point of this set. So it holds:

Theorem 3.4.16 *Every non-empty bounded closed set of $\mathbb{R}$ contains its infimum and its supremum, i.e., a smallest and a largest element.*

Lemma 3.4.17 *The set of accumulation points of a sequence in an ordered field is always closed.*

Proof Let x be a contact point of the set of accumulation points of the sequence (x_n). In every open ε-neighborhood of x, there is an accumulation point y of (x_n). Since this ε-neighborhood is also a neighborhood of y, it contains infinitely many members of the sequence (x_n). Therefore, x is an accumulation point of (x_n). □

Let (x_n) be a bounded sequence of real numbers. According to the Bolzano-Weierstrass theorem 3.4.3, the set of its accumulation points is non-empty and naturally also bounded. Due to Lemma 3.4.17, it is furthermore closed and therefore has a smallest and a largest element. These accumulation points (which coincide for convergent sequences) are called the **limit inferior** and the **limit superior** of the sequence and are denoted by

$$\liminf x_n \quad \text{resp.} \quad \limsup x_n \, .$$

By the way, the accumulation point that we constructed in the proof of the Bolzano-Weierstrass theorem 3.4.3 is the limit inferior.

For any sequence (x_n) in $\mathbb{R}$ we count $-\infty$ or ∞ to the accumulation points of (x_n), if a subsequence of (x_n) exists that converges to $-\infty$ or to ∞, if therefore (x_n) is not bounded below or not bounded above. *In this way, every sequence of real numbers in $\overline{\mathbb{R}} = \mathbb{R} \cup \{-\infty, \infty\}$ has a smallest and a largest accumulation point, i.e., a limit inferior and a limit superior.* For example, ∞ is the limit inferior and the limit superior of the sequence $n, n \in \mathbb{N}$, the natural numbers.

Exercises

Exercise 3.4.1 Determine all accumulation points as well as the limit inferior and limit superior of the sequence $(-1)^n/2 + (-1)^{n(n+1)/2}/3$ in $\mathbb{R}$.

Exercise 3.4.2 Provide a sequence in $\mathbb{R}$ for which the set of accumulation points is the set of natural numbers.

Exercise 3.4.3 Let $A \subseteq \mathbb{R}$, $A \neq \emptyset$. A is bounded from below if and only if the set $-A = \{-x \mid x \in A\}$ is bounded from above. In this case, $\operatorname{Inf} A = -\operatorname{Sup}(-A)$ applies.

Exercise 3.4.4 Let $A, B \subseteq \mathbb{R}$ be non-empty.

a) $A + B$ is bounded from above (or below) if and only if A and B are both bounded from above (or both from below). In this case,

$$\operatorname{Sup}(A + B) = \operatorname{Sup} A + \operatorname{Sup} B \quad (\text{resp. } \operatorname{Inf}(A + B) = \operatorname{Inf} A + \operatorname{Inf} B).$$

applies.
b) If $A \neq \{0\} \neq B$, then $A \cdot B$ is bounded if and only if A and B are both bounded.
c) If A and B are bounded and $A, B \subseteq \mathbb{R}_+$, then $\operatorname{Sup}(A \cdot B) = (\operatorname{Sup} A) \cdot (\operatorname{Sup} B)$. applies.

Exercise 3.4.5 Prove the following **Dedekind's Cut Theorem**: If A and B are non-empty subsets of $\mathbb{R}$ with $a<b$ for all $a \in A$ and all $b \in B$, then there exists a real number x with $a \leq x \leq b$ for all $a \in A$ and all $b \in B$. — If furthermore $A \cup B = \mathbb{R}$, then this real number x is uniquely determined and $x = \operatorname{Sup} A = \operatorname{Inf} B$ applies. (In this case, it is said that x is defined by the so-called Dedekind cut (A, B). Often, pairs (A, B), $A, B \subseteq \mathbb{Q}$, with $a<b$, $a \in A$, $b \in B$, $A \neq \emptyset \neq B$, $A \cup B = \mathbb{Q}$ are used instead of the Dedekind cuts in Remark 3.4.13. For $\alpha \in \mathbb{Q}$ the two pairs $(\mathbb{Q}_{<\alpha}, \mathbb{Q}_{\geq\alpha})$ and $(\mathbb{Q}_{\leq\alpha}, \mathbb{Q}_{>\alpha})$ have to be identified.)

Exercise 3.4.6 Let K be an ordered field that has a non-stationary null sequence. Furthermore, let $A \subseteq K$ and $x \in K$.

a) x is an accumulation point of A if and only if there is a sequence converging to x with pairwise different elements from A.
b) x is a contact point of A if and only if there is a sequence of elements from A converging to x.

Exercise 3.4.7 Let K be an ordered field and $A, B \subseteq K$.

a) It is $\overline{A \cup B} = \overline{A} \cup \overline{B}$, $(A \cap B)^\circ = \mathring{A} \cap \mathring{B}$, furthermore $\overline{A \cap B} \subseteq \overline{A} \cap \overline{B}$ and $(A \cup B)^\circ \supseteq \mathring{A} \cup \mathring{B}$. Show by examples that the last two inclusions can be strict.
b) The set of accumulation points of A is closed in K.

Exercise 3.4.8 Determine $\overline{A}$, $\mathring{A}$ and $\operatorname{Rd} A$ for the following subsets A of $\mathbb{R}$:

$\{1/n \mid n \in \mathbb{N}^*\}$, $\mathbb{N}$, $\mathbb{Q}$, $\mathbb{R} - \mathbb{Q}$; $[a, b]$, $]a, b[$, $[a, b[$, $]a, b]$ with $a, b \in \mathbb{R}$, $a < b$;

$$\{a/g^n \mid a \in \mathbb{Z}, n \in \mathbb{N}\} \text{ with } g \in \mathbb{N},\ g \geq 2 \text{ fixed.}$$

Exercise 3.4.9 Every infinite sequence of real numbers contains an infinite monotone subsequence.

Exercise 3.4.10 A sequence of real numbers is convergent if and only if it is bounded and has exactly one accumulation point. — Prove the Cauchy convergence criterion 3.4.8 again (using the theorem 3.4.3 of Bolzano-Weierstraß).

3.4 Consequences of Completeness

Exercise 3.4.11 An Archimedean ordered field K is (exactly then) a real number field when the interval nesting principle 3.3.5 (for K instead of $\mathbb{R}$) applies.

Exercise 3.4.12 Let (x_n) be a bounded sequence of real numbers. Then the following applies:

a) $\limsup x_n = \lim_{n \to \infty} \left(\mathrm{Sup}\{x_m \mid m \geq n\} \right)$,
$\liminf x_n = \lim_{n \to \infty} \left(\mathrm{Inf}\{x_m \mid m \geq n\} \right)$.
b) $\limsup x_n = \mathrm{Inf}\{x | x \geq x_n \text{ for almost all } n\} = \mathrm{Sup}\{x | x \leq x_n \text{ for an infinite number of } n\}$,
$\liminf x_n = \mathrm{Sup}\{x | x \leq x_n \text{ for almost all } n\} = \mathrm{Inf}\{x | x \geq x_n \text{ for an infinite number of } n\}$.

Exercise 3.4.13 Let (x_n) and (y_n) be bounded sequences of real numbers. It applies

$$\liminf x_n + \liminf y_n \leq \liminf (x_n + y_n) \leq \limsup x_n + \liminf y_n$$
$$\leq \limsup(x_n + y_n) \leq \limsup x_n + \limsup y_n.$$

These formulas also apply when the plus sign is replaced by the multiplication sign and moreover all x_n and all y_n are nonnegative.

Exercise 3.4.14 A subset of $\mathbb{R}$ is called **perfect** if it is equal to the set of its accumulation points. A perfect set is necessarily closed. Show that every non-empty perfect set is uncountable.[8] (One can conclude similarly as in Example 3.3.6.)

Exercise 3.4.15 Let $A \subseteq \mathbb{R}$. A point $x \in \mathbb{R}$ is called a **condensation point** or **point of density** of A, if in every neighborhood of x there are uncountably many elements of A.

a) Every uncountable subset A of $\mathbb{R}$ has at least one point of density. (One can reduce to the case that A is bounded, and then conclude as in Theorem 3.4.3.)
b) The set of points of density of A is perfect. It is particularly uncountable if A itself is uncountable, cf. exercise 3.4.14.
c) Every closed subset of $\mathbb{R}$ is the disjoint union of a countable and a perfect set. This decomposition is unique. (Every point of a perfect set in $\mathbb{R}$ is a point of density of this set.)

Exercise 3.4.16 A subset $A \subseteq \mathbb{R}$ is exactly an interval when A contains all numbers that lie between any two numbers of A.

Exercise 3.4.17 Let $A \subseteq \mathbb{R}$.

---

[8] It can be shown that a non-empty perfect subset $A \subseteq \mathbb{R}$ has the power of the continuum, cf. about [15], § 6, exercise 6.

a) Two points $a, b \in A$ may be called equivalent if the closed interval with a and b as endpoints belongs entirely to A. Show that this defines an equivalence relation on A, whose equivalence classes are intervals. These are called the **connected components** of A. (They coincide with the connected components of A as a topological subspace of $\mathbb{R}$, cf. Sect. 4.3, in particular Theorem 4.3.6.) If all these components of A are single-pointed, then A is called **totally disconnected**. For example, $\mathbb{R} - \mathbb{Q}$ and $\mathbb{Q}$ (as well as any other countable subset of $\mathbb{R}$) are totally disconnected. Any set $A \subseteq \mathbb{R}$ has at most countably many connected components with more than one point.
b) If A is open, then A is a disjoint union of countably many open intervals (namely the connected components of A).
c) If A is closed, then all connected components of A are closed intervals. (However, there can be uncountably many, cf. exercise 3.4.18.)

Exercise 3.4.18 (Cantor's Discontinuum) Let $C_0 := [0, 1]$ and $C_1 := C_0 -]1/3, 2/3[$. In general, C_{n+1} arises from C_n by removing the open middle third from each connected component of C_n, $n \in \mathbb{N}$, see Fig. 3.6. The intersection $C := \bigcap_{n=0}^{\infty} C_n$ is called the **Cantor's discontinuum** or the **Cantor's wipe set**. Show:

a) A number $x \in [0, 1]$ belongs to C if and only if x has a ternary expansion (see example 3.3.10) in which the digit 1 does not occur.[9]
b) C is a perfect (closed) totally disconnected subset of $\mathbb{R}$, which (like any non-empty perfect set) has the power of the continuum. (For a purely topological characterization of Cantor's discontinuum, we refer to exercise 4.4.27b).)

Exercise 3.4.19 Let A be a non-empty subset of $\mathbb{R}$, which is simultaneously open and closed. Then $A = \mathbb{R}$. (Consider a connected component of A. — The sets $\emptyset$ and $\mathbb{R}$ are thus the only borderless subsets of $\mathbb{R}$.)

Exercise 3.4.20
a) $\mathbb{R}$ cannot be represented as a disjoint union of countably many closed bounded intervals. (Let $\mathbb{R} = \biguplus_{n \in \mathbb{N}} [a_n, b_n]$. Then $\mathbb{R} - \biguplus_{n \in \mathbb{N}}]a_n, b_n[$ is perfect in contradiction to exercise 3.4.14. — For illustration, consider the following example: $\bar{I}_0, \bar{I}_1, \bar{I}_2, \ldots$ is the list of the fired shells of all open intervals that are removed in the construction of the Cantor discontinuum in exercise 3.4.18 from the sets $C_0, C_1, C_2, \ldots$ in order. Then $\bigcup_{n \in \mathbb{N}} \bar{I}_n$ is *not* the full (open) unit interval $]0, 1[$. Which numbers are missing?)
b) More general than a) is: $\mathbb{R}$ cannot be represented as a disjoint union of countably many closed and bounded (i.e., compact, see Sect. 3.9) subsets of $\mathbb{R}$. (This is traced back to a): Suppose $\mathbb{R} = \biguplus_{n \in \mathbb{N}} K_n$ with bounded closed sets K_n. It

---

[9] It is therefore allowed that from a certain point all digits of x are equal to 2, e.g. $1/3 = (0,1)_3 = (0,0222\ldots)_3 \in C$. The mapping $\{0,2\}^{\mathbb{N}^*} \tilde{\to} C$, $(z_n) \mapsto \sum_{n \in \mathbb{N}^*} z_n/3^n = \lim_{k \to \infty} \sum_{n=1}^{k} z_n/3^n$, is bijective. See exercise 1.8.14a).

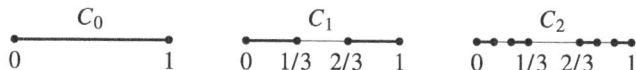

Fig. 3.6 Construction of Cantor's discontinuum

can be assumed that each $K_n \neq \emptyset$ is and lies in a connected component of the open set $\mathbb{R} - \bigcup_{k=0}^{n-1} K_k$, since in any case, according to the Bolzano-Weierstrass theorem, only finitely many of these connected components K_n meet. Let $a_n := \operatorname{Inf} K_n$ and $b_n := \operatorname{Sup} K_n$. Now construct closed bounded intervals I_n recursively in the following way: $I_0 = [a_0, b_0]$; $I_n := I_{n-1}$, if $K_n \subseteq \bigcup_{k=0}^{n-1} I_k$, or $I_n := [a_n, b_n]$ otherwise. Then $\mathbb{R}$ is the disjoint union of the different intervals in the sequence $I_0, I_1, I_2, \ldots$ — For a generalization see Example 4.3.5b).

Note The result of the present exercise has been used since Zeno of Elea as an argument against atomism. If one wants to describe the continuum atomistically, one necessarily has to allow uncountable sets. This is one of Cantor's great discoveries. He wrote in 1884 (in a letter to Mittag-Leffler): "I believe [...], that the totality of field atoms is of the first power, the totality of ether atoms is of the second power [...]." We owe this hint to J. Suck. In his letter, Cantor understands the first power to be the power $\aleph_0$ of $\mathbb{N}$ and the second to be the smallest uncountable cardinal number $\aleph_1$. That $\aleph_1$ coincides with the power $\aleph$ of $\mathbb{R}$ is the continuum hypothesis, cf. note 1.8.21.

Exercise 3.4.21
a) Let $x = a/b$ with coprime integers $a, b, b > 0$. Then the sequence $x_n := \{nx\} = nx - [nx]$, $n \in \mathbb{N}$, has exactly the b accumulation points $0, 1/b, \ldots, (b-1)/b$.
b) Let $x \in \mathbb{R}$ be irrational. Then the set of accumulation points of the sequence $x_n := \{nx\} = nx - [nx]$, $n \in \mathbb{N}$, is the closed unit interval $[0, 1]$. (Proceed in the following steps: (1) (x_n) has an accumulation point in $[0, 1]$. (2) (x_n) has 0 or 1 as an accumulation point. (3) (x_n) has every point from $]0, 1[$ as an accumulation point. — With the continued fraction development of x, cf. example 3.3.11, the exercise can even be solved constructively. For example, for the number $\varPhi = (1 + \sqrt{5})/2$ of the golden ratio with the continued fraction development $[1, 1, 1, 1, \ldots]$, provide the smallest possible $n \in \mathbb{N}$ with $|\{n\varPhi\} - \tfrac{1}{2}| \leq 10^{-6}$.)

Exercise 3.4.22 In the power set $\mathfrak{P}(\mathbb{N})$ of $\mathbb{N}$ there are uncountable chains, thus uncountable subsets, which are completely ordered (with respect to inclusion). (B. Kaup—This may be surprising and is certainly surprisingly easy to prove. Remember that $\mathbb{Q}$ like $\mathbb{N}$ is countable.) Furthermore, in $\mathfrak{P}(\mathbb{N})$ there are uncountable subsets whose elements are almost pairwise disjoint, where two sets are called **almost disjoint** if their intersection is finite.

Exercise 3.4.23 Every well-ordered subset of $\mathbb{R}$ is countable.

3.5 The complex numbers

Since in $\mathbb{R}$ as in any ordered field, the square of any element is non-negative, the equation $x^2 + 1 = 0$ has no real solution x. The polynomial $X^2 + 1 \in \mathbb{R}[X]$ is therefore a prime polynomial and has in the quadratic $\mathbb{R}$-algebra

$$\mathbb{C} = \mathbb{C}_\mathbb{R} = \mathbb{R}[i] = \mathbb{R}[\sqrt{-1}] = \mathbb{R}[X]/(X^2 + 1),$$

which is a field, the two zeros $\pm i := \pm X \bmod (X^2 + 1)$. We have already described this **field of complex numbers** in Example 2.9.35. With the **imaginary unit** $i = \sqrt{-1}$ every element has $z \in \mathbb{C}$ a unique representation

$$z = a + bi, \quad a = \Re z, \; b = \Im z \in \mathbb{R},$$

and the multiplication in $\mathbb{C}$ is already determined by the equation $i^2 = -1$:

$$(a + bi) \cdot (c + di) = ac + bdi^2 + adi + bci = (ac - bd) + (ad + bc)i,$$
$$a, b, c, d \in \mathbb{R}.$$

For computing with the computer, it should be noted that due to

$$(a + bi) \cdot (c + di) = (ac - bd) + ((a + b)(c + d) - ac - bd)i$$

the product can be calculated with three instead of four multiplications (and five instead of two additions or subtractions) of real numbers. The inverse to $z = a + bi \neq 0$ is

$$z^{-1} = \frac{1}{a + bi} = \frac{a - bi}{(a + bi)(a - bi)} = \frac{a}{a^2 + b^2} + \frac{-b}{a^2 + b^2}i.$$

Already here, the **complex conjugation**

$$z = a + bi \mapsto \bar{z} = a - bi, \quad a, b \in \mathbb{R},$$

and the **norm**

$$N(z) = z\bar{z} = a^2 + b^2$$

appear. Thus, $z^{-1} = \bar{z}/z\bar{z} = \bar{z}/N(z)$ for $z \neq 0$, and for any $z \in \mathbb{C}$ applies

$$\Re z = \frac{1}{2}(z + \bar{z}), \quad \Im z = \frac{1}{2i}(z - \bar{z}).$$

$z \in \mathbb{C}$ is thus real (i.e., it is $\Im z = 0$), when $z = \bar{z}$ is, and purely imaginary **purely imaginary** (i.e., it is $\Re z = 0$), when $z = -\bar{z}$ is. $2\Re z = z + \bar{z}$ is the **trace** $\mathrm{Tr}(z)$ of z. The complex conjugation is, besides the identity, the only $\mathbb{R}$-algebra endomorphism (= $\mathbb{R}$-algebra automorphism) of $\mathbb{C}$. Because such an endomorphism necessarily maps a zero of $X^2 + 1$ to a zero of $X^2 + 1$, i.e., i to i or to $-$i, and is thus the identity or the complex conjugation.

As elements of the 2-dimensional $\mathbb{R}$ vector space $\mathbb{C} = \mathbb{R} + \mathbb{R}i$ with the basis 1, i, the complex numbers can be identified by marking a Cartesian coordinate system in a plane of our perceptual space with the points of this plane (which we will discuss in more detail in Volumes 3 and 4). This is then referred to as the **complex**

3.5 The complex numbers

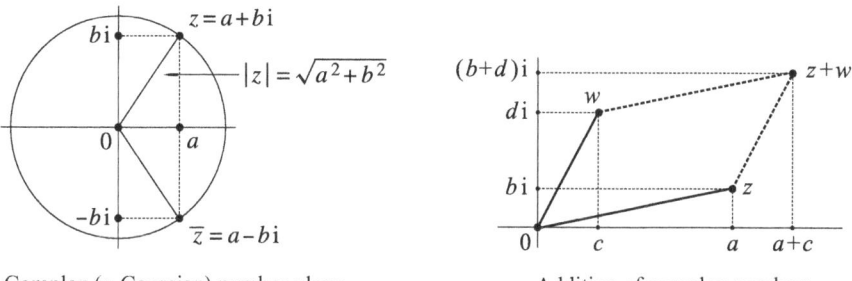

Fig. 3.7 Calculating in the Gaussian number plane

or **Gaussian number plane**, see Fig. 3.7. The real number a corresponds to the point with $(a, 0)$ as a pair of coordinates and the purely imaginary number bi to the point with $(0, b)$ as a pair of coordinates. In particular, the standard basis elements $e_1 = (1, 0)$ and $e_2 = (0, 1)$ of the $\mathbb{R}^2$ are identical to the coordinate pairs of the basis elements $1, i$ of $\mathbb{C}$. The addition of complex numbers can be easily illustrated with this. The sum $z + w = (a + c) + i(b + d)$ of two points $z = a + ib$ and $w = c + id \in \mathbb{C}$ is the fourth corner of the parallelogram determined by $0, z$ and w (if these three points do not lie on a straight line), see Fig. 3.7. We can also describe $z + w$ as the point that results from z by shifting by w (or from w by shifting by z).[10]

According to the Pythagorean theorem, the norm $N(z) = z \cdot \bar{z} = a^2 + b^2$ is the square of the distance of the point $z = a + bi = (a, b)$ from the origin $0 = (0, 0)$, cf. Fig. 3.7. The distance

$$|z| := \sqrt{z \cdot \bar{z}} = \sqrt{a^2 + b^2}$$

itself is called the **magnitude** of the complex number z. If $z = a$ is real, this magnitude $|z| = |a|$ coincides with the magnitude of z as a real number. The set of complex numbers z with a fixed magnitude r is the circle (i.e., here the periphery of the circle) with radius r around the origin. The interior of this circle is the set of $z \in \mathbb{C}$ with $|z| < r$. Accordingly, the circle with radius r around the point $z_0 \in \mathbb{C}$ or its interior is the set of $z \in \mathbb{C}$ with $|z - z_0| = r$ or with $|z - z_0| < r$. It is $|\bar{z}| = |z|$. Furthermore:

Proposition 3.5.1 *For any $z, w \in \mathbb{C}$ the following applies:*

(1) $|z| \geq 0$, and only for $z = 0$ is $|z| = 0$.
(2) $\text{Max}\left(|\Re z|, |\Im z|\right) \leq |z| \leq |\Re z| + |\Im z|$.
(3) $|zw| = |z||w|$.
(4) $|z + w| \leq |z| + |w|$ **(Triangle Inequality)**.
(5) $|z - w| \geq ||z| - |w||$.

---

[10] We describe the geometric interpretation of multiplication further below.

Fig. 3.8 Open and closed ε neighborhood in $\mathbb{C}$

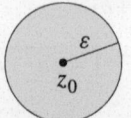
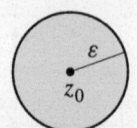

Proof (1) is trivial. (2) If $z = a + bi$, $a, b \in \mathbb{R}$, then $|z| = \sqrt{a^2 + b^2} \geq \sqrt{a^2} = |a|$ and analogously $|z| \geq |b|$ as well as

$$|z| = \sqrt{|a|^2 + |b|^2} \leq \sqrt{|a|^2 + 2|a||b| + |b|^2} = |a| + |b|.$$

(3) results from $|zw| = \sqrt{zw\overline{zw}} = \sqrt{z\overline{z}w\overline{w}} = \sqrt{z\overline{z}}\sqrt{w\overline{w}} = |z||w|$.

(4) The triangle inequality follows from

$$|z + w|^2 = (z + w)(\overline{z} + \overline{w}) = z\overline{z} + z\overline{w} + \overline{z}w + w\overline{w} = z\overline{z} + 2\Re(z\overline{w}) + w\overline{w}$$
$$\leq z\overline{z} + 2|z\overline{w}| + w\overline{w} = |z|^2 + 2|z||w| + |w|^2 = (|z| + |w|)^2.$$

From (4) follows (5) as in the real numbers, see 3.1.5. □

The triangle inequality 3.5.1 (4) is, by the way, a special case of the Minkowski inequality from exercise 3.1.26. For $z, w \in \mathbb{C}$ the $d(z, w) := |z - w| = |w - z| \in \mathbb{R}_+$ is called the **distance** between z and w. For a triangle $(z, w, v) \in \mathbb{C}^3$ in the complex plane, the triangle inequality

$$d(z, v) = |z - v| = |(z - w) + (w - v)| \leq |z - w| + |w - v| = d(z, w) + d(w, v),$$

applies, which motivates the term "triangle inequality": In a triangle of the Gaussian plane, the sum of the lengths of two sides is always at least as large as the length of the third side.

We can use the absolute value function on $\mathbb{C}$ to define limit concepts that we introduced in Sects. 3.2 and 3.3 for ordered and especially for real fields, also for complex numbers. If $z_0 \in \mathbb{C}$ and $\varepsilon \in \mathbb{R}$, $\varepsilon > 0$, then

$$B(z_0; \varepsilon) := \{z \in \mathbb{C} \mid |z - z_0| < \varepsilon\} \quad \text{resp.} \quad \overline{B}(z_0; \varepsilon) := \{z \in \mathbb{C} \mid |z - z_0| \leq \varepsilon\}$$

are called the **open** and the **closed ε-environment** of z_0, see Fig. 3.8. These are the circular disks with the center z_0 and the diameter 2ε, once without and the other time with periphery. The periphery is the 1-sphere (=circle)

$$S(z_0; \varepsilon) := \{z \in \mathbb{C} \mid |z - z_0| = \varepsilon\}.$$

A **neighborhood** of z_0 per se is a subset of $\mathbb{C}$, which contains such an (open or closed) ε neighborhood of z_0 with a (real) $\varepsilon > 0$.

A subset $A \subseteq \mathbb{C}$ is called **bounded**, if it lies entirely within a disk, i.e., if there is an $R \geq 0$ with $|z| \leq R$ for all $z \in A$, i.e., with $A \subseteq \overline{B}(0; R)$. The set $A \subseteq \mathbb{C}$ is exactly bounded when the sets $\{\Re z \mid z \in A\}$ and $\{\Im z \mid z \in A\}$ of the real and imaginary parts of the elements of A are bounded sets in $\mathbb{R}$. A sequence is called **bounded** if the set of its elements is bounded. As defined in 3.2.1, one defines:

3.5 The complex numbers

Fig. 3.9 Polar coordinates of $z \in \mathbb{C}^\times$

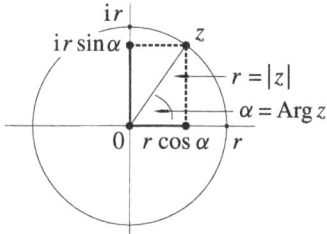

Definition 3.5.2 A sequence (z_n) in $\mathbb{C}$ is called **convergent** (in $\mathbb{C}$), if there is an $z \in \mathbb{C}$ with the following property: For every (no matter how small) $\varepsilon \in \mathbb{R}_+^\times$ there exists an $n_0 \in \mathbb{N}$ with $|z_n - z| \leq \varepsilon$ for all natural numbers $n \geq n_0$.

As in the real case, it is proven that the **limit** or **Limes**

$$z = \lim_{n \to \infty} z_n$$

of the convergent sequence (z_n) is uniquely determined. It is characterized by the fact that in each of its surroundings almost all sequence members z_n lie. By definition, the sequence (z_n) converges to z if and only if the sequence of distances $|z_n - z|$, $n \in \mathbb{N}$, is a null sequence in $\mathbb{R}$. If all z_n are real, then the sequence (z_n) can only be convergent in $\mathbb{C}$ if it is already convergent in $\mathbb{R}$, and for the determination of the limit it is then irrelevant whether it is considered as a sequence in $\mathbb{R}$ or as a sequence in $\mathbb{C}$. *Furthermore, the calculation rules 3.2.5 and 3.2.6 for limits (including their proofs verbatim) apply to the complex case.* They can, of course, also be formally derived from the calculation rules in the real case using the following criterion:

Proposition 3.5.3 *Let (z_n) be a sequence in $\mathbb{C}$, $z_n = a_n + ib_n$, $a_n, b_n \in \mathbb{R}$, and $z = a + ib \in \mathbb{C}$, $a, b \in \mathbb{R}$. Then the following holds: The sequence (z_n) converges to z if the sequence (a_n) of real parts converges to a and the sequence (b_n) of imaginary parts converges to b. — So for convergent sequences (z_n)*

$$\lim z_n = \lim(\Re z_n) + i \lim(\Im z_n).$$

Proof The assertion follows directly from the estimates

$$|a_n - a| = |\Re(z_n - z)| \leq |z_n - z| \quad \text{and analog} \quad |b_n - b| \leq |z_n - z|,$$

and conversely from $|z_n - z| = |(a_n - a) + i(b_n - b)| \leq |a_n - a| + |b_n - b|$. □

Cauchy Sequences are defined in the complex as in $\mathbb{R}$. Analogous to Proposition 3.5.3, it holds that *a sequence (z_n) of complex numbers is a Cauchy sequence exactly when the sequences of their real and imaginary parts are Cauchy sequences of real numbers*. This follows from Theorem 3.4.8:

Theorem 3.5.4 (Cauchy's Convergence Criterion for $\mathbb{C}$) *A sequence of complex numbers converges if and only if it is a Cauchy sequence.*

A point $z \in \mathbb{C}$ is called a **cluster point** of the sequence (z_n) in $\mathbb{C}$, if in every (no matter how small) neighborhood of z there are infinitely many members of the sequence, which is equivalent to a subsequence of (z_n) converging to z, cf. Lemma 3.4.2. The Theorem 3.4.3 of Bolzano-Weierstrass also applies in the complex:

Theorem 3.5.5 (Bolzano-Weierstrass Theorem in $\mathbb{C}$) *Every bounded sequence of complex numbers has a cluster point.*

Proof We show that every bounded sequence (z_n) in $\mathbb{C}$ has a convergent subsequence. With (z_n) the real sequences $(\Re z_n)$ and $(\Im z_n)$ are also bounded. According to 3.4.3, $(\Re z_n)$ therefore has a convergent subsequence. Thus, there is a subsequence of (z_n) for which the sequence of real parts converges. If we replace the sequence (z_n) with this subsequence, we can assume from the outset that $(\Re z_n)$ converges. If we then choose the subsequence so that the sequence of imaginary parts also converges, we obtain a convergent subsequence of (z_n). □

The definitions from 3.4.14 for the topological terms **touch point, accumulation point, inner point** and **closed** or **open subsets** are literally transferred to the complex. Then the statements of Proposition 3.4.15, if there K is replaced by $\mathbb{C}$. The **closed hull** $\overline{A}$ is again the set of contact points, the **open core** (or the **interior**) $\mathring{A}$ the set of inner points and $\operatorname{Rd} A = \overline{A} - \mathring{A}$ the **boundary** of $A \subseteq \mathbb{C}$. For a subset A of $\mathbb{R}$ one has to distinguish between the terms regarding $\mathbb{R}$ and regarding $\mathbb{C}$. With $A \subseteq \mathbb{R}$ of course always $\mathring{A} = \emptyset$ regarding $\mathbb{C}$; the closed hull $\overline{A}$ however is the same in both cases. If $A \subseteq A' \subseteq \overline{A}$ applies, then A is called **dense** in A'. Every point of A' is then a contact point of A.

Already in example 2.2.16 (2) we have extensively discussed the multiplicative group $\mathbb{C}^\times$ of $\mathbb{C}$ from an algebraic point of view and used the polar coordinate representation of complex numbers. The basis is the unique decomposition

$$z = ru, \quad r := |z| \in \mathbb{R}_+^\times, \; u \in U = \{z \in \mathbb{C}^\times \mid |z| = 1\}$$

of a complex number $z \in \mathbb{C}^\times$ as a product of a positive real number r and an element u of the **circle group** U of the complex numbers of magnitude 1. It provides the product representation $\mathbb{R}_+^\times \times U \xrightarrow{\sim} \mathbb{C}^\times$, $(r, u) \mapsto ru$, of the group $\mathbb{C}^\times$. It is $u = a + bi$ with $a^2 + b^2 = 1$. Therefore, there is exactly one angle $\operatorname{Arg} z = \alpha \in \mathbb{R}/\mathbb{Z}2\pi$ with $a = \cos\alpha$, $b = \sin\alpha$, and we obtain the **polar coordinate representation**

$$z = r(\cos\alpha + i\sin\alpha), \quad r = |z| \in \mathbb{R}_+^\times, \; \alpha = \operatorname{Arg} z \in \mathbb{R}/\mathbb{Z}2\pi,$$

of $z \in \mathbb{C}^\times$, cf. again example 2.2.16(2) and Fig. 3.9. The representative of $\operatorname{Arg} z$ in the interval $]-\pi, \pi]$ is called the **standard argument** of z. If $w = s(\cos\beta + i\sin\beta)$ is another complex number $\neq 0$ in polar coordinate representation, then

$$zw = rs(\cos(\alpha + \beta) + i\sin(\alpha + \beta)), \quad \text{therefore} \quad |zw| = |z||w|, \; \operatorname{Arg} zw = \operatorname{Arg} z + \operatorname{Arg} w.$$

3.5 The complex numbers

One multiplies complex numbers by multiplying their magnitudes in $\mathbb{R}_+$ *and adding their arguments in* $\mathbb{R}/\mathbb{Z}2\pi$. Multiplication with a complex number $z \neq 0$ is therefore in the complex number plane a rotation around the origin with rotation angle *Arg z*, followed by a stretching with stretching factor $|z|$ and 0 as the center of stretching. Of course, the rotation and the stretching can also be performed in reverse order. The exponentiation in $\mathbb{C}^\times$ yields in polar coordinate representation the so-called **Moivre's formulas**: For $z = r(\cos\alpha + i\sin\alpha) \in \mathbb{C}^\times$ and $n \in \mathbb{Z}$ applies

$$z^n = r^n(\cos n\alpha + i\sin n\alpha).$$

With the complex e-function

$$e^w = \exp w = e^a e^{ib} = e^a(\cos b + i\sin b), \quad w = a + bi,\ a,b \in \mathbb{R},$$

cf. example 2.2.16 (2), the polar coordinate representation of $z \neq 0$ is simply the representation

$$z = \exp(\ln|z| + i\,\text{Arg}\,z)$$

of z as an image of the e-function, whose pre-image in $\mathbb{C}$ is only determined up to an integer multiple of $2\pi i$. Each of the values $\ln|z| + i(\text{Arg}\,z + 2k\pi)$, $k \in \mathbb{Z}$, is called a **logarithm of** z and

$$\ln z := \ln|z| + i\,\text{Arg}\,z \quad \text{with} \quad \pi < \text{Arg}\,z \leq \pi$$

its **principal value**. For any $a \in \mathbb{C}^\times$ we define the **power function** a^z by

$$a^z := e^{z\ln a}, \quad z \in \mathbb{C}.$$

Example 3.5.6 (n-th roots) Let $n \in \mathbb{N}^*$ and $w \in \mathbb{C}$. If $w = 0$, so the equation $z^n = w$ only has the solution $z = 0$. Now let $w \neq 0$ with the polar coordinate representation $w = r(\cos\alpha + i\sin\alpha)$. Since the quotient of two solutions z of $z^n = w$ is obviously an n-th root of unity, *the n pairwise different numbers*

$$z_k := \zeta_n^k z_0 = \sqrt[n]{r}\left(\cos\frac{\alpha + 2k\pi}{n} + i\sin\frac{\alpha + 2k\pi}{n}\right), \quad k = 0,\ldots,n-1,$$

are all n-th roots of w, where $z_0 = \sqrt[n]{r}\bigl(\cos(\alpha/n) + i\sin(\alpha/n)\bigr)$ is a specific solution and

$$\zeta_n = \cos(2\pi/n) + i\sin(2\pi/n)$$

is the primitive n-th standard root of unity. The n-th roots of w form the vertices of a regular n-gon, whose circumcircle is the circle around 0 with radius $\sqrt[n]{r}$, and the polynomial $X^n - w$ has in $\mathbb{C}[X]$ the prime factor decomposition

$$X^n - w = \prod_{k=0}^{n-1}(X - \zeta_n^k z_0), \quad \text{in particular} \quad X^n - 1 = \prod_{k=0}^{n-1}(X - \zeta_n^k).$$

The function
$$r(\cos\alpha + i\sin\alpha) \longmapsto \sqrt[n]{r}(\cos(\alpha/n) + i\sin(\alpha/n)), \quad r > 0,\ -\pi < \alpha \le \pi,$$
on $\mathbb{C}^\times$ is referred to as the **principal value** $w \mapsto \sqrt[n]{w}$ of the n-th root. It maps $\mathbb{C}^\times$ bijectively onto the half-open sector·
$$\{s(\cos\beta + i\sin\beta) \mid\ > 0,\ -\pi/n < \beta \le \pi/n\}. \quad \diamond$$

Example 3.5.7 (Quadratic Equations) By definition of $\mathbb{C}$, the number -1 has the two square roots i and $-$i in $\mathbb{C}$. Following the previous example, now every complex number $z = a + bi$, $a, b \in \mathbb{R}$, has a square root and for $z \ne 0$ exactly two. For $z = r(\cos\alpha + i\sin\alpha)$, $r > 0$, $-\pi < \alpha \le \pi$, these are
$$\pm\sqrt{z} = \pm\sqrt{r}(\cos(\alpha/2) + i\sin(\alpha/2)).$$
One can also directly confirm by squaring that z has the square roots
$$\pm\sqrt{z} := \pm\frac{1}{\sqrt{2}}\left(\sqrt{|z|+a} + i\varepsilon\sqrt{|z|-a}\right), \quad \varepsilon := \begin{cases} \operatorname{Sign} b, & \text{if } b \ne 0, \\ 1, & \text{if } b = 0, \end{cases}$$

This allows to solve any quadratic equation in $\mathbb{C}$. The quadratic equation $x^2 - px + q = 0$, $p, q \in \mathbb{C}$, has exactly the solutions
$$x_{1,2} = \frac{1}{2}\left(p \pm \sqrt{p^2 - 4q}\right),$$
see Proposition 2.9.36. In $\mathbb{C}$, even every non-constant polynomial
$$X^n + c_{n-1}X^{n-1} + \cdots + c_1 X + c_0 \in \mathbb{C}[X], \quad n \ge 1,$$
has a root in $\mathbb{C}$, i.e. the field $\mathbb{C}$ is algebraically closed. This is the statement of the **Fundamental Theorem of Algebra**, see Theorem 3.9.7. In the case $n \ge 5$, however, no explicit formulas (which use only root expressions in addition to the basic arithmetic operations) can be given for the roots (**Abel-Ruffini theorem**). $\diamond$

Example 3.5.8 (Chebyshev Polynomials) For $\varphi \in \mathbb{R}$ and $n \in \mathbb{N}$,
$$\cos n\varphi + i\sin n\varphi = (\cos\varphi + i\sin\varphi)^n = \sum_{m=0}^{n} \binom{n}{m} i^m \sin^m\varphi \cos^{n-m}\varphi$$
$$= \sum_{k=0}^{[n/2]} \binom{n}{2k}(-1)^k \sin^{2k}\varphi \cos^{n-2k}\varphi + i \sum_{k=0}^{[(n-1)/2]} \binom{n}{2k+1}(-1)^k \sin^{2k+1}\varphi \cos^{n-2k-1}\varphi.$$

By comparing the real and imaginary parts, we obtain
$$\cos n\varphi = \sum_k (-1)^k \binom{n}{2k}(\cos^{n-2k}\varphi)(1 - \cos^2\varphi)^k,$$
$$\sin n\varphi = \sum_k (-1)^k \binom{n}{2k+1} \cos^{n-2k-1}\varphi \sin^{2k+1}\varphi.$$
In particular, $\cos n\varphi$ is a polynomial in $\cos\varphi$. We have the explicit representation

3.5 The complex numbers

$$\cos n\varphi = 2^{n-1} T_n(\cos \varphi),$$

with the so-called (for $n \geq 1$ **normalized**) **Chebyshev polynomials**

$$T_0 := 2 \quad \text{and} \quad T_n := \sum_{k=0}^{\lfloor n/2 \rfloor} \left(-\frac{1}{4}\right)^k \frac{n}{n-k} \binom{n-k}{k} X^{n-2k}, \quad n \geq 1.$$

For the *proof* of the last given representation of $\cos n\varphi$, we first note that

$$\cos(n+2)\varphi = \cos 2\varphi \cos n\varphi - 2\cos\varphi \sin\varphi \sin n\varphi$$
$$= (2\cos^2\varphi - 1)\cos n\varphi + 2\cos\varphi(\cos(n+1)\varphi - \cos\varphi \cos n\varphi)$$
$$= 2\cos\varphi \cos(n+1)\varphi - \cos n\varphi$$

is. Therefore, it is sufficient to show that the polynomials T_n satisfy the following recursion:

$$T_0 = 2, \quad T_1 = X, \quad T_{n+2} = XT_{n+1} - T_n/4, \; n \in \mathbb{N}.$$

This is easily confirmed by substitution. — Some authors call the polynomials

$$\widetilde{T}_n := 2^{n-1} T_n, \quad n \in \mathbb{N},$$

Chebyshev polynomials. It is $\cos n\varphi = \widetilde{T}_n(\cos \varphi)$. In particular, therefore, $\widetilde{T}_n(1) = 1$, applies. $\widetilde{T}_n(-1) = (-1)^n$ and $\widetilde{T}_n(0) = (-1)^{n/2}$, if n is even, and $\widetilde{T}_n(0) = 0$, if n is odd. The $\widetilde{T}_n$ satisfy the recursion

$$\widetilde{T}_0 = 1, \quad \widetilde{T}_1 = X, \quad \widetilde{T}_{n+2} = 2X\widetilde{T}_{n+1} - \widetilde{T}_n, \; n \in \mathbb{N}$$

and have integer coefficients. *Furthermore, in* $\mathbb{C}(X)$

$$\frac{X^n + X^{-n}}{2} = \widetilde{T}_n\left(\frac{X + X^{-1}}{2}\right), \quad n \in \mathbb{N}.$$

Proof Because of $(X + X^{-1})(X^{n+1} + X^{-(n+1)}) = (X^{n+2} + X^{-(n+2)}) + (X^n + X^{-n})$ the equations $X^n + X^{-n} = F_n(X + X^{-1})$ with polynomials $F_n \in \mathbb{C}[Y]$, which are determined by the recursion

$$F_0 = 2, F_1 = Y, \quad F_{n+2} = YF_{n+1} - F_n, \; n \in \mathbb{N},$$

apply. However, the polynomials $2\widetilde{T}_n(Y/2)$ obviously also satisfy this recursion. □

The identity $(X^n + X^{-n})/2 = \widetilde{T}_n\big((X + X^{-1})/2\big)$ can also be proven as follows: Let $\varphi \in \mathbb{R}$ and $z := \cos\varphi + i\sin\varphi$. Then, due to the Moivre's formula and $z^{-1} = \bar{z}$, the equation $(z^n + z^{-n})/2 = \cos n\varphi = \widetilde{T}_n(\cos \varphi) = \widetilde{T}_n\big((z + z^{-1})/2\big)$ applies, from which the assertion with the identity theorem for polynomials 2.9.31 results.

By definition, $\widetilde{T}_n\big(\cos((2k+1)\pi/2n)\big) = \cos((2k+1)\pi/2) = 0$ for $k \in \mathbb{Z}$, i.e., *the polynomials* $\widetilde{T}_n$ *and* T_n *of degree n have the n different real roots*

$$\cos \frac{(2k+1)\pi}{2n}, \quad k = 0, \ldots, n-1,$$

all of which lie in the interval $]-1, 1[$. For $n \geq 1$ follows,

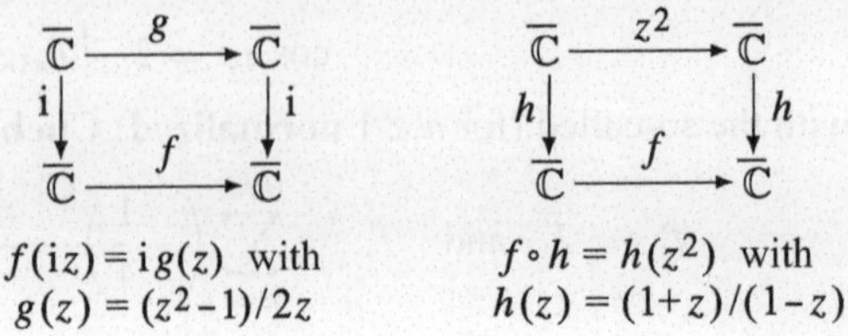

Fig. 3.10 Conjugacy of the functions $f(z) = (z^2 + 1)/2$, $g(z) = (z^2 - 1)/2$ and z^2

$$\begin{array}{ccc} \overline{\mathbb{C}} & \xrightarrow{g} & \overline{\mathbb{C}} \\ {\scriptstyle i}\downarrow & & \downarrow{\scriptstyle i} \\ \overline{\mathbb{C}} & \xrightarrow{f} & \overline{\mathbb{C}} \end{array}$$

$f(iz) = i g(z)$ with
$g(z) = (z^2-1)/2z$

$$\begin{array}{ccc} \overline{\mathbb{C}} & \xrightarrow{z^2} & \overline{\mathbb{C}} \\ {\scriptstyle h}\downarrow & & \downarrow{\scriptstyle h} \\ \overline{\mathbb{C}} & \xrightarrow{f} & \overline{\mathbb{C}} \end{array}$$

$f \circ h = h(z^2)$ with
$h(z) = (1+z)/(1-z)$

$$T_n = \prod_{k=0}^{n-1}\left(X - \cos\frac{(2k+1)\pi}{2n}\right). \quad \diamond$$

Example 3.5.9 (Joukowski Function) We consider the so-called **Joukowski function** (after N. Joukowski (1847–1921))

$$f: \mathbb{C}^{\times} = \mathbb{C} - \{0\} \longrightarrow \mathbb{C} \quad \text{with} \quad f(z) := \frac{1}{2}(z + z^{-1}) = \frac{z^2 + 1}{2z}$$

in more detail. If necessary, we extend f (like any rational function $R(z)$, $R \in \mathbb{C}(Z)$) in a canonical way to a mapping $f: \overline{\mathbb{C}} \to \overline{\mathbb{C}} = \mathbb{C} \uplus \{\infty\}$, in this case by $f(0) = f(\infty) = \infty$, see example 3.8.14 . For a complex number $z = r(\cos\varphi + i\sin\varphi), r > 0$, is $z^{-1} = r^{-1}(\cos\varphi - i\sin\varphi)$ and consequently

$$f(z) = \frac{1}{2}(r + r^{-1})\cos\varphi + \frac{i}{2}(r - r^{-1})\sin\varphi.$$

In particular, the f-image of the circle with radius r around the origin is the ellipse with 0 as the center and the semi-axis lengths

$$\frac{1}{2}(r + r^{-1}) \quad \text{resp.} \quad \frac{1}{2}|r - r^{-1}|,$$

which degenerates to the (twice traversed) line from $+1$ to -1 at $r = 1$. The circles with radii r and r^{-1} have the same image ellipse; at $r > 1$ it is traversed in the same sense as the circle, at $r < 1$ in the opposite sense. On the **upper half-plane**

$$\mathbb{H} := \{z \in \mathbb{C} \mid \Im z > 0\}$$

f is therefore injective. The image $f(\mathbb{H})$ is the doubly slit plane

$$\mathbb{C} - \{a \in \mathbb{R} \mid |a| \geq 1\}.$$

The value of the inverse function at the point $w \in f(\mathbb{H})$ is obtained by solving the quadratic equation $w = (z + z^{-1})/2$, i.e. $z^2 - 2zw + 1 = 0$ to

$$z = w + \sqrt{w^2 - 1} = w + i\sqrt{1 - w^2} = w + i\sqrt{1 - w}\sqrt{1 + w}.$$

For the two roots in the last expression, the principal value is to be chosen, see example 3.5.6. That this indeed provides the correct sign can be easily verified by the reader. By the way, for continuity reasons, it is sufficient to confirm this for a single value, and for $w = 0$, $z = i$ is obtained. — With the Joukowski function f, one also overlooks the function $g(z) := \frac{1}{2}(z - z^{-1}) = (z^2 - 1)/2z$.

3.5 The complex numbers

Because of $f(iz) = ig(z)$ for all $z \in \mathbb{C}^\times$ (and also for all $z \in \overline{\mathbb{C}}$), the diagram from Fig. 3.10 is commutative, where i denotes the multiplication with i on $\mathbb{C}^\times$ or on $\mathbb{C}$, i.e., the rotation around 0 with the angle $\pi/2$. For example, g on the **right half-plane** $\{z \in \mathbb{C} \mid \Re z > 0\}$ is injective with the double-slitted plane $\mathbb{C} - \{ai \mid a \in \mathbb{R}, |a| \geq 1\}$ as an image and the inverse function $w \mapsto -if^{-1}(iw) = w + \sqrt{1 - iw}\sqrt{1 + iw}$. Finally, we note the equation

$$f\left(\frac{1+z}{1-z}\right) = \frac{1+z^2}{1-z^2}, \quad \text{i.e.} \quad (f \circ h)(z) = h(z^2) \quad \text{with} \quad h(z) := \frac{1+z}{1-z},$$

with the help of which properties of the Joukowski function can often be traced back to properties of the square function. Note that the function $h \colon \overline{\mathbb{C}} \mapsto \overline{\mathbb{C}}$ is bijective with inverse function $z \mapsto (z-1)/(z+1)$. *The discrete dynamic systems $(\overline{\mathbb{C}}, f)$ and $(\overline{\mathbb{C}}, z^2)$ are therefore conjugated*, see exercise 2.4.16. As the n-th iterated $f_n = f \circ \cdots \circ f$ (n times) of the Joukowski function f directly results in the function $((z+1)^{2^n} + (z-1)^{2^n})/((z+1)^{2^n} - (z-1)^{2^n})$, see also exercise 3.5.45. ◊

Exercises

Exercise 3.5.1 Determine the real and imaginary part, the magnitude, and the conjugate complex number for

$$\frac{1}{1+i}; \quad \frac{2-i}{2+i}; \quad \frac{3+4i}{1+2i}; \quad (2+i)^n, n \in \mathbb{Z}; \quad \left(\frac{1-i}{1+i}\right)^n, n \in \mathbb{Z}.$$

Exercise 3.5.2
a) Determine the square roots of i, $8 - 6i$, $3 - 4i$.
b) Determine the fourth roots of $-i$, $-1 + i$.
c) Determine the complex roots of the polynomials $X^2 + (1 - 4i)X - 5 + i$ and $X^2 + (2i - 3)X + 1 + 2i$ as well as $X^2 - (2 + 2i)X + 3 - 2i$ and $X^2 + (2 - 6i)X + 3 - 18i = 0$.

Exercise 3.5.3 Provide all complex roots of the biquadratic polynomial $X^4 - pX^2 + q = 0$, $p, q \in \mathbb{C}$, on.

Exercise 3.5.4 Provide the complex roots of the so-called **self-reciprocal polynomial** of the fourth degree $F = X^4 + rX^3 + sX^2 + rX + 1, r, s \in \mathbb{C}$. (Write F/X^2 as a polynomial in $Y := X + X^{-1}$.—In general, for the roots of arbitrary polynomials of the third and fourth degree, there are explicit solution formulas by Tartaglia/Cardano or Ferrari, see for example 14, Theorem 36.6 and 36.7, or 16, § 54, exercise 41, 42.)

Exercise 3.5.5 Provide the prime factor decomposition of the following polynomials over $\mathbb{C}$ or $\mathbb{R}$:

$$X^3 + 1; \quad X^4 + 1; \quad X^4 - 5; \quad X^4 + 5; \quad X^4 + 4; \quad X^4 + X^2 + 1;$$
$$X^3 - 2X^2 + 2X - 1; \quad X^5 + 2X^4 + 2X^3 + 4X^2 + X + 1.$$

Also determine the prime factor decomposition of these polynomials over $\mathbb{Q}$. (The so-called **Sophie Germain identity** $X^4 + 4 = (X^2 + 2X + 2)(X^2 - 2X + 2)$ or in homogeneous form $X^4 + 4Y^4 = (X^2 + 2XY + 2Y^2)(X^2 - 2XY + 2Y^2)$ should be remembered.)

Exercise 3.5.6 Provide the prime factor decomposition of the polynomials $X^n + 1$, $n \in \mathbb{N}^*$, and $X^{2n} + a$, $n \in \mathbb{N}^*$, $a \in \mathbb{R}_+^\times$, over $\mathbb{C}$ or $\mathbb{R}$.

Exercise 3.5.7 Determine the minimal polynomial of a complex number $a + bi$, $a, b \in \mathbb{R}$, over $\mathbb{R}$. For each quadratic prime polynomial $\mu \in \mathbb{R}[X]$, $\mathbb{R}[X]/(\mu) \widetilde{\to} \mathbb{C}$ is.

Exercise 3.5.8 Let $n \in \mathbb{N}^*$. The polynomial $F \in \mathbb{C}[X]$ of degree $< n$, which for the n-th roots of unity $1, \zeta_n, \ldots, \zeta_n^{n-1}$ has the given values $b_0, \ldots, b_{n-1}$, is

$$F = a_0 + a_1 X + \cdots + a_{n-1} X^{n-1} \quad \text{with} \quad a_\nu := \frac{1}{n} \sum_{k=0}^{n-1} b_k \zeta_n^{-\nu k}.$$

In particular, for $\nu = 0, \ldots, n-1$ the estimate $|a_\nu| \leq \text{Max}\left(|b_0|, \ldots, |b_{n-1}|\right)$ is used. (For the calculation of the a_ν one might use the so-called fast Fourier transformation, which we will discuss in Vol. 6.)

Exercise 3.5.9 Let $F = a_0 + a_1 X + \cdots + a_{n-1} X^{n-1} + X^n$ be a normalized polynomial from $\mathbb{C}[X]$. Then for every root α of F in $\mathbb{C}$ the following estimates apply:

a) $|\alpha| \leq \text{Max}\left(1, |a_0| + \cdots + |a_{n-1}|\right)$.
b) $|\alpha| \leq \text{Max}\left(|a_0|, 1 + |a_1|, \ldots, 1 + |a_{n-1}|\right)$.
c) $|\alpha| \leq 2R$ with $R := \text{Max}\left(|a_\nu|^{1/(n-\nu)}, \nu = 0, \ldots, n-1\right)$. (**Cauchy's root estimate** — From $|\alpha| > 2R$ and $F(\alpha) = 0$ follows the contradiction

$$|\alpha|^n = \left|a_0 + \cdots + a_{n-1}\alpha^{n-1}\right| \leq \sum_{\nu=0}^{n-1} R^{n-\nu}|\alpha|^\nu = R\frac{|\alpha|^n - R^n}{|\alpha| - R} < |\alpha|^n.)$$

Exercise 3.5.10 Determine the partial fraction decompositions of the following rational functions over $\mathbb{C}$ or — insofar as they are real rational functions — over $\mathbb{R}$, cf. theorem 2.10.26 and the remarks on it as well as example 2.10.31:

$$\frac{2X^3 - X^2 - 10X + 19}{X^2 + X - 6}; \quad \frac{X^3 - 17X^2 - 39X - 15}{(X-1)(X-2)((X+2)^2 + 1)}; \quad \frac{1}{X^4 - X^3 - X + 1};$$

$$\frac{3X^4 - 9X^3 + 4X^2 - 34X + 1}{(X-2)^2(X+3)^2}; \quad \frac{X}{(X-1)(X^2+4)}; \quad \frac{2X^2 + 2X + 3}{2X^3 - 11X^2 + 18X - 9};$$

$$\frac{1}{X^6 + 2X^4 + X^2}; \quad \frac{X^5 - X^4 + X + 1}{X^3 + 2}; \quad \frac{X^2 - 2 + 2}{(X^4 - 1)^2}; \quad \frac{1}{X^5 + 3X^4 + 4X^3 + 2X^2};$$

$$\frac{1}{X^3+1}; \quad \frac{1}{X^4+1}; \quad \frac{1}{X^n-1}, n \in \mathbb{N}^*; \quad \frac{1}{X^n+1}, n \in \mathbb{N}^*; \quad \frac{1}{(X^2+1)^n}, n \in \mathbb{N}^*;$$

$$\frac{1}{(X-\alpha_1)\cdots(X-\alpha_r)}, \quad \alpha_1, \ldots, \alpha_r \in \mathbb{C} \text{ pairwise distinct,}$$

(for $\alpha_\rho := 1 - \rho, \rho = 1, \ldots, r$, cf. exercise 1.6.17a)).

Exercise 3.5.11 Sketch the following sets of points in the Gaussian number plane $\mathbb{C}$.

a) $\{z \in \mathbb{C} \mid |z+1| \leq |z-1|\}$.
b) $\{z \in \mathbb{C} \mid 1 < |z - 3\mathrm{i}| < 7\}$.
c) $\{z \in \mathbb{C} \mid |z^2 - z| \leq 1\}$.
d) $\{z \in \mathbb{C} \mid z\bar{z} + z + \bar{z} < 0\}$.
e) $\{z \in \mathbb{C} \mid z\bar{z} - 2z - 2\bar{z} = 2\}$.
f) $\{z \in \mathbb{C} \mid |z - \mathrm{i}| + |z + \mathrm{i}| \leq 3\}$.
Which of these are open or closed?

Exercise 3.5.12
a) Every complex number w of magnitude 1 has the form $w = z/\bar{z} = z^2/\mathrm{N}(z)$ with a $z \in \mathbb{C}^\times$, which is uniquely determined except for a factor $r \in \mathbb{R}^\times$, i.e., the form

$$w = \frac{s^2 - t^2}{s^2 + t^2} + \mathrm{i}\frac{2st}{s^2 + t^2}$$

with a pair of $(s, t) \neq (0, 0)$ real numbers (which is uniquely determined except for a factor from $\mathbb{R}^\times$). In other words: The mapping $\mathbb{C}^\times \to \mathrm{U}$, $z \mapsto z/\bar{z}$, is a surjective group homomorphism with $\mathbb{R}^\times$ as its kernel. Compare this with the homomorphism $z \mapsto z/|z|$, whose kernel is $\mathbb{R}_+^\times$.
b) Analogously, every rational complex number $w \in \mathbb{C}_\mathbb{Q} = \mathbb{Q}[\mathrm{i}]$ with $\mathrm{N}(w) = w\bar{w} = 1$ has a representation of the same form as in a), but the pair (s, t) is in $\mathbb{Q}^2 - \{(0, 0)\}$ and again is uniquely determined up to a factor from $\mathbb{Q}^\times$. One deduces the so-called **Indian formulas**: Every Pythagorean number triple $(a, b, c) \in (\mathbb{N}^*)^3$ with $a^2 + b^2 = c^2$ and $\mathrm{ggT}(a, b, c) = 1$ as well as $b \equiv 0 \bmod 2$ has a representation $a = s^2 - t^2$, $b = 2st$, $c = s^2 + t^2$ with uniquely determined $s, t \in \mathbb{N}^*$, $\mathrm{ggT}(s, t) = 1$. (See the remark (1) in exercise 1.7.13.)

Exercise 3.5.13 If the real parts of $z, w \in \mathbb{C}$ are both positive or both negative, then the inequality $|(z - w)/(z + \bar{w})| < 1$ applies.

Exercise 3.5.14 If $z, w \in \mathbb{C}$ with $|z| < 1, |w| < 1$, then also $|(z + w)/(1 + \bar{z}w)| < 1$. This last inequality also holds if z and w are both greater than 1 in absolute value.

Exercise 3.5.15 If $z_1, \ldots, z_n \in \mathbb{C}^\times$ with $|z_1 + \cdots + z_n| = |z_1| + \cdots + |z_n|$, then $z_i/z_j \in \mathbb{R}_+^\times$ for all $i, j = 1, \ldots, n$.

Exercise 3.5.16 Let (z_n) be a convergent sequence of complex numbers.

a) The sequence $(z_{n+1} - z_n)$ of differences between neighboring terms is a null sequence.
b) If $\lim z_n \neq 0$, then $z_n \neq 0$ for almost all n and the sequence (z_{n+1}/z_n) of quotients of neighboring terms converges to 1.
c) Show by examples that a sequence (z_n) does not have to be convergent if it fulfills the conditions from a) or from b) or even both.

Exercise 3.5.17 Let $z \in \mathbb{C}$.

a) (z^n) is a null sequence if and only if $|z| < 1$ is. (z^n) is convergent if and only if $|z| < 1$ or $z = 1$ is.
b) The sequence $\frac{1}{n+1}\sum_{k=0}^{n} z^k$, $n \in \mathbb{N}$, converges if and only if $|z| \leq 1$ is. If $|z| \leq 1$ and $z \neq 1$, then the limit is 0, for $z = 1$ the limit is 1.

Exercise 3.5.18 Determine the accumulation points or the limit of the sequences

$$\left(\frac{(n+\mathrm{i})^2}{n^2+\mathrm{i}}\right); \quad \left(\frac{\mathrm{i}^n}{1+\mathrm{i}n}\right); \quad ((-\mathrm{i})^n); \quad \left(\left(\frac{1+\mathrm{i}}{2+\mathrm{i}}\right)^n\right).$$

Exercise 3.5.19 A sequence (z_n) of complex numbers **converges** by definition **towards** ∞, if the sequence $(|z_n|)$ of magnitudes converges towards ∞ in the sense of Definition 3.2.8. Transfer exercise 3.3.2 to complex rational functions

$$R = F/G \in \mathbb{C}(T)^\times, \quad F, G \in \mathbb{C}[T]^*, \quad \gcd(F, G) = 1.$$

(If one includes improper convergence in $\mathbb{C}$, the limit of a convergent sequence of complex numbers thus lies in $\overline{\mathbb{C}} = \mathbb{C} \cup \{\infty\}$. See also Example 3.8.14.)

Exercise 3.5.20 Show that the closed disk $\overline{B}(z_0; \varepsilon)$, $z_0 \in \mathbb{C}$, $\varepsilon > 0$, is the closed hull of the open disk $B(z_0; \varepsilon)$.

Exercise 3.5.21 Construct a sequence of complex numbers that has every complex number as an accumulation point. Show in general: If $A \subseteq \mathbb{C}$ is any nonempty closed subset, then there exists a sequence (z_n) of points $z_n \in A$, for which the set of accumulation points is entirely A. (Every set $A \subseteq \mathbb{C}$ has a *countable* subset that is dense in A.)

Exercise 3.5.22 Let $A \subseteq \mathbb{C}$ be boundaryless (i.e., open and closed) and $\neq \emptyset$. Then $A = \mathbb{C}$, i.e., $\emptyset$ and $\mathbb{C}$ are the only boundaryless subsets of $\mathbb{C}$. (Let $a_0 \in A$. For each $z \in \mathbb{C}^\times$, the set $\{t \in \mathbb{R} \mid a_0 + tz \in A\}$ is open and closed in $\mathbb{R}$, thus equal to $\mathbb{R}$, cf. exercise 3.4.19.)

Exercise 3.5.23 The set $\{z \in \mathbb{C} \mid |z^2 - 1| \leq 1\}$ is closed. Sketch this set and show that $\{z \in \mathbb{C} \mid |z^2 - 1| < 1\}$ is its inner core. Consider more generally the set $\{z \in \mathbb{C} \mid |z^2 - 1| \leq r^2\}$ depending on $r \in \mathbb{R}_+^\times$ (distinguishing the cases $r < 1$, $r = 1$ and $r > 1$) and replace $z^2 - 1$ with any quadratic polynomial $z^2 - pz + q = (z-a)(z-b)$. (Note that the set $\{z \in \mathbb{C} \mid |z^2 - pz + q| = r^2\}$ is the

3.5 The complex numbers

Fig. 3.11 Polar coordinates of $z \pm w$

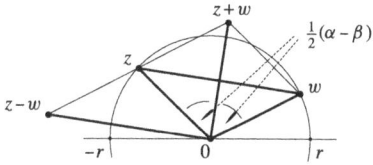

set of points in $\mathbb{C}$ for which the product of the distances from the roots a and b of the quadratic polynomial $Z^2 - pZ + q$ has the constant value r^2. Such sets are called **Cassini curves**.)

Exercise 3.5.24 Let $K \subseteq \mathbb{C}$ be bounded and closed. Then the set $\{|z| \mid z \in K\} \subseteq \mathbb{R}$ is also bounded and closed. In particular, if $K \neq \emptyset$ is, there are elements $u, v \in K$ with $|u| \leq |z| \leq |v|$ for all $z \in K$.—In general, $\{|z| \mid z \in A\}$ is closed when A is closed. Provide a closed set A in $\mathbb{C}$ for which neither the set $\{\Re z \mid z \in A\}$ nor the set $\{\Im z \mid z \in A\}$ is closed. (With the designations from example 4.4.26: *The absolute value function* $\mathbb{C} \to \mathbb{R}$, $z \mapsto |z|$, *is proper*.)

Exercise 3.5.25 An infinite bounded set $A \subseteq \mathbb{C}$ has at least one accumulation point.

Exercise 3.5.26 Let $z_n, n \in \mathbb{N}$ be a sequence of complex numbers. Then $\bigcap_{n \in \mathbb{N}} \overline{\{z_m \mid m \geq n\}}$ is the set of its accumulation points.

Exercise 3.5.27
a) Let $n \in \mathbb{N}^*$. The distinct n-th roots of unity are exactly the solutions x of the equation $x^{n-1} + \cdots + x + 1 = 0$. It follows $\sum_{\nu=0}^{n-1} X^\nu = \prod_{\nu=1}^{n-1} (X - \zeta_n^\nu)$.
b) Let $m, n \in \mathbb{N}^*$. For which $x \in \mathbb{C}$ is both $x^n = 1$ and $x^m = 1$.

Exercise 3.5.28 For the n-th primitive standard roots of unity ζ_n, $1 \leq n \leq 8$, $n \neq 7$, the following applies

$$\zeta_1 = 1, \quad \zeta_2 = -1, \quad \zeta_3 = \tfrac{1}{2}(-1 + i\sqrt{3}), \quad \zeta_4 = i,$$

$$\zeta_5 = \frac{1}{4}\sqrt{2}\left(\sqrt{3 - \sqrt{5}} + i\sqrt{5 + \sqrt{5}}\right) = \frac{1}{4}\left(\sqrt{5} - 1 + i\sqrt{2(5 + \sqrt{5})}\right),$$

$$\zeta_6 = \tfrac{1}{2}(1 + i\sqrt{3}) = \zeta_3 + 1 = \sqrt{\zeta_3}, \quad \zeta_8 = \tfrac{1}{2}\sqrt{2}(1 + i) = \sqrt{i}.$$

(To calculate ζ_5, use exercise 3.5.4. — Also determine ζ_{10}, ζ_{12} and ζ_{15}. For ζ_{15} see example 3.3.9. In general, describe how to calculate ζ_{mn} using ζ_m and ζ_n when $m, n \in \mathbb{N}^*$ are coprime.)

Exercise 3.5.29 Provide the real and imaginary parts of *all* 3-rd, 5-th and 6-th roots of unity. (Consider these roots of unity in the Gaussian plane.)

Exercise 3.5.30 Calculate $\sqrt{1+\sqrt{-3}}+\sqrt{1-\sqrt{-3}}$. (Exercise from Leibniz to Huygens — The roots are always the principal values.)

Exercise 3.5.31 Let $z, w \in \mathbb{C}$ with $|z|=|w|=r$, $\text{Arg}\, z = \alpha$, $\text{Arg}\, w = \beta$, see Fig. 3.11. It is

$$z \pm w = r\bigl((\cos\alpha + i\sin\alpha) \pm (\cos\beta + i\sin\beta)\bigr) = r(e^{i\alpha} \pm e^{i\beta})$$
$$= r(e^{i(\alpha-\beta)/2} \pm e^{-i(\alpha-\beta)/2})\, e^{i(\alpha+\beta)/2}$$
$$= \begin{cases} 2r\bigl(\cos(\alpha-\beta)/2\bigr)\bigl(\cos(\alpha+\beta)/2 + i\sin(\alpha+\beta)/2\bigr) & \text{for } z+w, \\ 2ri\bigl(\sin(\alpha-\beta)/2\bigr)\bigl(\cos(\alpha+\beta)/2 + i\sin(\alpha+\beta)/2\bigr) & \text{for } z-w. \end{cases}$$

Determine $|z \pm w|$ and $\text{Arg}(z \pm w)$. (The sign of $\cos \tfrac{1}{2}(\alpha - \beta)$ or $\sin \tfrac{1}{2}(\alpha - \beta)$ should be considered.)

In the case of $\beta = 0$, one obtains

$$(\cos\alpha + i\sin\alpha) + 1 = 2\cos\tfrac{1}{2}\alpha \bigl(\cos\tfrac{1}{2}\alpha + i\sin\tfrac{1}{2}\alpha\bigr),$$

$$(\cos\alpha + i\sin\alpha) - 1 = 2i\sin\tfrac{1}{2}\alpha \bigl(\cos\tfrac{1}{2}\alpha + i\sin\tfrac{1}{2}\alpha\bigr).$$

From this, derive formulas for the following sums ($n \in \mathbb{N}$):

$$\sum_{k=0}^{n} \binom{n}{k} \cos k\alpha,\quad \sum_{k=0}^{n} \binom{n}{k} \sin k\alpha,\quad \sum_{k=0}^{n} (-1)^{n-k} \binom{n}{k} \cos k\alpha,\quad \sum_{k=0}^{n} (-1)^{n-k} \binom{n}{k} \sin k\alpha.$$

Exercise 3.5.32

a) For $n \in \mathbb{N}$ show (by considering $(1+i)^n$)

$$\sum_{k=0}^{[n/2]} (-1)^k \binom{n}{2k} = \sqrt{2}^n \cos\frac{n\pi}{4},\quad \sum_{k=0}^{[(n-1)/2]} (-1)^k \binom{n}{2k+1} = \sqrt{2}^n \sin\frac{n\pi}{4}.$$

b) For $n \in \mathbb{N}$ is $(X+iY)^n = P_n + iQ_n \in \mathbb{C}[X,Y]$ with

$$P_n = \sum_{k \geq 0} (-1)^k \binom{n}{2k} X^{n-2k} Y^{2k} = \prod_{k=0}^{n-1} \left(X - Y \cot \frac{(2k+1)\pi}{2n} \right),$$

$$Q_n = \sum_{k \geq 0} (-1)^k \binom{n}{2k+1} X^{n-2k-1} Y^{2k+1} = nY \prod_{k=1}^{n-1} \left(X - Y \cot \frac{k\pi}{n} \right).$$

Exercise 3.5.33 For $\varphi \in \mathbb{R}$, $\varphi \neq k\pi$, $k \in \mathbb{Z}$, and $n \in \mathbb{N}^*$ applies

$$\sin n\varphi / \sin \varphi = 2^{n-1} U_{n-1}(\cos\varphi),$$

3.5 The complex numbers

where the normalized polynomials $U_n \in \mathbb{Q}[X]$ are defined by

$$U_n := \sum_{k=0}^{\lfloor n/2 \rfloor} \left(-\frac{1}{4}\right)^k \binom{n-k}{k} X^{n-2k}$$

They satisfy the recursion

$$U_0 = 1, \quad U_1 = X, \quad U_{n+2} = XU_{n+1} - \frac{1}{4}U_n, \quad n \in \mathbb{N},$$

which differs from that of the Chebyshev polynomials T_n only in the initial condition for $n = 0$. (The U_n are called **(normalized) Chebyshev polynomials of the second kind.**) Determine $U_0, \ldots, U_5$. It is

$$U_n = \prod_{k=1}^{n}\left(X - \cos\frac{k\pi}{n+1}\right) \quad \text{and} \quad U_{2n} = \prod_{k=1}^{n}\left(X^2 - \cos^2\frac{k\pi}{2n+1}\right).$$

One concludes (using $\sin\varphi \sim \varphi$ for $\varphi \to 0$):

$$n+1 = 2^n U_n(1) = 2^n \prod_{k=1}^{n}\left(1 - \cos\frac{k\pi}{n+1}\right),$$

$$2n+1 = 2^{2n} U_{2n}(1) = 2^{2n} \prod_{k=1}^{n} \sin^2 \frac{k\pi}{2n+1} = 2^{2n} \prod_{k=1}^{n} \sin^2 \frac{2k\pi}{2n+1} = 2^{2n} \prod_{k=1}^{2n} \sin \frac{k\pi}{2n+1}.$$

Exercise 3.5.34 For $\varphi \in \mathbb{R}$ and $n, m \in \mathbb{N}$ applies

$$\cos^n \varphi = \frac{1}{2^n} \sum_{k=0}^{n} \binom{n}{k} \cos(n - 2k)\varphi,$$

$$\sin^{2m} \varphi = \frac{1}{2^{2m}} \sum_{k=0}^{2m} (-1)^{m+k} \binom{2m}{k} \cos(2m - 2k)\varphi,$$

$$\sin^{2m+1} \varphi = \frac{1}{2^{2m+1}} \sum_{k=0}^{2m+1} (-1)^{m+k} \binom{2m+1}{k} \sin(2m + 1 - 2k)\varphi.$$

(With $z = \cos\varphi + i\sin\varphi$ is $\cos\varphi = (z + z^{-1})/2$, $\sin\varphi = (z - z^{-1})/2i$. See also example 3.5.9.)

Exercise 3.5.35 For $\varphi \in \mathbb{R}$, $\varphi \neq 2m\pi$, $m \in \mathbb{Z}$, and $n \in \mathbb{N}$ applies

$$\sum_{k=0}^{n} \cos k\varphi = \frac{\sin\frac{1}{2}(n+1)\varphi}{\sin\frac{1}{2}\varphi} \cos\frac{n\varphi}{2} = \frac{1}{2} + \frac{\sin\frac{1}{2}(2n+1)\varphi}{2\sin\frac{1}{2}\varphi},$$

$$\sum_{k=0}^{n} \sin k\varphi = \frac{\sin\frac{1}{2}(n+1)\varphi}{\sin\frac{1}{2}\varphi} \sin\frac{n\varphi}{2} = \frac{\cos\frac{1}{2}\varphi - \cos\frac{1}{2}(2n+1)\varphi}{2\sin\frac{1}{2}\varphi}.$$

Fig. 3.12 Square roots in $\mathbb{C}$

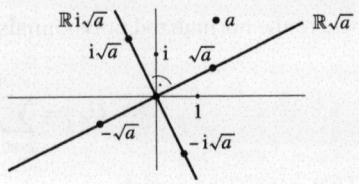

(Use $\sum_{k=0}^{n} z^k$ with $z := \cos\varphi + i\sin\varphi$ or $\cos k\varphi = (e^{ik\varphi} + e^{-ik\varphi})/2$ and $\sin k\varphi = (e^{ik\varphi} - e^{-ik\varphi})/2i$.)

Exercise 3.5.36 From the result of exercise 3.5.35, infer for $\varphi \in \mathbb{R}, \varphi \neq m\pi, m \in \mathbb{Z}$, and $n \in \mathbb{N}$ using $\sin^2 k\varphi + \cos^2 k\varphi = 1$ and $\cos^2 k\varphi - \sin^2 k\varphi = \cos 2k\varphi$ the equations

$$\sum_{k=0}^{n} \cos^2 k\varphi = \frac{n+1}{2} + \frac{\cos n\varphi \sin(n+1)\varphi}{2\sin\varphi} = \frac{2n+3}{4} + \frac{\sin(2n+1)\varphi}{4\sin\varphi},$$

$$\sum_{k=0}^{n} \sin^2 k\varphi = \frac{n+1}{2} - \frac{\cos n\varphi \sin(n+1)\varphi}{2\sin\varphi} = \frac{2n+1}{4} - \frac{\sin(2n+1)\varphi}{4\sin\varphi}.$$

Exercise 3.5.37 For $m, n \in \mathbb{N}^*$ is $\zeta_m \zeta_n$ a primitive k-th root of unity with $k := mn/\gcd(mn, m+n)$.

Exercise 3.5.38
a) Let $n \in \mathbb{N}, n \geq 2$. The sum of all n-th roots of unity is equal to 0.
b) Let $n \in \mathbb{N}, n \geq 1$. The product of all n-th roots of unity is equal to $(-1)^{n-1}$.

Exercise 3.5.39 For $n \in \mathbb{N}^*$ is $\sum_{k=0}^{n-1} \zeta_{2n}^k = 2/(1-\zeta_{2n})$.

Exercise 3.5.40 For $n \in \mathbb{N}$ is

$$n+1 = 2^n \prod_{k=1}^{n} \sin\frac{k\pi}{n+1} = 2^{2n} \prod_{k=1}^{n} \sin^2\frac{k\pi}{2(n+1)}, \quad 2n+1 = 2^{2n} \prod_{k=1}^{n} \sin^2\frac{k\pi}{2n+1}.$$

(Exercise 3.5.27a) and 3.5.31. — See also Exercise 3.5.33.)

Exercise 3.5.41 Determine the images of the circles $\{z \in \mathbb{C} \mid |z-a| = r\}, r > 0$, $a \in \mathbb{C}$, and the (real) lines $a + \mathbb{R}b$, $a, b \in \mathbb{C}, b \neq 0$, under the inverse formation $z \mapsto z^{-1}$ on $\mathbb{C}^\times$. Also consider the so-called **Inversion** or **Reflection on the unit circle** $z \mapsto \overline{z}^{-1} = z/|z|^2$.

Exercise 3.5.42 The images of the rays $r(\cos\varphi + i\sin\varphi)$, $r \in \mathbb{R}_+^\times$, φ constant, are under the Joukowski function $z \mapsto (z + z^{-1})/2$ are the hyperbolic loads

$$\frac{a^2}{\cos^2\varphi} - \frac{b^2}{\sin^2\varphi} = 1, \quad \text{Sign } a = \text{Sign } \cos\varphi,$$

3.5 The complex numbers

which degenerate at $\cos\varphi = 0$ to the imaginary axis $\mathbb{R}i$ and at $\cos\varphi = \pm 1$, i.e., $\sin\varphi = 0$, to the rays $\{a \in \mathbb{R} \mid a \leq 1\}$ or $\{a \in \mathbb{R} \mid a \geq -1\}$.

Exercise 3.5.43 The Joukowski function $z \mapsto (z + z^{-1})/2$ is injective on $\{z \in \mathbb{C} \mid |z| > 1\}$ with $D := \mathbb{C} - \{a \in \mathbb{R} \mid |a| \leq 1\}$ as the image. Specify the inverse function. (The problem is to determine the root in $z = w + \sqrt{w^2 - 1}$, $w \in D$, so that $|z| > 1$ is.)

Exercise 3.5.44

a) The function $T: u \mapsto (u - 1)/(u + 1)$ is injective on $\mathbb{C} - \{-1\}$ with image $\mathbb{C} - \{1\}$. The inverse function there is $w \mapsto (1 + w)/(1 - w)$. The image of the simply slit plane $\mathbb{C} - \mathbb{R}_-$ under T is the doubly slit plane $\mathbb{C} - \{a \in \mathbb{R} \mid |a| \geq 1\}$. (These are fractional linear functions, cf. Exercise 2.10.21.)

b) The function $h: z \mapsto (z^2 - 1)/(z^2 + 1)$ is injective on the right half-plane $\{z \in \mathbb{C} \mid \Re z > 0\}$ again with the doubly slit plane $\mathbb{C} - \{a \in \mathbb{R} \mid |a| \geq 1\}$ as an image. Its inverse function is $w \mapsto \sqrt{(1 + w)/(1 - w)}$ (with the principal value of the root function).

Exercise 3.5.45 Let $a \in \mathbb{C}^\times$, $\sqrt{a}$ be a square root of a (not necessarily its principal value, cf. example 3.5.6) and $g_a: \mathbb{C}^\times \to \mathbb{C}$ the function

$$g_a(z) := \frac{1}{2}\left(z + \frac{a}{z}\right) = \sqrt{a}\, f\left(\frac{z}{\sqrt{a}}\right),$$

where f denotes the Joukowski function $z \mapsto (z + z^{-1})/2$, cf. example 3.5.9. Then the iteration converges $z_{n+1} = g_a(z_n)$, $n \in \mathbb{N}$, of the Babylonian root extraction, see example 3.3.7, with any starting value $z_0 \in \mathbb{C}$, which is not on the straight line $\mathbb{R}i\sqrt{a}$, converges towards a square root of a, specifically towards the one that lies in the same half-plane determined by $\mathbb{R}i\sqrt{a}$ as z_0, see Fig. 3.12. In particular, for every $z_0 \in \mathbb{C}$ with $\Re z_0 > 0$ the recursively defined sequence $z_{n+1} = f(z_n)$, $n \in \mathbb{N}$, converges towards 1. (The general statement follows from this special case.) If you replace the starting value z_0 with $-z_0$, you get the sequence $(-z_n)$ and as a limit the second square root of a. For $a \notin \mathbb{R}_-$ and $z_0 := 1$ you always get the principal value $\sqrt{a}$ of the square root function. In this case, z_n, $n \in \mathbb{N}$, is the rational function $z_n = R_n(a)/S_n(a)$, where the polynomials

$$R_n = \sum_{k \geq 0} \binom{2^n}{2k} X^k, \quad S_n = \sum_{k \geq 0} \binom{2^n}{2k+1} X^k \in \mathbb{Z}[X], \quad n \in \mathbb{N},$$

through the equations $(1 \pm \sqrt{X})^{2^n} = R_n \pm S_n\sqrt{X}$, $n \in \mathbb{N}$, in the quadratic $\mathbb{Z}[X]$-algebra $(\mathbb{Z}[X])[\sqrt{X}]$ or through the recursion

$$R_0 = S_0 = 1; \quad R_{n+1} = R_n^2 + XS_n^2, \quad S_{n+1} = 2R_n S_n, \quad n \in \mathbb{N},$$

are determined. It is

$$R_n = \prod_{k=0}^{2^{n-1}-1}\left(X + \tan^2\frac{(2k+1)\pi}{2^{n+1}}\right), \quad S_n = 2^n \prod_{k=1}^{2^{n-1}-1}\left(X + \tan^2\frac{k\pi}{2^n}\right), \quad n \in \mathbb{N}^*.$$

Fig. 3.13 Overhang of stacked bricks

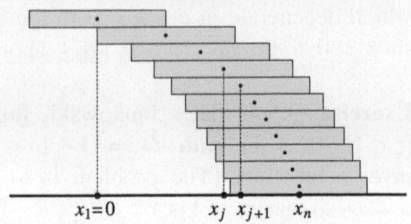

(For the determination of the zeros of R_n and S_n use for example Exercise 3.5.32b). What happens if the initial value z_0 lies on the critical line $\mathbb{R}\mathrm{i}\sqrt{a}$? Without restriction, let $a = 1$. The iteration is as chaotic as the iteration of the square function on the unit circle, see the end of exercise 2.4.16d). — The Babylonian root extraction for k-th roots, see the end of example 3.3.7, has a much more confusing convergence behavior in the complex case for $k \geq 3$ than in the above case $k = 2$.)

3.6 Series

In this section, the numbers considered are always real or complex numbers, unless otherwise stated. When comparing numbers in terms of size, they are always real numbers. On $\mathbb{C}$ the relation "$z \preceq w$ exactly when $|z| \leq |w|$" is a quasi-order, see exercise 1.4.2.

Series are simply sequences, viewed from a different perspective. Let be (x_n) is a sequence of real or complex numbers. To investigate how the sequence changes from term to term, one considers the differences

$$a_0 := x_0, \quad a_1 := x_1 - x_0, \quad \ldots, \quad a_n := x_n - x_{n-1}, \quad \ldots$$

For all n, then $x_n = a_0 + a_1 + \cdots + a_n$.[11] The sequence (x_n) is therefore the series $\sum_{k=0}^{\infty} a_k$ in the sense of the following definition.

Definition 3.6.1 Let (a_n) be a sequence of real or complex numbers. Then the sequence (x_n) of **partial sums**

$$x_n := \sum_{k=0}^{n} a_k = a_0 + a_1 + \cdots + a_n, \quad n \in \mathbb{N},$$

---

[11] This perspective can also be useful in other situations. For example, if you want to store the first n prime numbers $p_1 = 2, p_2 = 3, \ldots, p_n$ in a computer, it is less laborious to input the differences $d_1 := p_1 = 2, d_2 := p_2 - p_1 = 1, \ldots, d_n := p_n - p_{n-1}$ instead of the prime numbers themselves. (Since moreover all d_k except for d_2 are even, they can be conveniently halved. — Often one generates the prime numbers $\leq N$ for a given $N \in \mathbb{N}^*$ quickly with the sieve of Eratosthenes, cf. exercise 1.7.7.) Or: Because of $n^2 - (n-1)^2 = 2n - 1$ one generates the sequence, n^2, $n \in \mathbb{N}$, of square numbers conveniently, starting with 0, by successively adding the odd natural numbers: $n^2 = 1 + 3 + \cdots + (2n - 1)$, $n \in \mathbb{N}$.

3.6 Series

is the **(infinite) series** of a_n, $n \in \mathbb{N}$. It is denoted by

$$\sum_{k=0}^{\infty} a_k, \quad \text{short also with} \quad \sum_k a_k \quad \text{or} \quad \sum a_k,$$

If the sequence (x_n) converges, then the series is called **convergent** and its limit is the **sum** of the series. It is also denoted by $\sum_{k=0}^{\infty} a_k$.

The following *necessary* criterion for the convergence of series is trivial.

Proposition 3.6.2 *If the series converges $\sum_n a_n$, then (a_n) is a null sequence.*

Proof If (x_n) denotes the sequence of partial sums and if (x_n) is convergent, then $\lim a_n = \lim(x_n - x_{n-1}) = \lim x_n - \lim x_{n-1} = 0$. □

Conversely, the series $\sum a_n$ is by no means convergent when the sequence (a_n) is a null sequence.

Example 3.6.3 (Harmonic Series) *The harmonic series $\sum_{n=1}^{\infty} \frac{1}{n}$ is divergent.* Its partial sums, i.e., the harmonic numbers $H_k = \sum_{n=1}^{k} \frac{1}{n}$, are not bounded due to

$$H_{2^n} = 1 + \frac{1}{2} + \left(\frac{1}{3} + \frac{1}{4}\right) + \left(\frac{1}{5} + \frac{1}{6} + \frac{1}{7} + \frac{1}{8}\right) + \cdots + \left(\frac{1}{2^{n-1}+1} + \cdots + \frac{1}{2^n}\right)$$

$$\geq 1 + \frac{1}{2} + \left(\frac{1}{4} + \frac{1}{4}\right) + \left(\frac{1}{8} + \frac{1}{8} + \frac{1}{8} + \frac{1}{8}\right) + \cdots + \left(\frac{1}{2^n} + \cdots + \frac{1}{2^n}\right)$$

$$= 1 + \frac{1}{2} + \frac{1}{2} + \frac{1}{2} + \cdots + \frac{1}{2} = 1 + \frac{n}{2}$$

According to example 3.3.8 (2), even $H_k = \ln k + \gamma + \rho_k$ with a monotonically decreasing null sequence (ρ_k). — From the divergence of the harmonic series, the following result can be derived: *Similar bricks can be stacked on top of each other without mortar to achieve an arbitrarily large overhang.* To do this, we consider what overhang can be achieved with n stones using the construction method shown in Fig. 3.13: Each stone has a length of 2. We number the bricks from top to bottom. The x-coordinate of the center of the j-th stone is x_j, and let it be $x_1 = 0$. For the tower not to tip over, for each j the center of gravity of the top j stones must lie above the $(j+1)$-th stone, i.e., it must be

$$x_{j+1} - 1 \leq \frac{1}{j} \sum_{m=1}^{j} x_m, \quad j = 1, \ldots, n-1,$$

We show by induction over n that this implies $x_{j+1} \leq H_j$, $j = 1, \ldots, n-1$. The induction start $n = 1$ is trivial. When concluding from n to $n+1$, the first $n-1$ inequalities already apply. From $x_{n+1} - 1 \leq \frac{1}{n} \sum_{m=1}^{n} x_m$ it follows that

$$x_{n+1} \leq 1 + \frac{1}{n} \sum_{m=1}^{n} H_{m-1} = 1 + \frac{1}{n} \sum_{k=1}^{n-1} \frac{n-k}{k} = 1 + H_{n-1} - \frac{n-1}{n} = H_n.$$

Conversely, for $0 < \lambda \leq 1$, it follows from $x_{j+1} := \lambda H_j$, $j = 1, \ldots, n-1$:

$$\frac{1}{j}\sum_{m=1}^{j} x_m = \frac{\lambda}{j}\sum_{m=1}^{j} H_{m-1} = \frac{\lambda}{j}\sum_{k=1}^{j-1}\frac{j-k}{k} = \lambda H_j - \lambda = x_{j+1} - \lambda \geq x_{j+1} - 1,$$

so that with this choice of x_j the center of gravity conditions are fulfilled. Therefore, the overhang can be at most $x_n = H_{n-1}$ with n stones and becomes arbitrarily large as n increases. Due to $x_n - \ln n < \gamma < x_n - \ln(n-1)$, see example 3.3.8, $n - 1 < e^{x_n}/e^{\gamma} = e^{x_n} \cdot 0{,}561459\ldots < n$. Therefore, for an overhang ≥ 10, you need at least 12.367 and at most 12.368 stones of length 2.[12]   ◊

Example 3.6.4 (Telescope series) Because of

$$\frac{1}{k(k+1)} = \frac{1}{k} - \frac{1}{k+1}$$

it follows that

$$\sum_{k=1}^{n}\frac{1}{k(k+1)} = \left(1 - \frac{1}{2}\right) + \left(\frac{1}{2} - \frac{1}{3}\right) + \cdots + \left(\frac{1}{n-1} - \frac{1}{n}\right) + \left(\frac{1}{n} - \frac{1}{n+1}\right)$$

$$= 1 - \frac{1}{n+1}, \quad n \in \mathbb{N}^*,$$

and thus

$$\sum_{k=1}^{\infty}\frac{1}{k(k+1)} = \frac{1}{1 \cdot 2} + \frac{1}{2 \cdot 3} + \frac{1}{3 \cdot 4} + \cdots = 1,$$

since $1/(n+1)$, $n \in \mathbb{N}$, is a null sequence. In a similar way, the sum can often be easily calculated for other series. These are referred to as **telescope series**.   ◊

The geometric series are particularly important:

Theorem 3.6.5 (Geometric Series) *Let $x \in \mathbb{C}$. The geometric series $\sum_{n=0}^{\infty} x^n$ converges exactly when $|x| < 1$ is. It is then $\sum_{n=0}^{\infty} x^n = 1/(1-x)$.*

Proof If $|x| \geq 1$, then (x^n) is not a null sequence and therefore $\sum x^n$ is divergent. Now let $|x| < 1$. Then $\sum_{n=0}^{k} x^n = (1 - x^{k+1})/(1 - x)$ holds. Since (x^{k+1}) like $(|x|^{k+1})$ is a null sequence, see Lemma 3.3.3 (3), the assertion follows.   □

From the calculation rules 3.2.5 for limits, the following calculation rules for series are obtained directly.

---

[12] In fact, you need 12.368 stones. For this, the more precise representation of γ in Vol. 2, Sect. 3.8 is used. How many stones are needed for an overhang ≥ 20?

3.6 Series

Proposition 3.6.6 *For convergent series $\sum a_n$ and $\sum b_n$ and an arbitrary $\lambda \in \mathbb{C}$ are also $\sum (a_n + b_n)$ and $\sum \lambda a_n$ convergent, and it holds:*

$$\sum_{n=0}^{\infty}(a_n + b_n) = \sum_{n=0}^{\infty} a_n + \sum_{n=0}^{\infty} b_n, \quad \sum_{n=0}^{\infty} \lambda a_n = \lambda \sum_{n=0}^{\infty} a_n.$$

Proof The partial sums of the series $\sum (a_n + b_n)$ (or $\sum \lambda a_n$) are formed by adding the partial sums of the series $\sum a_n$ and $\sum b_n$ (or by multiplying the partial sums of $\sum a_n$ with λ). □

In the situation of theorem 3.6.6, the series $\sum a_n b_n$ is generally not convergent. For example, it follows from the Leibniz criterion 3.6.8 below that the series $\sum a_n$ with $a_n := (-1)^n/\sqrt{n}$ converges, whereas the series $\sum a_n^2$ is the divergent harmonic series. We first provide two simple convergence criteria for series with real terms.

Theorem 3.6.7 *Let (a_n) be a sequence of non-negative real numbers. The series converges exactly when $\sum a_n$, if the sequence of its partial sums is bounded.*

Proof Because of $a_n \geq 0$ the sequence of partial sums of $\sum a_n$ is monotonically increasing. Therefore, it is convergent if and only if it is bounded. □

Theorem 3.6.8 (Leibniz Criterion) *Let $(a_n)_{n \in \mathbb{N}^*}$ be a monotonically decreasing null sequence of real numbers. Then the alternating series $\sum_{n=1}^{\infty}(-1)^{n-1}a_n = a_1 - a_2 + a_3 - \cdots$ is convergent, and for its sum applies*

$$s_{2k+2} \leq \sum_{n=1}^{\infty}(-1)^{n-1}a_n \leq s_{2k+1}, \quad k \in \mathbb{N},$$

where $s_m = \sum_{n=1}^{m}(-1)^{n-1}a_n$, the m-th partial sum is. In particular, the error estimation $\left|\sum_{n=1}^{\infty}(-1)^{n-1}a_n - s_m\right| \leq a_{m+1}$ applies.

Proof By assumption, $a_n \geq 0$ holds for all $n \in \mathbb{N}^*$. We show that the sequences $(s_{2k+2})_{k \in \mathbb{N}}$ and $(s_{2k+1})_{k \in \mathbb{N}}$ form an interval nesting (which then necessarily defines the sum of the series), see theorem 3.3.5. However, it is

$$s_{2k+2} = s_{2k+1} - a_{2k+2} \leq s_{2k+1},$$

$$s_{2k+3} = s_{2k+1} + (a_{2k+3} - a_{2k+2}) \leq s_{2k+1},$$

$$s_{2k+4} = s_{2k+2} + (a_{2k+3} - a_{2k+4}) \geq s_{2k+2}.$$

Finally, $s_{2k+1} - s_{2k+2} = a_{2k+2}, k \in \mathbb{N}$, is a null sequence. □

The proof of theorem 3.6.8 shows that the Leibniz criterion is completely equivalent to the interval nesting principle 3.3.5: If $I_k := [b_k, c_k], k \in \mathbb{N}$, is an interval nesting $I_0 \supseteq I_1 \supseteq \cdots$ with $\lim_k (c_k - b_k) = 0$ and (without restriction) $b_0 = 0$, then the members of the sequence $b_0, c_0, b_1, c_1, b_2, c_2, \ldots$ are the partial sums s_m, $m \in \mathbb{N}$, of the alternating series $\sum_{n=1}^{\infty}(-1)^{n-1}a_n$ to the monotonically decreasing null sequence $(a_n)_{n \in \mathbb{N}^*}$ with $a_{2k+1} = c_k - b_k$, $a_{2k+2} = c_k - b_{k+1}$, $k \in \mathbb{N}$.

Example 3.6.9 The **alternating harmonic series** and the **Leibniz series**

$$\sum_{n=1}^{\infty}(-1)^{n-1}\frac{1}{n} = 1 - \frac{1}{2} + \frac{1}{3} - + \cdots \quad \text{resp.} \quad \sum_{n=1}^{\infty}(-1)^{n-1}\frac{1}{2n-1} = 1 - \frac{1}{3} + \frac{1}{5} - + \cdots$$

converge according to the Leibniz criterion. *The sum of the first series is* $\ln 2$, *cf. exercise 3.3.27a), that of the second* $\pi/4$. We will prove both results several times among others in Vol. 2. The convergence speed of these series is very poor; it is, as one can easily consider, no better than the estimates in theorem 3.6.8 indicate. ◊

The Cauchy convergence criterion 3.4.8, formulated for series, reads:

Theorem 3.6.10 (Cauchy's Convergence Criterion) *The series* $\sum a_k$ *of complex numbers converges if and only if, for every (no matter how small)* $\varepsilon > 0$ *there exists a* $n_0 \in \mathbb{N}$ *with* $\left|\sum_{k=m}^{n} a_k\right| \leq \varepsilon$ *for all* $n \geq m \geq n_0$.

Proof If x_n is the n-th partial sum of $\sum a_k$, then $x_n - x_{m-1} = \sum_{k=m}^{n} a_k$ for $n \geq m$. □

As a consequence, we immediately obtain:

Corollary 3.6.11 *Let* (a_k) *be a sequence of complex numbers. If the series* $\sum |a_k|$ *converges, then the series* $\sum a_k$ *also converges.*

Proof Because of $\left|\sum_{k=m}^{n} a_k\right| \leq \sum_{k=m}^{n} |a_k|$, the series $\sum a_k$ satisfies the Cauchy Convergence Criterion $\sum |a_k|$ just like the series 3.6.10. □

Definition 3.6.12 A series $\sum a_k$ is called **absolutely convergent**, if the series $\sum |a_k|$ converges.

Corollary 3.6.11 now states: *Absolutely convergent series are convergent*, and for the sum of the absolutely convergent series $\sum a_k$ naturally $|\sum_{k=0}^{\infty} a_k| \leq \sum_{k=0}^{\infty} |a_k|$ applies. As the alternating harmonic series $\sum(-1)^k/(k+1)$ from example 3.6.9 shows, a convergent series is not generally absolutely convergent. A simple criterion is:

Theorem 3.6.13 *Let there be* (a_k) *and* (b_k) *sequences of complex numbers. If the series* $\sum a_k$ *converges absolutely and the sequence* (b_k) *is bounded, then the series* $\sum a_k b_k$ *also converges absolutely.*

Proof Let $|b_k| \leq S$ for all $k \in \mathbb{N}$. Then $\sum_{k=0}^{n} |a_k b_k| \leq \sum_{k=0}^{n} |a_k| S \leq S \sum_{k=0}^{\infty} |a_k|$. The claim now follows from theorem 3.6.7. □

3.6 Series

Very often, the convergence of series is proven by comparison with known series.

Theorem 3.6.14 (Majorant Criterion) $\sum b_k$ *let there be a convergent series of non-negative real numbers b_k and $\sum a_k$ a series of complex numbers. If $|a_k| \leq b_k$ holds for all $k \in \mathbb{N}$, then $\sum a_k$ is also convergent, and even absolutely. It is $|\sum_{k=0}^{\infty} a_k| \leq \sum_{k=0}^{\infty} b_k$.*

Proof It is $\sum_{k=m}^{n} |a_k| \leq \sum_{k=m}^{n} b_k$, from which the assertion follows with theorem 3.6.10. □

For the convergence statement of theorem 3.6.14, it is sufficient that the estimate $|a_k| \leq b_k$ holds for almost all k. The series $\sum b_k$ is then called a **convergent majorant** for the series $\sum a_k$. Conversely, by comparing with a known divergent series, one can occasionally infer the divergence of a given series. If, for example, $0 \leq b_k \leq c_k$ holds for almost all k and the series $\sum b_k$ diverges, then the series $\sum c_k$ also diverges. In this case, $\sum b_k$ is called a **divergent minorant** to $\sum c_k$.

Example 3.6.15 Due to $1/k^2 < 1/(k-1)k$ for $k \geq 2$, the telescopic series converging according to example 3.6.4 $\sum_{k=2}^{\infty} 1/(k-1)k$ is a convergent majorant to $\sum_{k=1}^{\infty} 1/k^2$, and it holds

$$\sum_{k=1}^{\infty} \frac{1}{k^2} = 1 + \sum_{k=2}^{\infty} \frac{1}{k^2} < 1 + \sum_{k=2}^{\infty} \frac{1}{(k-1)k} = 1 + 1 = 2.$$

In Vol. 2 we show the formula found by Euler

$$\sum_{k=1}^{\infty} \frac{1}{k^2} = \frac{\pi^2}{6}.$$

From the divergence of the harmonic series $\sum 1/k$ follows due to $1/k \leq 1/\sqrt{k}$ the divergence of $\sum 1/\sqrt{k}$. ◊

The following important criterion is obtained by comparison with the geometric series:

Theorem 3.6.16 (Quotient Criterion) *Let $\sum_{k=0}^{\infty} a_k$ be a series of complex numbers with $a_k \neq 0$ for almost all k. Furthermore, let there be a real number q with $0 < q < 1$ and $|a_{k+1}/a_k| \leq q$ for almost all $k \in \mathbb{N}$. Then the series $\sum_{k=0}^{\infty} a_k$ is absolutely convergent. In particular, $\sum a_k$ converges absolutely if the sequence of quotients $|a_{k+1}/a_k|$ converges to a number < 1.*

Proof Let $|a_{k+1}| \leq q|a_k|$ for $k \geq k_0$. Then for $m \geq k_0$:

$$|a_m| \leq q|a_{m-1}| \leq \cdots \leq q^{m-k_0}|a_{k_0}|.$$

Since the geometric series $|a_{k_0}|q^{-k_0} \sum q^m$ converges for $0 < q < 1$ according to Example 3.6.5, the majorant criterion provides the assertion. □

In particular, a sequence (a_k), *which fulfills the conditions of* sentence 3.6.16 *is a null sequence.*—*If in the situation of* sentence 3.6.16 *the inequality applies for the quotients* $|a_{k+1}/a_k| \geq 1$ *for almost all k, then the series* $\sum a_k$ *is divergent.*

Example 3.6.17 The series $\sum_{k=0}^{\infty} k^n a^k$ are convergent for any $n \in \mathbb{N}$ and any $a \in \mathbb{C}$ with $|a| < 1$. For $a \neq 0$, the sequence

$$\left|\frac{(k+1)^n a^{k+1}}{k^n a^k}\right| = \left(1 + \frac{1}{k}\right)^n |a|, \quad k \in \mathbb{N}^*,$$

of quotients converges for $k \to \infty$ towards $|a| < 1$. *The convergence behavior of a series cannot always be determined with the quotient criterion.* The series $\sum 1/k$ and $\sum 1/k^2$ both do not meet the condition of the quotient criterion. Although the quotients

$$\frac{1}{k+1} \Big/ \frac{1}{k} = \frac{k}{k+1} \quad \text{resp.} \quad \frac{1}{(k+1)^2} \Big/ \frac{1}{k^2} = \frac{k^2}{(k+1)^2}$$

are less than 1, they converge towards 1. The first series diverges, the second converges. ◊

Example 3.6.18 (g-al-Fractions) Let $g \in \mathbb{N}$, $g \geq 2$. The convergence of a g-al-fraction

$$\sum_{n=1}^{\infty} \frac{z_n}{g^n}, \quad z_n \in \{0, 1, \ldots, g-1\},$$

cf. example 3.3.10, follows immediately from the convergence of the majorizing geometric series

$$\sum_{n=1}^{\infty} \frac{g-1}{g^n} = \frac{g-1}{g} \sum_{n=0}^{\infty} \frac{1}{g^n} = 1.$$

If the sequence of digits $(z_n)_{n \in \mathbb{N}^*}$ is periodic from a point $\mu + 1 \in \mathbb{N}^*$ with the (not necessarily minimal) period length $\lambda \geq 1$, i.e. is $z_{n+\lambda} = z_n$ for all $n \geq \mu + 1$, the value is the rational number

$$r = \sum_{n=1}^{\mu} \frac{z_n}{g^n} + \frac{z_{\mu+1} g^{\lambda-1} + \cdots + z_{\mu+\lambda}}{g^{\mu+\lambda}} \sum_{n=0}^{\infty} \left(\frac{1}{g^\lambda}\right)^n$$

$$= \frac{1}{g^\mu}\left(z_1 g^{\mu-1} + \cdots + z_\mu + \frac{1}{g^\lambda - 1}(z_{\mu+1} g^{\lambda-1} + \cdots + z_{\mu+\lambda})\right) = \frac{m}{g^\mu(g^\lambda - 1)},$$

$$m := c(g^\lambda - 1) + d, \quad c := z_1 g^{\mu-1} + \cdots + z_\mu < g^\mu,$$

$$d := z_{\mu+1} g^{\lambda-1} + \cdots + z_{\mu+\lambda} < g^\lambda - 1.$$

3.6 Series

If $r > 0$ and $r = a/b$ is the reduced representation of r with (uniquely determined) coprime natural numbers a, b, $0 < a < b$, and are $g = \prod_p p^{\alpha_p}$ and $b = \prod_p p^{\beta_p}$ the prime factor decompositions of g and b, then g^μ is a multiple of $b_1 := \prod_{p \mid g} p^{\beta_p}$ and $g^\lambda - 1$ is a multiple of $b_1 := \prod_{p \nmid g} p^{\beta_p}$, thus $\mu \geq \text{Max}(\lceil \beta_p/\alpha_p \rceil, p \mid g)$ and $\lambda \geq \text{Ord}_{b_2} g$, where $\text{Ord}_{b_2} g$ the smallest positive natural number n is with $b_2 \mid (g^n - 1)$, i.e. the order of g in the prime residue class group $\mathbf{A}_{b_2}^\times$. Thus, the sequence of digits $(z_1, z_2, z_3, \ldots)$ of the g-al development of the (reduced) fraction a/b has the periodicity type $\left(\text{Max}(\lceil \beta_p/\alpha_p \rceil, p \mid g), \text{Ord}_{b_2} g\right)$ (cf. exercise 1.7.38). The periodicity type thus does not depend on the numerator a. The algorithm for calculating the digits according to example 3.3.10 shows that the sequence of digits has the same periodicity type as the sequence g^n, $n \in \mathbb{N}^*$, in the minimal ring $\mathbf{A}_b = \mathbb{Z}/\mathbb{Z}b$. The sequence of digits is purely periodic if and only if $\text{ggT}(b, g) = 1$ is. In this case, the period length is the order of g in $\mathbf{A}_b^\times$ and in particular a divisor of $\varphi(b) = |\mathbf{A}_b^\times|$ or even of $\varepsilon(b) = \text{Exp}\, \mathbf{A}_b^\times$, cf. example 2.7.10. It has the maximum possible value $\varphi(b)$, if g is a primitive residue modulo b, which is only possible if b is of the form $2, 4, p^\alpha, 2p^\alpha$ is with $p \in \mathbb{P}$, $p \geq 3$, and $\alpha \geq 1$, see the remarks following theorem 2.7.14. All fractions $a/175 = a/5^2 \cdot 7$, $0 < a < 175$, $\text{ggT}(a, 175) = 1$, have a pre-period length of 2 and a period length of 6 in the decimal system $(= \varphi(7))$. What are the values for this in the hexadecimal system ($g = 16$)? Which fractions $1/b$ and how many fractions a/b with coprime natural numbers a, b, $0 < a < b$, in total are purely periodic in the decimal system with a minimum period length of 4? Also consider again exercise 3.3.22. ◊

Example 3.6.19 (Riemann Zeta Function) *The series $\sum_{n=1}^\infty 1/n^s$ converges for $s \in \mathbb{R}$, $s > 1$, and diverges for $s \in \mathbb{R}$, $s \leq 1$.* The harmonic series is a divergent minorant for all these series with $s \leq 1$, for $s > 1$ we have

$$\sum_{n=1}^{2^{k+1}-1} \frac{1}{n^s} = 1 + \left(\frac{1}{2^s} + \frac{1}{3^s}\right) + \cdots + \left(\frac{1}{(2k)^s} + \cdots + \frac{1}{(2^{k+1}-1)^s}\right)$$

$$\leq 1 + \left(\frac{1}{2^s} + \frac{1}{2^s}\right) + \cdots + \left(\frac{1}{(2^k)^s} + \cdots + \frac{1}{(2^k)^s}\right)$$

$$= 1 + 2 \cdot \frac{1}{2^s} + \cdots + 2^k \frac{1}{(2^k)^s} < \sum_{m=0}^\infty \left(\frac{1}{2^{s-1}}\right)^m = \frac{2^{s-1}}{2^{s-1} - 1}.$$

Consequently, the partial sums of $\sum 1/n^s$ for $s > 1$ are bounded, and the series is thus convergent. We set for $s > 1$

$$\zeta(s) := \sum_{n=1}^\infty \frac{1}{n^s}$$

and denote the function $s \mapsto \zeta(s)$ as **Riemann Zeta Function** (after B. Riemann (1826–1866)). The calculations used here with the powers n^s will be thoroughly justified in Sect. 3.10. The powers $1/n^s$ are even defined for all complex numbers

s, see example 2.2.16 (2). Because of $|1/n^s| = 1/n^{\Re s}$, the series $\sum 1/n^s$ converges for all $s \in \mathbb{C}$ with $\Re s > 1$ absolutely. For a further extension of the ζ-function, we refer to Vol. 2. ◊

The following lemma describes a frequently used summation trick.

Lemma 3.6.20 (Abelian Partial Summation) *Let $a_0, \ldots, a_n$ and $b_0, \ldots, b_n$ be elements of an arbitrary ring. If we then set $A_m := \sum_{k=0}^{m} a_k$ for $m = 0, \ldots, n$, then it is*

$$\sum_{k=0}^{n} a_k b_k = \sum_{k=0}^{n-1} A_k(b_k - b_{k+1}) + A_n b_n.$$

Proof With $A_{-1} := 0$ it follows that ·

$$\sum_{k=0}^{n} a_k b_k = \sum_{k=0}^{n} (A_k - A_{k-1}) b_k = \sum_{k=0}^{n} A_k b_k - \sum_{k=0}^{n} A_{k-1} b_k$$

$$= \sum_{k=0}^{n} A_k b_k - \sum_{k=0}^{n-1} A_k b_{k+1} = \sum_{k=0}^{n-1} A_k(b_k - b_{k+1}) + A_n b_n. \quad \square$$

Theorem 3.6.21 (Abel's Convergence Criterion) *Let $\sum a_k$ be a convergent series of complex numbers and (b_k) a monotonic and bounded sequence of real numbers. Then the series $\sum a_k b_k$ is also convergent.*

Proof With $A_m := \sum_{k=0}^{m} a_k$ is $\sum_{k=0}^{n} a_k b_k = \sum_{k=0}^{n-1} A_k(b_k - b_{k+1}) + A_n b_n$ according to Lemma 3.6.20. Since the sequence (A_n) of partial sums converges and $\sum(b_k - b_{k+1})$ due to the monotony of the sequence (b_k) absolutely converges, the series $\sum A_k(b_k - b_{k+1})$ also converges according to Theorem 3.6.13. Due to the convergence of the sequence $(A_n b_n)$, the sequence $\sum_{k=0}^{n} a_k b_k$, $n \in \mathbb{N}$, is then also convergent. $\square$

Theorem 3.6.22 (Dirichlet's Convergence Criterion) *Let (a_k) be a sequence of complex numbers. The sequence of partial sums $A_m = \sum_{k=0}^{m} a_k$, $m \in \mathbb{N}$, is bounded and the sequence (b_k) is a monotone null sequence of real numbers. Then the series $\sum a_k b_k$ is convergent.*

Proof According to Lemma 3.6.20, $\sum_{k=0}^{n} a_k b_k = \sum_{k=0}^{n-1} A_k(b_k - b_{k+1}) + A_n b_n$. The sequence $(A_n b_n)$ converges to 0, and the series $\sum A_k(b_k - b_{k+1})$ converges according to Theorem 3.6.13 (for the same reasons as in the proof of Theorem 3.6.21). This results in the assertion. $\square$

By the way, in integral calculus, the partial integration corresponds to the Abel's partial summation, which leads to analogous convergence criteria for integrals, see Vol. 2.

3.6 Series

Finally, we briefly touch on infinite products. The focus is essentially on the convergence in the multiplicative group $\mathbb{C}^\times$. Therefore, we write the terms of an infinite product in the form $1 + a_k$, $k \in \mathbb{N}$, with the deviations $a_k \in \mathbb{C}$ from the neutral element $1 \in \mathbb{C}^\times$.

Definition 3.6.23 Let (a_k) be a sequence of complex numbers. The **infinite product**

$$\prod_{k=0}^{\infty}(1 + a_k)$$

is called **convergent**, if there exists an $n_0 \in \mathbb{N}$ such that the sequence $\prod_{k=n_0}^{n}(1 + a_k)$, $n \geq n_0$, converges to a limit $\prod_{k=n_0}^{\infty}(1 + a_k)$ that is *different* from 0. In this case, the limit

$$\prod_{k=0}^{\infty}(1 + a_k) := \lim_{n \to \infty} \prod_{k=0}^{n}(1 + a_k) = \prod_{k=0}^{n_0-1}(1 + a_k) \prod_{k=n_0}^{\infty}(1 + a_k)$$

is also called the **infinite product** of the sequence $(1 + a_k)_{k \in \mathbb{N}}$.

The convergence and possibly the value $\prod_{k=0}^{\infty}(1 + a_k)$ of an infinite product obviously do not depend on the choice of n_0 with the property required in the definition. A convergent infinite product is (like a finite product) *exactly 0 when one of the factors is equal to 0*. In a convergent infinite product, only finitely many factors can be equal to 0. The sequence of partial products can converge (towards 0) without the infinite product being convergent in the sense of Definition 3.6.23 (for example in the case of $1 + a_k = 1/2$, i.e. $a_k = -1/2$, for all $k \in \mathbb{N}$). The Cauchy convergence criterion for infinite products is:

Theorem 3.6.24 (Cauchy's Convergence Criterion for Products) $\prod_{k=0}^{\infty}(1 + a_k)$, *converges if and only if for every $\varepsilon > 0$ there exists an $n_0 \in \mathbb{N}$ such that for all $n \geq m \geq n_0$ holds* $\left| \prod_{k=m}^{n}(1 + a_k) - 1 \right| \leq \varepsilon$.

Proof Let $x_n := \prod_{k=0}^{n}(1 + a_k)$, $n \in \mathbb{N}$. To prove the necessity of the given condition, we can immediately assume that $x = \lim x_n \neq 0$ is. Then the difference $\prod_{k=m}^{n}(1 + a_k) - 1 = (x_n - x_{m-1})/x_{m-1}$ is arbitrarily small for sufficiently large $n \geq m$.

Conversely, if the given condition is satisfied. Then, of course, the factors $1 + a_k$ are different from 0 for sufficiently large k. We can therefore immediately assume that all factors $1 + a_k \neq 0$ are. Then the assumption $|x_n - x_{m-1}| \leq \varepsilon |x_{m-1}|$ provides for $n \geq m \geq n_0$ and in particular because of $|x_n| \leq (\varepsilon + 1)|x_{m-1}|$ the boundedness of the sequence (x_m). Overall, with the Cauchy convergence criterion for sequences, the convergence of (x_n) is obtained. If $x = \lim x_n = 0$ were, then $|x_{m-1}| = |x - x_{m-1}| \leq \varepsilon |x_{m-1}|$ would result, which is a contradiction to $\varepsilon < 1$. ∎

This easily leads to the following convergence criterion for products:

Theorem 3.6.25 *If the series $\sum a_k$ converges absolutely, then the product $\prod(1 + a_k)$ converges.*

Proof The triangle inequality and the inequality from Exercise 3.1.14b) provide for all $n \geq m$ with $\sum_{k=m}^{n} |a_k| \leq \frac{1}{2}$ the estimates

$$\left|\prod_{k=m}^{n}(1+a_k) - 1\right| \leq \prod_{k=m}^{n}(1+|a_k|) - 1 \leq \frac{1}{1-\sum_{k=m}^{n}|a_k|} - 1 = \frac{\sum_{k=m}^{n}|a_k|}{1-\sum_{k=m}^{n}|a_k|}$$

$$\leq 2 \sum_{k=m}^{n} |a_k|,$$

from which the assertion follows with Theorem 3.6.24. □

We note that in general, the convergence of $\sum a_k$ does not imply the convergence of $\prod(1 + a_k)$ and vice versa, see Exercise 3.6.23. But if $\prod(1 + |a_k|)$ converges, so does $\prod(1 + a_k)$ due to $\infty > \prod(1 + |a_k|) \geq 1 + \sum |a_k|$ and Theorem 3.6.25.

The inequality $1 + a \leq \exp a$ for $a \in \mathbb{R}$ (cf. exercise 3.10.8h)) provides the often useful estimates $\prod_{k=1}^{n}(1 + a_k) \leq \exp\left(\sum_{k=1}^{n} a_k\right)$ for $a_k \in \mathbb{R}$, $a_k \geq -1$, as well as

$$a_k \geq -1$$

From $\left(1 + \frac{k}{n}\right)^n \leq e^k$ one obtains by the way

$$\binom{n}{k}\left(\frac{k}{n}\right)^k < e^k \quad \text{or} \quad \binom{n}{k} < \left(\frac{en}{k}\right)^k, \quad k, n \in \mathbb{N}^*, \ k \leq n.$$

The complex e-function $\mathbb{C} \to \mathbb{C}^\times$ with the logarithm functions as partial inverse functions provide a close connection between the additive group $\mathbb{C}$ and the multiplicative group $\mathbb{C}^\times$, cf. example 2.2.16 (2). This allows many problems about infinite products to be generally reduced to those about series. We will revisit this in Vol. 2.

Exercises

Exercise 3.6.1 Examine the following series for convergence or divergence:

$$\sum_{n=2}^{\infty} \frac{1}{\sqrt[3]{n^2-1}}; \quad \sum_{n=1}^{\infty} \frac{1}{n\sqrt[n]{n}}; \quad \sum_{n=1}^{\infty} \frac{n-1}{n(n+1)}; \quad \sum_{n=1}^{\infty} \sqrt[n+1]{a}\left(\sqrt[n(n+1)]{a}-1\right), \ a > 0;$$

$$\sum_{n=1}^{\infty} \frac{(2n)!}{2^n (n!)^2}; \quad \sum_{n=0}^{\infty} (-1)^n \frac{\sqrt{n}}{n+1}; \quad \sum_{n=1}^{\infty} \frac{n^n}{n!3^n}; \quad \sum_{n=0}^{\infty} (-1)^n \frac{(n+1)^{n-1}}{n^n}; \quad \sum_{n=0}^{\infty} \frac{(n!)^2}{(2n)!} 3^n;$$

3.6 Series

$$\sum_{n=1}^{\infty}\left(\frac{n}{n+1}\right)^{n^2}; \quad \sum_{n=1}^{\infty}(-1)^n\left(\sqrt{n+1}-\sqrt{n}\right); \quad \sum_{n=1}^{\infty}\frac{\sqrt{n}+(-1)^n}{n}; \quad \sum_{n=0}^{\infty}\frac{2^n+n}{3^n}.$$

Exercise 3.6.2 For which $z \in \mathbb{C}$ do the following series converge:

$$\sum_{n=1}^{\infty}\frac{z^n}{n^2}; \quad \sum_{n=0}^{\infty}n!z^n; \quad \sum_{n=1}^{\infty}\frac{n!}{n^n}z^n; \quad \sum_{n=0}^{\infty}\frac{z^n}{1+|z|^n}; \quad \sum_{n=0}^{\infty}\frac{z^n}{1+z^{2n}}; \quad \sum_{n=0}^{\infty}\binom{n}{k}\frac{z^n}{n!}, \ k \in \mathbb{N}.$$

Exercise 3.6.3 Calculate the sums of the following telescoping series:

$$\sum_{n=1}^{\infty}\frac{1}{4n^2-1}; \quad \sum_{n=0}^{\infty}\frac{1}{9n^2+15n+4}; \quad \sum_{n=0}^{\infty}\frac{1}{4n^2+8n+3}; \quad \sum_{n=1}^{\infty}\frac{2n+1}{n^2(n+1)^2};$$

$$\sum_{n=1}^{\infty}\frac{1}{n^2+kn}, \ k \in \mathbb{N}^*; \quad \sum_{n=0}^{\infty}\frac{n}{(n+1)!}; \quad \sum_{n=1}^{\infty}\frac{1}{n(n+1)(n+2)};$$

$$\sum_{n=1}^{\infty}\frac{n}{(n+1)(n+2)(n+3)}; \quad \sum_{n=1}^{\infty}\frac{1}{n(n+1)(n+2)(n+3)};$$

$$\sum_{n=1}^{\infty}\frac{4n+1}{(2n-1)2n(2n+1)(2n+2)}; \quad \sum_{n=1}^{\infty}\frac{1}{n(n+1)(n+2)\cdots(n+k)} = \frac{1}{k \cdot k!}, \ k \in \mathbb{N}^*.$$

For sums $\sum_{n=1}^{\infty}P(n)/n(n+1)(n+2)\cdots(n+k)$ with an arbitrary polynomial function P of degree $<k$, $k \in \mathbb{N}^*$, one writes

$$P(n) = a_k + a_{k-1}(n+k) + \cdots + a_1(n+2)\cdots(n+k), \quad a_1,\ldots,a_k \in \mathbb{C},$$

and obtains with the last equation

$$\sum_{n=1}^{\infty}\frac{P(n)}{n(n+1)(n+2)\cdots(n+k)} = \sum_{m=1}^{k}\frac{a_m}{m \cdot m!}.$$

As an example, determine

$$\sum_{n=0}^{\infty}\frac{3n^2-7n+1}{(n+2)(n+5)(n+6)(n+8)}.$$

Exercise 3.6.4 Prove the so-called **Root Criterion**: Let (a_n) be a sequence of complex numbers. If there is a real number q with $0 < q < 1$ and $\sqrt[n]{|a_n|} \leq q$ for almost all n, then the series $\sum a_n$ is absolutely convergent. (If the quotient criterion provides the convergence of a series, so does the root criterion, but generally not the other way around. Proof or counterexample!)

Exercise 3.6.5 Prove the so-called **Condensation criterion** by Cauchy: If (a_n) is a monotonically decreasing null sequence of real numbers, then $\sum a_n$ converges if and only if $\sum 2^n a_{2^n}$ converges. Investigate for which $s \in \mathbb{R}$ the series $\sum_{n=2}^{\infty} (\ln n)^s/n$ converges.

Exercise 3.6.6 Let (a_n) be a monotonically decreasing null sequence of real numbers. If the series $\sum a_n$ converges, then the sequence (na_n) is a null sequence.

Exercise 3.6.7 Let (a_n) be a sequence of complex numbers and $q \in \mathbb{R}$, $0 < q < 1$. Let $a_n \neq 0$ and $|a_{n+1}/a_n| \leq q$ hold for all $n \geq n_0$. If x is the sum of the series convergent according to the quotient criterion 3.6.16 $\sum a_n$, then the error estimation $\left| x - \sum_{k=0}^{n} a_k \right| \leq |a_{n+1}|/(1-q)$ applies for all $n \geq n_0 - 1$.

Exercise 3.6.8 Let (a_n) be a sequence of positive real numbers.

a) The series $\sum a_n$ converges if and only if $\sum a_n/(1+a_n)$ converges.
b) The series $\sum a_n/(1+n^2 a_n)$ is convergent.
c) If (a_n) is monotonically decreasing and $\sum a_n/(1+na_n)$ converges, then $\sum a_n$ also converges.
d) If (a_n) is monotonically increasing and bounded, then $\sum ((a_{n+1}/a_n) - 1)$ converges.
e) If $\sum a_n$ converges, then $\sum a_n/\sqrt[n]{a_n}$ also converges.

Exercise 3.6.9 Let (a_n) and (b_n) be sequences of complex numbers, where $b_n \neq 0$ holds for almost all n. If the sequence $(|a_n/b_n|)$ converges to a positive real number, then $\sum a_n$ is absolutely convergent if and only if $\sum b_n$ is absolutely convergent. Show with an example that the statement does not generally hold for simple convergence.

Note Let $a_n \neq 0$ and $b_n \neq 0$ for all n and moreover $\lim a_n/b_n = c \neq 0$. The series $\sum a_n$ converges absolutely. Then

$$\sum_n a_n = \sum_n (a_n - cb_n) + c \sum_n b_n = \sum_n \left(1 - cb_n/a_n\right) a_n + c \sum_n b_n.$$

If the sum $\sum b_n$ is known, the calculation of the sum $\sum a_n$ is reduced to the calculation of the sum $\sum_n (1 - cb_n/a_n)a_n$. But since $(1 - cb_n/a_n)$ is a null sequence, the last series generally converges faster than the original series. Carry out this **acceleration of convergence** with the series $\sum 1/n^2$ and the comparison series $\sum_{n=1}^{\infty} 1/n(n+1) = 1$ and then use $\sum_{n=1}^{\infty} 1/n(n+1)(n+2) = 1/4$, cf. Example 3.6.3.

Exercise 3.6.10 Let (a_n) and (b_n) be sequences of complex numbers.

3.6 Series

a) If the series $\sum |a_n|^2$ and $\sum |b_n|^2$ converge, then the series $\sum a_n b_n$ converges absolutely and the Cauchy-Schwarz inequality

$$\left|\sum_{n=0}^{\infty} a_n b_n\right| \leq \sum_{n=0}^{\infty} |a_n||b_n| \leq \left(\sum_{n=0}^{\infty} |a_n|^2\right)^{1/2} \left(\sum_{n=0}^{\infty} |b_n|^2\right)^{1/2}.$$

holds (To prove absolute convergence, $|a_n b_n| \leq \frac{1}{2}(|a_n|^2 + |b_n|^2)$ is sufficient.)

b) If the series $\sum_{n=1}^{\infty} |a_n|^2$ converges, then the series $\sum_{n=1}^{\infty} a_n/n$ converges absolutely.

c) For $c_n \in \mathbb{R}_+^\times$, $n \in \mathbb{N}^*$, one of the series $\sum_{n=1}^{\infty} c_n$ and $\sum_{n=1}^{\infty} 1/(n^2 c_n)$ is divergent.

Exercise 3.6.11 Let $s \in \mathbb{R}$, $s > 1$. Then the following applies:

a) $\sum_{n=1}^{\infty} (-1)^{n-1}/n^s = (1 - 2^{1-s})\zeta(s)$. (The left side even converges for all $s > 0$, and one obtains a definition of the ζ-function on $]0, 1[$, cf. Example 3.6.19.)
b) $\sum_{n=0}^{\infty} 1/(2n+1)^s = (1 - 2^{-s})\zeta(s)$.
c) Let $s = a + bi$ be a complex number with $a = \Re s \leq 1$. Then the series $\sum_{n=1}^{\infty} n^{-a} \cos(b \ln n)$ diverges. (For $0 < c < d$ there are at least $\left[e^d(1 - e^{c-d})\right]$ numbers $n \in \mathbb{N}^*$ with $c \leq \ln n \leq d$.) The divergence of the ζ-series follows

$$\sum_{n=1}^{\infty} \frac{1}{n^s} = \sum_{n=1}^{\infty} \frac{\cos(b \ln n) - i \sin(b \ln n)}{n^a}.$$

Exercise 3.6.12 The series $\sum_{n=0}^{\infty} z^n/n^n$ converges for all $z \in \mathbb{C}$. Calculate the sum for $z = i$ and $z = 1 + 2i$ up to an error, which is $\leq 10^{-4}$ in absolute value.

Exercise 3.6.13 Let a_n, $n \in \mathbb{N}$, be a sequence of complex numbers. Furthermore, let $0 = n_0 < n_1 < \cdots$ be a strictly monotonically increasing sequence in $\mathbb{N}$. For $b_k := \sum_{n=n_k}^{n_{k+1}-1} a_n$ and $c_k := \sum_{n=n_k}^{n_{k+1}-1} |a_n|$, $k \in \mathbb{N}$, the following applies:

a) If $\sum_{n=0}^{\infty} a_n$ is convergent, then so is $\sum_{k=0}^{\infty} b_k$ and both sums are equal.
b) If $\sum_{k=0}^{\infty} b_k$ is convergent and c_k, $k \in \mathbb{N}$, is a null sequence, then $\sum_{n=0}^{\infty} a_n$ also converges.

Exercise 3.6.14 Let $p, q \in \mathbb{N}^*$. The sequence a_n, $n \in \mathbb{N}^*$, arises from the harmonic sequence $1/n$, $n \in \mathbb{N}^*$, by multiplying each p consecutive terms with 1 and the next q terms with -1. $\sum_{n=1}^{\infty} a_n$ converges exactly when $p = q$. (Calculate the sum for $p = q = 2$ using example 3.6.9; for $p = q \geq 3$ see Vol. 2, exercise 3.2.24.)

Exercise 3.6.15 Let h_n, $n \in \mathbb{N}^*$, be a sequence of natural numbers ≥ 2.

a) It is

$$\sum_{n=1}^{\infty} \frac{h_n - 1}{h_1 \cdots h_n} = 1.$$

b) Let c_n, $n \in \mathbb{N}^*$, be a sequence in $\mathbb{N}$ with $c_n < h_n$ for all n and $c_n \neq h_n - 1$ for infinitely many n. Then

$$\sum_{n=1}^{\infty} \frac{c_n}{h_1 \cdots h_n}$$

is convergent, and the sum is a real number x with $0 \leq x < 1$.

c) For every real number x with $0 \leq x < 1$, there is a uniquely determined sequence (c_n) as in b), for which the given series converges to x. Almost all $c_n = 0$ are exactly when x is rational and x has a (not necessarily reduced) representation a/b with $a, b \in \mathbb{N}$, where b is of the form $b = h_1 \cdots h_n$. (If $h_n := g$ for all n with a $g \in \mathbb{N}$, $g \geq 2$, then we get the usual g-adic expansion, see example 3.3.10. The development given here corresponds to the representation of natural numbers according to the end of example 1.7.6. Another interesting example is the choice $h_n := n + 1$ for all $n \in \mathbb{N}^*$. In this case, x is rational exactly when almost all $c_n = 0$ are. Therefore, for example, $\sum_{n=1}^{\infty} 1/(n+1)!$ is irrational. As we will see in volume 2, this number is equal to $e - 2$.)

Exercise 3.6.16 Prove the following **Dubois-Reymond convergence criterion**: Let (a_k) and (b_k) be sequences of complex numbers. The series $\sum a_k$ converges and the series $\sum (b_k - b_{k+1})$ converges absolutely. Then $\sum a_k b_k$ also converges. (Use Abel's partial summation 3.6.20.)

Exercise 3.6.17 The series $\sum_{n=1}^{\infty} a_n$ of complex numbers, i.e., the sequence of their partial sums, is bounded. Then $\sum_{n=1}^{\infty} a_n/n$ is convergent.—If $\sum_{n=1}^{\infty} a_n/n$ converges, then $\left(\sum_{k=1}^{n} a_k\right)/n$, $n \in \mathbb{N}^*$, is a null sequence. (With Abel's partial summation 3.6.20 one obtains $\sum_{k=1}^{n} a_k = \sum_{k=1}^{n} (a_k/k) \cdot k = -\sum_{k=1}^{n-1} A_k + A_n n$, $A_k := \sum_{m=1}^{k} a_m/m$. Now use exercise 3.3.9a).—Generally, with exercise 3.3.11: If $(b_n)_{n \in \mathbb{N}^*}$ is a strictly monotonically increasing sequence in $\mathbb{R}_+^\times$ with $\lim b_n = \infty$ and $\sum_{n=1}^{\infty} a_n/b_n$ converges, then $\lim_{n \to \infty} \left(\sum_{k=1}^{n} a_k\right)/b_n = 0$.

Exercise 3.6.18 The series $\sum_{k=1}^{\infty} z^k/k$ or $\sum_{k=0}^{\infty} z^{2k+1}/(2k+1)$ converge exactly for all $z \in \mathbb{C}$ with $|z| \leq 1$, $z \neq 1$, or with $|z| \leq 1$, $z \neq \pm 1$. (For $|z| = 1$ use theorem 3.6.22.)

Exercise 3.6.19 Let $\varphi \in \mathbb{R}$. Prove the convergence of the following series:

$$\sum_{k=1}^{\infty} \frac{\sin k\varphi}{k}; \quad \sum_{k=1}^{\infty} \frac{\cos k\varphi}{k}, \quad \varphi \neq 2m\pi, \, m \in \mathbb{Z};$$

$$\sum_{k=0}^{\infty} \frac{\sin(2k+1)\varphi}{2k+1}; \quad \sum_{k=0}^{\infty} \frac{\cos(2k+1)\varphi}{2k+1}, \quad \varphi \neq m\pi, \, m \in \mathbb{Z}.$$

(For the sums of these series, we refer to Vol. 2, Sect. 3.7.)

3.6 Series

Exercise 3.6.20 Determine all $z \in \mathbb{C}$, for which the following series converge:

$$\sum_{n=1}^{\infty} \frac{1}{2n+1}\left(\frac{z-1}{z+1}\right)^{2n+1} \; ; \; \sum_{n=0}^{\infty} \frac{1}{n}\left(\frac{z-1}{z+1}\right)^{n}.$$

Exercise 3.6.21 If the a_k, $k \in \mathbb{N}$, are all real and all non-negative or all non-positive and the product $\prod(1 + a_k)$ converges, then the series $\sum a_k$ also converges.

Exercise 3.6.22
a) Examine the following infinite products for convergence:

$$\prod_{n=1}^{\infty}\left(1+\frac{z}{n}\right), z \in \mathbb{C}; \quad \prod_{n=1}^{\infty}\left(1-\frac{z}{n^3}\right), z \in \mathbb{C}; \quad \prod_{n=1}^{\infty}\sqrt[n]{a}, a \in \mathbb{R}_{+}^{\times}.$$

(In the first product, for $z = 1$, the n-th partial product is equal to $n + 1$. With theorem 3.6.25, this provides a new proof for the divergence of the harmonic series.)

b) Calculate the following telescope products:

$$\prod_{n=2}^{\infty}\left(1-\frac{1}{n^2}\right); \quad \prod_{n=2}^{\infty}\left(1-\frac{2}{n(n+1)}\right); \quad \prod_{n=1}^{\infty}\left(1+\frac{1}{n(n+2)}\right); \quad \prod_{n=2}^{\infty}\frac{n^3-1}{n^3+1}.$$

(See exercise 1.5.4. — By the way, $\prod_{n=2}^{\infty}\left(1+\frac{1}{n^2}\right) = (\sinh \pi)/2\pi$, see Vol. 2, Example 1.1.10.)

c) For $q \in \mathbb{C}, |q| < 1$, calculate $\prod_{n=0}^{\infty}(1 + q^{2^n})$. (Use exercise 1.5.5a); however, the product can also be interpreted directly as a geometric series by "multiplying out".)

Exercise 3.6.23 Let (ε_k) be a null sequence of real numbers with $\varepsilon_k \neq -1$ for all $k \in \mathbb{N}$ and $\sum \varepsilon_k^2 = \infty$ (like $\varepsilon_k := 1/\sqrt{k+1}$).

a) For $a_k := (-1)^k \varepsilon_{[k/2]}$, $k \in \mathbb{N}$, is $\sum_{k=0}^{\infty} a_k = 0$, but $\prod_{k=0}^{\infty}(1 + a_k)$ diverges.
b) For (a_k) with $a_{2m} := \varepsilon_m$, $a_{2m+1} := -\varepsilon_m/(1 + \varepsilon_m)$ is $\prod_{k=0}^{\infty}(1 + a_k) = 1$, but $\sum_{k=0}^{\infty} a_k$ diverges.

Note The prerequisite about the divergence of the series $\sum \varepsilon_k^2$ in the above examples is not random. It is indeed the case: If (a_k) is a sequence of complex numbers, for which $\sum |a_k|^2$ converges, then the product $\prod(1 + a_k)$ converges exactly when the series $\sum a_k$ converges, see again Vol. 2, exercise 2.2.28.

Exercise 3.6.24 Another illustration of the divergence of the harmonic series, see example 3.6.3, is the following problem: A snail moves during the day from one end of an arbitrarily stretchable rubber band of length $\ell > 0$ to the other, covering a unit of length per day, after which the band is stretched by ℓ_0 each following night. One should examine whether and possibly on which day the snail reaches the other end of the band. (As long as the snail is moving, its distance from the starting point on the evening of the n-th day is equal to nH_n. Consider in particular the case $\ell = \ell_0$.)

Exercise 3.6.25 Let (q_n) be any sequence of real numbers with $0 < q_n < 1$ for all n.

a) The (telescope) series $\sum_{n=0}^{\infty} q_0 \cdots q_n (1 - q_{n+1})$ converges.
b) Prove the following generalization of the quotient criterion: A series $\sum a_n$ of complex numbers is absolutely convergent if for (almost) all n: $a_n \neq 0$ and

$$\left| \frac{a_{n+1}}{a_n} \right| \leq q_n \frac{1 - q_{n+1}}{1 - q_n}.$$

c) For $a, c \in \mathbb{R}_+^{\times}$ with $c > a + 1$ the series

$$\sum_{n=0}^{\infty} \frac{(a)_{n+1}}{(c)_{n+1}} = \sum_{n=0}^{\infty} \frac{a(a+1) \cdots (a+n)}{c(c+1) \cdots (c+n)}.$$

converges.

3.7 Summability

Contrary to the situation with sequences, the convergence and possibly the sum of a series $\sum a_k$ generally depend significantly on the order of the terms a_k. If $\sigma \in \mathfrak{S}(\mathbb{N})$ is a permutation (with infinite range), then the partial sums of the series $\sum_k a_{\sigma k}$ are usually significantly different from those of the original series $\sum_k a_k$.

Example 3.7.1 Let $x \, (= \ln 2)$ be the non-zero sum of the alternating harmonic series $1 - \frac{1}{2} + \frac{1}{3} - \frac{1}{4} + - \cdots$. Then

$$x = \left(1 - \frac{1}{2} + \frac{1}{3} - \frac{1}{4}\right) + \left(\frac{1}{5} - \frac{1}{6} + \frac{1}{7} - \frac{1}{8}\right) + \cdots,$$

$$\frac{1}{2}x = \left(\frac{1}{2} - \frac{1}{4}\right) + \left(\frac{1}{6} - \frac{1}{8}\right) + \cdots,$$

and by addition one obtains

$$\frac{3}{2}x = \left(1 + \frac{1}{3} - \frac{1}{2}\right) + \left(\frac{1}{5} + \frac{1}{7} - \frac{1}{4}\right) + \cdots = 1 + \frac{1}{3} - \frac{1}{2} + \frac{1}{5} + \frac{1}{7} - \frac{1}{4} + - \cdots.$$

(One can obviously omit the brackets here, cf. Example 3.6.13). The last series, however, arises from the original series solely by rearranging the terms. By rearranging in a suitable way, it is even possible to make the new series converge to any given real number or even diverge, see Examples 3.7.13 and 3.7.14. ◊

To avoid the difficulties indicated by the last example, we introduce the concept of summability following N. Bourbaki (* 1935), where it is clear from the definition

3.7 Summability

that the summability and possibly the sum are independent of the order of the summands.

Let a_i, $i \in I$, be a family of complex numbers. As with any additive abelian groups, we set for a *finite* subset H of I

$$a_H := \sum_{i \in H} a_i$$

and call a_H the **partial sum** to the index set H. With $\mathfrak{E}(I)$ we denote as before the set of all finite subsets of I.

Definition 3.7.2 The family $a_i, i \in I$ of complex numbers is called **summable** if there is a complex number z with the following property: For every (no matter how small) $\varepsilon \in \mathbb{R}_+^\times$ there is an $H_0 \in \mathfrak{E}(I)$ with $|a_H - z| \leq \varepsilon$ for all $H \in \mathfrak{E}(I)$ with $H \supseteq H_0$.

The element z in Definition 3.7.2 is uniquely determined: If z' also has the analogous property as z and is $z \neq z'$, then there are finite subsets H_0 and H_0' of I with $|a_H - z| \leq \varepsilon$ and $|a_H - z'| \leq \varepsilon$ for $H := H_0 \cup H_0' \in \mathfrak{E}(I)$. This leads to the contradiction $|z - z'| \leq |z - a_H| + |a_H - z'| \leq \varepsilon + \varepsilon = \frac{2}{3}|z - z'|$.

If the family a_i, $i \in I$, is summable, then the uniquely determined number z according to 3.7.2 is called the **sum** of the a_i, $i \in I$, and is denoted by

$$\sum_{i \in I} a_i, \quad \text{short also with} \quad \sum_i a_i \quad \text{or} \quad \sum a_i,$$

In every (no matter how small) neighborhood of the sum $\sum_{i \in I} a_i$, all finite partial sums a_H are located, provided $H \in \mathfrak{E}(I)$ is large enough. As announced, summability and sum do not depend on the indexing. More precisely: *If $\sigma : J \to I$ is a bijective mapping, then the family a_i, $i \in I$, is exactly summable when the family $a_{\sigma(j)}$, $j \in J$, is summable, and then is $\sum_{i \in I} a_i = \sum_{j \in J} a_{\sigma j}$.* For finite I, the sum defined here is of course the ordinary sum of the a_i.

If a_n, $n \in \mathbb{N}$, is a summable sequence of complex numbers with sum a, then the series $\sum_{n=0}^\infty a_n$ is convergent and it holds $a = \sum_{n=0}^\infty a_n = \sum_{n \in \mathbb{N}} a_n$. Let $\varepsilon > 0$. Then there exists a $H_0 \in \mathfrak{E}(\mathbb{N})$ with $|a_H - a| \leq \varepsilon$ for all $H \in \mathfrak{E}(\mathbb{N})$ with $H \supseteq H_0$. If now $n_0 := \text{Max } H_0 \in \mathbb{N}$, then $\left|\sum_{k=0}^n a_k - a\right| \leq \varepsilon$ for all $n \geq n_0$. Conversely, the convergence of the series $\sum_{n=0}^\infty a_n$ does not imply the summability of a_n, $n \in \mathbb{N}$, as example 3.7.1 shows.

Similar to Proposition 3.6.6, the following **calculation rules for summable families** are proven: *If a_i, $i \in I$, and b_i, $i \in I$, summable families, then also $a_i + b_i$, $i \in I$, and λa_i, $i \in I$, for $\lambda \in \mathbb{C}$ summable and it holds*

$$\sum_{i \in I}(a_i + b_i) = \sum_{i \in I} a_i + \sum_{i \in I} b_i, \quad \sum_{i \in I} \lambda a_i = \lambda \sum_{i \in I} a_i.$$

If furthermore a_j, $j \in J$, is another summable family, whose index set J is disjoint to I, then a_k, $k \in I \uplus J$, is summable with

$$\sum_{k \in I \uplus J} a_k = \sum_{i \in I} a_i + \sum_{j \in J} a_j.$$

The family a_i, $i \in I$, of complex numbers is summable, if the families $\Re a_i$, $i \in I$, and $\Im a_i$, $i \in I$, of real numbers are summable. In this case, it is

$$\sum_{i \in I} a_i = \left(\sum_{i \in I} \Re a_i\right) + i\left(\sum_{i \in I} \Im a_i\right).$$

Let a_i, $i \in I$, be summable with sum z. For each $\varepsilon > 0$, then $|a_i| > \varepsilon$ is for only finitely many $i \in I$. There is indeed an $H_0 \in \mathfrak{E}(I)$ such that $|a_H - z| \leq \varepsilon/2$ is for all $H \in \mathfrak{E}(I)$ with $H \supseteq H_0$. For $i \notin H_0$, then

$$|a_i| = |a_{H_0 \uplus \{i\}} - a_{H_0}| \leq |a_{H_0 \uplus \{i\}} - z| + |z - a_{H_0}| \leq \varepsilon.$$

It follows, *that the set of indices $i \in I$ with $a_i \neq 0$ is countable*. This set is indeed the union of the countably many finite sets $I_n := \{i \in I \mid |a_i| > 1/n\}$, $n \in \mathbb{N}^*$. — The analogue to the Cauchy sequences are the Cauchy-summable families.

Definition 3.7.3 A family a_i, $i \in I$, of complex numbers is called a **Cauchy summable family**, if for every $\varepsilon > 0$ there is an $H_0 \in \mathfrak{E}(I)$ with $|a_E| \leq \varepsilon$ for all $E \in \mathfrak{E}(I)$ with $E \cap H_0 = \emptyset$.

Therefore, the Cauchy summability criterion is as follows:

Theorem 3.7.4 (Cauchy Summability Criterion) *A family of complex numbers is summable if and only if it is Cauchy summable.*

Proof Let initially a_i, $i \in I$, be summable with sum z and let $\varepsilon > 0$ be given. Then there is an $H_0 \in \mathfrak{E}(I)$ with $|a_H - z| \leq \frac{1}{2}\varepsilon$ for all $H \in \mathfrak{E}(I)$ with $H \supseteq H_0$. If then $E \cap H_0 = \emptyset$ for a $E \in \mathfrak{E}(I)$, it follows with $H := E \uplus H_0$

$$|a_E| = |a_H - a_{H_0}| \leq |a_H - z| + |z - a_{H_0}| \leq \frac{1}{2}\varepsilon + \frac{1}{2}\varepsilon = \varepsilon.$$

It is now to be shown in reverse that a Cauchy summable family a_i, $i \in I$, is summable. Let (ε_n) be a monotonically decreasing null sequence of positive real numbers. By assumption, there exists a sequence (H_n) of finite subsets H_n of I with $|a_E| \leq \varepsilon_n$, if $E \in \mathfrak{E}(I)$, $E \cap H_n = \emptyset$. By replacing H_n with $H_0 \cup \cdots \cup H_n$, $n \in \mathbb{N}$, we can assume that $H_0 \subseteq H_1 \subseteq H_2 \subseteq \cdots$ is. We show that the sequence a_{H_n}, $n \in \mathbb{N}$, is a Cauchy sequence and therefore converges to a number z. Let $\varepsilon > 0$ be given and $\varepsilon_{n_0} \leq \varepsilon$. For $n \geq m \geq n_0$ then $|a_{H_n} - a_{H_m}| = |a_{H_n - H_m}| \leq \varepsilon_{n_0} \leq \varepsilon$ applies because $(H_n - H_m) \cap H_{n_0} = \emptyset$. Finally, we show that a_i, $i \in I$, is summable with sum z. Let $\varepsilon > 0$ and n be chosen such that $\varepsilon_n \leq \frac{1}{2}\varepsilon$ and $|a_{H_n} - z| \leq \frac{1}{2}\varepsilon$ hold. Because of $a_H - a_{H_n} = a_{H-H_n}$ and $(H - H_n) \cap H_n = \emptyset$ then for all $H \supseteq H_n$:

$$|a_H - z| \leq |a_H - a_{H_n}| + |a_{H_n} - z| \leq \varepsilon_n + \frac{1}{2}\varepsilon \leq \varepsilon. \quad \square$$

Since a subfamily of a Cauchy summable family is trivially again a Cauchy summable family, it follows from theorem 3.7.4:

3.7 Summability

Corollary 3.7.5 *Every subfamily of a summable family of complex numbers is summable.*

Because of $|a_E| \leq \sum_{i \in E} |a_i|$ for $E \in \mathfrak{E}(I)$ with $|a_i|$, $i \in I$, also a_i, $i \in I$, itself a Cauchy summable family, and we obtain:

Corollary 3.7.6 *Let a_i, $i \in I$, be a family of complex numbers. If the family $|a_i|$, $i \in I$, is summable, then so is a_i, $i \in I$.*

We call a family a_i, $i \in I$, **absolutely** or **normally summable**, if $|a_i|$, $i \in I$, is summable. Corollary 3.7.6 states that *every absolutely summable family is also summable*. We will soon see that here (unlike in Corollary 3.6.11) the converse also holds. First, we obtain:

Theorem 3.7.7 (Majorant Criterion) *Let b_i, $i \in I$, be a summable family of non-negative real numbers and a_i, $i \in I$, a family of complex numbers. If $|a_i| \leq b_i$ holds for all $i \in I$, then a_i, $i \in I$, is absolutely summable and $|\sum a_i| \leq \sum |a_i| \leq \sum b_i$.*

The statement 3.6.7 about series with non-negative terms corresponds to:

Theorem 3.7.8 *A family a_i, $i \in I$, of non-negative real numbers is summable if and only if the family a_H, $H \in \mathfrak{E}(I)$, of finite partial sums is bounded. In this case, $\sum_{i \in I} a_i = \mathrm{Sup}\{a_H \mid H \in \mathfrak{E}(I)\}$.*

Proof Of course, in general, the family a_H, $H \in \mathfrak{E}(I)$, is bounded if the family a_i, $i \in I$, is summable.

Let now $a_i \geq 0$ for all $i \in I$ and $S := \mathrm{Sup}\{a_H \mid H \in \mathfrak{E}(I)\} \in \mathbb{R}$. For $\varepsilon > 0$ there is then an $H_0 \in \mathfrak{E}(I)$ with $a_{H_0} \geq S - \varepsilon$. For all $H \in \mathfrak{E}(I)$ with $H \supseteq H_0$ applies $S - \varepsilon \leq a_{H_0} \leq a_H \leq S$, thus $|a_H - S| \leq \varepsilon$. □

We can now easily prove the reversal to Corollary 3.7.6.

Theorem 3.7.9 *Every summable family of complex numbers is absolutely summable.*

Proof Let a_i, $i \in I$, initially be a summable family of *real* numbers. If we set $I_+ := \{i \in I \mid a_i \geq 0\}$ and $I_- := \{i \in I \mid a_i < 0\}$, then the subfamilies a_i, $i \in I_+$, and a_i, $i \in I_-$, are summable. Then also $|a_i| = -a_i$, $i \in I_-$, is summable and overall $|a_i|$, $i \in I_+ \uplus I_- = I$. Now let a_i, $i \in I$, be any summable family of complex numbers. Then the families $\Re a_i$, $i \in I$, and $\Im a_i$, $i \in I$, are summable and consequently also $|\Re a_i|$, $i \in I$, and $|\Im a_i|$, $i \in I$. Because of $|a_i| \leq |\Re a_i| + |\Im a_i|$ finally the family $|a_i|$, $i \in I$, is summable. □

For families of *real* numbers a_i, $i \in I$, the concept of improper summability is naturally explained. First, let $a_i \geq 0$ for all $i \in I$. Then we define

$$\sum_{i \in I} a_i := \infty,$$

if the family a_H, $H \in \mathfrak{E}(I)$, is not bounded. Similarly, let $\sum a_i = -\infty$, if all $a_i \leq 0$ are and $\{a_H \mid H \in \mathfrak{E}(I)\}$ is not bounded. Now, if a_i, $i \in I$, is any family of real numbers, we set (as in the proof of Theorem 3.7.9)

$$I_+ := \{i \in I \mid a_i \geq 0\}, \quad I_- := \{i \in I \mid a_i < 0\}$$

and say that a_i, $i \in I$, is **improperly summable** if one of the sums

$$a_+ := \sum_{i \in I_+} a_i \quad \text{resp.} \quad a_- := \sum_{i \in I_-} a_i$$

is in $\mathbb{R}$, and then define

$$\sum_{i \in I} a_i := a_+ + a_-.$$

In an obvious way, one extends the improper summability to families a_i, $i \in I$, with $a_i \in \overline{\mathbb{R}} = \mathbb{R} \cup \{\infty, -\infty\}$.

For summability in the narrower sense, we summarize once again: *If a_i, $i \in I$, is a family of complex numbers, then the following statements are equivalent:*

(1) a_i, $i \in I$, is summable.
(2) a_i, $i \in I$, is a Cauchy-summable family.
(3) a_i, $i \in I$, is absolutely summable, i.e. $|a_i|$, $i \in I$, is summable.
(4) The family of partial sums $\sum_{i \in H} |a_i|$, $H \in \mathfrak{E}(I)$, is bounded.
(5) The families $\Re a_i$, $i \in I$, and $\Im a_i$, $i \in I$, are summable.

The equivalence of (1) and (3) provides the following important relationship between summable sequences and convergent series:

Theorem 3.7.10 (Rearrangement theorem for absolutely convergent series) *A sequence a_n, $n \in \mathbb{N}$, in $\mathbb{C}$ is summable exactly when the series $\sum_{n=0}^{\infty} a_n$ is absolutely convergent.*

The following more general statement is of fundamental importance for calculating with summable families:

Theorem 3.7.11 (Great rearrangement theorem) *Let a_i, $i \in I$, be a summable family of complex numbers. Furthermore, let $I = \biguplus_{j \in J} I_j$ be a decomposition of the index set I into (pairwise disjoint) subsets $I_j \subseteq I$, $j \in J$. Then each of the subfamilies a_i, $i \in I_j$, is summable and with $s_j := \sum_{i \in I_j} a_i$ it holds: The family s_j, $j \in J$, is also summable, and it is*

$$\sum_{i \in I} a_i = \sum_{j \in J} s_j = \sum_{j \in J} \left(\sum_{i \in I_j} a_i \right).$$

Proof Let $s := \sum_{i \in I} a_i$, and let $\varepsilon > 0$. We are looking for an $F_0 \in \mathfrak{E}(J)$ with $|s_F - s| \leq \varepsilon$ for all $F \in \mathfrak{E}(J)$ with $F \supseteq F_0$. By assumption, there is an $H_0 \in \mathfrak{E}(I)$

3.7 Summability

with $|a_H - s| \leq \frac{1}{2}\varepsilon$ for all $H \in \mathfrak{E}(I)$ with $H \supseteq H_0$. Each of the finitely many elements of H_0 lies in an I_j. Therefore, there is a finite subset $F_0 \subseteq J$ such that $H_0 \subseteq \bigcup_{j \in F_0} I_j$ is.

Let $F \supseteq F_0$ be with $|F| = n \in \mathbb{N}^*$. Since s_j, $j \in F$, the sum of the family a_i, $i \in I_j$, there is an $H'_j \in \mathfrak{E}(I_j)$ with $|a_{H'_j} - s_j| \leq \varepsilon/2n$ and $H'_j \supseteq H_0 \cap I_j$. Then

$$|s_F - s| = \left| \sum_{j \in F} s_j - s \right| \leq \sum_{j \in F} |s_j - a_{H'_j}| + \left| \sum_{j \in F} a_{H'_j} - s \right| \leq n\frac{1}{2n}\varepsilon + \frac{1}{2}\varepsilon = \varepsilon,$$

applies since $H := \bigcup_{j \in F} H'_j$ is the disjoint union of the H'_j, so $a_H = \sum_{j \in F} a_{H'_j}$ applies, and since H includes H_0 by construction, so it is $|a_H - s| \leq \frac{1}{2}\varepsilon$. □

The statement 3.7.11 is often also called the **large associative law**. To be able to apply it conveniently, the following remark is useful:

Lemma 3.7.12 *Let the situation be the same as in theorem 3.7.11 with the exception that the summability of the family a_i, $i \in I$, is not assumed. Then a_i, $i \in I$, is exactly summable if each of the families $|a_i|$, $i \in I_j$, and also the family $t_j := \sum_{i \in I_j} |a_i|$, $j \in J$, is summable.*

Proof For every finite subset $H \subseteq I$ it is obviously $\sum_{i \in H} |a_i| \leq \sum_{j \in J} t_j < \infty$. The claim therefore follows from theorem 3.7.8 and corollary 3.7.6. □

A simple application of the major rearrangement theorem is:

Theorem 3.7.13 (Great Distributive Law) *Let a_i, $i \in I$, and b_j, $j \in J$, be summable families of complex numbers. Then the family of products $a_i b_j$, $(i,j) \in I \times J$, is summable, and it holds*

$$\sum_{(i,j) \in I \times J} a_i b_j = \left(\sum_{i \in I} a_i \right)\left(\sum_{j \in J} b_j \right).$$

Proof Every finite subset H of $I \times J$ is contained in a finite subset $F \times G$ with $F \in \mathfrak{E}(I), G \in \mathfrak{E}(J)$. Then

$$\sum_{(i,j) \in H} |a_i b_j| \leq \sum_{i \in F, j \in G} |a_i||b_j| = \left(\sum_{i \in F} |a_i| \right)\left(\sum_{j \in G} |b_j| \right) \leq \left(\sum_{i \in I} |a_i| \right)\left(\sum_{j \in J} |b_j| \right) < \infty,$$

is true since the families a_i, $i \in I$, and b_j, $j \in J$, are absolutely summable according to theorem 3.7.9. The family $a_i b_j$, $(i,j) \in I \times J$, is therefore summable according to 3.7.8 and 3.7.6. From theorem 3.7.11 it now follows

$$\sum_{(i,j) \in I \times J} a_i b_j = \sum_{i \in I} \left(\sum_{j \in J} a_i b_j \right) = \sum_{i \in I} \left(a_i \sum_{j \in J} b_j \right) = \left(\sum_{i \in I} a_i \right)\left(\sum_{j \in J} b_j \right). \quad \square$$

Specifically for sequences, we obtain the following:

Theorem 3.7.14 (Cauchy Product of Absolutely Convergent Series) *Let a_n, $n \in \mathbb{N}$, and b_n, $n \in \mathbb{N}$, be summable sequences of complex numbers (i.e., the series $\sum a_n$ and $\sum b_n$ are absolutely convergent). Then the sequence c_n, $n \in \mathbb{N}$, with*

$$c_n := \sum_{i=0}^{n} a_i b_{n-i} = a_0 b_n + a_1 b_{n-1} + \cdots + a_{n-1} b_1 + a_n b_0$$

*summable (i.e., the series $\sum c_n$ – called the **Cauchy Product** of the series $\sum a_n$ and $\sum b_n$ – is absolutely convergent), and it holds*

$$\sum_{n=0}^{\infty} c_n = \left(\sum_{n=0}^{\infty} a_n\right)\left(\sum_{n=0}^{\infty} b_n\right).$$

Proof It holds $\sum_{m,n} a_m b_n = (\sum_m a_m)(\sum_n b_n)$ according to the great distributive law 3.7.13. From theorem 3.7.11 follows $\sum_{m,n} a_m b_n = \sum_j \left(\sum_{m+n=j} a_m b_n\right) = \sum_j c_j$. □

In the following discussion of **infinite products**, we recall the following notations: For a family a_i, $i \in I$, and a finite subset $H \subseteq I$ let

$$a^H := \prod_{i \in H} a_i \quad \text{and} \quad (1+a)^H := \prod_{i \in H}(1 + a_i).$$

In analogy to Definition 3.6.23 we define:

Definition 3.7.15 The family $1 + a_i$, $i \in I$, of complex numbers is called **multiplicable**, if there is a finite subset $I_0 \subseteq I$ and a $u \in \mathbb{C}^\times$ with the following property: For every $\varepsilon > 0$ there exists a $H_0 \in \mathfrak{E}(I - I_0)$ such that for all $H \in \mathfrak{E}(I - I_0)$ with $H \supseteq H_0$ the following holds: $|(1+a)^H - u| \leq \varepsilon$. Then $(1+a)^{I_0} u$ is called the **product** $\prod_{i \in I}(1 + a_i)$ of the family $1 + a_i$, $i \in I$.

The Cauchy criterion, which is proven quite analogously to theorem 3.7.4 (see also theorem 3.6.24), reads here:

Theorem 3.7.16 (Cauchy's Multiplicability Criterion) *The family is exactly then $1 + a_i$, $i \in I$, of complex numbers multiplicable, if for every $\varepsilon > 0$ there is a $H_0 \in \mathfrak{E}(I)$ with $|(1+a)^E - 1| \leq \varepsilon$ for all $E \in \mathfrak{E}(I)$ with $E \cap H_0 = \emptyset$ exists.*

Unlike with series $\sum a_k$ with associated products $\prod(1 + a_k)$, see problem 3.6.23, there is the following simple relationship between summability and multiplicability.

Theorem 3.7.17 *The family $1 + a_i$, $i \in I$, is exactly then multiplicable, if the family a_i, $i \in I$, is summable.*

Proof Firstly, let a_i, $i \in I$, be summable. Then $\sum |a_i| < \infty$ is true, and the convergence of $\prod(1 + a_i)$ follows from the Cauchy criterion, just like in theorem 3.6.25.

Now conversely, let $1 + a_i$, $i \in I$, be multipliable. To prove the summability of a_i, $i \in I$, we use, for simplicity, the principal value of the logarithm function on $\mathbb{C} - \mathbb{R}_-$ and the exponential function in the complex with

3.7 Summability

$$\ln z = \ln|z| + i\operatorname{Arg} z, \quad \operatorname{Arg} z \in \,]-\pi,\pi[; \quad e^z = e^{\Re z}(\cos \Im z + i\sin \Im z),$$

cf. example 2.2.16 or Vol. 2, Sect. 2.2. Then $e^{\ln z} = z$ is true for all $z \in \mathbb{C} - \mathbb{R}_-$. Furthermore, we use the following statements:

(1) The addition theorem $\ln zw = \ln z + \ln w$ applies to complex numbers $z, w \in \mathbb{C} - \mathbb{R}_-$, for which $|\operatorname{Arg} z + \operatorname{Arg} w| < \pi$ is.
(2) For every $\varepsilon > 0$ there is a $\delta > 0$ with $|\ln z| \leq \varepsilon$, if only $|z - 1| \leq \delta$ is. (This is the continuity of ln at the point $z_0 = 1$.)
(3) It is $|e^z - 1| \leq e^{|z|} - 1 \leq |z|e^{|z|}$ for all $z \in \mathbb{C}$, see the remark on exercise 3.10.8.

To prove the summability of $|a_i|$, $i \in I$, we can assume that for *all* $H \in \mathfrak{E}(I)$ the following applies: $|(1+a)^H - 1| < 1$. For this H, according to (1), $\ln(1+a)^H = \sum_{i \in H} \ln(1+a_i)$ applies. Due to (2), the family $\ln(1+a_i)$, $i \in I$, therefore meets the Cauchy summability criterion and is summable (incidentally with $\ln \prod (1+a_i)$ as the sum). With (3), one obtains

$$|a_i| = |e^{\ln(1+a_i)} - 1| \leq |\ln(1+a_i)|e^{|\ln(1+a_i)|}.$$

From the summability of $|\ln(1+a_i)|$, $i \in I$, (and the boundedness of $e^{|\ln(1+a_i)|}$, $i \in I$) the summability of $|a_i|$, $i \in I$ now follows. □

It should be noted again that the formula $\ln \prod(1+a_i) = \sum \ln(1+a_i)$ generally does not hold for finite families a_i (not even when both sides are defined). Furthermore, theorem 3.7.17 for real families a_i, $i \in I$, can be easily proven by applying theorem 3.6.25 and exercise 3.6.21 each to the subfamilies with $a_i \geq 0$ or with $a_i < 0$. — The major rearrangement theorem 3.7.11 applies analogously to infinite products. We leave the formulation and proof to the reader.

Exercises

Exercise 3.7.1
a) Let $z_1, \ldots, z_r \in \mathbb{C}$ with $|z_i| < 1$ for $i = 1, \ldots, r$. Then the family $z^m = z_1^{m_1} \cdots z_r^{m_r}$, $m = (m_1, \ldots, m_r) \in \mathbb{N}^r$, is summable, and it holds

$$\sum_{m \in \mathbb{N}^r} z^m = \frac{1}{1-z_1} \cdots \frac{1}{1-z_r}.$$

b) Infer for $w \in \mathbb{C}, |w| < 1$, and all $r \in \mathbb{N}^*$:

$$\frac{1}{(1-w)^r} = \sum_{k=0}^{\infty} \binom{k+r-1}{r-1} w^k.$$

c) Use b) to calculate the sums of the following series:

$$\sum_{n=0}^{\infty} \left(\frac{2}{3}\right)^n; \quad \sum_{n=0}^{\infty} \frac{n}{(1+i)^n}; \quad \sum_{n=0}^{\infty} \frac{n^2}{2^n}; \quad \sum_{n=0}^{\infty} \frac{n^3}{(3i)^n}; \quad \sum_{n=0}^{\infty} \frac{n^4 i^n}{(\sqrt{2})^n}.$$

Exercise 3.7.2
a) The family $a_{m,n} := m^{-n}$, $m,n \in \mathbb{N} - \{0,1\}$, is summable with sum
$$\sum_{m,n\geq 2}\frac{1}{m^n} = \sum_{n=2}^{\infty}(\zeta(n)-1) = 1.$$

b) Let $Q := \{m^n \mid m,n \in \mathbb{N} - \{0,1\}\}$ be the set of true powers of natural numbers. Then the family $1/(q-1)$, $q \in Q$, is summable with sum 1.
(The results of a) and b) do not contradict each other. — For $\sum_{q\in Q} 1/q$ see exercise 3.7.11.)

Exercise 3.7.3 For $(m,n) \in \mathbb{N}^* \times \mathbb{N}^*$ let $a_{m,n} := \begin{cases} 1/(m^2-n^2), & \text{if } m \neq n, \\ 0, & \text{if } m = n. \end{cases}$ Then it holds: For each fixed $m \in \mathbb{N}^*$ the family $(a_{m,n})_{n\in\mathbb{N}^*}$ is summable, and for each fixed $n \in \mathbb{N}^*$ the family $(a_{m,n})_{m\in\mathbb{N}^*}$ is summable. Furthermore, the sums

$$\sum_{m\in\mathbb{N}^*}\left(\sum_{n\in\mathbb{N}^*} a_{m,n}\right) \quad \text{and} \quad \sum_{n\in\mathbb{N}^*}\left(\sum_{m\in\mathbb{N}^*} a_{m,n}\right),$$

exist but are different. Justify why the major rearrangement theorem is not applicable here.

Exercise 3.7.4 Examine the following families $a_{m,n}$, $(m,n) \in \mathbb{N}^2$, for summability and determine the sum if possible.
a) $a_{m,n} := 1/(m+n+1)$.
b) $a_{m,n} := mnw^{m+n}$, $w \in \mathbb{C}$ fixed.

Exercise 3.7.5 For $w \in \mathbb{C}$ with $|w| < 1$ and $g \in \mathbb{N}$, $g \geq 2$, applies
$$\prod_{n\in\mathbb{N}}\left(\sum_{k=0}^{g-1} w^{kg^n}\right) = \frac{1}{1-w}.$$

Exercise 3.7.6 The family $1 + a_i$, $i \in I$, is exactly then multipliable, when a^H, $H \in \mathfrak{E}(I)$, is summable. In this case applies
$$\prod_{i\in I}(1+a_i) = \sum_{H\in\mathfrak{E}(I)} a^H.$$

Exercise 3.7.7 Let a_i, $i \in I$, be a summable family of complex numbers. I_n, $n \in \mathbb{N}$, be a sequence of subsets of I with $I_0 \subseteq \cdots \subseteq I_n \subseteq I_{n+1} \subseteq \cdots$ and $\bigcup_{n=0}^{\infty} I_n = I$. (In this case, I_n, $n \in \mathbb{N}$, is called an **exhaustion** of I and writes $I_n \uparrow I$.) Then holds
$$\sum_{i\in I} a_i = \lim_{n\to\infty}\sum_{i\in I_n} a_i.$$

Fig. 3.14 Summation scheme for exercise 3.7.15a)

$$\begin{array}{|ccc|}
(1,1) & (1,2) & (1,3) \cdots \\
\hline
(2,1) & (2,2) & (2,3) \cdots \\
\hline
(3,1) & (3,2) & (3,3) \cdots \\
\vdots & \vdots & \vdots \\
\end{array}$$

Exercise 3.7.8

a) Let $A \subseteq \mathbb{P}$ be a *finite* subset of the set $\mathbb{P}$ of prime numbers and $N(A)$ the set of natural numbers $n \in \mathbb{N}^*$, whose prime factors all belong to A. Then for every $s \in \mathbb{R}$ with $s > 0$ (or every $s \in \mathbb{C}$ with $\Re s > 0$)

$$\prod_{p \in A}(1 - p^{-s})^{-1} = \sum_{n \in N(A)} n^{-s}.$$

b) For $s \in \mathbb{R}$, $s > 1$, (or more generally for $s \in \mathbb{C}$, $\Re s > 1$) the following applies:

$$\prod_{p \in \mathbb{P}}(1 - p^{-s})^{-1} = \zeta(s) \quad \text{(Formel von Euler)}.$$

In particular, the ζ-function has no zero for $\Re s > 1$.

c) The product $\prod_{p \in \mathbb{P}}(1 - p^{-1})^{-1}$ is divergent. It follows that the sum $\sum_{p \in \mathbb{P}} p^{-1}$ also diverges (**Euler's theorem**). (**Remark** $\sum_{p \in \mathbb{P}, p \leq n} p^{-1} - \ln \ln n$, $n \geq 2$, converges to the constant

$$\beta := \gamma + \sum_{p \in \mathbb{P}} \left(p^{-1} + \ln(1 - p^{-1})\right) = 0{,}261497\ldots \quad \text{(Satz von Mertens)}.$$

Compare this with the result $\lim_{n \to \infty} (H_n - \ln n) = \gamma$ from example 3.3.8. A proof of Mertens' theorem can be found, for example, in the highly recommended textbook Hardy, G. H.; Wright, E. M.: An Introduction to the Theory of Numbers, Oxford.)

d) One can deduce from b): For $s \in \mathbb{R}$, $s > 1$, (or more generally for $s \in \mathbb{C}$, $\Re s > 1$,)

$$1/\zeta(s) = \prod_{p \in \mathbb{P}}(1 - p^{-s}) = \sum_{n \in \mathbb{N}^*} \mu(n)n^{-s},$$

applies, where $\mu : \mathbb{N}^* \to \mathbb{Z}$ is the Möbius μ-function, see problem 2.1.23.

e) For $s \in \mathbb{R}$ with $s > 1$ (or more generally for $s \in \mathbb{C}$ with $\Re s > 1$) is

$$\zeta^2(s) = \prod_{p \in \mathbb{P}}(1 - p^{-s})^{-2} = \sum_{n \in \mathbb{N}^*} \frac{\tau(n)}{n^s}, \quad \frac{\zeta(s)}{\zeta(2s)} = \prod_{p \in \mathbb{P}}(1 + p^{-s}) = \sum_{n \in \mathbb{N}^*} \frac{|\mu(n)|}{n^s},$$

$$\frac{\zeta^2(s)}{\zeta(2s)} = \prod_{p \in \mathbb{P}} \frac{p^s + 1}{p^s - 1} = \sum_{n \in \mathbb{N}^*} \frac{2^{\omega(n)}}{n^s},$$

where $\tau(n)$ is the number of divisors $d \in \mathbb{N}^*$ of $n \in \mathbb{N}^*$ and $\omega(n)$ is the number of *different* prime divisors of n. (For $s = 2$, with $\zeta(2) = \pi^2/6$ and $\zeta(4) = \pi^4/90$ we get the formula

$$\prod_{p\in\mathbb{P}} \frac{p^2+1}{p^2-1} = \frac{5}{2}.$$

By the way, $\prod_{n\geq 2} \frac{n^2+1}{n^2-1} = (\sinh\pi)/\pi = (e^\pi - e^{-\pi})/2\pi = 3{,}67607791\ldots$ See exercise 3.6.22b).

Exercise 3.7.9 Let $f(n)$, $n \in \mathbb{N}^*$, be a sequence of complex numbers with $\sum_{n\in\mathbb{N}^*} 2^{\omega(n)}|f(n)| < \infty$, where $\omega(n)$ for $n \in \mathbb{N}^*$ is the number of different prime divisors of n. Furthermore, let $g(m) := \sum_{n\in\mathbb{N}^*} f(mn)$ for $m \in \mathbb{N}^*$. Then the inversion formula $f(1) = \sum_{m\in\mathbb{N}^*} \mu(m)g(m)$ applies. (Apply the major rearrangement theorem and the summation property of the Möbius µ-function given in exercise 2.1.23 to the summable family $a_{m,n} := \mu(m)f(n)$, $(m,n) \in \mathbb{N}^* \times \mathbb{N}^*$, $m \mid n$.)

Exercise 3.7.10 As a generalization of the major rearrangement theorem 3.7.11, prove the following **great sieve formula**: Let a_i, $i \in I$, be a family of complex numbers and let $I = \bigcup_{j\in J} I_j$ be a pointwise finite cover of I, i.e., for each $i \in I$, the set $N_i := \{j \in J \mid i \in I_j\}$ is (non-empty) and finite. Furthermore, let the family $2^{|N_i|}a_i$, $i \in I$, summable. For $H \in \mathfrak{E}(J)$ let $I_H := \bigcap_{j\in H} I_j$ and $s_H := \sum_{i\in I_H} a_i$. Then the family s_H, $H \in \mathfrak{E}(J)$, is summable, and it holds

$$\sum_{H\in\mathfrak{E}(J)} s_H = \sum_{i\in I} 2^{|N_i|} a_i \quad \text{and} \quad \sum_{i\in I} a_i = \sum_{H\in\mathfrak{E}(J), H\neq\emptyset} (-1)^{|H|-1} s_H.$$

(Consider the summable family A_k, $k \in K := \{(i,H) \in I \times \mathfrak{E}(J) \mid i \in I_H\} = \biguplus_{i\in I}\{(i,H) \mid H \in \mathfrak{E}(N_i)\}$ and $A_k := a_i$ for $k = (i,H) \in K$. — For finite I one obtains the so-called **(small) sieve formula**. Here, the a_i, $i \in I$, may be elements of any additive abelian group. If I is finite and $a_i = 1$ for all $i \in I$, then one gets the sieve formula from exercise 1.6.23.)

Exercise 3.7.11 Let $Q = \{m^n \mid m, n \in \mathbb{N}^* - \{1\}\}$, cf. exercise 3.7.2. Then the family $1/q$, $q \in Q$, is summable with sum

$$\sum_{q\in Q} \frac{1}{q} = -\sum_{n=2}^{\infty} \mu(n)(\zeta(n) - 1) = 0{,}87446436840\ldots$$

(Use the cover $Q = \bigcup_{p\in\mathbb{P}} Q_p$ of Q with $Q_p := \{m^p \mid m \in \mathbb{N} - \{0,1\}\}$, $p \in \mathbb{P}$, and the large sieve formula from exercise 3.7.10.—For a discussion of the set Q of true powers, we refer to [22], Section 6.B, Remark (2) to Exercise 4.)

Exercise 3.7.12 The following two limits exist, but are different:

$$\lim_{n\to\infty} \left(\sum_{k\in\mathbb{N}^*} \frac{n}{(n+k)(n+k+1)}\right) \quad \text{and} \quad \sum_{k\in\mathbb{N}^*} \left(\lim_{n\to\infty} \frac{n}{(n+k)(n+k+1)}\right).$$

Exercise 3.7.13 Let $(a_n)_{n\in\mathbb{N}}$ be a null sequence of real numbers, which is neither properly nor improperly summable. Then for every $r \in \mathbb{R}$ there is a permutation

3.7 Summability

n	0	1	2	3	4	5	6	7	8	9	10	11	12	13	14	15
Q(n)	1	1	1	2	2	3	4	5	6	8	10	12	15	18	22	27

Fig. 3.15 Number $Q(n)$ of partitions of n with odd natural numbers

σ of $\mathbb{N}$ with $\sum_{n=0}^{\infty} a_{\sigma n} = r$. Moreover, one can achieve that the series $\sum a_{\sigma n}$ converges improperly to ∞ or $-\infty$. (Riemann)

Exercise 3.7.14 Let (r_n) and (s_n) monotonically increasing sequences of natural numbers, $n \in \mathbb{N}^*$. Let $r_n + s_n = n$ hold for all $n \in \mathbb{N}^*$. The series $\sum_{k=1}^{\infty} a_k$ is such that among the first n terms exactly the first r_n terms of the series $\sum_{k=1}^{\infty} 1/(2k-1)$ and the first s_n terms of the series $\sum_{k=1}^{\infty} (-1/2k)$ occur. Then

$$\sum_{k=1}^{n} a_k = \sum_{k=1}^{r_n} \frac{1}{2k-1} - \sum_{k=1}^{s_n} \frac{1}{2k} = H_{2r_n} - \frac{1}{2}(H_{r_n} + H_{s_n})$$

is the n-th partial sum and $\sum_{k=1}^{\infty} a_k = \frac{1}{2} \ln 4t$, if $t := \lim r_n/s_n \in \overline{\mathbb{R}}_+$ exists. (Use example 3.3.8 and the continuity of the logarithm, cf. Sect. 3.10.) In particular,

$$\sum_{k=1}^{\infty} a_k = \frac{1}{2} \ln \frac{4p}{1-p},$$

if $\lim(r_n/n) = p$ holds, e.g. at $r_n := [pn]$, $p \in [0, 1]$. It follows (with $p = 1/2$)

$$\sum_{k=1}^{\infty} \frac{(-1)^{k-1}}{k} = \ln 2$$

for the alternating harmonic series, from which one can obtain any value $x \in \mathbb{R}$ by explicit rearrangement, e.g. ($p = 1/5$)

$$0 = \sum_{k=1}^{\infty} \left(\frac{1}{2k-1} - \frac{1}{2} \left(\frac{1}{4k-3} + \frac{1}{4k-2} + \frac{1}{4k-1} + \frac{1}{4k} \right) \right).$$

By the way, in example 3.7.1, $p = 2/3$.

Exercise 3.7.15 Let $z \in \mathbb{C}, |z| < 1$. Show:

a) The family z^{mn}, $m, n \in \mathbb{N}^*$, is summable, and it holds

$$\sum_{m,n \geq 1} z^{mn} = \sum_{n=1}^{\infty} \tau(n) z^n = \sum_{n=1}^{\infty} \frac{z^n}{1-z^n} = \sum_{n=1}^{\infty} \frac{1+z^n}{1-z^n} z^{n^2}.$$

($\tau(n)$ is the number of divisors $d \in \mathbb{N}^*$ of n.—For the last equation, sum according to the scheme indicated in Fig. 3.14.)

b) It is

$$z = \sum_{n=1}^{\infty} \mu(n) \frac{z^n}{1-z^n}, \quad \frac{z}{(1-z)^2} = \sum_{n=1}^{\infty} nz^n = \sum_{n=1}^{\infty} \varphi(n) \frac{z^n}{1-z^n},$$

$$\sum_{n=1}^{\infty} \sigma(n) z^n = \sum_{n=1}^{\infty} \frac{nz^n}{1-z^n}, \quad \frac{z}{1-z} = \sum_{n=0}^{\infty} \frac{z^{2^n}}{1+z^{2^n}}.$$

($\sigma(n) = \sum_{d|n} d$, $n \in \mathbb{N}^*$, is the sum of the divisors $d \in \mathbb{N}^*$ of n.)

Exercise 3.7.16 Let $n \in \mathbb{N}$. As in example 2.5.13, $P(n)$ denotes the number of **partitions** of n, i.e., the sequences $(v_1, \ldots, v_n) \in \mathbb{N}^n$ with $n = v_1 \cdot 1 + \cdots + v_n \cdot n$. For $z \in \mathbb{C}, |z| < 1$, the equation holds

$$\sum_{n \in \mathbb{N}} P(n) z^n = \left(\prod_{n \in \mathbb{N}^*} (1-z^n) \right)^{-1}.$$

Remark According to Euler, for $|z| < 1$, the representation

$$f(z) := \prod_{n \in \mathbb{N}^*} (1-z^n) = 1 + \sum_{n=1}^{\infty} (-1)^n \left(z^{(3n-1)n/2} + z^{(3n+1)n/2} \right)$$

$$= \sum_{n \in \mathbb{Z}} (-1)^n z^{(3n+1)n/2}.$$

applies. We refer to this so-called **Pentagonal number theorem** of Euler in Vol. 2, exercise 1.2.27c). By inverting this power series, the numbers $P(n)$ are easily calculated recursively, $n \in \mathbb{N}$. The first 16 values can be found in Fig. 2.15. — $P(n)$ can also be interpreted for $n \in \mathbb{N}$ as the number of ways to distribute n objects into n compartments, if two such distributions are identified when they arise from a permutation of both the objects and the compartments. For n objects and m compartments, the corresponding number is the number $p(n,m)$ of n-tuples $(v_1, \ldots, v_n) \in \mathbb{N}^n$ with $v_1 \cdot 1 + \cdots + v_n \cdot n = n$ and $v_1 + \cdots + v_n \leq m$. The numbers $p(n,m)$ fulfill the recursion $p(n,m) = p(n, m-1) + p(n-m, m)$ with the initial conditions $p(0,m) = 1$ for $m \in \mathbb{N}$, $p(n,0) = 0$ for $n \in \mathbb{N}^*$ and $p(n,m) = 0$ for $n < 0$, which makes them easy to calculate. Note that $p(n,m)$ also represents the number of m-tuples $(\mu_1, \ldots, \mu_m) \in \mathbb{N}^m$ with $\mu_1 \cdot 1 + \cdots + \mu_m \cdot m = n$ (proof!). For all $z \in \mathbb{C}$ with $|z| < 1$ (see Vol. 2, Ex. 1.2.27b))

$$\sum_{n \in \mathbb{N}} p(n,m) z^n = \frac{1}{(1-z)(1-z^2) \cdots (1-z^m)}.$$

Exercise 3.7.17 For all $z \in \mathbb{C}, |z| < 1$, applies

$$\prod_{n \in \mathbb{N}^*} (1+z^n) = \prod_{n \in \mathbb{N}^*} \frac{1-z^{2n}}{1-z^n} = \frac{f(z^2)}{f(z)} = \left(\prod_{m \in \mathbb{N}^*} (1-z^{2m-1}) \right)^{-1} = \sum_{n \in \mathbb{N}} Q(n) z^n,$$

where $f(z)$ has the same meaning as in exercise 3.7.16 and $Q(n)$ is the number of sequences $v_1, v_3, v_5, \ldots$ of natural numbers with $n = v_1 \cdot 1 + v_3 \cdot 3 + v_5 \cdot 5 + \cdots$. It follows, in the formal power series algebra $\mathbb{C}\ z$ calculating or the identity theorem for convergent power series, see Vol. 2, Theorem 1.2.13, anticipating: *For $n \in \mathbb{N}$ the number of partitions of n with exclusively odd natural numbers is equal to the number of representations of n as a sum of different positive natural numbers (not considering the order of the summands).* With the remark from exercise 3.7.16 the $Q(n)$, $n \in \mathbb{N}$, can be easily calculated recursively, see Fig. 3.15.

Exercise 3.7.18 Let X be a countable completely ordered set and r_x, $x \in X$, a summable family of *positive* real numbers with sum $S \in \mathbb{R}_+$.

a) The mapping $x \mapsto \sum_{y<x} r_y$ is a (injective) strictly monotonic mapping from X to $\mathbb{R}$. (Every countable total order can thus be realized as a subset of $\mathbb{R}$ up to isomorphism. It is easy to prove directly that this is even possible in $\mathbb{Q}$, see exercise 1.8.13a). Under no circumstances can *every* totally ordered set with a power $\leq \aleph$ be *order-faithfully* embedded in $\mathbb{R}$. Example?)
b) Let X even be well-ordered. $X \times [0, 1[$ carries the lexicographic order (see example 1.4.8). Then the mapping

$$X \times [0, 1[\xrightarrow{\sim} [0, S[, \quad (x,t) \mapsto \left(\sum_{y<x} r_y\right) + tr_x$$

is an isomorphism of ordered sets. (To prove surjectivity, consider the smallest element of the set $\{x \in X \mid \sum_{y \leq x} r_y > s\}$, where $s \in [0, S[$ is. — As already noted in exercise 3.4.23, every well-ordered subset of $\mathbb{R}$ is countable.)

3.8 Continuous Functions

Continuity is a general topological concept. Therefore, continuous functions could also be defined on any ordered fields K or more generally on subsets of K^n, $n \in \mathbb{N}$. We will discuss this in Chapter 4. Here we only consider functions on subsets D of $\mathbb{C}$ with values in $\mathbb{K}$ and use

$$\mathbb{K}$$

as a common designation for the fields $\mathbb{R}$ and $\mathbb{C}$. These functions form the $\mathbb{K}$-algebra $\mathbb{K}^D$. The completeness of $\mathbb{R}$ and thus also of $\mathbb{C}$ has serious consequences for continuous functions compared to general ordered fields. In the case of $\mathbb{K} = \mathbb{R}$, $\overline{\mathbb{K}} = \overline{\mathbb{R}} = \mathbb{R} \uplus \{\infty, -\infty\}$ is, and in the case of $\mathbb{K} = \mathbb{C}$, $\overline{\mathbb{K}} = \overline{\mathbb{C}} = \mathbb{C} \uplus \{\infty\}$ is.

Continuous functions are characterized by the fact that small changes in the arguments only lead to small changes in the function values, see Fig. 3.16. For example, in this sense, the area $f(x) = x^2$ of a square continuously depends on the length x of the side. If, for example, in the case of the area, a deviation from the target value $f(a) = a^2$, $a \in \mathbb{R}_+$, which is at most equal to $\varepsilon > 0$ in absolute value, this

Fig. 3.16 f is continuous at the point a

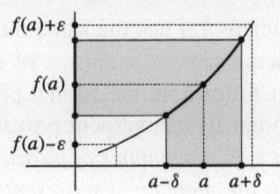

is achieved, for example, by the side deviating from the target value a by at most $\delta := \mathrm{Min}\left(1, \varepsilon/(2|a|+1)\right)$; because from $|x-a| \leq \delta$ follows $|x| \leq |a| + 1$ and

$$|f(x) - f(a)| = |x^2 - a^2| = |x-a||x+a| \leq \delta(|x| + |a|) \leq \delta(2|a|+1) \leq \varepsilon.$$

In this situation, it is said that the function f at the point a has the limit $a^2 = f(a)$ or that f is continuous at a. We first discuss general limits of functions.

In the following, let D always be a subset of $\mathbb{C}$ (in particular, D may also be a subset of $\mathbb{R}$).

Definition 3.8.1 Let $f: D \to \mathbb{K}$ be a function and $a \in \mathbb{K}$. The number $c \in \mathbb{K}$ is called **limit** or **Limes** of f at the point a, if for every (no matter how small) $\varepsilon > 0$ there exists a $\delta > 0$ such that $|f(x) - c| \leq \varepsilon$ is for all $x \in D$ with $|x - a| \leq \delta$.

If $a \notin \overline{D}$, then according to this definition, every point of $\mathbb{K}$ is a limit of f at a. But *if $a \in \overline{D}$, then the limit c of f at a, if it exists at all, is uniquely determined.* If also $c' \neq c$ is a limit of f at a, then there exists for $\varepsilon := |c' - c|/3$ a $\delta > 0$ and a $\delta' > 0$ such that for all $x \in D$ it holds: From $|x - a| \leq \delta$ it follows $|f(x) - c| \leq \varepsilon$ and from $|x - a| \leq \delta'$ follows $|f(x) - c'| \leq \varepsilon$. Since a is a point of contact of D, there is now an $x \in D$ with $|x - a| \leq \mathrm{Min}(\delta, \delta')$. For such an x, the contradiction

$$|c' - c| \leq |c' - f(x)| + |f(x) - c| \leq \varepsilon + \varepsilon = \frac{2}{3}|c' - c|.$$

arises. In the case of $a \in \overline{D}$, the uniquely determined limit c of f in a, if it exists, is denoted by

$$\lim_{x \to a, x \in D} f(x) \quad \text{or short with} \quad \lim_{x \to a} f(x).$$

If $a \in D$ and $c = \lim_{x \to a} f(x)$ exists, then necessarily $c = f(a)$. Why?

Theorem 3.8.2 *Let $f: D \to \mathbb{K}$ be a function, a a point of contact of D and c an element of $\mathbb{K}$. The following statements are equivalent:*

(i) *It is $\lim_{x \to a} f(x) = c$.*
(ii) *For every neighborhood $V \subseteq \mathbb{K}$ of c there exists a neighborhood U of a with $f(U \cap D) \subseteq V$.*
(iii) *It holds $\lim_{n \to \infty} f(x_n) = c$ for every sequence (x_n) in D with $\lim_{n \to \infty} x_n = a$.*

Proof From (i) follows (ii): Let V be a neighborhood of c. There is a ε-neighborhood of c that lies entirely within V. For this $\varepsilon > 0$, due to (i), there is by definition a $\delta > 0$ with $|f(x) - c| \leq \varepsilon$ for all $x \in D$, $|x - a| \leq \delta$. For the δ-neighborhood U of a, then $f(U \cap D) \subseteq V$ applies.

From (ii) follows (iii): Let (x_n) be a sequence with $x_n \in D$ and $\lim x_n = a$, and let V be a neighborhood of c. According to (ii), there is a neighborhood U of a with $f(U \cap D) \subseteq V$. Due to $\lim x_n = a$ and $x_n \in D$, almost all members of the sequence (x_n) lie in $U \cap D$ and thus almost all members of the sequence $(f(x_n))$ in V. This was to be shown.

From (iii) follows (i): Assume, (i) is false. Then there exists an $\varepsilon_0 > 0$, for which there is no δ in the sense of Definition 3.8.1, i.e., in particular: For every $n \in \mathbb{N}^*$ there is an $x_n \in D$, for which $|x_n - a| \leq 1/n$ is true, but $|f(x_n) - c| > \varepsilon_0$. The sequence thus obtained (x_n) in D converges to a, but $(f(x_n))$ does not converge to c. Contradiction.

From a topological standpoint, condition (ii) provides the best characterization of a limit. — The Cauchy convergence criterion for sequences also provides one for limits of functions:

Theorem 3.8.3 (Cauchy's convergence criterion for limits) *Let $f : D \to \mathbb{K}$ be a function and a a contact point of D. The limit $\lim_{x \to a} f(x)$ exists if and only if, for every $\varepsilon > 0$, there exists a $\delta > 0$ such that $|f(x) - f(x')| \leq \varepsilon$ for all $x, x' \in D$ with $|x - a| \leq \delta$ and $|x' - a| \leq \delta$.*

Proof $c := \lim_{x \to a} f(x)$ exists. For a given ε, there is a δ with $|f(x) - c| \leq \varepsilon/2$ for all $x \in D$ with $|x - a| \leq \delta$. If then $x, x' \in D$ with $|x - a|, |x' - a| \leq \delta$, then

$$|f(x) - f(x')| \leq |f(x) - c| + |c - f(x')| \leq \varepsilon/2 + \varepsilon/2 = \varepsilon.$$

Now let the Cauchy criterion be fulfilled and (x_n) be a sequence in D with $\lim x_n = a$. Then $(f(x_n))$ is a Cauchy sequence in $\mathbb{K}$ and thus convergent. If $\varepsilon > 0$ is given, then according to the assumption there is a $\delta > 0$ with $|f(x) - f(x')| \leq \varepsilon$ for all $x, x' \in D$ with $|x - a| \leq \delta$, $|x' - a| \leq \delta$. Since (x_n) in $\mathbb{K}$ converges, there is a $n_0 \in \mathbb{N}$ with $|x_n - a|, |x_m - a| \leq \delta$ for $m, n \geq n_0$. For these m, n, $|f(x_n) - f(x_m)| \leq \varepsilon$ also applies. The fact that the limit of the sequences $(f(x_n))$ is always the same and thus equal to the limit of f in a, is as follows: If (x_n) and (x'_n) are both sequences in D that converge to a, then one considers the mixed sequence $x_0, x'_0, x_1, x'_1, \ldots$, which also converges to a. Then the limit of the image sequence $f(x_0), f(x'_0), f(x_1), f(x'_1), \ldots$ is the common limit of the subsequences $(f(x_n))$ and $(f(x'_n))$.

Of course, one can formulate sentence 3.8.3 analogously to theorem 3.8.2 (2) also with environments: $\lim_{x \to a} f(x)$ exists exactly when for every environment V of 0 in $\mathbb{K}$ there is an environment U of a with $f(x) - f(x') \in V$ for all $x, x' \in U \cap D$. — The following calculation rules result directly from the corresponding calculation rules for limits of sequences with the sequence criterion 3.8.2 (3):

Theorem 3.8.4 (Calculation rules for limits of functions) *Let f and g $\mathbb{K}$-valued functions on D. If for $a \in \overline{D}$ the limits $\lim_{x \to a} f(x)$ and $\lim_{x \to a} g(x)$ exist, then the following applies:*

(1) *The sum $f+ g$ has a limit in a, and it is*
$$\lim_{x \to a}(f + g)(x) = \lim_{x \to a} f(x) + \lim_{x \to a} g(x).$$

(2) *The product fg has a limit in a, and it is*
$$\lim_{x \to a}(fg)(x) = \left(\lim_{x \to a} f(x)\right)\left(\lim_{x \to a} g(x)\right).$$

In particular, λf for each $\lambda \in \mathbb{K}$ has a limit, and it is $\lim_{x \to a}(\lambda f)(x) = \lambda \lim_{x \to a} f(x)$.

(3) *If $g(x) \neq 0$ for all $x \in D$ and $\lim_{x \to a} g(x) \neq 0$, then f/g has a limit in a, and it is*
$$\lim_{x \to a}\left(\frac{f}{g}\right)(x) = \frac{\lim_{x \to a} f(x)}{\lim_{x \to a} g(x)}.$$

According to (1) and (2), the functions *thus form $f \in \mathbb{K}^D$, for which the limit exists at the point $a \in \overline{D}$, a $\mathbb{K}$-subalgebra of $\mathbb{K}^D$.*—In addition to the (proper) limits, we define **improper limits** $\lim_{x \to a} f(x) = \pm\infty$ for real-valued functions f and $\lim_{x \to a} f(x) := \lim_{x \to a} |f(x)| = \infty$ for complex-valued functions f analogously to definition 3.2.8. Note that for real-valued functions the improper limit $\lim_{x \to a} |f(x)| = \infty$ can exist without $\lim_{x \to a} f(x)$ existing, for example for $f(x) := 1/x$ on $\mathbb{R}^\times$ and $a := 0$. The calculation rules 3.8.4 for limits apply to improper limits, where, as with sequences, one must exclude the case of a product in which one factor has the limit 0.

In addition to the limits of a function $f : D \to \mathbb{K}$, where the argument x approaches a fixed number $a \in \overline{D}$, the limit behavior of $f(x)$ for other movements of x is also interesting. Particularly important is the **behavior at infinity**: For an unbounded domain $D \subseteq \mathbb{R}$, we say that $\lim_{x \to \infty} f(x) = c \in \mathbb{K}$ if for every $\varepsilon > 0$ a $S \in \mathbb{R}$ exists with $|f(x) - c| \leq \varepsilon$ for all $x \in D$, $x \geq S$. Similarly, we define $\lim_{x \to -\infty} f(x)$ for downward unbounded $D \subseteq \mathbb{R}$. For an unbounded domain $D \subseteq \mathbb{C}$ we set $\lim_{x \to \infty} f(x) = \lim_{|x| \to \infty} f(x) = c \in \mathbb{K}$, if for every $\varepsilon > 0$ there is a $S \in \mathbb{R}$ with $|f(x) - c| \leq \varepsilon$ for all $x \in D$, $|x| \geq S$. The criteria 3.8.2 and 3.8.3 as well as the calculation rules 3.8.4 immediately apply to these situations. Finally, improper limits can also be defined for behavior at infinity. We leave it to the reader to carry this out.

Example 3.8.5 (Left and right-sided limits) Let $D \subseteq \mathbb{R}$. Furthermore, let $f : D \to \mathbb{K}$ be a function and $a \in \mathbb{R}$ a real number, which is a contact point (and thus even an accumulation point) of the set $D_{<a} := D \cap \,]-\infty, a[$. Then the limit of $f|D_{<a}$ at the point a is called the **left-hand limit** of f at a and is denoted by

$$\lim_{x \to a-} f(x) = \lim_{x \to a, x < a} f(x) \quad \text{or with} \quad f(a-)$$

Similarly, the **right-hand limit**

$$f(a+) = \lim_{x \to a+} f(x) = \lim_{x \to a, x > a} f(x),$$

is defined if a is an accumulation point of $D_{>a} := D \cap]a, \infty[$. If both values exist, the difference $f(a+) - f(a-)$ is called the **jump height** of f at a and a is a **jump discontinuity** of f (even if the jump height is 0).

Let a be an accumulation point of $D_{<a}$ and $D_{>a}$. $f(a-)$ (or $f(a+)$) exists exactly when $\lim_{n \to \infty} f(x_n)$ exists for every strictly monotonically increasing (or every strictly monotonically decreasing) sequence (x_n) in D with $\lim x_n = a$, cf. exercise 3.8.2. $\lim_{x \to a} f(x)$ exists exactly when both one-sided limits $f(a-)$ and $f(a+)$ exist and coincide, so the jump height at a is 0, and moreover $f(a-) = f(a+) = f(a)$ is, if a belongs to D. The sign function Sign x has a jump height of 0 at $a \neq 0$ and a jump height of 2 at $a = 0$, and the Gauss bracket $[x]$ has a jump height of 0 for all $a \notin \mathbb{Z}$ and a jump height of 1 for $a \in \mathbb{Z}$. ◊

To compare the limit behavior of two functions, the **Landau symbols**

O (to be read as : capital$-O$) and o (to be read as : small$-o$)

are used. If f and g are $\mathbb{K}$-valued functions on D and if $a \in \overline{D}$, then one writes

$$f = O(g) \quad \text{resp.} \quad f = o(g) \quad \text{for} \quad x \to a, \ x \in D,$$

if there is a neighborhood U of a and $f|(U \cap D) = hg|(U \cap D)$ with a bounded function $h: U \cap D \to \mathbb{K}$ resp. with a function $h: U \cap D \to \mathbb{K}$, for which $\lim_{x \to a} h(x) = 0$ is. The symbols O and o are defined similarly for $x \to \pm\infty$ at $D \subseteq \mathbb{R}$ and for $x \to \infty$ at $D \subseteq \mathbb{C}$. Finally, notations like $f = h + O(g)$ for $f - h = O(g)$ etc. are used.

Example 3.8.6 From $f = o(g)$ follows $f = O(g)$. The function f is exactly then limited to a neighborhood U of a on $U \cap D$, if $f = O(1)$ for $x \to a$ is, whereas $f = o(1)$ for $x \to a$ with $\lim_{x \to a} f(x) = 0$ is equivalent. So one can characterize $O(g)$ or $o(g)$ through $O(1)g$ or $o(1)g$ ◊

Example 3.8.7 (Asymptotic Equality) Let f and g be $\mathbb{K}$-valued functions on D and $a \in \overline{D}$. Then f and g are called **asymptotically equal in** a, if there is a neighborhood U of a such that g on $U \cap D$ does not vanish anywhere and $\lim_{x \to a} f/g = 1$ is. We then write

$$f \sim g \quad \text{for} \quad x \to a, \ x \in D.$$

The definition of asymptotic equality at infinity can be left to the reader. In this context, we often speak more precisely of *multiplicative* asymptotic equality. *Additive* asymptotic equality exists when the corresponding limit of $f - g$ is equal to 0. Compare the special case of sequences (i.e., $D = \mathbb{N}$ and $a = \infty$), which was already covered in exercise 3.2.9. For example, the famous **Prime Number Theorem** from 1896 by G. Hadamard (1865–1963) and Ch. de La Vallée Poussin (1866–1962) states that

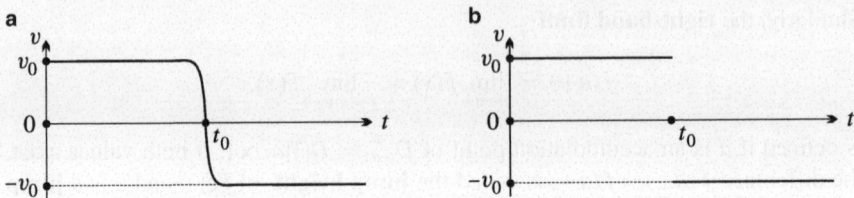

Fig. 3.17 Speed function $v(t)$ in elastic collision and its idealization with jump discontinuity

$$\pi(x) \sim \frac{x}{\ln x}$$

is for $x \to \infty$ and the prime number function π. We will prove this theorem in Vol. 5. For the n-th prime number p_n thus applies $n = \pi(p_n) \sim p_n/\ln p_n$, from which

$$p_n \sim n \ln p_n \sim n \ln n$$

results, since from the additive asymptotic equality of $\ln p_n$ and $\ln n + \ln \ln p_n$ the multiplicative asymptotic equality $\ln p_n \sim \ln n + \ln \ln p_n$ follows and since it also applies $\ln n + \ln \ln p_n \sim \ln n$, see exercise 3.2.9. (Note $\ln x = o(x)$ for $x \to \infty$.) For example, it is $p_{664.579} = 9.999.991$ is the largest prime number $\leq 10^7$, and it is $10^7/\ln 10^7 \approx 620.421$ or $664.579 \cdot \ln 664.579 \approx 8.909.950$.[13]   ◊

We now clarify the fundamental concept of continuity. Let again $D \subseteq \mathbb{C}$ and $f: D \to \mathbb{K}$ be a $\mathbb{K}$-valued function on D.

Definition 3.8.8 The function f is called **continuous** at the point $a \in D$, if

$$\lim_{x \to a, x \in D} f(x) = f(a)$$

f is called **continuous** (in D), if f is continuous at every point of D.

According to the definition of the limit $\lim_{x \to a} f(x)$ and with theorem 3.8.2, it follows:

Theorem 3.8.9 *For $f: D \to \mathbb{K}$ and $a \in D$ are equivalent:*

(i) *f is continuous at a.*
(i') *For every (no matter how small) $\varepsilon > 0$ there is a $\delta > 0$ with $|f(x) - f(a)| \leq \varepsilon$ for all $x \in D$ with $|x - a| \leq \delta$.*
(ii) *For every neighborhood $V \subseteq \mathbb{K}$ of $f(a)$ there is a neighborhood U of a with $f(U \cap D) \subseteq V$.*

---

[13] The additive deviations of p_n and $n \ln n$ are therefore considerable. In fact, $p_n - n \ln n, n \in \mathbb{N}^*$, is unbounded.

3.8 Continuous Functions

(iii) *It holds* $\lim_{n\to\infty} f(x_n) = f(a)$ *for every sequence* (x_n) *in* D *with* $\lim_{n\to\infty} x_n = a$.

The calculation rules 3.8.4 imply:

Theorem 3.8.10 *Let* $D \subseteq \mathbb{C}$ *and* $a \in D$.

(1) *The continuous a-valued functions in* $\mathbb{K}$ *form a* $\mathbb{K}$*-subalgebra of the* $\mathbb{K}$*-algebra* $\mathbb{K}^D = \text{Abb}(D, \mathbb{K})$ *of all* $\mathbb{K}$*-valued functions on* D.
(2) *The continuous D-valued functions form a* $\mathbb{K}$*-subalgebra of the* $\mathbb{K}$*-algebra* $\mathbb{K}$ *of all* $\mathbb{K}^D$*-valued functions on* D, *which are denoted with*

$$C_\mathbb{K}(D) \quad \text{or short with} \quad C(D).$$

It is $C_\mathbb{K}(D)^\times = C_\mathbb{K}(D) \cap (\mathbb{K}^D)^\times = C_\mathbb{K}(D) \cap (\mathbb{K}^\times)^D$.

If $f: D \to \mathbb{K}$ is continuous at the point $a \in D$, then for every $D' \subseteq D$ with $a \in D'$, the restriction $f|D': D' \to \mathbb{K}$ at the point a is also trivially continuous. In particular, the continuity of f over the entire D implies the continuity of the restriction $f|D'$ of f on D'. Conversely, the continuity of $f|D'$ in $a \in D'$ does not generally imply the continuity of the function $f: D \to \mathbb{K}$ at a. However, it is evident (see exercise 3.8.3):

Proposition 3.8.11 *Let* $f: D \to \mathbb{K}$ *be a function and* U *a neighborhood of* $a \in D$. *If* $f|(U \cap D)$ *is continuous at* a, *then f is also continuous at* a.

Because of Proposition 3.8.11, continuity is said to be a **local property**. Similarly elementary statements about continuous functions will often be used in the following without proof. We mention as an example the following consequence of the continuity of real-valued functions.

Proposition 3.8.12 *Let* $f: D \to \mathbb{R}$ *be a function continuous at point* $a \in D$ *and* c *a real number with* $f(a) > c$. *Then there exists a neighborhood* U *of* a *with* $f(x) > c$ *for all* $x \in U \cap D$.

Proof Due to the continuity of f at a, there exists a neighborhood $V :=]c, \infty[$ of $f(a)$ with a neighborhood U of a with $f(U \cap D) \subseteq V$.

Example 3.8.13

(1) Constant functions are obviously continuous on $\mathbb{C}$.
(2) The identity $x \mapsto x$ is continuous on $\mathbb{C}$.
(3) The absolute value function $x \mapsto |x|$ is continuous on $\mathbb{C}$. This follows from $\big||x| - |a|\big| \leq |x - a|$.
(4) The Gauss bracket $x \mapsto [x]$ on $\mathbb{R}$ is continuous exactly at the points $a \in \mathbb{R} - \mathbb{Z}$. For $a \in \mathbb{Z}$ applies $\lim_{n\to\infty}(a - \frac{1}{n}) = a$, but $\lim_{n\to\infty}[a - \frac{1}{n}] = a - 1 \neq a = [a]$.

(5) The sign function $x \mapsto \operatorname{Sign} x$ on $\mathbb{R}$ is not continuous exactly at the point $a = 0$.

(6) The so-called **Dirichlet function** on $\mathbb{R}$, that is the indicator function $e_\mathbb{Q}$ with

$$e_\mathbb{Q} : x \mapsto \begin{cases} 1, & \text{if } x \in \mathbb{Q}, \\ 0, & \text{if } x \notin \mathbb{Q}, \end{cases}$$

is not continuous at any point $x \in \mathbb{R}$. On $\mathbb{C}$ it is continuous at all points $x \in \mathbb{C} - \mathbb{R}$, since it is constantly equal to 0 on the open subset $\mathbb{C} - \mathbb{R}$ of $\mathbb{C}$.

(7) A ball strikes a wall perpendicularly and is elastically reflected. The speed v as a function of time is then a continuous function, which before the reflection has the value v_0 and after the reflection the value $-v_0$. v is then approximately represented by the graph in Fig. 3.17a. However, it is often useful to idealize the situation and describe the speed profile by the discontinuous function in Fig. 3.17b. This function has a jump discontinuity at the point t_0 with the jump height $-2v_0$, see example 3.8.5. ◊

Example 3.8.14 (Polynomial Functions — Rational Functions) Since the identity $x := (x \mapsto x)$ is continuous on $\mathbb{C}$, the subalgebra generated by x $\mathbb{C}[x] \subseteq \operatorname{Abb}(\mathbb{C}, \mathbb{C})$ of the polynomial functions on $\mathbb{C}$ is even a $\mathbb{C}$-subalgebra of the $\mathbb{C}$-algebra $C_\mathbb{C}(\mathbb{C})$ of the continuous functions $\mathbb{C} \to \mathbb{C}$. A rational function $R = F/G \in \mathbb{C}(X)$, $F \in \mathbb{C}[X]$, $G \in \mathbb{C}[X]^*$, defines a continuous function outside the (finite) set of zeros $\operatorname{NS}_\mathbb{C}(G)$ of the denominator G. If F and G have no common zero in $\mathbb{C}$ (which, according to the Fundamental Theorem of Algebra 3.9.7, is equivalent to F and G being coprime), then $\mathbb{C} - \operatorname{NS}_\mathbb{C}(G)$ is the largest set on which $R(x) = F(x)/G(x)$ is well-defined. For $c \in \operatorname{NS}_\mathbb{C}(G)$, then $\lim_{x \to c, x \neq c} R(x) = \infty$. If $R \neq 0$, then

$$\lim_{x \to \infty} R(x) = \begin{cases} 0 & , \text{ if } \deg F < \deg G \\ \operatorname{LK}(F)/\operatorname{LK}(G) & , \text{ if } \deg F = \deg G \\ \infty & , \text{ if } \deg F > \deg G, \end{cases}$$

see exercise 3.8.2. More precisely, $R(x) \sim \operatorname{LK}(F) x^{\operatorname{Grad} F - \operatorname{Grad} G} / \operatorname{LK}(G)$ holds for $x \to \infty$. With the given limits for $x \to \infty$ we always consider a rational function $R \in \mathbb{C}(X)$ as a mapping $\overline{\mathbb{C}} \to \overline{\mathbb{C}}$. — Discuss in a similar way the rational functions $R \in \mathbb{R}(X)$ as mappings $\mathbb{R} \uplus \{\infty\} \to \mathbb{R} \uplus \{\infty\}$. For a non-constant polynomial $F \in \mathbb{R}[X]^*$,

$$\lim_{x \in \mathbb{R}, x \to \infty} F(x) = \operatorname{Sign} \operatorname{LK}(F) \cdot \infty, \qquad \lim_{x \in \mathbb{R}, x \to -\infty} F(x) = (-1)^{\operatorname{Grad} F} \operatorname{Sign} \operatorname{LK}(F) \cdot \infty.$$

When can one also naturally consider a rational function $R \in \mathbb{R}[X]$ as a mapping $\mathbb{R} \uplus \{\pm\infty\} \to \mathbb{R} \uplus \{\pm\infty\}$? ◊

Compositions of continuous functions are again continuous, more precisely:

Theorem 3.8.15 *Let $f: D \to \mathbb{K}$ and $g: D' \to \mathbb{K}$ be functions with $f(D) \subseteq D'$. If f is continuous at $a \in D$ and g is continuous at $f(a)$, then the composition $g \circ f : D \to \mathbb{K}$ is also continuous at a. — In particular, $g \circ f$ is continuous on the entire D, if f is continuous on D and g is continuous on D'.*

Fig. 3.18 Successive Approximation

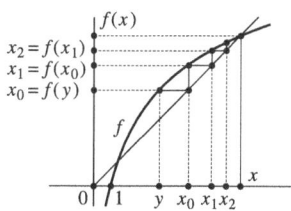

Proof We use the continuity criterion (ii) from theorem 3.8.9. Let W be a neighborhood of $(g \circ f)(a) = g(f(a))$. By assumption, there are neighborhoods V of $f(a)$ and U of a with $g(V \cap D') \subseteq W$ and $f(U \cap D) \subseteq V \cap D'$. Then $(g \circ f)(U \cap D) \subseteq W$ is true. — Of course, one can also conclude with theorem 3.8.9 (iii): Let (x_n) be a sequence in D with $\lim x_n = a$. From the continuity of f at a, it follows that $\lim f(x_n) = f(a)$, and from the continuity of g at $f(a)$, it follows that $\lim g(f(x_n)) = g(f(a))$.

We now prove a continuation theorem for continuous functions.

Theorem 3.8.16 *Let $f : D \to \mathbb{K}$ be a continuous function and $D \subseteq \widetilde{D} \subseteq \overline{D}$, i.e. D is dense in $\widetilde{D}$. At every point $x \in \widetilde{D} - D$ the limit exists*

$$\widetilde{f}(x) := \lim_{y \to x, y \in D} f(y).$$

If we set $\widetilde{f}(x) := f(x)$ for $x \in D$, then $\widetilde{f}$ is a continuous extension of f to $\widetilde{D}$.

Proof Let $a \in \widetilde{D}$ and $\varepsilon > 0$ be given. There then exists a $\delta > 0$ with $|f(x) - \widetilde{f}(a)| \leq \varepsilon/2$ for all $x \in D$ with $|x - a| \leq \delta$. For $x \in D$, $|x - a| \leq \delta$, thus already applies $|f(x) - \widetilde{f}(a)| \leq \varepsilon$. We show that this also applies for all $\widetilde{x} \in \widetilde{D}$ with $|\widetilde{x} - a| \leq \delta/2$. There exists for such a $\widetilde{x}$ a $x \in D$ with $|\widetilde{x} - x| \leq \delta/2$ and $|\widetilde{f}(\widetilde{x}) - f(x)| \leq \varepsilon/2$. Then $|x - a| \leq \delta$ is and consequently $|f(x) - \widetilde{f}(a)| \leq \varepsilon/2$, from which the assertion arises in the following way:

$$\left|\widetilde{f}(\widetilde{x}) - \widetilde{f}(a)\right| \leq \left|\widetilde{f}(\widetilde{x}) - f(x)\right| + \left|f(x) - \widetilde{f}(a)\right| \leq \varepsilon/2 + \varepsilon/2 = \varepsilon.$$

Example 3.8.17 (Fluctuation of a Function) The concept of fluctuation can be used to characterize the continuity of a function or, in a certain sense, to quantify its discontinuity. Let $f : D \to \mathbb{K}$ be any function ($D \subseteq \mathbb{C}$). Then

$$S(f) := S(f; D) := \mathrm{Sup}\left\{|f(x) - f(y)| \mid x, y \in D\right\} \in \overline{\mathbb{R}}_+ = \mathbb{R}_+ \uplus \{\infty\}$$

is called the **(global) fluctuation** of f on D. It is finite exactly when f is bounded on D. For $a \in \overline{D}$ is called

$$S_a(f) := S_a(f; D) := \mathrm{Inf}\,(S(f; U \cap D), U \in \mathcal{U}(a)),$$

where $\mathcal{U}(a)$ is the set of all neighborhoods of a, the **fluctuation of f at the point** a. The Cauchy criterion 3.8.3 states that *the fluctuation $S_a(f)$ is exactly 0 when the limit*

$\lim_{x \to a} f(x)$ exists. In particular, f at the point $a \in D$ is continuous exactly when $S_a(f) = 0$ is. The definition of fluctuation at infinity is left to the reader. ◊

Example 3.8.18 (Hölder and Lipschitz Continuous Functions) A function $f: D \to \mathbb{K}$ is called **Hölder continuous** with **exponent** $\alpha > 0$ (after O. Hölder (1859–1937)), if there is a constant $L > 0$ with

$$|f(x) - f(x')| \le L|x - x'|^\alpha \quad \text{for all} \quad x, x' \in D.$$

Here we use general powers, as defined in Sect. 3.10. Since these power functions for $\alpha > 0$ are continuous on $\mathbb{R}_+$ and vanish at the origin, Hölder continuous functions are continuous. Hölder continuous functions f with the exponent 1 are called **Lipschitz continuous** (after R. Lipschitz (1832–1903)). For Lipschitz continuous functions, there is therefore a so-called **Lipschitz constant** $L > 0$ for f with $|f(x) - f(x')| \le L|x - x'|$ for all $x, x' \in D$. Hölder exponents $\alpha > 1$ are of little interest, see Vol. 2, Ex. 2.3.2.

If there is a point $a \in \overline{D}$ and a neighborhood U of a such that $f|(U \cap D)$ is Hölder (or Lipschitz) continuous, then f at a is called **locally Hölder-(or locally Lipschitz-)continuous**. Finally, f is called **Hölder- (or Lipschitz-)continuous at the point** $a \in D$ if there exists a neighborhood U of a and a constant $L > 0$ such that $|f(x) - f(a)| \le L|x - a|^\alpha$ (or $\le L|x - a|$) holds for all $x \in U \cap D$, i.e., if $f(x) = f(a) + O(|x - a|^\alpha)$ (or $f(x) = f(a) + O(|x - a|)$) holds for $x \to a$. The case $\alpha > 1$ is also interesting. ◊

Example 3.8.19 (Contracting Functions) A function $f: D \to \mathbb{K}$ is called **contracting**, if $|f(x) - f(x')| < |x - x'|$ for all $x, x' \in D$, $x \ne x'$ applies. Contracting functions are in particular Lipschitz-continuous with the Lipschitz constant $L = 1$. If f is even Lipschitz-continuous with a Lipschitz constant $L < 1$, then f is called **strongly contracting**. L is also called a **contraction factor** of f. For strongly contracting functions f on a *closed* set $D (\subseteq \mathbb{C})$ with $f(D) \subseteq D$ the following fixed point theorem applies, which is a special case of the general **Banach's Fixed Point Theorem** 4.5.12 (after S. Banach (1892–1945)). ◊

Theorem 3.8.20 *Let* $f: D \to D$ *be a strongly contracting function of the nonempty closed subset* $D \subseteq \mathbb{C}$ *onto itself. Then f has exactly one fixed point x. If $x_0 \in D$ is any point in D, then the sequence*

$$x_0, x_1 = f(x_0), \ldots, x_n = f(x_{n-1}) = f^n(x_0), \ldots$$

(f^n *is the n-th iterate of f) converges to the unique fixed point x of f and with a contraction factor $L < 1$ of f is*

$$|x - x_n| \le \frac{1}{1-L}|x_{n+1} - x_n| \le \frac{L^n}{1-L}|x_1 - x_0| \quad \text{for all} \quad n \in \mathbb{N}.$$

Proof If x and x' are fixed points of f, then $|x - x'| = |f(x) - f(x')| \le L|x - x'|$, thus $|x - x'| = 0$ and $x = x'$. — Now let $x_0 \in D$ be arbitrary and $x_n := f^n(x_0)$, $n \in \mathbb{N}$. For $n \in \mathbb{N}^*$ holds

3.8 Continuous Functions

Fig. 3.19 Interval halving method for root finding

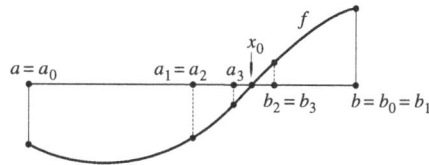

$$|x_{n+1} - x_n| = |f(x_n) - f(x_{n-1})| \leq L|x_n - x_{n-1}|$$

and consequently $|x_{n+1} - x_n| \leq L^n|x_1 - x_0|$ for $n \in \mathbb{N}$. We obtain

$$|x_n - x_m| \leq \sum_{i=0}^{n-m-1} |x_{m+i+1} - x_{m+i}| \leq L^m \frac{1 - L^{n-m}}{1 - L} |x_1 - x_0|$$

for all $m, n \in \mathbb{N}$ with $m \leq n$. Thus, x_n, $n \in \mathbb{N}$, is a Cauchy sequence in D and therefore convergent with a limit $x := \lim x_n$ in D (since D is closed). Because of

$$x = \lim x_n = \lim x_{n+1} = \lim f(x_n) = f(\lim x_n) = f(x)$$

x is a fixed point of f. Furthermore, we get $|x - x_0| = |\lim x_n - x_0| = \lim |x_n - x_0| \leq |x_1 - x_0|/(1 - L)$. If we apply this inequality with x_n instead of x_0 as the initial value, we obtain the first of the given inequalities and due to $|x_{n+1} - x_n| \leq L^n|x_1 - x_0|$ also the second.

Theorem 3.8.20 can be used to solve the equation $f(x) = x$ by **successive approximation**. This method is self-correcting: Any rounding or calculation errors do not propagate. If you want to determine the fixed point x up to an error $\leq \varepsilon$: $|x - x_n| \leq \varepsilon$, you can choose the **a posteriori estimation** $L|x_n - x_{n-1}|/(1 - L) \leq \varepsilon$ as the termination condition. But you can also limit the number of steps n with the **a priori estimation** $|x - x_n| \leq L^n|x_1 - x_0|/(1 - L)$ by the termination condition $L^n|x_1 - x_0|/(1 - L) \leq \varepsilon$ to guarantee an error $\leq \varepsilon$. To determine a Lipschitz constant L, the following statement, which directly follows from the mean value theorem of differential calculus (see Vol. 2, Theorem 2.3.4 or 2.3.7), often helps and should already be used by the reader when dealing with examples to avoid unnecessarily complicated estimates: *If $f : D \to \mathbb{C}$ is a differentiable function f on the interval $D \subseteq \mathbb{R}$ or on the (open) convex set $D \subseteq \mathbb{C}$ and has f a bounded derivative f', then f is Lipschitz continuous with the Lipschitz constant $L := \mathrm{Sup}\{|f'(t)| \mid t \in D\}$ (but with no smaller one).*

If for a given $y > e$ the equation

$$y = \frac{x}{\ln x}$$

is to be solved with $x > e$, it is the fixed point equation $x = y \ln x$. The function $f(t) = y \ln t$ has the derivative $f'(t) = y/t$ and is therefore strongly contracting on every interval $[a, \infty[$, $a > y$. Because of $f(y) = y \ln y > y$, f also maps such an interval onto itself, provided that a is close enough to y. With the starting value $x_0 = y \ln y = f(y)$, the limit of the (monotonically increasing) sequence $x_0 = y \ln y$,

Fig. 3.20 Regula falsi

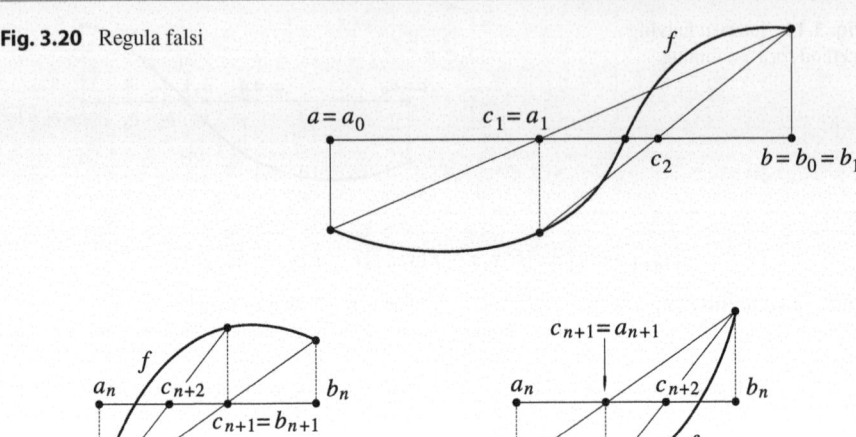

Fig. 3.21 Behavior of the secants in the Regula falsi

$x_1 = y \ln x_0 = y \ln y + y \ln \ln y$, $x_2 = y \ln x_1, \ldots$ results as the only solution according to theorem 3.8.20, see Fig. 3.18.

By the way, asymptotically $x(y) = x = y \ln x \sim y \ln y = x_0(y)$ for $y \to \infty$, as was already noted at the end of example 3.8.7. In the case of $y = 10^6$, one obtains for $x_n \cdot 10^{-6}$, $n = 0, \ldots, 4$, the approximation values 13,8155, 16,4413, 16,6153, 16,6258, 16,6265. According to the prime number theorem, see loc. cit., the x_n estimates for the millionth prime number p_{10^6}. The true value is $p_{10^6} = 15.485.863$.[14]

Example 3.8.21 (One-sided continuous functions) Let $D \subseteq \mathbb{R}$ and $a \in D$. Then a function $f: D \to \mathbb{K}$ is called **left-continuous** in a (or **right-continuous**) continuous, if $f|(D \cap]-\infty, a])$ (or $f|(D \cap [a, \infty[)$) is continuous at the point a. Exactly then is f continuous in a, when f in a is both left- and right-continuous. ◊

Example 3.8.22 (Monotonic functions) Let $I \subseteq \mathbb{R}$ be an interval with more than one point and $f: I \to \mathbb{R}$ monotonic. For an inner point a of I, both the left-sided limit $f(a-)$ and the right-sided limit $f(a+)$ exist. (If f is monotonically increasing, these are $\mathrm{Sup}\{f(x) \mid x \in I, x < a\}$ or $\mathrm{Inf}\{f(x) \mid x \in I, x < a\}$.) The jump height $f(a+) - f(a-)$ is always ≥ 0 for monotonically increasing functions and always ≤ 0 for monotonically decreasing functions. Its absolute value is equal to the fluctuation $S_a(f)$ of f in a. Exactly then is f continuous in a, if the jump height in a is equal to 0. Only in countably many points $a \in I$ can $S_a(f) \neq 0$ be, see problem 3.8.28, and *thus f has only countably many discontinuity points.* If one replaces $f(a)$ at each point $a \in \overset{\circ}{I}$ by $f(a-)$ (or by $f(a+)$), one obtains a left-continuous (or right-continuous) function on the interior $\overset{\circ}{I}$ of I. Finally, for the

---

[14] Efforts are constantly being made to improve the estimate by the prime number theorem.

3.8 Continuous Functions

boundary points $c<d$ of I in $\overline{\mathbb{R}}$ the (possibly improper) limits $f(c+)$ and $f(d-)$ exist. If a boundary point belongs to I, then f is continuous there if the function value matches the corresponding one-sided limit. ◊

Although continuity is a local property, it allows conclusions about the global course of a continuous function.[15] The following theorem was first proven by B. Bolzano in 1817 (but used without comment before).

Theorem 3.8.23 (Nullstellensatz) *Let $f: [a,b] \to \mathbb{R}$, $a < b$, be a continuous real-valued function on the closed interval $[a,b] \subseteq \mathbb{R}$. If $f(a)$ and $f(b)$ have different signs, then f has a root x_0 in the interval $[a,b]$, i.e., there exists an $x_0 \in [a,b]$ with $f(x_0) = 0$.*

Proof It suffices to consider the case $f(a) \leq 0$ and $f(b) \geq 0$. We use the interval having method and define an interval nesting $[a_n, b_n]$, $n \in \mathbb{N}$, in $[a,b]$ with the following properties:

$$[a_n, b_n]$$

Then the number defined by this interval nesting $x_0 \in [a,b]$ is a root of f: It is namely $x_0 = \lim a_n$ and due to the continuity of f consequently $f(x_0) = f(\lim a_n) = \lim f(a_n) \leq 0$. Similarly, $f(x_0) = f(\lim b_n) = \lim f(b_n) \geq 0$, so overall $f(x_0) = 0$. We now define a_n and b_n recursively by $a_0 = a$, $b_0 = b$ and

$$f(x_0) = f(\lim b_n) = \lim f(b_n) \geq 0$$

$$b_0 = b$$

see Fig. 3.19. Then conditions (1) and (2) are obviously fulfilled. □

An immediate consequence of the root theorem is:

Theorem 3.8.24 (Intermediate Value Theorem) *Let $f: [a,b] \to \mathbb{R}$ be a continuous real-valued function. For any value $c \in \mathbb{R}$ between $f(a)$ and $f(b)$, there exists an $x_0 \in [a,b]$ such that $f(x_0) = c$.*

Proof Like f, the function $g: [a,b] \to \mathbb{R}$ with $g(x) := f(x) - c$ is continuous. Since c lies between $f(a)$ and $f(b)$, 0 lies between $g(a) = f(a) - c$ and $g(b) = f(b) - c$. According to the Zero Theorem 3.8.23, g has a zero $x_0 \in [a,b]$. Then $f(x_0) = c$ is true. □

Example 3.8.25 If $F \in \mathbb{R}[X]^*$ is a real polynomial of odd degree, then according to example 3.8.14 the limits $\lim_{x \to \infty} F(x)$ and $\lim_{x \to -\infty} F(x)$ have different signs. Therefore, F has a root in $\mathbb{R}$, which we have already mentioned following theorem 2.10.30. ◊

---

[15] The conclusion from the local to the global is one of the basic motives for analysis.

Example 3.8.26 (Regula falsi) The interval halving method used in the proof of the root theorem provides a simple and reliable method for approximating the roots of a continuous function f, in which the length of the interval in which the root to be constructed lies is halved at each step. The **Regula falsi** (= rule of the false) described below provides a sequence that sometimes converges more quickly to a root of f. The basic idea is to interpolate f in the interval $[a,b]$, $a<b$, linearly through the secant

$$h(x) := \frac{f(b)-f(a)}{b-a}(x-a) + f(a)$$

to the points $(a, f(a))$ and $(b, f(b))$ of the graph of f and to take their root

$$a - \frac{b-a}{f(b)-f(a)}f(a) = \frac{af(b)-bf(a)}{f(b)-f(a)} = b - \frac{a-b}{f(a)-f(b)}f(b)$$

as an approximation of a root of f. With $f(a) < 0 < f(b)$, we obtain the following recursion scheme for the interval ends a_n and b_n and the roots c_n of the associated secants, which is indicated in Fig. 3.20:

$$a_0 = a, \quad b_0 = b; \quad c_{n+1} = \frac{a_n f(b_n) - b_n f(a_n)}{f(b_n) - f(a_n)},$$

$$a_{n+1} = \begin{cases} c_{n+1}, & \text{if } f(c_{n+1}) \leq 0, \\ a_n & \text{otherwise}, \end{cases} \quad b_{n+1} = \begin{cases} b_n, & \text{if } f(c_{n+1}) \leq 0, \\ c_{n+1} & \text{otherwise}. \end{cases}$$

We show: *The sequence (c_n) converges to a root of f.* By construction, always $f(a_n) \leq 0$, $f(b_n) \geq 0$, and the monotonic sequences (a_n) and (b_n) converge to values $\alpha, \beta \in [a,b]$ with $f(\alpha) \leq 0 \leq f(\beta)$. (In general, $\alpha < \beta$, so $[a_n, b_n]$, $n \in \mathbb{N}$, no interval nesting.) Also, the sequence (s_n) of secant slopes $s_n := \bigl(f(b_n) - f(a_n)\bigr)/(b_n - a_n)$ is obviously monotonically increasing, see the two sketches in Fig. 3.21. Therefore, the sequence $(1/s_n)$ is monotonically decreasing and positive, thus convergent. Thus, the sequence $c_{n+1} = a_n - f(a_n)/s_n = b_n - f(b_n)/s_n$, $n \in \mathbb{N}$, also converges, either towards α or β, say towards α. In the case $\alpha = \beta$, of course, $f(\alpha) = f(\beta) = 0$. Otherwise, $(1/s_n)$ is certainly not a null sequence, and the rules for limits again yield $f(\alpha) = 0$. — For a variant with an error estimate see Vol. 2, Ex. 2.6.10. ◊

Theorem 3.8.27 *Every continuous function f, which maps a closed interval $[a,b] \subseteq \mathbb{R}$, $a<b$, onto itself, has a fixed point, i.e., an $x_0 \in [a,b]$ with $f(x_0) = x_0$.*

Proof The auxiliary function $g: [a,b] \to \mathbb{R}$ with $g(x) := f(x) - x$ is also continuous. Because of $a \leq f(a)$ and $f(b) \leq b$ is $g(a) = f(a) - a \geq 0$ and $g(b) = f(b) - b \leq 0$. According to the intermediate value theorem, g thus has a root x_0 in $[a,b]$. This is a fixed point of f. □

For the calculation of a fixed point of f in special cases, we refer to Banach's fixed point theorem 3.8.20 and Exercise 3.8.34b).

3.8 Continuous Functions

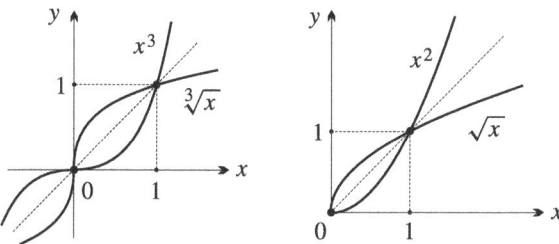

Fig. 3.22 Root functions

Example 3.8.28 Another conclusion from the intermediate value theorem is: *For every continuous function $f: S^1 \to \mathbb{R}$ on the unit circle $S^1 = U = \{z \in \mathbb{C} \mid |z| = 1\} \subseteq \mathbb{C}$ there are antipodal points $z_0, -z_0 \in S^1$ with $f(z_0) = f(-z_0)$*. *Proof.* With f also $g: [-1, 1] \to \mathbb{R}$, $x \mapsto f\left(x + i\sqrt{1 - x^2}\right) - f\left(-x - i\sqrt{1 - x^2}\right)$, is continuous, since $x \mapsto \sqrt{1 - x^2}$ is continuous on the interval $[-1, 1]$, see example 3.8.32. Moreover, $g(1) = -g(-1)$. According to theorem 3.8.23, g therefore has a zero $x_0 \in [-1, 1]$, and for $z_0 := x_0 + i\sqrt{1 - x_0^2} \in S^1$ applies $f(z_0) = f(-z_0)$.

For example, on every great circle of the Earth's surface, there are two antipodal points where the same temperature prevails at a given time (assuming that the temperature depends continuously on the location). An analogous statement as above about the existence of antipodal points with the same function value naturally applies to functions on any circular line in the complex number plane. *Therefore, there is no injective continuous mapping $D \to \mathbb{R}$ of a set $D \subseteq \mathbb{C}$, which includes a circular line.* On the other hand, there are even bijective mappings $\mathbb{C} \to \mathbb{R}$, since $\mathbb{R}$ and $\mathbb{C} = \mathbb{R} \times \mathbb{R}$ are equipotent, cf. example 1.8.14. According to what has been said, these mappings cannot be continuous. The above statement can be generalized to spheres of higher dimensions. We will discuss this so-called Borsuk-Ulam theorem in later volumes. ◊

We continue to discuss continuous real-valued functions on real intervals.

Lemma 3.8.29 *A real-valued continuous function on an interval $I \subseteq \mathbb{R}$ is injective if and only if it is strictly monotonic.*

Proof If f is strictly monotonic, then f is naturally injective (regardless of whether f is continuous or not). Conversely, let f be injective. We first consider the case where I is a closed interval $[a, b]$, $a < b$. Let's assume $f(a) < f(b)$. Then we have to show that f is strictly monotonically increasing. We first show that $f([a, b]) \subseteq [f(a), f(b)]$ is. Let $a < x < b$. If $f(x) < f(a) < f(b)$ (or $f(x) > f(b) > f(a)$), then according to the intermediate value theorem 3.8.24 in the interval $[x, b]$ (or $[a, x]$) there would be another point where f takes the value $f(a)$ (or $f(b)$) in contradiction to the injectivity of f. Now, if $x, y \in [a, b]$, $x < y$, are arbitrary, then initially according to the proven $f(a) \leq f(x) < f(b)$. If we apply this conclusion to the interval $[x, b]$, it follows that $f(x) < f(y) \leq f(b)$.

Let the interval I be arbitrary, and there are two points $x_1, x_2 \in I$, $x_1 < x_2$, for which $f(x_1) < f(x_2)$ is. Then f is strictly monotonically increasing. If $x, y \in I$, $x<y$, then there is a closed interval $[a, b] \subseteq I$, which contains the points x_1, x_2, x, y. According to the proven, $f|[a,b]$ is strictly monotonic and due to $f(x_1) < f(x_2)$ even strictly monotonically increasing. So $f(x) < f(y)$ is. □

For monotone functions, continuity is already characterized by the validity of the intermediate value theorem. We say a function $f: I \to \mathbb{R}$ on an interval $I \subseteq \mathbb{R}$ **satisfies the intermediate value theorem**, if f on every interval $[x, y] \subseteq I$ takes all values between $f(x)$ and $f(y)$.

Theorem 3.8.30 *Let $f: I \to \mathbb{R}$ be a real-valued monotone function on the interval $I \subseteq \mathbb{R}$, which satisfies the intermediate value theorem. Then f is continuous.*

Proof We can restrict ourselves to the case where f is monotonically increasing. Let $a \in I$ and $\varepsilon > 0$ be given. We assume that a is not a boundary point of I, and leave the modifications that are otherwise necessary to the reader. Let $x_1, y_1 \in I$ with $x_1 < a < y_1$. Then $f(x_1) \leq f(a) \leq f(y_1)$. If $f(x_1) < f(a) - \varepsilon$, then there is an $x_2 \in]x_1, a[$ with $f(x_2) = f(a) - \varepsilon$, since f satisfies the intermediate value theorem by assumption. In any case, there is therefore an $x \in I$, $x<a$, with $f(a) - \varepsilon \leq f(x) \leq f(a)$. Similarly, there is an $y \in I$, $a<y$, with $f(a) \leq f(y) \leq f(a) + \varepsilon$. Due to the monotonicity of f, then $f([x, y])$ is in the ε-neighborhood of $f(a)$. □

Note that a *monotone* function $f: I \to \mathbb{R}$ on the interval $I \subseteq \mathbb{R}$ satisfies the intermediate value theorem exactly when the image $f(I)$ is also an interval, cf. Example 3.4.16. — For theorem 3.8.30 also see example 3.8.22.

Theorem 3.8.31 (Inverse theorem) *Let f be a real-valued continuous and strictly monotone function on the interval $I \subseteq \mathbb{R}$. Then $J := f(I)$ is also an interval in $\mathbb{R}$, and the inverse function $f^{-1}: J \to I$ to the bijective function $f: I \to J$ is also continuous and strictly monotone (of the same type of monotonicity).*

Proof To show that J is an interval, it suffices to show that J with any two points $x' < y'$ contains the entire interval $[x', y']$ (see Example 3.4.16). This follows from the validity of the intermediate value theorem for the continuous function f.

With f, of course, f^{-1} is also strictly monotonic, and of the same type of monotonicity. Furthermore, the function f^{-1} satisfies the intermediate value theorem due to $f^{-1}(J) = I$: If $x', y' \in J$, $x' = f(x)$, $y' = f(y)$ with $x, y \in I$, and c is between $x = f^{-1}(x')$ and $y = f^{-1}(y')$, then $c \in I$ and $c = f^{-1}(f(c))$, where $f(c)$ is between $f(x) = x'$ and $f(y) = y'$. According to theorem 3.8.30, f^{-1} is therefore continuous. □

3.8 Continuous Functions

Example 3.8.32 (Root Functions) Let $n \in \mathbb{N}^*$. If n is odd, then the power function $f: x \mapsto x^n$ is on $\mathbb{R}$ strictly monotonically increasing. Since the values of f are not limited either upwards or downwards, theorem 3.8.31 provides: (1) $f(\mathbb{R}) = \mathbb{R}$. (2) $f^{-1}: \mathbb{R} \to \mathbb{R}$ is continuous and strictly monotonically increasing. The inverse function f^{-1} of f is the **root function** $\mathbb{R} \to \mathbb{R}$, $x \mapsto \sqrt[n]{x} = x^{1/n}$, see Fig. 3.22 on the left. — Now let $n \in \mathbb{N}^*$ be even. Then $f: x \mapsto x^n$ on $\mathbb{R}_+$ is strictly monotonically increasing. Since $f(0) = 0$ and the values of f are not limited upwards, theorem 3.8.31 provides: (1) $f(\mathbb{R}_+) = \mathbb{R}_+$. (2) $f^{-1}: \mathbb{R}_+ \to \mathbb{R}_+$ is continuous and strictly monotonically increasing. In this case, the inverse function f^{-1} is the **root function** $\mathbb{R}_+ \to \mathbb{R}_+$, $x \mapsto \sqrt[n]{x} = x^{1/n}$, see Fig. 3.22 on the right. □

Exercises

Exercise 3.8.1 Let $D := \{1/n \mid n \in \mathbb{N}^*\}$ and $f: D \to \mathbb{K}$ be a function. $\lim_{x \to 0} f(x)$ exists if and only if the sequence $f(1/n)$, $n \in \mathbb{N}^*$, converges. In this case, the limit of the sequence is equal to $\lim_{x \to 0} f(x)$. With this value at position 0, a continuous extension of f to $\overline{D} = D \uplus \{0\}$ is defined. A sequence $x_n \in \mathbb{K}$, $n \in \mathbb{N}^*$, therefore converges if and only if the function $D \to \mathbb{K}$, $1/n \mapsto x_n$, has a continuous extension $\overline{D} \to \mathbb{K}$.

Exercise 3.8.2 Let $D \subseteq \mathbb{R}$ and $a \in \overline{D}$. For a function $f: D \to \mathbb{K}$, $\lim_{x \to a} f(x)$ exists if and only if the limit $\lim_{n \to \infty} f(x_n)$ for every *monotone* sequence (x_n) in D with $\lim_{n \to \infty} x_n = a$ exists and is always the same.

Exercise 3.8.3 Let $D \subseteq \mathbb{C}$ and $a \in \overline{D}$. The limit of $f: D \to \mathbb{K}$ at point a exists if and only if there is a neighborhood U of a such that the restriction $f|(D \cap U)$ has a limit at a. In this case, both limits coincide.

Exercise 3.8.4 Let $D = D_1 \cup \cdots \cup D_n \subseteq \mathbb{C}$ and $a \in \overline{D}$ ($= \overline{D}_1 \cup \cdots \cup \overline{D}_n$). The limit of $f: D \to \mathbb{K}$ in a exists if and only if for all i with $a \in \overline{D}_i$ the limits of the restrictions $f|D_i$ exist and coincide.

Exercise 3.8.5 It is $\lim_{z \to 0, z \neq 0} 1/z = \infty$; $\Re(1/z)$ and $\Im(1/z)$ also do not have a limit in the proper sense at point 0. (Try to get an idea of the by no means trivial graphs of the functions $\Re(1/z) = (\cos \varphi)/r$ and $\Im(1/z) = -(\sin \varphi)/r$ of $\mathbb{C}^\times$ in $\mathbb{R}$ (with $z = r(\cos \varphi + i \sin \varphi)$ and $r > 0$).)

Exercise 3.8.6 Confirm the following calculation rules for the Landau symbols. (We leave the exact situation description to the reader here and in the next exercise.)

a) From $f_1 = O(g)$ and $f_2 = O(g)$ follows $f_1 + f_2 = O(g)$.
b) From $f_1 = O(g_1)$ and $f_2 = O(g_2)$ follows $f_1 f_2 = O(g_1 g_2)$.
c) From $f_1 = O(g_1)$ and $f_2 = o(g_2)$ follows $f_1 f_2 = o(g_1 g_2)$.
d) $f = O(g)$ (or $f = o(g)$) is equivalent to $|f| = O(|g|)$ (or $|f| = o(|g|)$).

e) From $f_1 = O(g_1)$ and $f_2 = O(g_2)$ follows $f_1 + f_2 = O(\text{Max}(|g_1|, |g_2|))$.
f) From $f = O(g)$ (or $f = o(g)$) follows $af = O(g)$ (or $af = o(g)$) for every $a \in K$.

Exercise 3.8.7 Confirm the following calculation rules for asymptotic equality:

a) Let $f \sim g$. Exactly then $h = o(g)$ applies, when $f + h \sim g$ is.
b) From $f_1 \sim g_1$ and $f_2 \sim g_2$ follows $f_1 f_2 \sim g_1 g_2$.
c) From $f \sim g$ follows $g \sim f$.
d) From $f \sim g$ and $g \sim h$ follows $f \sim h$.

Exercise 3.8.8 Investigate whether the following limits of functions exist, and if so, determine their values, where x in parts b) to e) each runs in $\mathbb{R}$:

a) $\lim_{x \to 1, x \in \mathbb{C} - \{1\}} (x^n - 1)/(x^m - 1)$, $m, n \in \mathbb{Z} - \{0\}$. What is the result for the limit transition $x \in \mathbb{C} - \{\zeta\}$, where ζ is any $|m|$-th root of unity in $\mathbb{C}$?
b) $\lim_{x \to 1} (\sqrt{x+3} - 2)/(x - 1)$.
c) $\lim_{x \to 1} \dfrac{1}{x-1} \left(\dfrac{3}{x^2 + 5} - \dfrac{1}{x^2 + 1} \right)$.
d) $\lim_{x \to \infty} \sqrt{x}(\sqrt{1+x} - \sqrt{x})$.
e) $\lim_{x \to \infty} \sqrt{x+1}(\sqrt{x+a} - \sqrt{x+b})$, $a, b \in \mathbb{R}$.

Exercise 3.8.9 Investigate at which points the following functions $\mathbb{R} \to \mathbb{R}$ are continuous:

a) $x \mapsto \begin{cases} -x, & \text{if } x < 0 \text{ or } x > 1, \\ x^2 & \text{otherwise.} \end{cases}$

b) $x \mapsto \begin{cases} x^2 + 2x + 1, & \text{if } -1 \leq x \leq 0, \\ 1 - x & \text{otherwise.} \end{cases}$

c) $x \mapsto \begin{cases} x, & \text{if } x \in \mathbb{Q}, \\ 1 - x & \text{otherwise.} \end{cases}$

d) $x \mapsto \begin{cases} 2x^2, & \text{if } x \in \mathbb{Q}, \\ x^3 + x & \text{otherwise.} \end{cases}$

Exercise 3.8.10 For which choice of $a, b \in \mathbb{C}$ are the following functions $\mathbb{R} \to \mathbb{C}$ continuous:

a) $x \mapsto \begin{cases} 1 + x^2, & \text{if } x \leq 1, \\ ax - x^3, & \text{if } 1 < x \leq 2, \\ bx^2 & \text{otherwise.} \end{cases}$

b) $x \mapsto \begin{cases} x^2 + a, & \text{if } x \leq 1, \\ bix + 1 & \text{otherwise} \end{cases}$

3.8 Continuous Functions

Exercise 3.8.11 Let $M \subseteq \mathbb{K}$. The indicator function $e_M : \mathbb{K} \to \mathbb{K}$ is exactly at the points of the boundary $\operatorname{Rd} M = \overline{M} - \overset{\circ}{M}$ of M in $\mathbb{K}$ not continuous.

Exercise 3.8.12 Let K be an ordered field. The continuity of a function $f : K \to K$ is defined as in the case $K = \mathbb{R}$. The following statements are equivalent: (i) K is complete, i.e., a real number field. (ii) $\emptyset$ and K are the only subsets of K that are both open and closed in K (i.e., K is connected as a topological space, cf. 4.3). (iii) For continuous functions $K \to K$ the intermediate value theorem applies (i.e., every continuous function $f : K \to K$ takes on all values between any two values). (Note: If $A \subseteq K$ is a non-empty set bounded above without an upper limit, then the indicator function $e_{\operatorname{OS}(A)}$ of the set $\operatorname{UB}(A) = \operatorname{UB}_K(A)$ of upper bounds of A in K is continuous.)

Exercise 3.8.13 Let $D \subseteq \mathbb{K}$ and let $a \in \overline{D}$, $a \notin D$. The limit $\lim_{x \to a} f(x)$ exists for a function $f : D \to \mathbb{K}$ if and only if f can be extended to a function $\overline{f} : D \uplus \{a\} \to \mathbb{K}$ that is continuous at a. In this case, necessarily $\overline{f}(a) = \lim_{x \to a} f(x)$.

Exercise 3.8.14 Which of the continuous functions

$$\mathbb{C}^\times \to \mathbb{C}, \; z \mapsto \overline{z}/z, \quad \text{resp.} \quad U - \{1\} \to \mathbb{C}, \; z \mapsto (\overline{z} - 1)/(z - 1),$$

can be continuously extended to $\mathbb{C}$ or to $U = \{z \in \mathbb{C} \mid |z| = 1\}$?

Exercise 3.8.15 The function $f : \mathbb{R} \to \mathbb{R}$ given by Riemann, with

$$f(x) := \begin{cases} 0, & \text{if } x \notin \mathbb{Q}, \\ 1/b, & \text{if } x = a/b, \; a,b \in \mathbb{Z}, \; b > 0, \; \gcd(a,b) = 1, \end{cases}$$

is continuous exactly at $\mathbb{R} - \mathbb{Q}$.

Exercise 3.8.16 The function $f : \mathcal{C} \to \mathbb{R}$, $x = \sum_{k=1}^\infty 2e_I(k)/3^k \mapsto f(x) := \sum_{k=1}^\infty e_I(k)/2^k$, $I \subseteq \mathfrak{P}(\mathbb{N}^*)$, on the Cantor's discontinuum $\mathcal{C}$, cf. exercise 3.4.18, is continuous. Its image is the full unit interval $[0, 1] \subseteq \mathbb{R}$. (Note: f is *not* injective. Also determine the fibers of f.) Every continuous mapping $[0, 1] \to \mathcal{C}$ is constant.

Exercise 3.8.17 Let D be the union of arbitrarily many open or finitely many closed sets $D_i \subseteq \mathbb{K}$, $i \in I$. A function $f : D \to \mathbb{K}$ is continuous if and only if the restrictions $f|D_i$, $i \in I$, are all continuous. (Cf. exercises 3.8.3 and 3.8.4. — Show with examples that for arbitrary subsets $D_1, D_2 \subseteq \mathbb{K}$ the function $f : D \to \mathbb{K}$ on $D = D_1 \cup D_2$ does not need to be continuous if $f|D_1$ and $f|D_2$ are continuous.)

Exercise 3.8.18 Let $D \subseteq \mathbb{R}$. A function $f : D \to \mathbb{K}$ is continuous at the point $a \in D$ if for every strictly monotonic sequence (x_n) with $x_n \in D - \{a\}$ and $\lim x_n = a$, $\lim f(x_n) = f(a)$ applies. (See exercise 3.8.2.)

Exercise 3.8.19 Let $f, g: D \to \mathbb{K}$ be continuous. Then $|f|: x \mapsto |f(x)|$ and for $\mathbb{K} = \mathbb{R}$ also the functions $\mathrm{Max}\,(f, g): x \mapsto \mathrm{Max}\,(f(x), g(x))$ or $\mathrm{Min}\,(f, g): x \mapsto \mathrm{Min}\,(f(x), g(x))$ are continuous on D.

Exercise 3.8.20 Let $f: D \to \mathbb{C}$ be a complex-valued function. Then the real-valued functions $x \mapsto \Re(f(x))$ and $x \mapsto \Im(f(x))$ are called the **real** and the **imaginary part** of f and are denoted by $\Re f$ and $\Im f$ respectively. Show: For $a \in \overline{D}$ $\lim_{x \to a} f(x)$ exists if and only if $\lim_{x \to a} \Re(f(x))$ and $\lim_{x \to a} \Im(f(x))$ exist. In this case,
$$\lim_{x \to a} f(x) = \lim_{x \to a} \Re(f(x)) + \mathrm{i} \lim_{x \to a} \Im(f(x)).$$
In particular, f is continuous at the point $a \in D$ if and only if $\Re f$ and $\Im f$ are continuous in a.

Exercise 3.8.21
a) Two continuous $\mathbb{K}$-valued functions on $D \subseteq \mathbb{C}$ already coincide if their restrictions on a dense subset $D' \subseteq D$ in D coincide. For example, two continuous functions $\mathbb{R} \to \mathbb{K}$ are already equal if their values on $\mathbb{Q}$ coincide. (This was already used in example 1.8.17.)
b) Let $f: I \to \mathbb{R}$ be a real-valued continuous function on the interval $I \subseteq \mathbb{R}$. Furthermore, let $I' \subseteq I$ be a dense subset of I (which is usually not an interval). If $f|I'$ is monotonic (or strictly monotonic), then so is f, and of the same type of monotonicity.

Exercise 3.8.22
a) The functions $f: D \to \mathbb{K}$ and $g: D \to \mathbb{K}$ are Hölder continuous with the exponent $\alpha > 0$, see example 3.8.18. Then also $f + g$ and λf, $\lambda \in \mathbb{K}$, are Hölder continuous with the exponent α. Furthermore, fg is locally Hölder continuous with the same exponent, as is f/g (if g does not vanish anywhere). The locally Hölder continuous functions $D \to \mathbb{K}$ with exponent α form a $\mathbb{K}$-subalgebra of the algebra $C_\mathbb{K}(D)$ of continuous $\mathbb{K}$-valued functions on D. Provide an example that the product of two Hölder continuous functions with exponent α does not need to be Hölder continuous with exponent α.
b) If $f: D \to \mathbb{K}$ and $g: D' \to \mathbb{K}$ are Hölder continuous with the exponents α and β respectively and if $f(D) \subseteq D'$, then the composition $g \circ f: D \to \mathbb{K}$ is Hölder continuous with the exponent $\alpha\beta$.

Exercise 3.8.23
a) Polynomial functions are locally Lipschitz continuous.
b) Let $0 < \alpha < 1$. Then the function $x \mapsto x^\alpha$ from $\mathbb{R}_+$ to $\mathbb{R}_+$ is not Lipschitz continuous at the origin. However, it is Lipschitz continuous on any interval $[a, \infty[$, $a > 0$.

Exercise 3.8.24 Let $0 \in D$ and $f: D \to \mathbb{K}$ be a Hölder continuous function at the origin with $f(0) = 0$ and the exponent $\alpha > 0$. For every summable family x_i, $i \in I$,

Fig. 3.23 x as a fixed point of $f(x) = a + \arctan x$ or $g(x) = \tan(x-a)$

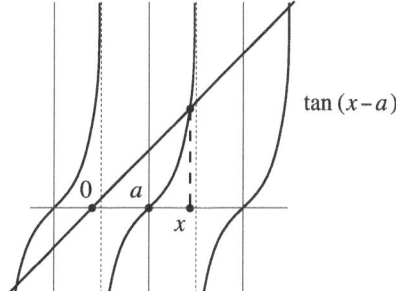

in D, the family $|f(x_i)|^{1/\alpha}$, $i \in I$, is also summable. In particular, $f(x_i)$, $i \in I$, with x_i, $i \in I$, is summable if f is Lipschitz continuous at the origin.

Exercise 3.8.25 Let $f: D \to \mathbb{K}$ and $g: D' \to \mathbb{K}$ be functions with $f(D) \subseteq D'$. Furthermore, for $a \in \overline{D}$ the limit $a' := \lim_{x \to a} f(x)$ exists. Let $a' \in D'$, and g be continuous in a'. Then $\lim_{x \to a} g(f(x))$ also exists and is equal to $g(a')$.

Exercise 3.8.26
a) Let $f: D \to \mathbb{K}$ be continuous on the closed set $D \subseteq \mathbb{K}$. For every closed set $A \subseteq \mathbb{K}$, the pre-image set $f^{-1}(A) \subseteq D$ is also closed in $\mathbb{K}$. In particular, the set of zeros of f is closed.
b) Let $f: D \to \mathbb{K}$ be continuous on the open set D in $\mathbb{R}$ (or $\mathbb{C}$). For every open set $A \subseteq \mathbb{K}$, the pre-image set $f^{-1}(A) \subseteq D$ is also open in $\mathbb{R}$ (or $\mathbb{C}$).

Exercise 3.8.27 A monotone function $f: I \to \mathbb{R}$ is left-(or right-) continuous at an inner point a of the interval $I \subseteq \mathbb{R}$ if and only if there is *one* monotonically increasing (or monotonically decreasing) sequence (x_n) in $I - \{a\}$ with $\lim x_n = a$ and $\lim f(x_n) = f(a)$.

Exercise 3.8.28 Let $f: I \to \mathbb{R}$ be a monotone function on the open interval $I \subseteq \mathbb{R}$. For all $a, b \in I$, $a<b$, it holds that $\sum_{x \in]a,b[} |h(x)| \leq |f(b) - f(a)|$, where $h(x)$ denotes the jump height of f at x. In particular, it follows that the set of true jump points $x \in I$ (with $h(x) \neq 0$) is countable.

Exercise 3.8.29 Let h_x, $x \in \mathbb{R}$, be a summable family of real numbers. Then the function $\mathbb{R} \to \mathbb{R}$ with $H(t) := \sum_{x<t} h_x$ is left-continuous and has a jump discontinuity at every point $x \in \mathbb{R}$ with jump height h_x. In particular, the points of discontinuity of H are exactly the points $x \in \mathbb{R}$ with $h_x \neq 0$.

Exercise 3.8.30 Let $f: D \to \mathbb{K}$ be a function.

a) For each $r \in \mathbb{R}_+$ the set of points $a \in \overline{D}$ with $S_a(f) \geq r$ is closed. (For the fluctuation $S_a(f)$ of f in a see example 3.8.17.)

Fig. 3.24 Eccentric anomaly x and area law

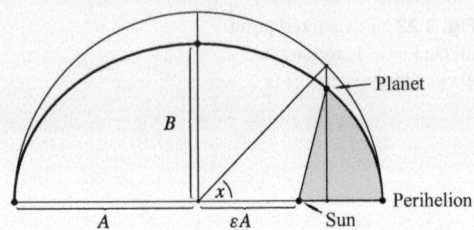

b) The set of points $a \in \overline{D}$, in which $\lim_{x \to a} f(x)$ does not exist, is the union of countably many closed sets. In particular, the set of discontinuity points of a function f on a closed set $D \, (=\overline{D})$ is always the union of countably many closed sets.

Exercise 3.8.31 Prove the so-called **Baire's Category Theorem**: Let U_n, $n \in \mathbb{N}$, be a sequence of dense open sets in $\mathbb{K}$. Then the intersection $\bigcap_{n \in \mathbb{N}} U_n$ is also dense in $\mathbb{K}$ and moreover uncountable. (Replacing U_n by $\bigcap_{k \leq n} U_k$, one can assume that (U_n) decreases monotonically. In the case of $\mathbb{K} = \mathbb{R}$, for each interval $[a,b]$, $a<b$, successive intervals $[a_n, b_n] \subseteq U_n$ are found with $a_n < b_n$ and $[a, b] \supseteq [a_0, b_0] \supseteq \cdots \supseteq [a_n, b_n] \supseteq [a_{n+1}, b_{n+1}] \supseteq \cdots$. In the case of $\mathbb{K} = \mathbb{C}$, one concludes similarly with closed disks. — For a generalization see theorem 4.5.18.)

Exercise 3.8.32 There is no function $f \colon \mathbb{R} \to \mathbb{R}$, whose points of discontinuity are exactly the irrational numbers. (This follows with exercise 3.8.30b) and exercise 3.8.31.)

Exercise 3.8.33 Let $f \colon D \to D$ be a continuous function. Does the sequence $x_0 \in D$ converge for a $x_n = f^n(x_0) \, (= f(x_{n-1})$, $n \in \mathbb{N}^*)$ for $n \to \infty$ to a point $x \in D$, then this x is necessarily a fixed point of f.

Exercise 3.8.34 Let $f \colon [a, b] \to [a, b]$, $a, b \in \mathbb{R}$, $a<b$, be a monotonically increasing function.

a) f has a fixed point (namely $\operatorname{Sup}\{x \in [a, b] \mid x \leq f(x)\}$). Provide an example that a monotonically decreasing function $[a, b] \to [a, b]$ generally does not have a fixed point.
b) If f is also continuous, then every sequence (x_n) with $x_0 \in [a, b]$ arbitrary and $x_{n+1} = f(x_n)$, $n \in \mathbb{N}$, converges to a fixed point of f. For $x_0 = a$, (x_n) converges to the smallest and for $x_0 = b$ to the largest fixed point of f. (For monotonically decreasing continuous functions $g \colon [a, b] \to [a, b]$, a fixed point (which exists according to theorem 3.8.27) cannot generally be determined with such recursion, as the trivial example $x \mapsto 1 - x$ on $[0, 1]$ shows. But if the composition $g \circ g$ (which is monotonically increasing) has only one fixed point (which then necessarily coincides with that of g), then the sequence (x_n) also now converges to this fixed point. Proof!)

3.8 Continuous Functions

Exercise 3.8.35 Let $f(x) := ax + b$, $a, b \in \mathbb{C}$, $|a| < 1$. What sequence does the method of successive approximation in theorem 3.8.20 yield for the fixed point $x = b/(1-a)$ of f, if one starts with $x_0 = b$?

Exercise 3.8.36 The following exercises illustrate the Banach fixed-point theorem 3.8.20 and partially use means of analysis that will be covered later. In particular, justify if necessary with the help of an estimation of the derivative, cf. the remark after theorem 3.8.20, the strong contractibility of the considered functions f, whose fixed point is to be determined.

a) $f(x) = \frac{1}{4}(1-x^3)$ on $[0,1]$. (The fixed point is a root of $x^3 + 4x - 1$.)
b) $f(x) = \frac{1}{2}(x - e^x) + 1$ on $[0, \frac{1}{2}]$. (The fixed point is the root of $e^x + x - 2$.)
c) $f(x) = \cos x$ on $[0, 1]$. (Repeatedly press the cos button of a just started calculator.)
d) $f(x) = (x+2)/(x+1)$ on $[1, 2]$. (The fixed point is $\sqrt{2}$.)
e) $f(x) = \sqrt{1+x}$ on $\mathbb{R}_+$. (The fixed point is the number Φ of the golden ratio. Therefore, one also writes

$$\Phi = \sqrt{1 + \sqrt{1 + \sqrt{1 + \cdots}}}.$$

More precisely: If one chooses $x_0 \in \mathbb{R}_+$ as the starting value for the iteration, then $x_n = f^n(x_0) = \sqrt{1 + \sqrt{1 + \sqrt{1 + \cdots + \sqrt{1 + x_0}}}}$ (n root signs), $n \in \mathbb{N}^*$. If $x_0 \in \mathbb{Q}_+^\times$, but $1 + x_0 \notin {}^2\mathbb{Q}_+^\times$, e.g. $x_0 = 1$, then x_n has degree 2^n over $\mathbb{Q}$, $n \in \mathbb{N}$. The minimal polynomials $P_n \in \mathbb{Q}[X]$ of the x_n can be determined recursively by $P_0 = X - x_0$, $P_{n+1} = P_n(X^2 - 1)$, $n \in \mathbb{N}$. (To prove this, show $x_{n+1} \notin \mathbb{Q}[x_n]$ by induction over n.) See also exercise 3.3.25b).
f) $f(x) = a + \arctan x = a + \frac{1}{2}\pi - \arctan x^{-1}$ to $[a, \infty[$, where $a > 0$. (The fixed point x satisfies the equation $x = \tan(x - a)$ with $x \in]a, a + \frac{1}{2}\pi[$, see Fig. 3.23.) Consider in particular the case $a = k\pi$, $k \in \mathbb{N}^*$, i.e. the equation $x = \tan x$. Use $x_0 := a + \frac{1}{2}\pi$ as the initial value.
(**Note** With the power series representation $\arctan x = \sum_{n=0}^{\infty}(-1)^n x^{2n+1}/(2n+1)$ for $x \in]-1, 1]$, cf. Vol. 2, Sect. 2.4, one obtains for $x_0 := a + \frac{1}{2}\pi$ for *large a* the representations

$$x_1 = x_0 - \arctan\frac{1}{x_0} = x_0 - \frac{1}{x_0} + \cdots, \quad x_2 = x_0 - \arctan\frac{1}{x_1} = x_0 - \frac{1}{x_0} - \frac{2}{3x_0^3} + \cdots,$$

$$x_3 = x_0 - \arctan\frac{1}{x_2} = x_0 - \frac{1}{x_0} - \frac{2}{3x_0^3} - \frac{13}{15x_0^5} + \cdots$$

etc. For *small a*, the fixed point x has a power series development in $a^{1/3}$. Cf. Vol. 2, Example 1.2.26.)
g) $f(x) = M + \varepsilon \sin x$ on $\mathbb{R}$, $M \in \mathbb{R}$, $\varepsilon \in [0, 1[$. Choose some exemplary values for M and ε (such as $M = \pi/4, \pi/2, 3\pi/4$ and $\varepsilon = 1/10, 1/4, 1/2, 3/4$) and start with $x_0 = M$.

(**Note** The equation $x = M + \varepsilon \sin x$ is called the **Kepler's equation**. When using the power series representation $\sin x = \sum_{n=0}^{\infty} (-1)^n x^{2n+1}/(2n+1)!$, cf. Vol. 2, Sect. 1.4, one obtains (in powers of ε):

$$x_0 = M, \quad x_1 = M + \varepsilon \sin M,$$

$$x_2 = M + \varepsilon \sin(M + \varepsilon \sin M) = M + \varepsilon \sin M + \varepsilon^2 (\sin 2M)/2 + \cdots,$$

$$x_3 = M + \varepsilon \sin x_2 = M + \varepsilon \sin M + \varepsilon^2 (\sin 2M)/2$$
$$+ \varepsilon^3 (3 \sin 3M - \sin M)/8 + \cdots$$

etc. As a closed expression for x we mention the representation found by F. W. Bessel (1784–1846) valid for all $M \in \mathbb{R}$ and $\varepsilon \in [0, 1[$

$$x = M + 2 \cdot \sum_{n=1}^{\infty} \frac{J_n(n\varepsilon)}{n} \sin nM,$$

in which the J_n, $n \in \mathbb{N}^*$, are Bessel functions, which will be discussed in one of the following volumes. *Proof.* The derivative dx/dM is (for fixed ε) an even periodic function with the period 2π and has the Fourier development

$$\frac{a_0}{2} + \sum_{n=1}^{\infty} a_n \cos nM, \quad a_n := \frac{1}{\pi} \int_0^{2\pi} \frac{dx}{dM} \cos nM dM$$

$$= \frac{1}{\pi} \int_0^{2\pi} \cos(nx - n\varepsilon \sin x) dx = 2 J_n(n\varepsilon),$$

from which, by integration due to $x(0) = 0$, the assertion follows.

The Kepler's equation is of great importance in the study of the motion of a planet around the sun. If its (elliptical) orbit has the semi-axis lengths A, B with $0 < B \leq A$ and the **eccentricity** $\varepsilon := \sqrt{A^2 - B^2}/A$ and t is the time since the last perihelion passage, then the ray Sun-Planet has swept over the area $AB\pi t/T$ in this period (area law = 2nd Kepler's law). Here, T is the orbital period of the planet. If x, as indicated in Fig. 3.24, is the so-called **eccentric anomaly**, then for $\varepsilon \leq \cos x$ and $0 \leq x \leq \pi$ this area is on the other hand equal to

$$\frac{1}{2}(A \cos x - A\varepsilon) \cdot B \sin x + \frac{B}{A}\left(\frac{A^2 x}{2} - \frac{A^2}{2} \cos x \sin x\right) = \frac{AB}{2}(x - \varepsilon \sin x).$$

The eccentric anomaly x thus fulfills the Kepler's equation $x - \varepsilon \sin x = M$ with the **mean anomaly** $M := 2\pi t/T$ at time t.[16] This also applies to the x with $\varepsilon > \cos x$ (and $0 \leq x \leq \pi$) as well as for $x > \pi$, as similar calculations show.)

---

[16] The angle between the rays Sun-Perihelion and Sun-Planet is called the **true anomaly** (at time t).

3.8 Continuous Functions

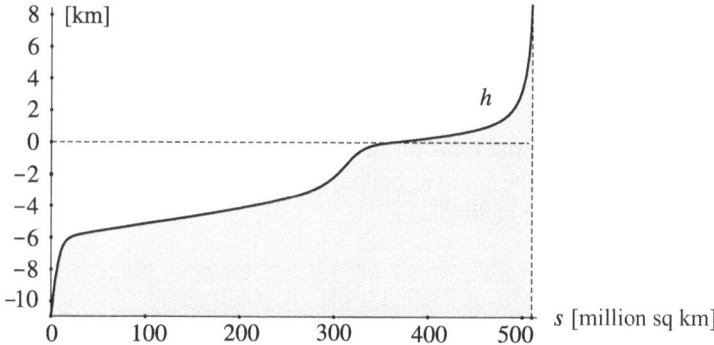

s [million sq km]	4.5	88.5	207.5	278.5	303	318	333.5	362
h(s) [km]	-6	-5	-4	-3	-2	-1	-0.2	0
s [million sq km]	443	470	494	504	507	509.5	510	
h(s) [km]	0.5	1	2	3	4	5	8	

Fig. 3.25 Value table and graph of the hypsographic curve h

Exercise 3.8.37 For $r \in [0, 4]$ is $f(x) := rx(1 - x)$ a function of $[0, 1]$ in itself, which is strongly contracting at $r \in [0, 1[$. For $r \in [0, 1[$ and every initial value $x_0 \in [0, 1]$ the iterated sequence (x_n) with $x_{n+1} = f(x_n)$ thus converges to 0. This also applies to $r = 1$. At $r \in]1, 2]$ f defines a monotonically increasing mapping of the interval $[0, \frac{1}{2}]$ into itself; for every initial value $x_0 \in]0, 1[$ the above sequence (x_n) then (from $n = 1$ monotonically increasing) converges to the non-zero fixed point of f, see also exercise 3.8.34b). At $r \in]2, 3[$ and an initial value x_0, which differs by less than $(3 - r)/2r$ from the fixed point $(r - 1)/r$ of f, the sequence (x_n) still converges to this fixed point. Finally, experimentally examine the behavior of the iterated sequence (x_n) for other starting values and for parameter values $r \in [3, 4]$. — For $r \in]1, 3]$ and $x_0 \in]0, 1[$ converges (x_n) always against $(r - 1)/r$. — The discrete dynamic systems defined by the functions $f : [0, 1] \to [0, 1]$ are discrete variants of the Verhulst differential equation (which models the growth of populations). For parameter values $r \in]3, 4]$ the purely periodic points x_0 with a period $k \geq 2$ are interesting. These are fixed points of the k-times iterated mapping $f^k = f \circ \cdots \circ f$. (See also exercise 2.4.16. — The discrete dynamic systems defined by $f(x) = rx(1 - x)$ $([0, 1], f)$ are prime examples of the behavior of dynamic systems depending on parameters (here r).)

Exercise 3.8.38 Provide a contracting function $[0, 1] \to [0, 1]$ that is not strongly contracting.

Exercise 3.8.39 Sketch the graph of a function $f : [-1, 1] \to \mathbb{R}$ that is continuous outside the zero point and discontinuous at the zero point, but does not have a jump discontinuity there.

Exercise 3.8.40 Prove the following generalization of Banach's fixed point theorem 3.8.20: Let $f: D \to D$ be a function on the non-empty closed set $D \subseteq \mathbb{C}$ with the iterates f^n, $n \in \mathbb{N}$. For each $n \in \mathbb{N}$ there exists a $L_n \in \mathbb{R}_+$ with $|f^n(x) - f^n(x')| \leq L_n |x - x'|$ for all $x, x' \in D$. Here, $M := \sum_{n \in \mathbb{N}} L_n < \infty$. (In theorem 3.8.20, $L_n = L^n$ with $0 \leq L < 1$.) Then f has exactly one fixed point x. If $x_0 \in D$ is any point in D, then the sequence $x_{n+1} := f^{n+1}(x_0) = f(x_n)$, $n \in \mathbb{N}$, converges to x and it is

$$|x - x_n| \leq \left(\sum_{i=n}^{\infty} L_i\right) |x_1 - x_0| \quad \text{resp.} \quad |x - x_n| \leq M|x_{n+1} - x_n| \leq L_n M |x_1 - x_0|.$$

Exercise 3.8.41 Calculate all the roots of the following functions (approximately up to an error $\leq 10^{-5}$) using the interval halving method or the Regula falsi and compare the quality of the two methods.

a) $x^3 - 3x + 1$ on $\mathbb{R}$.
b) $x^3 - 2x - 5$ on $\mathbb{R}$.
c) $\ln x - (x-2)^2$ on $\mathbb{R}_+^\times$.
d) $e^x + x - 2$ on $\mathbb{R}$.
e) $x - \tan x$ on $]\frac{1}{2}(2k-1)\pi, \frac{1}{2}(2k+1)\pi[$, $k = 1, 2, 3, \ldots$ (Compare exercise 3.8.36f).)

Exercise 3.8.42 A continuous function $f: [a,b] \to \mathbb{R}$, which has a root in $[a,b]$ at all, has both a smallest and a largest root there. (Exercise 3.8.26a).) In particular, f has a largest and a smallest root in $]a,b[$ if $f(a)f(b) < 0$ is.

Exercise 3.8.43 A function $f: D \to \mathbb{K}$ is called **locally constant**, if for every $a \in D$ there is a neighborhood U of a such that $f|(U \cap D)$ is constant. Show: If D is an interval in $\mathbb{R}$ and f is locally constant, then f is constant.

Exercise 3.8.44 A continuous function f on an interval $I \subseteq \mathbb{R}$, whose values are all rational, is constant. More generally: Every continuous function $I \to \mathbb{K}$, which only takes countably many values, is constant. □

Exercise 3.8.45 Let $f, g: [a,b] \to \mathbb{R}$ be continuous functions with $f(a) \leq g(a)$ and $f(b) \geq g(b)$. Then there is an $x_0 \in [a,b]$ with $f(x_0) = g(x_0)$. (Consider $f - g$.)

Exercise 3.8.46 Let $f: [a,b] \to \mathbb{R}$, $a < b$, be a continuous function with $f(a) = f(b)$. For every $n \in \mathbb{N}^*$ there is then an $x_n \in [a, b - (b-a)/n]$ with $f(x_n) = f(x_n + (b-a)/n)$. (The case $n = 2$ corresponds to the statement in Example 3.8.28.) If $f(x) \geq f(a) = f(b)$ for all $x \in [a,b]$, then for every $c \in [0, b-a]$ there is an $x_0 \in [a, b-c]$ with $f(x_0) = f(x_0 + c)$.

Exercise 3.8.47 Let $I \subseteq \mathbb{R}$ be an interval and $f: I \to \mathbb{R}$ continuous. Then $f(I) \neq \mathbb{R}^\times$.

3.8 Continuous Functions

Exercise 3.8.48 Let $I \subseteq \mathbb{R}$ be an interval and $f: I \to \mathbb{R}$ a continuous function. For $x_1, \ldots, x_n \in I$ and $t_1, \ldots, t_n \in \mathbb{R}_+$ with $t_1 + \cdots + t_n = 1$ there is an $x_0 \in I$ with

$$f(x_0) = t_1 f(x_1) + \cdots + t_n f(x_n).$$

Exercise 3.8.49 Let $f: [a, b] \to \mathbb{R}$ be a continuous function with $f(a) \leq 0$ and $f(b) > 0$. Prove the zero theorem 3.8.23 for f in the following way: The set of $x \in [a, b]$ with $f(x) \leq 0$ is non-empty and closed. Its greatest element is $< b$ and a zero of f.

Exercise 3.8.50 Let $W \subseteq \mathbb{R}$ be dense in $\mathbb{R}$, i.e. $\overline{W} = \mathbb{R}$. The function $f: I \to \mathbb{R}$ on the interval $I \subseteq \mathbb{R}$ satisfies the intermediate value theorem. For every convergent sequence (a_n) in I with a limit $a \in I$, for which $(f(a_n))$ is a constant sequence in W, let $f(a) = f(a_n), n \in \mathbb{N}$ apply. Then f is continuous.

Exercise 3.8.51 The function $f: I \to \mathbb{R}$ on the closed interval $I \subseteq \mathbb{R}$ satisfies the intermediate value theorem, and the fibers of f are closed. Then f is continuous. (Use the result of exercise 3.8.50.)

Exercise 3.8.52 Provide a function $f: I \to \mathbb{R}$ on an interval $I \subseteq \mathbb{R}$ that satisfies the intermediate value theorem, but is not continuous there.

Exercise 3.8.53 Prove the theorem 3.8.31 about the continuity of the inverse function using the theorem 3.4.3 of Bolzano-Weierstraß, without using theorem 3.8.30.

Exercise 3.8.54 Let $f:]a, b[\to \mathbb{R}$, $-\infty \leq a < b \leq \infty$, be a left-continuous and monotonically increasing function with $A := f(a+) = \lim_{t \to a, t > a} f(t)$, $B := f(b-) = \lim_{t \to b, t < b} f(t) \in \overline{\mathbb{R}}$. The function $h:]A, B[\to]a, b[$ is defined by

$$h(s) := \operatorname{Sup}\{t \in]a, b[\mid f(t) \leq s\} \in]a, b[.$$

a) h is monotonically increasing and right-continuous.
b) Exactly then h is continuous at the point $s_0 \in]A, B[$ when f takes the value s_0 at most once.
c) For all $t \in]a, b[$ and $s \in]A, B[$ applies: Exactly then is $t \leq h(s)$, when $f(t) \leq s$ is.
d) If f is strictly monotonically increasing and continuous, then h is the inverse function to f.

Note h in a way replaces the inverse function to f, if f does not bijectively map the interval $]a, b[$ onto $]A, B[$. Let $f: \mathbb{R} \to \mathbb{R}$ be, for example, the distribution of heights on the Earth's surface, i.e., $f(t)$, $t \in \mathbb{R}$, is the total area of points on the Earth's surface that have a height $<t$ relative to sea level. f is then (as a distribution function of a measure, see Vol. 6), left-continuous, and h is called the **hypsographic curve**, cf. Fig. 3.25. With its help, the distribution of heights on the Earth's surface is usually described.

3.9 Continuous Functions on Compact Sets

In this section, we deal with continuous functions on compact sets in $\mathbb{K}$. We call a subset $K \subseteq \mathbb{K}$ **compact** (in $\mathbb{K}$), if it is closed and bounded. A subset K of $\mathbb{R}$ is compact in $\mathbb{R}$ if and only if it is compact in $\mathbb{C}$. The reference to the field $\mathbb{R}$ or $\mathbb{C}$ is therefore superfluous. The decisive property of compact sets is described in the following theorem:

Theorem 3.9.1 *Let K be a subset of $\mathbb{K}$. Then K is compact if and only if every sequence of elements from K has a subsequence that converges to an element from K.*

Proof Let K be compact and (x_n) be a sequence in K. As K is also (x_n) bounded, it therefore has convergent subsequences according to the Bolzano-Weierstrass theorem 3.4.3. Since K is closed, their limit is in K. — Conversely, if the given condition for the sequences in K is met, then K is bounded. Otherwise, there would be a sequence (x_n) in K with $|x_n| \geq n$, $n \in \mathbb{N}$, and this sequence would not have a convergent subsequence. K is also closed. To show this, it must be shown that the limit x of any convergent sequence (x_n) with elements from K also lies in K. Since by assumption the limit of a subsequence of (x_n) lies in K and this coincides with x, x lies in K. □

In Sect. 4.4 we will discuss the concept of compactness in a more general context. The main theorem of this section is the following statement:

Theorem 3.9.2 *Let $f: K \to \mathbb{K}$ be a continuous function on the compact set K. Then also $\text{Im} f = f(K)$ is compact.*

Proof Let $f(x_n)$, $n \in \mathbb{N}$, with $x_n \in K$ a sequence of elements from $f(K)$. According to Theorem 3.9.1, there exists for (x_n) a convergent subsequence (x_{n_k}) with $x := \lim_{k \to \infty} x_{n_k} \in K$. Due to the continuity of f, then $\lim_{k \to \infty} f(x_{n_k}) = f(x) \in f(K)$ applies, i.e. the given sequence $(f(x_n))$ has a convergent subsequence $(f(x_{n_k}))$, whose limit lies in $f(K)$, and $f(K)$ is compact again according to Theorem 3.9.1. □

Theorem 3.9.2 particularly states that every continuous function on a compact set is bounded, i.e. its image is bounded. The following corollaries, which go back to K. Weierstraß, provide additional information beyond the boundedness of continuous functions on compact sets.

Corollary 3.9.3 *Let $f: K \to \mathbb{R}$ be a continuous real-valued function on the compact set $K \neq \emptyset$. Then there exist $x_1, x_2 \in K$ such that for all $x \in K$ the following holds: $f(x_1) \leq f(x) \leq f(x_2)$.*

Proof As a compact subset of $\mathbb{R}$ is $f(K)$ bounded. Since the set is also closed and non-empty, it contains its infimum $f(x_1)$ and its supremum $f(x_2)$. □

3.9 Continuous Functions on Compact Sets

We can also formulate Corollary 3.9.3 as follows: *A continuous real-valued function on a non-empty compact set attains its global maximum and its global minimum.*

Corollary 3.9.4 *Let $f : [a,b] \to \mathbb{R}$ be a continuous real-valued function on the closed interval $[a,b] \subseteq \mathbb{R}$, $a \le b$. Then $f([a,b])$ is also a closed and bounded interval in $\mathbb{R}$.*

Proof The interval $[a,b]$ is compact. Therefore, the function f attains its global maximum and its global minimum and, according to the intermediate value theorem, also all values in between. □

Corollary 3.9.5 *Let $f : K \to \mathbb{K}$ be a continuous function on the non-empty compact set K. Then there are elements $x_1, x_2 \in K$ with $|f(x_1)| \le |f(x)| \le |f(x_2)|$ for all $x \in K$.*

Proof The statement follows from Corollary 3.9.3, applied to the continuous function $x \mapsto |f(x)|$. □

Example 3.9.6 In the preceding statements, the compactness of the domain of the continuous function f is essential: The function $x \mapsto 1/x$ on the bounded interval $]0, 1[$ for example, is continuous, but its image is the unbounded interval $]1, \infty[$. The image of the continuous function $x \mapsto x$ on $]0, 1[$ is bounded, but the supremum 1 and the infimum 0 of the function values are not themselves function values. ◊

As a beautiful application of the properties of continuous functions on compact sets, we prove the Fundamental Theorem of Algebra already cited in Theorem 2.10.29.

Theorem 3.9.7 (Fundamental Theorem of Algebra) *The field $\mathbb{C}$ is algebraically closed, i.e., every non-constant polynomial $F \in \mathbb{C}[X]$ has a root in $\mathbb{C}$.*

Proof We can assume $F(0) \ne 0$. Because of $\lim_{z \to \infty} F(z) = \infty$, see example 3.8.14, there is an $R \ge 0$ such that $|F(z)| \ge |F(0)|$ is for all $z \in \mathbb{C}$ with $|z| \ge R$. In addition, the closed and bounded circular disc $\overline{B}(0; R) = \{z \in \mathbb{C} \mid |z| \le R\}$ is compact. According to corollary 3.9.5, there is therefore an $z_0 \in \overline{B}(0; R)$ with $|F(z_0)| \le |F(z)|$ for all $z \in \overline{B}(0; R)$. Then $|F(z_0)| \le |F(z)|$ is for all $z \in \mathbb{C}$ due to $|F(z_0)| \le |F(0)| \le |F(z)|$ for $|z| \ge R$.

We show that $F(z_0) = 0$ is. To do this, we assume that $b_0 := F(z_0) \ne 0$, and construct an $z \in \mathbb{C}$ with $|F(z)| < |F(z_0)|$, which contradicts the choice of z_0. Developing the polynomial F around z_0 (cf. the remark following Corollary 2.9.29) provides for each $z \in \mathbb{C}$ a representation

$$F(z) = b_0 + b_s(z - z_0)^s + \cdots + b_n(z - z_0)^n$$

Fig. 3.26 Shadow points in the Sunrise Lemma

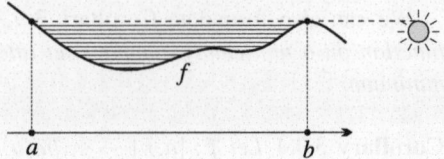

with constants $b_s, \ldots, b_n$, where $s \geq 1$ and $b_s \neq 0$ is. To determine z with $|F(z)| < |F(z_0)|$, we first choose an $x_0 \in \mathbb{C}$ with $x_0^s = -b_0/b_s (\neq 0)$, cf. Example 3.5.6. For $z := z_0 + rx_0, r \in \mathbb{R}_+^\times$, then

$$F(z) = b_0 + b_s r^s x_0^s + r^{s+1} G(r) = b_0(1 - r^s) + r^{s+1} G(r)$$

with a polynomial $G \in \mathbb{C}[X]$ and consequently

$$|F(z)| \leq |b_0|(1 - r^s) + r^{s+1}|G(r)| = |b_0| - r^s(|b_0| - r|G(r)|) < |b_0| = |F(z_0)|,$$

if moreover $r < 1$ and $r|G(r)| < |b_0|$ is, which can always be achieved because of $\lim_{r \to 0} rG(r) = 0$. □

The above proof of the Fundamental Theorem of Algebra essentially follows the proof by J. R. Argand (1768–1822) from the year 1806. The basic idea can already be found in 1748 by d'Alembert (1717–1783). For the algebraic consequences of the Fundamental Theorem of Algebra, we refer to Sect. 2.10, in particular to Theorem 2.10.30 on the prime factor decomposition of real polynomials.

Note 3.9.8 The above proof of the Fundamental Theorem of Algebra hardly provides a way to approximate the roots of a polynomial with reasonable effort. We will not provide any general methods for this difficult numerical problem later either. One will use the methods developed for arbitrary continuous or differentiable functions (such as the Newton method to be discussed in Vol. 2 and Vol. 5). To determine the real roots of real polynomials, one can also use the interval halving method from the proof of Theorem 3.8.23 or the Regula falsi according to Example 3.8.26. Rational roots of polynomials with integer coefficients can already be found with exercise 1.7.31a). ◊

Example 3.9.9 *The roots of a polynomial depend continuously on the coefficients.* We prove the following statement, which we will sharpen in Vol. 5. ◊

Theorem 3.9.10 *Let $n \in \mathbb{N}^*$. The normalized polynomial $F = a_0 + a_1 X + \cdots + X^n \in \mathbb{C}[X]$ has the roots $\alpha_1, \ldots, \alpha_n$ (multiple roots noted multiple times). Then for every $\varepsilon > 0$ there is a $\delta > 0$ such that the roots of all normalized polynomials $G = b_0 + b_1 X + \cdots + X^n \in \mathbb{C}[X]$ with $|b_i - a_i| \leq \delta$, $i = 0, \ldots, n-1$, lie in the union of the open ε-neighborhoods $B(\alpha_j; \varepsilon)$, $j = 1, \ldots, n$.*

Proof Let $\varepsilon > 0$ be given. We then initially choose $\delta \leq 1/n$, so every root β of G is obviously in absolute value $\leq c := |a_0| + \cdots + |a_{n-1}| + 1$, cf. exercise 3.5.9.

3.9 Continuous Functions on Compact Sets

Moreover, let δ be also $< \varepsilon^n/nc^{n-1}$. Then the assertion holds. If β were a root of G with $|\beta - \alpha_j| \geq \varepsilon$ for all $j = 1, \ldots, n$, one would obtain the contradiction

$$\varepsilon^n \leq |\beta - \alpha_1| \cdots |\beta - \alpha_n| = |F(\beta)| = |F(\beta) - G(\beta)| \leq \sum_{\nu=0}^{n-1} |a_\nu - b_\nu||\beta|^\nu$$

$$\leq n\delta c^{n-1} < \varepsilon^n. \quad \square$$

To conclude this section, we briefly touch on the concept of uniform continuity. The function $f:]0,1[\to \mathbb{R}$ with $f(x) := 1/x$ is continuous. For every point $a \in]0,1[$ and every $\varepsilon > 0$ there is therefore a $\delta > 0$ such that for all $x \in]0,1[$ with $|x - a| \leq \delta$ applies $|(1/x) - (1/a)| \leq \varepsilon$. This δ depends not only on ε (which is self-evident), but also significantly on the position a. The closer a is to 0, the smaller δ has to be chosen. Certainly, $\delta < a$ must be. In particular, for a given $\varepsilon > 0$ there is no $\delta > 0$, with which the continuity condition for *all* $a \in]0,1[$ is fulfilled. The function f is thus *not* uniformly continuous in the sense of the following definition.

Definition 3.9.11 A function $f: D \to \mathbb{K}$ is called **uniformly continuous** (in D), if for every (no matter how small) $\varepsilon > 0$ there is a $\delta > 0$ such that for all $x, y \in D$ with $|x - y| \leq \delta$ applies: $|f(x) - f(y)| \leq \varepsilon$.

Note that the uniform continuity of f is a global property that crucially also depends on the domain $D(\subseteq \mathbb{C})$. Of course, uniformly continuous functions are continuous. The converse also applies for compact domains:

Theorem 3.9.12 *Every continuous function $f: K \to \mathbb{K}$ on a compact set K is even uniformly continuous.*

Proof Assume the statement is false. Then there is an $\varepsilon_0 > 0$ and corresponding sequences x_n, $n \in \mathbb{N}^*$, and y_n, $n \in \mathbb{N}^*$, in K, for which $|x_n - y_n| \leq 1/n$ is true, but $|f(x_n) - f(y_n)| > \varepsilon_0$. Since K is compact, according to theorem 3.9.1, there is a convergent subsequence (x_{n_k}) of (x_n) with $x := \lim_{k \to \infty} x_{n_k} \in K$. Then $\lim_{k \to \infty} y_{n_k} = x$ is also true and due to the continuity of f

$$\lim_{k \to \infty} \left(f(x_{n_k}) - f(y_{n_k})\right) = \lim_{k \to \infty} f(x_{n_k}) - \lim_{k \to \infty} f(y_{n_k}) = f(x) - f(x) = 0$$

contradicts $|f(x_{n_k}) - f(y_{n_k})| > \varepsilon_0$ for all k. $\quad \square$

As an application of the concept of uniform continuity, we present the following extension theorem:

Theorem 3.9.13 *Every uniformly continuous function $f: D \to \mathbb{K}$ can be extended to a (uniquely determined) continuous function on the closure $\overline{D}$ of D. The extension $\overline{D} \to \mathbb{K}$ is also uniformly continuous.*

Proof According to Theorem 3.8.16, it suffices to show that for every $x \in \overline{D}$ the limit $\lim_{y \to x} f(y)$ exists. Due to the uniform continuity of f, the conditions of the Cauchy convergence criterion 3.8.3 for limits are met. We leave the proof of the addition to the reader. (See also Theorem 4.5.14.) □

For the existence of a continuous (not necessarily uniformly continuous) extension $\overline{D} \to \mathbb{K}$, it suffices in Theorem 3.9.13 to assume that for every $x \in \overline{D}$ there exists a neighborhood U of x such that $f|(D \cap U)$ is uniformly continuous.

Exercises

Exercise 3.9.1 There is no continuous surjective function $f : [0, 1] \to [0, 1[$.

Exercise 3.9.2 Let $n \in \mathbb{N}$, $n \geq 2$. There is no continuous real-valued function on the closed interval $[a, b] \subseteq \mathbb{R}$ that takes each of its values exactly n-times.

Exercise 3.9.3
a) There is no continuous bijective mapping $f : I \to U = \{z \in \mathbb{C} \mid |z| = 1\}$, where I is an open or a closed bounded interval in $\mathbb{R}$.
b) There is no continuous bijective mapping $f : \mathbb{R} \to U$. (There are continuous bijective mappings $\mathbb{R} \xrightarrow{\sim}]-1, 1[$.)

Exercise 3.9.4 Let $f : D \to \mathbb{R}$ be a continuous function on the compact subset $D \neq \emptyset$ of $\mathbb{C}$. Furthermore, let $E := \{\Re z \mid z \in D\} \subseteq \mathbb{R}$. Then E is compact, and for every $x \in E$ the set $D_x := \{z \in D \mid \Re z = x\}$ is also compact. Furthermore, the function $g : E \to \mathbb{R}$ with $g(x) := \text{Sup}\{f(z) \mid z \in D_x\}$ is **upper semi-continuous**, i.e. for every $a \in E$ and every $\varepsilon > 0$ there is a neighborhood U of a with $g(x) \leq g(a) + \varepsilon$ for all $x \in E \cap U$,[17] and it holds $\text{Sup}\{f(z) \mid z \in D\} = \{g(x) \mid x \in E\}$. If D is also convex (i.e. D contains the line segment connecting any two points), then $g : E \to \mathbb{R}$ is even continuous.

Exercise 3.9.5 Let D be compact and $f : D \to \mathbb{K}$ a continuous injective function. Then the inverse mapping $f^{-1} : f(D) \to D$ is also continuous. Show by examples that the statement is generally false for any $D \subseteq \mathbb{C}$.

Exercise 3.9.6 Let $f : D \to D$ be a contracting mapping of the compact set $D \neq \emptyset$ into itself, cf. example 3.8.19. Then f has exactly one fixed point. (Consider a point where $|f(x) - x|$ becomes minimal.) For every point $x_0 \in D$ the sequence $(x_n) = f^n(x_0)$, $n \in \mathbb{N}$, in D with $x_{n+1} = f(x_n)$, $n \in \mathbb{N}$, converges to this fixed point. Provide a closed set $D \neq \emptyset$ in $\mathbb{C}$ with a contracting mapping $f : D \to D$ without a fixed point, for example for $D = \mathbb{R}$.

---

[17] Similarly, **lower semi-continuity** is defined.

Exercise 3.9.7 Let $f: \mathbb{R} \to \mathbb{R}$ be continuous. A point $x \in \mathbb{R}$ is called a **shadow point for** f, if there is a $y > x$ in $\mathbb{R}$ with $f(y) > f(x)$. The points $a, b \in \mathbb{R}$, $a<b$, are not shadow points, but the open interval $]a, b[$ contains only shadow points for f. Then $f(x) < f(b)$ for all $x \in]a, b[$ and $f(a) = f(b)$, see Fig. 3.26. (Consider a point in $[x, b]$, where $f|[x, b]$ reaches the maximum — In the recommended book Spivak, M.: Calculus, New York 1967, this statement is called the **Sunrise Lemma**.)

Exercise 3.9.8 A uniformly continuous function $f: D \to \mathbb{K}$ on a bounded set $D \subseteq \mathbb{C}$ is bounded.

Exercise 3.9.9 Provide an example of a bounded continuous function on a bounded interval $I \subseteq \mathbb{R}$ that is not uniformly continuous. (Sketch the graph of such a function.)

Exercise 3.9.10
a) For $\alpha \in \mathbb{R}_+$ the function $x \mapsto x^\alpha$ on $\mathbb{R}_+$ is uniformly continuous if and only if $\alpha \leq 1$.
b) For $\alpha \in \mathbb{R}$ the function $x \mapsto x^\alpha$ on $[1, \infty[$ is uniformly continuous if and only if $\alpha \leq 1$.

Exercise 3.9.11
a) Every Hölder-continuous and in particular every Lipschitz-continuous function $f: D \to \mathbb{K}$ is uniformly continuous.
b) The functions $x \mapsto x^\alpha$ on $[0, 1]$ are uniformly continuous for $\alpha \geq 0$, but not Lipschitz-continuous for $0 < \alpha < 1$.

Exercise 3.9.12 Every bounded monotone continuous function $f: I \to \mathbb{R}$ on an interval $I \subseteq \mathbb{R}$ is uniformly continuous.

Exercise 3.9.13 A continuous function $f: D \to \mathbb{K}$ on a closed unbounded set $D \subseteq \mathbb{C}$ is uniformly continuous if $\lim_{x \to \infty} f(x) \in \mathbb{K}$ exists. If $D \subseteq \mathbb{R}$ is closed and D is unbounded above and below, the existence of the limits $\lim_{x \to -\infty} f(x)$ and $\lim_{x \to \infty} f(x)$ is sufficient for uniform continuity.

Exercise 3.9.14 A function $f: D \to \mathbb{C}$ is called **locally bounded** if for every point $x \in D$ there exists a neighborhood U of x for which $f|(D \cap U)$ is bounded. If $f: D \to \mathbb{C}$ is locally bounded and D is compact, then f is bounded.

Exercise 3.9.15 Let I_1, I_2 be intervals in $\mathbb{R}$ with $I_1 \cap I_2 \neq \emptyset$ and $I := I_1 \cup I_2$. A function $f: I \to \mathbb{C}$ is uniformly continuous if its restrictions $f|I_1$ and $f|I_2$ are uniformly continuous.

Exercise 3.9.16 Let K and L be compact subsets of $\mathbb{K}$. Then their Minkowski sum $K + L$ is also compact.

Exercise 3.9.17 Let $K \subseteq \mathbb{K}$ be a compact subset and U an open subset of $\mathbb{K}$ with $K \subseteq U$. Then there exists an $\varepsilon > 0$ such that the ε-tube

$$K_\varepsilon := \bigcup_{x \in K} \overline{B}_\mathbb{K}(x; \varepsilon) = K + \overline{B}_\mathbb{K}(0; \varepsilon)$$

around K (which is also compact according to exercise 3.9.16) still lies entirely within U. (Otherwise, there would be sequences x_n, $n \in \mathbb{N}^*$, and y_n, $n \in \mathbb{N}^*$, in $\mathbb{K}$ with $x_n \in K$ and $y_n \notin U$ and $|x_n - y_n| \leq 1/n$. For $x \in \mathbb{K}$ is $\overline{B}_\mathbb{K}(x; \varepsilon) = \{y \in \mathbb{K} \mid |y - x| \leq \varepsilon\}$.)

Exercise 3.9.18 Let $F: \mathbb{K} \to \mathbb{K}$, $t \mapsto F(t)$, $F \in \mathbb{K}[T]^*$, $\text{Grad } F \geq 1$, be a non-constant polynomial function on $\mathbb{K}$.

a) The F- pre image of any compact set $K \subseteq \mathbb{K}$ is also compact.
b) The F- image of any closed set $A \subseteq \mathbb{K}$ is also closed in $\mathbb{K}$. (This follows from a).)

Remark With the notations from Example 4.4.26 it follows: *Non-constant polynomial functions $\mathbb{K} \to \mathbb{K}$ are proper mappings.*

3.10 Real Exponential, Logarithm, and Power Functions

It is surprising, at least at first glance, that the additive group $(\mathbb{R}, +)$ and the multiplicative group $(\mathbb{R}_+^\times, \cdot)$ of positive real numbers are isomorphic.[18] In this section, the natural isomorphisms between the groups $\mathbb{R}$ and $\mathbb{R}_+^\times$ will be described.

Firstly, the multiplicative group $\mathbb{R}_+^\times$ (like the group $K_+^\times$ of positive elements of any ordered field K) is torsion-free. Since the exponentiation $a \mapsto a^n$ for each $n \in \mathbb{N}^*$ is surjective according to example 3.8.32 and thus an automorphism of $\mathbb{R}_+^\times$, $\mathbb{R}_+^\times$ is also a divisible group. The inverse of $a \mapsto a^n$ is the n-th root $a \mapsto a^{1/n}$. Overall, for every rational number $x = p/q$, $p, q \in \mathbb{Z}$, $q > 0$, the power mapping

$$a \mapsto a^x = a^{p/q} = (a^p)^{1/q} = (a^{1/q})^p, \quad a \in \mathbb{R}_+^\times,$$

is an endomorphism of $\mathbb{R}_+^\times$ and $x \mapsto (a \mapsto a^x)$ a ring homomorphism $\mathbb{Q} \to \text{End } \mathbb{R}_+^\times$, cf. exercise 2.8.17 (where the abelian group $(\mathbb{R}_+^\times, \cdot)$, on which $\mathbb{Q}$ operates, is however written multiplicatively).

This operation of the field $\mathbb{Q}$ on $\mathbb{R}_+^\times$ can be *naturally* extended to an operation of the field $\mathbb{R}$ on $\mathbb{R}_+^\times$. For this, the base $a \in \mathbb{R}_+^\times$ is first chosen firmly. On $\mathbb{Q}$, the exponential function $x \mapsto a^x$ (which is a group homomorphism $(\mathbb{Q}, +) \to (\mathbb{R}_+^\times, \cdot)$) is monotonous, and for $a > 1$ strictly monotonously increasing and for $0 < a < 1$ strictly monotonously decreasing as well as for $a = 1$ constantly equal to 1. This follows directly from $a^y > 1$ for $a > 1$, $y > 0$ and $a^y < 1$ for $a < 1$, $y > 0$.

---

[18] Compare this, for example, with the situation with $(\mathbb{Q}, +)$ and $(\mathbb{Q}_+^\times, \cdot)$.

3.10 Real Exponential, Logarithm, and Power Functions

Lemma 3.10.1 *Let $a \in \mathbb{R}_+^\times$. The exponential function $x \mapsto a^x$ on $\mathbb{Q}$ can be uniquely extended to a continuous group homomorphism $(\mathbb{R}, +) \to (\mathbb{R}_+^\times, \cdot)$.*

Proof The uniqueness and the homomorphism property of the continuous extension follow from the fact that $\mathbb{Q}$ is dense in $\mathbb{R}$ and $\mathbb{Q} \times \mathbb{Q}$ is dense in $\mathbb{R} \times \mathbb{R}$, see problem 3.8.21a). According to theorem 3.9.13, it is sufficient to show that the function $x \mapsto a^x$ is uniformly continuous on every $\mathbb{Q}$ interval of the form $\{x \in \mathbb{Q} \mid |x| \leq m\}$, $m \in \mathbb{N}^*$. We restrict ourselves to the case $a \geq 1$. For $x, y \in \mathbb{Q}$ with $-m \leq y \leq x \leq m$ and $|x - y| = x - y \leq 1/n$, $n \in \mathbb{N}^*$, then

$$|a^x - a^y| = a^y(a^{x-y} - 1) \leq a^m(a^{x-y} - 1) \leq a^m(a^{1/n} - 1).$$

applies. Since the sequence $a^{1/n} = \sqrt[n]{a}$, $n \in \mathbb{N}^*$, converges to 1, see problem 3.3.5, the assertion follows. □

We also refer to the continuation according to Lemma 3.10.1 as

$$x \mapsto a^x, \quad x \in \mathbb{R}.$$

It is called the (real) **exponential function to the base** a. The monotonicity properties of the exponential functions also transfer from $\mathbb{Q}$ to $\mathbb{R}$. Their graphs are shown in Fig. 3.27. For $a \neq 1$, the exponential function is strictly monotonic, and its image is (according to the intermediate value theorem 3.8.24) entirely $\mathbb{R}_+^\times$. So it follows:

Theorem 3.10.2 *The exponential homomorphism $\mathbb{R} \to \mathbb{R}_+^\times$, $x \mapsto a^x$, is for $a \in \mathbb{R}_+^\times$, $a \neq 1$, a continuous (and therefore strictly monotonic) group isomorphism.*

The inverse isomorphism, which is also continuous and strictly monotonic according to Theorem 3.8.31, $\mathbb{R}_+^\times \to \mathbb{R}$ is called the (real) **logarithm function** (or briefly the (real) **logarithm**) **to the base** a and is denoted by

$$\mathbb{R}_+^\times \to \mathbb{R}, \quad x \mapsto \log_a x,$$

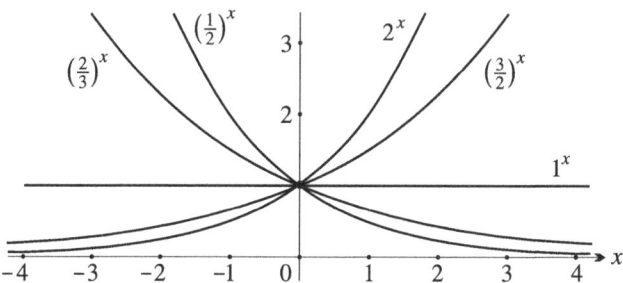

Fig. 3.27 Graphs of different exponential functions

Fig. 3.28 Graphs of various logarithmic functions

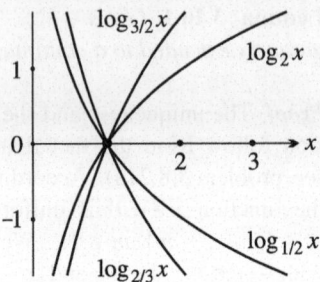

The graphs of the logarithm functions can be found in Fig. 3.28. By definition, the equation

$$y^x = a^{x \ln_a y}, \quad \text{d. h.} \quad \log_a y^x = x \ln_a y, \quad y \in \mathbb{R}_+^\times, \ x \in \mathbb{R},$$

applies, from which it follows in particular that *the function* $\mathbb{R} \times \mathbb{R}_+^\times \to \mathbb{R}_+^\times$ $(x,y) \mapsto y^x$, *even as a function in the two variables* (x,y) *on* $\mathbb{R} \times \mathbb{R}_+^\times (\subseteq \mathbb{C})$ *is continuous.* If $b := y \neq 1$ and one sets $u := y^x = b^x$, then $x = \log_b u$ and $\log_a u = x \log_a b$ result, thus

$$\log_a u = (\log_b u)(\log_a b), \quad \text{i.e.} \quad \log_b u = \frac{\log_a u}{\log_a b},$$

in particular $\log_{1/a} u = -\log_a u, \quad \log_b a = \dfrac{1}{\log_a b}, \quad u \in \mathbb{R}_+^\times,$

with which the logarithms to the base b can be calculated from the logarithms to the base a (and vice versa). The special exponential function with the Euler number e $= 2{,}718\ldots$ (cf. example 3.3.8) as a base plays a special role, as we will see in the next volume.[19] It is therefore often referred to as the **exponential function** per se. Often one writes

$$\exp x$$

for e^x. The inverse function $\log_e$ is called the **natural logarithm**. We usually denote it with

$$\ln x.$$

See Fig. 3.29. It is $a^x = e^{x \ln a} = \exp(x \ln a)$ for all $x \in \mathbb{R}$ and all $a \in \mathbb{R}_+^\times$.

Example 3.10.3 Let $a > 1$. By repeated squaring, we obtain the powers a^{2^ν}, $\nu \in \mathbb{N}$, and by repeated square rooting, the values $a^{1/2^\nu}$, $\nu \in \mathbb{N}$. If then $x \in \mathbb{R}_+$ and $(p_r \ldots p_0, q_1 q_2 \ldots)_2$ are the dual development of x (see Example 3.3.10), then (due to the continuity of the exponential function $a \mapsto a^x$)

---

[19] The function e^x on $\mathbb{R}$ coincides with its derivative. Also see exercise 3.10.8 for a first hint.

3.10 Real Exponential, Logarithm, and Power Functions

Fig. 3.29 Exponential function e^x and natural logarithm $\ln x$

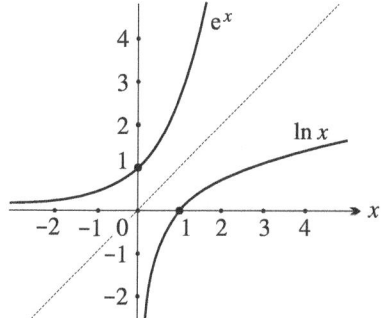

$$a^x = a^{[x]}a^{x-[x]} = \left(\prod_{\rho, p_\rho=1} a^{2^\rho}\right)\left(\prod_{\nu, q_\nu=1} a^{1/2^\nu}\right).$$

This calculation of a^x is at $x = [x] \in \mathbb{N}$ already in Remark (2) to Example 2.2.23 discussed method of rapid exponentiation. At $x \in]0,1[$, a^x can be approximated arbitrarily closely by suitable products of the iterated square roots $a^{1/2^\nu}$. For $n \in \mathbb{N}^*$ and $x = \sum_{\nu=1}^\infty q_\nu/2^\nu$, for example, $0 \leq x - \sum_{\nu=1}^n q_\nu/2^\nu < 1/2^n$ and consequently

$$0 \leq a^x - \prod_{\nu=1}^n a^{q_\nu/2^\nu} = \prod_{\nu=1}^n a^{q_\nu/2^\nu}\left(a^{x-\sum_{\nu=1}^n q_\nu/2^\nu} - 1\right) < a(a^{1/2^n} - 1).$$

In this way, the first exponential and logarithm tables to the base $a = 10$ were calculated. H. Briggs (1561–1630) used the roots $10^{1/2^\nu}$, $\nu = 1, \ldots, 54$, and was thus able to specify the values 10^x for all $x \in]0,1[$ with an error $< 10(10^{1/2^{54}} - 1) < 1{,}3 \cdot 10^{-15}$, i.e., obtain 14-digit logarithm tables; the first ones appeared in 1617. Through the work of Briggs, logarithms became popular. The calculation of logarithms (motivated by the desire to reduce the multiplication of real numbers to addition) was initiated by the Scot J. Napier (1550–1617) and the Swiss J. Bürgi (1552–1632). ◊

If one holds the exponent constant in the function $\mathbb{R} \times \mathbb{R}_+^\times \to \mathbb{R}_+^\times$, $(x,y) \mapsto y^x$, one obtains for $\alpha \in \mathbb{R}$ the **power function**

$$\mathbb{R}_+^\times \longrightarrow \mathbb{R}_+^\times, \quad x \mapsto x^\alpha = \exp(\alpha \ln x),$$

which is also continuous. For $\alpha > 0$, it can obviously be continued continuously to 0 by setting $0^\alpha := 0$. (0^0 is 1.)

Theorem 3.10.4 *The exponential functions $x \mapsto a^x$, $a \in \mathbb{R}_+^\times$, are the only continuous functions different from the zero function $f: \mathbb{R} \to \mathbb{R}$, which satisfy the following functional equation: $f(x+y) = f(x)f(y)$, $x, y \in \mathbb{R}$. (It then necessarily applies $f(0) = 1$, i.e. it is about homomorphisms of the additive group $(\mathbb{R}, +)$ into the multiplicative monoid $(\mathbb{R}, \cdot)$.)*

Proof Let $f(x_0) \neq 0$. From $f(x_0) = f(x_0 + 0) = f(x_0)f(0)$ follows $f(0) = 1$. Because of $1 = f(0) = f(x + (-x)) = f(x)f(-x)$, then $f(x) \neq 0$ for all $x \in \mathbb{R}$. From $f(x) = f((x/2) + (x/2)) = (f(x/2))^2$ even follows $f(x) > 0$. Let's say $a := f(1)$. We prove $f(x) = a^x$ for all $x \in \mathbb{R}$. Because of the continuity of f and the exponential function $x \mapsto a^x$, it is sufficient to show this for all $x \in \mathbb{Q}$.

For $n \in \mathbb{N}$ applies $f(n) = f(1 + \cdots + 1) = f(1) \cdots f(1) = a^n$. Furthermore, $f(-n) = f(n)^{-1} = a^{-n}$ is due to $f(n)f(-n) = f(n + (-n)) = f(0) = 1$. If finally p/q is a rational number with $p, q \in \mathbb{Z}$, $q > 0$, then $(f(p/q))^q = f((p/q) + \cdots + (p/q)) = f(p) = a^p$, thus as claimed $f(p/q) = a^{p/q}$. $\square$

It should be noted that there are discontinuous group homomorphisms $\mathbb{R} \to \mathbb{R}_+^\times$ (even bijective ones), see Example 3.10.6 and the remarks on it.

Example 3.10.5 From theorem 3.10.4 it follows that *the logarithm functions are the only non-trivial continuous group homomorphisms* $\mathbb{R}_+^\times \to \mathbb{R}$ *are*. If g is such a homomorphism and is $g(x_0) \neq 0$, then the values $g(x_0^n) = ng(x_0)$, $n \in \mathbb{Z}$, become arbitrarily large and arbitrarily small. According to the intermediate value theorem, there is an $a \in \mathbb{R}_+^\times$ with $g(a) = 1$. Due to $g(1) = 0$ $a \neq 1$. With the exponential isomorphism $h: \mathbb{R} \to \mathbb{R}_+^\times$, $x \mapsto a^x$, is then $f := hgh: \mathbb{R} \to \mathbb{R}_+^\times$ a continuous group homomorphism. According to theorem 3.10.4 is $f(x) = b^x$ with a $b \in \mathbb{R}_+^\times$. Because of $g(a) = 1$ is $b = f(1) = h(g(h(1))) = a^1 = a$, thus $hgh = f = h$ and consequently $hg = \mathrm{id}_{\mathbb{R}_+^\times}$ and $gh = \mathrm{id}_\mathbb{R}$, i.e. $g = h^{-1} = \log_a$. $\diamond$

Example 3.10.6 Important inequalities for the exponential functions are given by the following statements: $\diamond$

Lemma 3.10.7 *Let $a, b, r, s \in \mathbb{R}_+$ with $r + s = 1$. Then $a^r b^s \leq ra + sb$ applies.*

Proof We can assume $a, b, r, s \in \mathbb{R}_+^\times$ and use a trick from Cauchy for the proof. Due to the continuity of the exponential functions and since the rational numbers $u/2^n$, $n \in \mathbb{N}^*$, $0 < u < 2^n$, whose denominator is a power of 2, are dense in $]0, 1[$, it is sufficient to show the inequality for $r = u/2^n$, $s = (2^n - u)/2^n$, $n \in \mathbb{N}^*$, $0 < u < 2^n$, u odd. For $n = 1$, it is the inequality of the arithmetic and geometric mean, see Example 3.1.22. In the conclusion from n to $n + 1$, let

$$r = (2v + 1)/2^{n+1} = \frac{1}{2}(v/2^n + (v + 1)/2^n),$$

$$s = (2^{n+1} - 2v - 1)/2^{n+1} = \frac{1}{2}((2^n - v - 1)/2^n + (2^n - v)/2^n), \quad 0 \leq v < 2^n.$$

Then, by the induction hypothesis,

$$a^r b^s = \left(a^{v/2^n} a^{(v+1)/2^n}\right)^{1/2} \left(b^{(2^n - v - 1)/2^n} b^{(2^n - v)/2^n}\right)^{1/2}$$

$$\leq \left(\frac{va}{2^n} + \frac{(2^n - v)b}{2^n}\right)^{1/2} \left(\frac{(v + 1)a}{2^n} + \frac{(2^n - v - 1)b}{2^n}\right)^{1/2}.$$

3.10 Real Exponential, Logarithm, and Power Functions

If one applies the inequality of the arithmetic and geometric mean to the last product again, the assertion results. □

For another proof of Lemma 3.10.7 using differential calculus, we refer to Vol. 2, Ex. 2.5.2a).

Corollary 3.10.8 (Hölder's Inequality) *For any $a_i, b_i \in \mathbb{R}_+$, $i \in I$, and $p, q \in \mathbb{R}_+^\times$ with $p^{-1} + q^{-1} = 1$ the following holds*

$$\sum_{i \in I} a_i b_i \leq \left(\sum_{i \in I} a_i^p\right)^{1/p} \left(\sum_{i \in I} b_i^q\right)^{1/q}.$$

Proof We can assume $0 < \alpha := \left(\sum_i a_i^p\right)^{1/p} < \infty$ and $0 < \beta := \left(\sum_i b_i^q\right)^{1/q} < \infty$. According to Lemma 3.10.7 (applied with $a = \alpha^{-p} a_i^p$, $b = \beta^{-q} b_i^q$, $r = p^{-1}$, $s = q^{-1}$) $\alpha^{-1} \beta^{-1} a_i b_i \leq p^{-1} \alpha^{-p} a_i^p + q^{-1} \beta^{-q} b_i^q$, $i \in I$, holds, so

$$\alpha^{-1} \beta^{-1} \sum_i a_i b_i \leq p^{-1} \alpha^{-p} \sum_i a_i^p + q^{-1} \beta^{-q} \sum_i b_i^q = p^{-1} + q^{-1} = 1. \quad \square$$

For $p = q = 2$, Hölder's Inequality yields the **Cauchy-Schwarz Inequality**, which was already considered in Exercise 3.6.10a).

Remark 3.10.9 (Complex Exponential Function) With the real exponential function and the real trigonometric functions cos, sin, the complex exponential function exp: $\mathbb{C} \to \mathbb{C}$ can be explained by

$$e^{x+iy} = \exp(x + iy) = e^x(\cos y + i \sin y), \quad x, y \in \mathbb{R},$$

We have already discussed it in Example 2.2.16 (2) and in Sect. 3.5. It is a surjective group homomorphism $(\mathbb{C}, +) \to (\mathbb{C}^\times, \cdot)$ with $\mathbb{Z} 2\pi i$ as the kernel, which is also continuous (since in addition to exp, cos and sin are continuous). Its restriction to the strip $\mathbb{R} +]-\pi, \pi]i$ is bijective and provides as its inverse the (principal value of the) complex natural logarithm ln: $\mathbb{C}^\times \to \mathbb{C}$, cf. loc. cit., which, however, is only continuous on the slit plane $\mathbb{C} - \mathbb{R}_-$, but for example also on sectors of the form $\{z \in \mathbb{C}^\times \mid \alpha \leq \text{Arg } z \leq \pi\}$ with $-\pi < \alpha < \pi$. ◊

Exercises

Exercise 3.10.1 For $a \in \mathbb{R}_+^\times$, the following applies

$$\lim_{x \to \infty} a^x = \begin{cases} \infty, & \text{if } a > 1, \\ 0, & \text{if } a < 1; \end{cases} \qquad \lim_{x \to -\infty} a^x = \begin{cases} 0, & \text{if } a > 1, \\ \infty, & \text{if } a < 1. \end{cases}$$

Exercise 3.10.2 Let $a \in \mathbb{R}_+^\times$, $a \neq 1$. It follows that

$$\lim_{x \to \infty} \log_a x = \begin{cases} \infty, & \text{if } a > 1, \\ -\infty, & \text{if } a < 1; \end{cases} \qquad \lim_{x \to 0+} \log_a x = \begin{cases} -\infty, & \text{if } a > 1, \\ \infty, & \text{if } a < 1. \end{cases}$$

Exercise 3.10.3 Let $a \in \mathbb{R}$, $a > 1$, and $n \in \mathbb{N}^*$. Then $\lim_{x \to \infty} x^n/a^x = 0$ applies, i.e. $x^n = o(a^x)$ for $x \to \infty$. (Because of $x^n/a^x = \left(x/(\sqrt[n]{a})^x\right)^n$ it is sufficient to consider the case $n = 1$ and because of $a^x = 2^{(\log_2 a)x}$ the case $a = 2$. For $x \geq 4$, however, $2^x \geq 2^{[x]} \geq [x]^2 \geq x^2/2$ applies.)

Exercise 3.10.4 Let $a \in \mathbb{R}_+^\times$, $a \neq 1$, and $\alpha \in \mathbb{R}_+^\times$. It follows:

a) $\lim_{x \to \infty} (\log_a x)/x^\alpha = 0$, so $\log_a x = o(x^\alpha)$ for $x \to \infty$. (Use exercise 3.10.3.)
b) $\lim_{x \to 0+} x^\alpha \log_a x = 0$, so $\log_a x = o(x^{-\alpha})$ for $x \to 0$.
c) The function $f \colon [0, 1/2] \to \mathbb{R}$ with $f(x) := 1/\log_a x$ for $x > 0$ and $f(0) = 0$ is continuous, but not Hölder continuous at the zero point.

Exercise 3.10.5 For $a_i \in \mathbb{R}_+$, $i \in I$, and $p \geq 1$, $\sum_i a_i^p \leq \left(\sum_i a_i\right)^p$ applies. (One can assume that $0 < \alpha := \sum_i a_i < \infty$ is. Then $a_i/\alpha \leq 1$ is and consequently $(a_i/\alpha)^{p-1} \leq 1$, thus $a_i^p/\alpha^p = (a_i/\alpha)^p \leq a_i/\alpha$.)

Exercise 3.10.6 Let $f \colon \mathbb{R} \to \mathbb{R}$ be an additive function, i.e., an endomorphism of the additive groups. (It is also said that f satisfies the so-called **Cauchy functional equation** $f(x + y) = f(x) + f(y)$, $x, y \in \mathbb{R}$.) Examples of such additive functions are the homotheties $L_a \colon x \mapsto ax$ with a fixed $a \in \mathbb{R}$, i.e., the images of the Cayley representation $L \colon \mathbb{R} \to \operatorname{End} \mathbb{R}$, cf. example 2.6.19.

a) If f is continuous, then $f = L_a$ with $a \in \mathbb{R}$. (Conclude as in theorem 3.10.4.)
b) If f is monotonic, then $f = L_a$ with $a \in \mathbb{R}$.
c) If f is bounded in a neighborhood of 0, then $f = L_a$ with $a \in \mathbb{R}$.

(Part a) and b) follow from c). Conversely, c) is most conveniently proven by reducing it to a).)

Note There are (even bijective) non-continuous additive functions $f \colon \mathbb{R} \to \mathbb{R}$. This follows from the existence of Hamel bases of $\mathbb{R}$ over $\mathbb{Q}$, cf. Example 2.8.19. — From b) it follows: If $f \colon \mathbb{R}_+ \to \mathbb{R}_+$ is additive, then $f = L_a|\mathbb{R}_+$ with $a \in \mathbb{R}_+$ (because through $\widetilde{f}(x - y) := f(x) - f(y)$, $x, y \in \mathbb{R}_+$, there is an extension $\widetilde{f} \colon \mathbb{R} \to \mathbb{R}$ of f to (the Grothendieck group $G(\mathbb{R}_+) =) \mathbb{R}$ well-defined, *which is also monotonic*). This result justifies the brief indication "a Euro/kg" on price boards in shops, without having to presuppose the continuity of the price function in addition to its additivity.

Exercise 3.10.7
a) Prove again theorem 3.10.4 and the result of example 3.10.5 using exercise 3.10.6, where the requirement of continuity can also be replaced by a condition of monotonicity.
b) The only continuous or monotonic group endomorphisms of $\mathbb{R}_+^\times$ are the power functions $x \mapsto x^\alpha$, $\alpha \in \mathbb{R}$.

3.10 Real Exponential, Logarithm, and Power Functions

Exercise 3.10.8 The real exponential function exp can be introduced in the same way as its special value $\exp 1 = e = \lim_{n\to\infty} \left(1 + \frac{1}{n}\right)^n$ directly, namely by

$$e^x = \exp x = \lim_{n\to\infty} \left(1 + \frac{x}{n}\right)^n, \quad x \in \mathbb{R}.$$

This already illustrates its special character and is to be described in this exercise.

a) For $x \in \mathbb{R}, |x| < 1$, and $n \in \mathbb{N}^*$ applies

$$\left|\left(1 + \frac{x}{n}\right)^n - 1\right| \le \left(1 + \frac{|x|}{n}\right)^n - 1 \le \frac{1}{1-|x|} - 1 = \frac{|x|}{1-|x|}.$$

(The second estimate is very rough. The first even applies to *all* $x \in \mathbb{C}$.)

b) For $x \in \mathbb{R}^\times$ and $n_0 \in \mathbb{N}^*$ with $x + n_0 \ge 0$, the sequence $\left(1 + \frac{x}{n}\right)^n$, $n \ge n_0$, is strictly monotonically increasing. (One concludes as in example 3.3.8.—If $x > 0$ is an annual interest rate and the interest rate for a n-th of the year is x/n, $n \in \mathbb{N}^*$, then a capital $K_0 > 0$ grows after one year to $K_0\left(1 + \frac{x}{n}\right)^n$, if the already accrued capital is reinvested after each n-th of the year (compound interest). This makes the monotony for $x > 0$ "intuitive". For $x<0$, one finds a similar interpretation with a decay process.[20]

c) For $x > 0$ and $n \in \mathbb{N}^*$ is

$$\left(1 + \frac{x}{n}\right)^n \le \left(1 + \frac{\lceil x \rceil}{n}\right)^n \le \left(1 + \frac{\lceil x \rceil}{n\lceil x \rceil}\right)^{n\lceil x \rceil} = \left(1 + \frac{1}{n}\right)^{n\lceil x \rceil} < e^{\lceil x \rceil}.$$

d) For $x \in \mathbb{R}$ the sequence $\left(1 + \frac{x}{n}\right)^n$, $n \in \mathbb{N}^*$, converges. (One uses b) and c).) — We define $\exp^* x := \lim_{n\to\infty} \left(1 + \frac{x}{n}\right)^n$ (> 0).

e) For $x, y \in \mathbb{R}$ is $\exp^*(x+y) = \exp^* x \exp^* y$. (With the help of a), it is easy to see that

$$\left(1 + \frac{x}{n}\right)^n \left(1 + \frac{y}{n}\right)^n \Big/ \left(1 + \frac{x+y}{n}\right)^n$$

for $n \to \infty$ converges to 1.)

f) $\exp^*: \mathbb{R} \to \mathbb{R}^\times_+$ is strictly monotonically increasing and continuous. (Use e) and a).)

g) It is $\exp^* x = \exp x$ for all $x \in \mathbb{R}$. (According to example 3.3.8 is $\exp^* 1 = e = \exp 1$.)

h) It is $1 + x \le e^x$ for all $x \in \mathbb{R}$. (Use b).)

Note With Dini's theorem (see Vol. 2, Ex. 1.1.14) and b) it follows that $\left(1 + \frac{x}{n}\right)^n$ converges locally uniformly on $\mathbb{R}$ towards $\exp x$. This also applies in the complex

---

[20] Today, negative interest rates are also possible.

case, but there it is probably most quickly proven by direct comparison with the
exponential series

$$\sum_{k=0}^{\infty} \frac{x^k}{k!},$$

which represents the function exp both in the real and in the complex case. In particular, $|e^x - 1| \leq e^{|x|} - 1$ for all $x \in \mathbb{C}$, see a). — For the present exercise, also see Chap. 22 ("Algebra") from the "Feynman Lectures on Physics", Vol. I.

Exercise 3.10.9

a) Let a,b be real numbers >1. The sequences $q_i \in \mathbb{N}$, $c_i \in \mathbb{R}_+^\times$, $i \in \mathbb{N}$, are recursively determined by

$$c_{-2} = b, c_{-1} = a; \quad c_{i-1}^{q_i} \leq c_{i-2} < c_{i-1}^{q_i+1}, \quad c_i = c_{i-2}/c_{i-1}^{q_i}$$

where the procedure stops if $c_i = 1$ is. (c_i is always ≥ 1.) $\log_a b$ then has the continued fraction development (see Example 3.3.11)

$$\log_a b = [q_0, q_1, q_2, \ldots].$$

b) Calculate $q_0, \ldots, q_5$ for $\log_{10} 2, \log_2 10 = 1 + \log_2 5$ (these values are significant when comparing dual and decimal developments), $\ln 10 = \log_e 10$, $\log_{10} e = 1/\ln 10$, $\log_2 3 = 1 + \log_2(3/2)$. (The last value plays a role in the development of scale systems that (in addition to the octave 2:1) choose the pure fifth $3:2 = \frac{3}{2}:1$ as the guiding interval. The fourth approximation fraction $\log_2 \frac{3}{2} \approx [0, 1, 1, 2, 2] = \frac{7}{12}$ (so $2^{7/12} \approx \frac{3}{2}$ or $2^7 \approx (3/2)^{12}$) identifies 12 fifth jumps with 7 octave jumps and provides the classic circle of fifths. The third approximation $\log_2 \frac{3}{2} \approx [0, 1, 1, 2] (= [0, 1, 1, 1, 1]) = \frac{3}{5}$ identifies 5 fifths with 3 octaves. This happens in the pentatonic system, which in the classical system (e.g.) with the fifths c, g, d', a', e" and the faulty (small) sixth e", c'" is simulated. (Its tones are thus c, d, e, g, a.) In tempered pentatonic tuning, a fifth is the interval $2^{3/5}:1$ compared to the interval $2^{7/12}:1 (< 2^{3/5}:1)$ in classical tempered tuning. The pure fifth $\frac{3}{2}:1$ itself lies (of course) in between. What kind of scale systems do the fifth approximation fraction of $\log_2 \frac{3}{2}$ and its neighboring approximation fractions provide?) □

Topological Foundations 4

4.1 Metric Spaces

In the theory of fields $\mathbb{R}$ and $\mathbb{C}$ (for which we continue to use the common designation $\mathbb{K}$), the distances between points play a crucial role. With their help, we have introduced limits and other topological concepts for $\mathbb{K}$ in Chap. 3. The concept of the metric space generalizes these considerations and simultaneously motivates the examination of general topological spaces in this chapter.

Definition 4.1.1 Let X be a set. A mapping $d: X \times X \to \overline{\mathbb{R}}_+ = \mathbb{R}_+ \uplus \{\infty\}$ is called a **metric** on X, if for all $x, y, z \in X$ the following holds:

(1) It is $d(x, y) = 0$ if and only if $x = y$.
(2) It is $d(x, y) = d(y, x)$. (**Symmetry**)
(3) It is $d(x, z) \leq d(x, y) + d(y, z)$. (**Triangle Inequality**)

A set X together with a metric d on X is called a **metric space** $X = (X, d)$. The value $d(x, y)$, $x, y \in X$, is called the **distance** between x and y (with respect to d).

Induction over n immediately provides for a metric d on X: If $x_0, \ldots, x_n \in X$,

$$d(x_0, x_n) \leq d(x_0, x_1) + d(x_1, x_2) + \cdots + d(x_{n-1}, x_n).$$

Furthermore, for any $v, w, x, y \in X$ the inequality

$$|d(v, w) - d(x, y)| \leq d(v, x) + d(w, y),$$

applies, where we allow the difference $\infty - \infty = 0$. It is indeed $d(v, w) \leq d(v, x) + d(x, y) + d(y, w)$ and $d(x, y) \leq d(x, v) + d(v, w) + d(w, y)$.

If no misunderstandings are to be feared, we also denote different metrics with the same symbol d. Every subset $X' \subseteq X$ of a metric space (X, d) is also a metric space with the restriction $d|(X' \times X')$. Unless otherwise stated, a subset $X' \subseteq X$ is always considered as a metric space with this **induced metric**. If the values of the metric d are all finite, i.e., they all lie in $\mathbb{R}_+$ (and this is to be emphasized), we speak of a **real metric**. If d is any metric on X, then

$$x \sim_d y \iff d(x, y) < \infty$$

clearly defines an equivalence relation on X. The induced metric is real on each of the equivalence classes, and the distance between two points from different equivalence classes is ∞. Thus, every metric space can be viewed as a sum (i.e., as a disjoint union) of spaces with real metrics, where points from different summands have infinite distance.[1]

Occasionally, one has to weaken condition (1) for a metric to

(1′) It is $d(x, x) = 0$ for all $x \in X$.

One then speaks of a **pseudometric**. If d is a pseudometric on X, then by

$$x \equiv_d y \iff d(x, y) = 0$$

an equivalence relation on X is trivially defined, and on the set $\overline{X}$ of equivalence classes,

$$\overline{d}(\overline{x}, \overline{y}) := d(x, y), \quad x, y \in X,$$

a metric $\overline{d}$ with $\overline{d} \circ (\pi \times \pi) = d$ for the canonical projection $\pi : X \to \overline{X}$ is well-defined. In general, if $f : X \to Y$ is a mapping and d is a pseudometric on Y, then by $d(x, y) := d\bigl(f(x), f(y)\bigr)$, $x, y \in X$, a pseudometric on X is defined. In this way, every pseudometric is thus induced by a metric.

Let X be a metric space, $x \in X$ and $r \in \overline{\mathbb{R}}_+$. Then the sets

$$B(x; r) = B_{(X,d)}(x; r) := \{y \in X \mid d(x, y) < r\},$$

$$\overline{B}(x; r) = \overline{B}_{(X,d)}(x; r) := \{y \in X \mid d(x, y) \leq r\},$$

$$S(x; r) = S_{(X,d)}(x; r) := \{y \in X \mid d(x, y) = r\}$$

the **open (full) sphere** or the **closed (full) sphere** or the **sphere** with **center** x and **radius** r. Instead of "sphere" one also says "**ball**". $\overline{B}(x; r)$ is the disjoint union

---

[1] Consider the equivalence classes with respect to $\sim_d$ as islands in an insurmountable ocean not belonging to X.

4.1 Metric Spaces

Fig. 4.1 Open balls and complements of closed balls are open

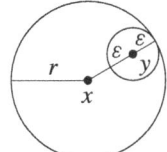
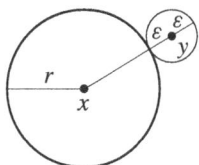

of $B(x; r)$ and $S(x; r)$. For $x \in X$ the open sphere $B_d(x; \infty)$ is the equivalence class of x with respect to the above introduced equivalence relation $\sim_d$ on X. The open spheres with radius ∞ thus form a partition of X. A subset $A \subseteq X$ is called **bounded** (with respect to the metric d), if there is an $r \in \mathbb{R}_+$ and an $x_0 \in X$ with $A \subseteq \overline{B}(x_0; r)$. A bounded set lies entirely within one of the equivalence classes with respect to $\sim_d$. With the spheres, as already in $\mathbb{R}$ and $\mathbb{C}$ Environments and open sets in arbitrary metric spaces are defined:

Definition 4.1.2 Let X be a metric space.

(1) A subset U of X is called an **environment** of the point $x \in X$, if there is an $\varepsilon > 0$ with $B(x; \varepsilon) \subseteq U$.
(2) A subset U of X is called **open** (in X), if for every $x \in U$ there is (depending on x) $\varepsilon > 0$ with $B(x; \varepsilon) \subseteq U$, i.e., if U is an environment of every point $x \in U$.
(3) A subset F of X is called **closed** (in X), if the complement $\complement_X F = X - F$ of F in X is open.

With $B(x; \varepsilon)$, every ball $\overline{B}(x; \varepsilon')$ for $0 < \varepsilon' < \varepsilon$ is a subset of U. Furthermore, see Fig. 4.1:

Lemma 4.1.3 *In a metric space X, the open balls are $B(x; r)$ open and the closed balls are $\overline{B}(x; r)$, $x \in X$, $r \in \mathbb{R}_+$, closed subsets of X.*

Proof We can assume without restriction $r < \infty$. Let $y \in B(x; r)$ and $\varepsilon := r - d(x, y) > 0$. Then $B(y; \varepsilon) \subseteq B(x; r)$ is due to $d(x, z) \leq d(x, y) + d(y, z) < d(x, y) + \varepsilon = r$ for all $z \in B(y; \varepsilon)$. It remains to show that the complement of $\overline{B}(x; r)$ in X is open. Let $y \notin \overline{B}(x; r)$. If $d(x, y) < \infty$, then $B(y; \varepsilon) \subseteq X - \overline{B}(x; r)$ is for $\varepsilon := d(x, y) - r > 0$ due to $d(x, z) \geq d(x, y) - d(y, z) > d(x, y) - \varepsilon = r$ for all $z \in B(y; \varepsilon)$. If $d(x, y) = \infty$, then $B(y; \varepsilon) \subseteq X - \overline{B}(x; r)$ is for every $\varepsilon \in \mathbb{R}_+^\times$. □

From Lemma 4.1.3, it immediately follows that a set $U \subseteq X$ is a neighborhood of $x \in X$ if and only if there is an open set U' with $x \in U' \subseteq U$. The following statements about open sets of a metric space are elementary but essential:

Proposition 4.1.4 *Let X be a metric space.*

(1) *$\emptyset$ and X are open subsets of X.*

Fig. 4.2 Separation of points in metric spaces by open balls

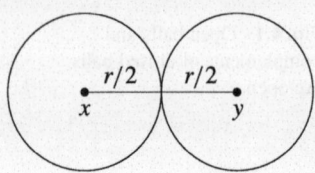

(2) If $U_i, i \in I$, is an arbitrary family of open subsets of X, then their union $\bigcup_{i \in I} U_i$ is also open in X.
(3) If $U_j, j \in J$, is a finite family of open subsets of X, then their intersection $\bigcap_{j \in J} U_j$ is also open in X.
(4) For any two points $x, y \in X$ with $x \neq y$ there exist neighborhoods U of x and V of y with $U \cap V = \emptyset$.

Proof (1) and (2) are trivial. To prove (3), let $J \neq \emptyset$ and $x \in U := \bigcap_j U_j$. There exists $\varepsilon_j > 0$ with $B(x; \varepsilon_j) \subseteq U_j, j \in J$. For $\varepsilon := \text{Min}\,(\varepsilon_j, j \in J) > 0$ is $B(x; \varepsilon) \subseteq U$. To prove (4), let's first consider $0 < r := d(x, y) < \infty$. Then $B(x; r/2) \cap B(y; r/2) = \emptyset$. With a z from this intersection, the contradiction $r = d(x, y) \leq d(x, z) + d(z, y) < r/2 + r/2 = r$ would arise, see Fig. 4.2. But if $d(x, y) = \infty$, then $B(x; s) \cap B(y; t) = \emptyset$ for all $s, t \in \mathbb{R}_+^\times$, i.e., it is $B(x; \infty) \cap B(y; \infty) = \emptyset$. □

By transitioning to the complements, we obtain from Proposition 4.1.4:

Corollary 4.1.5 *Let X be a metric space.*

(1) *X and $\emptyset$ are closed subsets of X.*
(2) *If $F_i, i \in I$, is an arbitrary family of closed subsets of X, then their intersection $\bigcap_{i \in I} F_i$ is also closed in X.*
(3) *If $F_j, j \in J$, is a finite family of closed subsets of X, then their union $\bigcup_{j \in J} F_j$ is also closed in X.*
(4) *For any two points $x, y \in X$ with $x \neq y$, there exist in X closed sets A and B with $y \notin A$, $x \notin B$ and $A \cup B = X$.*

Two metrics d and d' on the same set X are called **equivalent**, if there are numbers $\alpha, \beta \in \mathbb{R}_+^\times$ with $\alpha d \leq d' \leq \beta d$. Obviously, this defines an equivalence relation on the set of all metrics on X. For such equivalent metrics, the sets of open and thus also the sets of closed sets coincide due to $B_d(x; r) \subseteq B_{d'}(x; \beta r)$ and $B_{d'}(x; r) \subseteq B_d(x; \alpha^{-1} r)$ for all $x \in X$ and $r \in \overline{\mathbb{R}}_+$.

Example 4.1.6 (1) We already know $\mathbb{R}$ and $\mathbb{C}$ as metric spaces with the usual distance $d(x, y) = |y - x|$ as a metric. Here, $\mathbb{R}$ is a metric subspace of $\mathbb{C}$, but we emphasize that now *every* subset of $\mathbb{C}$ is considered as an independent metric space.

4.1 Metric Spaces

(2) On every set X, a metric is defined by

$$d(x, y) := \begin{cases} 1, & \text{if } x \neq y, \\ 0, & \text{if } x = y, \quad x, y \in X \end{cases}$$

with respect to which all subsets of X are open and closed. It is called the **discrete metric** on X. The constant distance 1 for $x \neq y$ can be replaced by any element from $\mathbb{R}_+^\times \cup \{\infty\}$. The natural choice for this is ∞. ◇

Example 4.1.7 (Metrics on Products) In addition to the subsets of metric spaces, the products of metric spaces also have canonical metrics, although there is a great deal of freedom of choice. — Let $X_i = (X_i, d_i)$, $i \in I$ be a family of metric spaces and $X := \prod_i X_i$ their product. Then let

$$d_p((x_i), (y_i)) := \left(\sum_{i \in I} d_i(x_i, y_i)^p \right)^{1/p}, \quad \text{if } p \in \mathbb{R}, \ p \geq 1, \text{ sowie}$$

$$d_\infty((x_i), (y_i)) := \mathrm{Sup}(d_i(x_i, y_i), i \in I), \quad (x_i), (y_i) \in X.$$

(Of course, $\infty^r := \infty$ for all $r \in \mathbb{R}_+^\times \cup \{\infty\}$.) *All of these are metrics on the product space X. d_1 is called the **sum metric**, d_2 the **Euclidean metric**, d_∞ the **Chebyshev-** or **supremum metric** and in general d_p the **p-metric** on X, $p \in [1, \infty]$.*

That the values of the given functions lie in $\overline{\mathbb{R}}_+$ and are only equal to 0 for $(x_i) = (y_i)$ is trivial, as is their symmetry in the arguments (x_i), (y_i). Furthermore, the triangle inequality is clear for the cases $p = 1$ and $p = \infty$. — For the *proof* of the triangle inequality for $1 < p < \infty$, let $q := p/(p-1)$, so $1/p + 1/q = 1$, and let $(x_i), (y_i), (z_i) \in X$. Because of $d_i(x_i, z_i) \leq d_i(x_i, y_i) + d_i(y_i, z_i)$, $i \in I$, it is sufficient for families $a_i, b_i \in \overline{\mathbb{R}}_+$, $i \in I$, to prove the inequality

$$\left(\sum_{i \in I} (a_i + b_i)^p \right)^{1/p} \leq \left(\sum_{i \in I} a_i^p \right)^{1/p} + \left(\sum_{i \in I} b_i^p \right)^{1/p}$$

where we can also assume that all a_i, b_i, $i \in I$, and even the sums $\alpha^p := \sum_i a_i^p$, $\beta^p := \sum_i b_i^p$ are finite as well as $\alpha, \beta > 0$. Then the sum $\sum_i (a_i + b_i)^p$ is also finite (e.g., because of $(a_i + b_i)^p \leq 2^p \mathrm{Max}\,(a_i, b_i)^p \leq 2^p(a_i^p + b_i^p)$). With the Hölder inequality 3.10.8 it now follows because of $(p-1)q = p$

$$\sum_{i \in I} a_i(a_i + b_i)^{p-1} \leq \left(\sum_{i \in I} a_i^p \right)^{1/p} \left(\sum_{i \in I} (a_i + b_i)^{(p-1)q} \right)^{1/q} < \infty,$$

$$\sum_{i \in I} b_i(a_i + b_i)^{p-1} \leq \left(\sum_{i \in I} b_i^p \right)^{1/p} \left(\sum_{i \in I} (a_i + b_i)^{(p-1)q} \right)^{1/q} < \infty, \quad \text{therefore}$$

$$\sum_{i\in I}(a_i+b_i)^p \le \left(\left(\sum_{i\in I}a_i^p\right)^{1/p}+\left(\sum_{i\in I}b_i^p\right)^{1/p}\right)\left(\sum_{i\in I}(a_i+b_i)^p\right)^{1/q}$$

and after shortening $\left(\sum_i(a_i+b_i)^p\right)^{1/q}$ the desired inequality. □

It is

$$d_p \ge d_q \quad \text{for } p,q \text{ with } 1 \le p \le q \le \infty.$$

This is trivial for $q = \infty$. For $q < \infty$ we have $(\sum_i a_i^p)^{1/p} \ge (\sum_i a_i^q)^{1/q}$ for all $a_i \in \overline{\mathbb{R}}_+, i \in I$, to show, where we can assume $\sum_i a_i^p < \infty$. According to exercise 3. 10.5 but $\sum_i a_i^q = \sum_i (a_i^p)^{q/p} \le (\sum_i a_i^p)^{q/p}$. For a finite index set I obviously applies

$$|I| \cdot d_\infty \ge d_1.$$

It follows: *If I is finite, then all metrics d_p, $1 \le p \le \infty$, on $X = \prod_{i\in I} X_i$ are equivalent.*

When we consider a finite product of metric spaces again as a metric space, this product space always carries one of the equivalent p-metrics d_p, $1 \le p \le \infty$, unless otherwise stated. Figure 4.3 shows some spheres in the $\mathbb{R}^2$ for different metrics. As **standard balls** and **standard spheres** one chooses

$$B^n := B_{\mathbb{R}^n}(0;1), \quad \overline{B}^n := \overline{B}_{\mathbb{R}^n}(0;1) \text{ resp. } S^{n-1} := S_{\mathbb{R}^n}(0;1)$$

in the $\mathbb{R}^n$ with the Euclidean metric d_2, $n \in \mathbb{N}$.

Let I be arbitrary again. In the product $X = \prod_{i\in I} X_i$ all factors X_i carry the discrete metric with $d(x_i, y_i) = 1$ for $x_i, y_i \in X_i$, $x_i \ne y_i$, cf. example 4.1.6 (2), so for $1 \le p < \infty$ the p-metric on X is equal to

$$d_p(x,y) = \begin{cases} n^{1/p}, & \text{if } x \text{ and } y \text{ are different in exactly } n \in \mathbb{N} \text{ components,} \\ \infty & \text{sonst.} \end{cases}$$

In particular, $d_1(x,y)$ is the number of components in which x and y differ. d_1 is called the **Hamming metric** on X. The sphere $B_{d_p}(x; \infty)$ is in any case the set of

Fig. 4.3 Spheres in the $\mathbb{R}^2$ for the metrics d_1, d_2, d_3, d_∞

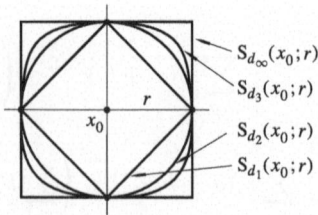

4.1 Metric Spaces

those I-tuples from X that differ from x only in finitely many components, which are thus almost equal to x. d_∞ is again the discrete metric on X.

Let $\eta_i \in \overline{\mathbb{R}}_+^\times$, $i \in I$ be a family with $\sum_i \eta_i^p < \infty$. Then, if $d_i \leq \eta_i$ applies to the metrics d_i on X_i, $i \in I$, then $d_p \leq \left(\sum_i \eta_i^p\right)^{1/p} < \infty$ applies to the p-metric d_p on the product X. ◊

If $X_i = \mathbb{K}$ applies to all $i \in I$ and $\mathbb{K}$ carries the standard metric, then we obtain on the $\mathbb{K}$-vector space $\mathbb{K}^I$ of the $\mathbb{K}$-valued functions on I the p-metrics

$$d_p(x, y) = \left(\sum_{i \in I} |y_i - x_i|^p\right)^{1/p}, \quad 1 \leq p < \infty, \quad \text{resp.}$$

$$d_\infty(x, y) = \mathrm{Sup}(|y_i - x_i|, \ i \in I),$$

$x = (x_i)$, $y = (y_i) \in \mathbb{K}^I$, see the preceding example. In particular, the distance from x to the origin is equal to

$$\|x\|_p := d_p(0, x) = \left(\sum_{i \in I} |x_i|^p\right)^{1/p}, \quad \text{and we have} \quad d_p(x, y) = \|y - x\|_p.$$

In addition to the metric properties

(1) $\|x\|_p \in \overline{\mathbb{R}}_+$ and $\|x\|_p = 0$ only when $x = 0$,
(2) $\|x + y\|_p \leq \|x\|_p + \|y\|_p$

for the equations still apply

(3) $\|ax\|_p = |a|\|x\|_p$ for all $a \in \mathbb{K}$.

Generally, a function $\|-\| : V \to \overline{\mathbb{R}}_+$ on a $\mathbb{K}$-vector space V with the properties (1), (2), (3) (for all $x, y \in V$, $a \in \mathbb{K}$ and with $\|-\|$ instead of $\|-\|_p$) is a **norm** on V. V, equipped with a norm, is called a **normed $\mathbb{K}$-vector space**. A norm on V defines a metric on V by

$$d(x, y) := \|y - x\|, \quad x, y \in V,$$

which is **translation invariant** (i.e., for which $d(x + z, y + z) = d(x, y)$ is) and for which moreover $d(ax, ay) = |a|d(x, y)$ applies, $x, y, z \in V$, $a \in \mathbb{K}$. A norm $\|-\|$ on V is determined by its (finite) values on the sphere

$$V_{<\infty} := B_V(0; \infty) = \{x \in V \mid \|x\| < \infty\},$$

that is a $\mathbb{K}$-subvector space of V is. For any $x \in V$, $B_V(x; \infty)$ is the coset $x + V_{<\infty}$. As a rule, for a given norm $\|-\|$ only this subspace $V_{<\infty}$, on which all values of the norm are finite, is considered.[2] Two norms $\|-\|$ and $\|-\|'$ on V are called **equivalent**, if their associated metrics are equivalent, i.e. if there are $\alpha, \beta \in \mathbb{R}_+^\times$ with $\alpha\|-\| \leq \|-\|' \leq \beta\|-\|$ exists. The norm is extreme with $\|x\| = \infty$ for all $x \in V - \{0\}$, which induces the discrete topology on V.

For the p-**norms** $\|-\|_p$, $1 \leq p < \infty$, on $\mathbb{K}^I$ one sets

$$\ell_\mathbb{K}^p(I) := \mathrm{B}_{d_p}(0; \infty) = \left\{(x_i) \in \mathbb{K}^I \,\bigg|\, \sum_i |x_i|^p < \infty\right\}.$$

This space is called the $\mathbb{K}$-ℓ^p-**space** to the index set I. Its elements are the so-called p-**summable** families $(x_i) \in \mathbb{K}^I$. The elements of $\ell_\mathbb{K}^1(I)$ are the summable families, and the elements of $\ell_\mathbb{K}^2(I)$ are also called the **square summable** families in $\mathbb{K}^I$. For the **Chebyshev-** or **Supremum norm** $\|-\|_\infty$ on $\mathbb{K}^I$ is

$$\ell_\mathbb{K}^\infty(I) := \mathrm{B}_{d_\infty}(0; \infty)$$

simply the space $\mathrm{B}_\mathbb{K}(I)$ of bounded $\mathbb{K}$-valued functions x on I with the Chebyshev norm $\|x\|_\infty = \mathrm{Sup}(|x_i|, i \in I)$. It is $\mathbb{K}^{(I)} \subseteq \ell_\mathbb{K}^p(I)$ for all p with $1 \leq p \leq \infty$ and in particular $\mathbb{K}^I = \ell_\mathbb{K}^p(I)$ for finite sets I. In the latter case, all p-norms on $\mathbb{K}^I$, $1 \leq p \leq \infty$, are equivalent.

We recommend to the reader to identify the metric space $\mathbb{R}^3$ with our **perceptual space** using a coordinate system. If the coordinate system is Cartesian, then the spheres $\overline{\mathrm{B}}_{d_2}(x; r)$ and $\mathrm{B}_{d_2}(x; r)$, $r \in \mathbb{R}_+$, in $\mathbb{R}^3$ with respect to the Euclidean norm $\|-\|_2$ correspond to the spheres with radius r and center P in perceptual space, where P has the coordinate triple $x \in \mathbb{R}^3$, once with and once without a boundary sphere, which corresponds to the sphere $\mathrm{S}_{d_2}(x; r) \subseteq \mathbb{R}^3$. What are the spheres and spheres with respect to the sum norm $\|-\|_1$ and the maximum norm $\|-\|_\infty$? Similarly, visualize the spaces $\mathbb{R}^2$ and $\mathbb{C}$ or $\mathbb{R} = \mathbb{R}^1$ using a plane or a line in perceptual space. This will be discussed in more detail in volumes 3 and 4.

If $v = (v_i)_{i \in I}$ is a basis of the $\mathbb{K}$ vector space V, then the norm transferred by $\mathbb{K}^{(I)}$ to V by means of the isomorphism $\mathbb{K}^{(I)} \xrightarrow{\sim} V$, $(x_i)_{i \in I} \mapsto \sum_{i \in I} a_i v_i$, is called

$$\left\|\sum_{i \in I} x_i v_i\right\|_p := \left(\sum_{i \in I} |x_i|^p\right)^{1/p}$$

the p-**norm** on V with respect to the basis v, $1 \leq p \leq \infty$. It is real-valued on the whole of V, but generally depends significantly on the chosen basis v.

The normalized $\mathbb{K}$ vector spaces and their generalizations play an important role in all of analysis. They are at the center of the so-called **functional analysis**. We will return to this topic several times in the following volumes.

---

[2] Basically, we allow the value ∞ for a norm. However, if this value is explicitly excluded, we speak of a **real norm** or a norm with values in $\mathbb{R}$.

4.1 Metric Spaces

The concept of convergence for sequences in $\mathbb{R}$ and $\mathbb{C}$ can be directly transferred to sequences in any metric spaces:

Definition 4.1.8 A sequence $(x_n)_{n \in \mathbb{N}}$ of elements of a metric space X is called **convergent** (in X), if an $x \in X$ exists with the following property: For every positive $\varepsilon > 0$ there is an $n_0 \in \mathbb{N}$ with $d(x_n, x) \leq \varepsilon$ for all $n \geq n_0$, $n \in \mathbb{N}$, i.e., almost all members of the sequence (x_n) lie within the sphere $\overline{B}(x; \varepsilon)$. Such an x is called **limit** or **Limes** of the sequence (x_n) and is denoted by

$$\lim x_n = \lim_{n \to \infty} x_n.$$

With the concept of neighborhood, convergence can be characterized as follows:

Lemma 4.1.9 *A sequence in the metric space X has exactly the limit $x \in X$, if almost all members of the sequence lie in every neighborhood of x.*

It follows that *the limit x of a convergent sequence (x_n) in the metric space X is uniquely determined*; because different points in X have disjoint neighborhoods according to Proposition 4.1.4 (4).

The concept of continuity also applies to mappings of metric spaces.

Definition 4.1.10 Let $f : X \to Y$ be a mapping of metric spaces.

(1) f is called **continuous at the point** $a \in X$, if for every $\varepsilon > 0$ there is a $\delta > 0$ such that $d(f(x), f(a)) \leq \varepsilon$ for all $x \in X$ with $d(x, a) \leq \delta$ applies, i.e., for every ball $\overline{B}(f(a); \varepsilon) \subseteq Y$, $\varepsilon > 0$, a ball $\overline{B}(a; \delta) \subseteq X$ with $\delta > 0$ and $f(\overline{B}(a; \delta)) \subseteq \overline{B}(f(a); \varepsilon)$ exists.
(2) The mapping f is called **continuous** (on all of X), if f is continuous at every point of X.

In analogy to Theorem 3.8.9 the following applies:

Proposition 4.1.11 *Let $f : X \to Y$ be a mapping of metric spaces.*

(1) *For a point $a \in X$ the following statements are equivalent:*
   (i) *f is continuous at a.*
   (ii) *For every neighborhood V of $f(a)$ in Y there exists a neighborhood U of a in X with $f(U) \subseteq V$.*
   (iii) *For every neighborhood V of $f(a)$ in Y the preimage $f^{-1}(V)$ is a neighborhood of a in X.*
   (iv) *For every convergent sequence (x_n) in X with $\lim x_n = a$ the image sequence $(f(x_n))$ is convergent in Y with $\lim f(x_n) = f(a)$.*

(2) *Exactly then is f continuous on X, when the preimage of every open set in Y is an open set in X.*

Proof (1) (i) $\Rightarrow$ (ii): For every environment V of $f(a)$ there is an $\varepsilon > 0$ with $\overline{B}(f(a); \varepsilon) \subseteq V$. By assumption, there is a $\delta > 0$ with $f(\overline{B}(a; \delta)) \subseteq \overline{B}(f(a); \varepsilon) \subseteq V$, and $U := \overline{B}(a; \delta)$ is an environment of a .— The implication (ii) $\Rightarrow$ (iii) is trivial.

To prove (iii) $\Rightarrow$ (iv) let $\lim x_n = a$ and V be an environment of $f(a)$. By assumption, $f^{-1}(V)$ is an environment of a, thus it contains almost all elements of the sequence (x_n). Then, almost all elements of the sequence $(f(x_n))$ are also in V.

The implication (iv) $\Rightarrow$ (i) is the least self-evident. Let $\varepsilon > 0$ be given. If there were no $\delta > 0$ with $f(\overline{B}(a; \delta)) \subseteq \overline{B}(f(a); \varepsilon)$, then for every $n \in \mathbb{N}^*$ there would be an $x_n \in X$ with $d(x_n, a) \leq 1/n$, but $d(f(x_n), f(a)) > \varepsilon$. The sequence (x_n) converges to a, but $(f(x_n))$ does not converge to $f(a)$. Contradiction!

To prove (2), we recall the characterization of open sets: A set is open if and only if it is a neighborhood of each of its points. Now let f be continuous and V an open subset of Y. Then, due to (i) $\Rightarrow$ (iii), the preimage $f^{-1}(V)$ is a neighborhood of each of its points and thus open in X. Conversely, if the preimages of open sets under f are always open and if V is a neighborhood of $f(a)$ for the point $a \in X$, then V contains an open neighborhood V' of $f(a)$. Then $f^{-1}(V')(\subseteq f^{-1}(V))$ is an open neighborhood of a, and f is continuous at a. $\square$

Since equivalent metrics define the same open sets and neighborhoods, one can switch to equivalent metrics when considering continuity (which occasionally simplifies calculations).

Definition 4.1.12 A mapping $f : X \to Y$ of metric spaces is called **distance-preserving** or **isometric**, if $d(f(x), f(y)) = d(x, y)$ holds for all $x, y \in X$. — f is called an **isomorphism of metric spaces** or an **isometry**, if f is bijective and distance-preserving.

A distance-preserving mapping is necessarily injective and continuous. If $f : X \to Y$ is an isometry, so is the inverse mapping $f^{-1} : Y \to X$. The isometries of a metric space X onto itself form a subgroup of the permutation group of X. If X carries the discrete metric, then the group of isometries of X is the full permutation group $\mathfrak{S}(X)$. A weakening of the isometries are the Lipschitz-continuous mappings of metric spaces, cf. Example 3.8.18.

Definition 4.1.13 A mapping $f : X \to Y$ of metric spaces is called **Lipschitz continuous** with **Lipschitz constant** $L \in \mathbb{R}_+$, if for all $x, x' \in X$ applies $d(f(x), f(x')) \leq Ld(x, x')$. If $X = Y$ and $L < 1$, then f is called **strongly contracting**. In this case, the Lipschitz constant $L < 1$ is also called a **contraction factor** of f. — f is called **Lipschitz continuous at the point** $a \in X$, if there is a neighborhood U of a in X and a $L \in \mathbb{R}_+$ with $d(f(a), f(x)) \leq Ld(a, x)$ for all $x \in U$.

4.1 Metric Spaces

Lipschitz continuous mappings are continuous. We also note that a self-mapping $f : X \to X$ of a metric space is called **contracting** if for all $x, x' \in X$ applies $\bigl(f(x), f(x')\bigr) < d(x, x')$. Contracting mappings are Lipschitz continuous with Lipschitz constant 1, but not necessarily strongly contracting, cf. exercise 3.8.38.

Exercises

Exercise 4.1.1 Let $f : \mathbb{R}_+ \to \mathbb{R}_+$ be a **subadditive function**, i.e. it is $f(x+y) \leq f(x) + f(y)$ for all $x, y \in \mathbb{R}_+$. Furthermore, let $f \neq 0, f(0) = 0$ and f monotonically increasing and continuous at the origin. (f is then continuous on all $\mathbb{R}_+$ and >0 on all $\mathbb{R}_+^{\times}$.) We extend f to a function $f : \overline{\mathbb{R}}_+ \to \overline{\mathbb{R}}_+$ by $f(\infty) := \lim_{x \to \infty, x < \infty} f(x)$. If d is a metric on X, then $f \circ d$ is also a metric on X and for both metrics exactly the same subsets of X are open. Frequently used examples of such functions f are Min (r, x), $r \in \mathbb{R}_+^{\times}$; $x/(1+x)$; $\arctan x$; x^{α}, $0 < \alpha < 1$. Without changing the set of open sets of a metric space X, one can therefore assume that in X all distances are $\leq r$ with fixed $r \in \mathbb{R}_+^{\times}$. (Continuous concave functions $f : \mathbb{R}_+ \to \mathbb{R}_+$ (see Vol. 2, Sect. 2.5) with $f \neq 0$ and $f(0) = 0$ meet the specified conditions.)

Exercise 4.1.2 Let d be the discrete metric on X. Characterize the convergent sequences in X.

Exercise 4.1.3 Let (x_n) be a convergent sequence in the metric space X with $\lim x_n = x$. The sequence (y_n) in X also converges to x if and only if $d(x_n, y_n)$, $n \in \mathbb{N}$, is a null sequence in $\mathbb{R}$.

Exercise 4.1.4 Two metrics on a set X define exactly the same open sets when the sets of convergent sequences with respect to the individual metrics coincide.

Exercise 4.1.5 Let $1 \leq p < q \leq \infty$ and I be an *infinite* set. Then the p-norm $\|-\|_p$ and the q-norm $\|-\|_q$ on $\mathbb{K}^I$ are not equivalent. It is $\ell_{\mathbb{K}}^p(I) \subset \ell_{\mathbb{K}}^q(I) \subset \mathbb{K}^I$. Also on $\mathbb{K}^{(I)}$ the norms $\|-\|_p$ and $\|-\|_q$ are not equivalent.

Exercise 4.1.6 Let $V = (V, \|-\|)$ be a normalized $\mathbb{K}$ vector space.

a) The addition $V \times V \to V$, $(x, y) \mapsto x+y$, and the negative formation $V \to V$, $x \mapsto -x$, are continuous.
b) If $V = V_{<\infty} = B_V(0; \infty)$, the norm $\|-\|$ is therefore real-valued, then the scalar multiplication $\mathbb{K} \times V \to V$, $(a, x) \mapsto ax$, is continuous. If there is a $x \in V$ with $\|x\| = \infty$, then the scalar multiplication is not continuous.

($V \times V$ or $\mathbb{K} \times V$ each carry one of the equivalent p metrics, such as d_1.)

Fig. 4.4 U and V separate x and y

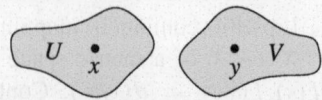

4.2 Topological Spaces and Continuous Mappings

The statements 4.1.9 and 4.1.11 show that convergence and continuity in metric spaces can be defined solely with the help of neighborhoods or also open sets. Therefore, it is advisable to make these concepts the basis of a general theory. The starting point for this is Proposition 4.1.4.

Definition 4.2.1 Let X be a set and $\mathcal{T}$ a subset of the power set $\mathfrak{P}(X)$. Then $\mathcal{T}$ is called a **topology** on X, if the following conditions are met:

(1) $\emptyset$ and X belong to $\mathcal{T}$.
(2) If U_i, $i \in I$, is an arbitrary family of elements in $\mathcal{T}$, then their union $\bigcup_{i \in I} U_i$ is also an element of $\mathcal{T}$.
(3) If U_j, $j \in J$, is a *finite* family of elements in $\mathcal{T}$, then their intersection $\bigcap_{j \in J} U_j$ is also an element of $\mathcal{T}$.

The elements of $\mathcal{T}$ are called the **open** sets of X (with respect to $\mathcal{T}$). — If $\mathcal{T}$ also fulfills the condition:

(4) For any two points $x, y \in X$ with $x \neq y$ there are open sets $U, V \in \mathcal{T}$ with $x \in U$, $y \in V$ and $U \cap V = \emptyset$;

then X is called a **Hausdorff space** (or a **separated topological space**). — A set $X = (X, \mathcal{T})$ with a topology $\mathcal{T}$ on X is called a **topological space**.

Condition (4) in the above definition is the so-called **Hausdorff separation axiom**: The open sets U and V separate the points x and y, see Fig. 4.4. Note that (1) follows from (2) and (3) for $I = \emptyset$.

According to Proposition 4.1.4 *every metric space is naturally a Hausdorff space*. We always consider a metric space (X, d) in this way as a topological space. Different metrics can define the same topology. For example, two equivalent metrics always define the same topology.[3] If there is a metric on a topological space X whose associated topology matches the given one, then X is called **metrizable**. Many important topological spaces are metrizable. However, proving this is not always straightforward. Furthermore, these metrics would often be artificial. *Therefore, it is generally much clearer to argue with the topology alone in metrizable spaces, if no naturally defined metrics are available.*

---

[3] But non-equivalent metrics can also define the same topology. Example? See Ex. 4.1.1.

4.2 Topological Spaces and Continuous Mappings

Definition 4.2.2 Let X be a topological space.

(1) A subset $U \subseteq X$ is called a **neighborhood of the point** $x \in X$, if there is an open set $U' \subseteq X$ with $x \in U' \subseteq U$. U is called a **neighborhood of the subset** $A \subseteq X$, if there is an open set $U' \subseteq X$ with $A \subseteq U' \subseteq U$.
(2) A subset $F \subseteq X$ is called **closed** (in X), if the complement $X - F$ of F in X is open in X.

For the closed sets, the following results from Definition 4.2.1 by passing to the complements:

Proposition 4.2.3 *Let X be a topological space. Then the following holds:*

(1) X and $\emptyset$ are closed in X.
(2) If F_i, $i \in I$, is an arbitrary family of closed sets in X, then their intersection is also $\bigcap_{i \in I} F_i$ closed in X.
(3) If F_j, $j \in J$, is a finite family of closed sets in X, then their union is also $\bigcup_{j \in J} F_j$ closed in X.

An open set $U \subseteq X$ is a neighborhood of each of its points, and conversely, a set $U \subseteq X$ is open if it is a neighborhood of each of its points. Then $U = \bigcup_{x \in U} U(x)$ is true, where $U(x)$ for $x \in U$ is an open set with $x \in U(x) \subseteq U$. A set $U \subseteq X$ is exactly a neighborhood of a subset $A \subseteq X$, if U is a neighborhood of each point $x \in A$ is. According to condition (3) in definition 4.2.1, the average of many environments of a point is finite $x \in X$ or a subset $A \subseteq X$ again an environment of x or A. — In a Hausdorff space X, every single-point and thus every finite subset $\{x\}$ *is closed.* If $x \in X$ is, then by definition every point $\neq x$ has an environment that does not contain x.[4]

The reader is advised to always consider simple examples of the following terms in our perceptual space or a plane of this space, see Fig. 4.5, and compare them with the corresponding term formations in Sects. 3.4 and 3.5. However, when examining general topological spaces, one should not rely too much on intuition.

Definition 4.2.4

---

[4] However, there are (necessarily infinite) spaces that are not Hausdorff, but in which every single-point set is closed. Any infinite space X, in which exactly the finite subsets of X and X itself are closed, is an example of this. It is generally shown that the following statements for a topological space X are equivalent: (i) The single-point subsets $\{x\} \subseteq X$ are closed in X. (ii) If $x, y \in X$, $x \neq y$, then there is an environment U of x with $y \notin U$. (iii) If $x \in X$, then $\{x\}$ is the intersection of all environments of x. Topological spaces that meet these conditions are also called T_1-**spaces**. Hausdorff spaces are also called T_2-**spaces**.

Fig. 4.5 Inner point and boundary point of the set A

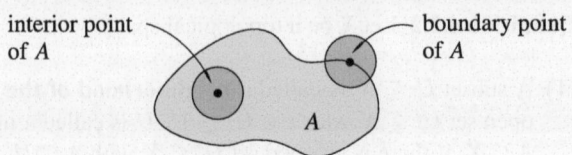

Let X be a topological space and $A \subseteq X$.

(1) A point $x \in X$ is called an **interior** point of A, if A is a neighborhood of x. — The set of all interior points is called the **interior** or the **open core** of A and is denoted by

$$\mathring{A}.$$

(2) A point $x \in X$ is called a **contact point** of A, if in every neighborhood of x there is an element of A. — The set of contact points of A is called the **closure** or the **closed hull** of A and is denoted by

$$\overline{A}.$$

If $\overline{A} = X$, then A is called **dense** (in X).

(3) A point $x \in X$ is called a **boundary point** of A, if in every neighborhood of x there is both an element of A and an element of the complement $X - A$. —The set of all boundary points of A is called the **boundary** of A and is denoted by

$$\partial A \quad \text{or} \quad \partial_X A.$$

(4) A point $x \in X$ is called an **accumulation point** of A, if in every neighborhood of x there is a point different from x in A.

In Fig. 4.5 the indicated boundary point is also an accumulation point of A. Apparently,

$$\mathring{A} \subseteq A \subseteq \overline{A}; \quad \mathring{A} \uplus \partial A = \overline{A}; \quad \partial A = \overline{A} \cap \overline{(X - A)} = \overline{A} - \mathring{A}.$$

A subset $A \subseteq X$ is exactly **boundary-less**, i.e. it is $\partial A = \emptyset$, when it is open and closed in X.[5] Every accumulation point of A is a contact point of A. Every contact point of A that does not belong to A is conversely an accumulation point of A. A is exactly dense in X when every non-empty open set in X has a non-empty intersection with A. An inner point of the complement $X - A$ of A in X is called an **outer point** of A.

Proposition 4.2.5 *Let A be a subset of the topological space X.*

---

[5] In English, boundary-less sets are called "clopen (= closed and open) sets".

4.2 Topological Spaces and Continuous Mappings

(1) *The open core $\mathring{A}$ of A is the largest open subset of X contained in A.*
(2) *The closed hull $\overline{A}$ of A is the smallest closed subset of X that includes A. An open set U in X intersects A exactly when it intersects $\overline{A}$.*
(3) *The boundary ∂A of A is closed in X.*

Proof (1) Let $x \in \mathring{A}$. Then there is an open neighborhood U of x with $U \subseteq A$. Since U is a neighborhood for all $y \in U$, it follows that $U \subseteq \mathring{A}$. Consequently, $\mathring{A}$ is open in X. Let $V \subseteq A$ and V be open in X. Then V is again a neighborhood for all $y \in V$ and thus $V \subseteq \mathring{A}$.

(2) Let $x \in X, x \notin \overline{A}$. There is an open neighborhood U of x with $U \cap A = \emptyset$. Then $U \cap \overline{A} = \emptyset$ is true, since U is a neighborhood for all $y \in U$. Consequently, $X - \overline{A}$ is open and $\overline{A}$ is closed in itself. Conversely, let B be closed in X with $A \subseteq B$. Then $\overline{A} \cap (X - B) = \emptyset$ is true, since $X - B$ is a neighborhood for every point $y \in X - B$ with $A \cap (X - B) = \emptyset$. Therefore, $\overline{A} \subseteq B$.

(3) follows from $\partial A = \overline{A} \cap \overline{(X - A)}$ due to (2). □

We leave the proof of the next statement to the reader, see exercise 4.2.4. We will frequently use similar simple set-theoretical relationships between the given topological terms in the following, without always proving them in detail. As already noted, the reader should not rely too much on intuition. See also the exercises for this section.

Proposition 4.2.6 *Let A be a subset of the topological space X.*

(1) *The following statements are equivalent:* (i) *A is open in X.* (ii) *It is $A = \mathring{A}$.* (iii) *It is $A \cap \partial A = \emptyset$.*
(2) *The following statements are equivalent:* (i) *A is closed in X.* (ii) *It is $A = \overline{A}$.* (iii) *It is $\partial A \subseteq A$.* (iv) *A contains all its accumulation points.*

Every subspace $X' \subseteq X$ of a topological space $X = (X, T)$ is naturally again a topological space with the so-called **induced** or **relative topology**

$$T \cap X' = \{U \cap X' \mid U \in T\}.$$

In the following, we always consider subsets of topological spaces with their relative topology as topological spaces (unless otherwise stated). If $A \subseteq X'$ is a subset of $X' \subseteq X$, one must carefully consider whether A is considered as a subset of the topological space X or of the topological space X' when dealing with the topological concepts introduced above and also later. For example, X' is always open in X', but not necessarily open in X. If X is Hausdorff, so is X' for every $X' \subseteq X$. If X is metric, the relative topology on X' coincides with the topology defined by the metric induced on X'.

In analogy to Definition 4.1.8 and taking into account Lemma 4.1.9, a sequence (x_n) of a topological space X is called **convergent** (in X) with **limit** or **Limes** $x \in X$, if almost all members of the sequence lie in every neighborhood of x. If X

is a *Hausdorff space*, a sequence in X has at most one limit in X. Because different points in X have disjoint neighborhoods, in which not simultaneously almost all members of a sequence can lie. A point $x \in X$ is called an **accumulation point** (or also **contact point**) of the sequence (x_n), if infinitely many members of the sequence (x_n) lie in every neighborhood of x. If (x_n) has a subsequence converging to x, then x is an accumulation point of (x_n).

Example 4.2.7 Let X be a set. We consider on X the two extreme topologies, namely first the **discrete topology**, where all subsets of X are open. This is exactly the case when every single-point subset of X is open in X. The discrete topology is induced by the discrete metric on X, cf. Example 4.1.6 (2). If (x_n) is a sequence in X and $x \in X$ is a point, then $\{x\}$ is a neighborhood of x and consequently x is only then a limit of (x_n), when the sequence (x_n) is stationary with limit x, i.e. when $x_n = x$ applies to all $n \geq n_0$ with a $n_0 \in \mathbb{N}$. The concept of convergence is therefore very restrictive in this case, but in the case of convergence, the sequence elements provide very precise information about the limit. The question of whether convergence exists at all and how to find the position n_0 can, of course, still be quite difficult. — A subspace X' of any topological space X is called **discrete**, when the induced topology on X' is the discrete topology. This is exactly the case when every point $x \in X'$ is a discrete point in X'. Here, $x \in X'$ is called a **discrete point** of X', if x has a neighborhood U in X with $U \cap X' = \{x\}$.[6] A point $x \in X$ is discrete in X exactly when $\{x\}$ is open in X.

In the extreme case opposite to the discrete topology, $\emptyset$ and X are the only open sets in X. This is then referred to as the **lump topology** on X. With respect to this topology, a point $x \in X$ only has X itself as a neighborhood. Consequently, every sequence in X converges to every point $x \in X$. The convergence is therefore completely meaningless in this case.

In both cases, the concept of convergence is quite trivial, albeit for different reasons. The choice of a "good" topology is often a compromise between too large and too small, which must be decided according to the pursued goal. For example, the natural topology of $\mathbb{R}$ has proven to be an excellent compromise for the set of real numbers, which underlies all considerations of analysis. ◊

Generally, a topology $\mathcal{T}$ on a set X is **finer** than the topology $\mathcal{U}$ (and $\mathcal{U}$ **coarser** than $\mathcal{T}$), if $\mathcal{U} \subseteq \mathcal{T}$ applies.[7] The discrete topology is thus the finest topology on X and the lump topology on X the coarsest. If $\mathcal{T}_i$, $i \in I$, is an arbitrary family of topologies on X, then the intersection $\bigcap_i \mathcal{T}_i$ is also a topology on X. It is the finest topology that is coarser than all $\mathcal{T}_i$. The union $\bigcup_i \mathcal{T}_i$ is generally not a topology on X. However, there is a coarsest topology on X that is finer than all $\mathcal{T}_i$, namely

---

[6] Some authors only call a subspace X' of X discrete in X when the topology induced on X' is discrete *and X' is moreover closed in X*. For example, the set of unit fractions $1/n$, $n \in \mathbb{N}^*$, is discrete but not closed in $\mathbb{R}$. Its closed hull $\{1/n \mid n \in \mathbb{N}^*\} \uplus \{0\}$ is not discrete.

[7] Note that these comparatives are also used when dealing with identical topologies.

the intersection of those topologies on X that are finer than all $\mathcal{T}_i$. The topologies thus form a complete lattice with respect to inclusion.

If $\mathcal{E} \subseteq \mathfrak{P}(X)$ is a set of subsets of X, there is a coarsest topology $\mathcal{T}$ on X with $\mathcal{E} \subseteq \mathcal{T}$, namely the intersection of all topologies on X that $\mathcal{E}$ include. It is called **the topology generated by $\mathcal{E}$ and is denoted by** $\mathcal{T}(\mathcal{E})$. $\mathcal{E}$ itself is called a **generating system** of $\mathcal{T}(\mathcal{E})$. For example, $\{\emptyset, U, X\}$ is the topology generated by a single subset $U \subseteq X$. Which sets belong to the topology generated by two sets $U, V \subseteq X$? In general, $\mathcal{T}(\mathcal{E})$ includes the set $\mathcal{B}(\mathcal{E})$ of *finite* intersections of elements from $\mathcal{E}$ and thus also *arbitrary* unions of elements from $\mathcal{B}(\mathcal{E})$. Since these unions obviously already form a topology, $\mathcal{T}(\mathcal{E})$ is equal to the set of these unions. $\mathcal{B}(\mathcal{E})$ is therefore a basis of the topology $\mathcal{T}(\mathcal{E})$ in the sense of the following definition.

Definition 4.2.8 Let $X = (X, \mathcal{T})$ be a topological space.

(1) A **base** of $\mathcal{T}$ is a family $\mathcal{B} = (U_i)_{i \in I}$ of open sets $U_i \in \mathcal{T}$ such that every open set $U \in \mathcal{T}$ is a union of elements of the family $\mathcal{B}$.
(2) Let $x \in X$ be. A **neighborhood base** of x (with respect to $\mathcal{T}$) is a family $\mathcal{B}(x) = (U_i)_{i \in I}$ of (not necessarily open) neighborhoods of x such that every neighborhood U of x contains an element of the family $\mathcal{B}(x)$.

A generating system of a topology $\mathcal{T}$ is also called a **subbase** of $\mathcal{T}$. A neighborhood base of $x \in X$ contains arbitrarily small neighborhoods of x. *A family $\mathcal{B} = (U_i)_{i \in I}$, of open sets $U_i \in \mathcal{T}$ is exactly a base of $\mathcal{T}$, if for every $x \in X$ the subfamily of those U_i, for which $x \in U_i$ is, forms a neighborhood base of x.* If $(U_i)_{i \in I}$ is a base of $\mathcal{T}$ and $x_i \in U_i$, $i \in I$, then obviously $\{x_i \mid i \in I\}$ is a dense subset of X. An important example of a base of a topology are the open balls $B_d(x; \varepsilon)$, $x \in X$, $\varepsilon \in \mathbb{R}_+^\times$, of a metric space $X = (X, d)$. It would even suffice if the ε is a null sequence $(\varepsilon_n)_{n \in \mathbb{N}}$ in $\mathbb{R}_+^\times$. The balls $B_d(x; \varepsilon_n)$, $n \in \mathbb{N}$, then form a neighborhood base of x in X. In particular, every point of a metric space has a countable neighborhood base, i.e. *a metric space fulfills the first countability axiom* in the sense of the following definition.

Definition 4.2.9 Let $X = (X, \mathcal{T})$ be a topological space.

(1) $(X, \mathcal{T})$ fulfills the **first countability axiom**, if every point $x \in X$ has a countable neighborhood base.
(2) $(X, \mathcal{T})$ fulfills the **second countability axiom** or has **countable topology**, if $\mathcal{T}$ has a countable base.[8]

If X satisfies the first or the second countability axiom, the same applies to every subspace $X' \subseteq X$. If $x \in X$ has a countable neighborhood base U_n, $n \in \mathbb{N}$,

---

[8] Note that this definition is somewhat misleading, as generally not the topology $\mathcal{T}$ itself is countable, but only a suitable base $\mathcal{B}$ of $\mathcal{T}$.

then also the countable neighborhood base $\mathring{U}_n$, $n \in \mathbb{N}$, from open neighborhoods of x. Furthermore, then $V_n := \bigcap_{k=0}^{n} U_k$, $n \in \mathbb{N}$, is a countable neighborhood base of x with $V_0 \supseteq V_1 \supseteq V_2 \supseteq \cdots$. If X satisfies the second countability axiom, then naturally also the first. The reverse is not generally true: For example, an uncountable topological space with the discrete topology satisfies the first but not the second countability axiom. A topological space already has countable topology if it has a countable subbase, cf. example 1.8.8. *A topological space with countable topology always has a countable dense subset.* For metric spaces, the reverse also applies:

Lemma 4.2.10 *For a metric space X, the following statements are equivalent:*

(i) *X has a countable topology.*
(ii) *X has a countable dense subset.*

Proof As already mentioned, (ii) follows from (i). — To prove (ii) $\Rightarrow$ (i), let $x_i \in X$, $i \in I$, be countably many points, for which $A = \{x_i \mid i \in I\}$ is dense in X. Then the open balls $B(x_i; 1/n)$, $(i, n) \in I \times \mathbb{N}^*$, a countable basis of the topology of X. Let U be any open set in X and $x \in U$. There is an $m \in \mathbb{N}^*$ with $B(x; 1/m) \subseteq U$. In the open ball $B(x; 1/2m)$ there is a point x_i of the dense subset A. Then $x \in B(x_i; 1/2m)$. Furthermore, $B(x_i; 1/2m) \subseteq B(x; 1/m)$ and thus $B(x_i; 1/2m) \subseteq U$ due to $d(x, y) \leq d(x, x_i) + d(x_i, y) < 1/2m + 1/2m = 1/m$ for $y \in B(x_i; 1/2m)$. Thus, U is a union of balls of the form $B(x_i; 1/n)$, $(i, n) \in I \times \mathbb{N}^*$. $\square$

Metric spaces that satisfy the equivalent conditions of Lemma 4.2.10 are also called **separable** metric spaces.

If x is the limit of a sequence (x_n) of elements of a subset A of the topological space X, then x is a contact point of A, and thus lies in $\overline{A}$. If the x_n, $n \in \mathbb{N}$, are also different from x, then x is even an accumulation point of A. If X fulfills the first countability axiom, then the reverse also applies:

Lemma 4.2.11 *Let A be a subset of the topological space X, which satisfies the first countability axiom.*

(1) *A point $x \in X$ is a contact point of A if and only if there is a sequence (x_n) of elements in A with $x = \lim x_n$ exists.*
(2) *A point $x \in X$ is an accumulation point of A if and only if there is a sequence (x_n) of elements in A with $x_n \neq x$ for all $n \in \mathbb{N}$ and $x = \lim x_n$ exists.*
(3) *A is closed in X if and only if the limit of every sequence in X that converges $x_n \in A$, $n \in \mathbb{N}$, even lies in A.*

Proof Let x be a contact or accumulation point of A and let U_n, $n \in \mathbb{N}$, be a neighborhood basis of x with $U_0 \supseteq U_1 \supseteq U_2 \supseteq \cdots$. For each $n \in \mathbb{N}$ there is an element $x_n \in U_n$, which can be chosen differently in the case of an accumulation

4.2 Topological Spaces and Continuous Mappings

point of x. Then $x = \lim x_n$. If U is a neighborhood of x, there is an $n_0 \in \mathbb{N}$ with $U_{n_0} \subseteq U$ and consequently with $x_n \in U$ for $n \geq n_0$. This proves (1) and (2). – (3) follows from (1). □

As already noted, a sequence (x_n) in X has the point $x \in X$ as an accumulation point if it has a subsequence converging to x. In spaces that satisfy the first countability axiom, the converse also holds:

Lemma 4.2.12 *Let X be a topological space that satisfies the first countability axiom. A point $x \in X$ is an accumulation point of the sequence (x_n) in X, if and only if (x_n) has a subsequence converging to x.*

Proof Let x be an accumulation point of the sequence (x_n) and $U_n, n \in \mathbb{N}$, a neighborhood base of x with $U_0 \supseteq U_1 \supseteq U_2 \supseteq \cdots$. Then for each $k \in \mathbb{N}$ there are infinitely many terms x_n of the sequence with $x_n \in U_k$. Therefore, there is an index sequence $n_0, n_1, \ldots$ with $n_0 < n_1 < \cdots$ and $x_{n_k} \in U_k, k \in \mathbb{N}^*$, and the subsequence $(x_{n_k})_k$ converges to x. □

The following simple statement is used repeatedly:

Proposition 4.2.13 *Let X be a topological space with countable topology. Then every open cover of X has a countable subcover.*

Proof By an **open cover** of X we naturally understand a cover of X with open sets. Let $U_i, i \in I$, be such an open cover. Furthermore, let $\mathcal{B} \subseteq \mathcal{T}$ be a countable basis of the topology $\mathcal{T}$ of X. Each U_i is the union of the elements of a subset $\mathcal{B}_i \subseteq \mathcal{B}, i \in I$. Let $\mathcal{B}'$ be the (countable) union set $\bigcup_{i \in I} \mathcal{B}_i \subseteq \mathcal{B}$. For each $V \in \mathcal{B}'$ let $i_V \in I$ be an index with $V \subseteq U_{i_V}$. If J denotes the countable subset $J := \{i_V \mid V \in \mathcal{B}'\} \subseteq I$, then

$$X = \bigcup_{V \in \mathcal{B}'} V \subseteq \bigcup_{V \in \mathcal{B}'} U_{i_V} = \bigcup_{i \in J} U_i.$$

applies. Thus, $U_i, i \in J$, is a countable subcover of $U_i, i \in I$. □

According to Proposition 4.1.11, the following definition of the continuity of mappings of topological spaces is not surprising:

Definition 4.2.14 Let X and Y be topological spaces and $f : X \to Y$ a mapping.

(1) The mapping f is called **continuous at the point** $a \in X$, if for every neighborhood V of $f(a)$ in Y there is a neighborhood U of a in X with $f(U) \subseteq V$.

(2) The mapping f is called **continuous** (on all of X), if f is continuous at every point of X.

$$C(X, Y)$$

denotes the set of continuous mappings $X \to Y$.

(3) The mapping f is called a **homeomorphism**, if f is bijective and both f and the inverse mapping f^{-1} are continuous.

Condition (1) is equivalent to the preimage $f^{-1}(V)$ of every neighborhood V of $f(a)$ being a neighborhood of a. Consider the continuous mappings as the homomorphisms of topological spaces and the homeomorphisms as their isomorphisms. The identity of a topological space is always a homeomorphism, as is the inverse mapping of a homeomorphism.—As with metric spaces, the following applies:

Theorem 4.2.15 *For a mapping $f: X \to Y$ between topological spaces X and Y the following statements are equivalent:*

(i) *f is continuous.*
(ii) *The preimage of every open set in Y is an open set in X.*
(iii) *The preimage of every closed set in Y is a closed set in X.*

Proof The equivalence of (i) and (ii) follows as in Proposition 4.1.11 (2) from the fact that a set is open if and only if it is a neighborhood of each of its points. The equivalence of (ii) and (iii) is obtained by passing to complements. □

Theorem 4.2.15 implies that a mapping $f: X \to Y$ of topological spaces is already continuous if for each element V of a subbasis of the topology of Y the preimage $f^{-1}(V)$ is open in X.—The following theorem deals with the continuity of concatenated mappings.

Theorem 4.2.16 *Let $f: X \to Y$ and $g: Y \to Z$ be mappings between the topological spaces X, Y, Z.*

(1) *If f is continuous at $a \in X$ and g is continuous at $f(a) \in Y$, then the composition $g \circ f: X \to Z$ is continuous at a.*
(2) *If f and g are continuous, then the composition $g \circ f$ is also continuous.*

Proof (1) For a neighborhood W of $(gf)(a) = g(f(a))$, $g^{-1}(W)$ is a neighborhood of $f(a)$, since g is continuous at $f(a)$, and thus $f^{-1}(g^{-1}(W)) = (gf)^{-1}(W)$ is a neighborhood of a, since f is continuous at a. — (2) follows from (1). □

Let $X' \subseteq X$ be a subspace of X, $\iota: X' \hookrightarrow X$ the canonical embedding, and $g: Z \to X'$ a mapping of a topological space Z into X'. Because of $(\iota \circ g)^{-1}(U) =$

4.2 Topological Spaces and Continuous Mappings

$g^{-1}(\iota^{-1}(U)) = g^{-1}(U \cap X')$ is, by definition of the relative topology of X', *the mapping $g : Z \to X'$ is continuous if and only if $\iota \circ g : Z \to X$ is continuous.* $h : Z \to X$ is called an **embedding** (of topological spaces) if h is injective and h induces a homeomorphism $Z \xrightarrow{\sim} h(Z)$. If $h(Z)$ is closed (or open) in X, we speak of a **closed** (or an **open**) **embedding**. Since the canonical embedding $\iota : X' \hookrightarrow X$ is continuous, for every continuous mapping $f : X \to Y$ its restriction $f|X' = f \circ \iota : X' \to Y$ to X' is continuous. Of course, $f|X'$ can be continuous at a point $a \in X'$ without f being continuous at a. However, if a is an interior point of X', then f is continuous at a if and only if $f|X'$ is continuous at a. Proof! Thus, $f : X \to Y$ is *continuous at a point only if $a \in X$ continuous, if there is a neighborhood U of a in X such that $f|U$ is continuous in a. Continuity is said to be a* **local property**.

A homeomorphism $f : X \to Y$ induces a bijection of the set of open sets (and also a bijection of the closed sets) from X to the set of open sets (or the closed sets) of Y. Furthermore, the composition of homeomorphisms is again a homeomorphism. *The homeomorphisms of a topological space onto itself thus form a subgroup of its permutation group.* It should be explicitly noted that the inverse mapping of a continuous bijective mapping is generally *not* continuous. For example, the identity of X is continuous, but not a homeomorphism, if X is provided with the discrete topology as the preimage area and the lump topology as the image area and X contains more than one point. For less coarse examples, see problem 4.2.23.

The continuity of the mapping $f : X \to Y$ at the point $a \in X$ can also be described as a limit relationship

$$f(a) = \lim_{x \to a} f(x)$$

if the following definition is used for the limit of mappings between topological spaces (see 3.8.8):

Definition 4.2.17 Let D be a subspace of the topological space X, $a \in X$ and $f : D \to Y$ a mapping from D into the topological space Y. Then $y \in Y$ is called a **limit** or **Limes** of f at the point a, symbolically

$$y = \lim_{x \to a, x \in D} f(x),$$

if for every neighborhood V of y in Y there exists a neighborhood U of a in X with $f(U \cap D) \subseteq V$.

If $a \notin \overline{D}$, then every point of Y is a limit of f at a. *If Y is Hausdorff and $a \in \overline{D}$, then the limit y (if it exists) is uniquely determined.* If $a \in D$ and f is continuous at a, then $f(a)$ is a limit of f at a. If $a \in D$ and Y is Hausdorff, then $\lim_{x \to a, x \in D}$ exists if and only if f is continuous at a (and the limit is necessarily $f(a)$). In the case of $a \notin D$, $y = \lim_{x \to a, x \in D} f(x)$ obviously holds if and only if the function defined on $D \cup \{a\}\,\widetilde{f}$ with

$$\widetilde{f}(x) := \begin{cases} f(x), & \text{if } x \in D, \\ y, & \text{if } x = a, \end{cases}$$

is continuous at a. — If X satisfies the first countability axiom, limits can be characterized with the help of sequences:

Lemma 4.2.18 *Let $f : D \to Y$ and $a \in X$ be as in Definition 4.2.17. Furthermore, let a in X have a countable neighborhood base. For a point $y \in Y$ the following are equivalent:*

(i) *It is $y = \lim_{x \to a, x \in D} f(x)$.*
(ii) *For every sequence (x_n) in D with $\lim x_n = a$ it holds $\lim f(x_n) = y$.*

Proof The implication (i) $\Rightarrow$ (ii) obviously holds for every topological space X. — To prove the converse (ii) $\Rightarrow$ (i), let U_n, $n \in \mathbb{N}$, be a countable neighborhood base of a in X with $U_0 \supseteq U_1 \supseteq U_2 \supseteq \cdots$. Suppose there is a neighborhood V of y in Y with $f(U \cap D) \not\subseteq V$ for all neighborhoods U of x in X. In particular, there are then points $x_n \in U_n \cap D$ with $f(x_n) \notin V$. Then $\lim x_n = a$ is true, but the sequence $f(x_n)$, $n \in \mathbb{N}$, does not converge to y. Contradiction! $\square$

As a corollary, we note:

Corollary 4.2.19 *Let $f : X \to Y$ be a mapping of topological spaces and $a \in X$. If a has a countable neighborhood base, then f is continuous in a if and only if $\lim f(x_n) = f(a)$ is for every sequence (x_n) in X with $\lim x_n = a$. In particular, if X satisfies the first countability axiom, then f is continuous if and only if for every convergent sequence in X, (x_n) holds*

$$\lim f(x_n) = f(\lim x_n).$$

Example 4.2.20 (Topologies on $\overline{\mathbb{R}}$) Since we often have to consider mappings with the image area $\overline{\mathbb{R}} = \mathbb{R} \uplus \{\pm\infty\}$, we want to make $\overline{\mathbb{R}}$ a topological space. Depending on the situation, we use one of the following two topologies: First, we define a metric on $\overline{\mathbb{R}}$ by $d(x, y) := |y - x|$, where we exceptionally allow the two differences $\infty - \infty = (-\infty) - (-\infty) = 0$. Then $\mathbb{R}$ is an open subspace of $\overline{\mathbb{R}}$, and on $\mathbb{R}$ the natural metric of $\mathbb{R}$ is induced. Furthermore, ∞ and $-\infty$ are open points in $\overline{\mathbb{R}}$. In other words: A subset $U \subseteq \overline{\mathbb{R}}$ is exactly open when $U \cap \mathbb{R}$ is open in $\mathbb{R}$. (The topology on $\overline{\mathbb{R}}$ is thus the image topology of $\mathbb{R}$ with respect to the embedding $\mathbb{R} \hookrightarrow \overline{\mathbb{R}}$, see example 4.2.24 below.) In this **metric topology**, the sequence of natural numbers does not converge to ∞. Rather, $\mathbb{R}$ is closed in $\overline{\mathbb{R}}$ with respect to the metric topology (which collides with the designation $\overline{\mathbb{R}}$). A mapping $f : X \to \overline{\mathbb{R}}$ of a topological

4.2 Topological Spaces and Continuous Mappings

space X into $\overline{\mathbb{R}}$ is continuous exactly when the two fibers $f^{-1}(\infty)$ and $f^{-1}(-\infty)$ are each open in X and the restriction $f|f^{-1}(\mathbb{R})$ is continuous.

For analysis, the following so-called **order topology** on $\overline{\mathbb{R}}$ is usually more important. This topology on $\overline{\mathbb{R}}$ is by definition generated by all open subsets of $\mathbb{R}$ and the intervals $]a, \infty]$, $[-\infty, a[$, $a \in \mathbb{R}$. (See exercise 4.2.35.) *$\overline{\mathbb{R}}$ is then homeomorphic to any closed interval $[a, b]$, $a, b \in \mathbb{R}$, $a < b$*. Obviously, for example, $\overline{\mathbb{R}} \xrightarrow{\sim} [-1, 1]$, $x \mapsto x/(1 + |x|)$ for $x \in \mathbb{R}$, $\pm\infty \mapsto \pm 1$, is a homeomorphism. On $\mathbb{R}$ also induces the order topology to the natural topology, and $\mathbb{R}$ is now dense in $\overline{\mathbb{R}}$, the sequences $(n)_{n \in \mathbb{N}}$ and $(-n)_{n \in \mathbb{N}}$ converge to ∞ or $-\infty$. The order topology is therefore coarser than the above metric topology. It is nevertheless also metrizable, since the interval $[-1, 1]$ is a metric space. A mapping $f : X \to \overline{\mathbb{R}}$ of a topological space X into $\overline{\mathbb{R}}$ is exactly continuous with respect to the order topology, when its restriction $f|f^{-1}(\mathbb{R})$ is continuous and for every point $x \in X$ with $f(x) = \infty$ (or $f(x) = -\infty$) and for every $a \in \mathbb{R}$ a neighborhood U of x exists with $f(x) > a$ (or $f(x) < a$) for all $x \in U$. If f is continuous with respect to the metric topology, it is also continuous with respect to the order topology on $\overline{\mathbb{R}}$. — Which mappings $\overline{\mathbb{R}} \to Y$ from $\overline{\mathbb{R}}$ into the topological space Y are continuous with respect to the metric or with respect to the order topology on $\overline{\mathbb{R}}$? ◇

Remark 4.2.21 (General Limits) In general topological spaces, limits cannot usually be characterized by the convergence of sequences. Instead, the concept of a convergent sequence must be generalized. As noted earlier, the order of natural numbers is not really used in the convergence of a sequence (x_n) in a topological space X. (x_n) converges to $x \in X$ if and only if for every neighborhood $U \in \mathcal{U}(x)$ of $x \in X$ there exists a finite set $E \subseteq \mathbb{N}$ with $x_n \in U$ for all $n \in \mathbb{N} - E$. The set $\mathcal{F}$ of complements of finite subsets of $\mathbb{N}$ has the following properties: (1) The intersection of finitely many elements of $\mathcal{F}$ again belongs to $\mathcal{F}$. (2) If $F \subseteq F' \subseteq \mathbb{N}$ holds and $F \in \mathcal{F}$ is true, then $F' \in \mathcal{F}$ is also true. Analogous properties hold for the set of complements of finite subsets of any set I. $\mathcal{F}$ is therefore a filter in the sense of the following definition:

Definition 4.2.22 Let I be a set. $\mathcal{F} \subseteq \mathfrak{P}(I)$ is called a **filter** on I, if the following holds:

(1) The intersection of finitely many elements of $\mathcal{F}$ again belongs to $\mathcal{F}$.
(2) If $F \subseteq F' \subseteq I$ holds and $F \in \mathcal{F}$ is true, then $F' \in \mathcal{F}$ is also true.

According to (1), in particular $I \in \mathcal{F}$ holds for every filter $\mathcal{F}$ on I. Condition (1) summarizes the following two conditions: (1_1) It is $I \in \mathcal{F}$, and (1_2) the intersection of two elements of $\mathcal{F}$ again belongs to $\mathcal{F}$. $\mathfrak{P}(I)$ is the largest filter on I and $\{I\}$ is the smallest. The filter $\mathfrak{P}(I)$ is defined by the condition $\emptyset \in \mathcal{F}$ characterizes it. The set of filters on I is (strictly) inductively ordered with respect to inclusion. An anti-atom in the set of filters is called an **Ultrafilter**. According to Zorn's Lemma, every filter $\neq \mathfrak{P}(I)$ on I is contained in an ultrafilter. The filter of complements of finite sets on I is called the **Fréchet filter** $\mathcal{F}r(I)$ on I. For every topological space X and every point $x \in X$, the set $\mathcal{U}(x)$ of neighborhoods of x is a filter on X, the so-called **neighborhood filter** of x. If $\mathcal{F}$ is a filter on I and $I' \subseteq I$, then $\mathcal{F} \cap I' = \{F \cap I' \mid F \in \mathcal{F}\}$ is a filter on I'. Usually, a filter is described by specifying a **filter base**: This is simply a subset $\mathcal{B} \subseteq \mathfrak{P}(I)$ such that the intersection of any finite number of elements from $\mathcal{B}$ always includes an element from $\mathcal{B}$. (In particular, $\mathcal{B} \neq \emptyset$ is therefore.) The associated filter $\mathcal{F}(\mathcal{B})$ with base $\mathcal{B}$ is then the set of all subsets of I that contain an element from $\mathcal{B}$ include. For example, a neighborhood base of a point x in the topological space X is by definition a filter base of the neighborhood filter $\mathcal{U}(x)$ of x. The sections $\mathbb{N}_{\geq n_0}$, $n_0 \in \mathbb{N}$, form a base of the Fréchet filter on $\mathbb{N}$. If $\mathcal{E}$ is any subset of $\mathfrak{P}(I)$, then the set $\mathcal{B}(\mathcal{E})$ of finite intersections of elements from $\mathcal{E}$ forms a filter base of the **filter generated by** $\mathcal{E}$ on I. — The convergence of a sequence and also Definition 4.2.17 are now special cases of the following general limit concept:

Definition 4.2.23 Let $I = (I, \mathcal{F})$ be a set with a filter $\mathcal{F}$ on I and x_i, $i \in I$, a family of elements in the topological space X. Then $x \in X$ is called the **limit** or **Limes** of x_i, $i \in I$, (with respect to $\mathcal{F}$), if for every neighborhood U of x there exists an $F \in \mathcal{F}$ with $x_i \in U$ for all $i \in F$. We then write

$$x = \lim_{i \in I} \mathcal{F} x_i = \lim_{i \in I} x_i.$$

If $\mathcal{F} = \mathfrak{P}(I)$, then every $x \in X$ is a limit of the family x_i, $i \in I$.[9] If X is *Hausdorff and $\mathcal{F} \neq \mathfrak{P}(I)$, then the limit x, if it exists, is uniquely determined.* If $\mathcal{F} = \{I\}$ and $x \in X$, then $x = \lim x_i$ applies exactly when $x_i \in \bigcap_{U \in \mathcal{U}(x)} U$ for all $i \in I$. The limits of a sequence (x_n) or the limits of a mapping $f : D \to Y$ in $a \in X$ as in Definition 4.2.17 are thus the limits $\lim_{n \in \mathbb{N}, \mathcal{F}_r(\mathbb{N})} x_n$ or $\lim_{x \in D, \mathcal{U}(a) \cap D} f(x)$. A mapping $f : X \to Y$ of topological spaces is exactly continuous in $a \in X$, when $f(a) = \lim_{x \in X, \mathcal{U}(a)} f(x)$ is the case. So, f is continuous in a exactly when the following applies: *If x_i, $i \in I$, a family in X with filtered index set $I = (I, \mathcal{F})$ and $a = \lim_{\mathcal{F}} x_i$, then $f(a) = \lim_{\mathcal{F}} f(x_i)$ applies.* Proof!

Many authors only consider the filters for inductively ordered sets $I = (I, \leq)$. We remind that an ordered set (or even just quasi-ordered set) I is inductively (quasi)ordered if for any two elements $i, j \in I$ a $k \in I$ with i,j $\leq k$ exists. Then the sections $I_{\geq i}$, $i \in I$, together with I form a filter base of a filter $\mathcal{F}$ on I. Families indexed with an inductively ordered set I $x_i \in X$, $i \in I$, are also called **nets** or **Moore-Smith sequences** (after E. Moore (1862–1932) and H. Smith (1892–1950)). If $I \neq \emptyset$, then obviously $x = \lim_{i \in I, \mathcal{F}} x_i$ applies exactly when *for every neighborhood $U \in \mathcal{U}(x)$ of x an $i_0 \in I$ exists with $x_i \in U$ for all $i \geq i_0$.* Since the Fréchet filter on $\mathbb{N}$ is the filter for the natural order on $\mathbb{N}$, the networks also prove to be canonical generalizations of sequences. General filters were introduced in

---

[9] Because of this pathology, many authors demand that the empty set does not belong to a filter $\mathcal{F}$, so that $\mathcal{F} \neq \mathfrak{P}(I)$ is the case.

1937 by H. Cartan (1904–2008). For an algebraic interpretation of the filters on a set I see exercise 4.2.31. ◊

As usual in mathematics, new structures are constructed from given ones, cf. the dictum of Kronecker at the beginning of Remark 1.5.7. In the following, we give some examples from the theory of topological spaces.

Example 4.2.24 (Image topologies—Quotient topologies) Let $X = (X, \mathcal{T})$ be a topological space and $f : X \to Y$ a mapping from X into an (arbitrary) set Y. Then

$$f_*\mathcal{T} := \{V \subseteq Y \mid f^{-1}(V) \in \mathcal{T}\}$$

is the finest topology on Y, with respect to which f is continuous. It is called the **image topology** of X with respect to f. $(Y, f_*\mathcal{T})$ has the following universal property:
For every topological space Z, a mapping $g : Y \to Z$ is continuous if and only if its composition $g \circ f : X \to Z$ is continuous.

A subset $V \subseteq Y$ belongs to $f_*\mathcal{T}$ if and only if $f^{-1}(V) = f^{-1}(V \cap \operatorname{Im} f)$ is open in X. The subspace $\operatorname{Im} f$ thus carries the image topology with respect to the surjective mapping $f : X \to \operatorname{Im} f$ and the complement $Y - \operatorname{Im} f$ the discrete topology. Furthermore, both $\operatorname{Im} f$ and $Y - \operatorname{Im} f$ are open with respect to $f_*\mathcal{T}$. *It is therefore sufficient to study image topologies for surjective mappings* $f : X \to Y$. In this case, we can identify Y with the quotient X/R_f, where R_f is the equivalence relation on X induced by f (with "$xR_f y \Leftrightarrow f(x) = f(y)$"). This is also referred to as a **quotient topology**. This is particularly common when f itself is already the canonical projection $\pi_R : X \to X/R$ from X onto the quotient set X/R with respect to an equivalence relation R on X. The open sets in X/R then correspond exactly to the R-saturated open sets in X, and the continuous mappings from X/R into a topological space Z to those continuous mappings $X \to Z$ that are constant on the equivalence classes with respect to R. The quotient topology is easy to define, but often difficult to oversee. A *surjective* continuous mapping $f : X \to Y$ of topological spaces is called a **quotient mapping**, if the topology of Y is the image topology of X with respect to f. If $f : X \to Y$ is a surjective continuous mapping, then f is certainly a quotient mapping if f is an open mapping or a closed mapping. A (not necessarily continuous) mapping $f : X \to Y$ of topological spaces is called **open** (or **closed**), if the f-images of open (or closed) sets in X are open (or closed) in Y. An equivalence relation R on X is called **open** (or **closed**), if the canonical projection $\pi_R : X \to X/R$ is open (or closed). Important examples of open equivalence relations arise in the following way: The group G operates on X as a group of homeomorphisms. *Then the canonical projection $X \to X \backslash G$ from X to the orbit space $X \backslash G$ (which carries the quotient topology, unless otherwise stated) is open.* If $U \subseteq X$ is open, then the G-invariant hull $GU = \bigcup_{g \in G} gU$ is also open.

For any family $f_i : (X_i, \mathcal{T}_i) \to Y$, $i \in I$, the **image topology** with respect to the f_i, $i \in I$, is characterized by the fact that a mapping $g : Y \to Z$ is continuous if and only if all compositions $g \circ f_i$, $i \in I$, are continuous. It is the average $\bigcap_{i \in I} f_{i*}\mathcal{T}_i$ of the image topologies of the individual f_i. In particular, all $f_i : X_i \to Y$, $i \in I$, are continuous with respect to this image topology. ◊

Example 4.2.25 (Sums of topological spaces) Let $X_i = (X_i, \mathcal{T}_i)$, $i \in I$, be a family of topological spaces and $X := \biguplus_{i \in I} X_i (= \bigcup_{i \in I} \{i\} \times X_i)$ their disjoint union with the canonical injections $\iota_i \colon X_i \to X$, $x_i \mapsto (i, x_i)$, with which we consider the X_i as subspaces of X. With respect to the image topology of the ι_i, $i \in I$, is a set $U \subseteq X$ open exactly when $U_i := \iota_i^{-1}(U) = U \cap X_i$ is open in X_i for all $i \in I$. This topology is called the **sum topology** on X. It is the topology generated by the $\mathcal{T}_i \subseteq \mathfrak{P}(X_i) \subseteq \mathfrak{P}(X)$ and the finest topology, with respect to which all ι_i are continuous. The topological space X, equipped with this topology, we also denote by $\coprod_{i \in I} X_i$. The X_i, $i \in I$, form an open decomposition of X. Together with the injections ι_i, $i \in I$, the **sum** $\coprod_{i \in I} X_i$ of the topological spaces X_i, $i \in I$, also called the **coproduct** of the X_i, has the following universal property: *For every topological space Z the mapping*

$$C\left(\coprod_{i \in I} X_i, Z\right) \xrightarrow{\sim} \prod_{i \in I} C(X_i, Z), \quad f \longmapsto (f \circ \iota_i)_{i \in I}$$

is bijective. The continuous mapping $\coprod_{i \in I} X_i \to Z$, which belongs to the tuple $(f_i)_{i \in I}$ of continuous mappings $f_i \colon X_i \to Z$, we denote with $\biguplus_i f_i$. If $X = \biguplus_{i \in I} X_i$ is an open decomposition of the topological space X, then X is obviously the sum of the open subspaces $X_i \subseteq X$, $i \in I$. For a family $f_i \colon X_i \to Y$ of mappings of the topological spaces X_i into a set Y, the image topology of the f_i, $i \in I$, is equal to the image topology of the one mapping $\biguplus_i f_i \colon \coprod_{i \in I} X_i \to Y$. If the $X_i = (X_i, d_i)$ are metric spaces, then $\coprod_{i \in I} X_i$ is also a metric space with respect to the metric d with $d|X_i = d_i$ and $d(x_i, x_j) = \infty$, if $x_i \in X_i$, $x_j \in X_j$, $i \neq j$. A given metric space (X, d) is the sum of the open equivalence classes with respect to the equivalence relation $\sim_d$ on X defined at the beginning of Sect. 4.1. ◊

Example 4.2.26 (Preimage Topologies) Let $Y = (Y, \mathcal{U})$ be a topological space and $f \colon X \to Y$ a mapping of an (arbitrary) set X into the space Y. Then

$$f^*\mathcal{U} := \{f^{-1}(V) \subseteq X \mid V \in \mathcal{U}\}$$

is the coarsest topology on X with respect to which f is continuous. It is called the **preimage topology** of Y with respect to f and has the following universal property:

For every topological space W, a mapping $g \colon W \to X$ is continuous if and only if its composition $f \circ g \colon W \to Y$ is continuous.

If $Y' \subseteq Y = (Y, \mathcal{U})$ is the case, then the topology induced by Y on Y' is the preimage topology $\iota_*\mathcal{U}$ with respect to the canonical embedding $\iota \colon Y' \hookrightarrow Y$. The **preimage topology** with respect to an arbitrary family $f_i \colon X \to (Y_i, \mathcal{U}_i)$, $i \in I$, is characterized by the fact that a mapping $g \colon W \to X$ is continuous if and only if all compositions $f_i \circ g \colon W \to Y_i$, $i \in I$, are continuous. It is the topology generated by the preimage topologies $f_i^*\mathcal{U}_i$, $i \in I$, on X. In particular, the $f_i \colon X \to Y_i$, $i \in I$, are continuous with respect to this preimage topology. ◊

4.2 Topological Spaces and Continuous Mappings

Example 4.2.27 (Product Topologies) Let $X_i = (X_i, \mathcal{T}_i)$, $i \in I$, be a family of topological spaces and $X := \prod_{i \in I} X_i$ their product with the canonical projections $p_i \colon X \to X_i$, $i \in I$. We are looking for a topology on X, for which the projections $p_i \colon X \to X_i$ are continuous and have the following typical universal property of a product: *A mapping $f \colon W \to X$ is exactly continuous when the compositions $p_i \circ f \colon W \to X_i$ are all continuous, i.e. the mapping*

$$C\left(W, \prod_{i \in I} X_i\right) \xrightarrow{\sim} \prod_{i \in I} C(W, X_i), \quad f \longmapsto (p_i \circ f)_{i \in I},$$

is bijective for every topological space W. We denote the continuous mapping $W \to \prod_i X_i$, which belongs to the I-tuple of continuous mappings $f_i \colon W \to X_i$, $i \in I$, as before also with $(f_i)_{i \in I}$. The universal property implies in particular: *A sequence $x_n = (x_{in})_{i \in I}$, $n \in \mathbb{N}$, in $X = \prod_i X_i$ converges exactly when for each $i \in I$ the component sequence x_{in}, $n \in \mathbb{N}$, converges; is $x_i = \lim_n x_{in}$ for $i \in I$, then $(x_i)_{i \in I} = \lim_n x_n$.* An analogous statement generally applies to limits of families with filtered index sets, see Remark 4.2.21 and Exercise 4.2.32.

According to example 4.2.26, the pre-image topology on X with respect to the family of projections $p_i \colon X \to X_i$, $i \in I$, has the desired universal property. The pre-image topology with respect to a single projection p_j contains exactly the open sets $\prod_{i \in I} U_i$ with $U_j \in \mathcal{T}_j$ and $U_i = X_i$ for $i \neq j$. The sought-after so-called **product topology** on $X = \prod_i X_i$ thus has the sets

$$\prod_{i \in I} U_i \quad \text{with} \quad U_i \in \mathcal{T}_i \text{ for all } i \in I \text{ and } U_i = X_i \text{ for almost all } i \in I$$

as a basis. It is sufficient that the U_i themselves each form a basis of the topology $\mathcal{T}_i$ of X_i. In particular, the projections $p_j \colon \prod_i X_i \to X_j$, $j \in I$, are not only continuous, but also open (but usually not closed, e.g., the projections $\mathbb{R}^2 \to \mathbb{R}$ are not closed). *Apparently, a product of non-empty topological spaces is Hausdorff if and only if each of its factors is Hausdorff.*—Further follows (see theorem 1.8.4 and example 1.8.8): ◊

Proposition 4.2.28 *Let X_i, $i \in I$, be a countable family of topological spaces. If each X_i, $i \in I$, fulfills the first or the second countability axiom, the same applies to the product space $\prod_i X_i$.*

Apparently, the product topology on X has the following universal property for the continuity of $f = (f_i) \colon W \to \prod_i X_i$ at a point $a \in W$: *Exactly then is f continuous at a, when all f_i are continuous at a, $i \in I$.* — If $f_i \colon W \to X_i$ is a family of mappings of the set W into the topological spaces X_i, $i \in I$, then the pre-image topology with respect to the family f_i, $i \in I$, is equal to the pre-image topology with respect to the one mapping $f = (f_i) \colon W \to \prod_i X_i$.

If the $X_i = (X_i, d_i)$, $i \in I$ are metric spaces, then the product topology on $\prod_i X_i$ is generally not induced by any of the p-metrics d_p, $1 \leq p \leq \infty$, from example 4.1.7. However, if I is finite, then the following applies:

Proposition 4.2.29 *If $X_i = (X_i, d_i)$, $i \in I$, is a finite family of metric spaces, then every p-metric d_p, $1 \leq p \leq \infty$, induces the product topology on the product space $X = \prod_{i \in I} X_i$.*

Proof Since the metrics are all equivalent according to example 4.1.7, it is sufficient to show this for the Chebyshev metric d_∞ on X. However, it is
$$B_{d_\infty}((x_i)_i; r) = \prod_{i \in I} B_{d_i}(x_i; r), \quad (x_i) \in X, \ r \in \overline{\mathbb{R}}_+.$$
The open balls of a metric space form a basis of its topology, and it holds $\prod_{i \in I} B_{d_i}(x_i; r_i) \supseteq \prod_{i \in I} B_{d_i}(x_i; \mathrm{Min}\,(r_i, i \in I))$ for $r_i \in \overline{\mathbb{R}}_+, i \in I$. □

For infinite products of metric spaces, see Ex. 4.2.15. — A slight generalization of product spaces are the so-called **fiber products** of topological spaces $X_i = (X_i, T_i)$, $i \in I$. For this, a family of continuous mappings $h_i \colon X_i \to Z$, $i \in I$, is given. The **fiber product** $\prod_{Z, i \in I} X_i$ includes a family of continuous mappings $p_{Z,i} \colon \prod_{Z, i \in I} X_i \to X_i$, $i \in I$, with $h_i \circ p_{Z,i} = h_j \circ p_{Z,j}$ for all $i, j \in I$, which has the following universal property: *For each family $f_i \colon W \to X_i$ of continuous mappings of topological spaces with $h_i \circ f_i = h_j \circ f_j$ for all $i, j \in I$ there is exactly one continuous mapping $f \colon W \to \prod_{Z, i \in I} X_i$ with $f_i = p_{Z,i} \circ f$ for all $i \in I$.* Apparently, the subspace
$$\prod_{Z, i \in I} X_i := \left\{ (x_i)_{i \in I} \in \prod_{i \in I} X_i \,\middle|\, h_i(x_i) = h_j(x_j) \text{ for all } i, j \in I \right\}$$
of the product $\prod_{i \in I} X_i$, equipped with the relative topology, and the restrictions $p_{Z,i} := p_i | \prod_{Z, i \in I} X_i$. If Z is single-pointed, then the fiber product is the ordinary product. The fiber product of two spaces X,Y over Z with continuous mappings $g \colon X \to Z$ and $h \colon Y \to Z$ is also written in the form $X \times_Z Y$.

A topological space X together with a continuous mapping $g \colon X \to Z$ is also called a **fiber space** (X, g) with base Z. The transition from $g \colon X \to Z$ to the fiber space $X \times_Z Y \to Y$ is then called the **base change** or the **pullback** from Z to Y by means of the continuous mapping $h \colon Y \to Z$. It is written as $g_{(Y)} \colon X_{(Y)} \to Y$. Its universal property is illustrated in Fig. 4.6. The pullback is the subspace
$$X_{(Y)} = X \times_Z Y = \{(x, y) \mid x \in X, y \in Y, h(x) = g(y)\} \subseteq X \times Y$$
of $X \times Y$, and $g_{(Y)} = p_Y | X_{(Y)}$ is the restriction to $X_{(Y)}$ of the projection $p_Y \colon X \times Y \to Y$. What is the pullback when the new base Y is single-pointed or more generally, when h is constant? ◊

Fig. 4.6 Fiber product = base change = pullback

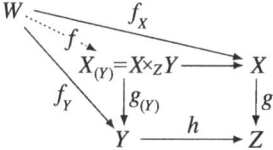

The field $\mathbb{K}$ always carries, unless otherwise stated, the natural topology given by the norm $|-|$. On $\mathbb{C}$ this norm is identical to the Euclidean norm $\|-\|_2$ on the product $\mathbb{C} = \mathbb{R} \times \mathbb{R}$.[10] In particular, $\mathbb{C}$ carries the product topology of $\mathbb{R} \times \mathbb{R}$. It is fundamental that the algebraic operations on $\mathbb{K}$ are continuous mappings. More precisely:

Proposition 4.2.30 *The following mappings are continuous:*

$$\mathbb{K} \times \mathbb{K} \to \mathbb{K}, \ (y, z) \mapsto y + z; \quad \mathbb{K} \times \mathbb{K} \to \mathbb{K}, \ (y, z) \mapsto y \cdot z;$$

$$\mathbb{K} \times \mathbb{K}^\times \to \mathbb{K}, \ (y, z) \mapsto y/z.$$

Proof Since continuity in metric spaces can be described with the convergence of sequences, see Proposition 4.1.11, the statement is merely a reformulation of the rules for limits. □

An important consequence is:

Theorem 4.2.31 *Let X be a topological space. Then the set*

$$C_{\mathbb{K}}(X) = C(X, \mathbb{K})$$

of continuous $\mathbb{K}$-valued functions on X is a $\mathbb{K}$-subalgebra of the $\mathbb{K}$-algebra $\mathbb{K}^X$ of all $\mathbb{K}$-valued functions on X with

$$C_{\mathbb{K}}(X)^\times = C_{\mathbb{K}}(X) \cap \left(\mathbb{K}^X\right)^\times = C_{\mathbb{K}}(X) \cap (\mathbb{K}^\times)^X,$$

i.e., $f: X \to \mathbb{K}$ and $g: X \to \mathbb{K}$ are continuous $\mathbb{K}$-valued functions on X as well as $a \in \mathbb{K}$, then the functions $f + g, f \cdot g, af$ and, if $g(x) \neq 0$ is for all $x \in X$, f/g is continuous on X.

Proof The mapping $(f, g): X \to \mathbb{K}^2$ with $x \mapsto (f(x), g(x))$ is continuous (since $\mathbb{K}^2$ carries the product topology). Then, according to theorem 4.2.16 also $f + g, fg$ or f/g

---

[10] Note that the norm $N_{\mathbb{R}}^{\mathbb{C}}(x)$, where $\mathbb{C}$ is considered as a square $\mathbb{R}$-algebra, is the square of $|x| = \|x\|_2$.

as a composition of (f, g) with the continuous functions $(y, z) \mapsto y + z$, $(y, z) \mapsto y \cdot z$ or $(y, z) \mapsto y/z$ on $\mathbb{K}^2$ or $\mathbb{K} \times \mathbb{K}^\times$ are continuous. Since constant mappings are continuous, $C_\mathbb{K}(X)$ is a $\mathbb{K}$-subalgebra of $\mathbb{K}^X$ with the specified unit group. □

Example 4.2.32 (Polynomial functions) The canonical projections $x_i \colon \mathbb{K}^I \to \mathbb{K}$, $i \in I$, are by definition of the product topology on $\mathbb{K}^I$ continuous. Therefore, the $\mathbb{K}$-subalgebra $\mathbb{K}[x_i, i \in I] \subseteq \mathbb{K}^{\mathbb{K}^I}$ generated by these functions of the polynomial functions on $\mathbb{K}^I$ is a subalgebra of the algebra $C_\mathbb{K}(\mathbb{K}^I)$ of continuous functions on $\mathbb{K}^I$. *Polynomial functions are therefore continuous.* Thus, all mappings $\mathbb{K}^I \to \mathbb{K}^J$, $x \mapsto (F_j(x))_{j \in J}$, are continuous, where the component functions $F_j(x)$, $j \in J$, are polynomial functions on $\mathbb{K}^I$. In particular, all linear mappings $\mathbb{K}^I \to \mathbb{K}^J$ are continuous, *if I is finite*, since then all linear forms $\mathbb{K}^I \to \mathbb{K}$, $x \mapsto \sum_{i \in I} a_i x_i$, $(a_i) \in \mathbb{K}^I$, are polynomial functions.[11] Similarly, every rational function $R(x) = F(x)/G(x)$, $F, G \in \mathbb{K}[X_i, i \in I]$, $G \neq 0$, is continuous except at the set of zeros $Z_\mathbb{K}(G) = G^{-1}(0) = \{x \in \mathbb{K}^I \mid G(x) = 0\} \subseteq \mathbb{K}^I$ of the denominator G. This set of zeros is closed as the fiber of the $\mathbb{K}$-valued polynomial function G over the closed point $0 \in \mathbb{K}$ according to 4.2.15 closed in $\mathbb{K}^I$. To keep this exception set $Z_K(G)$ as small as possible, one chooses the representation $R = F/G$ of the rational function $R \in \mathbb{K}(X_i, i \in I)$ reduced, i.e., with $\gcd(F, G) = 1$. From the identity theorem for polynomials 2.9.32 it follows directly, *that the set of zeros $Z_\mathbb{K}(G)$ is nowhere dense in $\mathbb{K}^I$ or — equivalently — that the complement $\mathbb{K}^I - Z_\mathbb{K}(G)$ is dense in $\mathbb{K}^I$*, see problem.4.2.12.

The continuity of polynomial functions provides important examples of open and closed sets. For example, for any real polynomial function $F \colon \mathbb{R}^I \to \mathbb{R}$ and any $a, b \in \mathbb{R}$, $a \leq b$, the set $\{x \in \mathbb{R}^I \mid a \leq F(x) \leq b\}$ as the preimage $F^{-1}([a, b])$ of the closed interval $[a, b] \subseteq \mathbb{R}$ is closed in $\mathbb{R}^I$. Similarly, the sets $\{x \in \mathbb{R}^I \mid F(x) > c\} = F^{-1}(]c, \infty[)$ are open in $\mathbb{R}^I$ for $c \in \mathbb{R}$. ◊

Example 4.2.33 (Natural topology of $\mathbb{K}$-vector spaces) We want to transfer the natural topologies of the $\mathbb{K}$-vector spaces $\mathbb{K}^n$ (i.e., the product topologies on $\mathbb{K}^n$), $n \in \mathbb{N}$, to any, initially finite-dimensional $\mathbb{K}$-vector spaces. So let V be an n-dimensional $\mathbb{K}$-vector space, $n \in \mathbb{N}$. A $\mathbb{K}$-basis $v = (v_1, \ldots, v_n)$ of V defines the $\mathbb{K}$-vector space isomorphism $f_v \colon \mathbb{K}^n \xrightarrow{\sim} V$, $(x_1, \ldots, x_n) \mapsto x_1 v_1 + \cdots + x_n v_n$. With the help of f_v we transfer the product topology from $\mathbb{K}^n$ to V, i.e., V carries the image topology with respect to f_v: A set $U \subseteq V$ is by definition open exactly when $f_v^{-1}(U)$ is open in $\mathbb{K}^n$, see example 4.2.24. f_v is therefore not only a linear isomorphism, but also a homeomorphism of topological spaces. This topology on V can also be defined by a p-norm $\|x_1 v_1 + \cdots + x_n v_n\|_p = (|x_1|^p + \cdots + |x_n|^p)^{1/p}$, $1 \leq p < \infty$, or $\|x_1 v_1 + \cdots + x_n v_n\|_\infty = \mathrm{Sup}(|x_1|, \ldots, |x_n|)$ with respect to the basis v, see Sect. 4.1. *This topology on V is independent of the choice of the basis v* and is called the **natural topology** on V. Let it be $w = (w_1, \ldots, w_n)$ another basis of V. Then the mapping

---

[11] For infinite I, linear forms on $\mathbb{K}^I$ are generally not continuous (with respect to the product topology on $\mathbb{K}^I$).

4.2 Topological Spaces and Continuous Mappings

$h := f_w^{-1} f_v \colon \mathbb{K}^n \to \mathbb{K}^n$ is a $\mathbb{K}$-linear automorphism, thus according to example 4.2.32 continuous. Since the inverse mapping $h^{-1} = f_v^{-1} f_w$ is also continuous, h is a homeomorphism of $\mathbb{K}^n$. Consequently, for $U \subseteq V$ the preimage $f_v^{-1}(U)$ is open in $\mathbb{K}^n$ exactly when $h(f_v^{-1}(U)) = f_w^{-1}(U)$ is open in $\mathbb{K}^n$. Since every linear mapping $\mathbb{K}^n \to \mathbb{K}^m$, $n, m \in \mathbb{N}$, is continuous, *this also applies to every $\mathbb{K}$-linear mapping between finite-dimensional $\mathbb{K}$-vector spaces with their natural topologies.*

If V is any $\mathbb{K}$-vector space, we define the **natural topology** on V as the image topology with respect to the canonical embeddings $W \hookrightarrow V$, where W is the finite-dimensional $\mathbb{K}$-subspaces of V are traversed. A subset $U \subseteq V$ is therefore open in V if and only if $U \cap W$ is open in W for every finite-dimensional $\mathbb{K}$-subspace W of V. If v_i, $i \in I$, is a $\mathbb{K}$-basis of V, then U in V is already open if $U \cap V_J$ is open in V_J for every finite subset $J \subseteq I$ and $V_J := \sum_{i \in J} \mathbb{K} v_i$. It is even sufficient if J traverses a cofinal subset of $\mathfrak{E}(I)$. Proof! It is often more convenient to argue with the closed sets in V. $F \subseteq V$ is closed with respect to the natural topology if and only if $F \cap W$ in W is closed for every finite-dimensional subspace W of V. The space $\mathbb{K}^{(\mathbb{N}^*)}$ with the natural topology is often also denoted by $\mathbb{K}^\infty$ or with $\mathbb{K}^{(\infty)}$. If I is infinite, then the natural topology on the product $\mathbb{K}^I = \mathrm{Abb}(I, \mathbb{K})$ is strictly finer than the product topology on K^I, cf. exercise 4.2.33f).

The natural topology of a $\mathbb{C}$-vector space V is identical to the natural topology of V, considered as a $\mathbb{R}$-vector space. Furthermore, we note here that on a finite-dimensional $\mathbb{K}$-vector space V all norms that only take finite values are equivalent, cf. Theorem 4.4.12, and thus *every real norm on V provides the natural topology of V*. On an infinite-dimensional $\mathbb{K}$-vector space there is no norm that defines the natural topology. *Every linear mapping between $\mathbb{K}$-vector spaces is continuous with respect to their natural topologies*, cf. exercise 4.2.33a).

Since $\mathbb{K}$ and therefore also $\mathbb{K}^n$ for every $n \in \mathbb{N}$ has a countable topology (e.g., the open intervals $]a, b[$, $a, b \in \mathbb{Q}$, $a < b$, form a basis of the topology of $\mathbb{R}$), *every finite-dimensional $\mathbb{K}$-vector space fulfills the second countability axiom with respect to its natural topology*. For infinite-dimensional $\mathbb{K}$-vector spaces see problem 4.2.33d).

The sentence 3.4.3 or 3.5.5 by Bolzano-Weierstraß immediately transfers from $\mathbb{R}$ and $\mathbb{C} = \mathbb{R}^2$ to any *finite-dimensional* $\mathbb{K}$ vector spaces V with their natural topology. A subset A of such an $\mathbb{K}$ vector space V is called **bounded**, if its preimage $f^{-1}(A)$ with respect to a $\mathbb{K}$ linear isomorphism $f \colon \mathbb{K}^n \xrightarrow{\sim} V$ lies in a poly-cylinder $\overline{B}_\mathbb{K}(0; R)^n \subseteq \mathbb{K}^n$, $R \in \mathbb{R}_+$. This condition is obviously independent of the choice of the isomorphism f. *A subset $A \subseteq V$ is exactly bounded when for each linear form $h \colon V \to \mathbb{K}$ the image $h(A)$ is bounded in $\mathbb{K}$.* A sequence in V is called bounded if the set of its members is bounded. With these designations, the following applies:

Theorem 4.2.34 (Bolzano-Weierstrass) *Let V be a finite-dimensional $\mathbb{K}$-vector space with its natural topology. Then every bounded sequence x_m, $m \in \mathbb{N}$, in V has a limit point in V, i.e., a convergent subsequence.*

Proof We can assume that $V = \mathbb{K}^n$ is and the sequence elements x_m all lie in the poly-cylinder $\overline{B}_\mathbb{K}(0; R)^n$. We conclude by induction over n. The case $n = 1$ is the classic Bolzano-Weierstrass theorem. In the conclusion from $n \geq 1$ to $n+1$, let $x_m = (x'_m, a_m)$

with $x'_m \in \overline{B}_{\mathbb{K}}(0, R)^n$ and $a_m \in \overline{B}_{\mathbb{K}}(0, R), m \in \mathbb{N}$. According to the induction hypothesis, the sequence (x'_m) has a convergent subsequence. If we replace (x_m) with the corresponding subsequence, we can assume that (x'_m) already converges. Since (a_m) has a convergent subsequence $(a_{m_k})_{k \in \mathbb{N}}$, then $x_{m_k} = (x'_{m_k}, a_{m_k}) \in \mathbb{K}^n \times \mathbb{K}, k \in \mathbb{N}$, is a convergent subsequence of (x_m). □

In Sect. 4.4 we will bring the Bolzano-Weierstrass theorem to the point with the concept of compact space, see the theorem 4.4.11 by Heine-Borel. ◊

Example 4.2.35 Natural topologies can be defined on vector spaces over any ordered field K. First, K carries a natural topology with the open intervals $]a, b[, a, b \in K, a < b$, as the basis of the topology. The open ε-neighborhoods $B(a; \varepsilon) =]a - \varepsilon, a + \varepsilon[$, $\varepsilon \in K_+^\times$, of a point $a \in K$ form a neighborhood basis of a. K obviously fulfills the first countability axiom exactly when K has a null sequence $\varepsilon_n \in K_+^\times, n \in \mathbb{N}$. Then already $]a - \varepsilon_n, a + \varepsilon_n[, n \in \mathbb{N}$, is a neighborhood basis of a for every $a \in K$. *The algebraic operations* $(x, y) \mapsto x + y, (x, y) \mapsto x \cdot y$ *as well as* $(x, y) \mapsto x/y$ *on K^2 respectively.* $K \times K^\times$ *are again continuous. Proof*. Since in general K does not satisfy the first countability axiom, it is not sufficient to argue with the calculation rules 3.2.5 for limits. So let ε be given with $0 < \varepsilon \leq 1$. Then the addition maps the closed square $\overline{B}(y_0; \varepsilon/2) \times \overline{B}(z_0; \varepsilon/2)$ into the closed ε-neighborhood $\overline{B}(y_0+z_0; \varepsilon)$ of $y_0 + z_0$. Therefore, the addition is continuous. Furthermore, the multiplication maps the closed rectangle $\overline{B}(y_0; \varepsilon/2\text{Max}\,(|z_0|, 1)) \times \overline{B}(z_0; \varepsilon/2(|y_0| + 1))$ into $\overline{B}(y_0z_0; \varepsilon)$. Therefore, the multiplication is continuous. For $z_0 \neq 0$, finally, $z \mapsto 1/z$ maps the closed interval $\overline{B}(z_0; \text{Min}\,(\varepsilon|z_0|^2, |z_0|)/2)$ into $\overline{B}(1/z_0; \varepsilon)$.

Thus, all polynomial functions $K^I \to K$ are continuous again on K^I and for finite I all K-linear mappings $K^I \to K^J$. As in the case of $K = \mathbb{R}$, the product topology on K^n then induces a **natural topology** on each n-dimensional K-vector space, $n \in \mathbb{N}$. Finally, the natural topology on any K-vector space is defined as the image topology with respect to the inclusions of its finite-dimensional subspaces. As in the case of $K = \mathbb{R}$, for infinite I, the natural topology on K^I must be carefully distinguished from the product topology on K^I, which is then genuinely coarser than the natural topology, cf. Exercise 4.2.33f). ◊

We conclude this paragraph with two significant existence theorems for continuous real-valued functions, which go back to H. Tietze (1880–1964) and P. Urysohn (1898–1924) and can be skipped on first reading.

Theorem 4.2.36 (Tietze-Urysohn Extension Theorem) *Let A be a non-empty closed subset of the metric space X and $f : A \to \mathbb{R}$ a continuous and bounded real-valued function on A. Then f can be extended to a continuous function $g : X \to \mathbb{R}$ that satisfies the following conditions:*

$$\text{Inf}\{g(x) \mid x \in X\} = \text{Inf}\{f(x) \mid x \in A\} \quad \text{and}$$

$$\text{Sup}\{g(x) \mid x \in X\} = \text{Sup}\{f(x) \mid x \in A\}.$$

4.2 Topological Spaces and Continuous Mappings

The extension g is also bounded, with the same bounds as f.

Proof ((according to [6], Sect. 4.5)) We can assume $\text{Inf}\{f(x) \mid x \in A\} = 1$ and $\text{Sup}\{f(x) \mid x \in A\} = 2$ as well as $d(x, y) \in \mathbb{R}_+$ for all $x, y \in X$.[12] Then the function $g: X \to \mathbb{R}$ with

$$g(x) := \begin{cases} f(x), & \text{if } x \in A, \\ \text{Inf}\{f(y)d(x, y) \mid y \in A\}/d(x, A), & \text{if } x \in X - A, \end{cases}$$

fulfills the required conditions. Here, $d(x, A) = \text{Inf}\{d(x, y) \mid y \in A\} \in \mathbb{R}_+$ is the distance from x and A, see exercise 4.2.27. The estimates $1 \leq g \leq 2$ are trivial. We leave the proof of the continuity of g on $X - \text{Rd } A$ to the reader, see exercise 4.2.36.

Let $x \in \text{Rd } A$ and $|f(y) - f(x)| \leq \varepsilon$ for all $y \in A \cap \overline{B}(x; \delta)$, where $\varepsilon > 0$ is given. For $z \in \overline{B}(x; \delta/4)$ and $y \in A - B(x; \delta)$ is $d(z, y) \geq 3\delta/4$, thus $f(y)d(z, y) \geq 3\delta/4$. Because of $f(x)d(z, x) \leq \delta/2$ is thus

$$g(z) = \frac{\text{Inf}\{f(y)d(z, y) \mid y \in A \cap \overline{B}(x; \delta)\}}{d(z, A)}$$

for $z \in \overline{B}(x; \delta/4) - A$. With $d(z, A) = \text{Inf}\{d(z, y) \mid y \in A \cap \overline{B}(x; \delta)\}$ for $z \in \overline{B}(x; \delta/4)$ and

$$(f(x) - \varepsilon)d(z, y) \leq f(y)d(z, y) \leq (f(x) + \varepsilon)d(z, y)$$

for $y \in A \cap \overline{B}(x; \delta)$ finally results in $f(x) - \varepsilon \leq g(z) \leq f(x) + \varepsilon$ for $z \in \overline{B}(x; \delta/4) - A$, thus overall $|g(z) - f(x)| \leq \varepsilon$ for all $z \in \overline{B}(x; \delta/4)$. □

Theorem 4.2.36 applies to any normal spaces X (**Urysohn's Theorem**). Here, a Hausdorff space X is called **normal** if any two disjoint closed subsets of X have disjoint neighborhoods in X. Metric spaces are normal. This follows from Theorem 4.2.36, but can be seen much more easily, see Ex. 4.2.28. For further examples, see Lemma 4.4.17. A Hausdorff space X is called **regular** if every closed set $A \subseteq X$ and every point $x \in X - A$ have disjoint neighborhoods in X. Normal spaces are regular. For normal spaces, we only prove the following frequently used special case of Urysohn's theorem here:

Theorem 4.2.37 (Urysohn's Separation Lemma) *Let X be a normal (Hausdorff) space and let $A, B \subseteq X$ be disjoint non-empty closed subsets of X. Then there exists a continuous function $g: X \to [0, 1] \, (\subseteq \mathbb{R})$ with $g(A) = \{0\}$ and $g(B) = \{1\}$.*

---

[12] It is sufficient for each $x_0 \in X$ to continue the restriction $f|(A \cap B_X(x_0; \infty))$ according to $B_X(x_0; \infty)$.

Proof The normality of X can obviously be expressed as follows: *For every open neighborhood V of a closed set $C \subseteq X$ there exists an open neighborhood U of C with $C \subseteq \overline{U} \subseteq V$.*—For the proof of 4.2.37 let now $y_n, n \in \mathbb{N}$, a sequence of pairwise different numbers in $]0,1[$, which form a dense subset of $]0,1[$. Following the preliminary remark, open sets $U_n \subseteq X$, $n \in \mathbb{N}$, can be chosen recursively with $A \subseteq U_n \subseteq \overline{U}_n \subseteq X - B$ and $\overline{U}_m \subseteq U_n$, if $y_m < y_n$ (which does not mean $m < n$). If $U_0, \ldots, U_n$ have already been chosen and $i_1, \ldots, i_r$ or $j_1, \ldots, j_s$ are the indices $\leq n$ with $y_{i_\rho} < y_{n+1} < y_{j_\sigma}$, then when choosing U_{n+1} besides $A \subseteq U_{n+1} \subseteq \overline{U}_{n+1} \subseteq X - B$, the condition

$$\overline{U}_{i_1} \cup \cdots \cup \overline{U}_{i_r} = \overline{U_{i_1} \cup \cdots \cup U_{i_r}} \subseteq U_{n+1} \subseteq \overline{U}_{n+1} \subseteq U_{j_1} \cap \cdots \cap U_{j_s}$$

must also be considered. Then $g: X \to [0,1]$ with

$$g(x) := \mathrm{Inf}\{y_n \mid x \in U_n\} \in [0,1]$$

is a function of the desired type: It is $g(x) = 0$ for $x \in A$ and $g(x) = 1$ for $x \in B$. For $y \in]0,1[$ are moreover

$$g^{-1}([0,y[) = \bigcup_{y_n < y} U_n \text{ and } g^{-1}([0,y]) = \bigcap_{y < y_n} U_n = \bigcap_{y < y_m} \overline{U}_m$$

open respectively closed in X, from which the continuity of g follows. □

Exercises

Exercise 4.2.1 For the following subsets A of the $\mathbb{R}^2$ (equipped with the natural topology) determine the interior $\mathring{A}$, the closed hull $\overline{A}$ and the boundary $\mathrm{Rd}\, A$:

a) $A := \{(x_1, x_2) \mid x_1 < |x_2|\} \subseteq \mathbb{R}^2$.
b) $A := \{(x_1, x_2) \mid x_1^2 \leq |x_2|\} \subseteq \mathbb{R}^2$.

Exercise 4.2.2 Let A, A_1, A_2 be subsets of the topological space X. Show:

a) Exactly then, $x \in X$ is not a point of contact of A, when x is an inner point of the complement $X - A$, i.e., it is $(X - A)^\circ = X - \overline{A}$ and thus $\overline{A} = X - (X - A)^\circ$. Furthermore, $\mathring{A} = X - \overline{(X - A)}$ results.
b) $\mathring{A} = X - \overline{(X - A)}$, $\mathring{A} = X - \overline{(X - A)}$.
c) $(A_1 \cup A_2)^\circ \supseteq \mathring{A}_1 \cup \mathring{A}_2$, $\overline{A_1 \cap A_2} \subseteq \overline{A}_1 \cap \overline{A}_2$. Provide examples that these inclusions can be genuine.
d) $\overline{(\overline{A})} = \overline{A}$, $(\mathring{A})^\circ = \mathring{A}$.
e) Let $A \subseteq X' \subseteq X$. The closed hull of A, considered as a subset of the subspace X', is $X' \cap \overline{A}$, where $\overline{A}$ is the closed hull of A in X. If A is dense in X and $U \subseteq X$ is open, then $U \cap A$ is dense in U and $\overline{U \cap A} = \overline{U}$.

Fig. 4.7 $B(x; r) \subset \overline{B}(x; r)$

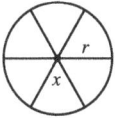

f) If $U_1, \ldots, U_n$ is open and dense in X, then $U_1 \cap \cdots \cap U_n$ is also open and dense in X.

Exercise 4.2.3 Let A, A_1, A_2 be subsets of the topological space X. Then the following holds:

a) $\partial(X - A) = \partial A$.
b) $\partial(A_1 \cup A_2) \subseteq (\partial A_1) \cup (\partial A_2)$.

Exercise 4.2.4 Prove Proposition 4.2.6.

Exercisse 4.2.5 Let X be a metric space. Then for every $x \in X$ and every $r > 0$ the inclusions $\overline{B(x; r)} \subseteq \overline{B}(x; r)$, $\partial B(x; r) \subseteq S(x; r)$ and $B(x; r) \subseteq (\overline{B}(x; r))^\circ$ apply. Show by examples that these inclusions can be strict, see Fig. 4.7.

Exercise 4.2.6

a) Determine the quotient topology of $\mathbb{R}^n/\mathbb{Q}^n$, $n \in \mathbb{N}^*$, where $\mathbb{R}^n$ carries the natural topology.
b) The composition $g \circ f : X \to Z$ of two quotient mappings $f : X \to Y$, $g : Y \to Z$ of topological spaces is also a quotient mapping. (However, the product $f_1 \times f_2 : X_1 \times X_2 \to Y_1 \times Y_2$ of two quotient mappings f_1, f_2 does *not* need to be a quotient mapping. Example?)

Exercise 4.2.7 X_i, $i \in I$, is a family of topological spaces. For any subsets $A_i \subseteq X_i$, $i \in I$, then in the product $\prod_{i \in I} X_i$ (equipped with the product topology)

$$\left(\prod_{i \in I} A_i\right)^\circ \subseteq \prod_{i \in I} \mathring{A}_i \quad \text{and} \quad \overline{\left(\prod_{i \in I} A_i\right)} = \prod_{i \in I} \overline{A_i}.$$

In particular, the product of closed sets is closed. If I is finite, then the equality sign also applies in the first formula. Provide an example of the true inclusion.

Exercise 4.2.8 For subsets A and B of the topological spaces X and Y respectively, the following applies:

$$\partial(A \times B) = \big((\partial A) \times \overline{B}\big) \cup \big(\overline{A} \times (\partial B)\big)$$
$$= ((\partial A) \times B) \cup (A \times (\partial B)) \cup ((\partial A) \times (\partial B)).$$

Exercise 4.2.9

a) The topological space X is a Hausdorff space if and only if the diagonal $\Delta_X = \{(x,x) \mid x \in X\}$ is closed in $X \times X$.
b) Let f and g be continuous mappings of the topological space W into the Hausdorff space X. Then the set $\{f = g\} := \{w \in W \mid f(w) = g(w)\}$ is closed and the set $\{f \neq g\} := \{w \in W \mid f(w) \neq g(w)\}$ is open in W. In particular, f and g coincide on all of W if they coincide on a dense subset of W.
c) Let $R \subseteq X \times X$ be an equivalence relation on the topological space X. If the quotient space X / R (with the quotient topology) is a Hausdorff space, then R is closed in $X \times X$. (The converse is generally not true, but see exercise 4.4.22.) The quotient space X / R is a Hausdorff space if and only if two different equivalence classes $A, B \subseteq X$ always have disjoint R-saturated neighborhoods.

Exercise 4.2.10 Let X be a topological space.

a) If X has a countable topology, then so does every subspace of X, and every basis of the topology of X contains a countable subfamily that is also a basis.
b) X has a countable open cover U_i, $i \in I$, where each U_i has a countable topology. Then X also has a countable topology.

Exercise 4.2.11 Let X be a space with a countable topology. Then the set $\mathcal{T}$ of open (and thus also of closed) sets of X has at most the power of the continuum. In particular, X itself has at most the power of the continuum if X is also Hausdorff.

Exercise 4.2.12 A subset A of a topological space X is called **nowhere dense**, if the complement of $\overline{A}$ in X is dense.

a) Subsets of nowhere dense sets are nowhere dense.
b) Finite unions of nowhere dense sets are nowhere dense.
c) The following statements are equivalent: (i) A is nowhere dense. (ii) $(X - A)°$ is dense in X. (iii) $\overline{A}$ is nowhere dense. (iv) $(\overline{A})° = \emptyset$. (v) $A \subseteq \mathrm{RD}\,\overline{A}$. (vi) For every open set U in X, $A \cap U$ is a nowhere dense subset in U. (vii) For every point $x \in X$ there is a neighborhood V of x such that $A \cap V$ is nowhere dense in V. (viii) For every non-empty open set $U \subseteq X$, $A \cap U$ is not dense in U. (Condition (viii) motivates the term "nowhere dense".)

Exercise 4.2.13 Let X be a topological space.

a) If X is discrete with countable topology, then X only has countably many points. In particular, a discrete subspace of a space with countable topology is countable. This applies especially to the spaces $\mathbb{K}^n$, $n \in \mathbb{N}$.
b) Let $A \subseteq X$. A point $x \in A$ is a discrete point of A if and only if x is a contact point, but not a limit point of A.

4.2 Topological Spaces and Continuous Mappings

c) A Hausdorff space with only finitely many points is discrete.
d) A subset A of a Hausdorff space X is discrete *and* closed if and only if every point $x \in X$ has a neighborhood U in which only finitely many points of A lie.
e) The union of finitely many *closed* discrete subspaces of X is discrete.
f) An infinite Hausdorff space X has an infinite discrete (not necessarily closed) subspace. (There is a point $x \in X$ with a neighborhood U such that $X - U$ is still infinite.)

Exercise 4.2.14 Let X be a set. A subset $\mathcal{B} \subseteq \mathfrak{P}(X)$ with $\bigcup_{U \in \mathcal{B}} U = X$ is exactly the basis of a topology on X when for any two elements $U, V \in \mathcal{B}$ and every $x \in U \cap V$ a $W \in \mathcal{B}$ with $x \in W \subseteq U \cap V$ exists.

Exercise 4.2.15 Let X_i, $i \in I$, be an arbitrary family of topological spaces and $X := \prod_{i \in I} X_i$ their product space. Furthermore, let $p \in [1, \infty[$.

a) If I is countable and the X_i, $i \in I$, are metric with metrics d_i, for which $d_i \leq \varepsilon_i \in \mathbb{R}_+^\times$ and $\sum_i \varepsilon_i^p < \infty$ apply (which can be assumed without further ado according to exercise 4.1.1), then the p-metric $d_p((x_i), (y_i)) = \left(\sum_{i \in I} d_i(x_i, y_i)^p\right)^{1/p}$, cf. example 4.1.7, induces the product topology on X. The same applies to the Chebyshev metric d_∞ on X, if $\lim_i \varepsilon_i = 0$ is. In particular, X with the product topology is metrizable.
b) If I is not countable and the spaces X_i, $i \in I$, all have more than one element, then X is not metrizable. (The intersection of countably many neighborhoods of a point $x \in X$ is always $\neq \{x\}$ in this case.) The sequence space $\mathbb{R}^\mathbb{N}$ is therefore metric, but the space $\mathbb{R}^\mathbb{R}$ is not.

Exercise 4.2.16 Let $X = (X, \mathcal{T})$ be a topological space and $A \subseteq X$ a subset.

a) A is closed in X if and only if, for every point $x \in X$, there exists a neighborhood U of x such that $A \cap U$ is closed in U.
b) A is called **locally closed** in X if, for every point $x \in A$, there exists a neighborhood U of x such that $A \cap U$ is closed in U. Show that: A is locally closed in X if and only if there exists a closed set $F \subseteq X$ and an open set $U \subseteq X$ with $A = F \cap U$. The intersection of finitely many locally closed sets is again locally closed.
c) A subset $C \subseteq X$ is called **constructible** (in X) if it is the union of finitely many locally closed sets. The locally closed sets form a subring $\mathcal{C}(X) = \mathcal{C}_\mathcal{T}(X)$ of the set ring $\mathfrak{P}(X)$, cf. exercise 2.6.1b), i.e., $\mathcal{C}(X)$ is closed with respect to finite unions, finite intersections, and complementation.

Exercise 4.2.17 Let $f : X \to Y$ be a mapping of topological spaces.

a) X or Y is discrete. When is f continuous?
b) X or Y carries the coarse topology. When is f continuous?

Exercise 4.2.18 The following mappings are continuous:

a) $f: \mathbb{R}^2 - \{0\} \to \mathbb{R}$ with $(x, y) \mapsto 1/\sqrt{x^2 + y^2}$.
b) $f: \mathbb{R} \to \mathbb{R}^2$ with $x \mapsto (e^x, \cos x)$.
c) $f: \mathbb{R}^2 \to \mathbb{R}^2$ with $(x, y) \mapsto \left(\sqrt{1 + \sin^2 xy}, xy/(1 + e^{xy})\right)$.

(Use that $\cos, \sin: \mathbb{R} \to \mathbb{R}$ are continuous.)

Exercise 4.2.19 Let f and g be continuous real-valued functions on the topological space X. Then the set $\{f < g\} := \{x \in X \mid f(x) < g(x)\}$ is open and the set $\{f \le g\} := \{x \in X \mid f(x) \le g(x)\}$ is closed in X.

Exercise 4.2.20 Investigate whether the following sets in $\mathbb{R}^2$ are open, closed, or neither open nor closed:

a) $\{(x, y) \mid x^2 < |y|\}$.
b) $\{(x, y) \mid \frac{1}{2} \le xy \le \cos xy\}$.
c) $\{(x, y) \mid y < x^2 \le 2y\}$.

Exercise 4.2.21 Let X,Y and Z be topological spaces. For all $y_0 \in Y$, $x_0 \in X$ the injections $\sigma_{y_0}: x \mapsto (x, y_0)$ and $\tau_{x_0}: y \mapsto (x_0, y)$ from X and Y into the product space $X \times Y$ are continuous. With a continuous mapping $f: X \times Y \to Z$ the partial mappings $f \circ \sigma_{y_0}: x \mapsto f(x, y_0)$ and $f \circ \tau_{x_0}: y \mapsto f(x_0, y)$ are also continuous for all $y_0 \in Y$ and $x_0 \in X$. (However, the reverse does not generally follow from this partial continuity of f, i.e., the continuity of f does not necessarily follow, see the following exercise.)

Exercise 4.2.22

a) The function $f: \mathbb{R}^2 \to \mathbb{R}$ with

$$f(x, y) := \begin{cases} xy/(x^2 + y^2), & \text{if } (x, y) \ne (0, 0), \\ 0, & \text{if } (x, y) = (0, 0), \end{cases}$$

is continuous in $\mathbb{R}^2 - \{0\}$, but not in 0. Each of the partial mappings $x \mapsto f(x, y_0)$ and $y \mapsto f(x_0, y)$ is however continuous on the whole $\mathbb{R}$.

b) The functions f and g from $\mathbb{R}^2$ in $\mathbb{R}$ with

$$f(x, y) := \begin{cases} xy^2/(x^2 + y^4), & \text{if } (x, y) \ne (0, 0), \\ 0, & \text{if } (x, y) = (0, 0); \end{cases}$$

$$g(x, y) := \begin{cases} 0, & \text{if } (x, y) = (0, 0) \text{ or } y \ne x^2, \\ 1 & \text{otherwise} \end{cases}$$

4.2 Topological Spaces and Continuous Mappings

are not continuous in 0; the restrictions on all lines $\mathbb{R}(x_0, y_0)$, $(x_0, y_0) \in \mathbb{R}^2 - \{0\}$, through the zero point are however continuous in 0. In addition, f is continuous in $\mathbb{R}^2 - \{0\}$. Where is g continuous?

Exercise 4.2.23 For a mapping $f : X \to Y$ of topological spaces, the following are equivalent: (i) f is continuous. (ii) For every subset $A \subseteq X$, $f(\overline{A}) \subseteq \overline{f(A)}$ applies. (iii) The (bijective) mapping $x \mapsto (x, f(x))$ from X to the graph $\Gamma_f \subseteq X \times Y$ of f is a homeomorphism. (Note that the inverse mapping $\Gamma_f \to X$, $(x, f(x)) \mapsto x$, is continuous for *any* $f : X \to Y$. The equivalence of (i) and (iii) therefore provides a wealth of examples of continuous bijective mappings that are not homeomorphisms.)

Exercise 4.2.24 Let A_i, $i \in I$, be any open or a finite closed cover of the topological space X. A mapping $f : X \to Y$ from X to the topological space Y is continuous if all restrictions $f|A_i$, $i \in I$, are continuous.

Exercise 4.2.25 Every bijective isometry between metric spaces is a homeomorphism.

Exercise 4.2.26 If $X = (X, d)$ is a metric space, then the metric $d : X \times X \to \overline{\mathbb{R}}_+$ is continuous with respect to the metric topology (and then also with respect to the order topology) on $\overline{\mathbb{R}}_+ \subseteq \overline{\mathbb{R}}$. (For the topologies on $\overline{\mathbb{R}}$ see example 4.2.20.) In particular, the sets $\Delta_{X,r} := \{(x, y) \in X \times X \mid d(x, y) < r\}$ or $\overline{\Delta}_{X,r} = \{(x, y) \in X \times X \mid d(x, y) \le r\}$ for $r > 0$ are open or closed neighborhoods of the diagonal $\Delta_X = \overline{\Delta}_{X,0} \subseteq X \times X$. (In general, the $\Delta_{X,r}$, $r > 0$, do not form a neighborhood basis of Δ_X, see Fig. 4.8. However, see exercise 4.4.10b).)

Exercise 4.2.27 Let X be a metric space. For subsets $A, B \subseteq X$ is called

$$d(A, B) := \operatorname{Inf}\{d(x, y) \mid x \in A, y \in B\} = d(B, A) \in \overline{\mathbb{R}}_+$$

the **distance** between A and B. If $A = \emptyset$ or $B = \emptyset$, then $d(A, B) = \infty$. (d is generally not a metric on $\mathfrak{P}(X) - \{\emptyset\}$, not even a pseudometric. But see example 4.5.29.)

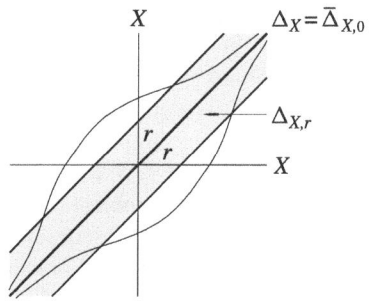

Fig. 4.8 Neighborhoods of the diagonal

Fig. 4.9 Fiber sum = Pushout

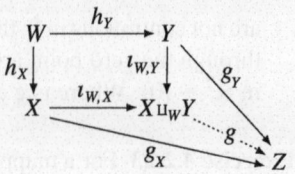

a) For every set A, $X \to \overline{\mathbb{R}}_+$ is $x \mapsto d(A,x) := d(A,\{x\})$ continuous on X with respect to the metric topology on $\overline{\mathbb{R}}$. It is exactly $d(A,x) = 0$ when $x \in \overline{A}$ is.
b) It is $d(A,B) = d(A,\overline{B}) = d(\overline{A},\overline{B})$.
c) It is $|d(A,x) - d(A,y)| \leq d(x,y)$ for all $x,y \in X$.
d) Provide closed sets $A, B \subseteq \mathbb{R}$ with $A \cap B = \emptyset$ and $d(A,B) = 0$.

Exercise 4.2.28 Let A, B be disjoint closed subsets of the metric space X. Then

$$U := \{x \in X \mid d(A,x) < d(B,x)\} \quad \text{and} \quad V := \{x \in X \mid d(B,x) < d(A,x)\}$$

are open and disjoint neighborhoods of A and B respectively. (Metric spaces are therefore normal.)

Exercise 4.2.29 (Fiber sums) The fiber sums are the objects dual to the fiber products. We first consider two continuous mappings $h_X : W \to X$ and $h_Y : W \to Y$ of topological spaces. We are looking for a topological space $X \sqcup_W Y$ and two continuous mappings $\iota_{W,X} : X \to X \sqcup_W Y$ and $\iota_{W,Y} : Y \to X \sqcup_W Y$, for which $\iota_{W,X} \circ h_X = \iota_{W,Y} \circ h_Y$ is, with the following universal property: If $g_X : X \to Z$ and $g_Y : Y \to Z$ are continuous mappings with $g_X \circ h_X = g_Y \circ h_Y$, then there is exactly one continuous mapping $g : X \sqcup_W Y \to Z$ with $g_X = g \circ \iota_{W,X}$ and $g_Y = g \circ \iota_{W,Y}$, see Fig. 4.9.

Show that the quotient space

$$X \sqcup_W Y := X \sqcup Y / R_{h_X = h_Y}$$

with the canonical mappings $\iota_{W,X} := \pi \circ \iota_X$ and $\iota_{W,Y} := \pi \circ \iota_Y$, where $R_{h_X = h_Y}$ is the smallest equivalence relation on $X \sqcup Y = X \uplus Y$ is, which contains the pairs $(\iota_X(h_X(w)), \iota_Y(h_Y(w)))$, and $\pi : X \sqcup Y \to X \sqcup Y / R_{h_X = h_Y}$ is the canonical projection, this universal property is possessed. $X \sqcup_W Y$ is called the **fiber sum** or the **Pushout** of X and Y with respect to the mappings h_X and h_Y. Since the pushout involves a quotient formation, it is often more confusing than, for example, the pullback, see example 4.2.27. A standard example of a pushout is the **blowing up of a subspace** $A \subseteq X$ to a point. In this case, $W = A$, $Y = \{P\}$ is single-pointed and h_X, h_Y are the canonical mappings $A \hookrightarrow X$ or $A \to \{P\}$. This fiber sum is also denoted with X/A. At least describe it set-theoretically. (The case $A = \emptyset$ is a special case with $X/\emptyset = X \sqcup \{P\}$. For examples see exercises 4.4.19 and 4.4.20.)

The reader formulates the universal property of the fiber sum $\bigsqcup_{i \in I, W} X_i$ for an arbitrary family of continuous mappings $h_i : W \to X_i$, $i \in I$, and shows that it

4.2 Topological Spaces and Continuous Mappings

can be obtained at $I \neq \emptyset$ and $i_0 \in I$ from the sum space $\coprod_{i \in I} X_i$ by identifying by means of the smallest equivalence relation on $\coprod_{i \in I} X_i$, which contains the pairs $\bigl(\iota_i(h_i(w)), \iota_{i_0}(h_{i_0}(w))\bigr)$, $i \in I$, $w \in W$. It can also be identified with the pushout of the two mappings $h' : \coprod_{i \neq i_0} \{i\} \times W \to \coprod_{i \neq i_0} W$, $(i, w) \mapsto \iota_i(h_i(w))$, and $h'' : \coprod_{i \neq i_0} \{i\} \times W \to X_{i_0}$, $(i, w) \mapsto h_{i_0}(w)$. (Many other constructions of topological spaces can often be traced back to fiber products and fiber sums.) □

Exercises 4.2.30 Let X be a Hausdorff space. Then the following are equivalent: (i) X is regular. (ii) A closed set in X is the intersection of its closed neighborhoods. (iii) Every point $x \in X$ has a neighborhood base of closed neighborhoods of x. (iv) For every closed set $A \subseteq X$ the space X/A resulting from blowing up A (cf. exercise 4.2.29) is also Hausdorf.

Exercise 4.2.31 Let I be a set and K a field. For a function $f \in K^I$, let $Z(f) = Z_K(f) = \{i \in I \mid f(i) = 0\}$ be the set of zeros of f.

a) The figures

$$\mathcal{F} \mapsto \mathfrak{a}(\mathcal{F}) := \{f \in K^I \mid Z(f) \in \mathcal{F}\} \quad \text{and} \quad \mathfrak{a} \mapsto \mathcal{F}(\mathfrak{a}) := \{Z(f) \mid f \in \mathfrak{a}\}$$

are inverse monotonically increasing mappings from the set of filters $\mathcal{F}$ on I to the set of ideals $\mathfrak{a}$ of the K-algebra K^I. The ultrafilters on I correspond to the maximal ideals of the K-algebra K^I. The principal ideal generated by $f \in K^I$ corresponds to the so-called **principal filter**

$$\mathcal{F}(Z(f)) := \{J \subseteq I \mid J \supseteq Z(f)\}$$

to the zero $Z(f)$ of f. A principal filter $\mathcal{F}(N)$, $N \subseteq I$, is exactly an ultrafilter when $N = \{x\}$, $x \in I$, is single-pointed. These filters correspond to the so-called **point ideals** $\mathfrak{m}_x = \{f \in K^I \mid f(x) = 0\}$, $x \in I$. (The existence of ultrafilters on non-empty sets thus also follows from Krull's theorem 2.7.18 about the existence of maximal ideals.—The algebra K^I can be replaced by a product ring $\prod_{i \in I} K_i$, where all factors K_i are fields.)

b) Every finitely generated ideal in K^I is a principal ideal. (It is sufficient to handle the case of two generators.) In particular, $f_1, \ldots, f_n \in K^I$ generate the unit ideal exactly when $Z(f_1) \cap \cdots \cap Z(f_n) = \emptyset$ is. If $|I| < |K| + 1$, then in this case there is even a linear combination $a_1 f_1 + \cdots + a_n f_n$ with coefficients $a_1, \ldots, a_n \in K$, which has no root.

Note The residue class fields $K^{\mathbb{N}}/\mathfrak{m}$ to a maximal ideal $\mathfrak{m} \subseteq K^{\mathbb{N}}$ with $K^{(\mathbb{N})} \subseteq \mathfrak{m}$ (i.e. $\mathfrak{m}$ is not a principal ideal) are called **ultraproducts** of K. They are true extension fields of K, but can be isomorphic to K. For example, the ultraproducts of $\mathbb{C}$ are all isomorphic to $\mathbb{C}$. (They are obviously algebraically closed like $\mathbb{C}$, have characteristic 0 and also the cardinal number $\aleph$. See Vol. 13 for this.) If K is an ordered field, then every ultraproduct of K is an ordered field, which is never archimedian ordered.

Its positive elements are represented by the sequences $(a_n) \in K^{\mathbb{N}}$ with $\{n \in \mathbb{N} \mid a_n > 0\} \in \mathcal{F} = \mathcal{F}(\mathfrak{m})$. Consider explicitly the case $K = \mathbb{R}$, which is studied in **Non-Standard Analysis**.

Exercise 4.2.32 Let $\mathcal{F}$ be a filter on the set I. This can be canonically associated with a topological space $I \uplus \{\infty\}$, such that $\mathcal{F} = \mathcal{U}(\infty) \cap I$ is, where $\mathcal{U}(\infty)$ is the neighborhood filter of ∞. The open sets with respect to this topology are all subsets of I as well as the sets of the form $F \uplus \{\infty\}$, $F \in \mathcal{F}$. (Apparently, this is a topology on $I \uplus \{\infty\}$, with respect to which I is an open discrete subspace. What space is obtained for the Fréchet filter $\mathcal{F} = \mathcal{F}r(\mathbb{N})$ of $I = \mathbb{N}$?) Show: If x_i, $i \in I$, is a family of points in a topological space X and x is another point in X, then $x = \lim_{i \in I} \mathcal{F} x_i$ holds if and only if the mapping $f : I \uplus \{\infty\} \to X$ with $i \mapsto x_i$, $\infty \mapsto x$, is continuous or — equivalently — continuous at the point ∞, i.e., if $x = \lim_{i \to \infty} f(i)$ holds. (General limits can thus be traced back to considerations of continuity. For example, it follows: If $x_i = (x_{ji})_{j \in J}$, $i \in I$ is a family of elements in the product space $X = \prod_{j \in J} X_j$ of the topological spaces X_j and is $x = (x_j)_{j \in J} \in X$, then $x = \lim_{i \in I} \mathcal{F} x_i$ applies exactly when component-wise $x_j = \lim_{i \in I} \mathcal{F} x_{ji}$ for all $j \in J$.)

Exercise 4.2.33 Let V,W be $\mathbb{K}$-vector spaces with their natural topologies, cf. Example 4.2.33.

a) Every $\mathbb{K}$-linear mapping $f : V \to W$ is continuous.
b) Every $\mathbb{K}$-subspace of V is closed in V. (It suffices to show this for finite-dimensional V.)
c) The topology induced on a $\mathbb{K}$-subspace V' of V is the natural topology of V'.
d) If V is infinitely dimensional, there is no norm on V that induces the natural topology of V. (One easily reduces to the case that all values of the norm $\|-\|$ are finite. Then there are $\mathbb{K}$-linear forms $f : V \to \mathbb{K}$ that are not continuous. If v_n, $n \in \mathbb{N}$, are linearly independent in V, one might consider a linear form f with $f(v_n) = n\|v_n\|$, $n \in \mathbb{N}$.)
e) If V has infinite dimension, then the natural topology of V does not satisfy the first countability axiom. (It is sufficient to consider the case $V = \mathbb{R}^\infty = \mathbb{R}^{(\mathbb{N}^*)}$. In this case, the cuboids $\sum_{i \in \mathbb{N}^*}[-r_i, r_i]e_i$, $(r_i) \in (\mathbb{R}_+^\times)^{\mathbb{N}^*}$, form a neighborhood base of the $0 \in \mathbb{R}^\infty$ from closed environments. If U is a neighborhood of 0, then recursively define the $r_i > 0$ such that $\sum_{i \in \mathbb{N}^*}[-r_i, r_i]e_i \subseteq U$ is, and use the Bolzano-Weierstrass theorem, cf. to this exercise 3.9.17 and Fig. 4.10.) If V has countable dimension, then V still has a countable dense subset. ($\mathbb{Q}^\infty = \mathbb{Q}^{(\mathbb{N}^*)}$ is dense in $\mathbb{R}^\infty$.) If the dimension of V is not countable, then V does not have a countable dense subset.
f) If I is infinite, then the natural topology on $\mathbb{K}^I$ is strictly finer than the product topology on $\mathbb{K}^I$. Every $\mathbb{K}$-linear form $f : \mathbb{K}^I \to \mathbb{K}$ with $f(e_i) \neq 0$ for infinitely many $i \in I$ — such linear forms exist — is not continuous with respect to the product topology.

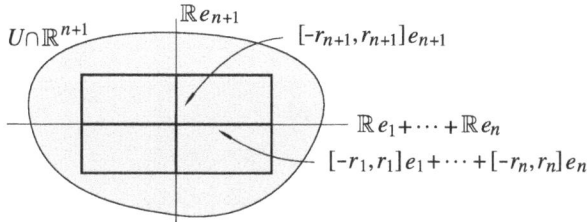

Fig. 4.10 Construction of a countable neighborhood base in $\mathbb{R}^\infty$

(These statements, with the exception of those in d) and e), also apply analogously to vector spaces with the natural topologies over any ordered field, cf. Example 4.2.35.)

Exercise 4.2.34 Let K be a subfield of the ordered field L (with the order induced by L). Then the natural topology of K is finer than the topology induced by the natural topology of L. If K is dense in L, then both topologies are the same. Provide an example where the natural topology on K is strictly finer than the induced topology.

Exercise 4.2.35 The natural topology on an ordered field is an example of a so-called order topology. Let $X = (X, \leq)$ be a completely ordered set. Then the open intervals $]a, b[$, $a, b \in X$, together with the intervals $[-\infty, b[$, $b \in X$, if $-\infty$ is a smallest element of X, and the intervals $]a, \infty]$, $a \in X$, if ∞ is a largest element of X, form a basis of a topology on X. (It is called the **order topology** defined by $\leq$ on X. It should also be noted that the order topology on a subset $X' \subseteq X$ generally does not coincide with the topology induced by the order topology on X.) Compare the order topology defined on $\mathbb{R}^2$ by the lexicographic order with the natural topology of $\mathbb{R}^2$.

Exercise 4.2.36 Prove the continuity of the function g in the proof of theorem 4.2.36 on $X - \partial A$.

4.3 Connected Spaces

A circle in a plane of our perceptual space is a connected topological space, but the union of two disjoint closed circles or two disjoint open circles is not, see Fig. 4.11. This fundamental phenomenon will be discussed in the present section.

Definition 4.3.1 A topological space X is called **connected**, if $X \neq \emptyset$ is and no decomposition $X = U_1 \uplus U_2$ of X into non-empty disjoint open subsets U_1, U_2 of

Fig. 4.11 Connected and disconnected topological space

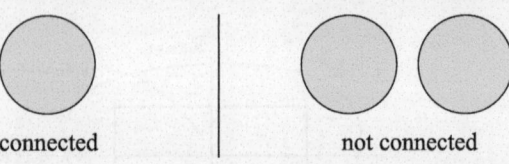

connected | not connected

X exists, if thus $X \neq \emptyset$ is and not the sum space of two non-empty subspaces, see example 4.2.25.[13]

A *non-empty* topological space X is connected if and only if it fulfills one of the following conditions:

(i) There is no decomposition $X = A_1 \uplus A_2$ of X into non-empty disjoint closed sets A_1, A_2 in X.
(ii) If $X = U \cup V$ with non-empty open subsets $U, V \subseteq X$, then $U \cap V \neq \emptyset$.
(iii) If U,V are non-empty open subsets of X with $U \cap V = \emptyset$, then $U \cup V \neq X$.
(iv) If $X = A \cup B$ with non-empty closed subsets $A, B \subseteq X$, then $A \cap B \neq \emptyset$.
(v) If A,B are non-empty closed subsets of X with $A \cap B = \emptyset$, then $A \cup B \neq X$.
(vi) $\emptyset$ and X are the only borderless subsets of X, i.e., X has exactly two borderless subsets.

A subset X' of a topological space X is called **connected**, if X', equipped with the topology induced by X, is connected. X is called **locally connected**, if every point $x \in X$ has a neighborhood base of connected neighborhoods. Continuous images of connected spaces are connected:

Theorem 4.3.2 *Let $f : X \to Y$ be a continuous mapping of topological spaces. If X is connected, then $\operatorname{Im} f = f(X) \subseteq Y$ is also connected.*

Proof We replace Y with $\operatorname{Im} f$ and can then assume that f is surjective. With X also $f(X)$ is non-empty. Let U,V be non-empty open sets in Y with $Y = U \cup V$. Then $X = f^{-1}(U \cup V) = f^{-1}(U) \cup f^{-1}(V)$ with the non-empty open sets $f^{-1}(U)$ and $f^{-1}(V)$. Consequently, $f^{-1}(U) \cap f^{-1}(V) = f^{-1}(U \cap V)$ and thus also $U \cap V = f(f^{-1}(U \cap V))$ is non-empty. Therefore, Y is connected. □

The following two lemmas are important:

Lemma 4.3.3 *Let A be a connected subset of the topological space X. Then every set $B \subseteq X$ with $A \subseteq B \subseteq \overline{A}$ is also connected.*

---

[13] Some authors also allow the empty space as a connected topological space.

4.3 Connected Spaces

Proof With A, B is also non-empty. Let U and V be open sets in X with $B \subseteq U \cup V$ and $U \cap B \neq \emptyset$ and $V \cap B \neq \emptyset$. Then $U \cap \overline{A} \neq \emptyset \neq V \cap \overline{A}$ is also true and consequently $U \cap A \neq \emptyset \neq V \cap A$. Since A is connected, it follows that $U \cap V \cap A \neq \emptyset$, and therefore $U \cap V \cap B \neq \emptyset$. □

Of course, $\overline{A}$ can be connected without A being connected. Example?

Lemma 4.3.4 *Let X_i, $i \in I$, be connected subsets of the topological space X with a non-empty intersection. Then the union $X' := \bigcup_{i \in I} X_i$ is also connected.*

Proof Let U,V be open sets in X with $U \cap X' \neq \emptyset \neq V \cap X'$ and $X' \subseteq U \cup V$. We have to show $U \cap V \cap X' \neq \emptyset$. Let's assume the opposite. Then $U \cap V \cap X_i = \emptyset$ is true for all $i \in I$. Furthermore, there are indices $j, k \in I$ with $U \cap X_j \neq \emptyset \neq V \cap X_k$. Since X_j and X_k are connected, $V \cap X_j = \emptyset$ is true, thus $X_j \subseteq U$, and $U \cap X_k = \emptyset$, thus $X_k \subseteq V$. It follows $X_j \cap X_k \subseteq U \cap V \cap X' = \emptyset$ in contradiction to the fact that the intersection of the X_i, $i \in I$, is not empty. □

Let X be a topological space and $x \in X$. From Lemma 4.3.4, it follows that the union of all connected subsets of X that contain the point x (to which the set $\{x\}$ belongs), is also connected. This set is the largest connected subset of X that contains x, and is called the **connected component** of x in X. If two connected components share a point, they are — again due to Lemma 4.3.4 — identical. *The connected components of X thus form a partition of X. X is connected if and only if X has exactly one connected component.* According to Lemma 4.3.3 *the connected components of X are closed in X.* Two points $x, y \in X$ are in the same connected component if and only if they are in a connected subset of X. This condition is therefore an equivalence relation on X. The space X is called **totally disconnected** if its connected components are singletons, i.e., if the single-point sets $\{x\}$, $x \in X$, are the only connected subsets of X. Every subspace of a totally disconnected space is also totally disconnected. — The following statement describes a situation where the connected components are easy to determine.

Proposition 4.3.5 *Let X be a topological space and $X = \biguplus_{i \in I} U_i$ a partition of X into pairwise disjoint open subsets U_i of X. Then each connected component of X lies in one of the sets U_i. In particular, the U_i, $i \in I$, are the connected components of X, if the U_i, $i \in I$, are connected.*

Proof Let $X' \subseteq X$ be connected. We have to show that $X' \subseteq U_{i_0}$ holds for a $i_0 \in I$. But let $X' \cap U_{i_0} \neq \emptyset$ for $i_0 \in I$. Then $X' = (X' \cap U_{i_0}) \uplus (X' \cap \bigcup_{i \neq i_0} U_i)$ is a decomposition of X' into open subsets of X'. Since the space X' is connected, $X' \cap \bigcup_{i \neq i_0} U_i = \emptyset$ and $X' \subseteq U_{i_0}$ are. □

Note that in the situation of Proposition 4.3.5 all U_i are also closed in X. — Nontrivial connected spaces result from the following fundamental theorem:

Theorem 4.3.6 *The connected subsets of $\mathbb{R}$ (with the natural topology) are exactly the non-empty intervals in $\mathbb{R}$.*

Proof We repeat the proof, see exercise 3.8.12. Let $I \subseteq \mathbb{R}$ be a non-empty interval and let $I = U_1 \uplus U_2$ be a partition of I into two non-empty disjoint open sets of I. Let $a \in U_1, b \in U_2$ and $a < b$. We consider the function $f : I \to \mathbb{R}$ with

$$f(x) := \begin{cases} -1, & \text{if } x \in U_1, \\ 1, & \text{if } x \in U_2. \end{cases}$$

f is continuous, since U_1 and U_2 are open in I. The restriction of f to the interval $[a, b] \subseteq I$ is also continuous, but does not satisfy the intermediate value theorem 3.8.24. Contradiction!

Conversely, let $M \subseteq \mathbb{R}$ be non-empty, but not an interval. Then there are points $a, b \in M$ and $c \in \mathbb{R} - M$ with $a < c < b$, see exercise 3.4.16, and $M = (M \cap]-\infty, c[) \uplus (M \cap]c, \infty[)$ is a partition into non-empty open subsets of M. $\square$

For theorem 4.3.6 see also exercise 3.4.19. — Theorem 4.3.6 together with theorem 4.3.2 give the following theorem:

Theorem 4.3.7 (General Intermediate Value Theorem) *Let $f : X \to \mathbb{R}$ be a continuous real-valued function on the connected topological space X. Then $\text{Im } f = f(X)$ is an interval in $\mathbb{R}$, i.e. f takes on all values between any two values.*

From theorems 4.3.6 and 4.3.2, it follows that the image of every continuous mapping $\gamma : [a, b] \to X$ of a nontrivial closed interval $[a, b] \subseteq \mathbb{R}$, $a < b$, into a topological space X is a connected subset of X. Such a continuous(!) mapping γ is called a **path** in X, see Fig. 4.12. The image $\gamma([a, b])$ of γ is called the **trajectory of** γ, which is thus connected. Furthermore, $\gamma(a)$ is called the **starting point** and $\gamma(b)$ the **endpoint of** γ. It is said that γ **connects** $\gamma(a)$ **and** $\gamma(b)$ **in** X. Since the trajectory of γ is connected, it always lies in the open ball $\text{B}(\gamma(a); \infty)$. *In particular, the starting and ending points of a path always have a finite distance.* The **return path** $\overleftarrow{\gamma} : [a, b] \to X$ with $\overleftarrow{\gamma}(t) := \gamma(a + b - t)$ then connects $\gamma(b)$ with $\gamma(a)$. Instead of the interval $[a, b]$, one can choose any other interval $[a', b']$, $a' < b'$. Instead of γ, one then considers the path

$$t \mapsto \gamma\left(a + \frac{b - a}{b' - a'}(t - a')\right)$$

on $[a', b']$. Often, the unit interval $[0, 1]$ is chosen as the definition interval for a path. In general, any continuous bijective mapping $\varphi : [a', b'] \xrightarrow{\sim} [a, b]$ (φ is thus a homeomorphism, cf. theorem 3.8.31) to the **reparametrized path** $\gamma \circ \varphi : [a', b'] \to X$ of the path $\gamma : [a, b] \to X$. If φ is monotonically increasing, one speaks of a **proper reparametrization**, otherwise of an **improper one**. In the

Fig. 4.12 Path from $x = \gamma(a)$ to $y = \gamma(b)$

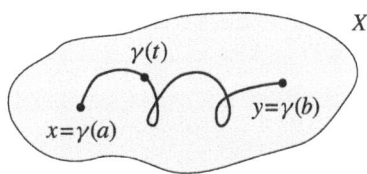

case of proper reparametrizations, the starting and ending points are preserved, in the case of improper ones, they are swapped. The trajectory does not change with reparametrizations. Often, two paths that arise from a *proper* reparametrization are identified.

If the path $\gamma \colon [a, b] \to X$ connects the points $x = \gamma(a)$ and $y = \gamma(b)$ and $\eta \colon [b, c] \to X$ the points $y = \eta(b)$ and $z = \eta(c)$ in X, then the **sum path** $\gamma\eta \colon [a, c] \to X$ with $\gamma\eta|[a, b] = \gamma$ and $\gamma\eta|[b, c] = \eta$ connects the points $x = (\gamma\eta)(a)$ and $z = (\gamma\eta)(c)$. One goes with γ from x to y and then with η further from y to z. Finally, for each $x \in X$ the **constant path** with $\gamma(t) = x$ for all $t \in [a, b]$ connects the point x with itself, it follows: For every topological space X is the relation $\sim$ with "$x \sim y$ exactly when x and y can be connected by a path in X", an equivalence relation on X. The equivalence classes with respect to this relation are called the **path-connected components** of X. *The path-connected components of a topological space X are connected.* Let Z be one of them and assume that there is a decomposition of Z into two disjoint (in Z) open non-empty subsets U and V. Then $x \in U$ and $y \in V$ in Z can be connected by a path γ, and the decomposition of Z would induce a similar decomposition of the trajectory of γ. However, this is — as already noted — connected. Contradiction! Of course, the statement also follows directly from Lemma 4.3.4. Each connected component of X is a union of certain path-connected components of X.

The topological space X is called **path-connected**, if it has exactly one path-connected component, i.e. if $X \neq \emptyset$ and any two points $x, y \in X$ can be connected by a path γ in X. Analogous to Theorem 4.3.2 applies: *Continuous images of path-connected spaces are again path-connected.* X is called **locally path-connected**, if every point $x \in X$ has a neighborhood basis of path-connected neighborhoods. Specifically, the following results:

Proposition 4.3.8 *Every path-connected topological space X is connected.*

Since the path-connected components of a locally path-connected space are open, they coincide with the connected components according to Proposition 4.3.5. — For product spaces, the following applies:

Theorem 4.3.9 *The product $X = \prod_i X_i$ of a family X_i, $i \in I$, of topological spaces is connected (or path-connected) if and only if all components X_i, $i \in I$, are connected (or path-connected).*

Fig. 4.13 Connection of $X_1 \times X_2$

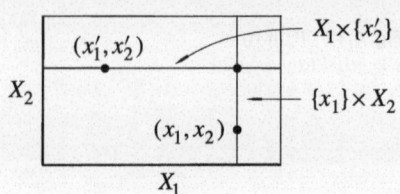

Proof With X the continuous images $X_i = p_i(X)$, $i \in I$, of X are also connected (or path-connected). — Conversely, let all X_i, $i \in I$, be connected. Then $X \neq \emptyset$ is (by the axiom of choice!). Let now initially I be finite. Induction over $|I|$ shows that it suffices to treat the case $I = \{1, 2\}$. If then $(x_1, x_2), (x_1', x_2') \in X_1 \times X_2$ are, both points lie in the connected subset $(X_1 \times \{x_2'\}) \cup (\{x_1\} \times X_2) \subseteq X_1 \times X_2$, see Fig. 4.13. — Let now I be arbitrary and $U \uplus V = X$ a decomposition of X into disjoint non-empty open sets. There is a *finite* subset $J \subseteq I$ and non-empty open sets $U', V' \subseteq X_J := \prod_{i \in J} X_i$ with $U' \times \prod_{i \notin J} X_i \subseteq U$ and $V' \times \prod_{i \notin J} X_i \subseteq V$. Then the decomposition $p_J(U) \uplus p_J(V) = X_J$ applies with the non-empty open sets $p_J(U), p_J(V) \subseteq X_J$, where $p_J = (p_i)_{i \in J} \colon X \to X_J$ is the canonical (open(!)) projection. Contradiction to the already proven case of finite index sets! — Now let all X_i, $i \in I$, be path-connected and $(x_i), (x_i') \in X$. If then $\gamma_i \colon [0, 1] \to X_i$ are paths in X_i from x_i to x_i', $i \in I$, then $\gamma := (\gamma_i) \colon [0, 1] \to X$ is a path in X from (x_i) to (x_i'). □

Let $V = (V, \|-\|)$ be a normalized $\mathbb{K}$-vector space. For two vectors $x, y \in V$ with *finite* distance $\|y - x\|$, the straight path $\gamma \colon t \mapsto x + t(y - x)$, $t \in [0, 1]$, from x to y is continuous, see exercise 4.1.6. Its trajectory is the **line segment**

$$[x, y] = \{sx + ty \mid s, t \in \mathbb{R}_+, s + t = 1\} \, (= [y, x]) \subseteq V.$$

In particular, every open ball $B(x_0; r) \subseteq V$, $r \in \overline{\mathbb{R}}_+$, is path-connected. More generally, every set that is star-shaped with respect to x_0 is path-connected $S \subseteq B(x_0; \infty)$. Here, a subset $S \subseteq V$ is called **star-shaped** with respect to a point $x_0 \in V$, if for all $x \in S$ the line segment $[x_0, x]$ belongs to S, see Fig. 4.14 on the left. Every ball $B(x_0; r)$ is even convex. Generally, a subset $K \subseteq V$ is called **convex**, if K is star-shaped with respect to every point. $x_0 \in K$ is. The empty set is convex, see Fig. 4.14 on the right.

For example, it follows:

Proposition 4.3.10 *Let V be a normed $\mathbb{K}$-vector space. Then V is locally path-connected, and consequently the connected components of any open set $G \subseteq V$ coincide with the path-connected components of G. These are open. In particular, G is connected if and only if G is path-connected.*

Proposition 4.3.11 *Every $\mathbb{K}$-vector space V is path-connected with respect to the natural topology and thus connected. — More generally, every non-empty star-shaped set and especially every non-empty convex set in V is connected.*

4.3 Connected Spaces

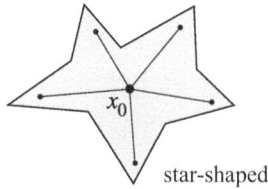
star-shaped

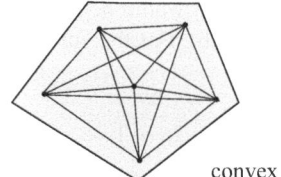
convex

Fig. 4.14 Star-shaped set and convex set

In the situation of Proposition 4.3.10, an open connected (and thus path-connected) subset $G \subseteq V$ is called a **domain** in V. In a domain $G \subseteq V$, any two points $x, y \in G$ can always be connected by a polygonal chain in G, see Ex. 4.3.18. It should be emphasized that in general a connected topological space X is not path-connected, so the connected components of X are not always the same as the path-connected components, see Ex. 4.3.12.

Example 4.3.12 (Length of a Path) Let $\gamma: [a, b] \to X$ be a path in the metric space $X = (X, d)$. If $a = t_0 \leq t_1 \leq \cdots \leq t_m = b$ is a subdivision of the interval $[a, b]$, we take the sum $\sum_{i=0}^{m-1} d(\gamma(t_i), \gamma(t_{i+1}))$ of the distances $d(\gamma(t_0), \gamma(t_1)), \ldots, d(\gamma(t_{m-1}), \gamma(t_m))$ as an approximation of the length of the path γ, cf. Fig. 4.15. If we refine the subdivision by adding another point $t \in [a, b]$ with $t_i \leq t \leq t_{i+1}$, then $d(\gamma(t_i), \gamma(t_{i+1})) \leq d(\gamma(t_i), \gamma(t)) + d(\gamma(t), \gamma(t_{i+1}))$ is according to the triangle inequality. The approximation of the path length is therefore not reduced when refining the subdivision. This motivates the following definition:

Definition 4.3.13 The **length** of the path $\gamma: [a, b] \to X$ is the supremum

$$L_a^b(\gamma) = L(\gamma)$$

of the sums $\sum_{i=0}^{m-1} d(\gamma(t_i), \gamma(t_{i+1}))$, where $a = t_0 \leq t_1 \leq \cdots \leq t_m = b$ runs through the finite subdivisions of the interval $[a, b]$. If $L(\gamma) < \infty$, then γ is called **rectifiable**.

There are non-rectifiable paths. Consider, for example, in $\mathbb{R}^2$ with the sum norm $\|(a, b)\|_1 = |a| + |b|$ the path $\gamma: [0, 1] \to \mathbb{R}^2$, $\gamma(t) := (t, th(1/t))$, $t > 0$,

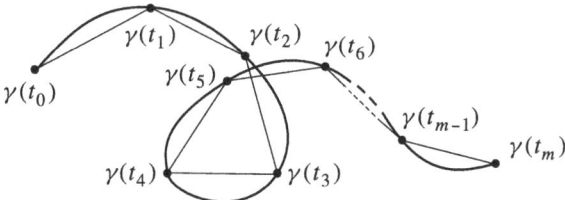

Fig. 4.15 Approximation of a path length

and $\gamma(0) = (0,0)$, where $h\colon \mathbb{R} \to \mathbb{R}$ is the distance to the nearest whole number, cf. exercise 1.2.1b). For the subdivision

$$0 < \frac{2}{2n+1} < \frac{1}{n} < \frac{2}{2n-1} < \frac{1}{n-1} < \cdots < \frac{2}{3} < 1$$

of the interval $[0, 1]$ the associated polyline has the length $1 + \sum_{\nu=1}^{n} 2/(2\nu + 1)$, and these lengths grow with $n \to \infty$ beyond all limits. *Every path in the trajectory of γ, connecting 0 and 1, has infinite length.*

Let again generally $\gamma\colon [a, b] \to X$ be a path in the metric space X. By definition, $L(\gamma) \geq d(x, y)$ applies, where $x := \gamma(a)$ and $y := \gamma(b)$ are the starting and ending point of γ, respectively, and $L(\gamma) = 0$ exactly when γ is constant. A reparametrization $\varphi\colon [a', b'] \xrightarrow{\sim} [a, b]$ changes the path length. $L(\gamma)$ not, since φ due to the strict monotony of φ the subdivisions of the interval $[a', b']$ are bijectively transformed into the subdivisions of the interval $[a, b]$ (possibly with a change of the order $\leq$ to $\geq$). *More generally, γ and $\gamma \circ \varphi$ have the same length, when $\varphi\colon [a', b'] \to [a, b]$ is a continuous and monotone surjective mapping*, cf. exercise 4.3.25a). If $L(\gamma) = d(x, y)(< \infty)$, then γ is called a **Geodesic** in X from x to y. In this case, $\gamma|[c, d]$ is also a geodesic in X for every subinterval $[c, d] \subseteq [a, b]$. The metric space X is called **geodesic**, if any two points $x, y \in X$ can be connected with a geodesic in X. (In particular, a geodesic space is path-connected, and the distance between any two points in it is finite.) If V is a normed $\mathbb{K}$-vector space and $x, y \in V$ are two points with finite distance $\|y - x\|$, then the straight path $t \mapsto x + t(y - x), t \in [0, 1]$, apparently the length $\|y - x\|$ and is thus a geodesic from x to y. So if all values of $\|-\|$ are finite, then V is a geodesic space. In general, however, there are geodesics with different trajectories for two points $x, y \in V$ that connect x and y. For example, if $V = \mathbb{R}^2$ is equipped with the sum norm $\|(a, b)\|_1 = |a| + |b|$, there are infinitely many essentially different geodesics from $(0, 0)$ to $(1, 1)$. The same applies to geodesics from $(0, 0)$ to $(0, 1)$ with respect to the Chebyshev norm $\|(a, b)\|_\infty = \mathrm{Max}\,(|a|, |b|)$, see Fig. 4.16. For the p-norms $\|-\|_p$ on $\mathbb{R}^2$ with $1 < p < \infty$, however, the path $[x, y]$ is the only geodesic that connects x and y.

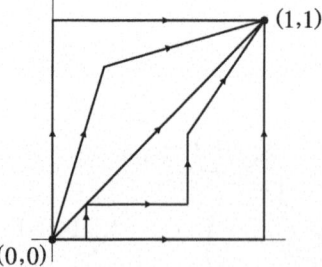 

Fig. 4.16 Geodesics with respect to $\|-\|_1$ or $\|-\|_\infty$

4.3 Connected Spaces

Path lengths are additive: If $\gamma: [a, b] \to X$ is a path and $c \in [a, b]$, then

$$L_a^b(\gamma) = L_a^c(\gamma) + L_c^b(\gamma)$$

applies (where we have denoted the restrictions of γ to the subintervals $[a, c]$ or $[c, b]$ again with γ for simplicity). Furthermore, the length function $L: [a, b] \to \mathbb{R}_+$, $t \mapsto L_a^t(\gamma)$, is monotonically increasing. If γ is rectifiable, then L is strictly monotonically increasing if and only if γ is not constant on any subinterval of $[a, b]$ with more than one point. It holds:

Lemma 4.3.14 *If $\gamma: [a, b] \to X = (X, d)$ is a rectifiable path of length $\ell := L(\gamma) < \infty$, then the (monotonically increasing) length function*

$$L: [a, b] \to [0, \ell], \quad t \mapsto L_a^t(\gamma),$$

is continuous.

Proof Due to the additivity of L, it is sufficient to show that L is (right-sided) continuous in a. (For the (left-sided) continuity in b, consider the return path.) Since L is monotonically increasing, the right-sided limit $\mu := L(a+) = \lim_{t \to a, t > a} L(t) \in \mathbb{R}_+$ exists, see example 3.8.22. Suppose it were $\mu > 0 = L(a)$. Then there is a $\delta > 0$ with $0 \leq L(d) - L(c) = L_c^d(\gamma) \leq \mu/3$ for all c,d with $a < c \leq d \leq a + \delta$. Since γ is continuous in a, we can choose δ so small that $d(\gamma(a), \gamma(t)) \leq \mu/3$ is for all t with $a \leq t \leq a + \delta$. By definition of $L(a + \delta) \geq \mu$ there is a subdivision $a = t_0 < t_1 \leq t_2 \leq \cdots \leq t_m = a + \delta$ of the interval $[a, a + \delta]$ with $\sum_{i=0}^{m-1} d(\gamma(t_i), \gamma(t_{i+1})) > 2\mu/3$. On the other hand, $d(\gamma(a), \gamma(t_1)) \leq \mu/3$ and

$$\sum_{i=1}^{m-1} d(\gamma(t_i), \gamma(t_{i+1})) \leq L_{t_1}^{a+\delta}(\gamma) = L(a + \delta) - L(t_1) \leq \mu/3,$$

so overall $\sum_{i=0}^{m-1} d(\gamma(t_i), \gamma(t_{i+1})) \leq 2\mu/3$. Contradiction! $\square$

Let $\gamma: [a, b] \to X$ be a rectifiable path of length ℓ as in Lemma 4.3.14. γ is then on the fibers of the surjective length function $L: [a, b] \to [0, \ell]$, $t \mapsto L_a^t(\gamma)$, is constant and therefore induces a mapping $\overline{\gamma}: [0, \ell] \to X$ with $\gamma = \overline{\gamma} \circ L$, which is also continuous and thus a path from $\gamma(a)$ to $\gamma(b)$, since L is a quotient mapping, cf. Corollary 4.4.6. However, we can also easily see this with the means proven so far: Let $A \subseteq X$ be closed in X. Then $\gamma^{-1}(A)$ is closed in $[a, b]$, thus compact, and consequently $\overline{\gamma}^{-1}(A) = L(L^{-1}(\overline{\gamma}^{-1}(A))) = L(\gamma^{-1}(A))$ is also compact and therefore closed according to Theorem 3.9.2. For all $s \in [0, \ell]$, $\overline{\gamma}(s) = L_0^s(\overline{\gamma}) = L_a^t(\gamma) = s$ is, where $t \in L^{-1}(s)$ is. Such a path $\overline{\gamma}$ is called **arc parameterized**. If γ is already not constant on any non-trivial subinterval of $[a, b]$, then L is bijective and $\overline{\gamma}$ is an arc parameterized reparameterization of γ.

Generally, it is difficult to determine the length of a path. In the next volume, we will obtain the following integral representation of its length for a continuously differentiable path $\gamma: [a, b] \to V$ in a $\mathbb{R}$-Banach space V:

$$L(\gamma) = \int_a^b \|\dot{\gamma}(t)\| dt,$$

where $\dot{\gamma}(t) = \lim_{s \to t, s \neq t}(\gamma(s) - \gamma(t))/(s-t)$ is the derivative (= speed) of γ. (See Vol. 2, Theorem 3.3.5. It is shown that $L: [a, b] \to \mathbb{R}, t \mapsto L_a^t(\gamma)$, is differentiable with derivative $\|\dot{\gamma}\|$.) This immediately gives, for example, that the once traversed unit circle $\gamma: t \mapsto (\cos t, \sin t), t \in [0, 2\pi]$, in the $\mathbb{R}^2$ with respect to the Euclidean norm $\|-\|_2$ has the length 2π; because it is $\dot{\gamma}(t) = (-\sin t, \cos t)$, thus $\|\dot{\gamma}(t)\|_2 = (\sin^2 t + \cos^2 t)^{1/2} \equiv 1$. Therefore, γ is an arc parameterization of the unit circle. (See also the end of Example 3.3.9.) ◊

Exercises

Exercise 4.3.1 Determine the connected components of the following sets, each with respect to the natural topology of the $\mathbb{R}^n$, $n \in \mathbb{N}$:

a) $\{(x, y) \mid x^2 + y^2 = 1\} \subseteq \mathbb{R}^2$.
b) $\{(x, y) \mid x^2 - y^2 = 1\} \subseteq \mathbb{R}^2$.
c) $\{(x_1, \ldots, x_n) \mid x_1 \cdots x_n \neq 0\} \subseteq \mathbb{R}^n$.
d) $\{(x_1, \ldots, x_n) \mid x_1 \cdots x_n = 0\} \subseteq \mathbb{R}^n$.
e) $\{(x, y) \mid x^2 < |y|\} \subseteq \mathbb{R}^2$.
f) $\{(x, y) \mid x^2 \geq y\} \subseteq \mathbb{R}^2$.

Exercise 4.3.2 Sketch the sets

$$A := \{(x, y) \mid 0 \leq 3x^2 + 3y^2 - 2xy < 8\}, \quad B := \{(x, y) \mid 0 \leq x^2 + y^2 - 6xy < 8\}$$

in the $\mathbb{R}^2$ and specify the sets $\mathring{A}, \overline{A}, \mathring{B}, \overline{B}$. Furthermore, determine the connected components of the sets $A, \mathring{A}, \overline{A}, B, \mathring{B}, \overline{B}, A \cup B, (A \cup B)^\circ, \overline{A} \cup \overline{B} = \overline{A \cup B}$.

Exercise 4.3.3

a) Determine the connected components of $\mathbb{Q}$ and $\mathbb{R} - \mathbb{Q}$. Show in general: A countable metric space is totally disconnected.
b) If $M \subseteq \mathbb{R}^2$ is a countable set, then $\mathbb{R}^2 - M$ is path-connected.
c) Let $m \in \mathbb{N}^*$. How many connected components does the complement of m pairwise different affine lines in $\mathbb{R}^2$ have at most, how many at least? What is the answer if the lines are replaced by circles (with positive radii)? Treat the analogous problem for the $\mathbb{R}^n$, $n \in \mathbb{N}^*$, replacing the affine lines with affine hyperplanes or the circles with (Euclidean) spheres.

4.3 Connected Spaces

Exercise 4.3.4 If the topological space X has only finitely many connected components, then these are simultaneously open and closed.

Exercise 4.3.5

a) The unit interval $[0, 1]$ does not possess a countable partition $[0, 1] = \biguplus_{i \in I} K_i$ into pairwise disjoint, closed sets $|I| > 1$. (This can be easily deduced from the result of exercise 3.4.20 b).
b) Let $X = \biguplus_{i \in I} A_i$ be a countable partition of the topological space X into pairwise disjoint, *closed* sets A_i. Then each path-connected component of X lies in one of the sets A_i. In particular, the A_i are the path-connected components of X if they are also path-connected. (Corresponding statements do not apply to connected subsets of X, not even when restricted to Hausdorff spaces.)

Exercise 4.3.6 Let X_i, $i \in I$, be a family of topological spaces. Determine the connected and path-connected components of the product space $\prod_i X_i$. (Cf. Theorem 4.3.9.)

Exercise 4.3.7 Let A and B be subsets of a topological space X. If A and B are both closed or both open and if $A \cup B$ and $A \cap B$ are connected, then A and B are also connected.

Exercise 4.3.8 A and B are proper subsets of the connected topological spaces X and Y respectively. Then the complement $(X \times Y) - (A \times B)$ of $A \times B$ in $X \times Y$ is also connected.

Exercise 4.3.9 $f: X \to Y$ is a surjective open or a surjective closed continuous mapping of topological spaces. If Y and the fibers $f^{-1}(y)$, $y \in Y$, are connected, then X is also connected. (Does a similar statement also apply to path-connected spaces?)

Exercise 4.3.10 $f: X \to Y$ is a mapping of the connected topological space X into the set Y. If f is **locally constant**, i.e., for every $x \in X$ there is a neighborhood U of x such that $f|U$ is constant, then f is already constant. If Y is a discrete topological space and f is continuous, then f is constant.

Exercise 4.3.11 Let A and B be connected subsets of a normed $\mathbb{K}$-vector space V. The Minkowski sum $A + B := \{x + y \mid x \in A, y \in B\} \subseteq V$ is also connected.

Exercise 4.3.12 Let $h: \mathbb{R} \to \mathbb{R}$ be the distance to the next whole number, cf. exercise 1.2.1b). The graph of the function $f: \mathbb{R} \to \mathbb{R}$ with $f(x) := h(1/x)$ for $x \neq 0$ and $f(0) := 0$ is connected, but not path-connected. Determine the three path-connected components of this graph.

Fig. 4.17 x and y are connectable with respect to $\mathcal{U}$

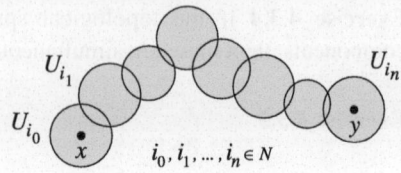

Exercise 4.3.13 An open set in a finite-dimensional $\mathbb{K}$-vector space with the natural topology has at most countably many connected components (which are all open according to Proposition 4.3.10 or 4.3.11).

Exercise 4.3.14 Let $\mathcal{U} = (U_i)_{i \in I}$ be a cover of the topological space X. The **nerve** N($\mathcal{U}$) of $\mathcal{U}$ is the graph with the vertex set I, where $\{i, j\} \in \mathfrak{E}_2(I)$ is an edge exactly when $U_i \cap U_j \neq \emptyset$ holds. Let $\mathcal{N} \subseteq \mathfrak{P}(I)$ be the set of connected components of the graph N($\mathcal{U}$). The sets $U_N := \bigcup_{i \in N} U_i$, $N \in \mathcal{N}$, form a partition of X. Two points $x, y \in X$ are called **connectable with respect to** $\mathcal{U}$, if there is a $N \in \mathcal{N}$ with $\{x, y\} \in U_N$, see Fig. 4.17. This $\mathcal{U}$ connectivity is an equivalence relation on X. Its equivalence classes U_N, $N \in \mathcal{N}$, $U_N \neq \emptyset$, we call the $\mathcal{U}$-**connectivity classes**.

a) If the U_i, $i \in I$, are connected, then every connectivity class with respect to $\mathcal{U}$ is connected.
b) If $\mathcal{U}$ is an open cover of X (i.e., all U_i, $i \in I$, are open), then every component of X lies in a connectivity class with respect to $\mathcal{U}$, see Proposition 4.3.5. In particular, any two points of X are connectable with respect to $\mathcal{U}$, if X is connected. (The last statement characterizes the connected spaces among the non-empty topological spaces.)
c) If $\mathcal{U}$ is an open cover of X and all U_i are connected, then the connectivity classes are the components of X. In particular, X is connected if and only if the nerve N($\mathcal{U}$) is a connected graph.
d) Statements analogous to b) and c) apply if $\mathcal{U}$ is a *finite* closed cover of X.

Exercise 4.3.15 Let X be a topological space.

a) If every point $x \in X$ has a connected neighborhood, then all connected components of X are open.
b) If every point $x \in X$ has a path-connected neighborhood, then all connected components of X are open and equal to the path-connected components of X.
c) The following statements are equivalent: (i) X is locally (path) connected. (ii) The (path-) connected components of open sets $U \subseteq X$ are open. (iii) Every point $x \in X$ has a neighborhood base of *open* (path) connected neighborhoods.

Exercise 4.3.16 Let X be a topological space. The space $[X]$ of the connected components of X (endowed with the quotient topology) is totally disconnected. (Show that every closed set $A \subseteq [X]$ with more than one point is not connected by finding

4.3 Connected Spaces

a decomposition for its preimage in X into non-empty closed subsets that are unions of connected components of X.)

Exercise 4.3.17 Let $X = (X, \mathcal{T})$ be a topological space.

a) The borderless sets form a subring $\mathcal{B} = \mathcal{B}_\mathcal{T}(X) \subseteq \mathfrak{P}(X)$ of the Boolean set ring $(\mathfrak{P}(X), \triangle, \cap)$. In the canonical isomorphism $\mathfrak{P}(X) \xrightarrow{\sim} F_2^X, A \mapsto e_A$, corresponds $\mathcal{B}$ the F_2-subalgebra $C_{F_2}(X) = C(X, F_2)$ of the continuous F_2-valued functions $X \to F_2$ (where F_2 carries the discrete topology). In particular, X is exactly connected when $X \neq \emptyset$ is and every continuous function $X \to F_2$ constant. Exactly then X has finitely many connected components, when the Boolean ring $\mathcal{B}$ is finite. In this case, $\mathcal{B} \cong F_2^s$, where s is the number of connected components of X. If X has infinitely many connected components, then for every $n \in \mathbb{N}, n \geq 2$, there is a decomposition of X into exactly n non-empty open subsets.

b) X is connected exactly when the $\mathbb{K}$-algebra $C_\mathbb{K}(X)$ (as a ring) is connected.

Exercise 4.3.18 Let G be a domain in the normed $\mathbb{K}$-vector space V. Then any two points $x, y \in G$ can be connected with a **polygonal chain**, i.e., there are points $x_0, x_1, \ldots, x_n \in G$ with $x_0 = x$, $x_n = y$ and $[x_0, x_1, \ldots, x_n] := \bigcup_{i=0}^{n-1} [x_i, x_{i+1}] \subseteq G$ for $n > 0$ (and $[x_0] := \{x_0\}$).

Exercise 4.3.19 Let I be a set and F a polynomial function different from 0 on $\mathbb{K}^I$ with the set of zeros $Z = Z_\mathbb{K}(F)$. ($\mathbb{K}^I$ carries the natural topology!) In the case of $\mathbb{K} = \mathbb{R}$, the (in $\mathbb{R}^I$ open) complement $\mathbb{R}^I - Z$ is usually not connected. (However, it only has finitely many connected components, as is stated without proof.) Quite different in the complex case: $\mathbb{C}^I - Z$ is always connected. (One can assume that I is finite. Then generally: If G is a domain in $\mathbb{C}^I \cong \mathbb{C}^{|I|}$, then $G - Z$ is also a domain.)

Exercise 4.3.20 Let V be a finite-dimensional $\mathbb{R}$-vector space of dimension ≥ 2 with the natural topology and $U_1, \ldots, U_r$ (affine) subspaces in V of codimension ≥ 2. If G is a domain in V, then $G - \bigcup_{\rho=1}^{r} U_\rho$ is also a domain. (It suffices to handle the case $r = 1$. If $x \in G - U_1$, then every point $y \in G$ can be connected by a path from x to y that lies entirely (possibly with the exception of the endpoint y) in $G - U_1$.)

Exercise 4.3.21 For a set X and a natural number n denote $\Delta_n = \Delta_n(X)$ the **generalized diagonal** of those n-tuples $(x_1, \ldots, x_n) \in X^n$, whose components are *not* pairwise different. Then $X^n - \Delta_n(X)$ is the set of n-tuples $(x_1, \ldots, x_n) \in X^n$, whose components are pairwise different. The spaces $\mathbb{K}^n$, $n \in \mathbb{N}$, will carry the natural topology in the following.

a) If G is a domain in $V := \mathbb{K}^n$, $n \geq 2$, then $G^n - \Delta_n(G)$ is also a domain (in V^n). (exercise 4.3.20.)

b) How many connected components do $\mathbb{R}^n - \Delta_n(\mathbb{R})$ and $T^n - \Delta_n(S^1)$ have, where $T^n = (S^1)^n$ is the n-dimensional torus? (The mapping $\mathbb{R} \to S^1, t \mapsto e^{2\pi i t}$, is a

Fig. 4.18 Θ-Space

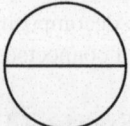

quotient mapping and induces a homeomorphism $\mathbb{T} = \mathbb{R}/\mathbb{Z} \xrightarrow{\sim} S^1 = \mathrm{U} = \{z \in \mathbb{C} \mid |z| = 1\}$. Therefore, the homeomorphism $\mathbb{T}^n \xrightarrow{\sim} T^n$ also applies, and the term "torus" used here does not conflict with the term introduced in Example 2. 2.16 (2).)

Remark The problem of determining the (path-)connected components of $X^n - \Delta_n(X)$ for a topological space X is a **shunting problem**: n (point-like) locomotives, which occupy the (pairwise different) points $x_1, \ldots, x_n \in X$, are to be moved without collisions so that they occupy the (pairwise different) points $y_1, \ldots, y_n \in X$. This means finding a path in $X^n - \Delta_n(X)$ that connects $(x_1, \ldots, x_n)$ and $(y_1, \ldots, y_n)$. From this perspective, determine for the topological space $X := S^1 \cup [-1, 1]e_1 \subseteq \mathbb{R}^2$ in Fig. 4.18 (the so-called Θ-space) the (number of) connected components ($=$ path-connected components) of $X^n - \Delta_n(X)$.

Exercise 4.3.22 $\mathbb{R}^n$ carries the natural topology. The intersection of two open connected sets in the $\mathbb{R}^n$ is usually not connected, even if this intersection is not empty. However, it applies: If $G_1, G_2 \subseteq \mathbb{R}^n$ are regions with $G_1 \cup G_2 = \mathbb{R}^n$, then $G_1 \cap G_2$ is also a region. (The proof for $n \geq 2$ is probably not quite simple. A suggestion for this can be found in Vol. 5. — The same applies to the spheres S^n (instead of $\mathbb{R}^n$), $n \geq 2$, but not for S^1. A general result can also be found in one of the further volumes.)

Exercise 4.3.23 $\mathbb{R}^n$ carries the natural topology in the following.

a) Let $H = x + U \subseteq \mathbb{R}^n$, $U \subseteq \mathbb{R}^n$ be a 1-codimensional subspace, an affine hyperplane in $\mathbb{R}^n$, $n \geq 1$. Then $\mathbb{R}^n - H$ has exactly two connected components, which are called the **open half-spaces** to H. (Their closed hulls are the closed half-spaces.)
b) Let F be an (affine) subspace of $\mathbb{R}^n$, $n \geq 1$, and $M \subseteq \mathbb{R}^n$ an open set with $\partial M = F$. Then M is one of the open half-spaces determined by F of the $\mathbb{R}^n$, if F is a hyperplane, and $M = \mathbb{R}^n - F$, if $\mathrm{codim}_{\mathbb{R}^n} F > 1$ is.
c) It is reminded that a subset $M \subseteq \mathbb{R}^n$ is called bounded if it lies within a cuboid $[-R, R]^n = \overline{B}_{d_\infty}(0; R)$, $R \in \mathbb{R}_+$. Let $G \subseteq \mathbb{R}^n$, $n \geq 2$, be an unbounded open set with a bounded boundary. Then G includes the complement of a cuboid $\overline{B}_{d_\infty}(0; R)$, $R \in \mathbb{R}_+$.

Exercise 4.3.24 Let V be an at least 2-dimensional $\mathbb{R}$ vector space (with the natural topology). If the mapping $f : V - \{0\} \to \mathbb{R}$ is continuous, then there exists $x \neq 0$ with $f(x) = f(-x)$. (Consider the difference $x \mapsto f(x) - f(-x)$. — More generally: If V

is at least $(n+1)$-dimensional and $f : V - \{0\} \to \mathbb{R}^n$ is continuous, then there exists a $x \neq 0$ with $f(x) = f(-x)$. This is the Borsuk-Ulam theorem, which is covered in Vol. 12.)

Exercise 4.3.25 Let $\gamma : [a, b] \to X$ be a path in the metric space $X = (X, d)$.

a) If $\varphi : [c, d] \to [a, b]$ is a monotone and surjective continuous function, then $L_c^d(\gamma \circ \varphi) = L_a^b(\gamma)$.
b) If $f : X \to Y$ is a mapping of metric spaces that is Lipschitz continuous with Lipschitz constant $\lambda \in \mathbb{R}_+$ on the trajectory $\gamma([a, b])$ of γ, i.e., if $d(f(x), f(y)) \leq \lambda d(x, y)$ for all $x, y \in \gamma([a, b])$, then $L_a^b(f \circ \gamma) \leq \lambda L_a^b(\gamma)$ applies.
c) If d' is a metric equivalent to d on X, then γ is rectifiable with respect to d if and only if γ is rectifiable with respect to d'.

Exercise 4.3.26 Let V be a finite-dimensional normed $\mathbb{R}$ vector space with a norm that induces the natural topology on V. (According to theorem 4.4.12, this applies exactly to the norms on V with $\|x\| < \infty$ for all $x \in V$.) A path $\gamma : [a, b] \to V$ is rectifiable if and only if for every $\mathbb{R}$ linear form $f : V \to \mathbb{R}$ the path $f \circ \gamma$ in $\mathbb{R}$ is rectifiable. This is already the case if this applies to a generating system of the $\mathbb{R}$ vector space $V^* = \mathrm{Hom}(V, \mathbb{R})$. (If $v_1, \ldots, v_n$ is a $\mathbb{R}$ basis of V, then the linear forms f on V are the functions $\sum_{i=1}^n a_i v_i \mapsto \sum_{i=1}^n a_i b_i$, where $b_1 = f(e_1), \ldots, b_n = f(e_n) \in \mathbb{R}$ are each fixed values.)

Exercise 4.3.27 Following the previous exercise, we are interested in the rectifiable paths $\gamma : [a, b] \to \mathbb{R}$ (where $\mathbb{R}$ is provided with the natural distance). The length of such a path is also called its variation. Thus, the graph $\Gamma_h \subseteq \mathbb{R}^2$ of a continuous function $h : [a, b] \to \mathbb{R}$ is rectifiable if and only if h has finite variation. The concept of **variation** is defined for arbitrary (not necessarily continuous) functions $h : [a, b] \to \mathbb{R}$ as the supremum over all sums $\sum_{j=0}^{m-1} |h(t_{j+1}) - h(t_j)|$, where $t_0, \ldots, t_m$ all finite decompositions $a = t_0 \leq \cdots \leq t_m = b$ of the interval $[a, b]$ are run through. Show: $h : [a, b] \to \mathbb{R}$ has finite variation if and only if $h = f - g$ with monotonically increasing functions $f, g : [a, b] \to \mathbb{R}$. For continuous h with finite variation, f and g can be chosen as continuous functions. (If h has finite variation, define $f(t)$ for $t \in [a, b]$ as the supremum over all sums $\sum_{j=0}^{m-1} \mathrm{Max}\left(h(t_{j+1}) - h(t_j), 0\right), a = t_0 \leq \cdots \leq t_m = t$.)

4.4 Compact Spaces

The sentence 3.4.3 or (more generally) the sentence 4.2.34 by Bolzano-Weierstraß is fundamental for the construction of analysis. With the concept of compactness, it is topologically founded.

Definition 4.4.1 A topological space X is called **quasi-compact**, if X fulfills the following condition: If U_i, $i \in I$, is an open cover of X (i.e., if U_i, $i \in I$, is a family of open sets in X with $\bigcup_{i \in I} U_i = X$), then there exists a *finite* subset $J \subseteq I$ with $\bigcup_{i \in J} U_i = X$. — The space X is called **compact**, if it is quasi-compact and Hausdorff.

The condition of quasi-compactness in the above definition is usually formulated as follows: Every open cover U_i, $i \in I$, of X has a finite subcover. Apparently, it is sufficient to verify this property for covers U_i, $i \in I$, to verify, whose elements U_i belong to a given basis $\mathcal{B}$ of the topology of X. For closed sets, the condition of quasi-compactness is formulated as follows: *If A_i, $i \in I$, is a family of closed sets in X with $\bigcap_{i \in I} A_i = \emptyset$, then there is a finite subset $J \subseteq I$ with $\bigcap_{i \in J} A_i = \emptyset$.*

A subspace X' of X is quasi-compact if and only if for every family U_i, $i \in I$, of open sets in X with $X' \subseteq \bigcup_{i \in I} U_i$ a finite subset $J \subseteq I$ with $X' \subseteq \bigcup_{i \in J} U_i$ exists. We also briefly say again: The open cover U_i, $i \in I$, of X' has a finite subcover. If X is Hausdorff and $X' \subseteq X$ is a quasi-compact subspace, then X' is also Hausdorff and therefore compact.

Proposition 4.4.2 *Let X be a Hausdorff space.*

(1) *If X' is a compact subspace of X, then X' is closed in X.*
(2) *If X is compact and $X' \subseteq X$ is closed in X, then X' is also compact.*

Proof (1) Let $x \in X$, $x \notin X'$. For every point $y \in X'$ there are open neighborhoods $U(y)$ of y and $V(y)$ of x with $U(y) \cap V(y) = \emptyset$. The cover $U(y)$, $y \in X'$, of X' has a finite subcover $U(y_1), \ldots, U(y_n)$. $V := V(y_1) \cap \cdots \cap V(y_n)$ is then an (open) neighborhood of x with $V \cap X' = \emptyset$. Therefore, $X - X'$ is open and X' is closed.

(2) In general, a closed subset X' of a quasi-compact space is again quasi-compact. For proof, let U_i, $i \in I$, be a cover of X' with sets U_i open in X. Then U_i, $i \in I$, together with $X - X'$ is an open cover of X, which by assumption has a finite subcover. From this, after omitting $X - X'$, one obtains a finite subcover U_i, $i \in J$, of X'. □

Continuous images of quasi-compact spaces are quasi-compact:

Theorem 4.4.3 *Let $f : X \to Y$ be a continuous mapping of topological spaces. If X is quasi-compact, then so is $\mathrm{Im} f = f(X)$. In particular: If X is quasi-compact and Y is Hausdorff, then $\mathrm{Im} f$ is compact and therefore closed in Y.*

Proof We replace Y with $\mathrm{Im} f$ and can assume that f is surjective. Let V_i, $i \in I$, be an open cover of Y. Then $U_i := f^{-1}(V_i)$, $i \in I$, is an open cover of X. Since X is quasi-compact, it contains a finite subcover U_i, $i \in J$, and $V_i = f(f^{-1}(V_i)) = f(U_i)$, $i \in J$, is a finite cover of $\mathrm{Im} f$. □

4.4 Compact Spaces

Corollary 4.4.4 *Let $f : X \to Y$ be a continuous mapping of Hausdorff spaces. If X is compact, then f is closed (i.e. f-images of closed subsets of X are closed in Y). In particular, $\operatorname{Im} f$ is closed in Y.*

Proof Let X be compact and $A \subseteq X$ be closed. Then A is compact according to Proposition 4.4.2 and $f(A)$ is closed according to Theorem 4.4.3. □

The following corollary is an important criterion for homeomorphisms:

Corollary 4.4.5 *Let $f : X \to Y$ be a bijective continuous mapping of Hausdorff spaces. If X is compact, then f is a homeomorphism.*

Proof We need to show that f^{-1} is also continuous, i.e., that $(f^{-1})^{-1}(A) = f(A)$ is closed in Y for every closed set $A \subseteq X$. This is stated by Corollary 4.4.4. □

Corollary 4.4.6 *Let $f : X \to Y$ be a continuous surjective mapping of Hausdorff spaces. X is compact. Then f is a quotient mapping, i.e., the topology on Y is the image topology of X with respect to f, and f induces a homeomorphism $X/R_f \overset{\sim}{\to} Y$, where R_f is the equivalence relation on X induced by f.*

Proof We have to show that the preimage $f^{-1}(B)$ of a subset $B \subseteq Y$ is closed in X if and only if B is closed in Y. If B is closed in Y, then $f^{-1}(B)$ is closed in X, since f is continuous. Conversely, if $f^{-1}(B)$ is closed in X, then $B = f(f^{-1}(B))$ is closed in Y according to Corollary 4.4.4. □

Corollary 4.4.6 fully corresponds to the isomorphism theorem $\operatorname{Im} f \cong X/\operatorname{ker} f$ for a homomorphism $f : X \to Y$ of groups, rings, modules, ...

Quasi-compactness can surprisingly already be tested with subbases:

Lemma 4.4.7 *Let $\mathcal{E} \subseteq \mathcal{T}$ be a subbase of the topological space $X = (X, \mathcal{T})$. X is already quasi-compact if every cover of X with elements from $\mathcal{E}$ has a finite subcover.*

Proof By definition, the set $\mathcal{B} := \mathcal{B}(\mathcal{E})$ of finite intersections of elements from $\mathcal{E}$ is a basis of the topology of X. We assume that every cover of X with elements from $\mathcal{E}$ has a finite subcover, but X is not quasi-compact. Then we consider the set $\mathfrak{M}$ of all those open covers $\mathcal{U} \subseteq \mathcal{B}$, which do not have a finite subcover. This set $\mathfrak{M}$ is not empty by assumption and is inductively ordered with respect to inclusion (even strictly). If $\mathcal{U}_i$, $i \in I$, is a non-empty chain in $\mathfrak{M}$, then it is $\bigcup_i \mathcal{U}_i$ apparently an upper limit of the considered chain. According to Zorn's Lemma 1.4.15 there is a maximum element $\mathcal{U}_0 \in \mathfrak{M}$. We show that $\mathcal{E} \cap \mathcal{U}_0$ is also a cover of X, which results in a contradiction. Let $x \in X$. If $x \notin U$ for each $U \in \mathcal{E} \cap \mathcal{U}_0$, then $x \in V = V_1 \cap \cdots \cap V_r$ would be for a $V \in \mathcal{U}_0, V_1, \ldots, V_r \in \mathcal{E}$. Because of $V_\rho \notin \mathcal{U}_0$ and the maximality of $\mathcal{U}_0$, $\mathcal{U}_0 \uplus \{V_\rho\}$ has a finite subcover for each ρ. So there is $U_1, \ldots, U_s \in \mathcal{U}_0$ such that $U_1, \ldots, U_s, V_\rho$

for each ρ covers the space X. But then $U_1, \ldots, U_s, V = V_1 \cap \cdots \cap V_r$ also covers all of X. Contradiction! So there is an $U \in \mathcal{U}_0$ with $x \in U$. □

A direct consequence of this lemma is the following important theorem.

Theorem 4.4.8 (Tychonoff's Theorem) *Let X_i, $i \in I$, be a family of non-empty topological spaces. The product $X = \prod_{i \in I} X_i$ is quasi-compact (or compact) if and only if each factor X_i, $i \in I$, is quasi-compact (or compact).*

Proof If X is quasi-compact, so is every image $X_i = p_i(X)$, $i \in I$. — Now let all X_i be quasi-compact. The products $\prod_i U_i$ with U_i open in X_i and $U_i = X_i$ except for *exactly one* index form a subbasis $\mathcal{E}$ of the topology of X. Due to Lemma 4.4.7 it is sufficient to show that every cover $W_j, j \in J$, of X with elements from $\mathcal{E}$ has a finite subcover. For each $i \in I$ let $\mathcal{U}_i$ be the set of open sets in X_i, which are $\neq X_i$ and as i-th factor of a W_j occur. If $U_i := \bigcup_{U \in \mathcal{U}_i} U \neq X_i$ were for every $i \in I$ and $x_i \in X_i - U_i$, then the I-tuple (x_i) would not be in any $W_j, j \in J$. So there is an $i_0 \in I$ with $U_{i_0} = X_{i_0}$. Because of the quasi-compactness of X_{i_0} there is a finite subset $J' \subseteq J$ such that the i-th components of the $W_j, j \in J'$, for $i \neq i_0$ are entirely X_i and their i_0-th components form a cover of X_{i_0}. Then the $W_j, j \in J'$, are but a finite subcover of the cover $W_j, j \in J$, of X. □

For finite index sets I, the theorem 4.4.8 of Tychonoff can easily be proven without the lemma 4.4.7 (and thus without Zorn's Lemma) by induction over $|I|$. We recommend the reader to carry this out. Another special case of Tychonoff's theorem is treated in theorem 4.5.7.

The following lemma describes the connection between the concept of quasi-compactness and the validity of the Bolzano-Weierstrass theorem:

Lemma 4.4.9 *Let X be a topological space.*

(1) *If X is quasi-compact, then every sequence (x_n) in X has a cluster point in X.*
(2) *Conversely, if every sequence (x_n) in X has a cluster point in X and the topology of X has a countable basis, then X is quasi-compact.*

Proof (1) Let X be quasi-compact. Assume that the sequence (x_n) in X has no accumulation point. Then every point $y \in X$ has an open neighborhood $U(y)$, in which only finitely many members of the sequence (x_n) lie. For $U(y), y \in X$, there is a finite subcover $U(y_1), \ldots, U(y_n)$. But all members of the sequence lie in $X = U(y_1) \cup \cdots \cup U(y_n)$. Contradiction!

(2) Since X has a countable basis, it suffices, according to the remark following Definition 4.4.1, to show that every countable open cover $U_n, n \in \mathbb{N}$, of X has a finite subcover (cf. also Proposition 4.2.13). But if $\bigcup_{i=0}^n U_i \neq X$ for all $n \in \mathbb{N}$ and $x_n \in X - \bigcup_{i=0}^n U_i$, then the sequence (x_n) due to $x_m \notin U_n$ for all $m, n \in \mathbb{N}, m \geq n$, would have no cluster point in X contrary to the assumption. □

Note that in a topological space that satisfies the first countability axiom, a sequence (x_n) has a limit point exactly when it has a convergent subsequence of (x_n), cf. Lemma 4.2.12. From Lemma 4.4.9 it follows:

Corollary 4.4.10 *A topological space X with countable topology is exactly quasi-compact when every sequence in X has a convergent subsequence in X.*

The fundamental theorem of Heine-Borel (after E. Heine (1821–1881) and E.Borel (1871–1956)) is now just a reformulation of the theorem 4.2.34 by Bolzano-Weierstraß and justifies the terminology introduced at the beginning of Sect. 3.9.

Theorem 4.4.11 (Heine-Borel Theorem) *Let V be a finite-dimensional $\mathbb{K}$-vector space with the natural topology. A subset $K \subseteq V$ is exactly compact when it is bounded and closed.*

Proof We recall that V is a Hausdorff space with countable topology. If now $K \subseteq V$ is compact, then K is closed according to Proposition 4.4.2. K is also bounded. If $h\colon V \to K$ is a $\mathbb{K}$-linear form, then $h(K) \subseteq \mathbb{K}$ is also compact and therefore bounded (since $h(K)$ is covered by finitely many of the open balls $B_{\mathbb{K}}(0; n)$, $n \in \mathbb{N}$). — If, conversely, K is bounded and closed, then every sequence in K has a convergent subsequence, whose limit lies in K, since K is closed, according to the Bolzano-Weierstraß theorem 4.2.34. According to Corollary 4.4.10, K is compact. □

According to the Heine-Borel theorem, for example, the closed balls $\overline{B}_{\|-\|_p}(x_0; r)$ and the spheres $S_{\|-\|_p}(x_0; r)$, $r \in \mathbb{R}_+$, are compact in $\mathbb{K}^n$, $n \in \mathbb{N}$, equipped with the p-norm, $1 \leq p \leq \infty$. This even applies to arbitrary (not necessarily real) norms on the $\mathbb{K}^n$:

Theorem 4.4.12 *Any two norms with values in $\mathbb{R}_+$ on a finite-dimensional $\mathbb{K}$-vector space V are equivalent. In particular, each such norm induces the natural topology on V.*

Proof It suffices to consider the case $V = \mathbb{K}^n$, $n \in \mathbb{N}^*$. Let $\|-\|$ be a norm on $\mathbb{K}^n$ with $M := \mathrm{Max}\left(\|e_1\|, \ldots, \|e_n\|\right) \in \mathbb{R}_+^\times$. We show that the norm $\|-\|$ is equivalent to the Chebyshev norm $\|-\|_\infty$. First, for an arbitrary $x = a_1 e_1 + \cdots + a_n e_n \in \mathbb{K}^n$

$$\|x\| = \|a_1 e_1 + \cdots + a_n e_n\| \leq |a_1|\|e_1\| + \cdots + |a_n|\|e_n\| \leq nM \|x\|_\infty.$$

So $\|-\| \leq nM \|-\|_\infty$. In particular, the identity $\mathrm{id}_{\mathbb{K}^n}$ is a continuous mapping $(\mathbb{K}^n, \|-\|_\infty) \to (\mathbb{K}^n, \|-\|)$ and consequently, by Corollary 4.4.4, the unit sphere $S_{\|-\|_\infty}(0; 1)$ is also compact and thus closed with respect to the topology defined by $\|-\|$. Therefore, there is a sphere $B_{\|-\|}(0; r)$, $r \in \mathbb{R}_+^\times$, with $B_{\|-\|}(0; r) \cap S_{\|-\|_\infty}(0; 1) = \emptyset$. If now $x \in \mathbb{K}^n - \{0\}$, then $x/\|x\|_\infty \in S_{\|-\|_\infty}(0; 1)$

and consequently $\|x/\|x\|_\infty\| = \|x\|/\|x\|_\infty \geq r$, i.e. $\|x\|_\infty \leq r^{-1}\|x\|$. Thus, the reverse is also true $\|-\|_\infty \leq r^{-1}\|-\|$, and $\|-\|$, $\|-\|_\infty$ are equivalent norms. □

Theorem 4.4.12 allows for topological considerations in finite-dimensional $\mathbb{K}$-vector spaces to always use a norm adapted to the problem (with values in $\mathbb{R}_+$).[14]— The following theorem generalizes Corollary 3.9.5:

Theorem 4.4.13 (Weierstrass Theorem) *Let $f : X \to \mathbb{R}$ be a continuous function on the quasi-compact space X. Then $f(X) \subseteq \mathbb{R}$ is bounded and closed. — In particular, there are points at $X \neq \emptyset$ points $x_1, x_2 \in X$ with $f(x_1) = \operatorname{Inf} f(X) = \operatorname{Min} f(X)$ and $f(x_2) = \operatorname{Sup} f(X) = \operatorname{Max} f(X)$. If X is also connected, then $f(X)$ is the compact interval $[\operatorname{Min} f(X), \operatorname{Max} f(X)]$.*

Proof According to Theorem 4.4.3, $f(X)$ is compact and then bounded and closed according to the Heine-Borel Theorem 4.4.11. The addition results from the Intermediate Value Theorem 4.3.7. □

The following theorem generally characterizes compact metric spaces by the validity of the Bolzano-Weierstrass theorem.

Theorem 4.4.14 *A metric space X is compact if and only if every sequence (x_n) in X has a limit point in X, i.e., a convergent subsequence in X. — The topology of a compact metric space has a countable basis.*

Proof According to Lemma 4.4.9, it is sufficient to show that the topology of a metric space X, in which every sequence has a limit point, has a countable base.

Let $m \in \mathbb{N}^*$. Then the open cover $B(x; 1/m)$, $x \in X$, has a finite subcover $B(x; 1/m)$, $x \in X_m$. Otherwise, there would be a sequence $y_0, y_1, \ldots$ in X with $y_{n+1} \notin \bigcup_{i=0}^{n} B(y_i; 1/m)$, and this sequence (y_n) would obviously have no limit point. The countable family $B(x; 1/m)$, $m \in \mathbb{N}^*$, $x \in X_m$, is now a base of the topology of X. Let $U \subseteq X$ be open and $y \in U$. There is an $m \in \mathbb{N}^*$ with $B(y; 1/m) \subseteq U$ and an $x \in X_{2m}$ with $y \in B(x; 1/2m) \subseteq B(y; 1/m) \subseteq U$. □

Remark 4.4.15 It should be noted that a topological space or a Hausdorff space X, in which every sequence has a limit point, is called **sequentially quasi-compact** or **sequentially compact**. Theorem 4.4.14 then states, *that a metric space is compact if and only if it is sequentially compact.* According to 4.4.9, every quasi-compact space is sequentially quasi-compact and every sequentially quasi-compact space, whose topology has a countable basis, is quasi-compact.

Quasi-compactness can always be characterized by the existence of limit points for families with filtered index sets. Let $\mathcal{F}$ be a filter on the index set I, cf. Definition

---

[14] Note that the values of a norm on the $\mathbb{K}$-vector space V already all lie in $\mathbb{R}_+$ when this is the case for the values on a $\mathbb{K}$-basis of V.

4.2.22, and x_i, $i \in I$, a family of points in the topological space X. A point $x \in X$ is called a **limit point** (or **contact point**) **of the family** x_i, $i \in I$, **with respect to**. $\mathcal{F}$, if for every neighborhood U of x and every $F \in \mathcal{F}$ there exists a $i \in F$ with $x_i \in U$, i.e. if $x \in \overline{\{x_i \mid i \in F\}}$ is for every $F \in \mathcal{F}$. For sequences, this corresponds (with respect to the Fréchet filter) exactly to the terminology introduced before Example 4.2.7. If $\mathcal{F} = \mathfrak{P}(I)$, then because of $\emptyset \in \mathcal{F}$ there is no limit point, but every point $x \in X$ is the limit of the family x_i, $i \in I$. If $\mathcal{F} \neq \mathfrak{P}(I)$, then each of these limits is also an accumulation point. It now applies: ◇

Lemma 4.4.16 *A topological space X is exactly quasi-compact, when every family x_i, $i \in I$, in X with a filter $\mathcal{F} \neq \mathfrak{P}(I)$ on the index set I has an accumulation point.*

Proof Let X be quasi-compact and x_i, $i \in I$, a family as in the lemma. Suppose it had no accumulation point. Then for each point $x \in X$ there would be an open neighborhood U_x of x and an $F_x \in \mathcal{F}$ with $U_x \cap F_x = \emptyset$. Since X is quasi-compact, there would be finitely many points $x_1, \ldots, x_n \in X$ with $U_{x_1} \cup \cdots \cup U_{x_n} = X$ and it would be $F := F_{x_1} \cap \cdots \cap F_{x_n} = \emptyset$. Since F belongs to $\mathcal{F}$, this is a contradiction to the assumption. $\mathcal{F} \neq \mathfrak{P}(I)$. — Conversely, let X satisfy the condition stated in the lemma. Let then A_j, $j \in J$, be a family of closed sets in X with $\bigcap_{j \in J} A_j = \emptyset$. Suppose it would be $A_K := \bigcap_{j \in K} A_j \neq \emptyset$ for all finite subsets K of J. Then the A_K would form the basis of a filter $\mathcal{F}$ on X, which is different from $\mathfrak{P}(X)$. So, by assumption, there would be an accumulation point $x \in X$ for this filter. Because of $A_i \in \mathcal{F}$, in particular, $x \in \overline{A_i} = A_i$ would then be for all $i \in I$. Contradiction! □ ◇

The following lemmas describe important properties of compact sets.

Lemma 4.4.17 *Let K, L be disjoint compact sets in the Hausdorff space X. Then K and L in X have disjoint neighborhoods. In particular, a compact space X is normal (i.e., disjoint closed sets in X have disjoint neighborhoods).*

Proof We first consider the case that $L = \{y\}$ is single-pointed. Since X is Hausdorff, for each $x \in K$ there are disjoint open neighborhoods U_x of x and V_x of y. Since K is compact, finitely many $U_{x_1}, \ldots, U_{x_m}$ cover all of K. Then $U_{x_1} \cup \cdots \cup U_{x_m}$ and $V_{x_1} \cap \cdots \cap V_{x_m}$ are open disjoint neighborhoods of K and y respectively. — Now let L be arbitrary (but compact). Then, according to the proven, for each point $y \in L$ there are disjoint neighborhoods U_y of K and V_y of y. Since L is compact, finitely many $V_{y_1}, \ldots, V_{y_n}$ cover all of L. Therefore, $U_{y_1} \cap \cdots \cap U_{y_n}$ and $V_{y_1} \cup \cdots \cup V_{y_n}$ are disjoint neighborhoods of K and L respectively. □

Therefore, Urysohn's separation lemma 4.2.37 can be applied to any compact spaces.

Lemma 4.4.18 (Lebesgue's Lemma (after H. Lebesgue (1875–1941))) *Let K be a compact subset of the metric space Y and U_i, $i \in I$, a cover of K by open sets in Y. Then there exists a $\lambda > 0$ such that every ball $\overline{B}(x; \lambda)$, $x \in K$, is in one of the sets U_i.*

Proof Every $x \in K$ lies in a set U_{i_x}, $i_x \in I$, of the cover. Since U_{i_x} is open, there is a ball $\overline{B}(x; r_x) \subseteq U_{i_x}$ with $r_x > 0$. Then $B(x; r_x/2)$, $x \in K$, an open cover of K, therefore has a finite subcover $B(x; r_x/2)$, $x \in E$, due to the compactness of K. We set $\lambda := \text{Min}\{r_x/2 \mid x \in E\}$. For each $y \in K$ there is then an $x \in E$ with $y \in B(x; r_x/2)$, thus $\overline{B}(y; \lambda) \subseteq B(x; r_x) \subseteq U_{i_x}$. □

A generalization of Lemma 4.4.18 can be found in exercise 4.4.9. — In generalization of 3.9.11 we define the concept of uniform continuity for mappings between any metric spaces:

Definition 4.4.19 A mapping $f : X \to Y$ of metric spaces X,Y is called **uniformly continuous**, if for every $\varepsilon > 0$ a $\delta > 0$ exists such that $d(f(x), f(y)) \leq \varepsilon$ is for all $x, y \in X$ with $d(x, y) \leq \delta$.

Every uniformly continuous mapping is naturally continuous. As a partial reversal applies:

Theorem 4.4.20 *Every continuous mapping $f : X \to Y$ of a compact metric space X into a metric space Y is uniformly continuous.*

Proof We provide a variant of the proof of theorem 3.9.12. Let $\varepsilon > 0$ be given. According to Lemma 4.4.18, there exists a $\delta > 0$ such that every sphere $\overline{B}(x; \delta)$ lies in one of the open sets $f^{-1}(B(f(x); \varepsilon/2))$. For $x, y \in X$ with $d(x, y) \leq \delta$, then $f(x), f(y) \in B(f(x_0); \varepsilon/2))$ holds for a $x_0 \in X$ and in particular $d(f(x), f(y)) \leq \varepsilon$. — One can also conclude: The preimage $(f \times f)^{-1}(\Delta_{Y,\varepsilon})$, $\Delta_{Y,\varepsilon} = \{(z, z') \in Y \times Y \mid d(z, z') < \varepsilon\}$, is a neighborhood of the diagonal $\Delta_X \subseteq X \times X$ and therefore includes according to exercise 4.4.10b) a set $\Delta_{X,\delta}$ with a $\delta > 0$. This also characterizes the uniform continuity. □

Compactness is often not given, but local compactness is.

Definition 4.4.21 A topological space is called **locally compact**, if it is Hausdorff and every point $x \in X$ has a compact neighborhood.

Finite-dimensional $\mathbb{K}$ vector spaces V with their natural topology are locally compact, but not compact when $V \neq 0$. *Every point of a locally compact space*

4.4 Compact Spaces

X even has a neighborhood basis of compact neighborhoods.[15] For the *proof*, one can assume that X is compact. If then U is an open neighborhood of $x \in X$, then according to Lemma 4.4.17 there are disjoint open neighborhoods U' of x and V of $X - U$. Then $U' \subseteq X - V \subseteq U$ and $X - V$ is a closed and therefore compact neighborhood of x. It follows: *Every open or closed subspace and thus every locally closed subspace of a locally compact space is again locally compact*, see exercise 4.4.8.

Example 4.4.22 (One-Point Compactification) To every locally compact space X, a compact space X' can be assigned in a canonical way such that X is an open subspace of X'. Thus, the locally compact spaces are precisely the open subspaces of the compact spaces. Let $X = (X, \mathcal{T})$ be locally compact and $X' := X \uplus \{\omega\}$ with a point ω, which does not belong to X. Then

$$\mathcal{T}' := \mathcal{T} \cup \{X' - K \mid K \subseteq X \text{ compact}\}$$

is obviously a topology on X', with respect to which X is an open subspace of X', whose topology coincides with the one given on X. X' is compact. *Proof.* The covering property of compactness is trivial.[16] X is Hausdorff: Two points $x, y \in X$ can be separated by disjoint neighborhoods even in X according to the assumption. But if $x \in X$ and K is a compact neighborhood of x in X, then K and $X' - K$ are neighborhoods of x and ω, which separate these points. □

$X' = X \uplus \{\omega\}$ is called the **one-point-** or **Alexandroff compactification** of the locally compact space X. ω is often also called the **point at infinity**. If X is already compact, then ω is a discrete point of X'. *The one-point compactification* $\mathbb{R}^n \uplus \{\omega\}$ *of* $\mathbb{R}^n$, $n \in \mathbb{N}$, *is homeomorphic to the sphere* $S^n = \{x \in \mathbb{R}^{n+1} \mid \|x\|_2 = 1\}$ (which is a priori compact according to the Heine-Borel theorem). A concrete homeomorphism is given, for example, by the **stereographic projection**, see Fig. 4.19. Each point $N := (0, 1) \in S^n \subseteq \mathbb{R}^{n+1} = \mathbb{R}^n \times \mathbb{R}$ different from the "North Pole" $(x, y) \in S^n$, $x \in \mathbb{R}^n, y \in \mathbb{R}, \|x\|_2^2 + y^2 = 1$, is associated with the intersection of the line through N and (x, y) with the $\mathbb{R}^n = \mathbb{R}^n \times \{0\} \subseteq \mathbb{R}^{n+1}$, i.e., $(x, y) \mapsto x/(1 - y)$. The inverse mapping is the mapping $z \mapsto \left(2z/(\|z\|_2^2 + 1), (\|z\|_2^2 - 1)/(\|z\|_2^2 + 1)\right)$ from $\mathbb{R}^n$ to $S^n - \{N\}$. If you apply this homeomorphism $S^n - \{N\} \longrightarrow \mathbb{R}^n$ using $N \mapsto \omega$, so one obtains the desired homeomorphism $S^n \xrightarrow{\sim} \mathbb{R}^n \uplus \{\omega\}$. Proof! (It is sufficient to show continuity.) ◇

A topological space X is called σ-**compact**, if it is locally compact and the point ω of the one-point compactification $X' = X \uplus \{\omega\}$ of X has a countable

---
[15] For **local quasi-compactness**, it is required that every point $x \in X$ has a neighborhood basis of quasi-compact neighborhoods.
[16] This also applies when X is (Hausdorff but) not necessarily locally compact; X' is always quasicompact.

Fig. 4.19 Stereographic projection of the sphere S^n from the "North Pole" N

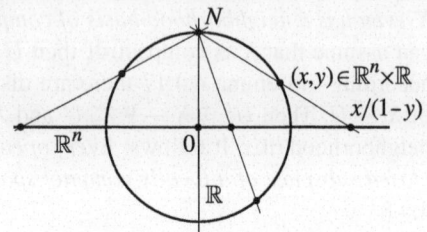

neighborhood basis. A locally compact and σ-compact topological space is also called **countably infinite**.

Proposition 4.4.23 *For a locally compact space X the following statements are equivalent:*

(i) *X is σ-compact.*
(ii) *X has a countable cover with compact sets.*
(iii) *There is a sequence K_k, $k \in \mathbb{N}$, of compact subsets $K_k \subseteq X$ with*

$$K_k \subseteq \mathring{K}_{k+1}, \ k \in \mathbb{N}, \ \text{and} \ \bigcup_{k \in \mathbb{N}} \mathring{K}_k = X.$$

Proof (i) ⇒ (ii): If U_n, $n \in \mathbb{N}$, is a countable neighborhood basis of ω with open sets U_n, then the complements $X' - U_k$, $k \in \mathbb{N}$, a cover of X with compact sets; because it is $\bigcap_{k \in \mathbb{N}} U_k = \{\infty\}$. — (ii) ⇒ (iii): Let L_k, $k \in \mathbb{N}$, be a cover of X with compact sets $L_k \subseteq X$. Then one defines the sequence K_k, $k \in \mathbb{N}$, with $L_0 \cup \cdots \cup L_n \subseteq \mathring{K}_n$ recursively. For K_0 one chooses a compact neighborhood of L_0. If the compact sets $K_0, \ldots, K_n$ with $K_k \subseteq \mathring{K}_{k+1}$, $k < n$, and $L_0 \cup \cdots \cup L_n \subseteq \mathring{K}_n$ are already defined, then one chooses for K_{n+1} a compact neighborhood of $K_n \cup L_{n+1}$. — (iii) ⇒ (i): If K_k, $k \in \mathbb{N}$, is a sequence of compact subsets of K as in (iii), then it is $(X - K_k) \uplus \{\omega\}$, $k \in K$, a neighborhood base of ω. For any arbitrary compact subset $K \subseteq X = \bigcup_{k \in \mathbb{N}} \mathring{K}_k$ there is indeed a $k_0 \in \mathbb{N}$ with $K \subseteq \mathring{K}_{k_0} \subseteq K_{k_0}$ and consequently $X' - K_{k_0} \subseteq X' - K$. □

Example 4.4.24 (Compactly Generated Topologies) Let X be a Hausdorff space. If A is a closed subset of X, then for every compact set $K \subseteq X$ the intersection $A \cap K$ is also closed and therefore compact. The topology of X is called **compactly generated**, if this property characterizes the closed sets in X, i.e., a subset $A \subseteq X$ is closed if and only if $A \cap K$ is compact for every compact set $K \subseteq X$. *The topology of a locally compact space is compactly generated*, cf. exercise 4.2.16a). Similarly, the topology of any Hausdorff space whose topology satisfies the first countability axiom is compactly generated. This follows from Lemma 4.2.11 (3) and the following trivial statement: If $(x_n)_{n \in \mathbb{N}}$ is a convergent sequence in the Hausdorff space X with limit x, then $\{x\} \cup \{x_n \mid n \in \mathbb{N}\} \subseteq X$ is compact.

4.4 Compact Spaces

Every Hausdorff space X can naturally be associated with a space with a compactly generated topology. To do this, one defines another, generally strictly finer, topology on X by the condition that a set $A \subseteq X$ should be closed if and only if $A \cap K$ is closed for every compact set $K \subseteq X$ with respect to the given topology. No additional *compact* sets are added, so the new topology is indeed compactly generated.

Important examples of spaces with compactly generated topology are the $\mathbb{K}$-vector spaces with their natural topology. The following proposition provides more details:

Proposition 4.4.25 *Let V be a $\mathbb{K}$-vector space with its natural topology. A subset is exactly compact $K \subseteq V$ if it is entirely within a finite-dimensional subspace $V' \subseteq V$ and is compact there. In particular, the topology of V is compactly generated.*

Proof The addition results from the fact that the finite-dimensional $\mathbb{K}$ vector spaces are compactly generated as locally compact spaces. — Now let $K \subseteq V$ be compact. We can assume that V is generated by K as a $\mathbb{K}$ vector space, and then have to show that V is finite-dimensional. Let $x_i \in K$, $i \in I$, be a $\mathbb{K}$ basis of V, cf. theorem 2.8.18. For each $i_0 \in I$, $B_\mathbb{K}(1;1) x_{i_0} + \sum_{i \neq i_0} B_\mathbb{K}(0;1) x_i$ is an (open) neighborhood of x_{i_0}, which does not contain any x_i, $i \neq i_0$. Since the set B of x_i, $i \in I$, (like any linearly independent set in a $\mathbb{K}$ vector space with respect to the natural topology) is also closed, B is a compact discrete subset of K and therefore finite $\square$ ◊

Example 4.4.26 (Proper Mappings) An important class of closed mappings of topological spaces are the proper mappings:

Definition 4.4.27 A continuous mapping $f: X \to Y$ of topological spaces is called **proper**, if f is closed and all fibers of f are quasi-compact.

For $X \neq \emptyset$ a constant mapping $X \to \{P\} \subseteq Y$ is therefore proper exactly when X is quasi-compact and $\{P\}$ is closed in Y. If $X' \subseteq X$ is a subspace of X, then the canonical inclusion $X' \hookrightarrow X$ is proper exactly when X' is closed in X, $X' \hookrightarrow X$ is therefore a closed embedding. — *If X is quasi-compact, then every continuous mapping $f: X \to Y$ into a Hausdorff space Y is proper.* Since the points in Y are closed, the fibers of f are therefore closed and consequently quasi-compact. If furthermore $A \subseteq X$ is closed, then A is quasi-compact and thus the image $f(A) \subseteq Y$ is compact, thus closed in Y, since Y is Hausdorff. — *If $f: X \to Y$ is proper and $Y' \subseteq Y$ as well as $X' := f^{-1}(Y')$, then the restriction $f|X': X' \to Y'$ is also proper.* If $A' = A \cap X'$ is a closed subset of X' with a closed subset $A \subseteq X$, then $f(A') = f(A) \cap Y'$ is closed in Y', since f is proper. — Conversely, if there is an open cover V_i of Y with $U_i := f^{-1}(V_i)$, $i \in I$, such that all restrictions $f|U_i: U_i \to V_i$, $i \in I$, are proper, then clearly f is also proper.

Not only the fibers, but the preimages of any quasi-compact sets are quasi-compact with respect to proper mappings:

Theorem 4.4.28 *Let $f: X \to Y$ be a proper mapping of topological spaces and let $L \subseteq Y$ be quasi-compact. Then the preimage $f^{-1}(L)$ is also quasi-compact.*

Proof By restricting f to the pre-image $f^{-1}(L)$, it suffices to show that X is quasi-compact if Y is quasi-compact. Let K_i, $i \in I$, be a family of closed sets in X such that every finite subfamily has a non-empty intersection. It must be shown that $\bigcap_{i \in I} K_i \neq \emptyset$ is. $f(K_i), i \in I$, is a family of closed sets in Y with the property that every finite subfamily has a non-empty intersection. Therefore, $\bigcap_{i \in I} f(K_i) \neq \emptyset$ is, since Y is quasi-compact. Let y be an element of this intersection. Then $f^{-1}(y) \cap K_i \neq \emptyset$ for each i and thus also $f^{-1}(y) \cap \bigcap_{i \in I} K_i \neq \emptyset$, since $f^{-1}(y)$ is quasi-compact. In particular, $\bigcap_{i \in I} K_i \neq \emptyset$ is. □

Corollary 4.4.29 *If $f: X \to Y$ and $g: Y \to Z$ are proper mappings, then the composition $g \circ f: X \to Z$ is also proper.*

For the next corollary, which we declare as a theorem, we recall the definition of a compactly generated topology: The topology of a Hausdorff space X is compactly generated if a subset $A \subseteq X$ is closed in X if and only if $A \cap K$ is closed (i.e., compact) for every compact set $K \subseteq X$, see example 4.4.24.

Theorem 4.4.30 *Let $f: X \to Y$ be a continuous mapping of topological spaces. Y is Hausdorff with compactly generated topology. Exactly then is f proper, when the f-preimages of any compact subsets of Y are quasi-compact in X.*

Proof The given condition is necessary according to theorem 4.4.28 for f to be proper. — Conversely, let the f-preimages of compact sets be quasi-compact. It remains to show that f is closed. But if $A \subseteq X$ is closed and $L \subseteq Y$ is compact, then $f(A) \cap L = f\bigl(A \cap f^{-1}(L)\bigr)$ as the image of the quasi-compact set $A \cap f^{-1}(L)$ is compact and $f(A)$ is therefore closed in Y, since the topology of Y is compactly generated. □

The requirement that the topology of Y is compactly generated is obviously necessary for the validity of theorem 4.4.30.

Proper mappings can be characterized as the universally closed mappings. Here, a continuous mapping $f: X \to Y$ of topological spaces is called **universally closed**, if for every topological space Z the product mapping $\mathrm{id}_Z \times f: Z \times X \to Z \times Y$, $(z, x) \mapsto (z, f(x))$, is closed. Then for every continuous mapping $g: Z \to Y$ the pullback mapping $f_{(Z)}: X_{(Z)} \to Z$ is also closed (see the end of example 4.2.27). Proof!

Theorem 4.4.31 *A continuous mapping $f: X \to Y$ of topological spaces is proper if and only if it is universally closed.*

4.4 Compact Spaces

Proof Let f be proper and Z a topological space. Furthermore, let $A \subseteq Z \times X$ be closed. It is to be shown that the image $B := \{(z, f(x)) \mid (z, x) \in A\}$ of A under $\widetilde{f} := \mathrm{id}_Z \times f$ is closed in $Z \times Y$. Let $(z_0, y_0) \notin B$ be given. Then $\widetilde{f}^{-1}(z_0, y_0) = \{z_0\} \times f^{-1}(y_0)$ is quasi-compact, since f is proper. Since $\big(\{z_0\} \times f^{-1}(y_0)\big) \cap A = \emptyset$ is given and since A is closed, there are, according to the Tube Lemma from Exercise 4.4.7, open neighborhoods W of z_0 in Z and U of $f^{-1}(y_0)$ in X with $(W \times U) \cap A = \emptyset$. Then $f(X - U)$ is closed in Y, since f is closed, and the open set $W \times \big(Y - f(X - U)\big)$ in $Z \times Y$ is disjoint from B. If $(z, f(x))$ with $(z, x) \in A$ were an element of $W \times \big(Y - f(X - U)\big)$, then $x \in U$ would be, which contradicts $(W \times U) \cap A = \emptyset$. Therefore, B is closed in $Z \times Y$.

If, conversely, f is universally closed. Then f is particularly closed. To show that the fibers of f are quasi-compact, we can assume that Y is single-pointed, and then have to show that X is quasi-compact, because according to the assumption, for every point $y_0 \in Y$ the projection $Z \times f^{-1}(y_0) \to Z$ as a pullback of the constant mapping $Z \to \{y_0\} \subseteq Y$ is closed. Now let A_i, $i \in I$, be a family of closed sets in X such that for each finite subset $J \subseteq I$ the intersection $\bigcap_{j \in J} A_j \neq \emptyset$ is. We have to show that $\bigcap_{i \in I} A_i \neq \emptyset$ is. Let $\mathcal{F}$ be the filter on X generated by the A_i and $Z := X \uplus \{\infty\}$ be the topological space associated with $\mathcal{F}$, whose open sets, in addition to the subsets of X, are the sets $F \uplus \{\infty\}$, $F \in \mathcal{F}$, see exercise 4.2.32. Then $X \subseteq Z$ is an open discrete subspace and ∞ is a contact point of X, as every neighborhood of ∞ intersects the subspace X because $F \neq \emptyset$ for all $F \in \mathcal{F}$. By assumption, the projection $p: Z \times X \to Z$ is closed. The closed hull A of the diagonal $\Delta_X \subseteq X \times X \subseteq Z \times X$ in $Z \times X$ thus meets the fiber $\{\infty\} \times X$, since the p-image of Δ_X is the subspace $X \subseteq Z$. Let (∞, x_0) be an element of A. For each A_i then $x_0 \in A_i$. Otherwise, $(A_i \uplus \{\infty\}) \times (X - A_i)$ would be a neighborhood of (∞, x_0) and $\big((A_i \uplus \{\infty\}) \times (X - A_i)\big) \cap \Delta_X = \emptyset$, thus (∞, x_0) would not be a point of contact of Δ_X. Therefore, $x_0 \in \bigcap_{i \in I} A_i \neq \emptyset$. □

Corollary 4.4.32 *Let $f_i: X_i \to Y_i$, $i \in I$, be a finite family of proper mappings of topological spaces. Then the product mapping $\prod_{i \in I} f_i: \prod_{i \in I} X_i \to \prod_{i \in I} Y_i$, $(x_i) \mapsto (f_i(x_i))$, is also proper.*

Proof To prove by induction over $|I|$ one can assume that $I = \{1, 2\}$ is. Then $f_1 \times f_2$ is the composition of the two proper mappings $\mathrm{id}_{X_1} \times f_2$ and $f_1 \times \mathrm{id}_{X_2}$ and thus also proper according to Corollary 4.4.29. □

We note without proof that Corollary 4.4.32 also applies to infinite families $f_i: X_i \to Y_i$, $i \in I$, of proper mappings of topological spaces. This is a generalization of Theorem 4.4.8 by Tychonoff. ◊

Exercises

Exercise 4.4.1 Let X be a topological space. A union of finitely many quasi-compact subsets of X is again quasi-compact.

Exercise 4.4.2 Let (x_n) be a convergent sequence in the topological space X with the limit x. Then the set $\{x_n \mid n \in \mathbb{N}\} \cup \{x\}$ is quasi-compact.

Exercise 4.4.3 Let A and B be compact subsets in a normed $\mathbb{K}$ vector space V. Then the Minkowski sum $A + B \subseteq V$ is also compact.

Exercise 4.4.4 Let V be a $\mathbb{R}$ vector space with the natural topology. The convex hull $\widehat{A}$ of a subset $A \subseteq V$ of V is by definition the smallest convex subset of V that includes A. ($\widehat{A}$ is the intersection of all convex sets in V that include A.)

a) Prove the so-called **Carathéodory's Lemma**: If $\mathrm{Dim}_{\mathbb{R}} V = n \in \mathbb{N}$ and $A \subseteq V$, then $\widehat{A}$ is the image under the (continuous) mapping

$$\Delta_n \times A^{n+1} \longrightarrow V, \quad ((a_0,\ldots,a_n),(x_0,\ldots,x_n)) \longmapsto \sum_{i=0}^{n} a_i x_i,$$

where $\Delta_n := \{(a_0,\ldots,a_n) \mid a_i \in \mathbb{R}_+, \sum_{i=0}^{n} a_i = 1\} \subseteq \mathbb{R}_+^{n+1}$ is the so-called (convex) **standard-n-simplex**. (It is to be shown that the image is convex. — If V is a $\mathbb{R}$-vector space and if $x_0,\ldots,x_n \in V$, then a linear combination of the form $a_0 x_0 + \cdots + a_n x_n$ with $(a_0,\ldots,a_n) \in \Delta_n$ is called a **convex linear combination** of the $x_0,\ldots,x_n$.)

b) If V is finite-dimensional and $K \subseteq V$ is compact, then $\widehat{K}$ is also compact

Exercise 4.4.5 Let V be a normalized $\mathbb{K}$-vector space.

a) If $V_{<\infty} = \mathrm{B}_V(0;\infty)$ is not finite-dimensional, then the unit sphere $\mathrm{S}(0;1)$ in V is not compact.
b) V is locally compact if and only if $V_{<\infty}$ is finite-dimensional.

Exercise 4.4.6 Let X be a Hausdorff space.

a) X is compact if and only if the following condition is met: For every family U_x, $x \in X$, where U_x is a neighborhood of x, there are finitely many points $x_1,\ldots,x_n \in X$ such that the neighborhoods $U_{x_1},\ldots,U_{x_n}$ already cover all of X.
b) If X is locally compact, then every compact subset of X has a compact neighborhood in X.

Exercise 4.4.7 (Tube Lemma) Let K and L be quasi-compact subsets of the topological spaces X and Y respectively. Furthermore, let W be a neighborhood of $K \times L$ in $X \times Y$. Then there are open sets U and V in X resp. Y with $K \times L \subseteq U \times V \subseteq W$. (First consider the case where L contains only one point. — The Tube Lemma implies Lemma 4.4.17. If $K, L \subseteq X$ are disjoint compact sets, then $K \times L \subseteq W := (X \times X) - \Delta_X$.)

4.4 Compact Spaces

Exercise 4.4.8 Let X be a locally compact topological space. A subspace $Y \subseteq X$ is locally compact if and only if Y is locally closed in X, cf. exercise 4.2.16b).

Exercise 4.4.9 (Variant of Lebesgue's Lemma) Let $f : X \to Y$ be a continuous mapping of the compact metric space X into the topological space Y. Furthermore, let U_i, $i \in I$, be an open cover of $f(X)$. Then there exists a $\lambda > 0$ such that for every $x \in X$ the image of $\overline{B}(x; \lambda)$ lies in one of the sets U_i, $i \in I$.

Exercise 4.4.10 Let $X = (X, d)$ be a metric space.

a) If $K \subseteq X$ is compact, then the open ε-tubes $B(K; \varepsilon) = \bigcup_{x \in K} B(x; \varepsilon)$, $\varepsilon > 0$, form a neighborhood base of K in X. (If U is an open neighborhood of K, consider the continuous function $x \mapsto d(X - U, x)$ on X.)
b) If X is compact, then the sets $\Delta_{X,\varepsilon} = \{(x, y) \in X \times X \mid d(x, y) < \varepsilon\}$, $\varepsilon > 0$, form a neighborhood base of the diagonal $\Delta_X \subseteq X \times X$.

Exercise 4.4.11 $X = (X, d)$ is a metric space. For subsets $A, B \subseteq X$ we adopt the concept of distance from exercise 4.2.27. Furthermore, let

$$\|A\| := \operatorname{Sup}\{d(x, y) \mid x, y \in A\} \; \left(\in \overline{\mathbb{R}}_+\right)$$

denote the so-called **diameter** of A. Exactly then is A is bounded (with respect to d), if $\|A\| < \infty$ is. In the following, the subsets $A, B \subseteq X$ are non-empty and A is compact.

a) $A \cap \overline{B} = \emptyset$ holds if and only if $d(A, B) > 0$ is. In addition, there is an $x \in A$ with $d(x, B) = d(A, B)$. If B is also compact, then there is $x \in A$ and $y \in B$ with $d(x, y) = d(A, B)$.
b) There is an $x, y \in A$ with $d(x, y) = \|A\|$. If the values of d are all finite, then $\|A\| < \infty$. In particular, the metric d is bounded if X is compact.

Exercise 4.4.12 Let $f : [a, b] \to \mathbb{R}$, $a, b \in \mathbb{R}$, $a < b$, be a function with the graph $\Gamma(f) \subseteq [a, b] \times \mathbb{R}$. The following statements are equivalent: (i) f is continuous. (ii) $\Gamma(f)$ is compact. (iii) $\Gamma(f)$ is closed and connected. (iv) $\Gamma(f)$ is closed and f satisfies the intermediate value theorem. (v) $\Gamma(f)$ is path-connected. (When concluding from (iii) to (iv), show with the help of exercise 4.4.7 that for every subinterval $[\alpha, \beta]$ of $[a, b]$ the graph of the restriction $f|[\alpha, \beta]$ is closed and connected.)

Exercise 4.4.13 Which of the following sets are compact (each with respect to the natural topology)?

a) $\{(x, y, z) \in \mathbb{R}^3 \mid x^n + y^n + z^n \leq 1\}$, $n \in \mathbb{N}^*$.
b) $\{(x, y) \in \mathbb{R}^2 \mid 0 \leq 3x^2 + 3y^2 - 2xy \leq 8\}$.
c) $\{(x, y) \in \mathbb{R}^2 \mid 0 \leq x^2 + y^2 - 6xy \leq 8\}$.

Exercise 4.4.14 A compact discrete space only has a finite number of points. A discrete *and closed* subset of a compact space is finite.

Exercise 4.4.15 Let X be a Hausdorff space. A subset $M \subseteq X$ is called **relatively compact**, if the closure $\overline{M}$ of M in X is compact.

a) $M \subseteq X$ is relatively compact if and only if M is contained in a compact subspace of X.
b) If X satisfies the first countability axiom, then $M \subseteq X$ is relatively compact if and only if every sequence (x_n) of elements in M has a limit point in X.
c) In a finite-dimensional $\mathbb{K}$-vector space with the natural topology, a subset is relatively compact if and only if it is bounded.

Exercisse 4.4.16 In a compact space, a sequence x_n, $n \in \mathbb{N}$, is convergent if and only if it has at most one limit point.

Exercise 4.4.17 A compact space X' is for every point $\omega \in X'$ the one-point compactification of $X := X' - \{\omega\}$. (Cf. Example 4.4.22.)

Exercise 4.4.18 Let X be a compact metric space $\neq \emptyset$ and $f: X \to X$ a contracting mapping of X into itself (i.e., $d(f(x), f(y)) < d(x, y)$ holds for all $x, y \in X$, $x \neq y$). Then f has exactly one fixed point x. For every initial value $x_0 \in X$, the sequence $x_n = f^n(x_0)$, $n \in \mathbb{N}$, converges to the fixed point x of f. (Cf. exercise 3.9.6.)

Exercise 4.4.19 Let $\overline{B}^n = \overline{B}_{\|-\|_2}(0; 1) \subseteq \mathbb{R}^n$ be the closed standard ball in $\mathbb{R}^n$ with boundary S^{n-1}, $n \in \mathbb{N}^*$.

a) By blowing up the edge $S^{n-1} \subseteq \overline{B}^n$ of $\overline{B}^n$ results in a compact space $\overline{B}^n/S^{n-1}$ that is homeomorphic to the sphere S^n, see Fig. 4.20.
b) By blowing together the base circle $S^1 \times \{0\}$ of the cylinder $S^1 \times [0, 1]$, the space $(S^1 \times [0, 1])/(S^1 \times \{0\})$ is created. This space is compact and homeomorphic to the cone $\{(x, y, z) \in \mathbb{R}^3 \mid z \in [0, 1], x^2 + y^2 = z^2\}$, see Fig. 4.21. This in turn is homeomorphic to the circular disc $\overline{B}^2$. What space is obtained when the union $(S^1 \times \{0\}) \uplus (S^1 \times \{1\})$ of the base and top circle is blown together to a point? (**Note** In general, for a topological space X, the space created by blowing together the base surface $X \times \{0\}$ of the cylinder $X \times [0, 1]$ over X is called the **cone over** X. If you also blow the top surface $X \times \{1\}$ together to a point, you get the so-called **suspension** of X, see Fig. 4.22.)
c) What spaces are obtained by blowing up finitely many points of the spheres S^1 or S^2, cf. Fig. 4.23? What space is created by blowing up the equator of the 2-sphere S^2? Consider further similar examples, such as the blowing up of finitely many points of the circular disc $\overline{B}^2$ (which can lie inside or on the edge).

4.4 Compact Spaces

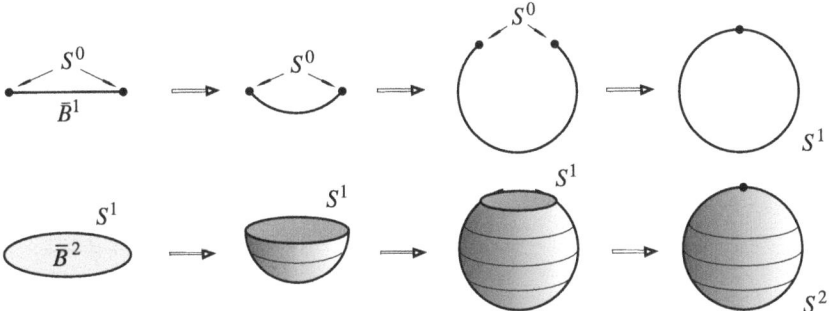

Fig. 4.20 Construction of the sphere S^n by blowing up the edge of $\overline{B}^n$

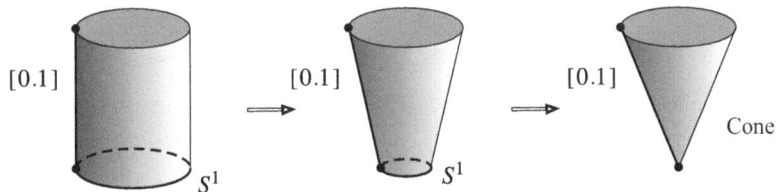

Fig. 4.21 Construction of a cone by blowing together the base surface of a cylinder

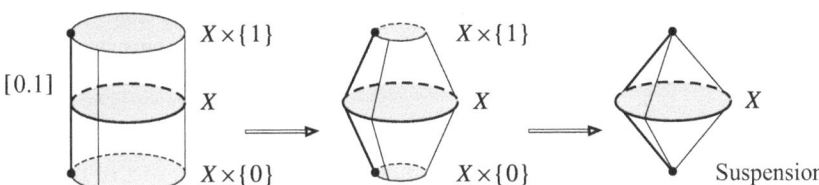

Fig. 4.22 Construction of the suspension of X from a cylinder over X

Exercise 4.4.20

a) Let $I \subseteq \mathbb{R}$ be the interval $[-1, 1]$ and Q the square $I^2 \subseteq \mathbb{R}^2$. If one identifies in Q the edges $I \times \{-1\}$ and $I \times \{1\}$ in the same direction, i.e. the points $(a, -1)$ and $(a, 1)$, $a \in I$, one obtains as quotient space (up to homeomorphism) the (compact) cylinder $I \times S^1$. If one also identifies the edges $\{-1\} \times I$ and $\{1\} \times I$ in the same direction, one obtains (up to homeomorphism) the torus $S^1 \times S^1$. But if one identifies the edges $I \times \{-1\}$ and $I \times \{1\}$ are in the same direction and the edges $\{-1\} \times I$ and $\{1\} \times I$ are in opposite directions, i.e. next to $(a, -1)$ and $(a, 1)$, $a \in I$, the points $(-1, b)$ and $(1, -b)$, $b \in I$, one obtains the (also

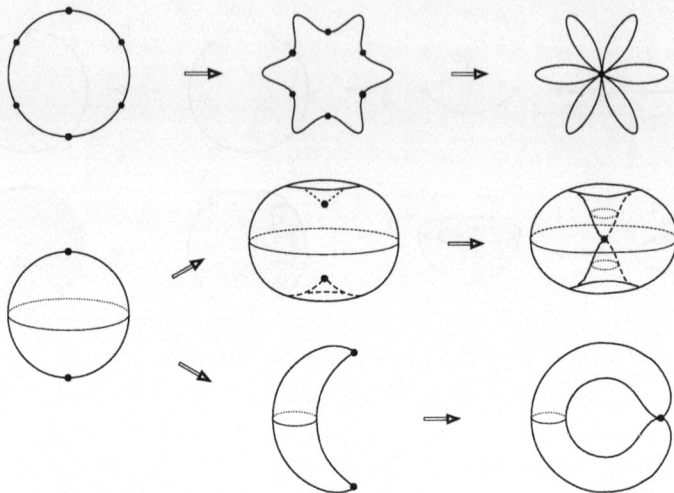

Fig. 4.23 Identification of points of a 1- or 2-sphere

compact) so-called **Klein bottle**, see Fig. 4.24.[17] Of course, one can also first identify the edges $\{-1\} \times I$ and $\{1\} \times I$ in opposite directions in Q. Then one obtains from Q the so-called (**compact**) **Möbius strip**. If one then identifies the edges $I \times \{-1\}$ or $I \times \{1\}$ in the same direction, one obtains the Klein bottle again from the Möbius strip, see Fig. 4.25. If one identifies the edges $I \times \{-1\}$ or $I \times \{1\}$ also in opposite directions, one obtains from the Möbius strip the so-called **projective plane** $\mathbb{P}^2(\mathbb{R})$. Fig. 4.25 may give an idea of these identification processes.

b) Let Z be the annulus $\{z \in \mathbb{C} \mid 1 \leq |z| \leq 2\}$, which is homeomorphic to the compact cylinder $I \times S^1$. If one identifies antipodal points $z, -z$ of the inner circle $\{|z| = 1\}$, one obtains again a Möbius strip, which in this representation is called a **cross cap**. If one also identifies antipodal points of the outer circle $\{|z| = 2\}$, one obtains a Klein bottle, cf. a).

Exercise 4.4.21 Let $A_1, \ldots, A_n$ be finitely many pairwise disjoint compact sets of the Hausdorff space X or pairwise disjoint closed sets of a normal space X. Then the space $\overline{X}$, which is obtained by identifying the points of the sets $A_1, \ldots, A_n$, is Hausdorff.

---

[17] The Klein bottle cannot be realized as a subspace of the perceptual space. The sketch in Fig. 4.24 uses self-penetrations and can only provide a vague idea. The same applies to the projective plane $\mathbb{P}^2(\mathbb{R})$, which will be introduced further below.

4.4 Compact Spaces

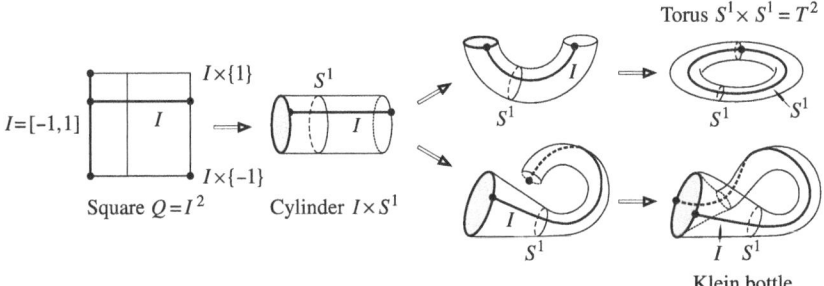

Fig. 4.24 Construction of torus and Klein bottle from a square

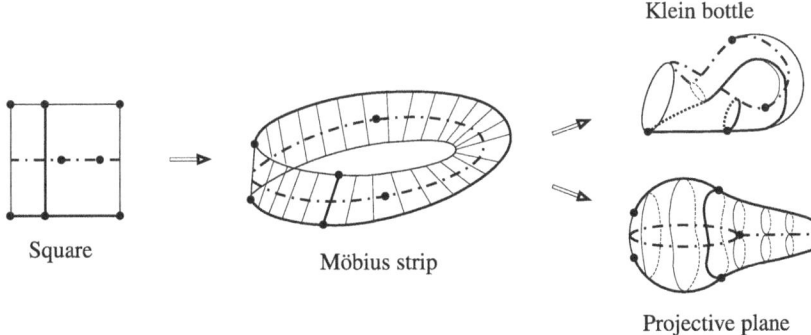

Fig. 4.25 Construction of Klein bottle and projective plane from the Möbius strip

Exercise 4.4.22 Let $R \subseteq X \times X$ be an equivalence relation on the compact space X. The quotient space X / R is Hausdorff, i.e., compact, if and only if R is closed in $X \times X$, cf. exercise 4.2.9c).

Exercise 4.4.23 Let $K_i, i \in I$, be a family of closed connected subsets of the compact space X, which contains the intersection of any finite subfamily. Then $K := \bigcap_i K_i$ is also connected. (It is $K \neq \emptyset$. If $K = L \uplus M$ with disjoint non-empty compact sets L and M, there would be disjoint open sets U, V in X with $L \subseteq U$ and $M \subseteq V$. Then $K_i \subseteq U \uplus V$ (and $K_i \cap U \neq \emptyset \neq K_i \cap V$, $K_i \cap U \cap V = \emptyset$) for suitable $i \in I$, otherwise $\bigcap_i (K_i \cap (X - U \cup V)) \neq \emptyset$ would be.) In particular, the intersection $\bigcap_{n=0}^{\infty} K_n$ of a descending sequence $K_0 \supseteq K_1 \supseteq K_2 \supseteq \cdots$ of compact connected subsets K_n of a Hausdorff space X is also connected (and compact).

Exercise 4.4.24 Let X_i, $i \in I$, be a family of Hausdorff spaces and $A \subseteq X := \prod_{i \in I} X_i$. A is relatively compact in X if and only if for every $i \in I$ the projection $p_i(A) \subseteq X_i$ is relatively compact in X_i. A is compact if and only if A is closed in X and $p_i(A)$ is compact in X_i, $i \in I$.

Exercise 4.4.25 Let X be a compact space and $x \in X$. The connected component of x in X is the intersection of the borderless environments K_i, $i \in I$, of x. (It is to be shown that this intersection $K := \bigcap_i K_i$ is connected. As in exercise 4.4.23 one introduces a representation $K = L \uplus M$, $x \in L$, to a contradiction $K_i \subseteq U \uplus V$, since then $K_i \cap U$ is an open and as complement of $K_i \cap V$ in K_i also closed environment of x.) If X is also totally disconnected, then the borderless sets in X form a basis of the topology of X. (A compact, totally disconnected topological space is also called a **boolean space**, cf. exercise 4.4.30.)

Exercise 4.4.26 Let X be a compact space. The space $\overline{X}$ of the connected components of X (equipped with the quotient topology) is a boolean space. (Cf. exercise 4.3.16.)

Exercise 4.4.27 $\mathcal{C} \subseteq [0, 1]$ is the Cantor discontinuum from exercise 3.4.18.

a) $\mathcal{C}$ is a Boolean space with countable topology and without discrete points. The mapping $\chi : \{0, 1\}^{\mathbb{N}^*} \xrightarrow{\sim} \mathcal{C}$, $(a_n) \mapsto \sum_{n=1}^{\infty} 2a_n/3^n$, is a homeomorphism.
b) Every Boolean space Y with countable topology and without discrete points is homeomorphic to $\{0, 1\}^{\mathbb{N}^*}$ and thus to $\mathcal{C}$. (Every such space Y is called a **Cantor's discontinuum** (or also a **Cantor space**). — For the proof, note that according to exercise 4.4.25 the borderless subsets $K \subseteq Y$ form a basis of the topology of Y. Thus, there is a sequence $K_1, K_2, \ldots$ of borderless sets that form a basis of the topology of Y. (Y has only countably many borderless sets.) With this sequence, construct non-empty borderless sets $Y(i_1, \ldots, i_n), n \in \mathbb{N}, i_1, \ldots, i_n \in \{0, 1\}$, with the following properties: (1) $Y(\emptyset) = Y$. (2) $Y(i_1, \ldots, i_n) = Y(i_1, \ldots, i_n, 0) \uplus Y(i_1, \ldots, i_n, 1)$. (3) $Y(i_1, \ldots, i_n, 0) = K_{n+1} \cap Y(i_1, \ldots, i_n)$, if K_{n+1} neither includes $Y(i_1, \ldots, i_n)$ nor is disjoint from it. — For every point $y \in Y$ there is then a uniquely determined 0–1 sequence i_n, $n \in \mathbb{N}^*$, with $y \in Y(i_1, \ldots, i_n)$ for all $n \in \mathbb{N}^*$, and the mapping $y \mapsto (i_n)$, $n \in \mathbb{N}^*$, is a homeomorphism from Y to $\{0, 1\}^{\mathbb{N}^*}$.)

Remark According to b), all products $\prod_{n=1}^{\infty} X_n$ are Cantor discontinua, where the X_i are finite discrete spaces with more than one point. Likewise, products of countable non-empty families of Cantor discontinua are again Cantor discontinua. As an application, we give the following construction of **Peano curves**. These are by definition surjective paths $[0, 1] \to [0, 1]^n$, $n \in \mathbb{N}^*$, $n \geq 2$. They were first given by Peano. For this purpose, let $f : \mathcal{C} \to [0, 1]$ be the surjective continuous mapping from exercise 3.8.16 with $f \circ \chi : \{0, 1\}^{\mathbb{N}^*} \to [0, 1]$, $(a_n) \mapsto \sum_{n=1}^{\infty} a_n/2^n$. For each $n \in \mathbb{N}^*$ is then also $f_n := f \times \cdots \times f : \mathcal{C}^n \to [0, 1]^n$ is surjective and continuous. Furthermore, there is a homeomorphism $g : \mathcal{C} \xrightarrow{\sim} \mathcal{C}^n$. Then $h := f_n \circ g : \mathcal{C} \to [0, 1]^n$ is a continuous surjective mapping that can be extended to a Peano curve $[0, 1] \to [0, 1]^n$. For this, one does not need to use the extension theorem 4.2.36 by Tietze-Urysohn. If $]a, b[$ is a connected component of $[0, 1] - \mathcal{C}$, one simply interpolates linearly on $]a, b[$: $h(sa + tb) := sh(a) + th(b)$, $s, t \in \mathbb{R}_+^{\times}$, $s + t = 1$. Since $\mathcal{C}^{\mathbb{N}^*}$ is also a Cantor

4.4 Compact Spaces

discontinuum, there are also continuous surjective paths $[0, 1] \to [0, 1]^{\mathbb{N}^*}$ that fill the so-called **Hilbert cube** $[0, 1]^{\mathbb{N}^*}$.

Exercise 4.4.28 A topological space X is called **Noetherian**, if the set of open sets of X (with respect to inclusion) is Noetherian ordered or — equivalently — if the set of closed sets of X is Artinian ordered.

a) X is Noetherian if and only if every open subspace of X is quasi-compact.
b) If X is Noetherian, then *every* subspace $X' \subseteq X$ is Noetherian.
c) A Noetherian Hausdorff space X is finite. (Every subspace of X is compact.)
d) A Noetherian space X has only finitely many connected components and is locally connected. (One concludes by Artinian induction over the closed subsets of X.)

(For examples, see exercise 4.4.29f).)

Exercise 4.4.29 (Zariski Topologies) Let A be a commutative ring. We recall the notations Spec A for the set of prime ideals in A and Spm A for the set of maximal ideals of A, cf. Sect. 2.7. For a family $a_i, i \in I$, $V(a_i, i \in I) = \bigcap_{i \in I} V(a_i) \subseteq \operatorname{Spec} A$ denotes the set of prime ideals in A that contain all $a_i, i \in I$, and $D(a_i, i \in I) = \bigcup_{i \in I} D(a_i)$ the complement Spec $A - V(a_i, i \in I)$. It is $V(a_i, i \in I) = V(\mathfrak{a})$ and $D(a_i, i \in I) = D(\mathfrak{a})$, where $\mathfrak{a} := \sum_{i \in I} Aa_i$ is the ideal generated by the a_i in A. It is exactly when $V(\mathfrak{a}) = \operatorname{Spec} A$, when the ideal $\mathfrak{a} \subseteq A$ contains only nilpotent elements, so $\mathfrak{a} \subseteq \mathfrak{n}_A$ is, see exercise 2.7.16. $V(\mathfrak{a}) = \emptyset$ applies if and only if $\mathfrak{a} = A$ is.

a) The sets $D(\mathfrak{a})$, $\mathfrak{a}$ ideal in A, form a topology on Spec A, whose closed sets are thus the sets $V(\mathfrak{a})$, $\mathfrak{a} \subseteq A$ ideal. The sets $D(a), a \in A$, are a basis of this topology. (This topology is called the **Zariski topology** on Spec A (after O. Zariski (1899–1986)). Unless otherwise stated, we always consider Spec A as a topological space with the Zariski topology and Spm A with the induced topology.)
b) For every homomorphism $\varphi \colon A \to B$ of commutative rings, the (contravariant) associated mapping Spec $\varphi \colon \operatorname{Spec} B \to \operatorname{Spec} A$, $\mathfrak{q} \mapsto \varphi^{-1}(\mathfrak{q})$, is continuous, see exercise 2.7.14. — If φ is surjective, then Spec φ is a closed embedding with $V(\ker \varphi)$ as the image. So then it is Spec B is homeomorphic to $V(\ker \varphi) \cong \operatorname{Spec}(A/\ker \varphi)$. In particular, Spec $A = \operatorname{Spec} A_{\text{red}}$, where $A_{\text{red}} = A/\mathfrak{n}_A$ is the reduction of A. — If $S \subseteq A$ is a submonoid of the multiplicative monoid of A and $\iota_S \colon A \to A_S$ is the canonical homomorphism with $a \mapsto a/1$, then the associated mapping Spec $\iota_S \colon \operatorname{Spec} A_S \to \operatorname{Spec} A$ is an embedding of topological spaces, whose image is the subspace $\operatorname{Spec}_S A = \{\mathfrak{p} \in \operatorname{Spec} A \mid \mathfrak{p} \cap S = \emptyset\} \subseteq \operatorname{Spec} A$, and in particular for $S := S_f = \{f^n \mid n \in \mathbb{N}\}, f \in A$, an open embedding with image $D(f) \cong \operatorname{Spec} A_f$, cf. Ex. 2.7.15.
c) Spec A is exactly connected when the ring A is connected. (One can assume $A = A_{\text{red}}$, cf. Ex. 2.7.17.)
d) Spec A and Spm A are quasi-compact. (If $a_i, i \in I$, are elements in A with $\sum_{i \in I} Aa_i = A$, then there is a finite subset $J \subseteq I$ with $\sum_{i \in J} Aa_i = A$.) Spec A is Hausdorff (i.e., compact) if and only if Spec $A = \operatorname{Spm} A$ is. In this case, Spec A is

also totally disconnected, thus a Boolean space. (Let $\operatorname{Spm} A = \operatorname{Spek} A$. It suffices to show: If $\mathfrak{m}, \mathfrak{n} \in \operatorname{Spm} A$, $\mathfrak{m} \neq \mathfrak{n}$, then there is an idempotent element $e \in \mathfrak{m} - \mathfrak{n}$. Then $\operatorname{Spm} A = D(e) \uplus D(1-e)$ is a decomposition of $\operatorname{Spm} A$ into open sets with $\mathfrak{m} \in D(1-e)$ and $\mathfrak{n} \in D(e)$. For this, one can assume that A is reduced, see problem 2.7.17. Then all localizations $A_\mathfrak{m} \xrightarrow{\sim} A/\mathfrak{m}$, $\mathfrak{m} \in \operatorname{Spm} A$, are fields, and it applies $Aa = Aa^2$ for all $a \in A$, because all residue class rings $A/\mathfrak{a}$, $\mathfrak{a} \subseteq A$ ideal, are also reduced, see exercise 2.7.16. If now $b \in \mathfrak{m} - \mathfrak{n}$ and $b = rb^2$, $r \in A$, then $e := rb$ is an idempotent element of the desired kind.)
e) Let $A = \prod_{i \in I} A_i$ be the product of the commutative rings A_i, $i \in I$. If I is finite, then $\operatorname{Spec} A = \bigsqcup_{i \in I} \operatorname{Spec} A_i$ applies. If the set of $i \in I$ with $A_i \neq 0$ is infinite, then there is no commutative ring whose spectrum is homeomorphic to the sum space $\bigsqcup_{i \in I} \operatorname{Spek} A_i$.
f) If A is Noetherian, then $\operatorname{Spec} A$ is Noetherian. ($\operatorname{Spec} A$ can be Noetherian without A being Noetherian. Example!) Describe the Zariski topology of the spectrum of a principal ideal domain, e.g., of $\operatorname{Spec} \mathbb{Z}$ or of $\operatorname{Spec} K[X]$, K field.

Exercise 4.4.30

a) Let B be a Boolean ring with the spectrum $X := \operatorname{Spec} B = \operatorname{Spm} B$, which is a Boolean space, cf. Ex. 4.4.29d). For each $\mathfrak{m} \in \operatorname{Spm} B$, $B/\mathfrak{m}$ is the only Boolean field F_2. Each element $b \in B$ therefore defines the function $f_b : X \to F_2$, $\mathfrak{m} \mapsto b \bmod \mathfrak{m}$. Show that: $b \mapsto f_b$ is an injective homomorphism $B \to F_2^X$ of Boolean rings, whose image is the Boolean subalgebra $C_{F_2}(X) = C(X, F_2) \subseteq F_2^X$ of continuous F_2-valued functions on X (which is canonically isomorphic to the Boolean algebra $\mathcal{B}(X)$ of borderless sets of X, cf. Ex. 4.3.17) (**Stone's representation theorem for Boolean rings** after M. Stone (1903–1983)). For an arbitrary commutative ring A, the Boolean functional algebra $C_{F_2}(\operatorname{Spec} A) \xrightarrow{\sim} \mathcal{B}(\operatorname{Spec} A)$ is canonically isomorphic to the Boolean ring $\operatorname{Idp}(A)$ of idempotent elements of A, cf. Ex. 2.6.6 a).
b) Let X be a Boolean space and $B := C_{F_2}(X)$ the Boolean ring of continuous F_2-valued functions on X. For a point $x \in X$, let $\mathfrak{m}_x = \{f \in B \mid f(x) = 0\} \in \operatorname{Spm} B$ be the point ideal for x in X. Then the mapping $x \mapsto \mathfrak{m}_x$ is a homeomorphism $X \xrightarrow{\sim} \operatorname{Spm} B$ of Boolean spaces. (Use, among other things, exercise 4.4.25.)

Exercise 4.4.31 Let X be a compact space and $A := C_\mathbb{K}(X)$ the $\mathbb{K}$-algebra of continuous $\mathbb{K}$-valued functions on X. For a point $x \in X$, let $\mathfrak{m}_x \in \operatorname{Spm} A$ be the point ideal $\mathfrak{m}_x := \{f \in A \mid f(x) = 0\}$ of those continuous $\mathbb{K}$-valued functions on X that vanish at x.

a) It is $\operatorname{Spm} A = \{\mathfrak{m}_x \mid x \in X\}$. (Let $\mathfrak{m} \in \operatorname{Spm} A$. Suppose it would be $\mathfrak{m} \not\subseteq \mathfrak{m}_x$ for all $x \in X$. Then there would be for each $x \in X$ a $f_x \in \mathfrak{m}$ with $f_x(x) \neq 0$. Since X is compact, there would be finitely many $x_1, \ldots, x_n \in X$ with $X = (X - Z(f_{x_1})) \cup \cdots \cup (X - Z(f_{x_n}))$. Then $|f_{x_1}|^2 + \cdots + |f_{x_n}|^2 = \overline{f}_{x_1} f_{x_1} + \cdots + \overline{f}_{x_n} f_{x_n} \in \mathfrak{m}$ would have no zero in X. Contradiction!)

b) The mapping $X \to \operatorname{Spm} A$, $x \mapsto \mathfrak{m}_x$, is a homeomorphism. In particular, $\operatorname{Spm} A$ is compact. (Use the Urysohn's separation lemma 4.2.37. The open sets $X - Z(f)$, $f \in A$, form a basis of the topology of X. For further discussion of the algebras $C_{\mathbb{K}}(X)$, X compact, see Vol. 2, Sect. 1.1. See also L. Gillman, M. Jerison, Rings of Continuous Functions, New York [2]1976.)

Exercise 4.4.32 Let X and Y be locally compact spaces and $f: X \to Y$ continuous. The canonical extension $f': X \uplus \{\omega_X\} \to Y \uplus \{\omega_Y\}$ of f to the one-point compactifications of X and Y (see Example 4.4.22) with $f'(\omega_X) := \omega_Y$ is continuous if and only if f is proper.

Exercise 4.4.33 A locally compact space X with a countable topology is σ-compact. Conversely, if X is metric and σ-compact, then X has a countable topology.

Exercise 4.4.34 For a topological space X, the following are equivalent. (i) X is metric and compact. (ii) X is compact, and there is a countable family f_i, $i \in I$, of continuous functions $f_i: X \to \mathbb{R}$, which separates the points of X, i.e. for $x, y \in X$, $x \neq y$, there is an $i \in I$ with $f_i(x) \neq f_i(y)$. (iii) X is homeomorphic to a closed subspace of the Hilbert cube $[0, 1]^{\mathbb{N}^*}$. (iv) X is compact with countable topology. (For the proof of (i) $\Rightarrow$ (ii) note: If Y is a dense subset of the metric space X, then the distance functions $x \mapsto \operatorname{Min}(1, d(x, y))$, $y \in Y$, separates the points of X. For the proof of the implication (iv) $\Rightarrow$ (ii) use the Urysohn's separation lemma 4.2.37. — By transitioning to the one-point compactification, see example 4.4.22, we obtain the following important **Metrizability criterion** : *A locally compact space X with countable topology is always metrizable*. Note that the one-point compactification of X also has a countable topology, see the preceding exercise.)

4.5 Complete Metric Spaces and Uniform Convergence

In this section, we will investigate, among other things, to what extent the Cauchy convergence criterion 3.4.8 or 3.5.4 can be transferred to arbitrary metric spaces.

Definition 4.5.1 Let X be a metric space.

(1) A sequence (x_n) in X is called a **Cauchy sequence**, if for every $\varepsilon > 0$ there is an $n_0 \in \mathbb{N}$ with $d(x_m, x_n) \leq \varepsilon$ for all $m, n \geq n_0$.
(2) X is called **complete**, if every Cauchy sequence in X is convergent in X.

Of course, every convergent sequence in X is a Cauchy sequence. A sequence (x_n) in X is already a Cauchy sequence, if for every $\varepsilon > 0$ there exists an $n_0 \in \mathbb{N}$ with $d(x_{n_0}, x_n) \leq \varepsilon$ for all $n \geq n_0$. In particular, for every Cauchy sequence (x_n) there exists an $n_0 \in \mathbb{N}$ such that the sequence $(x_n)_{n \geq n_0}$ is bounded. If (x_n) is a

Cauchy sequence in X and (y_n) is another sequence in X with $\lim_n d(x_n, y_n) = 0$, so (y_n) is also a Cauchy sequence. The completeness of a metric space X depends very much on the metric and not just on its topology. For example, the discrete space of proper fractions $1/n$, $n \in \mathbb{N}^*$, with respect to the metric induced by $\mathbb{R}$ is not complete, but it is with respect to the discrete metric, see also exercise 4.5.1 for another such example. With equivalent metrics, however, X is complete with respect to one metric if and only if it is with respect to the other. $\mathbb{R}$ and $\mathbb{C}$ are complete metric spaces. X is complete if and only if the balls $B(x; \infty)$, $x \in X$, (which are open and closed in X) are complete. As Lemma 3.4.7 proves:

Lemma 4.5.2 *Let X be a metric space. Every Cauchy sequence in X with a limit point in X is convergent. X is therefore complete if and only if every Cauchy sequence in X has a limit point in X.*

The following statement deals with subspaces of metric spaces in relation to completeness.

Proposition 4.5.3 *Let X be a metric space and $X' \subseteq X$ a subspace.*

(1) *If X' is complete, then X' is closed in X.*
(2) *If X is complete and X' is closed in X, then X' is also complete.*

Proof (1) Let (x_n) be a sequence in X', which converges in X. We have to show that $x := \lim x_n \in X'$ is. Since (x_n) is a Cauchy sequence in X', (x_n) converges in X'. Due to the uniqueness of the limit, $x \in X'$.

(2) Let (x_n) be a Cauchy sequence in X'. Then (x_n) is also a Cauchy sequence in X and therefore has a limit, which lies in X', since X' is closed in X. □

Example 4.5.4 Let $X_i = (X_i, d_i)$, $i \in I$, be a family of metric spaces.

(1) Apparently, *the sum $\coprod_{i \in I} X_i$ is complete with respect to* the metric specified in Example 4.2.25 *if and only if all X_i, $i \in I$, are complete.* In particular, $\overline{\mathbb{R}} = \mathbb{R} \sqcup \{\infty\} \sqcup \{-\infty\}$ is a complete metric space, see Example 4.2.20.

(2) *The product $X := \prod_i X_i$ is complete with respect to each p-metric d_p, $1 \le p \le \infty$, if all factors X_i are complete. In particular, for each set I and each complete metric space Y, the space Y^I of mappings from I to Y is a complete metric space with respect to each p-metric.* For the *proof*, let's first consider $p = \infty$ and $x_n = (x_{in})_{i \in I} \in X$, $n \in \mathbb{N}$, a Cauchy sequence in X. Apparently, for each $i \in I$ the sequence $(x_{in})_{n \in \mathbb{N}}$ is also a Cauchy sequence in X_i. Let $x_i \in \mathbb{K}$ be its limit, $i \in I$. Then $(x_i)_{i \in I} \in X$ is the limit of the output sequence x_n, $n \in \mathbb{N}$. Now let $p < \infty$ and again $x_n = (x_{in})_{i \in I} \in X$, $n \in \mathbb{N}$, a Cauchy sequence with respect to the metric d_p. Then also now x_{in}, $n \in \mathbb{N}$, for each $i \in I$ a Cauchy sequence in X_i, which converges to a $x_i \in \mathbb{K}$. Now we have to show that the family $x = (x_i)_{i \in I}$ is limit of the sequence x_n, $n \in \mathbb{N}$, with respect to d_p. We leave this to the reader. □

4.5 Complete Metric Spaces and Uniform Convergence   561

If all $X_i \neq \emptyset$ are, then conversely all X_i are complete, if (X, d_p) is complete for a $p \in [1, \infty]$. ◇

With Lemma 4.4.9 (1) and Lemma 4.5.2 we get:

Theorem 4.5.5 *Every compact metric space is complete.*

The converse is:

Theorem 4.5.6 *Let X be a complete metric space. If there are for every $\varepsilon > 0$ a finite number of balls $B(y_1; \varepsilon), \ldots, B(y_r; \varepsilon)$, $y_1, \ldots, y_r \in X$, that cover X, then X is a compact metric space.*

Proof Let $(x_n)_{n \in \mathbb{N}}$ be a sequence in X. We have to show that it has a convergent subsequence, cf. Theorem 4.4.14. Based on the assumption, we recursively define a descending sequence $N_1 \supseteq N_2 \supseteq N_3 \supseteq \cdots$ of infinite subsets $N_k \subseteq \mathbb{N}$, $k \in \mathbb{N}^*$, with $d(x_\nu, x_\mu) \leq 1/k$ for all $\mu, \nu \in N_k$. Then every subsequence $(x_{n_k})_{k \in \mathbb{N}^*}$ of (x_n) with $n_k \in N_k$ for all $k \in \mathbb{N}^*$ is a Cauchy sequence and thus convergent. □

Metric spaces that possess the covering property assumed in Theorem 4.5.6 are called **precompact**. With 4.5.6, we prove very clearly the following special case of Tychonoff's Theorem 4.4.8, which we use, however, for *finite* products:

Theorem 4.5.7 *Let X_i, $i \in I$, be a countable family of compact metric spaces. Then the product space $X := \prod_{i \in I} X_i$ is also metric and compact.*

Proof We can assume $I = \mathbb{N}$ and $X_i \neq \emptyset$ for all $i \in \mathbb{N}$. Then X is metric (e.g.) with the metric

$$d((x_i), (y_i)) = \sum_{i=0}^{\infty} 2^{-i} \operatorname{Min}(d_i(x_i, y_i), 1),$$

cf. exercise 4.2.15a). According to example 4.5.4 (2), (X, d) is complete. According to Theorem 4.5.6, it remains to show that X is precompact. Let $\varepsilon > 0$ be given and let $i_0 \in \mathbb{N}$ be chosen so large that $\sum_{i=i_0+1}^{\infty} 2^{-i} \leq \varepsilon/2$ is. Since the finite product $\prod_{i=0}^{i_0} X_i$ is compact, there are points $z'_1, \ldots, z'_r \in \prod_{i=0}^{i_0} X_i$ such that the balls $B(z'_\rho; \varepsilon/2)$, $\rho = 1, \ldots, r$, cover the space $\prod_{i=0}^{i_0} X_i$, where we rely on $\prod_{i=0}^{i_0} X_i$ choose the metric $\sum_{i=0}^{i_0} 2^{-i} \operatorname{Min}(d_i(x_i, y_i), 1)$. If we then extend the sequences z'_ρ (in any way) to sequences $z_\rho \in X$, $\rho = 1, \ldots, r$, then the spheres $B(z_\rho; \varepsilon)$, $\rho = 1, \ldots, r$, completely cover X. □

Let V be a finite-dimensional normed $\mathbb{K}$-vector space. If the values of the norm are all finite, then the given norm induces the natural topology on V according to theorem 4.4.12. Since every Cauchy sequence in V is then bounded, it has an accumulation point according to the Bolzano-Weierstraß theorem 4.2.34 and is therefore convergent. We have proven:

Theorem 4.5.8 *A finite-dimensional normed $\mathbb{K}$-vector space is complete.*

Let V be any normed $\mathbb{K}$ vector space and $V_{<\infty} = B(0; \infty)$ the subspace of vectors from V with finite norm. V is complete exactly when $V_{<\infty}$ is complete. For each coset $x_0 + V_{<\infty} \subseteq V$, $V_{<\infty} \xrightarrow{\sim} x_0 + V_{<\infty} \subseteq V$, $x \mapsto x_0 + x$, is an isometry. Based on this remark, we give the following standard definition:

Definition 4.5.9 A normed $\mathbb{K}$ vector space $V = (V, \|-\|)$ is called a $\mathbb{K}$-**Banach space**, if $V = V_{<\infty} = B_{\|-\|}(0; \infty)$ is (i.e., if its norm $\|-\|$ only takes finite values) and if $(V, \|-\|)$ is complete.

If the norms $\|-\|$ and $\|-\|'$ are equivalent on V, then V is a Banach space with respect to $\|-\|$ if and only if V is a Banach space with respect to $\|-\|'$. According to theorem 4.5.8, *a finite-dimensional $\mathbb{K}$-vector space is a Banach space with respect to any norm that only takes finite values, and its topology coincides with the natural topology.*

Example 4.5.10 For a set I and an $p \in \mathbb{R}$ with $p \geq 1$, let $\ell_{\mathbb{K}}^{p}(I)$ be the space of p-summable families $(x_i) \in \mathbb{K}^I$, and $\ell_{\mathbb{K}}^{\infty}(I)$ be the space of bounded families $(x_i) \in \mathbb{K}^I$, see section 4.1. The result of example 4.5.4 (2) implies: ◊

Proposition 4.5.11 *For every set I and every p with $1 \leq p \leq \infty$ the space $\ell_{\mathbb{K}}^{p}(I)$ with respect to the p-norm $\|-\|_p$ is a $\mathbb{K}$-Banach space.*

Proposition 4.5.11 is a special case of a much more general theorem from measure and integration theory, see Vol. 6. ◊

Many applications of the concept of completeness are based on the following simple but important Banach fixed-point theorem, which is proven verbatim like the special case 3.8.20, only replacing $|z - y|$ each time with the distance $d(y, z)$:

Theorem 4.5.12 (Banach's Fixed Point Theorem) *Let $f : X \to X$ be a strongly contracting mapping of the non-empty complete metric space X into itself. Then f has exactly one fixed point x. If $x_0 \in X$ is any point in X, then the sequence*

$$x_0, x_1 = f(x_0), \quad x_2 = f(x_1) = f^2(x_0), \quad \ldots, \quad x_n = f(x_{n-1}) = f^n(x_0), \quad \ldots$$

converges to the unique fixed point x of f. If $L < 1$ is a contraction factor of f, then

$$d(x_n, x) \leq \frac{1}{1-L} d(x_n, x_{n+1}) \leq \frac{L^n}{1-L} d(x_0, x_1), \quad n \in \mathbb{N}.$$

4.5 Complete Metric Spaces and Uniform Convergence

The method described in sentence 4.5.12 for solving the fixed point equation

$$f(x) = x$$

in X is the method of **successive approximation**. It is self-correcting: Any rounding or calculation errors do not propagate. As mentioned in exercise 3.8.40, theorem 4.5.12 has a generalization, see exercise 4.5.8b).

Every metric space $X = (X, d)$ can naturally be considered as a subspace of a complete metric space $\widehat{X}$ (in which X is dense). We proceed as in the construction of $\mathbb{R}$ as the completion of the space $\mathbb{Q}$ of rational numbers in sentence 3.4.12. So let

$$X_{\mathrm{CF}}^{\mathbb{N}} \subseteq X^{\mathbb{N}}$$

be the set of Cauchy sequences in X. The relation "$(x_n) \sim (y_n)$ if and only if $d(x_n, y_n)$, $n \in \mathbb{N}$, is a null sequence (in $\overline{\mathbb{R}}_+$)", is (due to the triangle inequality) an equivalence relation on $X_{\mathrm{CF}}^{\mathbb{N}}$ (even on $X^{\mathbb{N}}$). We denote the associated quotient space with

$$\widehat{X} := X_{\mathrm{CF}}^{\mathbb{N}}/\sim.$$

We embed X in $\widehat{X}$ by assigning to each $x \in X$ the equivalence class of the constant sequence (x). So we always consider X as a subset of $\widehat{X}$ with respect to this canonical inclusion. If X is complete, then obviously $X \hookrightarrow \widehat{X}$ is bijective. In general, the metric d on X can be extended to a metric $\widehat{d}$ on $\widehat{X}$. To do this, set

$$\widehat{d}(\widehat{x}, \widehat{y}) := \lim_{n \to \infty} d(x_n, y_n), \quad \widehat{x} = [(x_n)], \ \widehat{y} = [(y_n)] \in \widehat{X}.$$

Initially, the sequence $d(x_n, y_n)$, $n \in \mathbb{N}$, is convergent in $\overline{\mathbb{R}}_+$. It is either stationary with limit ∞ or a Cauchy sequence in $\mathbb{R}_+$ from a certain point n_0 onwards. If namely $d(x_n, y_n) \in \mathbb{R}_+$ for $n \geq n_0$, so there is for a given $\varepsilon > 0$ a $n_1 \geq n_0$ with $d(x_n, x_{n_1}), d(y_n, y_{n_1}) \leq \varepsilon/2$ for all $n \geq n_1$ and consequently $|d(x_n, y_n) - d(x_{n_1}, y_{n_1})| \leq d(x_n, x_{n_1}) + d(y_n, y_{n_1}) \leq \varepsilon$ for all $n \geq n_1$ (see the comment on Definition 4.1.1). Similarly, one proves the independence of the limit $\lim d(x_n, y_n)$ from the choice of representatives $(x_n), (y_n)$. The triangle inequality for $\widehat{d}$ results from the triangle inequality for d and the rules for limits. For $x, y \in X$ is $\widehat{d}(x, y) = d(x, y)$. X is even dense in $\widehat{X}$; because it is $[(x_n)] = \lim x_n$ for all $[(x_n)] \in \widehat{X}$. Finally, $\widehat{X}$ is complete. This immediately follows from the following lemma:

Lemma 4.5.13 *Let X be a dense subspace of the metric space Y. If every Cauchy sequence in X has a limit in Y, then Y is complete.*

Proof Let (y_n) be a Cauchy sequence in Y. Since X is dense in Y, for every $n \in \mathbb{N}$ there is a $x_n \in X$ with $d(x_n, y_n) \leq 1/(n+1)$. Then (x_n) is also a Cauchy sequence and, by assumption, has a limit $y \in Y$, and it is $\lim_n y_n = \lim_n x_n = y$. □

We summarize:

Theorem 4.5.14 *For every metric space $X = (X, d)$ there exists a complete metric space $\widehat{X} = (\widehat{X}, \widehat{d})$, which contains X as a dense metric subspace. (So it is $d = \widehat{d}|X$.) $(\widehat{X}, \widehat{d})$ has the following universal property: If $f : X \to Y$ is a uniformly continuous mapping from X into a complete metric space Y, then there exists exactly one continuous extension $\widehat{f} : \widehat{X} \to Y$ of f. The extension $\widehat{f}$ is also uniformly continuous. The space $(\widehat{X}, \widehat{d})$ is uniquely determined up to isometry by X and is called the* **completion** *or* **completeness** *of X. As a concrete model for $\widehat{X}$ one can choose the space $X_{CF}^{\mathbb{N}}/\sim$ of equivalence classes of Cauchy sequences in X.*

Proof The universal property still needs to be proven, which also implies the uniqueness of $\widehat{X}$. So let $f : X \to Y$ be uniformly continuous with a complete metric space Y and $\hat{x} \in \widehat{X}$. Since X is dense in $\widehat{X}$, there is a sequence (x_n) in X with $\lim_n x_n = \hat{x}$. In particular, (x_n) is a Cauchy sequence. Since f is uniformly continuous, $(f(x_n))$ is also a Cauchy sequence in Y and therefore convergent. Necessarily, $\widehat{f}(\hat{x}) := \lim_n f(x_n)$ must be set. This definition is independent of the choice of the sequence (x_n) converging to $\hat{x}$. If $\lim_n u_n = \hat{x}$ is also for a sequence (u_n) in X, then $(d(x_n, u_n))$ is a null sequence and thus also $\big(d(f(x_n), f(u_n))\big)$ — again due to the uniform continuity of f.[18] So $\lim_n f(x_n) = \lim_n f(u_n)$ is. This extension $\widehat{f}$ of f is uniformly continuous like f. For proof, let $\varepsilon > 0$ be given. Since f is uniformly continuous, there exists a $\delta > 0$ with $d(f(x), f(u)) \leq \varepsilon$ for all $x, u \in X$ with $d(x, u) \leq \delta$. If now $\hat{x}, \hat{u} \in \widehat{X}$ with $\widehat{d}(\hat{x}, \hat{u}) \leq \delta/2$, there are sequences $(x_n), (u_n)$ in X with $x_n \to \hat{x}$ and $u_n \to \hat{u}$ and $d(x_n, u_n) \leq \delta$, thus $d(f(x_n), f(u_n)) \leq \varepsilon$ for all $n \in \mathbb{N}$. Then $d(\widehat{f}(\hat{x}), \widehat{f}(\hat{u})) = d\big(\lim_n f(x_n), \lim_n f(u_n)\big) = \lim_n d(f(x_n), f(u_n)) \leq \varepsilon$ is due to the continuity of the metric $d : Y \times Y \to \overline{\mathbb{R}_+}$. □

With Proposition 4.5.3 (2) follows:

Corollary 4.5.15 *Let $X \subseteq Z$ be any subspace of the complete metric space Z. Then the closed hull $\overline{X}$ in Z is the completion of X.*

---

[18] Note: If $f : X \to Y$ is a mapping of metric spaces, then f is *uniformly continuous* if and only if for any sequences $x_n, u_n, n \in \mathbb{N}$, in X, for which $d(x_n, u_n), n \in \mathbb{N}$, is a null sequence, also $d(f(x_n), f(u_n)), n \in \mathbb{N}$, is a null sequence. Proof!

4.5 Complete Metric Spaces and Uniform Convergence

Example 4.5.16 Let (X_i, d_i), $i \in I$, a family of non-empty metric spaces with the completions $(\widehat{X}_i, \widehat{d}_i), i \in I$, and the products $X = \prod_{i \in I} X_i$ or $Y := \prod_{i \in I} \widehat{X}_i$. Regarding each p-norm, $1 \leq p \leq \infty$, is then $(X, d_{X,p})$ is a metric subspace of $(Y, d_{Y,p})$, and the latter is complete according to example 4.5.4 (2). Therefore, according to corollary 4.5.15, the completion $(\widehat{X}, \widehat{d}_p)$ of X is the closed hull $\overline{X}$ of X in $(Y, d_{Y,p})$. For finite I or — more generally — for any I, but $p = \infty$ is naturally $\overline{X} = \widehat{X} = Y$. Now let $p < \infty$. If $X_i \neq \widehat{X}_i$ for uncountably many $i \in I$, then certainly $\overline{X} \subset Y$. For example, an element $(y_i) \in Y$ with $y_i \notin X_i$ for uncountably many i is certainly not contained in $\overline{X}$. But if I is countable, $(y_i) \in Y$ arbitrary and $\varepsilon > 0$ given, then there is $\varepsilon_i > 0$ with $\sum_{i \in I} \varepsilon_i^p \leq \varepsilon^p$ as well as $x_i \in X_i$ with $\widehat{d}_i(x_i, y_i) \leq \varepsilon_i$. Then $d_{Y,p}((x_i), (y_i)) \leq \varepsilon$ is, i.e. X is dense in Y and $\widehat{X} = \overline{X} = Y$. So it follows: *If I is countable, then* $(\widehat{X}, \widehat{d}_p) = \left(\prod_{i \in I} \widehat{X}_i, d_p\right)$ *for every* $p \in [1, \infty]$. ◊

An important special case is the completion of normed $\mathbb{K}$ vector spaces. Let $\widehat{V}$ be the completion of the normed $\mathbb{K}$ vector space $(V, \|-\|)$. Then the equivalence relation $\sim$ on $V_{\mathrm{CF}}^{\mathbb{N}}$ is obviously the congruence relation modulo of the space $\mathfrak{n}_V$ of null sequences in V and $\widehat{V} = V_{\mathrm{CF}}^{\mathbb{N}}/\mathfrak{n}_V$ as a quotient space is also a $\mathbb{K}$ vector space. $\|\widehat{x}\| = \widehat{d}(0, \widehat{x})$ is a norm on $\widehat{V}$, whose associated metric $\widehat{d}$ is. Obviously, $(V_{<\infty})^\wedge = \widehat{V}_{<\infty} \subseteq \widehat{V}$, $V_{<\infty} = \widehat{V}_{<\infty} \cap V$ and the canonical injection $V/V_{<\infty} \to \widehat{V}/\widehat{V}_{<\infty}$ is an isomorphism of discrete $\mathbb{K}$ vector spaces. We note:

Theorem 4.5.17 *Let V be a normed $\mathbb{K}$-vector space with $V = V_{<\infty} = B_V(0; \infty)$. Then the completion $\widehat{V} = \widehat{V}_{<\infty}$ is a $\mathbb{K}$-Banach space, which contains V as a dense subspace and has the following universal property: If $f : V \to W$ is a continuous $\mathbb{K}$-linear mapping from V into a $\mathbb{K}$-Banach space W, then there exists a unique continuous $\mathbb{K}$-linear extension $\widehat{f} : \widehat{V} \to W$ of f.*

Proof According to theorem 4.5.14, it only remains to show *that a continuous $\mathbb{K}$-linear mapping $f : V \to W$ is uniformly continuous*. (The linearity of the continuous extension $\widehat{f}$ of f is then clear.) So let $\varepsilon > 0$ be given. Due to the continuity of f at the zero point, there is a $\delta > 0$ with $f\left(\overline{B}(0; \delta)\right) \subseteq \overline{B}(0; \varepsilon)$. If then $x, x' \in V$ with $\|x' - x\| \leq \delta$, it follows $\|f(x') - f(x)\| = \|f(x' - x)\| \leq \varepsilon$. □

For example, the $\mathbb{K}$-Banach space $\ell_{\mathbb{K}}^p(I)$, $1 \leq p \leq \infty$, cf. Proposition 4.5.11, is the completion of the subspace $(\mathbb{K}^{(I)}, \|-\|_p)$.

An important property of complete metric spaces, which they share with locally compact spaces, is described by the following theorem, which goes back to R. Baire (1874–1932) and for which a special case was already given in exercise 3. 8.31 :

Theorem 4.5.18 (Baire's Category Theorem) *Let X be a complete metric space or a locally compact topological space. Then the intersection $\bigcap_{i \in I} U_i$ of a countable family of open dense subsets $U_i \subseteq X$, $i \in I$, is also dense in X.*

Proof Without loss of generality, let $I = \mathbb{N}$ and the sequence U_i, $i \in \mathbb{N}$, be monotonically decreasing. (Otherwise, replace U_i with $\bigcap_{j \leq i} U_j$, $j \in \mathbb{N}$.) Let $D := \bigcap_{i \in \mathbb{N}} U_i$.

Let X initially be a complete metric space. We have $\overline{B}(x; \varepsilon) \cap D \neq \emptyset$ for all spheres $\overline{B}(x; \varepsilon)$, $x \in X$, $\varepsilon > 0$, to show. Since the U_i are open and dense, one can easily construct inductively a sequence of spheres $\overline{B}(x_i; \varepsilon_i) \subseteq U_i \cap \overline{B}(x; \varepsilon)$ with $\varepsilon_i > 0$, $\overline{B}(x_i; \varepsilon_i) \supseteq \overline{B}(x_{i+1}; \varepsilon_{i+1})$, $i \in \mathbb{N}$, and $\lim \varepsilon_i = 0$. Then, however, x_i, $i \in \mathbb{N}$, is a Cauchy sequence in X and $x := \lim x_i \in \bigcap_i \overline{B}(x_i; \varepsilon_i) \subseteq \overline{B}(x; \varepsilon) \cap \bigcap_i U_i = \overline{B}(x; \varepsilon) \cap D$.

If X is locally compact, then one replaces $\overline{B}(x; \varepsilon)$ with a compact set K with $\mathring{K} \neq \emptyset$ and constructs instead of the $\overline{B}(x_i; \varepsilon_i)$ a sequence K_i of compact sets with $K_i \subseteq U_i \cap K$, $\mathring{K}_i \neq \emptyset$, $K_i \supseteq K_{i+1}$, $i \in \mathbb{N}$. Then $\emptyset \neq \bigcap_{i \in I} K_i \subseteq K \cap \bigcap_i U_i = K \cap D$ is true. $\square$

Example 4.5.19 (1) As a simple consequence of theorem 4.5.18, we find: If X is a non-empty complete metric space or a locally compact topological space, each without discrete points, then the complement of any countable subset of X is dense in X. In particular, X itself is uncountable.

(2) Let V be an $\mathbb{K}$-Banach space of infinite dimension and x_n, $n \in \mathbb{N}$, a sequence of elements in V. The subspace generated by these elements $W := \sum_{n \in \mathbb{N}} \mathbb{K} x_n$ is the union of the countably many finite-dimensional subspaces $W_i := \sum_{n=0}^{i} \mathbb{K} x_n$, $i \in \mathbb{N}$, all of which are complete and thus closed and moreover nowhere dense in W, i.e., $U_i := V - W_i$ is open and dense in V. According to theorem 4.5.18, $\bigcap_{i \in \mathbb{N}} U_i = V - W$ is also dense in V. In particular, $V \neq W$. A $\mathbb{K}$-*Banach space, which is not finite-dimensional, therefore always has an uncountable vector space dimension.* ◊

We will discuss spaces of mappings between topological spaces and limit ourselves to Hausdorff spaces, which ensures the uniqueness of limits. Let Y be a Hausdorff space. The set of mappings of a set X into Y is the product space

$$Y^X = \text{Abb}(X, Y).$$

The product topology on Y^X is the so-called **topology of pointwise convergence**: A sequence $f_n \colon X \to Y$, $n \in \mathbb{N}$, converges exactly when $f \in Y^X$ with respect to the product topology, if it converges pointwise to f, i.e., if for each $x \in X$ the sequence $f_n(x)$, $n \in \mathbb{N}$, converges in Y, see example 4.2.27. However, one usually considers finer topologies on Y^X than the rather coarse product topology. An important special case is given by the Chebyshev metric on Y^X when Y is a metric space. The convergence with respect to this metric is the so-called **uniform**

4.5 Complete Metric Spaces and Uniform Convergence 567

convergence. A sequence $f_n \colon X \to Y$, $n \in \mathbb{N}$, of mappings from X to Y converges uniformly to $f \in Y^X$, if it converges with respect to the Chebyshev metric d_∞ to f, i.e., if for each $\varepsilon > 0$ there is an $n_0 \in \mathbb{N}$ with $d_\infty(f_n, f) \leq \varepsilon$ for all $n \geq n_0$, if thus $d(f_n(x), f(x)) \leq \varepsilon$ is for all $n \geq n_0$ and *all* $x \in X$. Therefore, the topology defined by the metric d_∞ on Y^X is called the **topology of uniform convergence**. The uniform convergence of $f_n \colon X \to Y$, $n \in \mathbb{N}$, implies pointwise convergence, as the topology of uniform convergence on Y^X is finer than the product topology on Y^X. If Y is complete, then Y^X is also complete with respect to the Chebyshev metric, see example 4.5.4 (2), i.e., every Cauchy sequence in $Y^X = (Y^X, d_\infty)$ converges. Therefore, the following applies:

Proposition 4.5.20 (Cauchy's Criterion for Uniform Convergence) *A sequence (f_n) of mappings $f_n \colon X \to Y$ from the set X into a complete metric space Y converges uniformly if and only if for every $\varepsilon > 0$ there exists an $n_0 \in \mathbb{N}$ such that for all $m, n \geq n_0$ and all $x \in X$ the following holds $d(f_m(x), f_n(x)) \leq \varepsilon$.*

In the case of uniform convergence, the continuity of mappings is transferred to the limit. More precisely:

Theorem 4.5.21 *Let Y be a metric space and X a topological space. Then the space $C(X, Y)$ of continuous mappings from X into Y is a closed subspace of $Y^X = (Y^X, d_\infty)$. In particular, the limit $f = \lim_n f_n$ of a uniformly convergent sequence of continuous mappings $f_n \colon X \to Y$, $n \in \mathbb{N}$, is continuous. If Y is complete, then so is $(C(X, Y), d_\infty)$.*

Proof Let $f \in Y^X$ be a point of contact of $C(X, Y) \subseteq Y^X$. We show that f is also continuous. Let $a \in X$ and $\varepsilon > 0$. Then there is an $g \in C(X, Y)$ with $d(f(x), g(x)) \leq \varepsilon/3$ for all $x \in X$. Since g is continuous, there exists a neighborhood U of a with $d(g(x), g(a)) \leq \varepsilon/3$ for all $x \in U$. For these x, the following holds

$$d(f(a), f(x)) \leq d(f(a), g(a)) + d(g(a), g(x)) + d(g(x), f(x)) \leq \varepsilon. \qquad \square$$

Since continuity is a local property, it follows:

Corollary 4.5.22 *Let $f_n \colon X \to Y$, $n \in \mathbb{N}$, be a locally uniformly convergent sequence of continuous mappings of the topological space X into the metric space Y. Then $\lim f_n$ is also continuous.*

The sequence $f_n \colon X \to Y$ is called **locally uniformly convergent**, if for every $x \in X$ there is a neighborhood U of x in X exists in such a way that the sequence $f_n|U$, $n \in \mathbb{N}$, converges uniformly on U. To provide the appropriate topological basis for this convergence, we define for an arbitrary family $\mathcal{R} = (A_i)_{i \in I}$ of subsets of X the **topology of $\mathcal{R}$-uniform convergence** on Y^X as the smallest topology, with respect to which all restriction mappings $p_{A_i} \colon Y^X \to (Y^{A_i}, d_\infty)$, $f \mapsto f|A_i$,

$i \in I$, are continuous, or — which is the same — as the pre-image topology of $(p_{A_i})_{i \in I} \colon Y^X \to \prod_i Y^{A_i}, f \mapsto (f|A_i)_{i \in I}$. The space Y^X, equipped with this topology, and also every subspace $F \subseteq Y^X$, equipped with the induced topology, we denote by $Y^X_{\mathcal{R}}$ or with $F_{\mathcal{R}}$. A sequence $f_n \colon X \to Y$, $n \in \mathbb{N}$, is therefore convergent in $Y^X_{\mathcal{R}}$ with limit $f \in Y^X$, i.e., $\mathcal{R}$-converges uniformly to $f \colon X \to Y$ if and only if for every $i \in I$ the sequence $f_n|A_i$, $n \in \mathbb{N}$, on A_i converges uniformly to $f|A_i$. In particular, (f_n) converges locally uniformly to f if and only if there is a cover $\mathcal{U} = (U_i)_{i \in I}$ of X with $\bigcup_{i \in I} \mathring{U}_i = X$ such that (f_n) converges $\mathcal{U}$-uniformly to f. In general, if $\mathcal{R}$ is a cover of X, then $(p_{A_i})_{i \in I} \colon Y^X_{\mathcal{R}} \to \prod_i Y^{A_i}$ is injective. In this case, we identify $Y^X_{\mathcal{R}}$ with a subspace of $\prod_{i \in I}(Y^{A_i}, d_\infty)$. This subspace is even closed:

Lemma 4.5.23 *Let $\mathcal{R} = (A_i)_{i \in I}$ be a cover of the set X and Y a metric space. Then $Y^X_{\mathcal{R}}$ is a closed subspace of $\prod_{i \in I} Y^{A_i} - \prod_{i \in I}(Y^{A_i}, d_\infty)$.*

Proof Let $f = (f_{A_i})_{i \in I}$ be a contact point of $Y^X_{\mathcal{R}}$. We have to show that for all $i, j \in I$ applies $f_{A_i}|(A_i \cap A_j) = f_{A_j}|(A_i \cap A_j)$. Suppose there is a $x \in A_i \cap A_j$ with $\varepsilon := d(f_{A_i}(x), f_{A_j}(x)) > 0$. Since f is a contact point of $Y^X_{\mathcal{R}}$, there is a $g \in Y^X_{\mathcal{R}}$ with $g \in \mathrm{B}_{Y^{A_i}}(f_{A_i}; \varepsilon/2) \times \mathrm{B}_{Y^{A_j}}(f_{A_j}; \varepsilon/2) \times \prod_{k \neq i,j} Y^{A_k}$. This results in the contradiction $d(f_{A_i}(x), f_{A_j}(x)) \leq d(f_{A_i}(x), g(x)) + d(g(x), f_{A_j}(x)) < \varepsilon/2 + \varepsilon/2 = \varepsilon$. □

If $\mathcal{R} = (A_i)_{i \in I}$ is a *countable* family of subsets of X, then $\prod_{i \in I} Y^{A_i}$ is metrizable, see exercise 4.2.15a), and $Y^X_{\mathcal{R}}$ fulfills the first countability axiom. In particular, the topology on $Y^X_{\mathcal{R}}$ is determined by the set of convergent sequences, cf. Corollary 4.2.19. If $\mathcal{R}$ is even a countable cover of X, then $Y^X_{\mathcal{R}} \subseteq \prod_{i \in I} Y^{A_i}$ is also metrizable. If $\mathcal{R} = (A_i)_{i \in I}$ is a *finite* cover of X, then obviously the topology of $\mathcal{R}$-uniform convergence is equal to the topology of uniform convergence, i.e., the topology of the subspace $Y^X_{\mathcal{R}} \subseteq \prod_{i \in I} Y^{A_i}$ is the topology of uniform convergence on Y^X. This can be generalized. To formulate this more general result, we introduce the following terminology: If $\mathcal{R} = (A_i)_{i \in I}$ and $\mathcal{S} = (B_j)_{j \in J}$ are families of subsets of X, then $\mathcal{R}$ is called **finer** than $\mathcal{S}$, if for every $i \in I$ there exists a finite subset $J' \subseteq J$ with $A_i \subseteq \bigcup_{j \in J'} B_j$. We then write $\mathcal{R} \preceq \mathcal{S}$. With these designations, the following applies:

Lemma 4.5.24 *Let Y be a metric space and X a set. Furthermore, let $\mathcal{R} = (A_i)_{i \in I}$ and $\mathcal{S} = (B_j)_{j \in J}$ be families of subsets of X. If $\mathcal{R} \preceq \mathcal{S}$ holds, then the topology of the $\mathcal{S}$-uniform convergence is finer than the topology of the $\mathcal{R}$-uniform convergence, i.e., the identity $Y^X_{\mathcal{S}} \to Y^X_{\mathcal{R}}$ is continuous. In particular, the topologies on $Y^X_{\mathcal{R}}$ and $Y^X_{\mathcal{S}}$ coincide when $\mathcal{R} \preceq \mathcal{S}$ and $\mathcal{S} \preceq \mathcal{R}$ hold.*

Proof We have to show that for every $i \in I$ the canonical mapping $Y^X_{\mathcal{S}} \to Y^{A_i}$, $f \mapsto f|A_i$, is continuous. However, by assumption, $A_i \subseteq B_{J'} := \bigcup_{j \in J'} B_j$ holds with

4.5 Complete Metric Spaces and Uniform Convergence

a finite subset $J' \subseteq J$. Then $Y^{B_{J'}} \to Y^{A_i}$, $g \mapsto g|A_i$, is continuous (even Lipschitz-continuous with Lipschitz constant 1), and the canonical mapping $Y_S^X \to Y^{B_{J'}} \subseteq \prod_{j \in J'} Y^{B_j}$ is also continuous. □

From now on, let X be a topological space (and Y still a metric space), and we focus on continuous mappings $X \to Y$. The following theorem generalizes theorem 4.5.21 and also corollary 4.5.22.

Theorem 4.5.25 *Let Y be a metric space and X a topological space. Furthermore, let $\mathcal{U} = (U_i)_{i \in I}$ be a cover of X with $\bigcup_{i \in I} \mathring{U}_i = X$. Then the space $C(X,Y)_\mathcal{U} = Y_\mathcal{U}^X \cap \prod_{i \in I} C(U_i, Y)$ of continuous mappings from X to Y is a closed subspace of $\prod_{i \in I} C(U_i, Y)$ and of $Y_\mathcal{U}^X \subseteq \prod_{i \in I} Y^{U_i}$, thus also of $\prod_{i \in I} Y^{U_i}$.*

Proof According to Theorem 4.5.21 and Exercise 4.2.7, $\prod_{i \in I} C(U_i, Y)$ is a closed subspace of $\prod_{i \in I} Y^{U_i}$. From this, the claims follow, the last one with Lemma 4.5.23. □

Let X be a Hausdorff space and $\mathcal{K}$ the set of compact subsets of X. Then the topology on $Y_\mathcal{K}^X$ is called the **topology of compact convergence** and the convergence with respect to this topology is the **compact convergence**. This topology is—like every topology of the $\mathcal{R}$-uniform convergence on Y^X with respect to a cover $\mathcal{R}$ of X—finer than the topology of pointwise convergence. If X is discrete, it is identical with it. For compact X, it is the topology of uniform convergence. If X is locally compact, the topology on $Y_\mathcal{U}^X$ for each cover $\mathcal{U} = (U_i)_{i \in I}$ with $\bigcup_{i \in I} \mathring{U}_i = X$ according to Lemma 4.5.24 is finer than the topology on $Y_\mathcal{K}^X$. Since also $\bigcup_{K \in \mathcal{K}} \mathring{K} = X$ is, the topology of compact convergence is the coarsest among the topologies on $Y_\mathcal{U}^X$ for the above covers $\mathcal{U}$. In particular, *for a locally compact space X a sequence $f_n \colon X \to Y$, $n \in \mathbb{N}$, converges compactly if and only if it converges locally uniformly.*

The following Arzelà-Ascoli theorem (named after C. Arzelà (1847–1912) and G. Ascoli (1887–1957)) characterizes the compact subsets of the complete metric space $C(X,Y) = (C(X,Y), d_\infty) \subseteq (Y^X, d_\infty)$, where X is compact and Y is a complete metric space. It can be seen as a generalization of the theorem 4.4.11 by Heine-Borel (for which X is finite and $Y = \mathbb{K}$ is). First, we introduce the following terminology: A set F of mappings $X \to Y$ is called **uniformly continuous**, if for every point $x \in X$ and every $\varepsilon > 0$ there exists a neighborhood U of x with $d(f(x), f(y)) \leq \varepsilon$ for all $y \in U$ and all $f \in F$ exists. If F is uniformly continuous, then so is the closed hull $\overline{F}$ of F in $C(X,Y)$. Then the following holds:

Theorem 4.5.26 (Arzelà-Ascoli Theorem) *Let X be a compact topological space and Y a complete metric space. The subset $F \subseteq C(X,Y)$ of continuous mappings satisfies the following conditions:*

(1) *For each $x \in X$ the set $\{f(x) \mid f \in F\} \subseteq Y$ is relatively compact.*
(2) *F is uniformly continuous.*

Then F is relatively compact in $C(X, Y)$ *(with respect to the topology of compact (= uniform) convergence).*

Proof We can assume that $F = \overline{F}$, so F is closed in $C(X, Y)$, and then we have to prove that F is compact. Since F is complete like $C(X, Y)$, it suffices according to Theorem 4.5.6 to show that F is precompact, i.e., that for each $\varepsilon > 0$ there are finitely many balls $\overline{B}(f_1; \varepsilon), \ldots, \overline{B}(f_r; \varepsilon), f_1, \ldots, f_r \in F$, that cover the set F.

Let $\varepsilon > 0$ be given. For each $x \in X$, due to condition (2), there is an open neighborhood $U(x)$ of x in X with $d(f(x), f(y)) \leq \varepsilon/4$ for all $f \in F$ and all $y \in U(x)$. Since X is compact, already finitely many environments $U(x_1), \ldots, U(x_m)$ completely cover X. Since according to assumption (1) the sets $F(x_i) := \{f(x_i) \mid f \in F\} \subseteq Y$ are relatively compact, there are also finitely many spheres $\overline{B}(y_1; \varepsilon/4), \ldots, \overline{B}(y_n; \varepsilon/4)$ in Y, which cover $F(x_1) \cup \cdots \cup F(x_m)$. Then F is the union of the finitely many sets

$$F_{j_1\ldots j_m} := \{f \in F \mid f(x_i) \in \overline{B}(y_{j_i}; \varepsilon/4), \ i = 1, \ldots, m\}, \quad 1 \leq j_1, \ldots, j_m \leq n,$$

and it holds $d_\infty(f, g) \leq \varepsilon$ for all $f, g \in F_{j_1\ldots j_m}$, i.e. it is $F_{j_1\ldots j_m} \subseteq \overline{B}(f; \varepsilon)$ for each $f \in F_{j_1\ldots j_m} \subseteq F$. If $x \in X$, say $x \in U(x_i)$, then

$$d(f(x), g(x)) \leq d(f(x), f(x_i)) + d(f(x_i), g(x_i)) + d(g(x_i), g(x))$$
$$\leq \varepsilon/4 + \varepsilon/2 + \varepsilon/4 = \varepsilon.$$

Thus, each set $F_{j_1\ldots j_m}$ is entirely within an ε-sphere $C(X, Y)$. □

Corollary 4.5.27 *In the situation of* Proposition 4.5.26 *a subset* $F \subseteq C(X, Y)$ *is compact if and only if it fulfills the following conditions:*

(1) *For each* $x \in X$ *the set* $\{f(x) \mid f \in F\} \subseteq Y$ *is relatively compact (and then even compact).*
(2) *F is uniformly continuous.*
(3) *F is closed in* $C(X, Y)$.

Proof According to the theorem 4.5.26 of Arzelà-Ascoli (and theorem 4.4.14), the given conditions are sufficient for F to be compact.

Conversely, let F be compact. Then F is closed in $C(X, Y)$ according to Proposition 4.4.2 (1). Furthermore, $\{f(x) \mid f \in F\}$ for each $x \in X$ according to theorem 4.4.3 as the image of F under the continuous mapping $f \mapsto f(x)$ compact and in particular relatively compact. To prove that F is uniformly continuous, let $x \in X$ and $\varepsilon > 0$. Since F is compact, there are finitely many functions $f_1, \ldots, f_n \in F$ such that the balls $\overline{B}(f_i; \varepsilon/3)$ in $C(X, Y)$, $i = 1, \ldots, n$, completely cover F. There is then an environment U of x with $d(f_i(x), f_i(y)) \leq \varepsilon/3$ for all $y \in U$ and all $i = 1, \ldots, n$. If now $f \in F$ is arbitrary and $f \in \overline{B}(f_{i_0}; \varepsilon/3)$, it follows for all $y \in U$

$$d(f(x), f(y)) \leq d\big(f(x), f_{i_0}(x)\big) + d\big(f_{i_0}(x), f_{i_0}(y)\big) + d\big(f_{i_0}(y), f(y)\big) \leq \varepsilon \quad \square$$

4.5 Complete Metric Spaces and Uniform Convergence

The theorem 4.5.26 by Arzelá-Ascoli can easily be transferred to locally compact spaces. Let X be locally compact and $\mathcal{K}$ the set of compact subsets $K \subseteq X$ and Y still a complete metric space. Since $C(X, Y)_{\mathcal{K}}$ according to sentence 4.5.25 is a closed subspace of $\prod_{K \in \mathcal{K}} C(K, Y)$, a subset $F \subseteq C(X, Y)$ is relatively compact in $C(X, Y)_{\mathcal{K}}$ if and only if for every compact set $K \in \mathcal{K}$ the set $F|K := \{f|K \mid f \in F\}$ is relatively compact in $C(K, Y)$ (with respect to the topology of uniform convergence), and is compact if and only if F is closed in $C(X, Y)_{\mathcal{K}}$ and the sets $F|K$ are compact in $C(K, Y)$, $K \in \mathcal{K}$, see exercise 4.4.24. The statements 4.5.26 and 4.5.27 imply:

Theorem 4.5.28 *Let X be a locally compact topological space and Y a complete metric space. A subset $F \subseteq C(X, Y)_{\mathcal{K}}$ is relatively compact if and only if it satisfies the following conditions:*

(1) *For every $x \in X$ the set $\{f(x) \mid f \in F\}$ is relatively compact in Y.*
(2) *F is equicontinuous.*

F is compact if and only if F additionally satisfies the following condition:

(3) *F is closed in $C(X, Y)_{\mathcal{K}}$.*

The topology of $C(X, Y)_{\mathcal{K}}$ can be described very clearly if the locally compact space X is even σ-compact, i.e., it has a countable cover of compact sets. Let $\mathcal{K}' = (K_k)_{k \in \mathbb{N}}$ be an exhaustion sequence of X as in Proposition 4.4.23 (iii). Then $C(X, Y)_{\mathcal{K}} = C(X, Y)_{\mathcal{K}'}$ is according to Lemma 4.5.23. In particular, $C(X, Y)_{\mathcal{K}}$ is metrizable and a sequence of functions $f_n : X \to Y$, $n \in \mathbb{N}$, is locally uniformly convergent if and only if the sequence (f_n) converges uniformly on each K_k, $k \in \mathbb{N}$. A subset $F \subseteq C(X, Y)_{\mathcal{K}}$ is relatively compact in $C(X, Y)_{\mathcal{K}}$ according to Theorem 4.4.14 if and only if every sequence (f_n) in F has a subsequence that converges in $C(X, Y)_{\mathcal{K}}$.

We leave it to the reader to formulate the theorem 4.5.26 of Arzelà-Ascoli and its consequences specifically for the function spaces $C_{\mathbb{K}}(X) = C(X, \mathbb{K})$. Note that the relatively compact subsets of $Y = \mathbb{K}$ are exactly the bounded subsets.

Example 4.5.29 (Hausdorff Distance) As a first application of the recently discussed theory of uniform convergence, we discuss the Hausdorff distance for closed sets of a metric space.

Let X be a metric space. With $\mathcal{F} = \mathcal{F}(X)$ we denote the set of closed subsets of X. For each $A \in \mathcal{F}$ the distance function is $d_A : X \to \overline{\mathbb{R}}_+$, $x \mapsto d(A, x) = \text{Inf}(d(z, x), z \in A)$, a continuous function on X with $A = d_A^{-1}(0)$, where $\overline{\mathbb{R}}_+$ is equipped with the metric $d(x, y) = |y - x|$, with respect to the $\overline{\mathbb{R}}_+$ is a complete metric space, see example 4.2.20. It is $d_\emptyset \equiv \infty$. The mapping $A \mapsto d_A$ is an embedding of $\mathcal{F}$ into the complete metric space $C(X, \overline{\mathbb{R}}_+)$, equipped with the Chebyshev metric, i.e., with the topology of uniform convergence. With respect to the embedding $A \mapsto d_A$ we consider $\mathcal{F}$ as a metric subspace of $C(X, \overline{\mathbb{R}}_+)$. To avoid misunderstandings, we denote this metric on $\mathcal{F}$ with ∂. For $A, B \in \mathcal{F}$ is therefore $\partial(A, B) = \|d_B - d_A\|_{X,\infty}$. If $A \neq \emptyset$, then $\partial(A, \emptyset) = \infty$. The empty set is in particular an isolated point of $\mathcal{F}$.[19] Furthermore, the important representation

$$\partial(A, B) = \text{Max}\,(\text{Sup}(d(A, y), y \in B), \text{Sup}(d(x, B), x \in A))$$
$$= \text{Sup}(\{d(x, B) \mid x \in A\} \cup \{d(A, y) \mid y \in B\}).$$

If one denotes with $\partial'(A, B)$ the right side of this equation, then a simple consideration for each $z \in X$ results in the inequality $d(A, z) \leq d(B, z) + \partial'(A, B)$ and thus due to $\partial'(B, A) = \partial'(A, B)$ also the inequality $d(B, z) \leq d(A, z) + \partial'(A, B)$, from which $\partial(A, B) \leq \partial'(A, B)$ follows. The inequality $\partial'(A, B) \leq \partial(A, B)$ is trivial. One calls $\partial(A, B) = \partial'(A, B)$ the **Hausdorff distance** between A and B.[20] Note that $x \mapsto \{x\}$ defines an isometric embedding of X into $\mathcal{F}(X) \subseteq C(X, \overline{\mathbb{R}}_+)$. For a more geometric interpretation of $\partial(A, B)$, refer to exercise 4.5.12. The following lemma, whose simple proof we leave to the reader, shows that when studying the Hausdorff distance, one can essentially limit oneself to spaces with real metrics. ◇

Lemma 4.5.30 *If $X = (X, d) = \coprod_{i \in I}(X_i, d_i)$ is the sum of the metric spaces $X_i = (X_i, d_i)$ (with $d|(X_i \times X_i) = d_i$ and $d(X_i \times X_j) \subseteq \{\infty\}$ for $i \neq j$), then the mapping $\mathcal{F}(X) \xrightarrow{\sim} \prod_{i \in I} \mathcal{F}(X_i)$, $A \mapsto (A \cap X_i)_{i \in I}$, is an isometric homeomorphism (where the metric on $\prod_{i \in I} \mathcal{F}(X_i)$ is the Chebyshev metric).*

A direct consequence of the theorem 4.5.26 by Arzelà-Ascoli is the following theorem:

Theorem 4.5.31 *If X is a complete metric space, then the space $\mathcal{F} = \mathcal{F}(X)$ of the closed subsets of X with respect to the Hausdorff distance is complete. If X is even compact, then also $\mathcal{F}$.*

Proof We can assume according to Lemma 4.5.30 that the metric d on X is real. Since $C(X, \overline{\mathbb{R}}_+) \subseteq (\overline{\mathbb{R}}_+)^X$ is complete according to Theorem 4.5.21 like $\overline{\mathbb{R}}_+$, it is necessary to show for the completeness of $\mathcal{F}$ that $\mathcal{F}$ is closed in $C(X, \overline{\mathbb{R}}_+)$. Let A_n, $n \in \mathbb{N}$, be a sequence of closed subsets of X, for which the sequence $f_n := d_{A_n}$, $n \in \mathbb{N}$, *uniformly* converges to a (continuous) function $f : X \to \overline{\mathbb{R}}_+$. We have to show $f = d_A$ for $A := f^{-1}(0)$. If infinitely many of the A_n are empty, then f is constantly equal to ∞ and $f = d_\emptyset$. So we can assume that the A_n, $n \in \mathbb{N}$, are not empty. Then $f(x) < \infty$ is true for all $x \in X$.

We first prove $f(x) \geq d(A, x)$ for each $x \in X$. For a given $\varepsilon > 0$ there is a $k \in \mathbb{N}$ with $d(A_n, z) = |d(A_n, z) - d(A_m, z)| \leq \varepsilon$ for all $z \in A_m$ and $n \geq m \geq k$. Now let ε_i, $i \in \mathbb{N}$, be a sequence in $\mathbb{R}_+^\times$ with $\sum_i \varepsilon_i \leq \varepsilon$. Furthermore, let $k_0 \in \mathbb{N}$ be an

---

[19] For this reason, the empty set is often excluded from considerations.
[20] The Hausdorff distance $\partial(A, B)$ should not be confused with the distance $d(A, B)$ from exercise 4.2.27.

4.5 Complete Metric Spaces and Uniform Convergence

index with $|d(A_{k_0}, x) - f(x)| \leq \varepsilon_0$ for all $x \in X$ and $d(A_n, z) \leq \varepsilon_1$ for all $z \in A_m$, $n \geq m \geq k_0$. There is a $y_0 \in A_{k_0}$ with $d(y_0, x) \leq d(A_{k_0}, x) + \varepsilon_0$.

One now recursively selects natural numbers k_i and points $y_i \in A_{k_i}, i \in \mathbb{N}^*$, with $k_0 < k_1 < k_2 < \cdots$, $d(y_i, y_{i-1}) \leq 2\varepsilon_i$ and $d(A_m, y_i) \leq \varepsilon_{i+1}$ for $m \geq k_i$. Because of

$$d(y_j, y_k) \leq \sum_{i=j+1}^{k} d(y_i, y_{i-1}) \leq 2 \sum_{i=j+1}^{k} \varepsilon_i \quad \text{for } j \leq k,$$

(y_k) is a Cauchy sequence in X and thus convergent in X. Let $y := \lim_{i \to \infty} y_i$. Then $f(y) = \lim_i f(y_i) = \lim_i f_{k_i}(y_i) = \lim_i d(A_{k_i}, y_i) = 0$, so $y \in A$. In addition,

$$d(y, y_0) = \lim_i d(y_i, y_0) \leq \sum_{i=1}^{\infty} d(y_i, y_{i-1}) \leq 2 \sum_{i=1}^{\infty} \varepsilon_i \leq 2\varepsilon - 2\varepsilon_0 \quad \text{as well as}$$

$$f(x) \geq d(A_{k_0}, x) - \varepsilon_0 \geq d(y_0, x) - 2\varepsilon_0 \geq d(y, x) - d(y, y_0) - 2\varepsilon_0$$
$$\geq d(y, x) - (2\varepsilon - 2\varepsilon_0) - 2\varepsilon_0 = d(y, x) - 2\varepsilon \geq d(A, x) - 2\varepsilon.$$

For each $\varepsilon > 0, f(x) \geq d(A, x) - 2\varepsilon$ therefore applies. Therefore, $f(x) \geq d(A, x)$ is as claimed. In particular, $A \neq \emptyset$.

To prove $f(x) \leq d(A, x)$, let $a \in A$ and $\varepsilon > 0$ be given. There exists $b_n \in A_n$ with $f_n(a) = d(A_n, a) \geq d(b_n, a) - \varepsilon$. Because of $a \in A$, i.e. $\lim f_n(a) = f(a) = 0$, $d(b_n, a) \leq f_n(a) + \varepsilon \leq 2\varepsilon$ is for $n \geq n_0$. For these $n, f_n(x) = d(A_n, x) \leq d(b_n, x) \leq d(b_n, a) + d(a, x) \leq 2\varepsilon + d(a, x)$ results. It follows $f(x) \leq d(a, x)$. This applies for every $a \in A$, i.e. it is $f(x) \leq d(A, x)$.

To prove the addition, we have to show that the set completed in $C(X, \overline{\mathbb{R}}_+)$ also fulfills conditions (1) and (2) of corollary 4.5.27 for a compact X. For every closed and thus compact set $A \subseteq X$ and every $x \in X$, however, $d_A(x) = d(A, x)$ is in the compact set $d(X \times X) \subseteq \mathbb{R}_+$. That $\mathcal{F}$ is uniformly continuous follows directly from the fact that every function $d_A \in \mathcal{F}$ is Lipschitz continuous with the Lipschitz constant 1: $|d(A, x) - d(A, y)| \leq d(x, y)$ for all $x, y \in X$, see exercise 4.2.27c). □

Let X continue to be compact, $A \in \mathcal{F}$ and $\varepsilon > 0$. Finally, many of the ε-spheres $\overline{B}(x; \varepsilon)$, $x \in A$, cover A. If these are the spheres with the centers $x_1, \ldots, x_n$, then A lies in the ε- sphere $\overline{B}(E; \varepsilon) \subseteq \mathcal{F}$, $E := \{x_1, \ldots, x_n\}$. In other words: *The finite subsets of X form a dense subset of $\mathcal{F}$.* Somewhat more generally, it follows:

Theorem 4.5.32 *If D is a dense subset of the compact metric space X, then the set $\mathfrak{E}(D)$ of finite subsets of D is a dense subset with respect to the Hausdorff distance of the set $\mathcal{F}$ of closed subsets of X.*

Let's consider, for example, the space $\mathbb{R}^n$, equipped with the Euclidean metric d_2, and within it a compact sphere $X := \overline{B}(0; R) \subseteq \mathbb{R}^n$, $R \in \mathbb{R}_+^\times$. Then *the set of closed convex subsets of X is a closed and thus compact subset of the* theorem 4.5.31 *compact* space $\mathcal{F} = \mathcal{F}(X)$. Proof! From theorem 4.5.32 it follows: *The (convex) polytopes in X (these are the convex hulls of finite subsets of X) form a dense subset with respect to the Hausdorff distance in the set of all closed convex sets in X. The same even applies to the polytopes whose vertices lie in $\mathbb{Q}^n \cap X$*. With this result, the investigation of compact convex sets can often be reduced to the investigation of polytopes. If a stretch is not important, it can even be assumed that the vertices of the polytopes are points of the standard grid $\mathbb{Z}^n \subseteq \mathbb{R}^n$. ◊

To conclude this chapter, the concept of summability for families of real or complex numbers from Sect. 3.7 will be generalized. The concept of summability used the fact that $(\mathbb{R}, +)$ or $(\mathbb{C}, +)$ are commutative topological groups. Similarly, for multiplicability it is crucial that $(\mathbb{R}^\times, \cdot)$ or $(\mathbb{C}^\times, \cdot)$ are commutative topological groups. We first define in general:

Definition 4.5.33 Let G be a (multiplicatively written) group, endowed with a topology. G is called a **topological group**, if the multiplication $G \times G \to G$ and the inverse formation $G \to G$ are continuous (where $G \times G$ carries the product topology). A mapping $\varphi: G \to H$ of topological groups is called a **homomorphism (of topological groups)**, if φ is a continuous group homomorphism. If φ is a group isomorphism and moreover a homeomorphism, then φ is called an **isomorphism (of topological groups)**.

Let G be a topological group. Subgroups of G are also topological groups with the induced topology. For each $g \in G$ the translations $L_g: x \mapsto gx$ and $R_g: x \mapsto xg$ are homeomorphisms of G with $L_{g^{-1}}$ and $R_{g^{-1}}$ as inverses. The formation of inverses is an involutory homeomorphism. For each open set $U \subseteq G$ and any $A \subseteq G$ the complex products $AU = \bigcup_{a \in A} aU$ and $UA = \bigcup_{a \in A} Ua$ are open. The fact that multiplication and the formation of inverses are continuous can be summarized in the condition that the quotient formation $f: G \times G \to G$, $(x, y) \mapsto xy^{-1}$, is continuous. The fiber $f^{-1}(e_G)$ is the diagonal Δ_G. Since G is Hausdorff if and only if Δ_G is closed in $G \times G$, see exercise 4.2.9a), it follows:

Proposition 4.5.34 *A topological group is Hausdorff if and only if $\{e_G\}$ is closed in G.*

If N is a normal subgroup in G, then the quotient group G/N with the quotient topology is also a topological group and the canonical projection $\pi_N: G \to G/N$ is an open homomorphism of topological groups; because for every open set $U \subseteq G$ the saturated hull NU is open in G. The product mapping $\pi_N \times \pi_N: G \times G \to G/N \times G/N$ is open and thus a quotient mapping. From this and from the commutative diagram in Fig. 4.26 it follows that — as claimed — the multiplication of G/N is continuous. The continuity of the inverse mapping is trivial. According to Proposition 4.5.34, G/N is Hausdorff if and only if N is closed in G. G/N is discrete if and only if N is open in G. G/N has the following universal property: If $\varphi: G \to H$ is a homomorphism of topological groups with $N \subseteq \ker \varphi$, there is

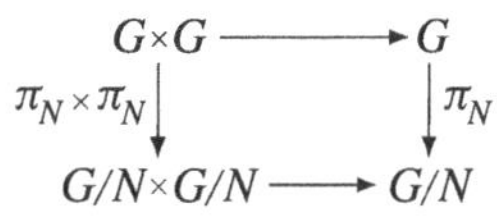

4.5 Complete Metric Spaces and Uniform Convergence

$$
\begin{array}{ccc}
G \times G & \longrightarrow & G \\
\pi_N \times \pi_N \downarrow & & \downarrow \pi_N \\
G/N \times G/N & \longrightarrow & G/N
\end{array}
$$

Fig. 4.26 Connection of the multiplications in the topological groups G and G/N, $N \subseteq G$ normal divisor

exactly one homomorphism of topological groups $\overline{\varphi} \colon G/N \to H$ with $\varphi = \overline{\varphi} \circ \pi_N$. Unlike with groups in general, $\overline{\varphi}$ does not have to be an isomorphism of topological groups when φ is surjective and $N = \operatorname{Ker}\varphi$ is. Otherwise, every bijective homomorphism of topological groups would be a homeomorphism. Rather, the surjective homomorphism $\varphi \colon G \to H$ induces an isomorphism $\overline{\varphi} \colon G/\operatorname{ker}\varphi \xrightarrow{\sim} H$ of topological groups, φ is therefore a quotient mapping, if φ is open.

Example 4.5.35 (1) Products of topological groups are (with the product topology) again topological groups.

(2) Every group is a topological group with the discrete topology. This example is simple, but infinite products of discrete groups and their subgroups are quite interesting as topological groups (and only in trivial cases discrete).

(3) $(\mathbb{R}, +)$ and $(\mathbb{C}, +)$ as well as $(\mathbb{R}^\times, \cdot)$ and $(\mathbb{C}^\times, \cdot)$ are Hausdorff topological groups. In general, for every ordered field K with the order topology, the groups $(K, +)$ and $(K^\times, \cdot)$ are topological groups, see example 4.2.35. Generally, a division area K with a topology is called a **topological division area**, when the addition, the multiplication and the inverse formation $K^\times \to K^\times$ are continuous.

(4) The circle group $U = \{z \in \mathbb{C} \mid |z| = 1\}$ is a compact subgroup of $\mathbb{C}^\times$. The canonical isomorphism $\mathbb{T} = \mathbb{R}/\mathbb{Z} \to U$, $t + \mathbb{Z} \mapsto e^{2\pi i t} = \cos 2\pi t + i \sin 2\pi t$, is an isomorphism of topological groups. Consequently, all torus groups $\mathbb{T}^n \xrightarrow{\sim} U^n$ are compact topological groups, which are moreover isomorphic to $\mathbb{R}^n/\mathbb{Z}^n$, $n \in \mathbb{N}$. The exponential mapping $\exp \colon \mathbb{C} \to \mathbb{C}^\times$ is an open surjective homomorphism of topological groups, which induces an isomorphism $\mathbb{C}/2\pi i \mathbb{Z} \xrightarrow{\sim} \mathbb{C}^\times$ of topological groups. Moreover, $\mathbb{R}_+^\times \times U \xrightarrow{\sim} \mathbb{C}^\times$, $(r, u) \mapsto ru$, is an isomorphism of topological groups.

(5) The additive group of a normed $\mathbb{K}$ vector space V is a Hausdorff topological group, see exercise 4.1.6a). If the values of the norm on V are all finite, then the scalar multiplication $\mathbb{K} \times V \to V$ is also continuous, see Ex. 4.1.6b). In general, a $\mathbb{K}$-vector space V with a topology is a **topological $\mathbb{K}$-vector space**, if $(V, +)$ is a Hausdorff topological group and its scalar multiplication is continuous.[21] Every normed $\mathbb{K}$ vector space V with $V = V_{<\infty}$ and in particular every $\mathbb{K}$ Banach space is therefore a topological $\mathbb{K}$ vector space. ◇

---

[21] Some authors do not require a topological vector space to be Hausdorff.

Let G again be a (multiplicative) topological group with neutral element $e = e_G$. If U is a neighborhood of e, then U^{-1} is also a neighborhood of e and thus also $\tilde{U} := U \cap U^{-1}$. It is $\tilde{U}^{-1} = \tilde{U}$. Neighborhoods of e with this property are called **symmetric**. *The symmetric neighborhoods of e thus form a neighborhood basis of e.* Let $n \in \mathbb{N}^*$. Since the multiple product formation $G^n \to G$, $(x_1, \ldots, x_n) \mapsto x_1 \cdots x_n$, is continuous (induction over n), there is for every neighborhood V of e a neighborhood U of e with $U^n = U \cdots U \subseteq V$. Note that U^n is also a neighborhood of e. *The neighborhoods U^n, U neighborhood of e, thus form a neighborhood basis of e.* If the neighborhoods V of e form a neighborhood basis of e, then for each $g \in G$ both the translates gV and Vg form a neighborhood basis of g. *In particular, G satisfies the first countability axiom,* cf. definition 4.2.9, *if e has a countable neighborhood basis.*

We only define summable or multiplicable families for abelian and Hausdorff topological groups. We prefer the additive notation and leave the corresponding formulations for the multiplicative case to the reader. *In the following, let $G = (G, +)$ be a commutative Hausdorff topological group with neutral element 0 and a_i, $i \in I$, a family of elements from G.* For a finite subset $J \in \mathfrak{E}(I)$, as before, a_J denotes the **partial sum** $a_J = \sum_{i \in J} a_i$.[22] Analogous to definition 3.7.2 we define:

Definition 4.5.36 The family a_i, $i \in I$, in $G = (G, +)$ is called **summable in G**, if there is an $c \in G$ with the following property: For every neighborhood U of c there is an $J_0 \in \mathfrak{E}(I)$ with $a_J \in U$ for all $J \in \mathfrak{E}(I)$ with $J \supseteq J_0$. — If c exists, then c is uniquely determined (since G is Hausdorff) and is called the **sum** of the a_i, $i \in I$. It is denoted by

$$\sum_{i \in I} a_i.$$

In multiplicative notation, we naturally speak of **multiplicable families**. The sum $\sum_{i \in I} a_i$ is obviously equal to the limit $\lim_{\mathcal{F}(I)} a_J$ of the partial sums a_J, $J \in \mathfrak{E}(I)$, with respect to that filter $\mathcal{F}(I)$ on $\mathfrak{E}(I)$ for which the sets $\{J \in \mathfrak{E}(I) \mid J \supseteq J_0\} \in \mathfrak{P}(\mathfrak{E}(I))$, $J_0 \in \mathfrak{E}(I)$, form a basis, cf. remark 4.2.21. This immediately results in:

Proposition 4.5.37 *Let $\varphi \colon G \to H$ be a homomorphism of commutative Hausdorff topological groups. If the family a_i, $i \in I$, is summable in G, then the image family $\varphi(a_i)$, $i \in I$, is summable in H and it holds*

$$\sum_{i \in I} \varphi(a_i) = \varphi\left(\sum_{i \in I} a_i\right).$$

---

[22] In multiplicative notation, $a^J = \prod_{i \in J} a_i$ is the corresponding **partial product**.

4.5 Complete Metric Spaces and Uniform Convergence

If this is used for the negative formation $x \mapsto -x$ and the sum homomorphism $(x, y) \mapsto x + y$ of G, the following calculation rules are obtained: If a_i, $i \in I$, and b_i, $i \in I$, are summable families in G, then the families $-a_i$, $i \in I$, and $a_i + b_i$, $i \in I$, are also summable, and the following applies

$$\sum_{i \in I}(-a_i) = -\sum_{i \in I} a_i, \quad \sum_{i \in I}(a_i + b_i) = \sum_{i \in I} a_i + \sum_{i \in I} b_i.$$

Furthermore, for a decomposition $I = I' \uplus I''$ of I: If a_i, $i \in I'$, and a_i, $i \in I''$, are summable families in G, then a_i, $i \in I$, is also summable and the following applies

$$\sum_{i \in I} a_i = \sum_{i \in I'} a_i + \sum_{i \in I''} a_i.$$

Note that a subfamily of a summable family does *not* need to be summable. Example? But compare the corollary 4.5.42 and the remark 4.5.45.

The great rearrangement theorem 3.7.11, also known as the **great associative law**, can be generally formulated as follows.

Theorem 4.5.38 (Major Rearrangement Theorem) *Let G be a Hausdorff commutative topological group and a_i, $i \in I$, a summable family in G. Furthermore, let $I = \biguplus_{j \in J} I_j$ be a decomposition of the index set I (with pairwise disjoint subsets $I_j \subseteq I$, $j \in J$) such that each of the subfamilies a_i, $i \in I_j$, is also summable (which is always the case in the situation of* Theorem 4.5.41*) with*

$$s_j := \sum_{i \in I_j} a_i.$$

Then the following holds: The family s_j, $j \in J$, is also summable, and it is

$$\sum_{i \in I} a_i = \sum_{j \in J} s_j = \sum_{j \in J} \left(\sum_{i \in I_j} a_i \right).$$

Proof The proof proceeds analogously to the proof of theorem 3.7.11. Let $s := \sum_{i \in I} a_i$ and U be a neighborhood of 0 in G. We are looking for an $L_0 \in \mathfrak{E}(J)$ with $s_L \in s + U$ for all $L \in \mathfrak{E}(J)$ with $L \supseteq L_0$. Let V be a neighborhood of 0 in G with $V + V \subseteq U$. By assumption, there is an $K_0 \in \mathfrak{E}(I)$ with $a_K \in s + V$ for all $K \in \mathfrak{E}(I)$ with $K \supseteq K_0$. Each of the finitely many elements of K_0 is in an I_j. Therefore, there is a finite subset $L_0 \subseteq J$ such that $K_0 \subseteq \bigcup_{j \in L_0} I_j$ is.

Let $L \supseteq L_0$ be with $|L| = n \in \mathbb{N}^*$ and W a symmetric environment of 0 with $nW = W + \cdots + W \subseteq V$. Since s_j, $j \in L$, the sum of the family a_i, $i \in I_j$, exists, there is an $K'_j \in \mathfrak{E}(I_j)$ with $a_{K'_j} \in s_j + W$ and $K'_j \supseteq K_0 \cap I_j$. Then

$$s_L - s = \sum_{j \in L} s_j - s = \sum_{j \in L}\left(s_j - a_{K'_j}\right) + \left(\sum_{j \in L} a_{K'_j} - s\right) \in nW + V \subseteq V + V \subseteq U,$$

applies since $K := \biguplus_{j \in L} K'_j$ is the disjoint union of the K'_j, so $a_K = \sum_{j \in L} a_{K'_j}$ applies, and since K includes K_0 by construction, so it is $a_K \in s + V$. □

In order to be able to formulate a Cauchy criterion for summability, we need the concept of the Cauchy summable family and the concept of the complete topological commutative group. We can directly transfer the first of these concepts from $\mathbb{K}$, see Definition 3.7.3.

Definition 4.5.39 A family $a_i, i \in I$, in G is called **Cauchy summable**, if for every neighborhood U of 0 in G there is an $J_0 \in \mathfrak{E}(I)$ such that $a_E \in U$ is for all $E \in \mathfrak{E}(I)$ with $E \cap J_0 = \emptyset$.

The Cauchy summability of a family $a_i, i \in I$, in G can also be expressed as follows: *For every neighborhood U of 0 there is a $J_0 \in \mathfrak{E}(I)$ with $a_J - a_{J'} \in U$ for all $J, J' \in \mathfrak{E}(I)$ with $J, J' \supseteq J_0$.* Subfamilies of Cauchy summable families are obviously again Cauchy summable. — A summable family $a_i, i \in I$, is Cauchy summable. Let $c := \sum_{i \in I} a_i$, U be a neighborhood of 0 in G and V a symmetric neighborhood of 0 with $V + V \subseteq U$. By assumption, there is a $J_0 \in \mathfrak{E}(I)$ with $a_J \in c + V$ for all $J \in \mathfrak{E}(I)$ with $J \supseteq J_0$. For $E \in \mathfrak{E}(I)$ with $E \cap J_0 = \emptyset$ then $a_E = a_{J_0 \uplus E} - a_{J_0} \in (c + V) - (c + V) = V - V \subseteq U$.

The reversal applies when G is complete. As mentioned several times, sequences are not sufficient for limit considerations if the topological space does not satisfy the first countability axiom. Therefore, Cauchy sequences are also generally not sufficient to define completeness. However, this is possible if G satisfies the first countability axiom, i.e., if 0 in G has a countable neighborhood base U_n, $n \in \mathbb{N}$, which we can also assume to be open and symmetric and the sequence (U_n) decreases monotonically: $U_0 \supseteq U_1 \supseteq U_2 \supseteq \cdots$. We call G **metrizable (as a topological group)**, if G has a *translation-invariant* metric d that generates the topology of G. Then $d(x, y) = d(y, x) = d(0, x - y) = d(-x, -y)$ for $x, y \in G$. The formation of the negative is thus (like any translation) an isometry. A metrizable group naturally satisfies the first countability axiom, and the neighborhoods $U_n := \{x \in G \mid d(0, x) < 1/(n+1)\}, n \in \mathbb{N}$, of $0 \in G$ meet the above conditions.[23]

Definition 4.5.40 Let $G = (G, +)$ be a Hausdorff commutative topological group.

(1) A sequence $x_n, n \in \mathbb{N}$, in G is called a **Cauchy sequence**, if for every neighborhood U of 0 in G there exists an $n_0 \in \mathbb{N}$ with $x_n - x_m \in U$ for all $n, m \geq n_0$.

---

[23] Note that the metrics induced by the natural metrics on $\mathbb{R}$ and $\mathbb{C}$ are not translation-invariant for the topological groups $\mathbb{R}^\times$ and $\mathbb{C}^\times$. Specify translation-invariant metrics that define the topology of these groups.

4.5 Complete Metric Spaces and Uniform Convergence

(2) G is called **sequentially complete**, if every Cauchy sequence in G converges. If G satisfies the first countability axiom, then G is called **complete**, if G is sequentially complete.

If the topological group G is metrizable with metric d, then a sequence x_n, $n \in \mathbb{N}$, is a Cauchy sequence with respect to d if and only if it is a Cauchy sequence according to Definition 4.5.1. G is in this case exactly a complete topological group when (G, d) is a complete metric space. *In particular, a complete normed $\mathbb{K}$-vector space $V = (V, \|-\|)$ and especially a $\mathbb{K}$-Banach space is a complete topological group with respect to addition.*

Every Hausdorff commutative group G, which satisfies the first countability axiom, has a canonical **completion** (or **completion**) $\widehat{G}$, which is also Hausdorff and commutative, satisfies the first countability axiom and contains G as a dense subgroup. For the construction, we now consider the group $G_{\mathrm{CF}}^{\mathbb{N}}$ of Cauchy sequences in G and within it the subgroup $\mathfrak{n}_G$ of null sequences. Then such a completion is the quotient group

$$\widehat{G} = G_{\mathrm{CF}}^{\mathbb{N}}/\mathfrak{n}_G.$$

The elements of G are represented in $\widehat{G}$ by the constant sequences. If U is a neighborhood of 0 in G, then the set $\widetilde{U}$ of elements from $\widehat{G}$, which are represented by a Cauchy sequence with elements from U, is a neighborhood of 0 in $\widehat{G}$. — The Cauchy summability criterion for topological groups is now as follows:

Theorem 4.5.41 (Cauchy's Summability Criterion) *Let G be a complete Hausdorff commutative topological group that satisfies the first countability axiom. A family a_i, $i \in I$, in G is summable if and only if it is Cauchy-summable.*

Proof The proof proceeds analogously to the proof of Theorem 3.7.4. As already noted, every summable family is Cauchy summable. Now let, conversely, $a_i, i \in I$, be Cauchy summable and $U_0 \supseteq U_1 \supseteq U_2 \supseteq \cdots$ a basis of symmetric neighborhoods of 0 in G. By assumption, there is a sequence $J_n \in \mathfrak{E}(I)$, $n \in \mathbb{N}$, with $a_E \in U_n$, if $E \in \mathfrak{E}(I)$ and $E \cap J_n = \emptyset$. By replacing J_n with $\bigcup_{k=0}^{n} J_k$, we can assume that $J_0 \subseteq J_1 \subseteq J_2 \subseteq \cdots$ holds. We show that the sequence a_{J_n}, $n \in \mathbb{N}$, is a Cauchy sequence and therefore converges to an $c \in G$. Let U be a neighborhood of 0 in G. There is an $n_0 \in \mathbb{N}$ with $U_{n_0} \subseteq U$. For $n \geq m \geq n_0$, then $a_{J_n} - a_{J_m} = a_{J_n - J_m} \in U_{n_0} \subseteq U$ holds because of $(J_n - J_m) \cap J_{n_0} = \emptyset$.

Finally, we show that $c = \sum_{i \in I} a_i$ is. Let U be a neighborhood of 0 in G and V be a neighborhood of 0 with $V + V \subseteq U$. Furthermore, let n be chosen such that $U_n \subseteq V$ and $a_{J_n} \in c + V$ is. For each $J \in \mathfrak{E}(I)$ with $J \supseteq J_n$ then due to $a_J - a_{J_n} = a_{J - J_n}$ and $(J - J_n) \cap J_n = \emptyset$

$$a_J - c = (a_J - a_{J_n}) + (a_{J_n} - c) \in U_n + V \subseteq V + V \subseteq U. \qquad \square$$

Since subfamilies of Cauchy summable families are again Cauchy summable, we obtain:

Corollary 4.5.42 *Let G be as in Theorem 4.5.41. Then every subfamily of a summable family in G is also summable.*

Example 4.5.43 (1) Let G_k, $k \in K$, be Hausdorff commutative groups. A family $a_i = (a_{ik})_{k \in K}$, $i \in I$, in the product $G = \prod_{k \in K} G_k$ (endowed with the product topology) is summable by definition of the product topology if and only if each of the component families a_{ik}, $i \in I$, is summable in G_k, $k \in K$. In this case, $\sum_{i \in I} a_i = \left(\sum_{i \in I} a_{ik}\right)_{k \in K}$. For example, in a discrete group H, a family b_i, $i \in I$, is Cauchy summable exactly when almost all b_i vanish. H is therefore complete. A family in a product $\prod_{k \in K} H_k$ of discrete groups is summable exactly when almost all members of the family vanish in each component. In particular, $h = \sum_{k \in K} h_k$ for each element $h = (h_k)_{k \in K} \in \prod_{k \in K} H_k$ (where H_k is identified in a canonical way with a subgroup of the product). This convention is often used without explicitly mentioning the topological background. For example, this applies to the formal monoidal algebras $A[\![M]\!] = A^M$, where A is an S-algebra and M is a monoid, in which each element ρ has only finitely many product representations $\rho = \sigma\tau$ with $\sigma, \tau \in M$, see the end of example 2.9.6. The multiplication of $A[\![M]\!]$ is also continuous. $A[\![M]\!]$ is therefore a topological ring. In general, a ring R with a topology is called a **topological ring**, if the addition and multiplication of R are continuous. (Then the negation is also continuous, $(R, +)$ is therefore a topological group.)

(2) Let $V = (V, \|-\|)$ be a normalized $\mathbb{K}$-vector space. A family x_i, $i \in I$, is apparently summable exactly when there is a $I_0 \in \mathfrak{E}(I)$ such that $x_i \in V_{<\infty}$ is for all $i \in I - I_0$ and x_i, $i \in I - I_0$, in $V_{<\infty}$ is summable. In this case, $\sum_{i \in I} x_i = x_{I_0} + \sum_{i \in I - I_0} x_i \in x_{I_0} + V_{<\infty}$. We therefore want to assume that $V = V_{<\infty}$ and moreover V is complete, thus a Banach space. Then the following proposition is useful:

Proposition 4.5.44 *Let x_i, $i \in I$, be a family in the $\mathbb{K}$-Banach space V. If $\|x_i\|$, $i \in I$, is summable (in $\mathbb{R}$), then x_i, $i \in I$, is summable in V.*

Proof According to theorem 4.5.41 it is sufficient to show that the family x_i, $i \in I$, is Cauchy summable. Let $\varepsilon > 0$ be given. Then there exists a $J_0 \in \mathfrak{E}(I)$ with $\sum_{i \in E} \|x_i\| \leq \varepsilon$ for all $E \in \mathfrak{E}(I)$ with $E \cap J_0 = \emptyset$. For this E also applies $\|x_E\| \leq \sum_{i \in E} \|x_i\| \leq \varepsilon$. □

A family x_i, $i \in I$, as in the proposition with $\sum_{i \in I} \|x_i\| < \infty$ is called **normally summable**. *Normally summable families in Banach spaces are therefore summable.* We expressly note that a summable family does not necessarily have to be normally summable. For $x = (1/(n+1))_{n \in \mathbb{N}} \in \ell^2_{\mathbb{R}}(\mathbb{N})$ for example, $x = \sum_{n \in \mathbb{N}} e_n/(n+1)$ (proof?), but $\sum_{n \in \mathbb{N}} \|e_n/(n+1)\|_2 = \sum_{n \in \mathbb{N}} 1/(n+1) = \infty$. However, if V is a finite-dimensional Banach space, then a family x_i, $i \in I$, is exactly summable when *it is normally summable.* Proof! See theorem 3.7.9, to which the general case can be traced back.

(3) Let K be an ordered field, equipped with the order topology. As already noted in example 4.5.35 (3), then K is a topological field, i.e. $(K, +)$ and $(K^\times, \cdot)$ are topological groups. In analogy to $\mathbb{R}$ and $\mathbb{C}$ the multiplicability of a family $a_i \in K$, $i \in I$, is defined by the following condition: There is a finite set $I_0 \in \mathfrak{E}(I)$ such that a_i, $i \in I - I_0$, a multiplicable family in $(K^\times, \cdot)$. In this case, we set $\prod_{i \in I} a_i = a^{I_0} \prod_{i \in I - I_0} a_i$. We now assume that the topology of K satisfies the first countability axiom, i.e., that there is a non-stationary null sequence in K. One must first distinguish between Cauchy sequences in $(K, +)$ and those in $(K^\times, \cdot)$. By definition, K is called **complete** if the additive group $(K, +)$ is complete. Then the multiplicative group $(K^\times, \cdot)$ is also complete. For a sequence x_n, $n \in \mathbb{N}$, in $K^\times$, the following conditions are equivalent: (i) (x_n) is a Cauchy sequence in $K^\times$. (ii) There is a $s \in K_+^\times$ with $|x_n| \geq s$ for (almost) all $n \in \mathbb{N}$ and (x_n) is a Cauchy sequence in $(K, +)$. See exercise 4.5.17. Since the null sequences form not only a subgroup, but apparently even a maximal ideal $\mathfrak{n}_K$ in the K-algebra $K_{\mathrm{CF}}^{\mathbb{N}}$ of Cauchy sequences, the completion

$$\widehat{K} = K_{\mathrm{CF}}^{\mathbb{N}} / \mathfrak{n}_K$$

is not only a complete additive topological group, but even an ordered field. The elements of the positivity area $\widehat{K}_+$ of $\widehat{K}$ are represented by those Cauchy sequences whose members almost all lie in K_+. As an example, consider for an ordered field k the rational function field $K = k(X) = \mathrm{Q}(k[X])$ in an indeterminate X with the order given in example 3.1.2 (1), in which X^{-n}, $n \in \mathbb{N}$, is a strictly monotonically decreasing null sequence. ◊

Remark 4.5.45 Let G again be any Hausdorff commutative additive topological group. To characterize completeness in general, one has — as already emphasized — to consider general families a_i, $i \in I$, in G over filtered index sets I, cf. example 4.2.21. Let $\mathcal{F}$ be a filter on I. Then $a_i, i \in I$, a **Cauchy family** with respect to $\mathcal{F}$, if $\mathcal{F} \neq \mathfrak{P}(I)$ is and for every neighborhood U of 0 in G there exists a $F \in \mathcal{F}$ with $a_i - a_j \in U$ for all $i, j \in F$. For example, a family a_i, $i \in I$, is exactly Cauchy summable when the family s_J, $J \in \mathfrak{E}(I)$, of the finite partial sums is a Cauchy family with respect to the filter $\mathcal{F}(I)$ on $\mathfrak{E}(I)$ specified following Definition 4.5.36. If $\lim_{\mathcal{F}} a_i$ exists, then a_i, $i \in I$, is naturally a Cauchy family with respect to $\mathcal{F}$. If, conversely, every Cauchy family is convergent (each with respect to the given filter), then G is called **complete**. In generalization of Theorem 4.5.41, one shows: *If G satisfies the first countability axiom and is G sequentially complete, then every Cauchy family in G converges, i.e. G is complete.* This justifies Definition 4.5.40 (2). *Every compact topological group G is complete.* One proves this like Theorem 4.5.5: If $a_i, i \in I$, is a Cauchy family in G, then according to Lemma 4.4.16 it has an accumulation point and is therefore convergent, since every accumulation point of a Cauchy sequence is a limit of the family. The topological group $(\mathbb{C}^\times, \cdot)$ is isomorphic to the product $\mathbb{R}_+^\times \times \mathrm{U}$, where U is the compact circle group, and thus like $\mathbb{R}_+^\times \cong \mathbb{R}$ and U is complete. Due to $U \cong \mathbb{T} = \mathbb{R}/\mathbb{Z}$, the completeness of U also directly follows from that of $\mathbb{R}$. Similarly, the isomorphism $\mathbb{C}^\times \cong \mathbb{C}/2\pi i\mathbb{Z}$ provides the completeness of $\mathbb{C}^\times$ from that of $(\mathbb{C}, +)$, cf. the proof of Theorem 3.7.17. — We mention that for every topological group G as at the beginning of this remark, a completion $\widehat{G}$ can be constructed and we refer to the literature, for example to [2]. ◇

Exercises

Exercise 4.5.1 Let $V = (V, \|-\|)$ be a normalized $\mathbb{K}$-vector space with $V = V_{<\infty}$. Then the mapping $x \mapsto x/(1 + \|x\|)$ is a homeomorphism from V to the open ball $B_V(0; 1)$. If V is a Banach space $\neq 0$, then V is complete, $B_V(0; 1)$ but not. (The completeness of a metric space thus depends significantly on the metric itself and not just on the topology.)

Exercise 4.5.2 Let X be a metric space with the following property: There is a (fixed) $\varepsilon > 0$ such that all balls $\overline{B}(x; \varepsilon)$, $x \in X$, are compact. Then X is complete. (A locally compact metric space is not generally complete.)

Exercise 4.5.3 A metric space X is precompact if and only if its completion is compact.

Exercise 4.5.4 Let x_{mn}, $(m, n) \in \mathbb{N} \times \mathbb{N}$, be a double sequence in a finite-dimensional $\mathbb{K}$-Banach space V with the following property: For each $m \in \mathbb{N}$ the sequence $(x_{mn})_{n \in \mathbb{N}}$ is bounded. Then there exists a sequence $0 \le n_0 < n_1 < \cdots$ of indices such that the sequences $(x_{mn_k})_{k \in \mathbb{N}}$ converge for all $m \in \mathbb{N}$. (Cf. theorem 4.5.7.)

Exercise 4.5.5 A locally compact or metric space with countably many points is totally disconnected. (For metric spaces see already exercise 4.3.3a). — There are countably infinite, connected Hausdorff spaces. Example?)

Exercise 4.5.6 Let $X = (X, d)$ be a complete metric space.

a) Let x_n, $n \in \mathbb{N}$, be a sequence in X with $\sum_{n=0}^\infty d(x_n, x_{n+1}) < \infty$. Then (x_n) is a Cauchy sequence, and for $x := \lim x_n$ and every $n \in \mathbb{N}$ $d(x_n, x) \le \sum_{i=n}^\infty d(x_i, x_{i+1})$ holds.
b) Let $d(x, y) < \infty$ for all $x, y \in X$. The n-th iterate $f^n \colon X \to X$ of the mapping $f \colon X \to X$ has the Lipschitz constant L_n, $n \in \mathbb{N}$. Let $M := \sum_{n=0}^\infty L_n < \infty$. Then f has exactly one fixed point x, and for every point $x_0 \in X$ the sequence $x_n := f^n(x_0)$, $n \in \mathbb{N}$, converges to x, and the estimates $d(x_n, x) \le \left(\sum_{i=n}^\infty L_i\right) d(x_0, x_1)$ or $d(x_n, x) \le M d(x_n, x_{n+1}) \le L_n M d(x_0, x_1)$ apply. (See exercise 3.8.40.) The sequence f^n, $n \in \mathbb{N}$, converges *uniformly* to the constant x on any non-empty subset of X with finite diameter.

4.5 Complete Metric Spaces and Uniform Convergence

Exercise 4.5.7 There is no bijective continuous mapping $f : \mathbb{R} \to \mathbb{R}^n$, $n \geq 2$. (Such an f would, according to corollary 4.4.5, induce a homeomorphism $I \to f(I)$ for every compact interval I, and $f(I) \subseteq \mathbb{R}^n$ would contain no interior points, cf. example 3.8.28. Now use Baire's category theorem. — Note that for all $n > m \geq 1$ there are surjective continuous mappings $\mathbb{R}^m \to \mathbb{R}^n$. This follows easily from the existence of Peano curves, cf. the remark on exercise 4.4.27. More generally than the result of the exercise, however, it holds that: For $n \neq m$ there is no bijective continuous mapping from $\mathbb{R}^m$ to $\mathbb{R}^n$. This follows, for example, from the Borsuk-Ulam theorem, see exercise 4.3.24.)

Exercise 4.5.8 Let X be a topological space and $F \subseteq C_\mathbb{K}(X)$ a uniformly continuous set of $\mathbb{K}$-valued continuous functions on X. For each quasi-compact subset $K \subseteq X$, the set $\{\|f\|_K \mid f \in F\} \subseteq \mathbb{K}$ is bounded.

Exercise 4.5.9 Let $V \neq 0$ be a normalized $\mathbb{R}$-vector space and $\overline{B}(x_n; R_n)$, $n \in \mathbb{N}$, a sequence of closed balls in V with centers x_n and radii $R_n \in \mathbb{R}_+$. When does this sequence converge with respect to the Hausdorff metric, and what set is then the limit?

Exercise 4.5.10 Let X be a metric space, A, B closed in X and $\varepsilon \in \overline{\mathbb{R}}_+^\times$.

a) From $\partial(A, B) < \varepsilon$ follows $A \subseteq B(B; \varepsilon)$ and $B \subseteq B(A; \varepsilon)$. (Here, $B(Y; \varepsilon)$ denotes for $Y \subseteq X$ and $\varepsilon > 0$ the open ε-tube $\bigcup_{y \in Y} B(y; \varepsilon)$ around Y and $\partial(A, B)$ the Hausdorff distance, see example 4.5.29.)
b) From $A \subseteq B(B; \varepsilon)$ and $B \subseteq B(A; \varepsilon)$ follows $\partial(A, B) \leq \varepsilon$.
c) $\partial(A, B)$ is the infimum of the set of $\varepsilon \in \overline{\mathbb{R}}_+^\times$ with $A \subseteq B(B; \varepsilon)$ and $B \subseteq B(A; \varepsilon)$.

Exercise 4.5.11 Let X be a metric space and $\mathcal{F}$ the space of closed subsets of X with the Hausdorff distance. Does the sequence (A_n) with $A_n \neq \emptyset$ for all $n \in \mathbb{N}$ in $\mathcal{F}$ converge to $A \in \mathcal{F}$, then A is the set of $x \in X$, for which a sequence of points $x_n \in A_n$ with $\lim x_n = x$ exists.

Exercise 4.5.12 Let X be a metric space and $\mathcal{K}$ the set of compact subsets in the space $\mathcal{F}$ of closed subsets of X, equipped with the Hausdorff distance.

a) If X is complete, then $\mathcal{K}$ is closed in $\mathcal{F}$ and in particular also complete.
b) If X is locally compact, then $\mathcal{K}$ is open in $\mathcal{F}$ and locally compact. (See Theorem 4.5.31.)

Exercise 4.5.13 $(a_n) \mapsto [a_0 + 1, a_1 + 1, \ldots] = \lim_{n \to \infty}[a_0 + 1, a_1 + 1, \ldots, a_n + 1]$ (with the continued fraction development from example 3.3.11) is a homeomorphism from $\mathbb{N}^\mathbb{N}$ to the space of irrational numbers >1. Specifically, it follows: The space $X := \mathbb{R} - \mathbb{Q}$ of all irrational numbers has countable topology and possesses a metric defining its topology, with respect to which it is *complete*. (Such spaces X are called

Polish.) Is $\mathbb{Q}$ also a Polish space? A countable product of Polish spaces is also a Polish space. □

Exercise 4.5.14 Let H be a subgroup of the topological group G.

a) The closed hull $\overline{H}$ is also a subgroup of G. If H is a normal divisor in G, then so is $\overline{H}$.
b) If H is open in G, then H is also closed in G.
c) If H is abelian and G is Hausdorff, then $\overline{H}$ is also abelian.
d) The connected component of G, which contains e_G, is a closed normal divisor in G.

Exercise 4.5.15 Let G be a topological group with neutral element e.

a) The closed environments of e form an environment basis of e. In particular, G is a regular topological space when G is Hausdorff. (If U is an environment of e and V is an environment of e with $V \cdot V^{-1} \subseteq U$, then $\overline{V} \subseteq U$.)
b) $\overline{\{e\}}$ is the intersection of all environments of e.
c) The Hausdorff topological group $\overline{G} := G/\overline{\{e\}}$ with the canonical projection $\pi: G \to \overline{G}$ has the following universal property: If $\varphi: G \to H$ is a homomorphism of topological groups and H is Hausdorff, then there is exactly one homomorphism $\overline{\varphi}: \overline{G} \to H$ of topological groups with $\varphi = \overline{\varphi} \circ \pi$.

Exercise 4.5.16 Let a_i, $i \in I$, be a summable family in a complete Hausdorff commutative topological group that satisfies the first countability axiom, and let n_i, $i \in I$, be a bounded family of integers. Then the family $n_i a_i$, $i \in I$, is also summable. (This statement generalizes Theorem 3.7.9.)

Exercise 4.5.17 Let K be an ordered field. In parts b) to e), K has a non-stationary null sequence.

a) For a sequence x_n, $n \in \mathbb{N}$, in $K^\times$ the following conditions are equivalent: (i) (x_n) is a Cauchy sequence in $(K^\times, \cdot)$. (ii) There is $s \in K_+^\times$ with $|x_n| \geq s$ for (almost) all $n \in \mathbb{N}$ and (x_n) is a Cauchy sequence in $(K, +)$.
b) Exactly then is K complete, i.e. the additive group $(K, +)$ is complete, when the multiplicative group $(K^\times, \cdot)$ is complete.
c) K is complete. A family a_i, $i \in I$, in K is exactly summable when the family $|a_i|$, $1 \in I$, of amounts is summable. (See theorem 3.7.9 and exercise 4.5.16.)
d) K is complete. A family $1 + a_i$, $i \in I$, in K is exactly then multiplicable when the family a_i, $i \in I$, is summable. (See theorem 3.7.17.)
e) The null sequences form a maximal ideal $\mathfrak{n}_K$ in the K-algebra $K_{CF}^\mathbb{N}$ of all Cauchy sequences, and the quotient field $\widehat{K} = K_{CF}^\mathbb{N}/\mathfrak{n}_K$ is in a natural way an ordered field, which is also complete.

Literature

1. Aigner, M.: A Course in Enumeration. Springer, Berlin Heidelberg (2007)
2. Bourbaki, N.: General Topology. Springer, Berlin Heidelberg (2008)
3. Cohn, P.M.: Basic Algebra. Springer, London (2003)
4. Conway, J.: A Course in Point Set Topology. Springer, (2014)
5. Deiser, O.: Einführung in die Mengenlehre, 2nd edn. Springer, Heidelberg (2004)
6. Dieudonné, J.: Grundzüge der modernen Analysis, 3rd edn. Vol. 1. Vieweg, Braunschweig (1986)
7. Ebbinghaus, H.-D.: Einführung in die Mengenlehre, 4th edn. Spektrum Akademischer Verlag, Heidelberg (2003)
8. Ebbinghaus, H.-D., et al.: Zahlen, 3rd edn. Springer, Berlin Heidelberg (1992)
9. Fischer, G.: Lehrbuch der Algebra, 3rd edn. Springer Spektrum, Heidelberg (2013)
10. Gerritzen, L.: Grundbegriffe der Algebra. Springer, Wiesbaden (1994)
11. Kunz, E.: Algebra. Vieweg Teubner, (1991)
12. Laures, G., Szymik, M.: Grundkurs Topologie, 2nd edn. Springer Spektrum, Heidelberg (2015)
13. Munkres, J.: Topology. Pearson Educational, London (2013)
14. Schafmeister, W., Wiebe, H.: Grundzüge der Algebra. B. G. Teubner, Stuttgart (1978)
15. Scheja, G., Storch, U.: Lehrbuch der Algebra, Teil 1, 2nd edn. B. G. Teubner, Stuttgart (1994)
16. Scheja, G., Storch, U.: Lehrbuch der Algebra, Teil 2. B. G. Teubner, Stuttgart (1988)
17. Stanley, R.P.: Enumerative Combinatorics, 2nd edn. Vol. 1. Cambbridge University Press, Cambridge (2011)
18. Storch, U., Wiebe, H.: Lehrbuch der Mathematik, 3rd edn. Vol. 1. Spektrum Akademischer Verlag, Heidelberg (2010)
19. Storch, U., Wiebe, H.: Lehrbuch der Mathematik, 2nd edn. Vol. 2. Spektrum Akademischer Verlag, Heidelberg (2010)
20. Storch, U., Wiebe, H.: Lehrbuch der Mathematik Vol. 3. Spektrum Akademischer Verlag, Heidelberg (2010)
21. Storch, U., Wiebe, H.: Lehrbuch der Mathematik Vol. 4. Spektrum Akademischer Verlag, Heidelberg (2011)
22. Storch, U., Wiebe, H.: Arbeitsbuch zur Analysis einer Veränderlichen. Springer Spektrum, Heidelberg (2014)
23. Storch, U., Wiebe, H.: Arbeitsbuch zur Linearen Algebra. Springer Spektrum, Heidelberg (2015)

The manufacturer's authorised representative in the EU is Springer Nature Customer Service Centre GmbH, Europaplatz 3, 69115 Heidelberg, Germany. If you have any concerns regarding our products, please contact ProductSafety@springernature.com

Printed and bound by CPI Group (UK) Ltd, Croydon, CR0 4YY

23/03/2026

02076398-0011